Vehicle Compatibility in Automotive Crashes

PT-102

Edited by
Stanley H. Backaitis

INTERNATIONAL®

Published by:
Society of Automotive Engineers, Inc.
400 Commonwealth Drive
Warrendale, PA 15096-0001
USA
Phone: (724) 776-4841
Fax: (724) 776-5760
April 2005

For permission and licensing requests contact:

SAE Permissions
400 Commonwealth Drive
Warrendale, PA 15096-0001-USA
Email: permissions@sae.org
Fax: 724-772-4028
Tel: 724-772-4891

SAE INTERNATIONAL®

For multiple print copies contact:

SAE Customer Service
Tel: 877-606-7323 (inside USA and Canada)
Tel: 724-776-4970 (outside USA)
Fax: 724-776-1615
Email: CustomerService@sae.org

ISBN 0-7680-1645-2

SAE/PT-102

Printed in USA

INTRODUCTION

Traffic-related fatalities and injuries remain a major problem throughout the world. According to a research working paper #3035, published by the World Bank in 2003, worldwide traffic fatalities are estimated to climb from 723,000 in 2000 to more than 1.2 million in 2020. Current annual fatality estimates in the United States are slightly more than 42,000 and well over 45,000 in the expanded European Community. It is estimated that for each fatality, there are at least 40 injuries. The total in social and economic costs amounts to hundreds of billions of dollars/euros. It is a serious economic burden to most advanced economies and is particularly severe to the economies of developing countries.

Vehicle safety experts worldwide agree that significant reduction in traffic fatalities and injuries can be realized through implementation of improved active and passive safety systems, driver education, traffic flow management, road and roadside architecture, and so forth.

Passive crash safety measures already have a proven track record in reducing road accident casualties through the introduction of safety belts; air bags; improvements in crashworthiness and energy absorption provisions in vehicle frontal, side, and rear structures; passenger compartment structural integrity; and energy absorption features within the occupant compartment. Passive safety measures still have a great potential in further reducing fatalities and injuries. None of these, however, will be of great significance unless disparities in crashworthiness among vehicles of different masses, sizes, and structural characteristics in mixed crash environments are successfully taken into account. This has been a research issue for many years, and it recently has gained much more momentum in view of rapidly increasing SUV, van, and light-truck populations relative to the number of passenger cars, and due to significant improvements in technologies that facilitate a better understanding of the interaction dynamics among widely differing size vehicles. The complexity of the subject requires the development of clear definitions, convergence of procedural directions, involvement of stakeholders from passenger car and heavy-vehicle manufacturers, research institutions, infrastructure suppliers, insurers, and governments at the global level.

The first part of this Compendium presents oral comments of a panel of seven experts, who were invited at the 2003 SAE Congress's Compatibility session to address a set of seven prepared questions as a way of providing a rough framework and a broad range of views of where compatibility research is at the present time and the direction it is going. This section was supplemented by subsequent comments by four additional compatibility researchers to the same set of questions. Inasmuch as each panelist rendered an unaudited opinion on the subject matter, by prior agreement, the commenters' names are listed but are not related to any specific comment.

The commenters were:

- Dr. Sayeed Barbat—Ford Motor Co., Detroit, Michigan*
- Prof. Kennerly Digges—George Washington University, automotive crash research
- Dr. Adrian Hobbs—TRL of UK, compatibility research for the European Union
- Mr. Mathew Jerinsky—General Motors*
- Mr. Hisaaki Kato—Toyota Co., Japan*
- Mr. Robert Lange—Alliance of Automobile Manufacturers
- Dr. Adrian Lund—Insurance Institute for Highway Safety
- Prof. Koji Mizuno—Japan Automobile Manufacturer Association and Nagoya University
- Mr. George Neat—Volpe National Transportation System Center*
- Dr. Dietmar Otte—Medical University of Hannover, Germany
- Mr. Allan Thebert—Chair of the SAE Impact and Rollover Test Procedures Standards Committee

* Did not participate as a panel member but provided subsequent written comments.

Pages following this introduction provide the seven questions, a short background related to each question, and the responses of the eleven commenters. It is also pointed out that the numbered sequence of the responses to each question does not imply the same commenter for the same repeated number.

In the second part of the Compendium, we have provided a collection of papers that a select panel of experts judged to be the best technical presentations and discussions on the entire spectrum of vehicle crash compatibility. The selection process was conducted in two review cycles, initially involving 179 published papers that were found in the SAE data bank. The first review cycle yielded a recommendation for further review of 97 papers. The second review cycle reduced the selection to 43 papers. This final selection was divided into 9 topical sections. Rear impact compatibility and that related to pedestrians were judged by the reviewers as not having reached sufficient maturity for dealing with this issue.

Although other sources have published technical papers on compatibility, copyright and other considerations limited this selection to SAE and ESV publications. They cover papers published from the early 1970s through 2004.

The reviewers for selected papers included the following persons:

- Dr. Saeed Barbat—Ford Motor Co.
- Mr. Pascal Delannoy—Teuchos, Snecma Group, UTAC, France
- Prof. Kennerly Digges—George Washington University
- Prof. Clay Gabler—Virginia Polytechnic Institute
- Prof. Kashiro Mizuno—Nagoya University, Japan
- Dr. Craig Newland—Australian Transport Ministry and IHRA Section Chairman on Side Impact Compatibility
- Mr. Satoshi Takizawa—Honda Co., Japan
- Mr. Guy Nusholtz—Daimler/Chrysler Corp. (participated in only the first review cycle)

The editor of this Compendium is grateful to all panelists for their comments and to the reviewers for their patience and hard work in selecting from the voluminous publications the most meaningful and insightful sets of papers on vehicle crash compatibility. In addition, many thanks to Dr. Roger Saul of the National Highway Transportation Safety Administration (NHTSA) for help in the design of questions to the compatibility panel and for chairing the Compatibility session at the 2003 SAE Congress.

PART 1. QUESTIONS AND INTRODUCTORY REMARKS POSED TO THE PANEL OF EXPERTS ON COMPATIBILITY AND THEIR RESPONSES

1. Definition of Vehicle Crash Compatibility

Increasing diversity in vehicle sizes, shapes, and masses in the U.S. traffic environment is creating concerns about the collision compatibility between large and small vehicles on the road today and the safety of their occupants. Vehicle structural developments, based on self-protection advocated during the last decade, are being questioned for adequacy to protect vehicle occupants. Partner protection is now being explored as a means of assuring occupant safety in collisions of any vehicle combination. The term "vehicle collision compatibility" has been used recently by some to describe the collision properties of any two vehicles in frontal crashes, and by others in front to side impacts and in rear impacts. But compatibility in the context of vehicle impacts with any exterior object that may be encountered in the collision environment may also encompass roadside objects, cyclists, pedestrians, and so forth. In addition, other control systems may, in the future, eliminate vehicle crashes or minimize their effects. This leads to the question of what do we mean by the word "compatibility" as is used in the automotive safety community. Webster defines "compatible" as capable of existing together in harmony with something else. Perhaps it is easier to define incompatibility than compatibility. At the 1973 ESV conference in Kyoto, Fiat presented a film showing a small Fiat car being totaled by an average-size American car, promptly followed by a scene in which a big Fiat truck tore the average American car to pieces.

Is there more to crash compatibility than fairness? Is compatibility achieved if the likelihood of survival or death in corresponding crashing vehicles were equal? Does decreasing the occupant protection provided by one vehicle or increasing the protection provided by the other attain compatibility? Does this ultimately increase or decrease crash incompatibility? We hope a reasonably clear understanding will emerge of what vehicle crash compatibility is, defined for optimum effectiveness, so that everyone can direct their efforts toward a common objective.

Respondent #1

The issue of collision compatibility between vehicles has to be seen as one of the aspects of automotive safety and not as an isolated issue. This leads to the definition of "improved compatibility" as "reduced injuries and fatalities in vehicle-to-vehicle crashes." While further elaboration of this equivalency will be time and region dependent (e.g., in the United States, the total number of fatalities associated with LTV to car collisions has increased over the last few years, and most harmful of these impacts are the side collisions between these vehicles), certain aspects are evident and universal. Compatibility is not confined to the design of the larger vehicle alone. Compatibility should not be measured by the ratios of the probability of survival in a larger vehicle versus those in a smaller vehicle, but by continuously reducing the probability of injuries in all vehicle-to-vehicle crashes along with reducing the total number of injuries and fatalities in all automobile crashes.

As a matter of principle, this needs to be achieved without degrading the safety performance of any one type or model of vehicle (which would change the ratios but not reduce the total number of injuries). Protection of vehicle occupants in single-vehicle crashes is a major part of automotive safety, and this needs to be kept in mind in devising any solutions for collision compatibility.

Respondent #2

1. *Compatibility is not fairness, because fairness is not achievable.*
2. *Compatibility is making sure that no passenger vehicles are unnecessarily aggressive in their collisions with other passenger vehicles.*
3. *Compatibility is both aggressivity and vulnerability.*
4. *Compatibility does not reduce self-protection.*
5. *Compatibility, as a concept, should not include roadside objects, as that would dilute the meaning and usefulness of the concept.*

Respondent #3

Historically, compatibility has been used to describe the problems caused by large cars hitting small cars. It has been attributed to mass ratio and was considered intractable. More recent research has shown that mass in itself is not the major problem; it is the effect that mass has had on making heavy cars stiffer than light cars. It has also become clear that in virtually all impacts, the crash performance is degraded because of poor structural interaction between impacting cars. This aspect of compatibility affects virtually all crashes and is not mass related. Because of this, compatibility covers so-called self-protection as well as partner protection. My definition of compatibility, which has been adopted by the IHRA Working Group, is: "A compatible car is one that protects not only its own occupants but other road users as well."

Respondent #4

Compatibility should be defined as the standard of injury outcome that does exist under the same impact conditions for different vehicle sizes, vehicle shapes, and vehicle weights. This definition could be expanded to all possible collision configurations, that is, that the injury outcome of the occupants of a vehicle should be the same in every accident event.

Respondent #5

My understanding is that compatibility is the ability of a vehicle not only to protect its occupants but also other road users as well. However, it is not realistic to expect the injury risk to be the same between small and large cars. It is impossible if you consider the mechanics where the acceleration is higher for the lighter car. But it does not mean that we cannot do anything for compatibility. When investigating the car-to-car crash between different sizes of cars, there are several possible improvements. The most important thing is to improve vehicle compatibility without sacrificing self- or partner protection. I mean that self-protection should be improved without reducing the level of partner protection, and vice versa. One example is the improvement of structural interactions that are effective for both self- and partner protection. We also should think that compatibility improvement should not increase the risk to other road users as well. For example, the lower frame will be useful for preventing underride, but this kind of structure should not increase the injury risk to the lower extremities of the pedestrian.

Respondent #6

Three types: mass, stiffness, and geometry.

Mass differences are inevitable; increased mass generally is accompanied by increased stiffness and increased height of load paths.

Frontal compatibility should match geometry and stiffness of two vehicles to maximize the combined crash energy absorption before intrusion occurs in either vehicle. Applicable to all frontal directions and extent of overlaps.

Front to side compatibility would require that the side structure be engaged so that the lower structure of the struck vehicle does not lag behind the upper structure as intrusion progresses. In turn, the side structure should be strong enough to limit and control the geometry of the intrusion, and added safety systems should provide for occupant self-protection.

Respondent #7

Compatibility can be understood as the similarities among differing objects or things. In the area of vehicle crash compatibility, discovering these similarities and understanding which aspects can be controlled will help to create harmony among them. This harmony is where one or more vehicle crash events results in little or no injuries to the occupants involved. Vehicle compatibility is not a self-protection goal but a balance among all vehicles involved in any crash condition.

Respondent #8

The major concern should be to develop techniques/guidelines/regulations, etc. that will mitigate serious injuries and fatalities in motor vehicles. Fairness does not seem to be the issue. For example, the occupants of the struck vehicle in a side-impact crash will never have the chance of survival that the occupants of the striking vehicle have. Even in frontal crashes, the emphasis on equal chance of survival could result in less total reduction in fatalities and serious injuries. To maximize total safety, it may be necessary to slightly decrease the occupant protection in what are now the most aggressive vehicles. If this is done properly, the ultimate effect will be to decrease incompatibility.

Respondent #9

The present vehicle fleet differs in mass, stiffness, and geometry, which are the main factors affecting compatibility in vehicle-to-vehicle crashes. Light truck vehicles (LTVs), sport utility vehicles (SUVs), and vans collectively make up a growing segment of the total automotive fleet sales, particularly in the United States. As a group, these vehicles are generally heavier, stiffer, and have higher ground clearance compared to those of passenger cars due to their functional requirements such as load-carrying capacity, towing capacity, and off-road capability. Heavier vehicles may have a stiffer front-end structure compared to lighter vehicles because all vehicles are designed to the same regulatory and consumer crash test speeds aimed at demonstrating self-protection to their own occupants. The structural stiffness of the front end of the LTVs/SUVs/vans group is, to some extent, dictated by these regulatory and consumer crash tests and may be potentially higher than the stiffness of the front end of passenger cars.

In a head-on collision between two cars of equal masses and comparable restraint systems, it is expected that the injury outcome to the occupants of the two cars will be the same and that the two cars are compatible. However, conservation of momentum dictates that the lighter vehicle undergoes a higher change in velocity; therefore, the odds of injury or fatality of the lighter vehicle's occupants will always be higher than that in the heavier vehicle. Even with identical masses and as observed from real-world accident data, loss of compatibility occurs due to differences in stiffness and geometry between the two cars.

In LTV/SUV-to-passenger car collisions, the potential injury risk to the occupants in the passenger car is generally higher. This is due to the larger velocity change and higher intrusions experienced by the generally lighter, smaller, and less stiff passenger cars.

Compatibility is a fleet-wide issue, and design actions to a small segment of vehicles can adversely affect the total.

2. What are the objectives of vehicle crash compatibility?

The U.S. vehicle passenger-carrying population changed during the last decade primarily from cars to where by now SUVs, LTVs, and light trucks compose nearly 50% of sales. This group of vehicles is significantly different in size, mass, and energy at collisions than passenger cars. Self-protection was the primary focus in many of these vehicle designs. But compatibility seems to imply that self-protection in the diverse vehicle population cannot be sought at the expense of occupants in collision partners and must consider the effects of collisions with roadside objects, pedestrians, and so forth.

Is there a logical upper and lower limit of exposures and interactions that would serve to bound and limit the compatibility objective? The safety community needs to define and quantify the objectives that would provide a framework of parameters for the design of crash-compatible vehicles.

Respondent #1

Reducing injury disparities between collision pairs while increasing the total societal benefit should be the governing objective of vehicle designs. The upper and lower limits of the interactions where vehicle designs might be effective are defined by the laws of physics and by the variables that are under the control of vehicle designers. As an example, in a collision between a locomotive and a small car, infrastructure changes such as separating the paths and reducing the likelihood of collision are more effective than any design changes to the automobile. The impact between a pedestrian and a heavy vehicle may set the other bound of potential solutions where the vehicle design changes are likely to result in smaller benefits than other measures aimed at reducing the frequency of these events.

Respondent #2

The objectives of vehicle crash compatibility should be to maximize the likelihood that the self-protection countermeasures in every vehicle are brought into play in collisions with other vehicles. Principally, this means that vehicle structures must interact predictably.

Respondent #3

The objectives of compatibility are to minimize the overall injuries occurring in any accident. It should improve both self-protection and partner protection. Compatible cars should interact in an impact in a way that is both more predictable and more similar to the performance achieved in crash tests. For frontal impact, this requires good structural interaction, throughout the impact, matching of the frontal force that each car imposes on its partner, and a strong passenger compartment. This should overcome the greatest compatibility problem, which is compartment intrusion. This will assure a safe environment in which the restraint systems can function properly. It also should improve the cars' deceleration pulses and allow further tuning of the cars' structures and restraint systems to reduce deceleration-induced injuries.

Respondent #4

The accident situation has to be the basis for the objectives. There are existing compatibility problems in size, shape, and weight differences. Regarding the physical basis of momentum exchange, the different mass relations are resulting in different crash severity delta-V distributions. The load transmission to the occupants could be influenced by stiffness and geometrical construction for energy absorption, as well as by safety protection provisions within the passenger car compartment. Concerning the fact that in an accident event, the collision configuration cannot be known before it occurs, compatibility has to be defined for all vehicle classes from the smallest cars to the heaviest LTVs and SUVs. Compatibility should be separately discussed for other vulnerable road users (i.e., pedestrians).

Respondent #5

Compatibility can include so many items—passenger cars, trucks, road obstacles, and pedestrians. At present, we cannot solve all issues together. So there need to be limitations or priorities for compatibility problems.

The compatibility target may differ among countries, based on accident analysis. In Japan, the number of collisions between passenger cars occupies about half of vehicle collisions. Thus, I think that the first target for compatibility should be passenger cars, including SUVs. The truck also is important, considering the fatality risk to car occupants. But the countermeasure should be done by the truck attaching the underrun protection system to prevent underride.

If the bounds or limits are applied for the compatibility objective, the countermeasures will be beneficial, at least if the partner protection to other road users remains at the same level.

Respondent #6

For frontal compatibility, the prime objective should be to prevent the heavier vehicle from defeating the self-protection system of the lighter vehicle. Failure to fully engage the lighter vehicle's self-protection structure may result in mis-engagement of the lighter vehicle's self-protection systems, yielding incompatibilities that may be detrimental to the occupants of both vehicles.

For front to side compatibility, improved side protection for the head and chest in the struck vehicles is the initial compatibility requirement. For the striking vehicle, some control of the stiffness and geometry over the initial 125 mm of crush needs to be exercised so that the side protection systems are not defeated.

Respondent #7

Similarities in vehicles can include, but are not limited to, the following: most travel on wheels, have a stiff structure, carry occupants, and have an engine. Coupling these similarities with consumer preferences such as performance, occupant load, cargo load, and vehicle size creates the constraints to which safety regulators and industry engineers are limited in creating crash-compatible vehicles. It is the safety regulators' goal to determine which vehicle aspects need to be constrained, and it is the industry engineers' goal to meet the constraints so that they are invisible to the consumer preferences.

Once those aspects of vehicle design that affect vehicle compatibility are understood, the next step is selection of what objects to include. This can be passenger vehicles, heavy trucks, pedestrians, motorcycles, roadway devices, etc. Often, this selection is further complicated by global landscapes. While the basic principles of compatible vehicles can be similar globally, consumer preferences can vary widely from one region to another. This limits what tests can be shared among countries. One possibility is to have a test or test series that can be scaled to meet the demands of different fleets. If a test that measures height or vehicle stiffness is used, it can be scaled to match the average height and vehicle stiffness for the fleet in question. Eventually over time, the scaled requirements can be more closely aligned. However, this burden is either put on the consumer or the automobile manufacturers' ability to meet compatibility requirements while being invisible to consumers' preferences.

Respondent #8

There does not appear to be a logical upper limit of exposure that needs to be considered. For a lower limit, there are considerations, such as pedestrians, but one could never say pedestrians and motor vehicles are compatible in a crash scenario. However, motorcycles, bicycles, and some of the smallest European cars (e.g., three-wheelers) are simply not compatible with full-size motor vehicles in a collision.

To define and quantify the objectives that would provide a framework of parameters for the design of crash-compatible vehicles has been actively investigated for several years by the National Highway Traffic Safety Administration (NHTSA) Office of Vehicle Safety Research, an International Harmonization of Research Activities (IHRA) Working Group on Vehicle Compatibility, the European Enhanced Safety of Vehicles Committee (EEVC) Working Group 15 on Vehicle Compatibility, and the motor vehicle industry. The general consensus is that the objective is to design vehicles so that the total number of fatalities and serious injuries in motor vehicles is minimized. How this is measured is a subject that is ongoing for resolution.

Respondent #9

In the United States, the media, regulatory agencies, automotive industry, and the Insurance Institute of Highway Safety have been focused on the injury/fatality outcomes of passenger car occupants in LTV/SUV-to-passenger car collisions. The main objective of vehicle crash compatibility is to maximize societal impact in providing partner protection without compromising self-protection and further improve the survivability of passenger car occupants in crashes with LTVs/SUVs in front, side, and rear collision modes. Another objective is to develop technologies that reduce override/underride in frontal impact and reduce intrusions in side impact during vehicle-to-vehicle collisions.

3. How does one recognize if a vehicle is compatible or incompatible?

What assessment mechanisms should be used to quantify vehicle collision compatibility? Would it be the dynamics of structural interactions, geometry of crush interfaces, stiffness of involved structures, equality of collision forces and/or pressure distributions, comparative crash performance, levels of risk of occupant and pedestrian injuries, or mitigation of injury potential? What needs to be measured? Are some measurements more important than others? Can laboratory measurements sufficiently predict the real-world outcome?

Respondent #1

The ultimate evaluation of improved compatibility should be based on the potential improvement in the protection of occupants in a vehicle. This biomechanical criterion can be translated into sets of mechanical criteria that may be different for different crash events. As an example, for the cases where a large vehicle impacts the side of a smaller vehicle, the improvement of compatibility may require the measurement of numbers generated by an appropriate ATD in the vehicle when a crash barrier representing an "average" vehicle on the road is crashed into it. Translated into mechanical terms, this may mean targets for parameters such as maximum door or B-pillar intrusion and/or maximum deceleration levels on the car.

While laboratory tests cannot totally predict real-world outcome, we believe it is possible to devise laboratory tests that will improve real-world performance. This requires synthesis of knowledge from crash data, of principles of physics involved in crashes, and of injury mechanisms.

The auto industry and others involved in the area have created two work groups of technical experts that will propose what needs to be measured and how collision compatibility is to be improved.

Respondent #2

This is work in progress. Work looking at force distributions on load cell barriers appears promising, at least for determining if vehicle structures are likely to line up for effective interaction. It may also be possible to specify maximum force generated.

Respondent #3

There are no cars that could be described as compatible. It is much easier to identify the ways and extent that they are incompatible. The frontal stiffness inhomogeneity common to all cars creates problems in engaging their structures when they impact. This compromises the ability of the car front to absorb the impact energy and adversely affects the deceleration pulse shape. Static misalignment and localized areas of high stiffness within the car front all indicate that good structural interaction is unlikely. Beyond that, current frontal stiffness is almost directly related to mass, as all cars are encouraged to stop in the same distance in crash tests, and heavy cars have more kinetic energy to absorb. Consequently, stiffness incompatibility, due to mass difference, is endemic across the fleet. For the fleet to become compatible, the frontal force each car imposes on its partner must be harmonized. To do this, it will be necessary to set a target force for all cars in the fleet.

Respondent #4

For the assessment of compatibility, the injury outcome has to be measured. There are two possibilities in principle: one is NCAP, and another is accident analysis. These two methods have to be combined as is proposed in the German NCAP proposal. A ranking system based on different impact configurations and vehicle classes based on different sizes, shapes, and weights should be developed within a research project.

The assessment of "good compatibility" should be pointed out on the measurements of a vehicle class-related comparison of stiffness values and bumper/frame heights, together with the results of a special defined crash test for compatibility.

Respondent #5

The main concept of compatibility control is that after good structural interaction of the car, the front structure absorbs energy while maintaining the compartment integrity.

Based on this concept, the test procedures should include the evaluation of structural interaction, energy absorption of front structures, and the compartment strength.

Structural interaction can be assessed by force measurements or barrier deformations in impact tests. Energy absorption in front structures will be assessed by the barrier force in ODB tests. Compartment strength can be measured by the barrier forces or car accelerations in overload tests. However, further research will be necessary to clarify the criteria that can predict the real-world car-to-car crashes because there are so many parameters in real-world crashes.

Respondent #6

FARS data provide ratios of people killed in cars versus people killed in SUVs, vans, and pickup trucks. However, it is unclear from this data a determination of other parameters affecting vehicle frontal incompatibilities other than mass.

In current NASS/CDS data, it is difficult to clarify stiffness and geometric incompatibility issues:

- *Pictures and measurements of POV or a struck object may not be available.*
- *Documentation does not show structure interaction.*
- *Effects of structural geometry and stiffness on intrusion are difficult to determine from case data.*

NCAP barrier test data provide a start but have numerous inadequacies:

- *Force concentrations and stiffnesses may be higher in a crash than measured on the barrier.*
- *The number of load cells is inadequate to discriminate load concentrations—particularly true for side impact compatibility, where the heights of the lowest load path are critical.*
- *The center of force is a first cut at measuring frontal geometric compatibility. However, the stiffness at various heights and horizontal positions may be a better measure.*
- *Front to side compatibility should consider both the self-protection of the struck vehicle and the geometric force distribution of the front of the striking vehicle.*

Respondent #7

Another way to tackle vehicle compatibility is to create new similarities among vehicles. For example, bumper beams have changed since they were first mandated. Today's beams have a stiff section usually covered with a homogeneous energy absorber. The designs have moved away from a centralized impact point. Manufacturers have found it necessary to distribute the load for effective bumper design. In the future, it may be found that stiff occupant compartments surrounded with homogeneous material on all sides will be the answer. This homogeneous material can be regulated across fleets to minimize the damage that one vehicle causes to another, while the stiff occupant compartment will protect the occupant. Unfortunately, until manufacturers find a way to do this and meet consumer preferences, it will remain only theoretical and a long-term objective.

Respondent #8

New crash tests have been proposed to evaluate compatibility, but there are still differing opinions on the test specifications to accomplish this. Average height of force for frontal collisions, which could be established with the proper choice of compatibility test requirements, is a parameter that should be strongly considered. A new crash test is essential to rate vehicles for compatibility, but a separate test is most likely still required to measure the degree of occupant protection provided by the vehicle design.

The dynamics of structural interactions, geometry of crush interfaces, stiffness of involved structures, comparative crash performance, levels of risk of occupant and pedestrian injuries, and mitigation of injury potential are all important parameters for recognition, with the exception of asking for equality of forces or pressures.

Leading candidates are average height of force, pressure distribution on crash surface, maximum force, and statistical data from the crash databases, FARS, GES, etc. Crash results from the databases, which are available only after the fact, will provide the best measure.

Laboratory measurements to predict the real-world outcome are not as sufficient as we would like, but that is what is available in the short term. A tool based on Ford Motor Company's System Optimization Model (SSOM), from more than 20 years ago, and further developed by the University of Virginia for NHTSA, was more recently resurrected for NHTSA as the Fleet Systems Model (FSM) at the Volpe Center. The FSM is a statistical model that utilizes computer models of representative motor vehicles in one-car and two-car crashes to represent the serious crashes in the U.S. fleet in a given year. This model, when completed, should provide the answer to the critical issue, which is total harm to the occupants.

Respondent #9

Today's U.S. cars and LTVs are designed to 56-kph (35-mph) rigid barrier impact (NCAP) for belted occupants. Therefore, low injury rates are expected in real-world collisions up to 56-kph Barrier Equivalent Velocity (BEV). For compatible vehicles, in the vehicle-to-vehicle test conditions, the occupants in both the bullet and the target will have injury responses no greater than those in a 56-kph BEV.

In the short term, geometrical alignment of the front-end energy-absorbing structures between cars and LTVs/SUVs is a first step toward improving compatibility in frontal collisions. Providing head protection to the occupants of passenger cars by introducing side inflatable or comparable effective technologies in passenger cars is a first step toward improving compatibility in side impacts. The ultimate measurements for compatibility in vehicle-to-vehicle collisions are occupant responses or injury indices as they are used in evaluating vehicle performance in single-vehicle accidents in the current regulations through FMVSS208, FMVSS214, NCAP, and SINCAP. The Alliance of Automotive Manufacturers has already announced Phase 1 performance criteria in the short term toward improving vehicle compatibility in front and side collisions.

Measuring and controlling force levels (stiffness) in the crash interaction zones between cars and trucks during a load cell barrier test while measuring occupant responses to ensure no degradation in self-protection may serve as a reasonable metric for compatibility.

4. Which collision configurations should be considered within the compatibility discussion?

Are there certain test configurations that would yield significant synergistic benefits, versus others that would be minimally beneficial or even detrimental? Could mathematical modeling be successfully used for compatibility development and optimization, considering the diversity of vehicle sizes, designs, and collision environments? What would be the maximum benefit in terms of lives saved and injuries reduced of a vehicle fleet that had an ideal compatibility solution? What would be the expected benefit in terms of lives saved and injuries reduced of a fleet that has been optimized for compatibility within the framework of real-world solutions?

Respondent #1

The answer to this question is time and region dependent. It needs to be established for each region of the world, based on its fleet composition and its traffic patterns. For the U.S. fleet, the data analysis shows that side impact of LTVs into cars is the most frequent mode of these crashes and also accounts for about 57% of the total harm associated with these crashes. Obviously, addressing this configuration is likely to produce larger societal benefits. Frontal crashes between LTVs and cars account for 33% of the total harm. We have looked at further dividing into subcategories. For example, of the 33% harm that is associated with frontal collisions between LTVs and cars, more than 12% can be attributed to full frontal crashes.

Respondent #2

We should consider all issues of compatibility, but we should focus on those where we think we understand what to do. Ultimately, it is difficult to assess how many lives would be saved by improved compatibility because there would be improvements in self-protection effectiveness as well. However, when considering the special issues associated with the different geometry of SUVs and cars, side impacts appear to be the area with the most potential for improvement.

Respondent #3

For good compatibility, all configurations where cars impact one another would require addressing. However, just as for self-protection, the situations that are addressed need to be based on accident need and our understanding and ability to bring about effective change. Currently, the principles of what is required are now understood for frontal impact, and there is great justification for action. Side impacts are also a priority, particularly in markets where SUVs and LTVs are prevalent. Unfortunately, there is currently limited knowledge about what may be needed to improve side impact compatibility. Those actions that are aimed at improving frontal impact compatibility should help. Otherwise, the major opportunity for action is within the target vehicle. This is not to say that improvements could not be gained by removing obviously dangerous features such as stiff frontal structures impacting high on the car's side, although at this stage we are not in a position to address other types of impacts, such as those with roadside obstacles, two-wheelers, and pedestrians. It is important to consider these requirements and ensure that, at the least, the situation is not made worse.

Respondent #4

The accident in-depth investigations are showing the following relevant issues for compatibility. The most frequent impact configurations in real accidents of all car classes including SUV and LTVs are with 64% the frontal impact, mainly front to side (24 %), and head on (21 %) on 1/3 overlaps (8%). Secondly, the side impact is important. Here the oblique impact into the compartment is with 8% the most frequent one. These configurations seem to be happening for all car classes, from small cars to SUVs and LTVs, with the same frequency.

Respondent #5

I think that the combination of tests by full frontal, offset frontal, and overload tests can control the compatibility performance effectively. MDB tests will be useful to predict the compatibility performances that reproduce car-to-car crashes in real-world accidents. However, this test strongly depends on MDB mass.

There have already been full frontal tests and ODB tests in regulations and NCAP. We can get the useful information based on these tests if the barrier force distributions and vehicle accelerations are also measured in the tests. Thus, to improve compatibility in the short term, it is useful to evaluate the compatibility based on the current test procedures such as full and offset frontal tests. For the long term, the MDB test and the PDB test will be useful.

If you optimize compatibility by vehicle fleet, you have to ride the same car. Mathematical simulation can clarify the countermeasures for each vehicle class in a fleet.

I think compatibility can be improved without changing the vehicle fleet. Vehicle mass distributions will also be difficult to change. The main target should be the structure modification. Test procedures should assess the structure, not the vehicle mass.

In Japan, minicars are increasing in numbers. Considering the environment and efficiency, it is beneficial to society that the number of minicars is increasing. I think this trend should not be changed for the purpose of compatibility optimization.

Respondent #6

The incompatibility between the front and side of vehicles is the greatest challenge. This crash configuration has the most unfavorable ratio of people killed in the struck vehicle to people killed in the striking vehicle. This is true when passenger cars strike passenger cars, and it is worse when the striking vehicle is a light truck. The issue lends itself to the study of vehicle models. However, such studies should be done in conjunction with studies of accident data and crash test data. Such studies will be required in order to assess alternative countermeasures and associated benefits.

The front-to-front incompatibility between light trucks and passenger cars represents the second greatest challenge in the United States. This compatibility issue may be more complex than the front-to-side compatibility issue. Incompatibility exists not only in head-on crashes, but also in offset and angled offset frontal crashes. All these crash modes need to be studied. The consequence of frontal-offset barrier tests on vehicle frontal compatibility needs continued review.

Respondent #7

In the short term, various safety regulators and industry engineers have shown that vehicle structure, mass, speed, and height can affect the balance in a crash event. Of these, manufacturers have shown that they can control a vehicle's height and structure. Mass can be controlled to a lesser extent and is often limited by consumer preference. Vehicle speed can be selected after careful study of recent crash events. Therefore, the test to ensure vehicle compatibility today will help to bound a vehicle's height and stiffness at various energy levels. If the stiffness can be balanced with homogeneity, then the height will play a lesser role. Until global fleets are more similar, this height and stiffness will need to be scaled to what their global fleets can meet.

Respondent #8

While there may be test configurations that might be minimally beneficial or even detrimental, these would be eliminated in the process of finding the best configurations to implement. Mathematical modeling and simulation could be successfully used for this process, as noted in the discussion of the Fleet Systems Model in Question 3.

Maximum benefit in terms of lives saved and injuries reduced of a vehicle fleet that had an ideal compatibility solution, considering more than 40,000 annual fatalities, it is reasonable to assume that more than 1,000 lives or approximately 2% could be saved. The fleet systems model will soon be able to make this kind of prediction by simulating a historical year and adjusting the results of incompatible crashes to be more compatible.

Respondent #9

Front and side impact test configurations should be considered within the compatibility discussion. Full frontal and 50% offset collinear frontal impact crash modes are to be considered to cover most of the real-world accidents. A perpendicular IIHS-type barrier side impact is also to be considered for compatibility.

Mathematical modeling can successfully be used for compatibility development and optimization, considering the diversity of vehicle and occupant sizes, designs, and collision environments.

5. Worldwide harmonization issues.

Europe is focusing on frontal crash conditions to address compatibility among cars of different sizes. In the United States, the issue is arguably a major challenge related to car/truck/SUV collisions in which the car is struck by the collision partner, both in front and side impacts. In Japan, the SUV population is relatively small, while the MPV and minicar numbers are large. Is it possible to devise solution sets that are common and at least not mutually exclusive? Are there common issues of compatibility across the world regions?

Respondent #1

There are many aspects that can be and should be harmonized worldwide. One of them is the definition of dummies used in crash tests (such as side impacts). This is crucial and basic to achieving any further harmonization in crash test procedures.

We have been studying a unified test procedure that will reduce the number of barriers used for crashworthiness testing and will address issues of self-protection and partner protection. It also will look at the harmonization in a more fundamental way—harmonizing the test protocol while permitting differences to account for regional variations in fleet composition.

The fundamental physics of the vehicle-to-vehicle crash is the same, but the specifics are different for different parts of the world. Even for the same part of the world, the situation changes from time to time as the public likes and dislikes change affecting the vehicles that people buy. It is therefore important that all safety criteria—regulated and non-regulated—be evaluated and adjusted at regular intervals.

Respondent #2

The issues are more common than is generally recognized, and the vehicle fleets are becoming more similar as SUV sales pick up in Europe.

Respondent #3

Although the composition of fleets varies across the world, with a higher proportion of SUVs and LTVs in North America, the principles of protection are the same, and the vulnerability of the road users is similar. With manufacturers wishing to sell similar types of cars across the different markets and with the inevitable cascading down of technology from the developed world to the developing one, there is a need for harmonization of approach. The only justifiable reason for adopting different approaches is to address different priorities. Where knowledge is good, such as with frontal impact, different approaches should be geared to the common aim, even if the initial steps are different. Where little is known, such as with side impact, there may be a political imperative to take an initial step. Such steps should be taken with care, addressing only those aspects about which there is some confidence, and with the acceptance that changes may be needed when greater knowledge has been gained.

Respondent #4

The importance of collision configurations could be different for the worldwide continents. For finding a worldwide acceptable harmonized solution for compatibility tests, a linkage of international research activities should be followed.

Respondent #5

In Japan, there are wide variations of vehicle fleet from SUVs to minicars. The number of compact sedans is decreasing every year, but the number of MPVs is increasing. Therefore, it will be difficult to apply the U.S. situation to Japan. In the United States, the aggressivity of SUVs will be important, whereas in Japan, the self-protection of small cars should also be a consideration. We also must consider that the vehicle fleet is changing year by year.

Although the vehicle fleet differs among countries, crash test procedures can be harmonized. For example, the United States and Japan have the full frontal impact tests, and barrier force distributions are important information on vehicle compatibility. However, the vehicle application and requirements may be different among regions. In Japan, we do not have as many SUVs, and that number does not increase. The AHOF is important for the SUV in the United States, but in Japan, we have to examine the requirements and their effectiveness. IHRA is taking an important role to harmonize the crash test procedures and their measurement.

Respondent #6

The priorities in the United States differ from those in Europe:

- *Europe has provided leadership in car-to-car frontal compatibility.*
- *Manufacturers in Japan have passenger vans that appear to have compatibility considerations built in, based on barrier test results.*
- *The United States needs to learn from these advancements but should not lose sight of its different priorities.*

Respondent #7

The solutions reached in various parts of the world will have much in common, and continued communication is essential to gain the benefits of the work in other countries. However, the United States is unique, at this time in the massive influx of SUVs, pickup trucks, and vans. The U.S. solution must address this difference in determining the best solution in terms of crash tests and parameter measurements.

There is general agreement that mass is not the only issue. Stiffness and geometry are also important parameters in improving compatibility.

Respondent #8

Compatibility test procedures and evaluating metric or performance criteria can be harmonized. The current Alliance compatibility working groups have members who are involved in developing compatibility test procedures and metrics for Europe, Japan, etc. The Phase II ongoing research at the

Alliance to develop test procedures and performance metrics for evaluating vehicle compatibility does entertain, to certain extent, the European requirements. Most of the test procedures and metrics under development are, to some extent, considering European and Japanese requirements. WG members who are active members in the IHRA have currently introduced some proposals.

6. Could alternate vehicle-based countermeasures obviate the search for compatibility solutions?

Respondent #1

I am unsure what this question refers to. It is possible to improve restraint systems further. However, these changes may be incremental. In some cases, these already may have reached their limit because of the possibility of causing unwanted injuries to unrestrained occupants.

In the future, collision avoidance technology may advance to vehicle-to-vehicle communications, thus making it possible to avoid these crashes. This will require large amounts of data transfer and onboard communications and is something that is likely in the distant future. Other similar concepts such as intelligent roadways or infrastructure changes that reduce the likelihood of crashes will also help. However, in the absence of such advances, we are all governed by the principles of physics. When two vehicles are involved in a crash, their kinetic energy has to be translated into structural deformation and motions of the vehicles and their contents. The key to improve occupant safety is first, to dissipate the maximum possible portion of this energy before it reaches the occupant, and second, to provide restraint systems that minimize injuries by putting more loads into anatomical regions capable of sustaining such loads without injury.

Respondent #2

Sorry, I have no idea what this question means. I hope it is clarified for the panel discussion.

Respondent #3

The argument that one should take primary safety action to prevent accidents from happening, rather than deal with the crash when it occurs, is very compelling. It has obvious appeal to the naive politicians, policy makers, and members of the public, who believe that all accidents are preventable, and manufacturers can see marketing advantages in adding primary safety systems to their cars. Unfortunately, there is virtually no evidence to show that significant numbers of accidents can be prevented by the primary safety systems that are currently being considered. It is clear that research should continue with such systems and their performance on the road be evaluated. However, until their benefits have been proved, it would be unwise to delay actions to improve passive safety. A field where there is very good evidence of the safety benefits.

Respondent #4

There are other types of countermeasures available via implementation of occupant protection devices (i.e., special air bags positioned in relevant places), further equipment of protection elements in the car interior (e.g., footwall pads and intelligent restraint systems), and expanded side protection (e.g., stiff or soft frames for energy absorption, adoption of height for force transmission).

Respondent #5

In accidents, many drivers step on the brake pedal but cannot use the brake performance until its maximum limitation. By the help of the antilock braking system and the brake assist system, drivers may avoid crashes and reduce the impact velocity. They will reduce the number of crashes as well as the severity of crashes. In the current situation, this will be the realistic way to avoid crashes. However, I think much room still remains for the structures to be improved for compatibility.

Respondent #6

Carl Clark and IIHS worked on a bumper air bag. I will leave that to Adrian Lund for comment.

There is no substitute for self-protection. However, overly aggressive vehicles should not be allowed to defeat self-protection systems.

Respondent #7

Another possible help to vehicle compatibility is driver education. Avoiding the crash in the first place is the best possible outcome. Driver education does not have to stop after someone receives his or her first license. Vehicles, roadways, and driving behavior change over the years. Consumers can benefit from understanding these changes. A better understanding of their vehicle's performance capabilities can help. A 1950s sedan does not perform the way a recently purchased modern SUV or sedan will perform, and the consumer may not even understand the differences. These differences can be voiced to the consumer through government consumer metrics and manufacturer marketing, but will the consumer listen?

However, it is not always practical to teach and educate every consumer to understand his or her vehicle and physics to the same level. Crash avoidance has become a hot topic to address this. Again, this is a harmony between various consumers' driving abilities, so that they are forced to drive similarly through computer-enhanced vehicle dynamics. It has yet to be determined if this technique of computer-enhanced vehicles from a simple traction control device to a complete computer-driven vehicle will answer this education need by the consumer. It poses even more concerns for the safety regulator, but as the roads get more crowded with even more complicated vehicles, it may be the answer to avoiding crashes before they can even start. Unfortunately, crash avoidance cannot be 100% effective, and until that day, crashworthiness and the tests needed to ensure crash compatibility will be needed.

Respondent #8

With more than 40,000 fatalities per year, any and all means should be used to improve safety. Of course, structural design controls and occupant protection devices will be valuable. However, the benefit gained should not preclude every other known technique to improve compatibility.

Respondent #9

Development of pre-crash sensing (radar/video based) to enable active geometry changes of vehicle structure will significantly help in improving vehicle compatibility between cars and trucks. Adaptive front end in terms of geometry and stiffness can be an alternative countermeasure. However, the reliability of the sensing system and the feasibility of its availability on all vehicles is not there yet.

7. Could insurance through premiums and publicity, more informative rating systems, and/or consumer-based pressures drive manufacturers toward the development of compatible vehicles?

In lieu of achieving physical compatibility, is there a place and role for insurance, various vehicle rating systems, and/or consumer-based initiatives to assure the development of vehicles with either absolute or scaled lower levels of aggressivity and significantly improved compatibility? Is there a place for industry, NGO, and governments to work together to assure that every consumer metric promotes and improves vehicle compatibility and human safety in the traffic environment?

Respondent #1

The use of insurance premiums is an intriguing concept and worth more thought before reaching any conclusions.

Automotive manufacturers need to find solutions that improve safety as well as improve consumers' perception of safety. One of the reasons for the popularity of larger vehicles is that consumers consider them to be safer and, given other factors of affordability and societal trends, will prefer them. All OEMs understand the importance of safety and have continuously attempted to improve it. If you look at data today and plot the trend over the past several years, the fatality rate per million registered vehicle years has declined for passenger cars as well as light trucks and LTVs.

We do not believe additional regulations are required at present. What is required is a willingness to look at the existing regulations and consumer metrics and adjust them to maximize safety. The automotive industry has taken the initiative to search for other solutions that will improve safety in this arena, and that is more likely to provide the desired outcome sooner. The cost to the industry may be substantial, but we are determined to move forward and seek the solutions.

Respondent #2

Vehicle ratings have a major disadvantage in moving people toward less aggressive vehicles. People tend to view these ratings to get the best protection they can for their families. Vehicle aggressivity is a secondary issue in the process of buying new vehicles. One important step that can be made is to point out to buyers that the very heaviest vehicles do not provide improved real-world results commensurate with their greater operational costs. Consumer ratings are likely to be helpful in building a more compatible fleet only to the extent that they encourage people to buy vehicles with higher levels of self-protection, including if they help to eliminate the very smallest and lightest vehicles from the fleet.

Respondent #3

Despite many years of concern about compatibility, only recently has frontal impact compatibility been understood. Side impact compatibility is still not understood. Although the principles behind frontal impact compatibility are now well accepted, much is still unknown about what performance levels should be set for frontal force, passenger compartment strength, and frontal homogeneity. Even if we knew the optimum values at which to set these requirements, it is unknown to what extent cars could be developed to meet them. Consequently, the setting of fixed standards in legislation would be difficult in the near future. For consumer information programs, this is much less of a problem. They could be developed to encourage the characteristics of cars to move in the direction of improved compatibility without setting fixed targets. With regard to frontal stiffness, cars with very weak or very stiff fronts could be rated less well than those closer to the current median force. As knowledge grew, the target force could be set more carefully and the corridor of acceptability tightened. With regard to structural interaction, a consumer information program could simply reward cars more as their stiffness homogeneity improved. Again, no fixed target would be required.

Respondent #4

The word "compatibility" must go into the mind of the consumer as it happened with the term "safety" that meanwhile is related to the NCAP test procedures. NCAP is a good example of a ranking system for the probability of injury outcome and gives the manufacturers the chance of public relations. Achieving satisfactory compatibility standards for vehicles, self-protection and partner protection have to be considered. Based on these contradictory demands, the realization of effective measures will be difficult by achieving a self-regulated agreement of the manufacturers. I guess the problem of compatibility could be achieved with initiatives of insurance but needs support by governmental regulations for defined accident configurations and vehicle classes, too. The existing NCAP tests could be the starting level for achieving the process for compatibility standards, for they have to be expanded with further test rules and combined with a newly developed accident controlling system on the basis of in-depth investigations.

Respondent #5

Dr. Tay from Queensland University of Technology interviewed college students. He found that in automobile accidents, students were more concerned about the likelihood of injuring or killing someone else than they were of injuring or killing themselves. If this attitude is an indication of their general preferences, providing information to consumers on vehicle aggressivity in real-world accidents will be useful. By this kind of information, the number of vehicles that are so aggressive can be reduced from voluntary activities by consumers.

If the insurance industry were to change the premiums based on not only self-protection but also on partner protection, the number of aggressive vehicles may decrease. There may be a lot of ways that can improve compatibility.

Respondent #6

There may be a role for consumer information to address vehicles that are outliers—those that have excessive kill ratios. However, a moderate kill ratio may be a selling feature. Many buyers of SUVs cite self-defense as a powerful motivation in their purchase decision. Compatibility, like environmental control, requires all to participate if it is to be useful. Standards are the best way to achieve uniformity for public benefits that are not market driven. Improved side impact protection to improve self-protection and thereby improve compatibility could be a market-driven feature.

Respondent #7

This seems to be more of a political question than a technical one. The general answer should be no, but any information in terms of rating systems could be beneficial. The auto industry appears to have turned the corner on attempting to improve compatibility in significant ways. More government interference or direction could be counterproductive.

Is there a place for industry, NGO, and governments to work together to assure that every consumer metric promotes and improves vehicle compatibility and human safety in the traffic environment? Absolutely, yes.

Respondent #8

Euro-NCAP among others is investigating methods to assess and reward compatible designs that promote vehicle compatibility. Reducing insurance premiums on trucks, which have compatible designs, may attract more buyers, and this can help the industry to implement new technologies.

Table of Contents

Chapter One – Crash Compatibility Definitions

Chapter Two – Accident Data

Chapter Three – Self Protection

Chapter Four – Frontal Impact

Chapter Five – Side Impact

Chapter Six – Factors Influencing Compatibility

Measuring Compatibility

Stiffness Effects and Geometry Effects

CHAPTER ONE – CRASH COMPATIBILITY DEFINITIONS

2004-01-1167

Feasible Steps towards Improved Crash Compatibility

Dietmar Haenchen, Thomas Schwarz, Gareth Thomas and Robert Zobel
Volkswagen AG

ABSTRACT

Compatibility has been a research issue for many years now. It has gained more importance recently due to significant improvements in primary and secondary safety. Using a rigorous approach, combining accident research and theoretical scientific considerations, measures to improve vehicle-vehicle compatibility, with an emphasis on feasibility, were discussed.

German accident research statistics showed that frontal impacts are of higher statistical significance than side impacts. Based on this and the high potential for improvement due high available deformation energy, the frontal impact configuration was identified as the most appropriate collision mode for addressing the compatibility issue. In side impacts, accident avoidance was identified as the most feasible and sensible measure.

For frontal vehicle-vehicle impacts, both trucks and passenger cars were identified as opponents of high statistical significance. Based on the bulkhead concept, a theoretical range of survivable collision velocities for both impact configurations was calculated and the limitations discussed. A pre-requisite in each case was structural interaction. Structural interaction was identified as necessary between all collision partners including car - truck. This means the addition of adequate under-run protection devices.

To evaluate compatibility, several test procedures have been proposed. The limitations of each were discussed, with a particular focus on the ability of each test to assess structural interaction potential. Finally, feasible steps towards achieving compatibility were discussed and illustrated in a "Roadmap for Compatibility".

INTRODUCTION

To improve the passive safety of the vehicle fleet we must consider two viewpoints:

Self-protection: the ability of a vehicle to protect its own occupants both in vehicle-to-vehicle collisions and collisions with other objects in the traffic environment.

Partner-protection: the ability of a vehicle to protect the occupants of the opponent vehicle in vehicle-to-vehicle crashes.

Compatibility aims at finding an optimum for self-protection <u>and</u> partner-protection. It is generally accepted that this should take place without compromising self- protection. Partner-protection is often referred to as low aggressivity towards other traffic participants.

The primary goal remains to prevent accidents through active safety measures. Significant improvements have already been achieved over the past few years. Electronic Stabilization Program (ESP), for example, has a significant influence, particularly in the reduction of single vehicle accidents. It will be much more difficult to prevent vehicle-to-vehicle collisions with active safety measures.

The **Compatibility** of a vehicle is understood as a combination of self- and partner protection in such a way that optimum overall safety is achieved. This means: Compatibility seeks to minimize the number of fatalities and injuries, regardless of the vehicle in which the injuries or fatalities occur. Additionally, customers expect further improvements in the level of self-protection. It will not be acceptable to reduce today's high levels of self-protection.

To investigate compatibility in an individual case, it is possible to impact two vehicles and evaluate the injury values and deformations in both vehicles. In the real accident world, infinite collision configurations are possible. It is impossible to test all these combinations. This means that it is not feasible to evaluate compatibility through vehicle-to-vehicle crash testing. Rather it is necessary to define a meaningful and repeatable vehicle-to-barrier test.

ACCIDENT RESEARCH

Statistical analysis of accident data is essential in predicting the real world benefit of proposed safety measures. A hierarchical analysis, beginning with few restrictions on the sample set, is sensible. Utilizing this methodology, the relevance of certain collision configurations, based on the injury risk and the frequency of the crash configuration, can be determined. However, statistical results should not be considered in isolation, as they neither take into account the potential for improvement nor do they consider the feasibility of implementing passive safety measures for each crash configuration. A subjective assessment of the achievable benefits is therefore necessary as well.

The following hierarchical analysis was carried out with respect to front and side compatibility, based on German accident data.

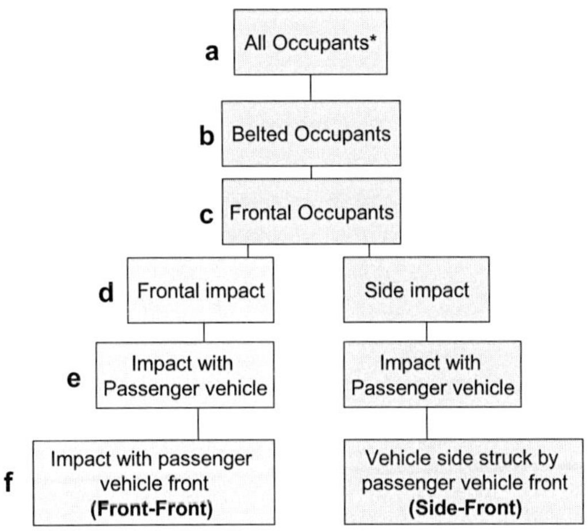

Figure 1: Overview – Statistical analysis

As a first step, assessing the influence of imposed data-set restrictions on accident numbers is appropriate. Through this methodology, the statistical relevance of each accident configuration can be determined. The data-set was initially restricted to consider only frontal belted occupants. These three initial data set restrictions (frontal belted occupants-a,b,c) were common for both the front-front and front-side compatibility analyses, Figure 2 and Figure 10.

FRONTAL-FRONTAL VEHICLE-VEHICLE ACCIDENTS

Frontal impacts are of high statistical relevance. For frontal belted occupants, they constitute nearly 50% of both the total number of collisions and accidents in which a MAIS 3+ injury occurs, Figure 2 (Columns c and d).

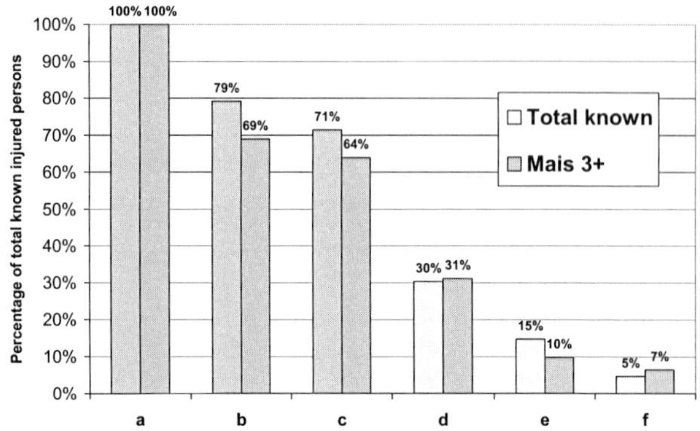

Figure 2: Statistical relevance of the vehicle-vehicle front-front impact configuration

In the frontal impact configuration, opponent passenger vehicles cause significant harm. One half of the total number of injured occupants and one third of occupants receiving a MAIS 3+ injury occur in collisions with another vehicle, Figure 2 (Columns d-e). The remaining two-thirds of serious injuries occur during collisions with other opponents, as shown below.

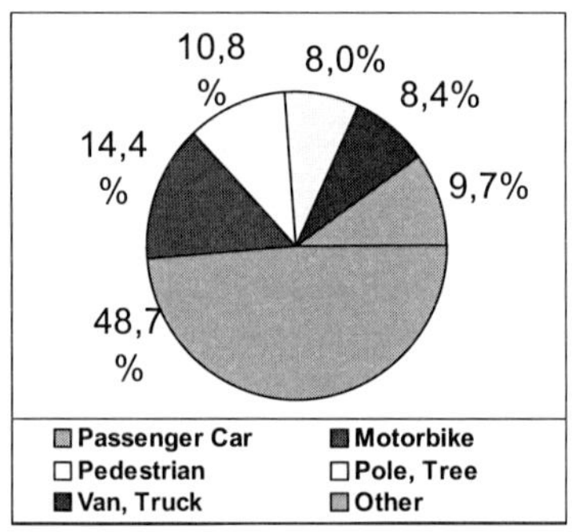

Figure 3: Distribution of opponent type for all injured belted front occupants in frontal collisions[1]

---

* Data extracted from the VW GIDAS (German-In-Depth-Accident-Study) database. Accidents including 1991 and thereafter-irrespective of vehicle build year and make-involving at least one (MAIS 1+) injured person.

[1] Throughout this paper, Van/Truck refers to LTV's (light trucks and vans) as well as heavy trucks.

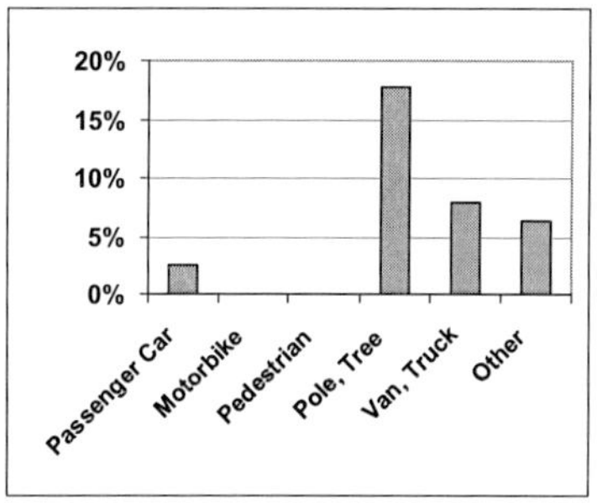

Figure 4: Risk of a MAIS 3+ injury for all injured belted front occupants in frontal collisions based on opponent type

Figure 3 and Figure 4 show the high statistical significance of frontal impacts with poles/trees. Although the number of injured occupants in pole/tree collisions is relatively low (8,0%, Figure 3) the injury risk is very high (About 17%, Figure 4). Impacts with Trucks/Vans are equally common (8,4%, Figure 3) and also pose a high injury risk (around 7%, Figure 4). For all ijured occupants, frontal impacts with passenger cars are very common (49%, Figure 3) but pose a comparatively low injury risk (around 2%, Figure 4).

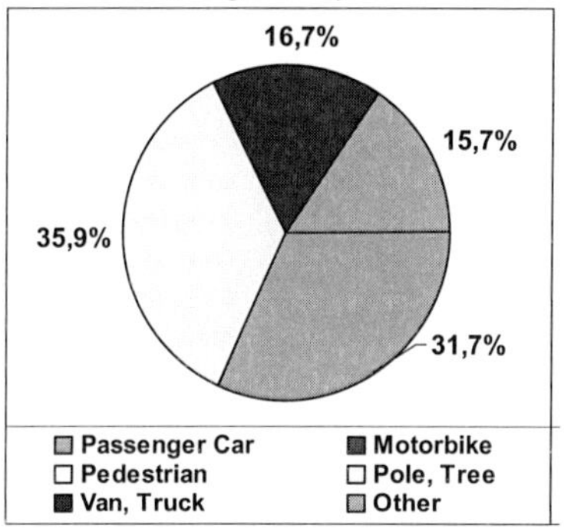

Figure 5: Distribution of opponent type for belted front occupants in frontal collisions with a MAIS 3+ injury

In general, the vehicle-vehicle and vehicle-pole/tree configurations have a similar statistical significance for MAIS 3+ injured occupants, (32%, 36%, Figure 5). Frontal collisions with Vans/Trucks are also significant, (16,7%, Figure 6).

The opponent front was struck in one third of cases for injured occupants in frontal impacts involving another passenger vehicle yet the most of the serious injuries (MAIS 3+) occur when struck by the opponent front (Figure 2).

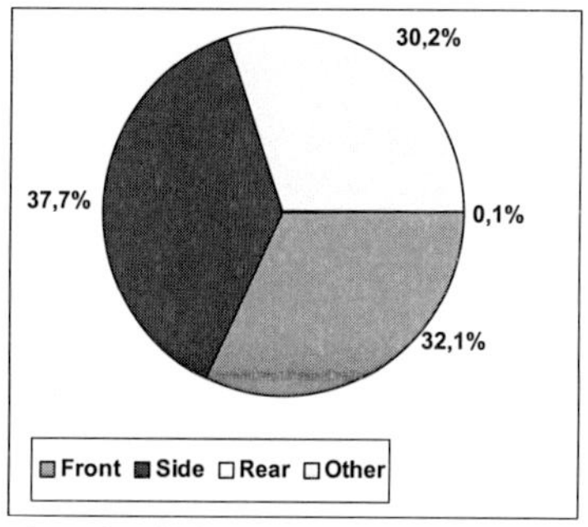

Figure 6: Distribution of collision area of opponent vehicle for belted front occupants in frontal collisions

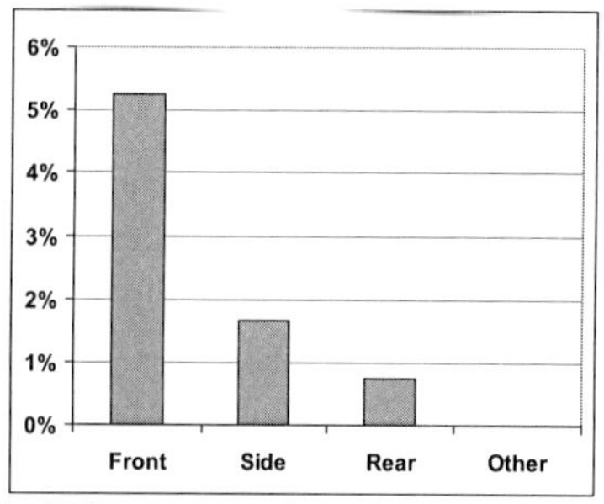

Figure 7: Risk of a MAIS 3+ injury for all injured belted frontal occupants in frontal collisions based on collision area of opponent vehicle

To further investigate the frontal-frontal vehicle-vehicle configuration, the influence of significant intrusion was isolated. "Intrusion" and "No Intrusion" represent significant intrusion (where intrusion threatens the survival space of the occupant) and insignificant intrusion (where intrusion itself is not considered to be the main cause of occupant injury)[2].

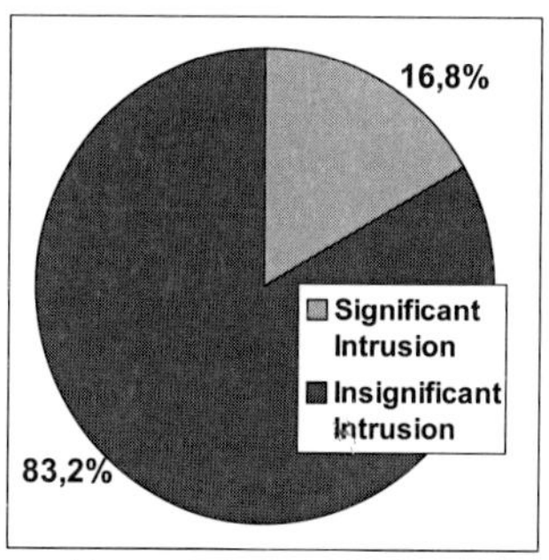

Figure 8: Distribution of injured, front, belted occupants with respect to significant intrusion for all injured occupants in the frontal-frontal vehicle-vehicle configuration

[2] Significant Intrusion defined as VDI6≥4. This was assumed to represent all cases in which threatening passenger compartment intrusion occurred.

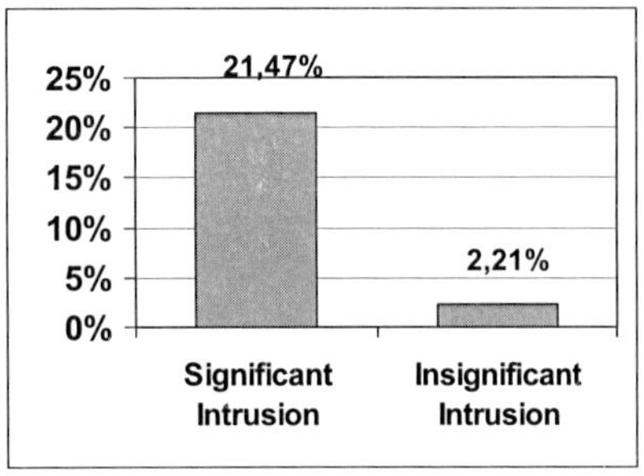

Figure 9: Risk of a MAIS 3+ injury risk for injured occupants in the frontal-frontal vehicle-vehicle configuration based on intrusion

Figure 8 shows that intrusion occurs in relatively few frontal-frontal vehicle-vehicle accidents (16,8%). However, when intrusion does occur (Figure 9) the risk of serious injury is very high (21,5%). Survival space is a clear requirement in frontal-frontal vehicle-vehicle collisions. Additional information on this topic can be found in [1].

FRONT-SIDE VEHICLE-VEHICLE ACCIDENTS

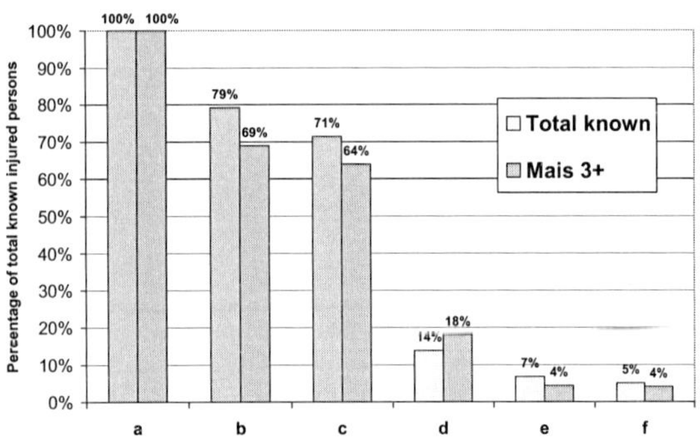

Figure 10 : Statistical relevance of the vehicle-vehicle front-side impact configuration

Figure 10 shows that side impacts are of a lower statistical significance than frontal impacts. The total number injured occupants and the risk of a MAIS 3+ injury for these occupants is significantly lower in the side configuration (14%, 18%, columns c and d) than in the front configuration (30%, 31%, Figure 2, columns c and d).

Side impacts with other vehicles are of lower statistical significance relative to the frontal configuration, Figure 10. Although half of the injured belted front occupants are involved in impacts with another vehicle, only one quarter of serious injuries occur when struck by another vehicle, Figure 10 columns d and e. Other collision partners dominate serious collisions in the side impact configuration as shown below.

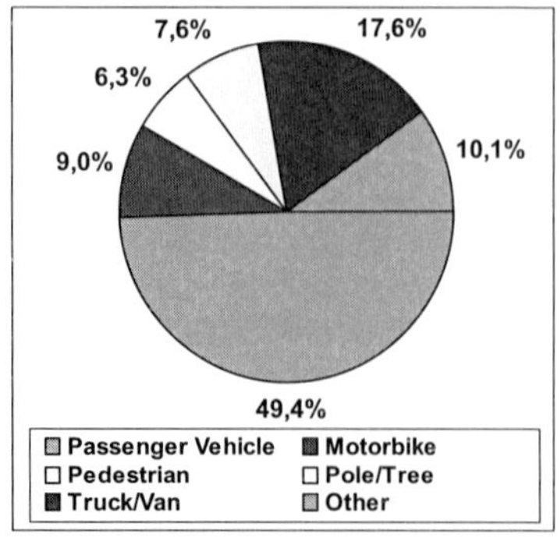

Figure 11: Distribution of opponent types for all injured belted front occupants in side collisions

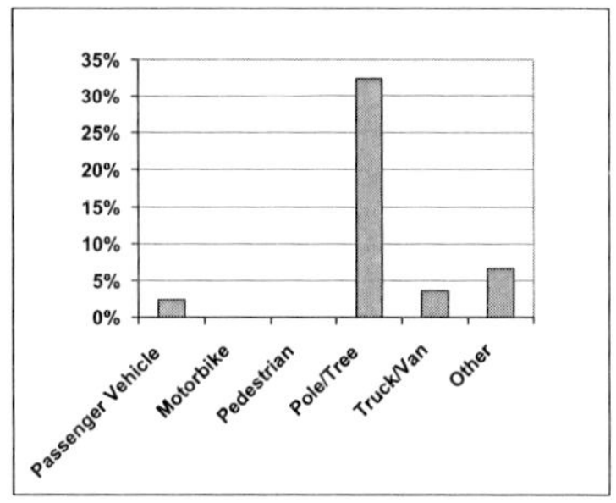

Figure 12: Risk of a MAIS 3+ injury for injured frontal belted occupants in side collisions based on opponent type

Figure 11 and Figure 12 show the dominance of pole/tree impacts for seriously injured occupants (MAIS 3+) in the side impact configuration. The number of injured occupants involved in side impacts with poles/trees is relatively low (8%, Figure 11) but the risk of serious injury is disproportionately high (over 30%, Figure 12). Poles/trees are an extremely high-risk accident opponent in the side impact configuration. This is also shown below in Figure 13. Poles/Trees are a dominant collision opponent for MAIS 3+ injured occupants (49%) with passenger vehicles also significant crash opponents (25%), Figure 13 . Trucks are also a relatively dangerous collision partner, with respect to serious injury, when an injured, frontal, belted passenger is involved in a side impact (13%, Figure 13).

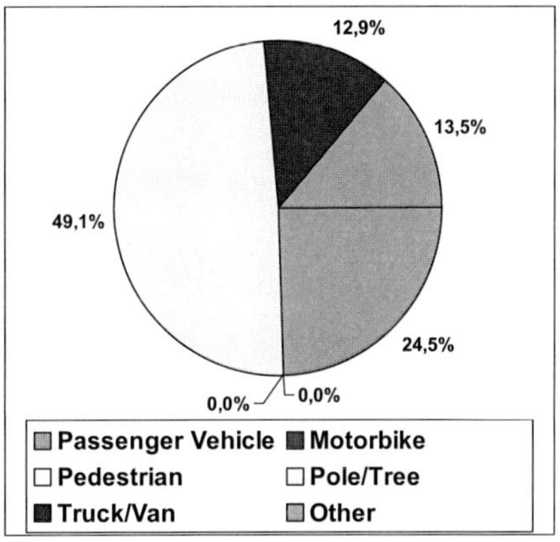

Figure 13: Distribution of opponent type for belted front occupants in side collisions with an MAIS 3+ injury

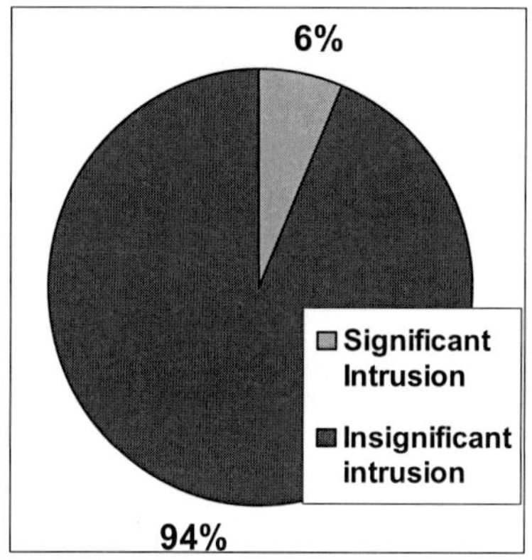

Figure 15: Frequency of intrusion for all injured occupants involved in side impacts and struck by the opponent vehicle front

The opponent front end is the most aggressive part the striking vehicle in the front-side impact configuration. Almost all of the MAIS 3+ injured occupants involved in side impacts with another vehicle occurred were struck by the opponent's front-end (Figure 14). As with the front-front vehicle configuration, the aggressivity of the opponent's frontal structural is high, though by no means unexpected.

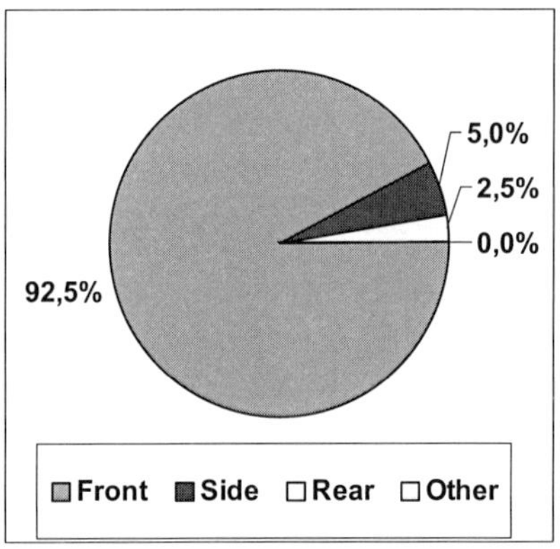

Figure 14: Distribution of the struck region of the opponent car for frontal, belted occupants sustaining a MAIS 3+ injury

To further investigate the front-side vehicle-vehicle configuration, the influence of significant intrusion was isolated. Again significant intrusion refers to the case where intrusion could be the primary mode of injury causation[3].

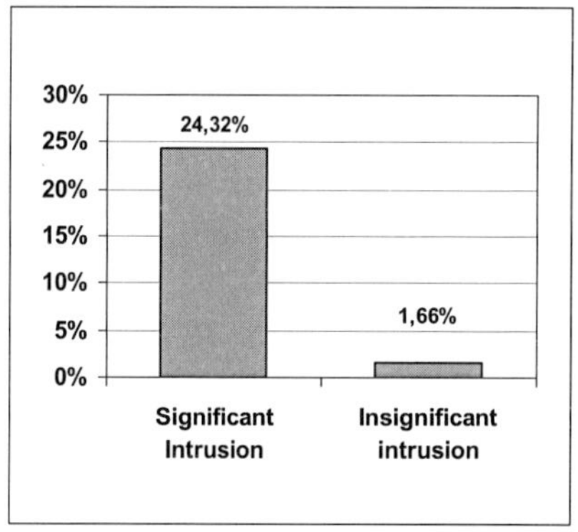

Figure 16: Risk of a MAIS 3+ injury for all injured, front, belted occupants involved in side impacts and struck by the opponent vehicle front

Figure 15 and Figure 16 show the high risk of sustaining serious injury due to intrusion in the front-side, vehicle-vehicle impact configuration. The number of cases in which intrusion occurs is low (6%), Figure 15, though the risk of injury when intrusion does occur, Figure 16, is alarmingly high. Intrusion in the front-side configuration can mean direct contact between the occupant and the intruding parts of the vehicle interior. Survival space is also a basic necessity in side collisions.

GENERAL STATEMENTS BASED ON ACCIDENT RESEARCH

Due to the available deformation travel, the frontal impact configuration offers the highest potential for injury prevention through passive safety measures. Front impacts are also of high statistical relevance with respect to both the total number of injured occupants and the number of seriously injured occupants (MAIS 3+). To maximise injury benefit in frontal impacts, increased compatibility is required in the vehicle-vehicle

[3] Intrusion defined as VDI6≥2. This represents significant intrusion into the passenger compartment.

and vehicle-truck configuration. To prevent under-riding of trucks, better structural interaction is needed. Truck under-ride prevention is essential. Increased structural interaction is also needed between two vehicles in the frontal-frontal, vehicle-vehicle configuration. Increased structural interaction would cause higher energy dissipation in the front-end, consequently lowering compartment loadings. A region for structural interaction that maximises structural interaction in both the vehicle-vehicle and vehicle truck configurations therefore needs to be defined. Furthermore, increased horizontal homogeneity may also be of benefit in single vehicle collisions against trees and poles, to increase energy dissipation in the front-end through better structural interaction, where the risk of serious injury is high. This is a topic for further research.

The potential for injury mitigation in side impacts is lower, due to the lack of deformation space. Furthermore, the statistical significance is also lower than that for frontal impacts. Preventing intrusion, although the risk to the occupant when intrusion occurs is very high, is difficult due to the lack of available space (deformation travel). This could be partly achieved through a combination of stiffer side structures and force homogeneity in the striking vehicle's front-end structure. However, preventing intrusion is only effective up to a certain accident severity level, above which excessive deceleration or contact with internal objects will be fatal even with perfectly stiff side structures. The most appropriate strategy for all side impacts is accident avoidance through active safety measures (e.g. ESP).

FUNDAMENTALS OF COMPATIBILITY IN CAR-TO-CAR COLLISIONS

Crash tests against fixed barriers force the manufacturer to provide sufficient energy dissipation in the vehicle front-end. If the barrier impact has to be conducted with an impact speed vB for a vehicle with mass m, then the required deformation energy is approximately equivalent to the kinetic energy of the barrier impact:

$$E = \frac{1}{2} * m * v_B^2$$

In a vehicle-vehicle front-front collision, two vehicles, vehicle 1 and vehicle 2 with masses m_1 and m_2 and velocities v_1 and v_2 can be considered. The kinetic energy before the crash is:

$$E_{before} = \frac{1}{2} * m_1 * v_1^2 + \frac{1}{2} * m_2 * v_2^2$$

The velocity after the crash is assumed to be the same for both vehicles (V_{after}). The rebound is neglected. So v_{after} is computed from the theory of conservation of momentum by

$$(m_1 + m_2) * v_{after} = m_1 * v_1 + m_2 * v_2$$

This results in a kinetic energy after the crash:

$$E_{after} = \frac{1}{2} * (m_1 + m_2) * v_{after}^2$$

The lost energy is equal to the deformation energy D of this collision

$$D = E_{before} - E_{after}$$

$$= \frac{1}{2} * \frac{m_1 * m_2}{m_1 + m_2} * (v_1 - v_2)^2$$

Thus, this well-known computation shows that the deformation energy depends on the two masses and the closing velocity $v_c = v_1 - v_2$ of the two colliding vehicles.

If there is a collision between two vehicles of different masses and a closing velocity of $v_c \leq 2 * v_B$ and if each vehicle is designed for a barrier impact with v_B, then sufficient deformation energy is present to withstand the collision without significant intrusion into one of the passenger compartments.

In this case, the available deformation energy of the two vehicles is

$$E = \frac{1}{2} * m_1 * v_B^2 + \frac{1}{2} * m_2 * v_B^2 = \frac{1}{2} * (m_1 + m_2) * v_B^2$$

For this collision the energy absorption will be

$$D = \frac{1}{2} * \frac{m_1 * m_2}{m_1 + m_2} * (v_1 - v_2)^2$$

$$= \frac{1}{2} * \frac{m_1 * m_2}{m_1 + m_2} * v_c^2 \leq \frac{1}{2} * \frac{m_1 * m_2}{m_1 + m_2} * (2 * v_B)^2$$

For two vehicles with 56 km/h design speed colliding with a closing velocity less than 112kmh, compartment collapse, even of the smaller vehicle, can be avoided, because sufficient energy absorption through deformation is possible. This finding holds, regardless of the mass ratio. This restricts the problem of compatibility to the question: Are we able to force both vehicles to deform?

Looking at real world crashes, various different types of deformation can be seen: Sometimes both vehicles deform similarly and sometimes one of the vehicles deforms only marginally, while the compartment of the other vehicle is completely destroyed. The second case is typical for a crash between a very stiff vehicle and a very soft vehicle. The question that arises; what is a stiff vehicle and what is a soft vehicle? Was the marginally deformed vehicle too stiff or was the destroyed vehicle too soft?

The bulkhead-concept means that we must define a maximum deformation force level for the front-end. A compartment has to be built that is able to withstand this maximum force level. This would avoid a compartment collapse, as long as one of the vehicles deforms.

The bulkhead-concept would force both vehicles to deform. This concept takes advantage of the fact that sufficient deformation energy is available. It will decide the question of too stiff and too soft. Forces up to the maximum force-level are acceptable.

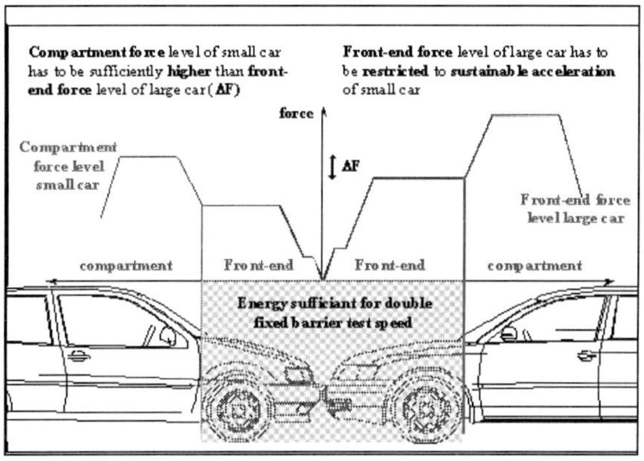

Figure 17: Bulkhead concept

When two vehicles collide, and if their closing velocity is less than twice their design speed, then there is sufficient deformation energy available for this particular crash. This concept holds, regardless of the mass-ratio of the two vehicles.

LIMITATIONS OF THE BULKHEAD-CONCEPT

When a (small) car with mass m_s collides with another vehicle with deformation force F, then for the deceleration a_s, the following equation applies:

$$m_s * a_s = F$$

For the large vehicle with mass m_l, which is designed for a barrier impact speed of v_B, we need a level of deformation energy D, which is sufficient to compensate for the kinetic energy at barrier impact test speed:

$$D = \frac{1}{2} * m_l * v_B^{\ 2}$$

On the other hand, deformation energy is computed by

$$D = \int F(s)ds = F * s_l$$

with a deformation travel sl of the large vehicle. The equation only holds for the average force level F of the force deflection curve of the large vehicle's front structure.

$$\frac{1}{2} * m_l * v_B^{\ 2} = D = F * s_l$$

$$\left(\frac{1}{2} * m_l * v_B^{\ 2}\right) \Big/ s_l = F$$

When this large vehicle and the small vehicle mentioned above collide, based on the principle of action and reaction, the forces are equal. This can be assumed by neglecting dynamic effects that produce oscillation for which the static approach generates the average behaviour. Thus the static computation is the approximate lower limit of the dynamic computation. It leads to

$$\left(\frac{1}{2} * m_l * v_B^{\ 2}\right) \Big/ s_l = F = m_s * a$$

or

$$\frac{1}{2 * s_l} * \frac{m_l}{m_s} * v_B^{\ 2} = a$$

This shows a relationship between barrier impact speed, mass ratio and the available deformation travel of the large vehicle. It is clear that limits of feasibility for the deformation travel exist. The length of vehicles is limited for several reasons (including environmental). The acceleration of the vehicles must be limited because the restraint system is able to protect the occupant in an acceptable manner only if vehicle deceleration occurs within certain defined limits.

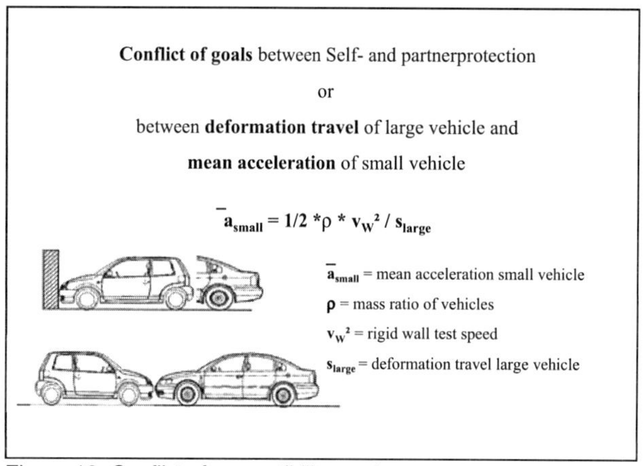

Figure 18: Conflict of compatibility goals

There is a relationship between these principle vehicle parameters, which prompts several questions. One question is; Which average deceleration limit is appropriate for a vehicle? We should bear in mind that we are speaking about the average deceleration of the passenger compartment. A compartment deceleration of 30 g is already an upper limit in terms of acceptable dummy load. The 30 g level is probably not acceptable from the viewpoint of older vehicle occupants.

It is easy to compute an upper limit for v_B to ensure compatibility. If we use a mass ratio of up to $\mu = \frac{m_l}{m_s} = 1{,}6$, approximately 80-90% of all frontal collisions in real-world accidents in Germany are covered. A deformation travel of $s_l = 0{,}7\ m$ is greater than the available deformation travel in the current vehicle fleet. If we accept a deceleration of 30 g that is higher than current restraint systems allow, then we find

$$30\ g = \frac{1}{2} \cdot 1{,}6 \cdot \frac{v_B^2}{0{,}7}$$

$$v_B = 16\ m/s = 57{,}8\ km/h$$

When we take into account that the 30 g level is an upper limit and that a mass ratio of 1,6 does not cover the entire vehicle fleet, then we must accept the fact, that the current barrier test speed is already at its upper limit. 56 km/h, the US NCAP test speed, is already equivalent to the EES that occurs when a 64 km/h test is conducted against a deformable barrier. An increase of barrier impact speed will decrease the potential for compatibility. If test speed would be increased to 60 km/h, then the mass ratio, for which compatibility measures are possible, would decrease according to the formula:

$$30\ g = \frac{1}{2} \mu \cdot \frac{(60\ km/h)^2}{0{,}7}$$

$$\mu = 1{,}48$$

If the test speed was increased to 64 km/h, then the analogous computation leads to a mass ratio of $\mu = 1{,}30$. This means that compatibility would only be possible, for example, in a range from 1000 kg to 1300 kg. This is clearly not sufficient. When environmental concerns dictate that small, fuel efficient vehicles will have an increasing market share, then we must allow for a great compatibility range.

The maximum front-end resistance of the large car determines the maximum deceleration of the small car. The higher the mass ratio between the colliding vehicles, the higher the deceleration in the lighter vehicle will be. As described above, a mass-ratio of 1:1.6 appears to be a reasonable limit for compatible car-to-car frontal-frontal collisions. Such a mass-ratio covers about 85% of the frontal car-to-car collisions in Europe.

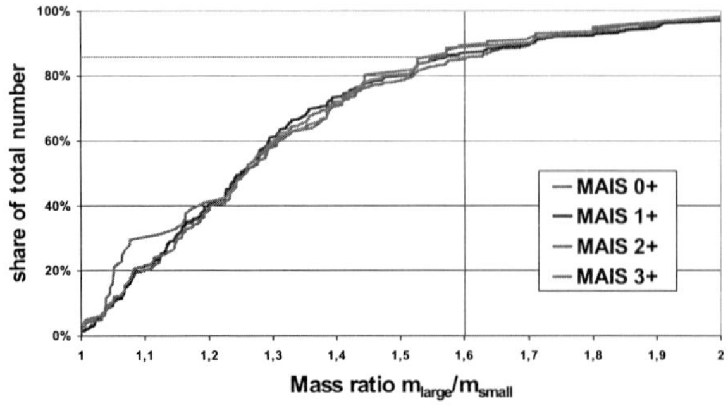

Figure 19: Mass ratio in front-front car-to-car collisions for accidents of different severities (source VW/GIDAS accident data base)

As a first step it seems reasonable to address the most likely car-to-car collisions. Therefore a lower and upper mass limit can be defined that includes volume models and excludes niche vehicles. As an example these mass limits can be set to 800 kg and 2000 kg (Figure 20). If we now follow the basic rules of the bulkhead principle it is possible to calculate the necessary front-end forces and passenger compartment forces for such vehicles.

To keep the maximum deceleration in the opponent vehicle below 40g, the maximum front-end force of the 1.6 times heavier vehicle is limited by the mass of the smaller opponent vehicle. Therefore the maximum front-end force from 800kg to 1280kg can be 320 kN (40g times 800kg). From there, the force increases with vehicle mass up to 2000kg. At this point the computation results in a maximum force of approximately 500 kN (2000kg/1.6*40g). This means the actual deceleration in the small vehicle will be always below or equal 40 g. As shown in [1], this still leads to acceptable occupant loads, if modern restraint systems are used. Softer front-end designs are not prohibited, because these limits are upper limits.

While the front-end force of the heavy opponent must take the deceleration of the lighter vehicle into account, the passenger compartment resistance of the lighter vehicle must withstand the forces of the front-end of the heavier vehicle. This means the same principal calculations apply to the minimum compartment force. For the heaviest vehicle, it is the same as the maximum front-end force plus a safety margin. It remains the same up to 1.6 for the lighter vehicle. From there the necessary minimum compartment force decreases to that of the lightest vehicle.

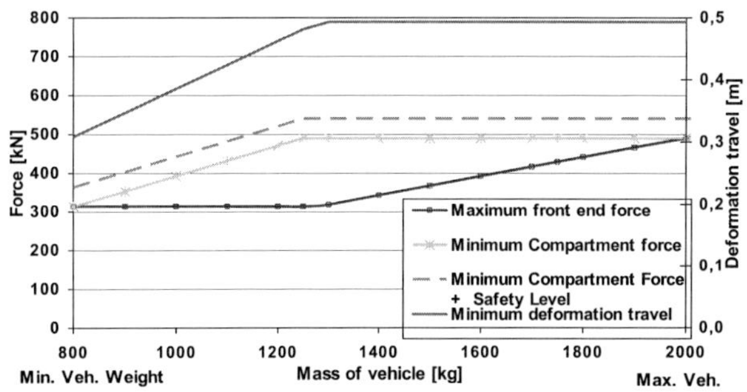

Figure 20: Calculated forces for a maximum deceleration of 40 g in small vehicle (theoretical calculations)

Assuming a rectangular force deflection characteristic, the minimum deformation travel of the front-end is calculated and shown in Figure 20. It should be noted that these values are highly theoretical. Real-world vehicles have an increasing force-deflection characteristic, which would result in much greater deformation lengths. For this reason, the maximum deceleration of the small vehicle will be 40g for a very limited time. Mean deceleration will be much lower and lower levels are acceptable.

SPECIAL CONSIDERATIONS FOR CAR-TO-TRUCK COLLISIONS

Is car-to-truck compatibility possible? Accidents statistics show clearly that this is a dangerous configuration.

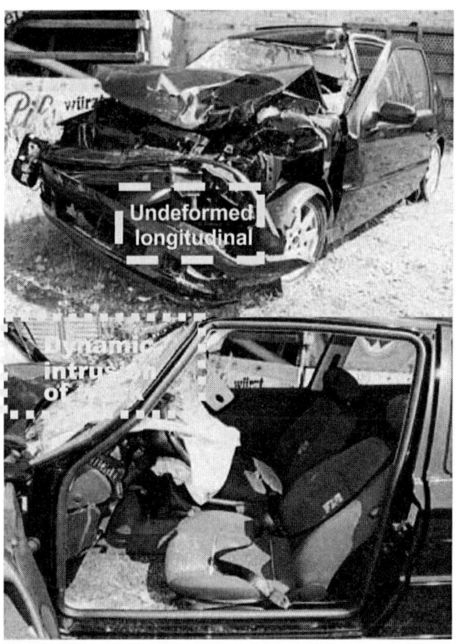

Figure 21 **Passenger Car with undeformed front structure after rear impact with a truck.** No truck photo is available because the truck was not damaged and was removed from the scene immediately after accident. A-pillar and roof rail of passenger car were directly hit. Thus passenger compartment intrusion occurred, although the closing velocity was in the range of 30...40 km/h (20...25 mph). Direct occupant contact was prevented by the airbag.

Figure 22 For the Van the result does not look much better

Thus theoretical computations made for vehicle-to-vehicle collisions must be applied to trucks. There is a first and inevitable pre-requisite for all further examinations: **If there is no support for the structures of the passenger car, then there is virtually no chance that the passenger car can protect its occupants**

The opposite question is: Due to the high mass ratio in car-to-truck collisions, is there any practical way to protect occupant in the passenger car?

The formula, describing the deformation energy needed for a car-to-truck accident is the same as derived in the last section.

$$D = E_{before} - E_{after} = \frac{1}{2} * \frac{m_1 * m_2}{m_1 + m_2} * (v_1 - v_2)^2$$

For velocity change dv, it is easily computed:

$$dv_1 = v_{after} - v_1 =$$

$$\frac{m_1 * v_1 + m_2 * v_2}{m_1 + m_2} - \frac{m_1 + m_2}{m_1 + m_2} v_1$$

$$= \frac{m_2 * (v_2 - v_1)}{m_1 + m_2} = -\frac{m_2}{m_1 + m_2} * (v_1 - v_2)$$

In the case of a truck of mass m_2, m_1 « m_2, or $m_1 + m_2$ • m_2.

This means for the formulae:

$$D = E_{before} - E_{after} = \frac{1}{2} * \frac{m_1 * m_2}{m_1 + m_2} * (v_1 - v_2)^2$$

$$\approx \frac{1}{2} * m_1 * (v_1 - v_2)^2$$

In other words, the deformation energy depends only on the mass of the passenger car and the closing velocity. The velocity change of the small vehicle is:

$$dv_1 = v_{after} - v_1 = -\frac{m_2}{m_1 + m_2} * (v_1 - v_2) \approx -(v_1 - v_2)$$

The passenger car has to change its velocity to the velocity of the truck and thus needs the energy for this change in deformation energy.

The good news is that the passenger car needs energy that is independent of the mass of the truck. The bad news is that it has a velocity change that equals the closing velocity between passenger car and truck.

Three cases may now be considered:

NO UNDER-RIDE PROTECTION

If there is no under ride protection available, then no deformation energy is available. The truck body will directly impact the occupant compartment. Even accidents at low closing velocities are very dangerous, depending on the question, whether the truck directly strikes the occupant.

RIGID UNDER-RIDE PROTECTION

If there is an under ride protection that is **rigid** and, if the passenger car structures interact appropriately with the truck under-ride structures, then the front-end structures of the passenger car can deform as designed. The deformation energy, available in car-to-barrier impact, is now available for car-to-truck collisions. As many cars are designed for a 56 km/h (35 mph) barrier impact, the available deformation energy is sufficient up to a closing velocity of 56 km/h (35 mph) in the case of a car-to-truck crash. This means that the large portion of low speed accidents, especially rear impacts, where a car hits a truck after braking and the closing velocity is thus low, would no longer be dangerous. This finding holds for the entire perimeter of the truck. It means that car-to-truck collisions, particularly in urban areas, could be significantly improved.

DEFORMABLE UNDER-RIDE PROTECTION

If there is an under-ride protection which is **deformable** and, if the passenger car structures interact appropriately with the truck under-ride protection, then the front-end structures of the passenger car can deform as designed and can also deform the truck under-ride protection. This dissipates additional deformation energy. But now for the bad news for the computation noted above; Even if the dissipation energy from the truck is sufficient to mitigate accidents at higher closing velocities, the closing velocity in this case will be the velocity change of the passenger car. The question to be answered is, whether the occupant, sitting in its now well-protected compartment, will survive the high velocity change.

A short example may help to clarify this question. Assuming that the truck provides as much deformation energy as the passenger car, the following deformation energy is available:

$$D_{available} = 2 * \left(\frac{1}{2} * m_1 * (56km/h)^2 \right)$$

What does this mean for the closing velocity? Up to which closing velocity is this sufficient?

$$\frac{1}{2} * m_1 * v_c^2 = D \le D_{available} = 2 * \frac{1}{2} * m_1 * (56km/h)^2$$

$$v_c \le \sqrt{2} * 56km/h = \sqrt{2} * 35mph = 79km/h = 49mph$$

$$dv_1 = 79km/h = 49mph$$

Further research is required to determine whether restraint systems, designed for such high velocity changes, can save people or whether they threaten even younger more resilient occupants at lower speeds. There are researchers who claim that the stiffening of restraint systems to accommodate the 56km/h = 35mph test speed has negative side effects at lower velocities.

Concluding this study on trucks, we have to state:

- If a **rigid under-ride protection device** is available and its height corresponds to the deformable structures of passenger vehicles, then, for collisions between passenger cars and trucks up to a closing speed of **56 km/h,** sufficient energy dissipation is available.
- If the **under-ride protection device is deformable** and provides similar dissipation energy as the passenger car, then even up to a closing speed of **79 km/h** sufficient energy dissipation is available.
- It has to be taken into account that the **closing velocity of a car-to-truck accident nearly equals the velocity change experienced by the passenger car**, since the truck, with its

much higher mass, essentially does not change its velocity. Thus, careful research must be carried out to determine, to what extent a deformable under-ride protection device will help reduce the risk of injury to passenger car occupants.

- These findings hold for **all directions of impact**, as long as under ride protection structures are available in the area of impact.

- A prerequisite for these considerations is that the deformable structures of the passenger car also deform in car-to-truck collisions. This means that **the truck under ride protection and the deformable structures of the passenger car have to be of the same height.** If part of the deformation energy of the passenger car is only able to be dissipated at a lower ground clearance, then this will not be available for car-to-truck collisions and thus the degree of protection for the occupant of the passenger car will decrease accordingly.

- There must be an agreement between passenger car manufacturers and truck manufacturers on an adequate height for structural interaction.

STRUCTURAL INTERACTION

Structural interaction is a prerequisite for compatible collisions and therefore also for all of the theoretical calculations mentioned above. Structural interaction means that the structural components of the front-end of the passenger car can deform at the same force level as in the barrier crash. That means that a similar amount of energy is absorbed without greater intrusions than in a barrier crash.

This is only necessary for crashes of high severity. In low severity crashes it might be even an advantage if there is poor structural interaction because this will result in a softer deceleration pulse. That is the reason why the influence of structural interaction is very difficult to detect in real world accident data. Unfortunately it is not known when designing a car which type of accident severity it will experience in its lifetime.

STRUCTURAL HEIGHTS TODAY

The principle energy absorbing structures of today's vehicles are the longitudinals, which support the transverse structural members. These structures, because of the existing bumper test requirements, lie at similar heights above the ground. Figure 23 shows the projection of the mean values of longitudinal heights of several vehicles of different sizes.

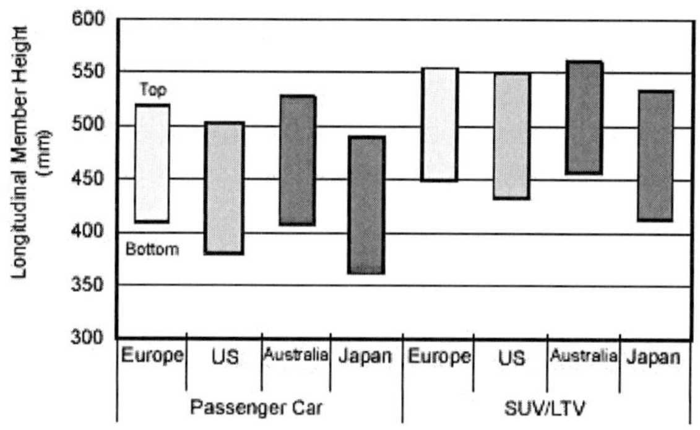

Figure 23: Mean values of structural heights [IHRA Working Group Compatibility]

It is a challenge for the future to design a test procedure that enforces the coherence of main structures to enable deformation similar to that in a barrier crash.

Convergence of frontal structures is required, between all crash partners, to maximize energy dissipation through structural interaction, in all collision configurations. The size of such a region is an open question for further research.

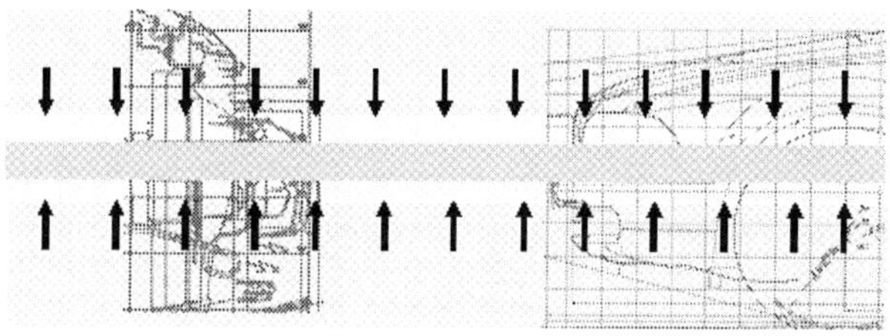

Figure 24: Convergence of frontal structures

POSSIBLE COMPATIBILITY TEST PROCEDURES

There are several test procedures that offer the potential to further investigate compatibility issues, each of which has inherent advantages and disadvantages:

U.S. GENERAL COMMENTS

In the US, incompatibilities between cars and LTV's (Light trucks and vans) have been made clear through analysis of accident statistics. LTV's have been shown to be up to 4 times as aggressive as passenger cars, in vehicle-vehicle collisions [2].The problem is based on a lack of structural interaction due to geometrical incompatibility between LTV's and passenger vehicles. As the US vehicle fleet consists of around 35% SUV's (more than 50% of new vehicle registrations), the problem of incompatibility is highly significant [2]. To compound the problem, a general lowering of the structures of large vehicles will be sought. The two proposals that are currently being discussed are:

- Static measurement of energy absorbing structures
- An evaluation of the AHOF (Average Height of force) based on load cell force measurements. US NCAP configuration.

OPEN QUESTIONS - US-NCAP CONFIGURATION WITH AN AHOF EVALUATION

Full overlap tests are much less demanding than offset tests for testing front-end structures. With respect to compatibility, however, full-overlap tests have the advantage of considering the entire front-end, which is important when rating the potential for structural interaction and considering the presence of supporting structural forces.

A particular disadvantage of the current US-NCAP configuration is the unrealistically high loading of single load cells by impacting mechanical parts (e.g. engine, transmission). These parts create very high peaks when impacting the infinitely stiff barrier; this would not occur in vehicle-to-vehicle crashes [3, 5]. Due to the rigid and flat interface, horizontal and vertical connections (eg. crossbeam) are mostly undetectable as the interface activates load pathways primarily in the x-direction. Shear forces created through vertical and horizontal connections go largely undetected.

The proposed AHOF evaluation determines the average height of all recorded load cell forces, over all previously recorded time intervals, for each discreet time point. In addition to this, a weighting factor with the total force is implemented to facilitate a more stable measurement.

The advantage of the assessment is its ease of implementation and its relative simplicity.

<u>Convergence of structures</u>

The AHOF evaluation encourages a general lowering of structures yet does not ensure structural convergence. The AHOF can be lowered either by a general lowering of frontal structures or by the addition of lower load paths. These measures increase the possibility for structural interaction yet do not enforce it.

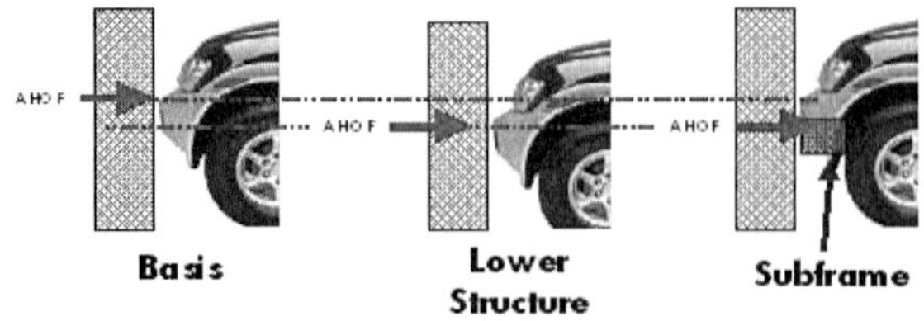

Figure 25: Approaches to lower the AHOF

<u>Force homogeneity</u>

The AHOF evaluation method does not evaluate force homogeneity. 1 load path, 2 load paths or a completely homogenous front end could all receive the same AHOF rating.

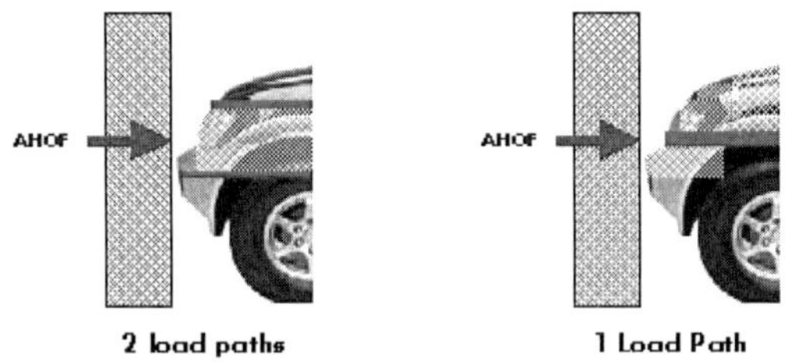

Figure 26: AHOF and measuring force homogeneity

Force homogeneity within a defined vertical region is considered an important step in promoting structural interaction in vehicle-vehicle collisions.

EUROPE – GENERAL COMMENTS

The European vehicle fleet has a distinctly different composition to that of the US fleet. Small passenger vehicles are much more common in Europe with SUV's comprising a relatively small percentage of the fleet. In Europe, two principle tests are currently being discussed:

- FWB, (Full Width Barrier) with a deformable face
- PDB (Progressive Deformable Barrier) in the offset configuration

OPEN QUESTIONS - FWB WITH DEFORMATION ELEMENT AND HOMOGENEITY EVALUATION

The FWB proposal is implemented in the US NCAP configuration with the primary difference being the inclusion of a deformation element with two homogenous layers of constant pressure. The deformation element has the advantage of preventing unrealistic peak loading that occur when only a rigid barrier interface is present. The deformation element also has the advantage of inducing shear forces in the front end and thus has the capacity to detect vertical and horizontal connections. The extent to which this can actually be achieved is unclear. The other potential drawback of the deformation element could be a further reduction in test severity due to energy dissipation in the barrier. Experience to date, however, suggests that the crash severity is still high enough for the optimization of restraint systems.

The FWB proposal occurs in the 100% overlap configuration. A deformation element composed of two sections is located in front of the barrier each of which is 150mm thick with a uniform and discreet pressure (0,34 and 1,71 Mpa). The homogeneity of front end forces is calculated based on the standard error between a target load level and the measured loads in each cell within a standard footprint area. A fixed size and location of the footprint area has not yet been finalized.

Mass dependency

A mass dependency exists in the current configuration as a consequence of the homogeneity assessment formula.

$$H_c(orig) = \frac{\sum_{i=1}^{n_c} (L - fi)^2}{n_c}$$

Figure 27: Overall „cell" homogeneity

The above formula describes the "in-homogeneity" based on the standard deviation between measured cell values (fi) and an average "target" load level (L). (nc) is a constant representing the number of cells in the footprint.

The calculations below show that a factor "k" larger force distribution is penalized by a factor k² in the homogeneity assessment.

$$H_c(new) = \frac{\sum_{i=1}^{n_c} (kL - kfi)^2}{n_c} = k^2 \frac{\sum_{i=1}^{n_c} (L - fi)^2}{n_c}$$

$$\therefore \; H_c(new) = k^2 * H_c(original)$$

It is clear that higher forces will be penalized in the proposed assessment. As there is a strong dependency between vehicle mass and front-end forces, a mass dependency is present.

Heavier vehicles will generally cause higher forces (although this increase in force is not linearly related to mass due to the discreet pressure changes in the barrier). This is attributable to the fact that the available deformation travel is similar for all vehicles. In order to provide the same level of self-protection, a heavier vehicle will exhibit higher front-end deformation forces. Vehicles of heavier mass will thus be penalized.

Force homogeneity should be considered irrespective of the magnitude of the forces. Force homogeneity and front-end stiffness should not be confused and should be considered separately in the evaluation procedures.

Sensitivity to impact location

Due to the resolution of the load cell wall, a sensitivity due to small vertical and horizontal movements of front-end load members may be present. For concentrated loadings, it may be advantageous for concentrated loadings to hit the junction between multiple cells rather than the middle of one cell. This would effectively "spread" concentrated forces over more load cells and result in a more homogeneous force distribution.

A smoothing process is proposed to remove this sensitivity. It involves the smoothing of the entire data set by taking the average of each combination of four adjacent load cells, for each point in time. A new smoothed data set results and the inhomogeneity, based on the standard deviation formula, Figure 27, is calculated.

Even with this smoothing, the sensitivity is not removed to a significant extent, only the magnitude of the average errors decreases. A point load hitting one cell will be smoothed over 9 cells. A point load hitting to junction of 4 cells will be smoothed over 16. It is therefore still beneficial to hit the junction between four cells.

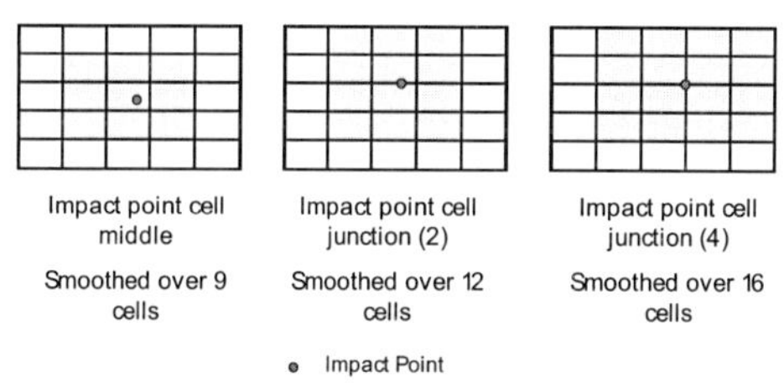

Figure 28: Load cells considered in smoothing process based on contact point of a concentrated load

In Figure 29 the influence of smoothing is further illustrated for a simplified 4*4 footprint. Even with smoothing, the resulting homogeneity values are strongly dependent on the contact point of the concentrated load. If the smoothing process were completely effective, the inhomogeneity should be similar in each case.

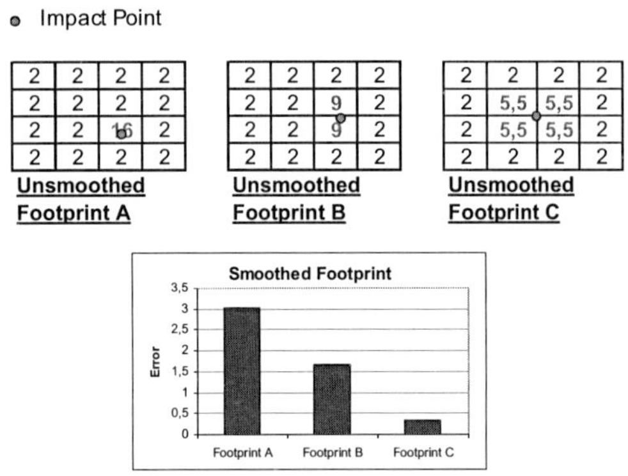

Figure 29: Influence of smoothing in removing the sensitivity of the FWB Assessment to load contact points

The sensitivity to load contact points and the appropriateness of applying smoothing routines in the proposed form is an open question for which more research is needed.

Repeatability

Due to the sensitivity to load contact mentioned above, the reproducibility of the FWB assessment procedure is unclear. It is also unknown if the honeycomb layer provides a stable transmission of forces to the load cell wall. The sensitivity of the honeycomb layer should also be addressed in repeatability testing.

Influence of discreet pressure changes in the deformation element

The discreet pressure changes in the barrier may have implications for the homogeneity evaluation. For a longitudinal member with a [125mm * 125mm] cross section, the following force ranges, based on the degree of barrier penetration, result.

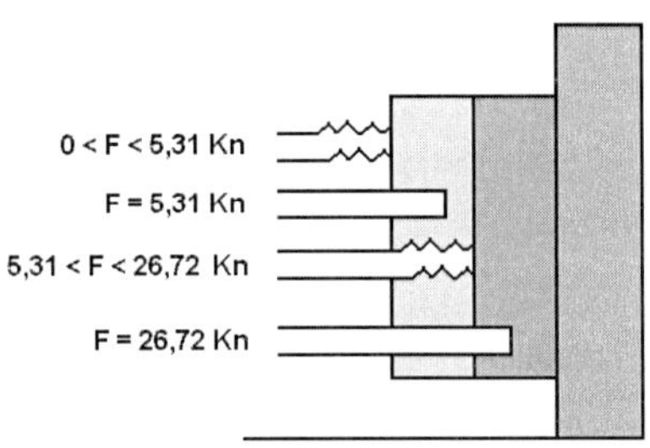

Figure 30: Influence of the discreet pressure changes in the barrier on resulting forces

Within deformation layers, the force provided is discreet (for a constant cross section area) irrespective of the depth of deformation. If the layers are too deep, then energy will be dissipated until the vehicle comes to rest, at a force level determined by the pressure of the barrier and not by the front end strength of the vehicle. A vehicle that comes to rest in the second honeycomb layer would appear perfectly homogeneous and stiffer load members may go undetected. It seems clear, that the depth of the layers should be thin enough, so that front end profiles are not disguised within them as a result of the constant pressure in the deformation element. This is also a subject for future research.

Assessing rounded front-end profiles

Rounded front-end profiles may be penalized in the force homogeneity evaluation because of non-planar contact with the deformation element.

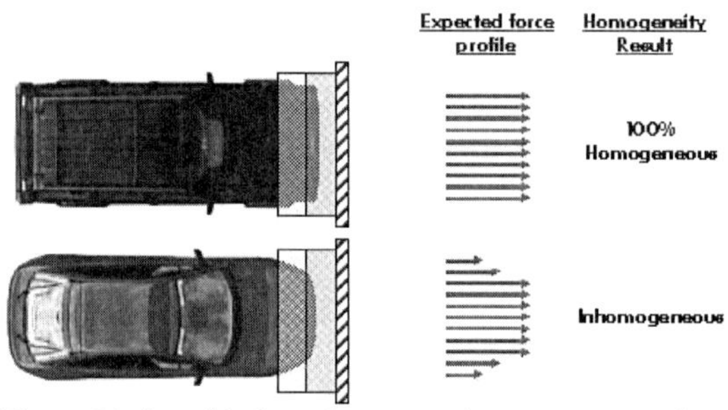

Figure 31: Possible force homogeneity assessment of rounded front-end profiles.

Rounded profiles may also offer distinct advantages in vehicle-vehicle collisions. Rounded profiles can enforce lateral force transmission through horizontal members and hence offer increased stability under impact loading. Round front ends may also help to induce a glance-off effect and shorten the crash episode in vehicle-vehicle frontal-frontal crashes.

Evaluating row and column homogeneity

The homogeneity evaluation algorithm considers row and column homogeneity separately in addition to overall or "cell" homogeneity. The overall evaluation is shown below.

$$Homogeneity = Q_{overall(cell)} + Q_{between\ rows} + Q_{between\ columns}$$

Mathematically, however, the row and column homogeneity values are already included in the overall "cell" evaluation.

$$Q_{cell} = Q_{within(Rows/Columns)} + Q_{between(Rows/Columns)}$$

This is illustrated below for a simplified 4x4 footprint.

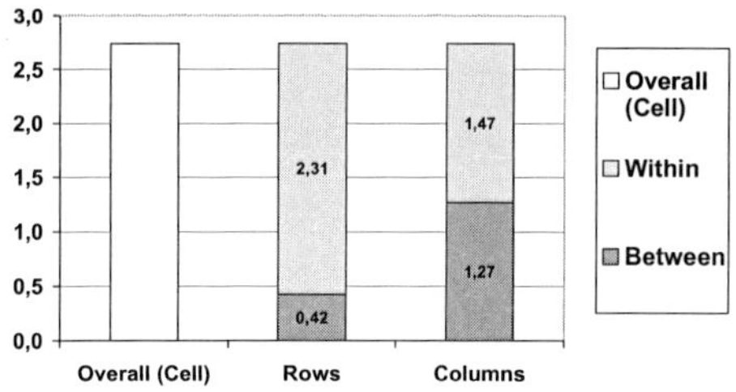

4	6	4	5
2	4	7	2
3	6	5	1
4	6	6	5

Figure 32: Relationship between cell, row and column homogeneity

In the case of the FWB evaluation, the row and column homogeneity values relate to the homogeneity between the rows and columns, based on the average value of each.

The reason behind isolating homogeneity between rows and columns and considering components in addition to overall cell homogeneity is unclear.

OFFSET DEFORMABLE BARRIER TEST INCLUDING LOAD CELL WALL

The additional load cell wall in the current offset deformable barrier test is of minimal benefit. There is a strong bridging effect, due to the deep deformable face, with the same stiffness over the full deformation depth. Most current cars fully deform the barrier and produce unrealistic force peaks due to impacting mechanical parts. It appears very difficult to gain additional information for a compatibility evaluation with this test configuration. It uses only the impact side longitudinal frame member and it also generated higher loading of the impacted side and detects soft compartments. It has to be determined, whether the compartment resistance derived from this test can be used for compatibility evaluation. [5]

OPEN QUESTIONS - FIXED PDB

The fixed PDB proposal occurs in the offset configuration at a proposed velocity of 60 kph. A deformable element placed in front of the barrier is illustrated below.

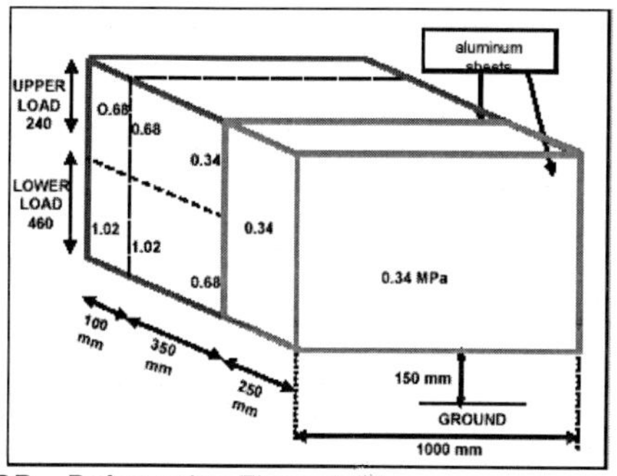

Figure 33 PDB – Deformation Element [6]

A large proportion of the PDB consists of an increasing stiffness profile into the depth of the barrier. There is also a significant difference between the upper and lower sections. After impact, the barrier is digitised resulting in a 2–dimensional picture with different colours representing distinct ranges of deformation. Based on this 2-dimensional picture, a PDB-Assessment value is calculated to represent vehicle aggressivity.

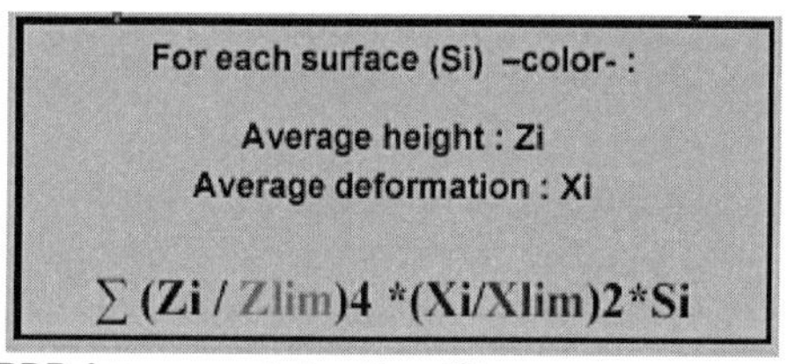

Figure 34: PDB Assessment algorithm [6]

The PDB Assessment algorithm evaluates vehicle aggressivity based on the average height (Z) and surface area (Si) of each respective colour (X) (representing deformation). The “lim” terms are constants and have little influence on the formula itself.

Influence of vehicle rotation

Vehicle rotation occurs in the offset configuration causing uneven deformation of the deformable element. This is more likely to occur at the end of the crash and the outer edge of the barrier may be deformed as a result of the rotation itself. This has consequences for the PDB assessment which uses a digitised picture based on the imprint on the barrier. The proposed solution is to cut the edge of the 2-dimensional picture to remove the influence of vehicle rotation before applying the assessment formula. It is an open question whether this “trimming” of the digitised picture is sufficient.

Assessing horizontal homogeneity

The PDB assessment method does not directly assess force homogeneity in the horizontal direction. There is no component in the PDB assessment formula which directly considers the horizontal direction. In the section of the barrier with progressive stiffness, (the main section of the barrier from 250 to 600mm), the assessment appears to be insensitive to horizontal homogeneity (based on VW calculations).

Whether it is feasible to measure homogeneity in the offset configuration is doubtful.

Softening of the front-end

The PDB assessment demands a general softening of the front-end of all vehicles. Limiting the barrier deformation, (ie. Limiting X), will result in a better assessment as the X/Xlim term will decrease accordingly (to the inverse power of two). Vehicle deformation is thus preferential to barrier deformation and softening of the front end is an expected consequence. The softening of the front-end for small vehicles is inappropriate for both self-protection and compatibility. Softening of the front-end of larger vehicles is not feasible and leads to a mass dependency in the proposed PDB Assessment.

Mass Dependency

Vehicle deformation is preferential to barrier deformation and thus lower front-end forces are desirable. This, however, is infeasible for larger vehicles due to the available length of the deformation zone. For the same self-protection level (self-protection should not be compromised through compatibility measures) larger vehicles are correspondingly stiffer. Larger vehicles thus require higher front-end deformation forces, penetrate the barrier to a larger extent and are penalized because of the X/Xlim term. A clear mass dependency is present in the PDB Assessment.

Lowering of frontal forces and front-end homogeneity

The PDB Assessment algorithm encourages a general lowering of front-end forces as a consequence of the Z/Zlim term. In order to receive a lower rating under the PDB Assessment, forces are required to converge toward ground level. The Z/Zlim is also to the power of 4 and thus has a dominant influence on the PDB Assessment value. If a particular colour in the 2d picture doubles its average height, it will be penalised by a factor four.

The PDB Assessment demands the following tendency in frontal structures. The following frontal force characteristics would be desirable.

Figure 35 Desirable force characteristics for the PDB assessment

Convergence of force toward ground level may be appropriate for large vehicles in order to increase geometrical compatibility with smaller opponents. The convergence of forces toward ground level for smaller vehicles is seen as inappropriate. It may lead to an increased possibility of under riding crash opponents including both larger vehicles and trucks.

The other implication of the convergence of forces toward ground level is that homogeneity is no longer considered. In order to improve structural interaction, force homogeneity within a distinct structural interaction zone is necessary. A degree of convergence is considered appropriate.

Further Comments

The PDB Assessment in its current form seeks to consider both front end stiffness and force homogeneity. Force homogeneity appears to be an important requirement to improve structural interaction and should be evaluated independently.

GENERAL REQUIREMENTS OF A TEST PROCEDURE

As mass limits are infeasible, enforcing force limits and measures to improve structural interaction is necessary. A compatibility procedure should evaluate both in a separate manner.

Force homogeneity within a defined corridor is considered the principle approach to improve the potential for structural interaction. Homogeneity should be evaluated independent of the magnitude of the force levels present which infers without a mass dependency. Force homogeneity relates only to the distribution of the force.

Enforcing force limits for a certain mass ratio is also important and should be enforced separately. The first requirement for improved crash compatibility is that support forces, within an interaction zone, must be present. In this way, energy dissipation in the front ends of both partners can be increased and the crash outcomes improve.

FUTURE RESEARCH - PROGRESSIVE DEFORMABLE BARRIER WITH A MASS DEPENDANT IMPACT SPEED

One-way of addressing this problem could be a mass dependent impact speed. Lower impact speeds for heavier vehicles would leave the possibility open to develop a softer front-end without compromising the self-protection level for these heavier vehicles. For lighter vehicles, the higher test severity would encourage high compartment forces that are beneficial in car-to-car collisions. Although, from a scientific point of view, this approach might be reasonable, it seems to be not communicable to the public. It will be very difficult to explain, especially to buyers of larger and often more expensive vehicles, why their vehicles are tested at a lower impact speed.

The change in barrier deformation attributable to the vehicle rotation and rebound also remains a problem for this test configuration.

FUTURE RESEARCH - OFFSET PROGRESSIVE DEFORMABLE MOVING BARRIER

Another way to address the problems mentioned above might be a Moving Deformable Barrier (MDB) with a deformable barrier face incorporating an increasing force deflection characteristic.

That way the impact speed could be kept the same for vehicles of different masses. Heavier vehicles would be tested at a lower severity and this would allow them to retain a reasonably soft front-end. The lighter vehicles would be tested with a higher test severity, forcing these vehicles to have sufficient compartment resistance under these conditions. That is especially important for lighter vehicles in vehicle-to-vehicle collisions.

The vehicle rotation during the rebound might be of less relevance because the MDB also starts to rotate.

However, it must be noted that such a test configuration places much higher demands on repeatability and reproducibility than fixed barrier tests. Before moving in such research directions, the ability of fixed barriers to address these questions should be further investigated. In a test series with this test configuration, practically no vehicle impacted the barrier with the overlap exactly as planned [7].

NEXT FEASIBLE STEPS

Since there is no full width test in Europe to date, of either a consumer or regulatory type, it may be more appropriate to move in this direction rather than adding another offset test. It must also be determined if such a test will be able to ensure sufficient structural interaction and be able to detect design measures that prevent over-/under-riding through interlocking of the front-end structures.

Any test with a mass-dependent test speed appears to be incommunicable to the public. For non-experts it will not be acceptable that a larger and, in most cases more expensive car, has, at least at a first glance, a lower self-protection level than a small car.

For the offset test using the ECE-barrier, the main disadvantage appears to be the early bottoming out of the barrier. There is a possibility that a deeper barrier with an increasing force deflection characteristic could have an advantage for predicting the behavior in vehicle-to-vehicle collisions because its behavior is closer to that of an opponent vehicle. This item must be discussed as a method for the optimization of frontal impact test procedures and is at this time a long-term question with several issues requiring further research.

The remaining possibility of using a Moving Deformable Barrier requires further investigation. Although such a test requires higher efforts to achieve reproducibility, it might have significant advantages compared with fixed barrier tests. With such a test, the rotation of the tested vehicle in offset collision configuration, which blurs the deformation picture of any deformable barrier, might be at least partially compensated. This may be of further interest as a long-term research approach. For this test configuration, a lot of open questions render it an option only feasible in the long-term and not available for mid-term application.

Taking all these pros and cons into account, it is not surprising that there is no "industry" position on compatibility available at this phase of research. There is a relevant part of industry that supports the attempt to enhance the full frontal impact by using layers that provide information about the front-end force distribution during a crash. Vehicle manufacturers, operating on a world-wide basis, appreciate the harmonization potential this test provides because it is close to FMVSS 208 test requirements. NHTSA (National Highway and Transport and Safety Association) expressed its goal to implement a first step of compatibility via an average height of force evaluation (or a similar approach) within the short term. A full width barrier with a shallow layer could support this goal because this barrier also has the potential also to be used for future compatibility evaluations. On the other hand, it appears to be rather clear for all manufacturers that a rigid barrier, even with load cells, is not able, for example, to measure the capabilities of transverse structural members. A rigid barrier already provides an infinitely stiff cross member, even for a car that has two unconnected longitudinals.

Harmonization under the umbrella of compatibility appears possible, as shown in the official IHRA report. However, it needs the support of all parties, including the U.S., at least to accept a full width frontal impact with a shallow layer as the equivalent of the rigid barrier. With regard to self-protection, the two tests are so similar that it makes no sense to conduct them both.

Figure 36 shows a possible roadmap for the next steps towards improved compatibility. It appears logical to take initial steps that are easier to accomplish and promise a measurable improvement in the current situation.

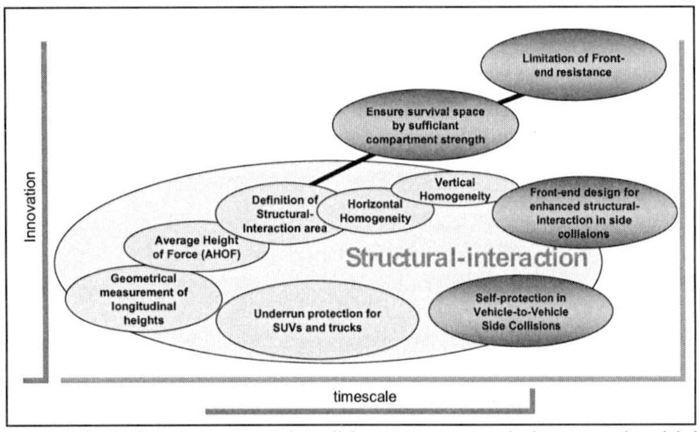

Figure 36: Roadmap for next feasible steps towards improved vehicle crash compatibility

CONCLUSIONS

- No compatibility test procedure exists today.
- A FWB with a deformable element has substantial potential to identify a homogeneous front structure. The reproducibility of this test is still questionable and requires further study because the grid of the load cells may lead to a sensitivity resulting from slight horizontal and vertical movements.
- The PDB generates more shear forces in the front structure. Thus there is a higher possibility to identify main structural elements and their connections (fork effects). The PDB, however, only evaluates the final deformation and does not identify local peak loads, as long as they do not result in end deformation. The influence of test vehicle rotation makes it difficult to identify homogeneity. The influence of the PDB Assessment, making cars softer, is questionable for small cars and for passenger compartment stiffness of all cars.
- Lowering the height of structural elements is not a goal in itself. Optimum structural height is a trade-off involving the demands of small passenger cars, large passenger cars, SUVs, MPVs, pick-ups and the under-ride protection systems of trucks. There will be a minimum height (that is acceptable in the context of under-ride protection systems of large vehicles) and there will be a maximum height (acceptable for passenger cars and especially for small passenger cars). It is not clear if these contradicting demands will allow a common range to be defined as a structural interaction zone. The areas defined by the U.S. (Part 581) and European bumper tests may become a basis for further discussions. The question of interacting forces is much more complicated. Taking into account that deformation travel is limited and self-protection cannot be compromised, up to a mass ratio of 1.6 and up to a closing velocity of 100 km/h, partner protection appears feasible.
- Good structural interaction and high compartment stiffness will have a positive benefit in a wide-range of accident scenarios. Most fatalities in vehicle-to-vehicle collisions are related to the fact that one compartment is significantly compromised. This creates a list of priorities for compatibility measures:

  1. Good structural interaction

  2. High compartment stiffness

  3. Control of interacting forces

  These are listed in order of increasing level of complexity relating to the fulfilment of each.
- Geometrical measurement of structural elements and restrictions of their height or dynamic measurements, such as average height of force (AHOF), lead the vehicle fleet in the right direction but do not necessarily enforce structural interaction. For a quick implementation of measures to achieve compatibility into regulation, there is no alternative, as all tests procedures continue to raise unanswered questions and require further research.

- Under-ride protection systems on trucks (even if they are rigid) provide a measurable improvement in overall safety on the roads. Without under-ride protection systems, even minor accidents (in terms of velocity) can be fatal.

- Side impact measures usually relate to self-protection through addressing stiffness of the side structure of the struck vehicle. A lowering of vehicle front structures in such a way that the sill of the struck vehicle will be involved does not appear to be feasible. A feasible contribution to front-side vehicle-vehicle compatibility may be a more homogeneous front-end especially in the horizontal direction. This will distribute forces to the A- and B-pillars of the struck vehicle and will reduce the intrusion into the side of the passenger compartment. Due to the high risk of this accident configuration, however, accident avoidance remains the most appropriate goal.

REFERENCES

[1] Schwarz, T./ Busch, S./ Zobel, R.: Influence of deceleration pulse on driver injury levels in vehicle-to-vehicle collisions, IMechE 2002, London

[2] Summers, S. Hollowell, T. Prasad, A. NHTSA'S research program for vehicle compatibility, ESV 2003

[3] Zobel, R./ Schwarz, T.: Development of criteria and standards for vehicle compatibility, ESV 2001

[4] Zobel, R.: Demands for compatibility of passenger vehicles, Society of Automotive Engineers: SAE technical paper series, SAE 98-S3-0-10, 1998

[5] Schwarz, T.: Selbst- und Partnerschutz bei frontalen Pkw-Pkw-Kollisionen (Kompatibilität), Fortschr.-Ber. VDI Reihe 12 Nr. 502, Düsseldorf: VDI Verlag 2002

[6] Delanoy, P. Faure, J.: Compatibility close from real life assessment. ESV 2003

[7] Seyer, K.: Report on crashtests within IHRA working group Compatibility, 2002

[8] Zobel, R. Principles for the development of a passenger car safety information system for consumers, based on real-life accident evaluation. Crash-Tech special '98, München, 1998

[9] Zobel, R. Accident Analysis and Measures to Establish Compatibility, 1999-01-0065, SAE-Conference, Detroit, Michigan, March 1-4, 1999

[10] Zobel, R. - Schwarz, T. Determination of compartment stiffness for compatible design of passenger vehicle front structures, Crash Tech 2000, München, 2000

2000-01-1378

Crash Compatibility Issue in Perspective

Adrian K. Lund, Brian O'Neill, Joseph M. Nolan and Janella F. Chapline
Insurance Institute for Highway Safety

ABSTRACT

Despite extensive media coverage to the contrary, mismatches among cars, utility vehicles, and pickups in crashes is not a big problem from a societal perspective. On the other hand, if you are riding in a small car that is about to be hit by a big utility vehicle, then the problem looms large. Crash compatibility has attracted a lot of attention lately because utility vehicles have become so popular. The concern is that their designs pose a threat to people riding in smaller cars. But the fact is, two-vehicle collisions between cars (including passenger vans) and utility vehicles or pickups account for only about 15 percent of all car occupant deaths. As a result, countermeasures that focus on making utility vehicles and pickups more crash compatible, however appropriate, can have only small effects on crash injuries and fatalities. On the other hand, improvements in crashworthiness not only reduce crash incompatibilities but also protect across a wider spectrum of crashes.

WHAT IS CRASH INCOMPATIBILITY?

It refers to vehicle design differences and similarities that can adversely affect occupant injury risk in two-vehicle crashes. Weight differences are one example, because in crashes between light and heavy vehicles the people in the light vehicles have higher injury risks than the people riding in heavy vehicles. The growth in sales of utility vehicles has led to a new focus on design characteristics besides weight, such as vehicle ride height and front-end stiffness, that may exacerbate crash incompatibility.

DATA FROM ON-THE-ROAD CRASHES

Assessing the influence of vehicle weight, ride height, and stiffness on occupant injury risk using real-world crash data is complex because of the interrelationships among the three characteristics. Utility vehicles and pickups are stiffer and ride higher than cars, for example, and on average they are heavier. Heavy vehicles often are larger, and size as well as weight affects injury risk.

Weight mismatches between vehicles contribute to incompatibility, but size mismatches need not if the vehicles weigh about the same. Factors such as ride height or front-end stiffness are harder to characterize, making their influence on crash compatibility more difficult to assess.

It is important to use real-world data to study crash incompatibility, particularly to help put the whole issue in perspective, because crashes involving two passenger vehicles in which compatibility is an issue are not the cause of most occupant deaths. Even people riding in light cars (less than 1,135 kg)—the occupants at greatest potential risk from crash incompatibility—are more likely to be killed in single-vehicle crashes or in crashes with other cars (including passenger versions of vans typically referred to as minivans, none of which weigh less than 1,135 kg) than in collisions with heavier utility vehicles or pickups. Using Fatality Analysis Reporting System (FARS) data, only 16 percent of occupant deaths in 1990-96 model light cars occurred in crashes with utility vehicles or pickups during calendar years 1991-97 (Figure 1). In contrast, 38 percent of light car occupant deaths occurred in single-vehicle crashes, and another 23 percent occurred in two-vehicle crashes with other cars.

Two-vehicle crashes are somewhat more important in accounting for nonfatal injuries. In Germany, it is reported that more than half of nonfatal car occupant injuries occur in two-vehicle crashes [1].* The same is true in the United States, where two-vehicle crashes accounted for 58 percent of moderate and more severe nonfatal car occupant injuries using 1991-97 National Automotive Sampling System (NASS) data (Figure 2). However, as with fatalities, collisions with other cars—not utility vehicles or pickups—resulted in 66 percent of these nonfatal injuries. Only 23 percent of the injuries occurred in collisions with utility vehicles or pickups. In fact for light cars, collisions with other cars is the leading cause of moderate and more severe nonfatal injury (47 percent) and 1 in 4 fatalities.

Why are car-to-car crashes so much of the problem for the lightest cars? Their small size and light weight make their occupants more vulnerable in collisions with any other object, even vehicles as light as themselves. The

*Numbers in brackets designate references at the end of the paper.

influence of weight is particularly great in this regard. By comparison with light cars, heavy cars (1,590-1,814 kg) experience only 28 percent of their nonfatal injuries in collisions with other cars and only 1 in 5 of their fatalities in such collisions.

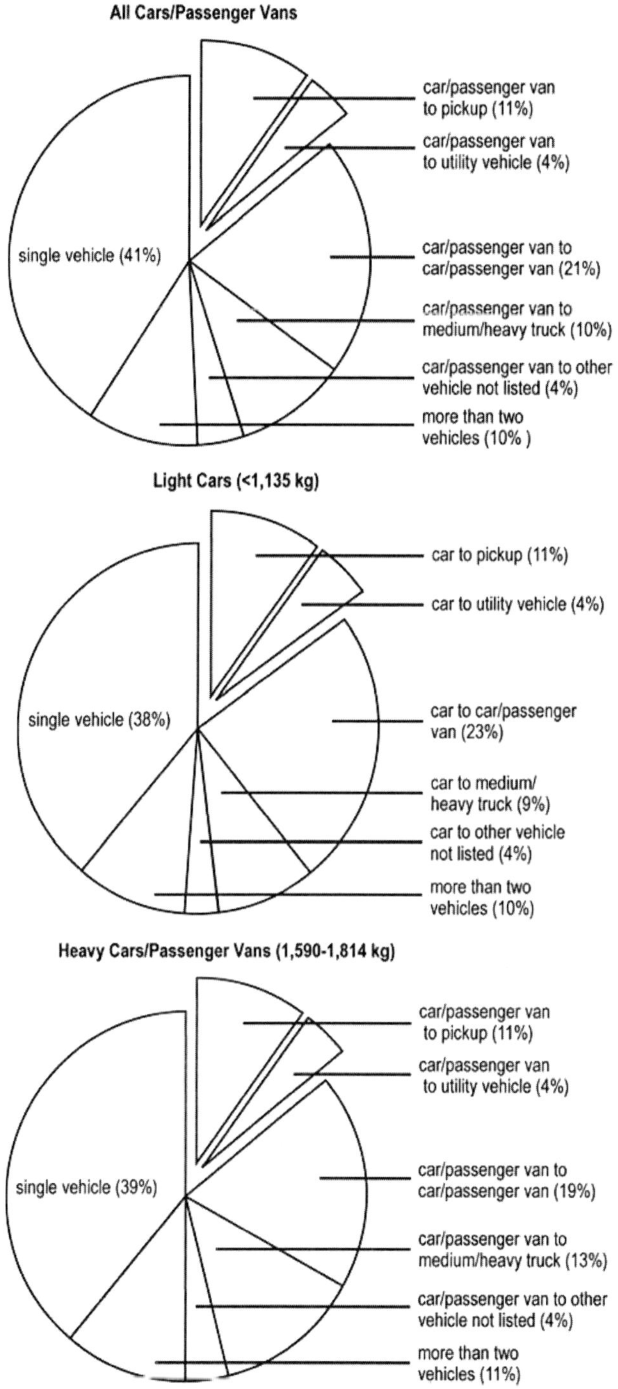

Figure 1. Distribution of Occupant Deaths in 1990-96 Model Cars/Passenger Vans by Crash Type

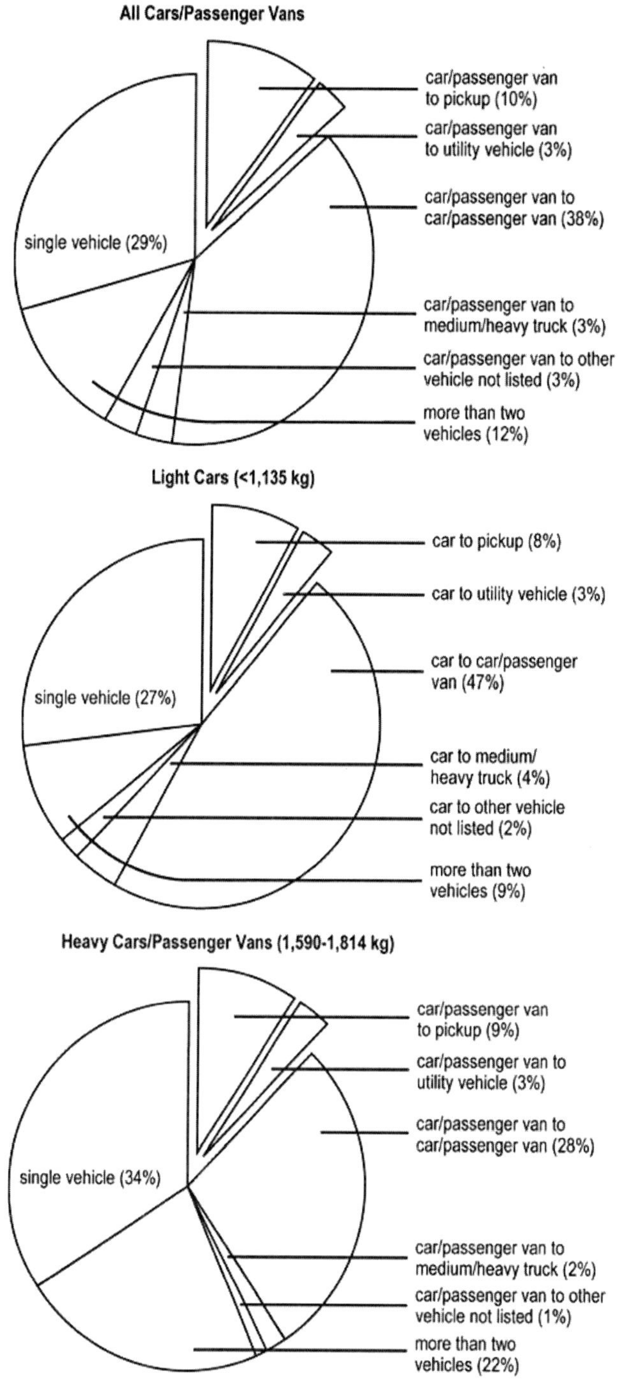

Figure 2. Distribution of Moderate to Severe (MAIS 2+) Nonfatal Occupant Injuries in 1990-96 Model Cars/Passenger Vans by Crash Type

EFFECT OF VEHICLE WEIGHT – Weight so strongly influences occupant death and injury risk that it is important to isolate this effect (to the extent possible) in assessments of other potential compatibility factors. Again using 1991-97 FARS data, as cars get lighter their death rates per million registered go up in two-vehicle crashes (Figure 3). Light cars also have the highest death rates per million in single-vehicle crashes (Figure 4). The same is true for utility vehicles and pickups—the lighter they are, the higher their occupant death rates. This is dramatically true for the very lightest utility vehicles and pickups.

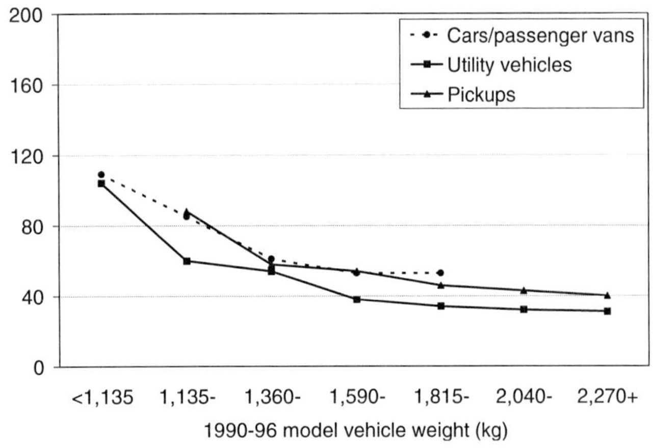

Figure 3. Occupant Death Rates for Two-Vehicle Crashes Involving 1990-96 Model Passenger Vehicles, Deaths per Million Vehicles per Year

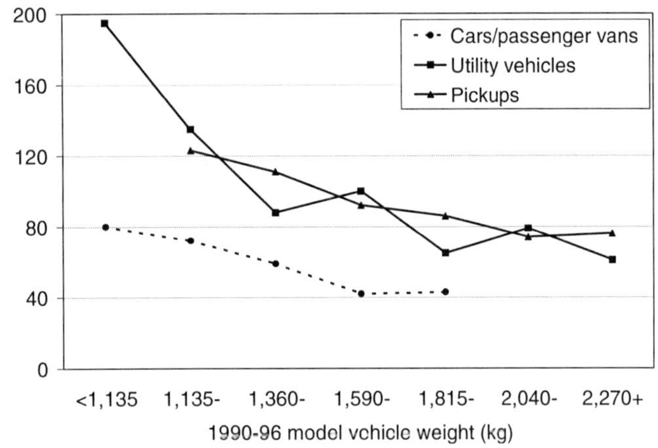

Figure 4. Occupant Death Rates for Single-Vehicle Crashes Involving 1990-96 Model Passenger Vehicles, Deaths per Million Vehicles per Year

Within weight classes, however, cars have lower overall occupant death rates than utility vehicles or pickups, mainly because cars have much lower death rates in single-vehicle crashes. In collisions between two vehicles, cars and pickups of similar weight tend to have similar occupant death rates, somewhat higher than utility vehicles.

COMPATIBILITY FACTORS OTHER THAN VEHICLE WEIGHT – Although crash incompatibilities exist among cars as well as between cars and utility vehicles or pickups, the most important compatibility factors other than weight—vehicle ride height and stiffness—vary most in collisions between cars and utility vehicles or pickups.

So a key question is, do car occupant deaths occur relatively more often in collisions with utility vehicles or pickups of a given weight than in collisions with other cars of the same weight? The answer is yes. For every million registered 1990-96 model cars weighing 1,590-1,814 kg, 45 deaths occur in the other cars (all model years) they collide with (Figure 5). In comparison, for every million utility vehicles in the same weight class, 76 deaths occur in the cars they collide with. The corresponding rate for pickups is 87. Car death rates are higher for every weight class of 1990-96 model utility vehicle and pickup, compared with 1990-96 model cars. The highest rates always involve pickups.

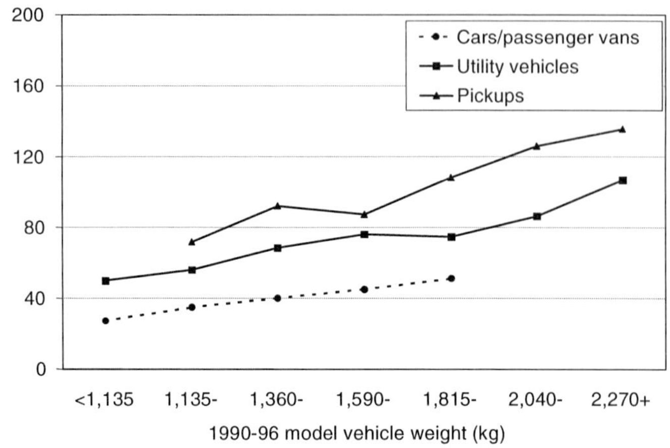

Figure 5. Occupant Death Rates for All Other Cars/ Passenger Vans in Two-Vehicle Crashes with 1990-96 Model Passenger Vehicles, Deaths per Million Vehicles per Year

These results indicate that utility vehicle and pickup characteristics beyond weight increase the risks to occupants in cars that collide with these vehicles. However, it is not possible from real-world crash data to identify the specific vehicle characteristics that contribute to the increased risks.

DATA FROM CRASH TESTS AND INJURY RISK

To partially address issues related to vehicle characteristics concerning side impact crashes, the Insurance Institute for Highway Safety (IIHS) conducted a series of six 90-degree front-to-side crash tests with both vehicles moving [2]. Front-to-side impacts were studied because real-world crash data show even larger incompatibility differences between cars and utility vehicles or pickups in side impacts than in other impacts [3]. The side struck vehicle, a Mercury Grand Marquis with a BioSID (biofidelic side impact dummy) in the driver position, was moving at 24 km/h in all tests. The striking vehicles, moving at 48 km/h in all tests, included a Lincoln Town Car and five Ford F-150 pickups. The baseline pickup had greater front-end stiffness than the car but was of comparable mass and ride height. The four other pickups systematically varied with respect to weight, front-end stiffness, and ride height (Table 1). Nolan et al. [2] reported that increased mass, stiffness, and ride height each had the predicted effect of increasing the extent of damage to the side struck car. However, injury risk for the driver dummy seated in the car was not so predictable (Table 2). Increased vehicle height clearly affected injury risk in terms of both the likelihood of the head contacting the striking vehicle and thoracic injury as measured by thoracic trauma index (TTI), thoracic deflection, and thoracic viscous criterion. Increased mass further increased the risk of a head strike but affected only the TTI measure of thoracic injury.

Table 1. Striking Vehicle Characteristics in Side Impact Crash Tests with Mercury Grand Marquis

Striking vehicle	Vehicle weight (kg)	Front-end stiffness	Ride height
1995 Lincoln Town Car	2,067	least stiff	normal
1997 Ford F-150 4x2 Normal weight, height	2,084	stiffer	normal
1997 Ford F-150 4x2 Raised height	2,085	stiffer	high
1997 Ford F-150 4x2 Added weight	2,402	stiffer	normal
1997 Ford F-150 4x2 Added weight, raised height	2,400	stiffer	high
1997 Ford F-150 4x4	2,401	stiffest	high

1992-93 Mercury Grand Marquis 2,007 ±2 kg

Table 2. Mercury Grand Marquis BioSID Injury Measures

Striking vehicle	Head injury criterion	Thoracic trauma index	Thoracic deflection (mm)	Thoracic viscous criterion (m/s)	Pelvis accel. (g)
Injury reference values	1000	85	42	1.00	130
1995 Lincoln Town Car	75	50	41	0.40	48
1997 Ford F-150 4x2 Normal weight, height	182	68	37	0.40	105
1997 Ford F-150 4x2 Raised height	227	**86**	**60**	**1.51**	105
1997 Ford F-150 4x2 Added weight	249	**101**	32	0.37	103
1997 Ford F-150 4x2 Added weight, raised height	no data	**111**	**62**	**1.54**	114
1997 Ford F-150 4x4	717	**87**	**48**	**0.44**	109

Most surprising was the unclear effect of vehicle stiffness. Although dummy injury measures were lower in the test with the vehicle struck by the car than by the similarly weighted but stiffer pickup, the stiffest vehicle (the 4x4 version of the pickup) did not cause the highest injury ratings. In fact, thoracic injury risk was lower than measured for the driver dummy in the car struck by a similar weight and height but less stiff 4x2 version of the pickup.

The explanation for this finding appears to be that the degree to which some areas of a vehicle's front end are stiffer than others—and where the stiffer areas are located—can affect aggressivity in side impacts. In the case of the 4x4 pickup, a pair of tow hooks happened to be located where they contacted the side of the struck car at a low level. This very stiff but low engagement with the car's side structure resulted in an intrusion profile of the struck car that accelerated the dummy's pelvis earlier than its thorax (Figure 6). The benefits of such a vertical intrusion profile have been discussed by others, as well [4]. These findings indicate homogeneity of stiffness may be more important than overall stiffness in improving vehicle compatibility. The inherently aggressive effects of front ends in side impacts may be mitigated by designs that spread the load of the striking vehicle both horizontally and vertically.

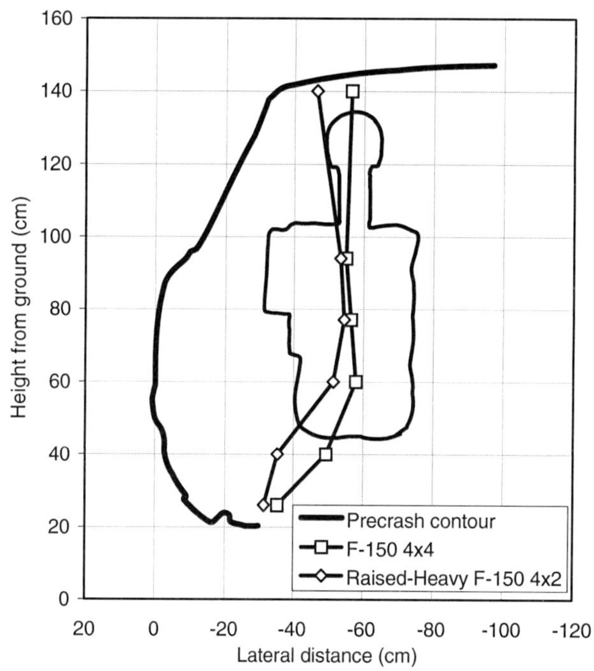

Figure 6. Vertical Crush Profile of 1992-93 Mercury Grand Marquis at BioSID Location, Raised-Heavy 1997 Ford F-150 4x2 and 1997 Ford F-150 4x4

Results of the crash tests also suggest that a major incompatibility factor associated with utility vehicles and pickups is their higher hoods, which are more likely to strike occupants' heads in side impacts. This is consistent with data from the NASS (1992-97), which show the likelihood of Abbreviated Injury Scale (AIS) 3+ head or face injuries increases sharply for occupants in side struck cars (passenger vans not included) when the striking vehicle is a utility vehicle or pickup rather than another car (Figure 7).

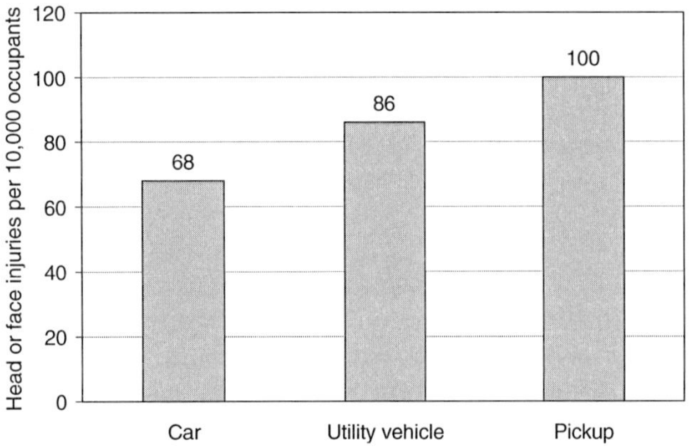

Figure 7. Occupant Head or Face Injuries (AIS 3+) in Side Struck Cars in Two-Vehicle Crashes by Type of Striking Vehicle

Clearly, countermeasures to reduce the risk of these head strikes are appropriate. However, how likely is it that practical modifications to utility vehicle and pickup front ends can achieve this? After all, every one of the pickup side impacts posed a threat to the struck driver's head. It is far more likely that meaningful reductions of head injury risk in such crashes will be achieved by the addition of side airbags that deploy themselves between the head and any intruding striking vehicle. Moreover, such airbags will benefit car occupants in single-vehicle side impacts, as well, where head injury from contact with external objects also is a risk.

This observation points out two basic issues. First, crash incompatibility is not just about the so called aggressiveness of one vehicle but also the vulnerability (or crashworthiness) of the other. Crash incompatibility can be reduced either by reducing an aggressive characteristic (mass or stiffness, for example) or by improving crashworthiness (the addition of a head airbag for side impacts, in this case, or increasing the mass of the lightest vehicles). Second, improvements in crashworthiness are likely to have more benefit than reductions in vehicle aggressiveness because it addresses injury risk in single-vehicle as well as two-vehicle crashes.

CHANGING THE VEHICLE WEIGHT MIX

To further explore the effects of vehicle weight for all impact modes, IIHS estimated the likely effects of changing the mix of weights of the vehicle fleet [3]. Because the effect of weight differences is so ubiquitous, limiting the weight of the largest vehicles (and hence limiting mass incompatibility) often is touted as a way to improve the safety of motor vehicle travel. However, *increasing* the size and weight of the smallest, lightest vehicles also reduces mass incompatibility in crashes, and it is informative to compare the overall safety benefits of these two strategies.

First, using 1991-97 FARS data IIHS estimated the number of deaths that would have been expected during 1997 if the fleet had been limited to vehicles from the 1990 model year or later. This makes for a fleet with a relatively modern mix of vehicle types and compatibility characteristics. The hypothetical mix of vehicle weights was then modified. In one scenario, occupants in the lightest cars (less than 1,135 kg) were moved to the next heavier group of cars (1,135-1,359 kg). In another scenario, occupants in the heaviest utility vehicles and pickups (1,815 kg or more) were put into lighter (1,590-1,814 kg) corresponding vehicles. In all, five scenarios were analyzed (Table 3).

What these scenarios reveal is that getting rid of the lightest cars would have a positive effect on highway safety. An estimated 714 lives would be saved each year. Eliminating heavy utility vehicles and pickups would have only a slightly positive effect of saving 159 lives, and putting occupants of heavier cars (1,590 kg or more) into lighter cars (1,360-1,589 kg) would theoretically lead to 423 more deaths.

Table 3. Changes in Passenger Vehicle Occupant Deaths from Changing the Vehicle Weight Mix

	Single-vehicle crashes	**Two-vehicle crashes with medium or heavy truck**	**Two-vehicle crashes with another passenger vehicle**	**Total**
Baseline				
Expected deaths in 1997 if all passenger vehicles had been 1990-96 models	12,675 deaths	2,723 deaths	11,032 deaths	26,429 deaths
Percent change in deaths, five scenarios:				
1. Eliminate lightest cars	-1.9	-2.6	-3.7	-2.7
2. Eliminate lightest cars, utility vehicles	-2.2	-2.6	-3.5	-2.8
3. Eliminate heaviest cars/ passenger vans	+2.8	-0.2	+0.6	+1.6
4. Eliminate heaviest cars/ passenger vans, utility vehicles, pickups	+4.8	+0.2	-3.0	+1.1
5. Eliminate heaviest utility vehicles, pickups	+2.0	+0.4	-3.7	-0.6

What these results demonstrate is that improving the crashworthiness of the most vulnerable vehicles results in far more benefits because those improvements can affect occupant protection across the spectrum of crashes to which those vehicles are exposed. In contrast, limiting the aggressiveness characteristics of certain vehicles affects only a subset of two-vehicle crashes involving those vehicles. For example, only 21 percent of occupant deaths in the lightest cars occur in two-vehicle crashes with utility vehicles, pickups, and heavy cars—the crashes in which incompatibility is likely to be greatest. So although it may be possible to improve overall safety by improving vehicle compatibility, the effect is likely to be smaller than expected. If compatibility improvements to utility vehicles, pickups, and heavy cars could reduce by 10 percent the death risk for people in the light cars with which they collide, it would mean an overall light car occupant fatality reduction of about 2 percent (10 percent of the 21 percent). A relatively huge improvement of 20 percent would mean only a 4 percent overall fatality reduction.

Are not these reductions worth seeking? Yes, but such improvements are difficult given design constraints on vehicles. The highest priority should continue to be improving the protection vehicles offer to their own occupants. Such improvements benefit people in most kinds of crashes, including single-vehicle crashes, and would have a much bigger effect on highway safety than compatibility improvements.

The bottom line is that vehicle occupants should pay close attention to their own vehicles. The best thing occupants can do to protect themselves is to travel in passenger vehicles that offer good crashworthiness.

ACKNOWLEDGMENTS

The authors would especially like to thank Maria Penny of the Insurance Institute for Highway Safety and Sara Jones for their computer programming contributions and dedication to this study. This work was supported by the Insurance Institute for Highway Safety.

REFERENCES

1. Zeidler, F.; Knöchelmann, F.; and Scheunert, D. 1999. Possibilities and limits in the design of compatible cars for real-world accidents (SAE 1999-01-0068). *Vehicle Aggressivity and Compatibility in Automotive Crashes* (SP-1442), 55-58. Warrendale, PA: Society of Automotive Engineers.
2. Nolan, J.M.; Powell, M.R.; Preuss, C.A.; and Lund, A.K. 1999. Factors contributing to front-side compatibility: a comparison of crash test results (SAE 99SC02). *Proceedings of the 43rd Stapp Car Crash Conference* (P-350), 13-24. Warrendale, PA: Society of Automotive Engineers.
3. Lund, A.K. and Chapline, J.F. 1999. Potential strategies for improving crash compatibility in the U.S. vehicle fleet (SAE 1999-01-0066). *Vehicle Aggressivity and Compatibility in Automotive Crashes* (SP-1442), 33-42. Warrendale, PA: Society of Automotive Engineers.
4. Hobbs, C.A. 1995. Dispelling the misconceptions about side impact protection. SAE Technical Paper Series 950879. Warrendale, PA: Society of Automotive Engineers.

2001-01-1167

NHTSA's Compatibility Research Program Update

Stephen Summers, Aloke Prasad and William T. Hollowell
National Highway Traffic Safety Administration

ABSTRACT

This paper provides an update of NHTSA's research activities in vehicle compatibility and aggressivity. This paper presents newly initiated efforts underway to develop test assessment methodologies intended to evaluate vehicle compatibility. The rigid barrier load cell data collected from 18 years of the agency's New Car Assessment Program testing are reviewed to evaluate potential test measures that may be used to evaluate a vehicle's compatibility in vehicle-to-vehicle crashes. These parameters are then evaluated using a series of vehicle-to-vehicle and moving deformable barrier (MDB)-to-vehicle tests. In these tests, the face of the MDB has been instrumented with an array of load cells to compute test measures. This study is part of NHTSA's ongoing compatibility research program and is being coordinated with the IHRA compatibility group.

INTRODUCTION

At the Fifteenth Enhanced Safety of Vehicles Conference held in Melbourne, Australia, during May 1996, a program of coordinated research was agreed upon that would be undertaken under the worldwide banner of International Harmonized Research Activities (IHRA). The research is comprised of six high priority areas in which nations would collaborate over a 5-year period. Among these is research in vehicle compatibility. The aim of the IHRA Vehicle Compatibility Working Group is to develop test procedure(s) designed to improve the compatibility of vehicle structures in front-to-front and front-to-side crashes, thus enhancing occupant protection in these crash modes. A secondary aim is to consider protection in vehicle impacts with pedestrians, heavy vehicles, and other obstacles

Despite the fundamental differences in the fleets found in the participating members' regions of the world, the IHRA Working Group remained focused on developing test procedures that could be used anywhere. While there were diverse views as to the types of tests that could be used, the members agreed to a work program that addresses the overall goals of the program. The program plan included efforts in using real world crash data to define the safety problem, using the findings from the review of crash data in combination with results obtained from crash testing and computer modeling to determine the key vehicle characteristics affecting vehicle compatibility, and the development of test protocols that evaluate the key characteristics which when used would ensure that vehicles become more compatible.

The Working Group members remained focused on developing test procedures that would be suitable anywhere (despite the identified differences in fleets among the participating regions). The consensus was that the test procedure(s) should include an evaluation of the force footprint imparted by a vehicle and, in parallel, evaluate the capability of the vehicle to sustain an overload condition on the occupant compartment. Recognizing that heavy and light vehicle collisions are inevitable, the overload condition is intended to establish performance levels for maintaining occupant compartment integrity. The partIculars regarding these evaluations remain under further development and debate.

In the U.S. fleet, light trucks, vans, and sport utility vehicles (LTV's) currently account for over one third of light vehicles, but over half of all light vehicle-to-vehicle fatalities occur in collisions between cars and LTV's[1, 2]. Eighty one percent of these vehicle-to-vehicle fatalities occur to passenger car occupants. NHTSA's compatibility research program is investigating vehicle-to-vehicle compatibility issues and the implications of changing U.S. fleet composition upon fleetwide occupant safety. Previous studies undertaken in NHTSA have evaluated FARS and NASS GES data in terms of fatality ratios between striking and struck drivers to identify the safety problems related to the compatibility issues [1,2,3,4]. There have been numerous test programs conducted by NHTSA and others in an effort to identify vehicle characteristics that are associated with observed vehicle incompatibility in real world crashes. These studies have shown the relative significance of vehicles mass, stiffness and geometry in front-front and front-side crashes. This paper will focus on NHTSA's recent research efforts to develop test methods for evaluating vehicle compatibility.

EVALUATION OF NCAP LOADCELL DATA

Before any new testing was conducted, it was essential to review the existing test data to see if it can provide some insights toward vehicle-to-vehicle compatibility. A review of the existing NCAP load cell data was conducted to see if the distribution of forces across the fixed rigid barrier during an NCAP test can provide an estimate of vehicle compatibility. Out of the 684 frontal NCAP tests currently stored in NHTSA's crash test database, 478 tests had acceptable load cell data. These tests were performed from 1982 to present and were conducted at the six facilities listed in Table 1. The number of tests indicate only those tests that passed quality control.

Test Performer	Number of Tests
Veridian / Calspan	232
TRC	99
MGA	63
Mobility Systems	58
KARKO	19
Dynamic Sciences	7

Five of the six test facilities use similar load cell barriers, as shown in Figure 1 below. This barrier configuration measures a 4 by 9 grid of force data.

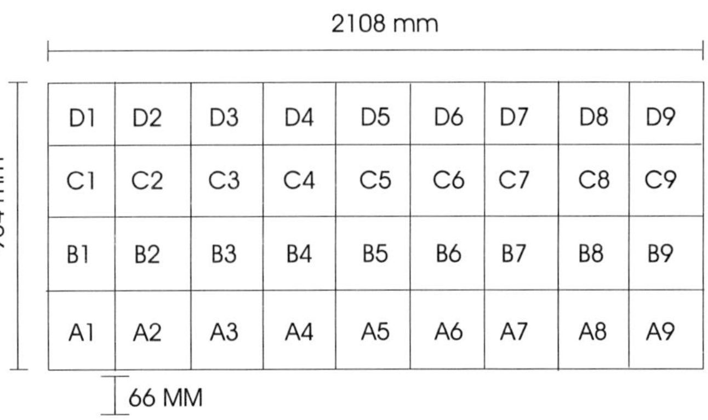

Figure 1: Loadcell barrier configuration

MGA Research, the sixth facility, uses a different load cell barrier that is longer and higher, but only measures a 2 by 3 array of force data. Each load cell "grid element" is measured using 5 load cells, but only the total force for each grid is reported. The geometry of the MGA barrier is shown in Figure 2. The loadcell data collected in MGA tests, although of reduced resolution, was otherwise acceptable and was used in this analysis.

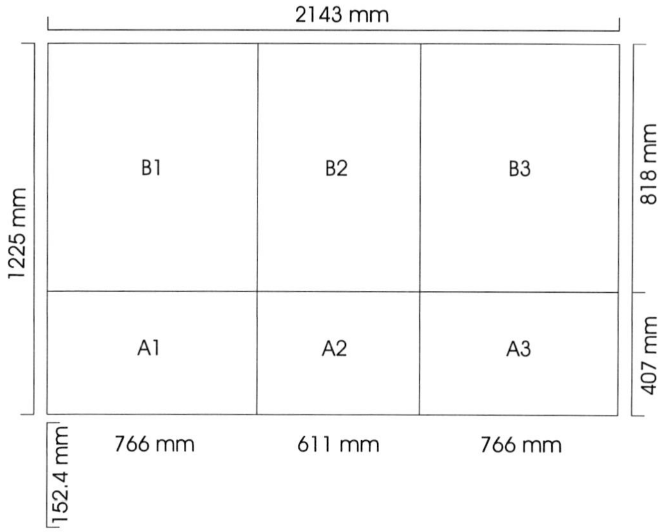

Figure 2: MGA Barrier configuration

The first step in evaluating the load cell measurements was to look at some measures of the vehicle stiffness. A reference accelerometer was identified for each test vehicle, where possible it was chosen near the drivers seating position. The load cell and accelerometer data were evaluated for consistency and combined to produce a force deflection signal. This signal was interpolated in 1 mm intervals from time zero to the first inflection point in the deflection. The force-deflection profiles were averaged by body style and weight categories as shown in Figure 3. The car test data were grouped based on test vehicle weight using the NCAP weight categories. The test weight averages 275 kg greater than the curb weight, but the weight difference varies considerably among vehicles. Also, the force deflection data were averaged at 1mm intervals. Since there is a significant difference in the crush distances, the number of tests averaged at each deflection interval is not the same. The number of tests averaged for each interval decreases as the deflection increases. Figures 4-10, show the average and standard deviations for each vehicle category. The number of tests averaged for each interval is shown in the top curve and relates to the axis along the right side. The area under the force-deflection signals is controlled by the mass of the vehicle and can be seen by comparing the profiles for the car groups.

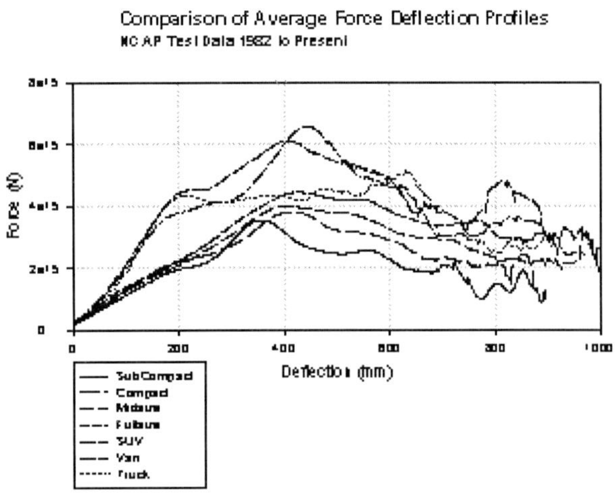

Figure 3: Average F-D profiles

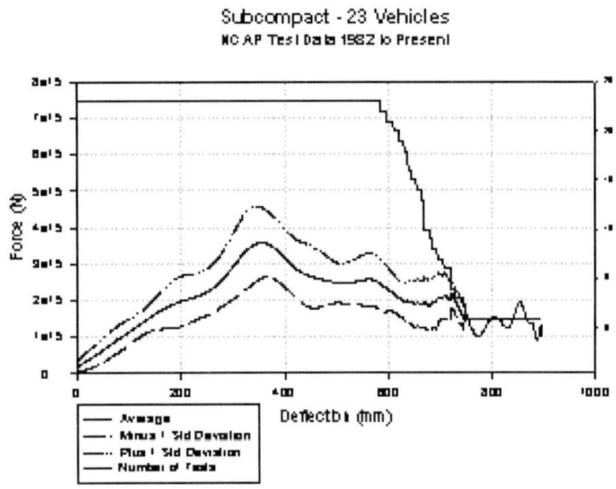

Figure 4: Test Weight < 1133.98 kg (2500 lbs)

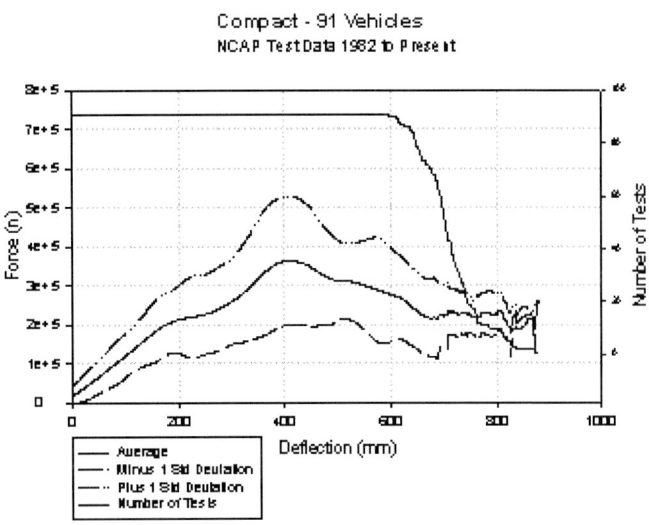

Figure 5: Test Weight < 1360.7 kg (3000 lbs)

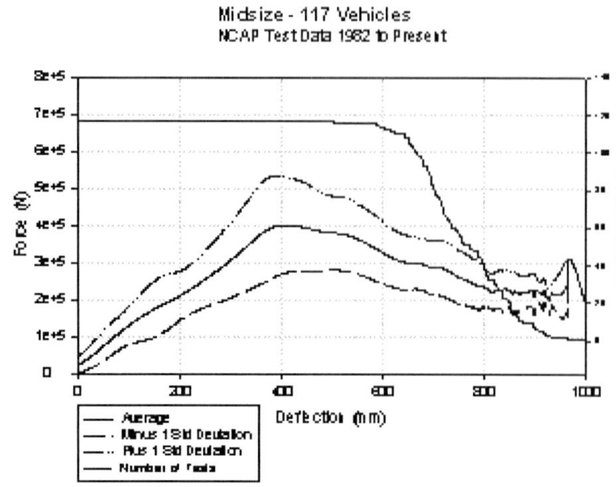

Figure 6: Test Weight < 1587.6 kg (3500 lbs)

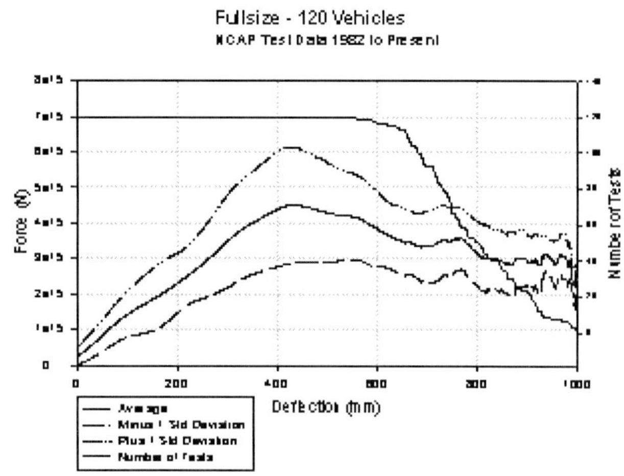

Figure 7: Test Weight > 1587.6 kg (3500 lbs)

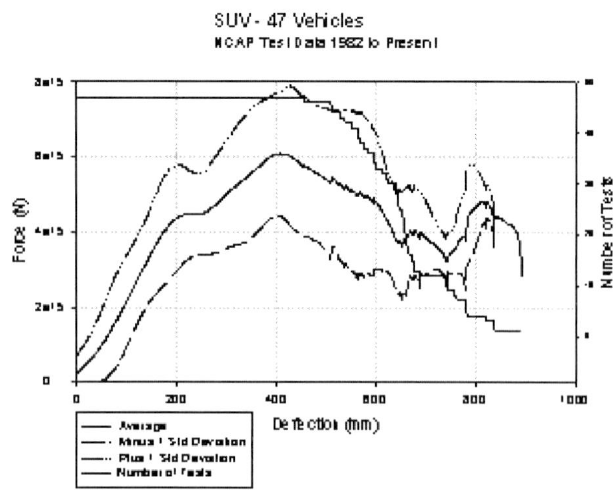

Figure 8: SUV

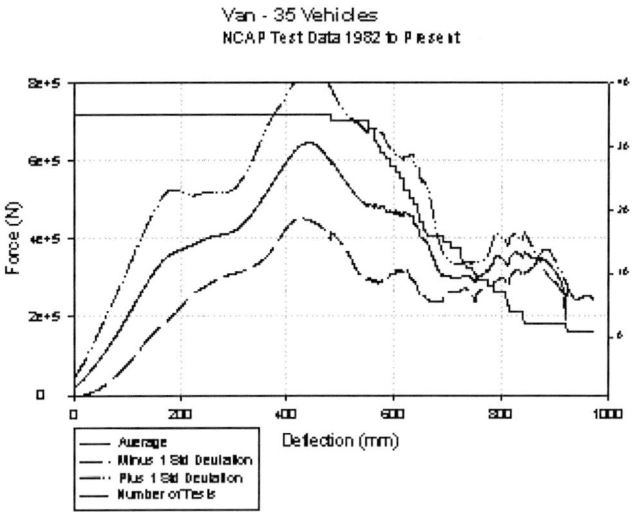

Figure 9: Van

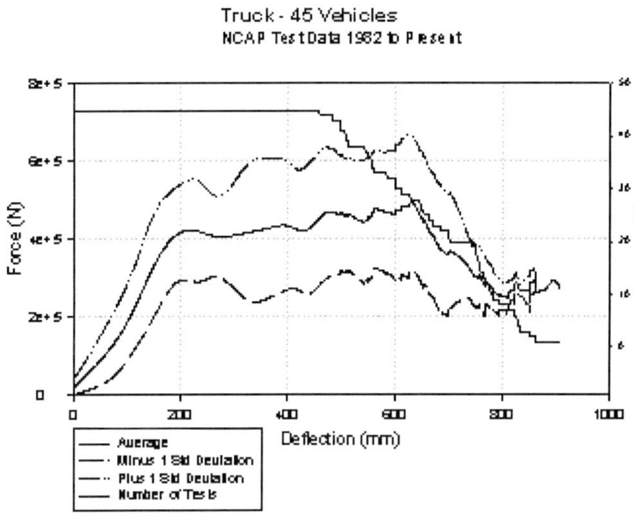

Figure 10: Trucks

Many of the NCAP tests measured significant forces prior to the indicated time zero causing the deviations in force levels at time zero as shown in Figures 4-10. However the data submitted by the test labs were not modified to correct these deviations. Examining the Force deflection profiles shows a striking consistency of the initial stiffness or slope of the force deflection curve among the car groups and within the SUV, truck, and van body styles. This dichotomy in initial stiffness may have some implications for compatibility in side impact collisions. The area under the force deflection signal is a function of the weight of the test vehicle, therefore it is not surprising to observe the increase in force between weight categories and the wide variability among the LTV categories, which cover a large range in weight.

Several methods were examined to try and characterize the initial stiffness for each test vehicle. These all involved using linear regression of the force deflection data over variable length deflections. After reviewing several alternatives, the initial stiffness for each test was determined by computing a linear fit over the longest crush distance with an $R^2 > 0.95$. The region of the fit was constrained to, start within the first 200 mm of deflection to reflect the initial stiffness of the vehicle. Only 9 tests out of 478 did not display linear behavior within the first 200 mm. The initial stiffness measurements for the remaining 469 vehicles will be evaluated against field crash data and presented in a future report.

Having established the trends in vehicle stiffness, it was desired to evaluate the load distribution across the barrier faces. The NCAP test data were analyzed to determine the average height at which the vehicles transferred force to the rigid barrier, as described in Reference 6. At any instant in time, the average height of force is computed as follows.

$$\frac{\sum_{1}^{N} F_i H_i}{\sum_{1}^{N} F_i}$$

Where F_i are the individual force measurements and H_i is the height of the load cell. This average height of force is computed for each time step and an overall average is computed using the force-time signal as a weighting function. This methods weights the average height of force to height at which the higher loads were transferred. The average height of force is shown in Figure 11 for the seven vehicle groups This figure shows the average height increases with car weight and a higher average height for the SUV and truck categories.

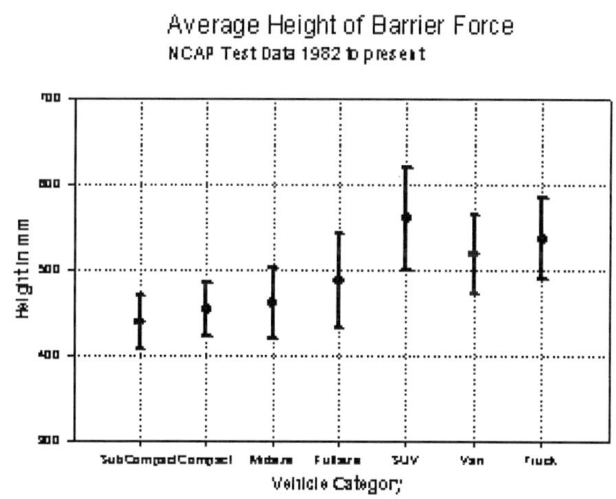

Figure 11: Average Height of Force

The power for the crash is computed by multiplying the force by the instantaneous velocity of the vehicle's center of gravity. The peak power was also computed for each of the NCAP tests. Figure 12 shows the average peak power for each of the vehicle categories.

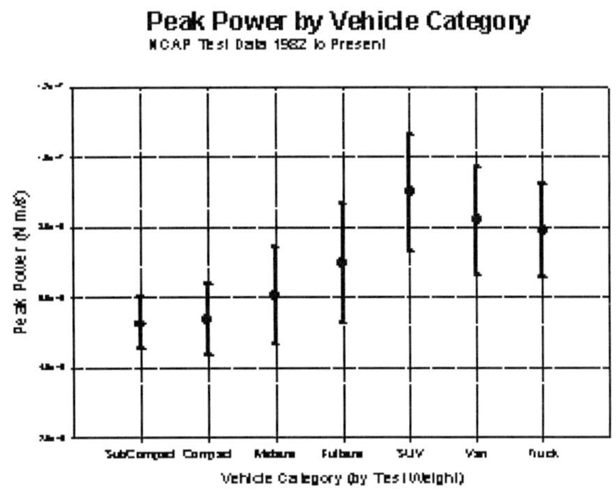

Figure 12: Peak Power

The peak power shows a general increase with car weight and the SUV, truck, and van categories are significantly above the passenger car categories.

CORRELATION OF TEST METRICS TO CRASH DATA

The significance of the test derived metrics can only be established by comparison to the field crash data. A recent study by UMTRI, see Reference 7, compared preliminary vehicle metrics derived from NCAP test data to real world crash performance as measured in FARS and NASS GES. In this study, it was necessary to decode the VIN data in order to map the field crash data to the corresponding NCAP test results. Not all NASS regions provide VIN information, so considerable effort was necessary to develop national estimates. Not surprisingly, there was a strong correlation between vehicle weight and aggressivity measures based on struck driver fatalities rates. An equivalent stiffness was used, based on post test deformation (X), where stiffness (k) was defined as shown below. (m = mass, v = velocity)

$$k = \frac{(mv^2)}{X^2}$$

These stiffness values were not found to have any significant correlation to the vehicle's aggressivity, even when the data were corrected for the dominant effect of the weight ratio of the vehicles. The study also evaluated unweighted averages for the height of force. The average height of force was shown to have a mild correlation with aggressivity, but only after correcting for the weight ratio of the vehicles involved in the crash. This study is currently being updated to include the vehicle metrics previously discussed, initial stiffness, weighted average height of force, and peak power. Additionally, the study is being augmented with an additional two years of FARS and NASS GES data and will try to separate aggressivity data for frontal and side impact collisions. The updated study is expected to be released towards the end of 2001.

EVALUATION OF COMPATIBILITY TEST PROCEDURES

The NCAP test derived metrics may have some potential for estimating a vehicle's compatibility. However, the NCAP test procedure does not account for the weight ratio of the vehicles involved in a vehicle to vehicle test. An NCAP test simulates a vehicle running into itself at 56 kmph. This is useful for measuring a vehicle's load distribution, but it does not fully evaluate occupant protection in real world vehicle-to-vehicle crashes where the change in velocity is a function of the weight of the collision partner. It is desirable to investigate test methods that can account for the weight ratio between the test vehicle and an average or expected collision partner. A second concern with rigid barrier testing is that the rigid wall will fully engage all of the structure of the test vehicle and does not allow for any override or underride that may occur in a real world crash.

NHTSA is investigating the use of a moving deformable barrier, MDB, with load cell instrumentation behind the honeycomb face. The goal would be to design an MDB that represents an average vehicle and to require occupant survivability in the struck vehicle, or self-protection. The test procedure must also place limits on the amplitude and distribution of the forces measured on the face of the MDB, or partner-protection. As a preliminary test of this concept, a test program was devised using a load cell moving deformable barrier, LCMDB, developed for side impact testing[8]. A preliminary test series was conducted to see if the load cell instrumentation can provide sufficient data to distinguish between aggressive and non aggressive vehicles. A full frontal test was used to provide the maximum coverage of the load cells and to provide for some limited correlation with NCAP test data. Using a fixed mass for the LCMDB, the vehicle weight ratio and the resulting change in velocity for the struck vehicle is controlled by weight of the subject vehicle. Thus, the severity of the crash will depend upon the mass of the test vehicle.

This preliminary test series was also conducted to evaluate the load distribution in LCMDB-to-vehicle crashes and to compare damage and injury measurements in an equivalent vehicle-to-vehicle crash. Three tests were conducted and are shown in Table 2 below.

Test	Vehicle 1	Vehicle 2
3362	LCMDB, 56.2 kmph, 2051 kg	1996 Plymouth Neon, 55.9 kmph, 1382 kg
3413	LCMDB, 56.2 kmph, 1377 kg	1997 Dodge Caravan, 56.8 kmph, 2138 kg
3414	1997 Dodge Caravan, 56.5 kmph, 2060 kg	1996 Plymouth Neon, 55.9 kmph, 1378 kg

The baseline test was a full frontal collinear crash involving a 1996 Plymouth Neon and a 1997 Dodge Caravan, each moving 56 kmph. In the second test, the weight of the LCMDB was adjusted to match the Dodge Caravan, and the LCMDB was fitted with an FMVSS No. 214 barrier face. This LCMDB was impacted into the Plymouth Neon. In the third test, the LCMDB weight was adjusted to match the

Plymouth Neon and the Dodge Caravan was impacted. This test series was intended to see if an MDB impact could produce vehicle acceleration, crush, and occupant responses similar to that in vehicle-to-vehicle tests.

INJURY CRITERIA

All vehicles had a belted 50th percentile male dummy in the drivers seat and a belted 5th percentile female in the passenger seat. The computed injury criteria for the dummies are shown in Tables 3 and 4 below. The injury criteria for the Neon driver struck by the LCMDB are generally severe, and considerably higher than measured for the baseline driver. The Neon passenger injury measures are significantly less severe and are generally consistent between the baseline and LCMDB tests. For all passengers, except the baseline Neon, the head data had numerous significant spikes that prevented computing HIC values. The Caravan passenger, when struck by the LCMDB, had an exceptionally high neck extension which led to the high maximum Nij value. Additionally, the chest injuries were higher for both the driver and passenger struck by the LCMDB. The HIC and right femur injury criteria for both the Caravan driver and the passenger were generally consistent between the baseline and the LCMDB test.

Driver 50th male	Baseline Caravan	LCMDB Caravan	Baseline Neon	LCMDB Neon
15 ms Hic	638	621	521	1414
Max Nij	0.58	0.47	0.54	0.86
Chest G's	44.6	51.3	69.0	100.4
Chest Disp.	29	45	40	39
L Femur	2839	5545	5299	9113
R Femur	5596	6384	6717	12053

Passenger 5th female	Baseline Caravan	LCMDB Caravan	Baseline Neon	LCMDB Neon
15 ms Hic			411	
Max Nij	0.70	3.34	0.89	0.90
Chest G's	34.9	48.3	55.8	55.3
Chest Disp.	29	33	35	34
L Femur	4770	5937	4862	6304
R Femur	3242	3533	3262	3589

INTRUSION

The collisions with the LCMDB produced significantly more deformation and intrusion in the Neon than the baseline test. The Caravan had slightly higher intrusion in the baseline test than in the LCMDB test, but the change was small compared to the Neon tests. The maximum exterior crush was unchanged for the Caravan tests.

Max in mm	Baseline Caravan	LCMDB Caravan	Baseline Neon	LCMDB Neon
Toepan Intrusion	274	212	189	521
Exterior Crush	588	586	528	606

CRASH PULSE

One of the goals for this test series was to evaluate the potential for using an MDB to approximate a vehicle-to-vehicle crash environment. The acceleration of the Neon center of gravity is plotted in Figure 13 for the Neon-Caravan and the Neon-MDB tests. The acceleration profiles are generally close in shape and timing. The peak acceleration when struck by the MDB is slightly higher. The corresponding acceleration plots for the Caravan-Neon and Caravan-MDB tests are shown in Figure 14. Here there is a significant difference in pulse duration, an odd drop in acceleration around 40 ms, and a higher peak acceleration for the Caravan-MDB test. The overall crash pulse shape and timing are generally similar.

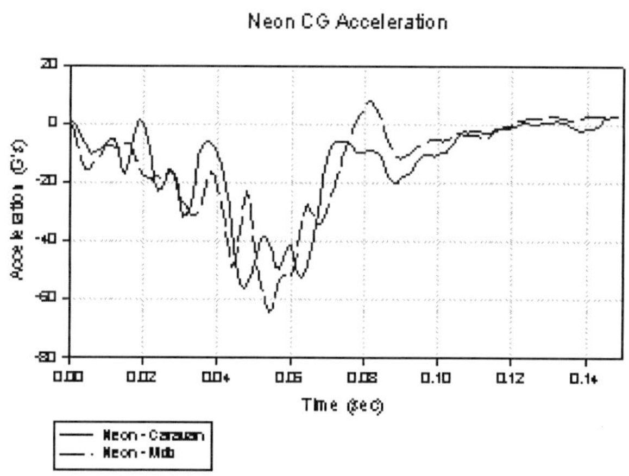

Figure 13: Comparison of Neon Acceleration

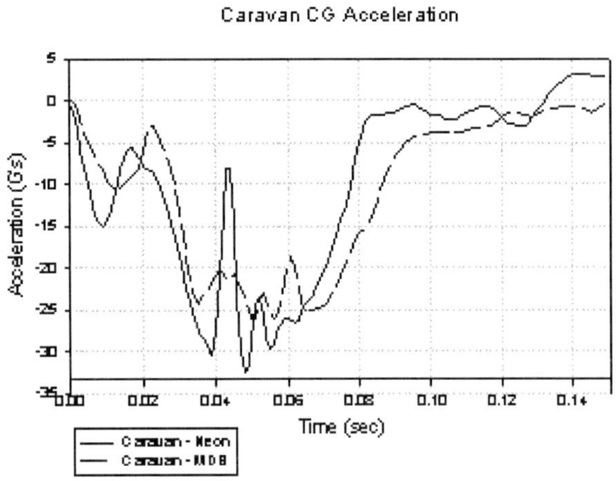

Figure 14: Comparison of Caravan Acceleration

FORCE DISTRIBUTION

Another goal for this test series was to see if the LCMDB can distinguish, based on load cell measurements, between striking the Neon and the Caravan. Before comparing force data, it is important to evaluate the accuracy of the force measurements. Assuming a constant mass for the LCMDB, the total measured force, divided by the mass should approximate the acceleration-time history of the LCMDB. Figures 15 and 16 compare the predicted and measured velocity profiles for the Neon-LCMDB and Caravan-LCMDB tests respectively. The two curves starting at 56 kmph and slowing down represent the measured and loadcell estimated velocity profile for the LCMDB, Approximately 40 percent of the change in velocity is not accounted for in the Neon-LCMDB test. The Caravan LCMDB test is better; however, still 15 percent of the change in velocity is not accounted for. This type of discrepancy is not generally observed in the NCAP test data and this simple impulse / velocity check was one of the primary factors used for quality control in reviewing the NCAP data.

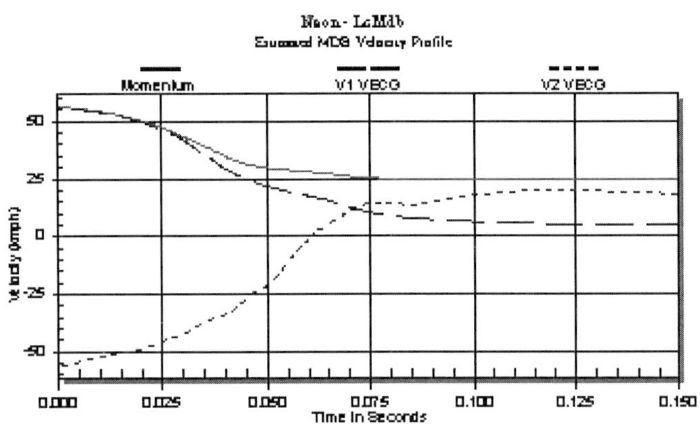

Figure 15: Evaluation of Force Data, Neon Test

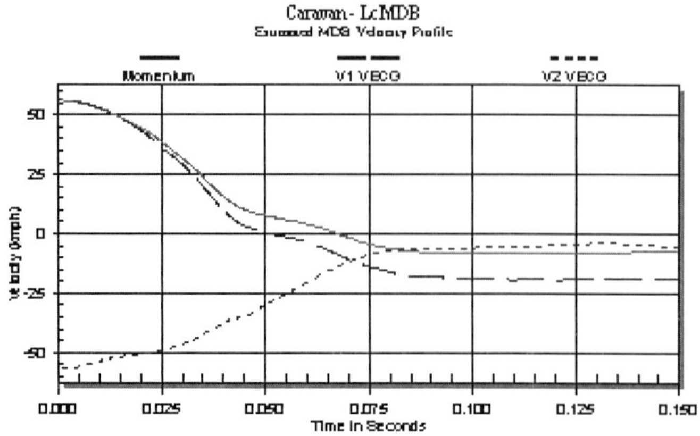

Figure 16: Evaluation of Force Data, Caravan Test

The missing force data may be understood by examining the crush patterns of the MDB face from the Neon test, shown in Figures 17 and 18. Figure 17 shows a two inch mounting surface below the bottom edge of the honeycomb face. Corresponding to this bottom edge in Figure 18, there is a noticeable crease in the deformed honeycomb. While 40 percent is a lot of force to be transferred through this small surface, this is clearly a design issue that should be addressed if an LCMDB is to be used for compatibility testing. A similar, but less severe folding pattern was also observed in the Caravan-LCMDB test.

Figure 17: Crushed MDB Face, Neon Test

Figure 18: Crushed MDB Face, Neon Test

The evaluation of the NCAP test data focused on the distribution of the force data across the surface of the load cell matrix. Figures 19 and 20 compare the impulse or time integral of the force data as measured in the NCAP and LCMDB tests respectively. These two plots cannot be directly compared because of the different geometry of the load cells. Figure 21 shows a comparison of the load cell layouts. The LCMDB effectively measures the locations corresponding to the middle two rows of the rigid barrier wall. Over this region, the LCMDB measurements are more widely distributed than for the rigid barrier. This distribution of force may be attributed to bumper element on the MDB face which was designed to spread out the load across the side of the vehicle. The bumper element is working against the goal of estimating the compatibility of the striking vehicle from the distribution of force on the load cell face. Because a significant percentage of the force was not measured in the Neon test, no strong conclusions can be reached about the distribution of force measurements from this test series.

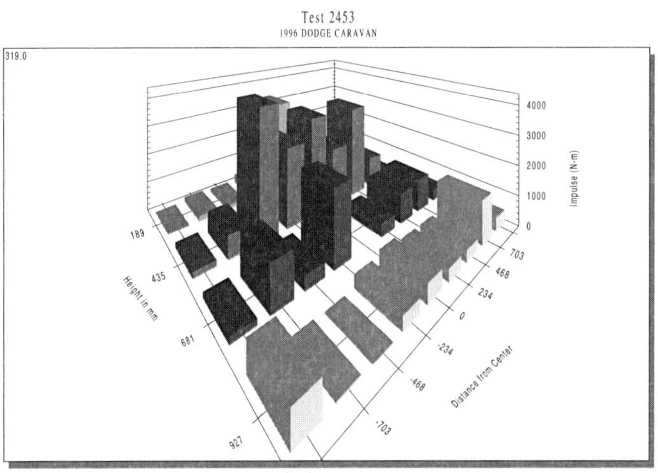

Figure 19: Distribution of Force - NCAP test

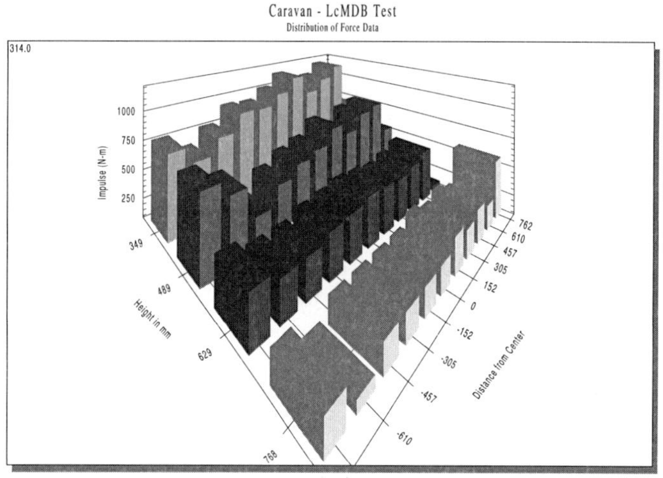

Figure 20: Distribution of Force - LcMDB

Comparison between Load cell Barrier and LcMDB

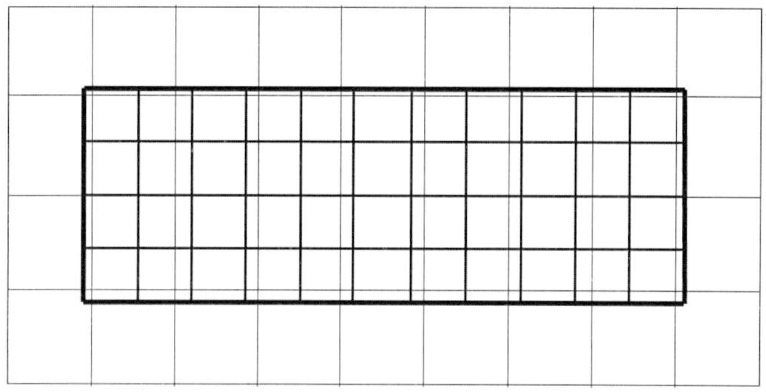

Figure 21: Overlay of NCAP and LcMDB measurements

CONCLUSION

NHTSA's compatibility research program is attempting to develop and evaluate compatibility test measures from new and existing test data. These measures should show some correlation to real world crash data and should consider both the self and partner protection aspects of vehicle compatibility. Initial testing using an existing load cell MDB produced a crash environment that reasonably replicated a vehicle to vehicle crash, but the load cell data did not produce a satisfactory measure of the striking vehicles compatibility. Additional research will be necessary to see if an MDB face can be designed that will meet the goals or approximating the crash performance of an average vehicle and providing an estimate of the compatibility of the striking vehicle.

ACKNOWLEDGMENTS

The authors would like to acknowledge to contributions of the IHRA Compatibility research group for their advice and feedback on this research program.

REFERENCES

1. H. Gabler, W.T. Hollowell, "The Aggressivity of Light Trucks and Vand in Traffic Crashes", SAE Paper 980908, March 1998

2. H. Gabler, W.T. Hollowell, "The Crash Compatibility of Cars and Light Trucks", Journal of Crash Prevention and Injury Control

3. H. Joksch, "Vehicle Aggressivity: Fleet Characterization Using Traffic Collision Data"",DOT HS 808 679, February 1998

4. H. Joksch, ""Fatality Risks in Collisions between Cars and Light Truck",DOT HS 808 802, October 1998

5. S. Summers, A. Prasad, W.T. Hollowell, "NHTSA's Compatibility Research Program", SAE Paper 01B-257, March 1999

6. K. Digges, A. Eigen, J. Harrison, "Application of Load Cell Barrier Data to Assess Vehicle Crash Performance and Compatibility", SAE Paper 1999-01-0720, March 1999

7. H. Joksch, "Vehicle Design versus Aggressivity", *To be published*

8. Trella, Samaha, Fleck, Strassburg, "A FMVSS 214 Moving Deformable Barrier with Dynamic Force and Deflection Spatial Measurement Capabilities for Full Scale Tests", SAE Paper 2000-01-0637, March 2000

1999-01-0068

Possibilities and Limits in the Design of Compatible Cars for Real World Accidents

Falk Zeidler, Fred Knöchelmann and Dieter Scheunert
DaimlerChrysler AG,71059 Sindelfingen, Germany

ABSTRACT

Up to now, incompatibility of cars was mainly derived from theoretical considerations due to differences with regard to mass, stiffness, and geometry. Numerous accident analyses and reports only showed, that mass is a dominant factor for the injury outcome in car-to-car collisions. The same analyses have difficulties to assess the influence of stiffness and geometry. One reason is the wide scattering of the injury outcome due to other more important variables than incompatibility. A second reason is the fact that stiffness correlates more or less with the mass. A third reason is the effect of the non-linearity of the geometry influence that means a positive influence of incompatible weak structures causes a reduced acceleration, but with an increased risk of intrusion into the passenger compartment. This can be observed at high impact speeds and/or low overlap degrees. These effects can be seen from accident analysis. Nevertheless, on the basis of theoretical considerations, the aspect of compatibility can be implemented in the design of new cars. This is true for a certain range of impact severity and for a limited number of car size groups only. A realistic approach in this regard is detailed in the paper.

INTRODUCTION

Looking at Germany's accident statistics, about two million road accidents occur each year resulting in nearly 500.000 injured persons and approx. 9.000 fatalities (fig.1). For the year 1996, for example, 2673 out of 5622 car occupants (47.5%) were killed in single vehicle accidents (fig. 1) against trees or rigid obstacles, rollovers, etc. suggesting that compatibility is not a very big issue. In contrast to this result, the number of injured car occupants in car-to-car collisions, namely 159,308 of 313,831 (50.8%) is very dominant (fig. 2), concluding that compatibility is more important in the case of a car-to-car collision. The question concerning the right design (self protection / partner protection) rises up immediately. The answer, found years ago, was and is still a stiff passenger cell with sufficient crumple zones. To understand in more detail the problems of compatibility, it is necessary to have a closer look at the distribution of the collision types and the mass of the other car. Most frequent is the frontal impact (55%, see fig. 3). 63.8% of the car-to-car collisions are distributed in the mass class 801 kg - 1200 kg (fig. 4). According to figure 5, a 30% frequency is given for incidents of "mid-class-car" versus "executive car".

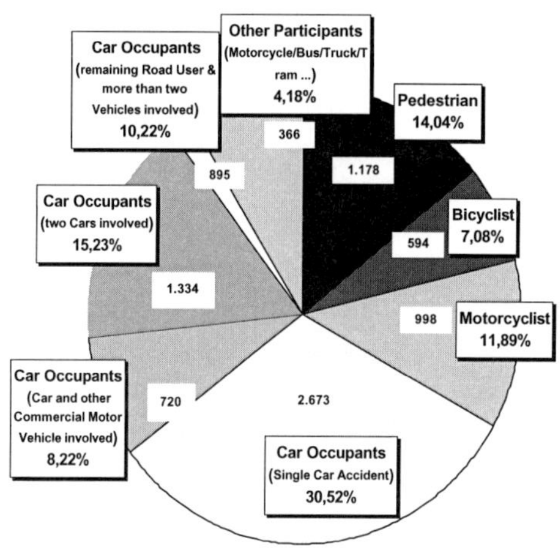

Figure 1. 8.758 Fatalities in Germany 1996 (Stat. Bundesamt)

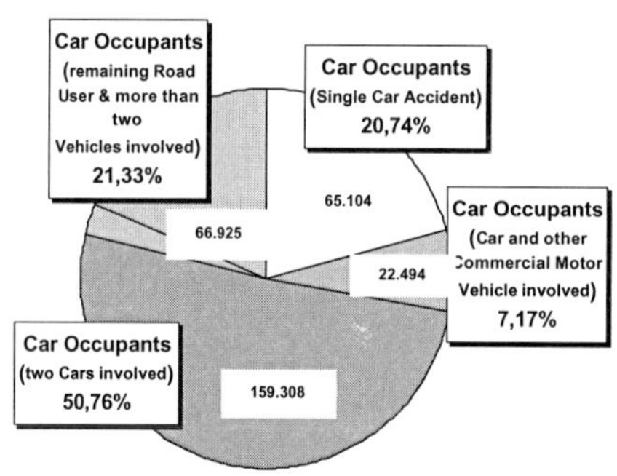

Figure 2. 313.831 injured occupants in Germany 1996 (Stat. Bundesamt)

The accident data shows that two thirds of the car-to-car collisions are in the same mass class which could be a hint that for compatibility reasons more parameters must be taken into account like geometry and stiffness. Mass plays a role in a few cases of "David" versus "Goliath", only (4%). A third effect can be observed in real world crashes.

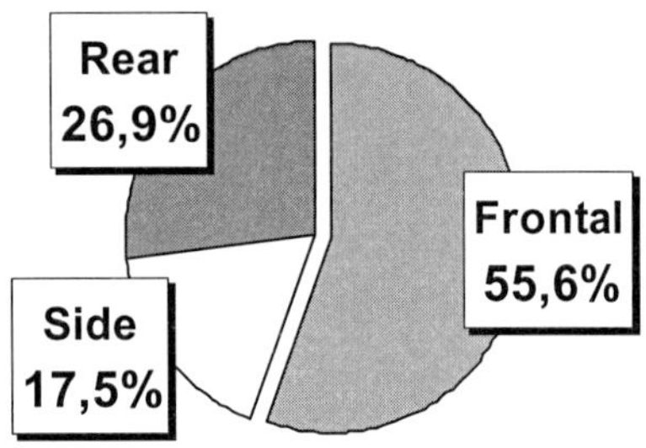

Figure 3. Collision configuration of 16.007 car-to-car accidents

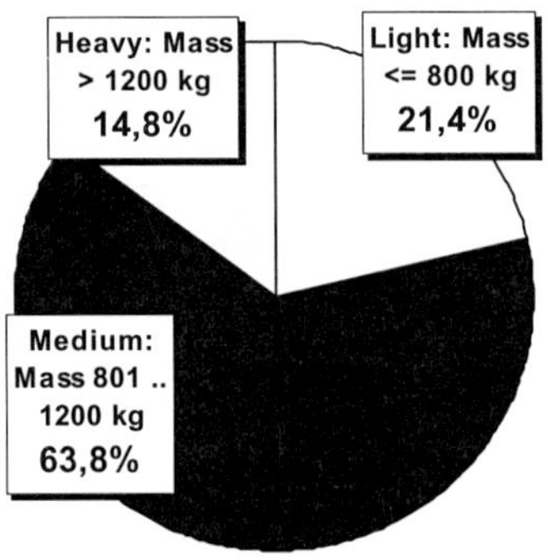

Figure 4. Mass classes of 16.007 car-to-car accidents

Small versus small	0.9%
Small versus midsize	11.3%
Small versus large	3.9%
Midsize versus midsize	49.0%
Midsize versus large	30.0%
Large versus large	4.9%

Figure 5. Frequency of car-to-car accidents in different car sizes (BASt)

REAL WORLD 'FORK' EFFECT

It is obvious that in many crashes where a car hits another car, the longitudinal side rails are not directly involved in absorbing energy. Activating as designed, the front end is mostly damaged in weaker zones with the result of high amounts of deformation, and a low deceleration of the car. This principle is best described in the term 'incompatibility'. In cases where the impact severity is only minor or moderate, the injury outcome of the passengers is minor even at a low level. If in this cases the design of the front end would be more compatible, which means homogenous according to the front end height and width, the load on the occupants would increase, but not necessarily so high that the occupants would suffer severe injuries. The big problem of such a 'by chance' design is that the rigid areas are starting at a point where no more deformation is acceptable because it would mean the beginning collapse of the passenger cell. Therefore, the injury outcome can differ significantly due to some centimeter more or less deformation and intrusion, this means the crash behaviour is non-linear. For example, if a truck impacts a passenger car and the overhanging structures of the truck directly hit the A-pillar and the windshield, no deformation energy is absorbed in the front end, but the occupant may be killed due to contact with parts of the truck. All physical parameters which describe the impact severity are very low. The usual physical parameters which describe the impact severity such as ΔV or EES may be relatively low in this case.

As a consequence of the non-linearity of the geometrical influence, this kind of incompatibility should never be a design concept in spite of its positive contribution in some cases.

COMPATIBILITY MEASURES

What measures should be used to determine compatibility? If a manufacturer is willing to design compatible structures, it is necessary to balance the relative structural stiffness of the active crumple zones of light and heavy cars.

STIFFNESS – Stiffness - in this case, more accurately expressed as the resistance to deformation of the front end of the vehicle - is an essential input variable for describing the changes of state during car-to-car collisions.

In the descriptions that follow, first the structural forces are assumed to increase in direct linear proportion to the deformation path. This assumption corresponds approximately to the behaviour of today's weight-optimized steel structures with the engine/gearbox unit and engine packaging in the front end.

Second, in the theoretical analyses, it is initially assumed, for the sake of simplicity, that the front end forces are distributed over an ideal surface area, both against the crash barrier and against the opposing car in a car-to-car collision.

This latter assumption corresponds with varying accuracy to actual car structures.

Figure 6 shows the design that has been adopted up to now, with all car front end structures permitted to have roughly the same deformation length of about 500 mm in the 50 km/h barrier test, regardless of their weight. This specification enables today's cars to remain within the legal biomechanical load limits relatively safely by means of the latest occupant restraint systems. The fact that this crumpling length is independent of vehicle mass leads to the generation of structural forces proportional to mass. As a result of this state of affairs, the deformation patterns that have come to light in actual accidents have eventually contributed to conceding a new priority to the theme of compatibility.

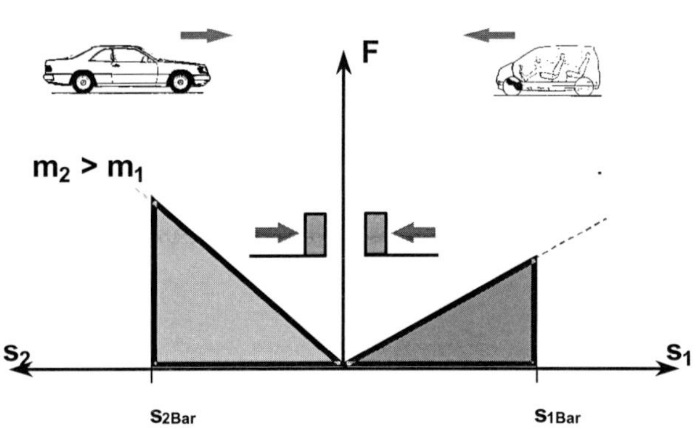

$DF_1; DF_2$: the slope is proportional to the mass and dependent on the deformation

Deformationlength in

car-to-car collision $s_{2car} = \sqrt{m_1/m_2} * s_{1car}$

Barrier impact $s_{2Bar} = s_{1Bar}$

Figure 6. Existing Deformation Characteristic

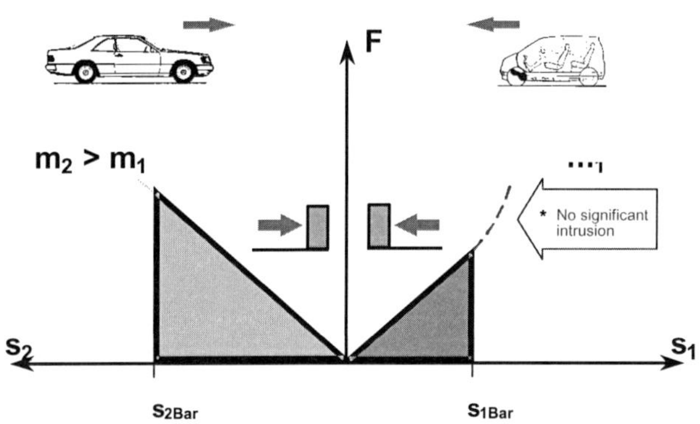

$DF_1; DF_2$: linear increasing with the deformation

Deformationlength in

car-to-car collision*): $s_{1car} = s_{2car}$

Barrier impact $s_{1Bar} = \sqrt{m_1/m_2} * s_{2Bar}$

Figure 7. Theoretically Ideal Deformation Characteristic

As a logical consequence, and as an essential step in achieving compatibility between the structural stiffness of vehicles of different weight classes, a specification conforming to Figure 7 is recommended. This formulation is characterized by the same linear increase in front end stiffness in proportion to deformation length, up to a given load condition, e.g. 50 km/h barrier test (Appel et al, 1994). As a result, in a car-to-car collision, the deformation of the two cars will be identical, while in a barrier impact the deformation characteristics correspond to the square root of the mass ratio.

DEFORMATION LENGTH – What criteria should be adopted in determining the stiffness or the increase in this characteristic, and what reasonable framework conditions should be considered?

Figure 8 shows the best method: minimum and maximum deformation lengths are introduced in addition to the deformation characteristic described above (figure 7). The minimum deformation length of 330 mm is determined from the objective of ensuring success in the 50 km/h barrier crash test with the currently valid biomechanical load values.

With the assumptions that:

- **Deformation forces increase linearly with the deformation**
- **Force-deflection characteristics of the front structures are appr. equal**
- **A deformation length of minimum 330 mm is required for "Minis"**
- **Vehicle masses** (for useful consideration) **600 < m < 1800** (kg)

the following relationship is valid to determine the required deformation length:

$$Def_{(req.)} = Def_{(min)} * \sqrt{m_{(exist.)} / m_{(min)}}$$

and approximating with the above mentioned value for the deformation length:

$$Def_{(req.)} \approx 13{,}5 * \sqrt{m_{(exist.)}}$$

Def in (mm)
m in (kg)

in the **theoretical range**

$$330 < Def_{(req)} < 570 \quad \text{(without bumper)}$$

Figure 8. Determination of Deformation Length (Flat barrier impact at 50 km/h)

It should also be noted that with decreasing vehicle mass, then the structural strength of the passenger compartment must increase.

An effective deformation length of more than 570 mm is not feasible due to the conflict of interests with other vehicle design requirements.

The result is shown in Figure 10 is an ideal compatibility curve. This curve provides sufficient protection for the small car in a car-to-car collision with a large car. Nevertheless, due to physics, the large car has – for this case – the minor impact severity, and consequently better protection.

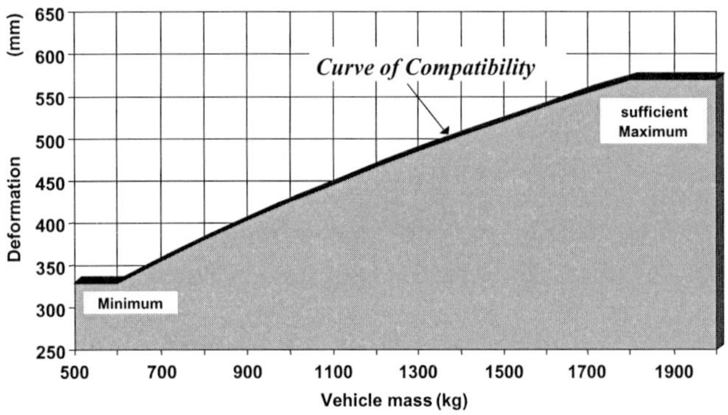

Figure 9. Ideal Curve of Compatibility (Flat barrier impact at 50 km/h)

SMALL VERSUS LARGE

During the MCC development process, all compatibility aspects were considered. The front end is homogenous in height and width so that the assumptions are very promising to withstand a car-to-car collision with a heavier and compatible car.

Therefore, Daimler-Benz carried out a crash test for a MCC against an E-class both at 50 kph with a mass ratio 1:1.8 and an overlap degree of 50% for the MCC (fig. 10). The results show that both vehicles offer enough survival space for the occupants. Even, at this high impact severity the loads for the dummies in the MCC remained significantly below the protection criteria specified in FMVSS 208 as well as ECE-Regulation 94.

Figure 10. Small versus large

DISCUSSION

As mentioned above, it is possible to design compatible cars within certain limits. Both, the heavy and the light car can benefit from this concept due to the reduction of the mean acceleration in the compression phase of the crash. The principle 'homogenous front end' is appropriate for car-to-car collisions where the mass ratio of the colliding cars are nearly 2:1. Considering and handling a greater mass ratio would mean more stiffness in the front end for the small car and would lead to more aggressivity if the car hits another car in a side impact. Therefore, when considering today's vehivles the mass ratio of 2:1 is a reasonable limit.

REFERENCES

1. Appel, H.; Ni, J.; Deter, T.: Grenzen der Passiven Sicherheit von leichten Fahrzeugen, VDI-Berichte 1994

2001-06-0060

DEVELOPMENT OF CRITERIA AND STANDARDS FOR VEHICLE COMPATIBILITY

Robert Zobel

Thomas Schwarz

Volkswagen AG

Germany

Paper Number: 140

ABSTRACT

This paper summarises the results of a project on vehicle crash compatibility, run by European automotive industry together with some research institutes. The project was funded by the European commission as BE97-4049.

There are three main issues that can be detected in real world accidents, influencing vehicle compatibility. These issues are mass differences, compartment integrity with regard to frontal car-to-car impact, and differences in bumper and sill height in side impact. Longitudinal mismatch in frontal impact, front end stiffness and other items which are from a theoretical point of view responsible for vehicle aggressiveness are not seen influential from the view point of real world accidents.

On the other hand, compartment collapse occurs, when there is not sufficient deformation energy available in vehicle front-end. And deformation energy is available, when it is provided by vehicle structures and when these structures interact. So compartment collapse can only be avoided, as long as sufficient deformation energy is available and is effective within the car-to-car collision.

So the paper will provide results with regard to the questions of deformation management. It will give guidelines for future research on compatibility, which has to focus on the problem of structural interaction.

GOAL

In vehicle-to-vehicle crashes, two vehicle safety viewpoints have to be considered (Figure 1):

- **Self-protection**, the ability of a vehicle to protect its own occupants, both in vehicle-to-vehicle accidents and against other objects in the traffic environment.
- **Partner-protection**, the ability of a vehicle to protect the occupants of the opponent vehicle in vehicle-to-vehicle crashes.

The intention of this project was to minimise fatalities and injuries in the vehicle fleet by taking into account self and partner protection. It was aimed to develop common design rules to achieve an optimum structural interaction of vehicles. This goal should be reached by requirements on implications of vehicle structure like restricted force levels, dummy loads etc.; it is not understood as a restriction of the design possibilities. It should be achieved as far as possible by computer simulation with finite element models (FEM). The knowledge should be summarised in a suggestion for a procedure for an enhancement of compatibility of the vehicle fleet that could lead to a European standard. The **Compatibility** of a vehicle is understood as a combination of self- and partner protection in such a way that optimum overall safety is achieved, this is regardless of in which vehicle the injuries or fatalities occur. It is not acceptable to compromise today's self-protection level.

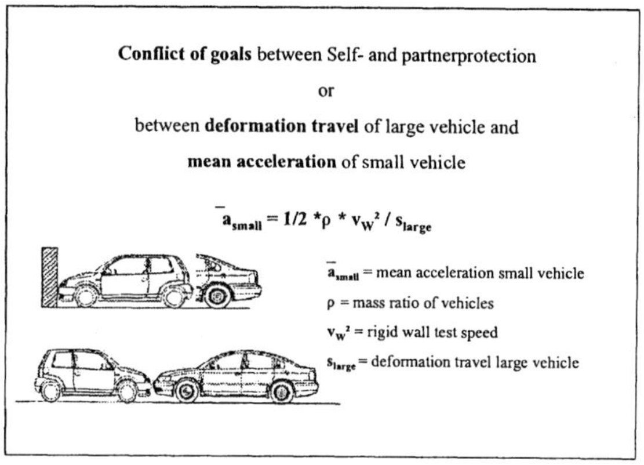

Figure 1: Conflict of goals in compatibility

ACTIONS

Two steps were taken to reach this goal:

- **Accident analysis** was conducted using accident databases from Germany, France, Great Britain, Sweden and Finland. Variables that might be relevant with regard to self and partner protection were considered.
- **Car-to-Car and barrier impact analysis** including
  - Crash tests were conducted to study the findings of accident analysis and derive criteria for the evaluation of vehicles.
  - Computer simulation was conducted to achieve a deeper understanding of the effect of variables considered and to study, whether it is possible to substitute crash testing for the evaluation of crash compatibility.

ACCIDENT ANALYSIS

The factors were rated as compulsory when they have a clear influence on compatibility. They were rated as possible, when an influence was not seen or not clearly seen in accident analysis but when other reasons were found to study this effect more deeply by crash tests and computer simulation (Table 1).

Variable	**Importance** based on accident analysis	**Importance** taking into account accident analysis, crash tests and computer simulation
Mass	**Compulsory**	**Frontal impact: Not considered, unchanged** **Side impact: Minor relevance, unchanged**
Front-end stiffness	Possible	Enable sufficient deformation energy up to a defined force-level
Compartment stiffness	**Compulsory**	**Compulsory proved**
Lateral fork effect	No effect	→ Structural interaction
Vertical fork effect	No effect	→ Structural interaction
Lateral longitudinal mismatch	No effect	→ Structural interaction
Vertical longitudinal mismatch	Possible	→ Structural interaction
Override of sills	**Compulsory**	**Compulsory proved**
Homogenous front-end	Possible	→ Structural interaction
Engine orientation	Possible	Unchanged
Corner impact	Special case	Unchanged

Table 1: Compatibility factors derived from accident analysis

The accident databases that were studied consist of mainly older vehicles. These vehicles have older generation restraint systems, a significant number have no airbags or pretensioners and are unlikely to be have been designed for the offset deformable crash test.

Only three of the issues that were highlighted in the list were found to be clearly relevant from the viewpoint of accident analysis.

Vehicle mass is a well-known influence factor in vehicle-to-vehicle crashes. It was decided that it was infeasible to affect vehicle mass with regards to compatibility. A mass ratio of 1.6 covers 80% of the European passenger vehicle-to-vehicle accidents and defines a range within which vehicle compatibility can be improved, if a self protection level of a rigid barrier impact speed of approximately 56 km/h is considered.

Vehicle-to-vehicle frontal accidents with fatalities or serious occupant injuries were mainly accidents with severe **deformation of the passenger compartment**. In severe accidents involving older vehicles high occupant decelerations and severe compartment intrusions result from loading of the compartment. These two effects cannot be discriminated by accident analysis.

Accident analysis found that **front-end stiffness** was not relevant but from a theoretical understanding of vehicle impacts it was rated as a 'possible' variable. The fact that this variable was not found in accident analysis is surprising and from a theoretical point of view questionable. This observation might change, when a fleet of stiffer vehicles exists and can be analyzed.

Fork effects - undeformed longitudinals intruding into an opponent vehicle - were observed only in a very small number of cases in accident statistics. However, cases of structural **mismatch** - different vertical or lateral position of load paths of colliding vehicles - were observed. In some statistics, it was found that mismatch in a vertical direction is an advantage or at least no disadvantage. This finding was not expected. A possible explanation is that in the majority of less severe accidents a mismatch is an advantage because it leads to lower force levels and thus to lower decelerations. In more severe accidents mismatch may be a disadvantage because it might lead to higher deformations. Due to the smaller number of more severe accidents in the databases this would explain the finding from accident analysis.

IMPACT ANALYSIS: FRONTAL IMPACT

Compartment Stiffness (Resistance)

Accident analysis results showed that **compartment stiffness** (resistance) was one of the main compatibility issues. Crash tests were performed to understand compartment **overload**. High-speed crash tests (so called destruction tests) were conducted with various small cars to overload the passenger compartment and investigate its deformation resistance well beyond the elastic limit. Time histories of acceleration and force measurement on a load cell wall were studied. Depending on the vehicle structure and engine position there were significant differences between the vehicles.

One of these small cars was used to study repeatability. In three similar tests against an offset deformable barrier (ODB) with a high EES (~74km/h) different intrusion patterns and structural forces occurred (Figure 2).

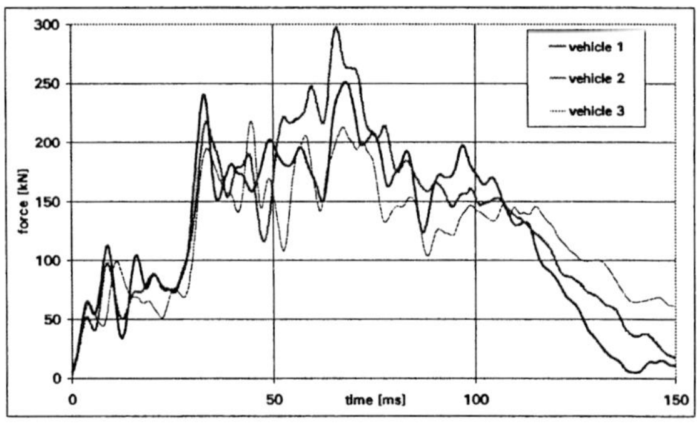

Figure 2: Structural forces from destruction tests with identical cars

Two further tests with this car were conducted at 64 km/h against an ODB. The force deflection curves derived from compartment acceleration for these tests are similar (Figure 3). This kind of test reflects an energy equivalent speed (EES) of 55...60 km/h. In this test most today's vehicles retain structural integrity. The maximum force detected in both of these tests was 250 kN. This is also the average maximum force level for the series of the three destruction tests with this car, although the values show a high scattering. In a vehicle-to-vehicle crash test with this car, the force level of 250 kN was also achieved.

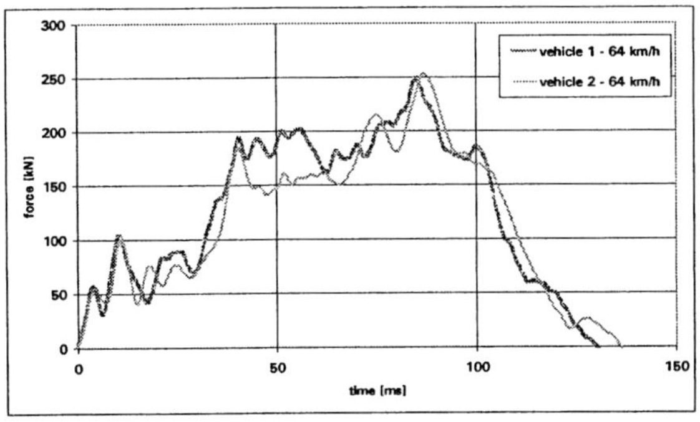

Figure 3: Structural forces (mean B-pillar acceleration * affiliated mass) from 64 km/h ODB test for identical cars as in Figure 1

The conclusion is that within the severity range of 64 km/h ODB test the maximum reproducible force level of vehicle structure of this car and possible some other vehicles can already be detected. This force level is identified with **compartment stiffness,** which in fact is a **compartment resistance (strength).** The maximum compartment resistance will be detected in different tests and at different speeds for each car. The destruction test is a valid tool for a research study of the compartment resistance, but it is not suitable for compliance or rating purposes.

Front-end Resistance

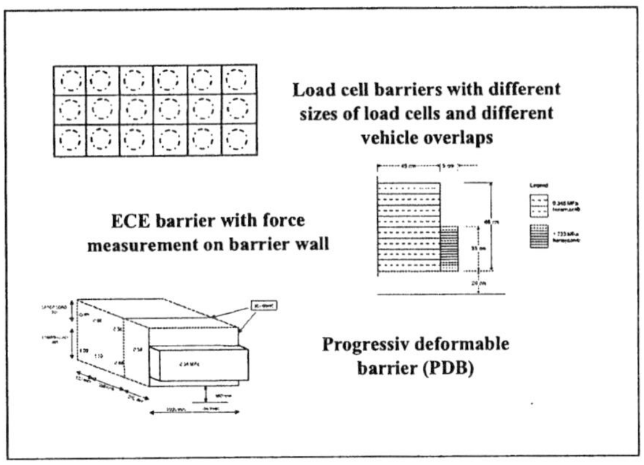

Figure 4: Different barrier types to measure front-end stiffness (resistance)

A series of tests was conducted against the Progressive Deformable Barrier [PDB], which has a progressively increasing force level and an upper and lower difference in stiffness. This barrier has a depth of 700 mm and is designed to avoid bottoming out. The test speed is 60 km/h to avoid reaching self-protection force and the overlap with the vehicle is 750mm to generate always the same force resistance. This barrier and the chosen test condition avoid bottoming out. The final deformation of PDB and the force-deflection curves can be analyzed to see whether the front-end is homogenous or if local differences in stiffness exist. Further research has to provide information which barrier type provides sufficient information with regard to front-end resistance (Figure 4).

VEHICLE-TO-VEHICLE TESTS

A **prediction of the deformation** in frontal vehicle-to-vehicle crash tests was made in accordance with the following rules: ① The force deflection curves of both vehicles from 64 km/h barrier impacts were compared. When theoretically computing the outcome of a vehicle-to-vehicle collision in most cases one vehicle has to provide more deformation energy than in a 64 km/h ODB test. This vehicle may be overloaded, which means that more significant intrusion into the passenger compartment occurs when compared to the 64 km/h ODB test. ② The PDB barrier test maximum force measured on the wall behind the deformable barrier was used to predict the loading on the passenger compartment of the opponent vehicle. The homogeneity of the barrier at the end of crash was generally evaluated under the viewpoint of structural interaction. The predictions held as shown in the following Table 2.

Test	64 km/h ODB test Prediction	PDB Prediction	Outcome	Structural interaction
Rover 75 vs. Renault Clio	Clio overloaded	Clio overloaded	Clio overloaded	Rover overrode Clio
Rover 75 vs. Volvo S80	Rover more intrusion	N/A	Balanced impact (S80 more intrusion than Rover)	Good interaction but high loading on A-pillar of S 80
Rover 75 vs. MCC Smart	Balanced impact	N/A	Smart overloaded (higher test speed than intended with theoretically 18% higher energy absorption)	Poor structural interaction. Rover longitudinal directly loaded Smart A-pillar
Rover 75 vs. VW Polo	Polo overloaded	Polo overloaded	Polo overloaded	No over-/underring
Rover 75 vs. Opel Astra	Astra overloaded	Balanced impact	Balanced impact (Astra and Rover had more intrusion than in the 64 ODB test)	Good structural interaction, Astra subframe was incorrectly modified.
Opel Astra vs. Renault Clio	Clio overloaded	Balanced impact	Balanced impact	Good structural interaction

Table 2: Predictions and results of frontal vehicle-to-vehicle tests

Two findings were made from the analysis of these vehicle-to-vehicle tests:

- The total force measured behind the deformable element can not be used as an indication how the vehicle will perform in vehicle-to-vehicle tests. It is necessary to distinguish between the mechanical parts forces and the structural forces (Figure 5). The force measured on the barrier is the sum of mechanical parts forces (mainly engine/transmission impact) and the structural forces (Figure 6). The mechanical parts and structural forces can be calculated by multiplication of the acceleration of these parts with the associated masses. Deformation management has to consider the front end and structural forces of both impact partners.
- The result of the car to car tests show that the outcome predicted from an assessment of deformation management can be wrong due to a lack of structural interaction. The PDB approach might be able to predict this.

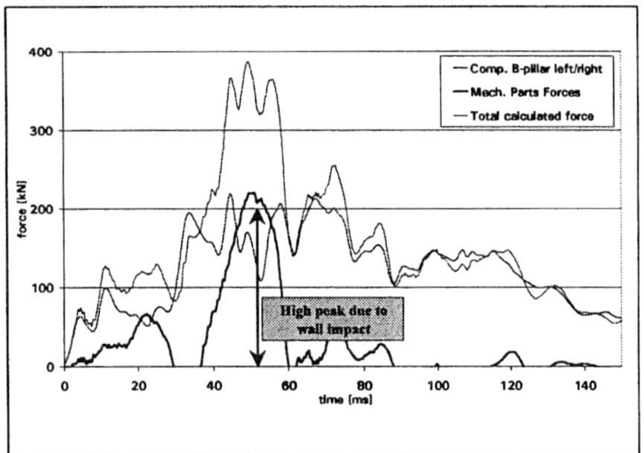

Figure 5: Comparison of structural forces (compartment) and mechanical parts forces

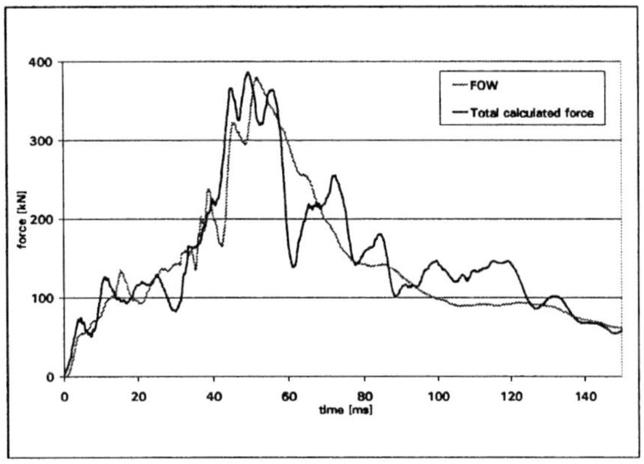

Figure 6 Comparison of measured wall force and total calculated force

Structural Interaction

Mismatch of load paths and 'Homogenous front end' initially 'excluded' or rated as 'possible' have to be combined and reassessed due to the importance of **structural interaction** found in severe frontal vehicle-to-vehicle crashes.

The outcome of CAE and test investigation emphasized that importance. Identical impact partners were used. Although they have optimal interaction conditions (same mass, same geometry and the same stiffness due to the alignment of longitudinals), they react differently depending on the vertical mismatch of longitudinals compared with the case of matching longitudinals. In these cases not all designed load paths were activated.

Due to this fact, the requirement can be deduced that the energy absorbing structural parts of the impact partners have to interact in a controlled and predictable manner. That is the prerequisite for every requirement dealing with stiffness and energy absorption.

Avoiding overriding completely, as an important part of structural interaction, will be difficult because 30mm height difference of the centerline of longitudinal cross sections already leads to partial over-/underriding. Therefore additional front-end measures (devices) to avoid over- and underriding in severe impacts have to be investigated (Figure 7, Figure 8).

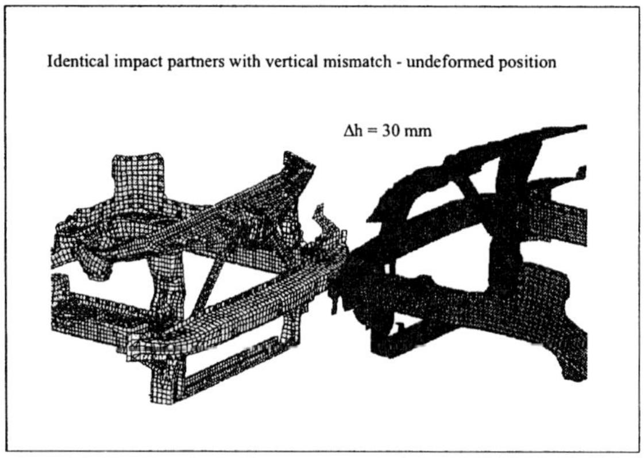

Figure 7: Identical impact partners with height difference before crash

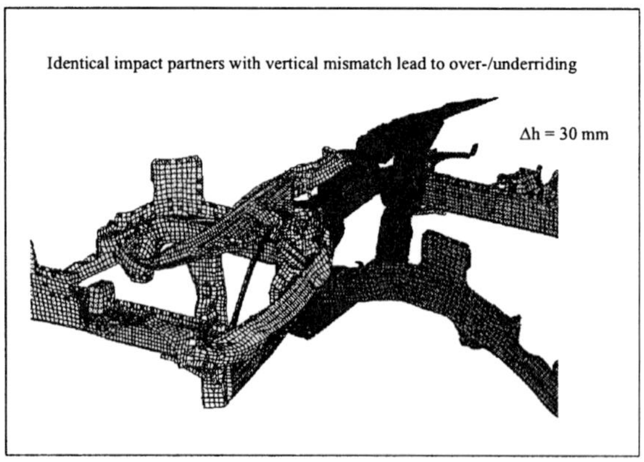

Figure 8: Over-/Underriding due to vertical mismatch

IMPACT ANALYSIS: SIDE IMPACT

Regarding side impact various parameter variations were conducted. In vehicle-to-vehicle side impacts the mass of the bullet vehicle is of less concern but the height difference between the struck vehicle sill and the bullet vehicle bumper is a more relevant issue. This is fully in line with the findings from accident analysis. Crash tests and simulations proved that when the longitudinals hit the sill of the target car substantially lower dummy loads are generated, although this might be difficult to achieve, which have to be further investigated.

In identical side impact tests with the mass of the bullet vehicle increased by 300 & 600kg there was no clear effect of increasing severity for the occupants of the struck vehicle.

Two tests were conducted with an identical target vehicle and two bullet vehicles with different initial front-end stiffness' derived from AZT (Test defined by German insurance companies). The crash with the lower initial front-end stiffness resulted in lower dummy loads in the target vehicle. This finding might be influenced by side effects like structure and geometry, which were not identical in both bullet vehicles.

A comparison was made of a lateral 90° impact and an angled impact into the side of the same struck vehicle. The results showed that the increase in severity in the case of an angled impact was not as great as expected. However, the bullet vehicle struck the target vehicle at the 'A' pillar and there was good interaction between the structures which mitigated the severity of the impact.

COMPUTER SIMULATION

As part of the project, computer simulation for compatibility investigations was used. To describe the possibilities of the Finite Element Method (FEM) the following impact modes must be distinguished:

- Barrier tests in the range where the vehicle retains its structural integrity
- Barrier tests in the range where the vehicle loses its structural integrity
- Vehicle-to-vehicle impacts

In general, FEM is a valid tool to predict vehicle behavior, although it is not yet able to simulate material failures like tearing of spot welds or detachment of structural parts.

Barrier tests are mainly used to design vehicles. As long as the vehicle deformed in a range where it retains its structural integrity, computer simulations are used as a tool to predict the vehicle performance. A validated model is necessary as a basis of such work.

In barrier tests that overload the structure (destruction tests) more material overload occurs. So, the general weakness of FEM prevents its use in destruction tests. The lack of repeatability in the overload tests reinforced that statement because whenever crash tests are not reproducible, computer simulation, which produces 100% reproducible results, cannot predict the scattering, and of course cannot predict a result of the single crash test, which is unique. Simulations done within the project proved that statement. Even when a vehicle is not overloaded in **vehicle-to-vehicle impacts** more material overload and sliding effects occur than in barrier impacts that are used to design the vehicles. The usage of FEM is, therefore, limited. The work done in the project shows that FEM computer simulation is able to reproduce vehicle-to-vehicle testing with respect to:

- the major effects of the crash and
- the amount of deformation and its characteristics.

Validation by a base line test is necessary. Good correlative quality of deceleration traces requires a great deal of effort and depends on the occurrence of sliding effects and local overload. They can vary from case to case and this limits the prediction quality.

The general weaknesses of FEM can not be solved as part of a compatibility project. They have to be addressed by methodological work like the EU funded project "Virtual testing", coordinated by TNO.

CONCLUSION

Frontal Impact

- The following variables were found to be relevant for frontal vehicle compatibility
  - Mass
  - Structural interaction (containing fork effects, mismatch of structural parts, homogeneity of the front end)
  - Deformation management (passenger compartment and front end stiffness)

  Structural interaction and deformation management have to be investigated further with respect to their potential for standards and requirements. Mass differences within the vehicle fleet are a consequence of customer demands. Therefore changing the mass differences of the vehicles is not possible. A mass ratio of 1.6 covers 80 to 90 % of European vehicle-to-vehicle accidents.
- Deformation management of the front-end and the passenger compartment should be described by resistance (deformation forces) rather than by stiffness.
- To avoid overcrushing of the opponent vehicle, the front-end resistance (deformation force) of the striking vehicle should be lower than the compartment resistance of the struck vehicle.
- High-speed 'destruction' tests do not necessarily provide more information regarding compartment resistance than the 64kph ODB test. Due to possible irreproducibility they might provide misleading results.

- Structural interaction and deformation management are required for compatibility. Deformation management is covered by the bulkhead concept, which means (Figure 9):
  - Sufficiently high force level of the compartment to ensure survival space
  - Restriction of the front-end resistance to force levels sustainable by the compartment resistance of the opponent vehicle and taking into account opponent vehicle acceleration
  - Ensuring this relationship up to a mass ration of 1.6, which covers 80 to 90 % of European vehicle-to-vehicle accidents

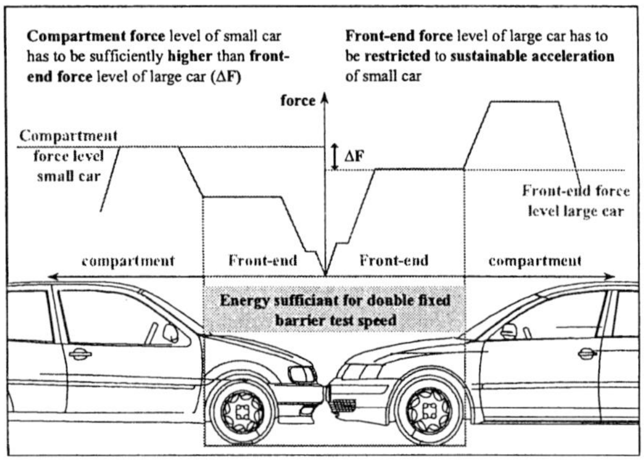

Figure 9: Bulkhead concept

- Within the severity range of EES 55...60km/h, computer simulation is an adequate predictor of vehicle crash behavior against barriers. It is not able to predict tearing of material or welding joints/spots. Until the CAE tools are improved, their use in vehicle-to-vehicle tests is limited.
- Force deflection curves of frontal impacts of a vehicle, which retains its structural integrity, provide some information about compartment resistance and the deformation energy available for partner protection. This information is valid only if load paths interact. The force deflection curves separated for mechanical parts and the passenger compartment in frontal impact ODB crash test with the vehicle retaining its structural integrity provide some information about the compartment resistance and the deformation energy of the vehicle front-end which is available for partner protection (Figure 10). This energy is only available and thus valid for evaluating partner protection ability, if the geometrical pre-assumptions hold and if load paths interact properly.

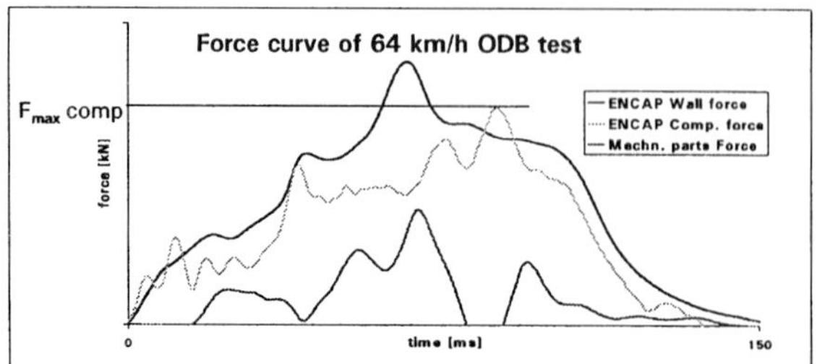

The higher F_{max} of compartment the higher self protection in car-to-car crash.

If for certain vehicles max. force level of compartment is not detected in 64 kph test, a higher crash speed can be choosen for this vehicle in accordance with vehicle design.

Figure 10: Compartment resistance derived from 64 km/h ODB test

- Geometrically, a structural fit of load paths should be created during the crash. Structural interaction must be ensured before high local loads on the compartment occur with the risk of high intrusion into the compartment. The PDB shows the final deformation and its local differences and is a candidate to assess the compatibility of front ends. Further research is needed to fully understand structural interaction.
- Although computer simulation was not able to replace hardware tests it might be valid to study structural interaction if validated models for the appropriate test modes exist. It can be a very useful tool to investigate the details of deformation.

Side Impact

- An override of sills by the bullet vehicle is a disadvantage for the occupant of the struck vehicle.
- The mass ratio is of minor influence.
- The stiffness of the longitudinals as observed in repair cost tests, might influence the load of the occupant of the struck vehicle.

Override of sills and the stiffness of the front end will be influenced by the measures to improve front impact compatibility. These effects have to be considered. When defining the standards and criteria for front impact the implications for side impact compatibility should be studied.

RECOMMENDATIONS

Some of the car-to-car crashes that were performed followed the predictions derived from barrier tests. These predictions were derived from an understanding of deformation management, the compartment resistance of both vehicles and the available deformation energy in the crash. The main reason that both types of prediction failed in particular cases is due to the lack of structural interaction in those impacts.

This leads to the following recommendations:

- Investigation of structural interaction and methods to assess the capability of two vehicles to interact in a frontal and lateral vehicle-to-vehicle crashes.

- Compatible structures need to have sufficient deformation energy available at a 'compatible' force level. Up to a specific severity passenger compartments have to resist the loading that occurs in a vehicle-to-vehicle impact. The front-end and compartment forces that ensure efficient deformation management have to be defined and the appropriate measurement of these forces has to be established in detail.
- Compatibility between the smallest and largest vehicles is probably not achievable due to the contradictory demands of self- and partner-protection. Up to a mass ratio of 1.6, compatibility can be established that holds for 80-90% of the European passenger vehicle fleet.
- Changes to the vehicle front may have implications for side impact. This must be investigated.

REFERENCES

Angulo, T. Car to Car Offset Compatibility Crash Test, European Automotive Congress, Paper STA99C107, Barcelona, 1999

Appel, H. - Deter, T. Crash compatibility of passenger cars achievable, but how? International Conference on Vehicle Safety 2000, 7-9 June 2000, IMechE HQ, London, UK, IMechE Conference Transactions 2000-2, pp. 55-65

Appel, H. - Deter, Th. Kollisions-Kompatibilität bei Personenkraftwagen erreichen, aber wie? VDI Berichte Nr. 1471, 1999, pp. 3-42

Steyer, C. - Delannoy, P. Proposal to improve compatibility in head on collisions. 16th international conference on the enhanced safety vehicles. Paper 98-S3-O-05. Windsor, 1998

Zeidler, F. - Knöchelmann, F., The Influence of Frontal Crash Test Speeds on the Compatibility of Passenger Cars in Real World Accidents. International Symposium on Real World Crash Injury Research, Leicestershire, UK, June 1997

Zeidler, F. - Knöchelmann, F. - Scheunert, D., Possibilities and Limits in the Design of Compatible Cars for Real World Accidents,1999-01-0068, SAE-Conference, Detroit, Michigan, March 1-4, 1999

Zobel, R. Principles for the development of a passenger car safety information system for consumers, based on real-life accident evaluation. Crash-Tech special '98, München, 1998

Zobel, R. Demands for compatibility of passenger vehicles. 16th international conference on the enhanced safety vehicles. Paper 98-S3-O-10. Windsor, 1998

Zobel, R. Accident Analysis and Measures to Establish Compatibility, 1999-01-0065, SAE-Conference, Detroit, Michigan, March 1-4, 1999

Zobel, R. - Schwarz, T. Determination of compartment stiffness for compatible design of passenger vehicle front structures, Crash Tech 2000, München, 2000

ACKNOWLEDGEMENT

This paper reflects the main conclusions of the project: Development of Safety Criteria and Standards for Vehicle Compatibility. It was conducted in the years 1998 to 2000 by European automotive industry and some research institutes and funded by the European Commission (Directorate Generale Research) under BE-97-4049. The project was co-ordinated by

Robert Zobel Volkswagen AG.

Contributions to this final text were provided by

Andersson, Johan	Volvo
Angulo, Theresa	Seat
Boening, Steffen	Ford AG
Delannoy, Pascal	Renault S.A.
Deter, Thomas	TU Berlin
Ernvall, Timo	University of Oulu
Krueger, Hans-Joachim	TUEV Rheinland
Monnet, Sebastien	PSA Peugeot Citroen
Provensal, Jacques	ACEA
Rogers, Ian	Landrover
Schwarz, Thomas	Volkswagen AG
Wrobel, Bert	Adam Opel AG
Wrobel, Michael	GdV
Zeidler, Falk	DaimlerChrysler AG

The authors thank all partners for ideas and results contributed to the project. The authors felt it worthwhile to present these findings as agreed to the safety community and so used the agreed text unchanged without attributing special findings to the partner, who provided it.

CHAPTER TWO – ACCIDENT DATA

980908

The Aggressivity of Light Trucks and Vans in Traffic Crashes

Hampton C. Gabler and William T. Hollowell
U.S. National Highway Traffic Safety Administration

ABSTRACT

Light trucks and vans (LTVs) currently account for over one-third of registered U.S. passenger vehicles. Yet, collisions between cars and LTVs account for over one half of all fatalities in light vehicle-to-vehicle crashes. Nearly 60% of all fatalities in light vehicle side impacts occur when the striking vehicle is an LTV. This paper will examine this apparent incompatibility between cars and LTVs in traffic crashes. An analysis of U.S. crash statistics will be presented to explore the aggressivity of LTVs in impacts with cars and identify those design imbalances between the cars and LTVs, e.g., mass, stiffness, and geometry, which lead to these severe crash incompatibilities.

INTRODUCTION

During the past decade, a profound shift in the composition of the passenger vehicle fleet has been realized in the U.S.. Fueled by the growing popularity of pickup trucks, minivans, and, more recently, by sports utility vehicles, the demographics of the U.S. fleet are characterized by a growing population of light trucks and vans (LTVs). As a group, LTVs are heavier, of more rugged construction, and have higher ground clearance than the passenger cars with which they share the road. The concern is that these design features, introduced to allow specialized functions e.g. off-road driving, may make LTVs fundamentally incompatible with cars in highway crashes, and in some cases dangerous to the occupants of cars struck by LTVs.

NHTSA has initiated a research program to investigate the problem of aggressive vehicles in multi-vehicle crashes. The near term objective of this program is to identify and demonstrate the extent of the problem of incompatible vehicles in vehicle-to-vehicle collisions. The focus of this goal is to identify and characterize compatible vehicle designs with the expectation that improved vehicle compatibility will result in large reductions in crash related injuries. Specifically, the objective is to identify those vehicle structural categories, vehicle models, or vehicle design characteristics which are aggressive based upon crash statistics and crash test data. LTV-to-car collisions are one specific, but growing, aspect of this larger problem.

The goal of this paper is to examine LTV aggressivity in vehicle-to-vehicle crashes. The specific objectives are to define the nature of the problem through examination of crash statistics, and to explore the relationships between crash aggressivity and vehicle design characteristics.

THE DEMOGRAPHICS OF LTV AGGRESSIVITY

As shown in Figure 1, registrations of LTVs currently account for over 1/3 of all light vehicle registrations (Polk, 1980-1996), and are a growing component of the U.S. fleet. During the period from 1980 to 1996, LTV vehicle registrations increased from 20 percent to 34 percent. Although LTVs only account for 1/3 of all registered vehicles, traffic crashes between an LTV and any other light vehicle now account for the majority of fatalities in vehicle-to-vehicle collisions. As shown in Table 1, in 1996 LTV-car crashes accounted for 5,259 fatalities while car-car crashes led to 4,013 deaths and LTV-LTV crashes resulted in 1,225 fatalities.

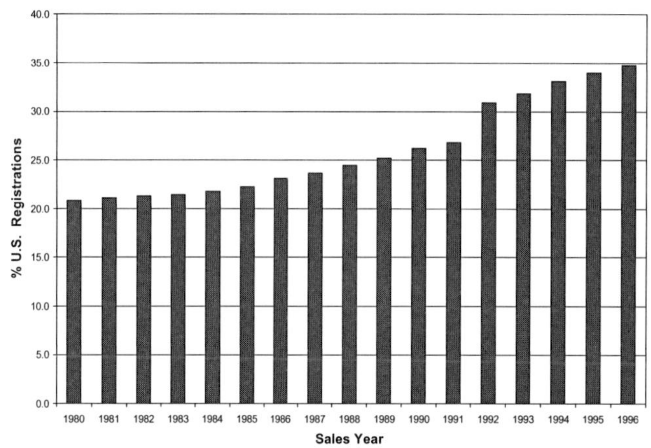

Figure 1. U.S. Light Truck and Van Registrations by Model Year from 1980-1996 as a fraction of light vehicle registrations (R.L. Polk Co., 1980-96).

A disproportionate number of the fatalities in LTV-car crashes are incurred by the car occupants. Of the 5259 fatalities in LTV-car crashes in 1996, 81 percent of the fatally-injured were occupants of the car. As shown in Figures 2 and 3 and tabulated in Table 2, side impacts in which an LTV was the bullet vehicle led to 56.9 percent of all fatalities in side struck vehicles. As illustrated in Figure 4 and tabulated in Table 3, frontal impacts in which an LTV was involved accounted for 59.7 percent of all fatalities in frontal-frontal impacts. These statistics suggest that LTVs and passenger cars are incompatible in traffic crashes, and that LTVs are the more aggressive of the two vehicle classes. In particular, crashes with an LTV cause a disproportionate number of vehicle-to-vehicle fatalities.

Table 1. Fatalities in Light Vehicle-to-Vehicle Crashes

Year	All car-car	All car-LTV	All LTV-LTV	Total
1980	6506	3580	510	10596
1981	6510	3292	482	10284
1982	5437	3452	556	9445
1983	5157	3408	505	9070
1984	5340	3540	593	9473
1985	5174	3608	635	9417
1986	5450	3895	660	10005
1987	5489	4277	788	10554
1988	5320	4676	802	10798
1989	5175	4730	861	10766
1990	4726	4719	867	10312
1991	4482	4297	873	9652
1992	4208	4421	804	9433
1993	4364	4451	977	9792
1994	4219	4972	1059	10250
1995	4097	5238	1183	10518
1996	4013	5259	1225	10497

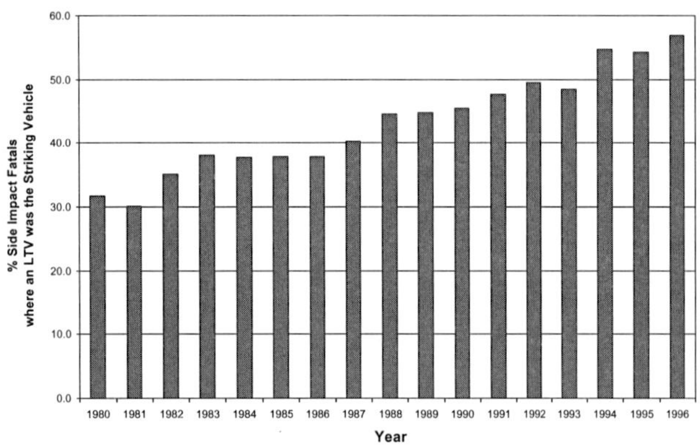

Figure 2. Side Impact Fatalities resulting from LTVs striking passenger cars and other LTVs as a fraction of total side impact fatalities (FARS 1980-96).

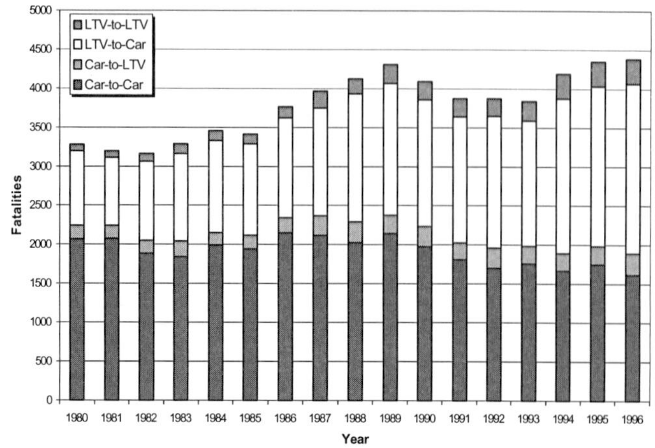

Figure 3. Distribution of Side Impact Fatalities (FARS 1980-96).

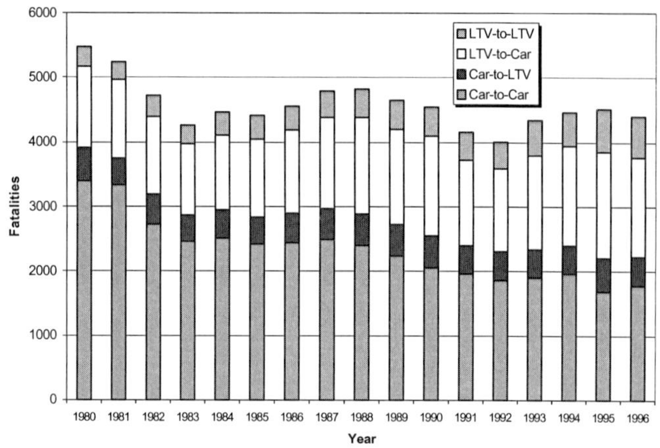

Figure 4. Distribution of Frontal-Frontal Impact Fatalities (FARS 1980-96).

Table 2. Light Vehicle-to-Vehicle Side Impacts: Fatalities in Side-Struck Vehicle

Year	Car into Car	Car into LTV	LTV into Car	LTV into LTV	Total
1980	2071	170	962	78	3281
1981	2077	161	876	87	3201
1982	1881	174	1015	102	3172
1983	1848	190	1134	118	3290
1984	1996	153	1186	122	3457
1985	1943	178	1168	130	3419
1986	2149	192	1285	147	3773
1987	2121	246	1382	216	3965
1988	2026	262	1645	194	4127
1989	2144	231	1697	238	4310
1990	1976	255	1628	234	4093
1991	1812	216	1614	232	3874
1992	1705	252	1698	223	3878
1993	1759	224	1609	256	3848
1994	1667	229	1983	318	4197
1995	1749	237	2049	316	4351
1996	1613	276	2181	314	4384

Table 3. Fatalities in Light Vehicle-to-Vehicle Frontal Impacts

Year	Car-Car	Car-LTV (LTV Fatals)	Car-LTV (car Fatals)	LTV -LTV	Total
1980	3395	521	1254	303	5473
1981	3331	422	1208	278	5239
1982	2731	463	1197	330	4721
1983	2465	400	1115	283	4263
1984	2515	434	1159	351	4459
1985	2420	415	1217	362	4414
1986	2440	461	1289	373	4563
1987	2493	483	1407	404	4787
1988	2398	487	1498	437	4820
1989	2238	481	1478	446	4643
1990	2056	500	1538	450	4544
1991	1964	437	1321	432	4154
1992	1857	448	1282	418	4005
1993	1899	439	1460	543	4341
1994	1956	442	1548	521	4467
1995	1692	510	1658	656	4516
1996	1776	453	1529	651	4409

The incompatibility of LTVs and cars is a growing problem. As shown in Figure 5, LTV market share has risen steadily from 1980 to 1996 [1]. LTVs captured over 43 percent of all light vehicle sales in 1996. Comparison of LTV registrations and LTV-caused fatalities over the same period show that LTV impacts have always caused a disproportionate number of vehicle-to-vehicle fatalities. For example in 1980, LTVs accounted for 20 percent of the registered light vehicle fleet, but side impacts in which an LTV was the bullet vehicle led to 31 percent of all fatalities in side struck vehicles. The magnitude of this problem then is not only due to the aggressivity of LTVs in crashes, but also the result of the dramatic growth in the LTV fraction of the U.S. fleet.

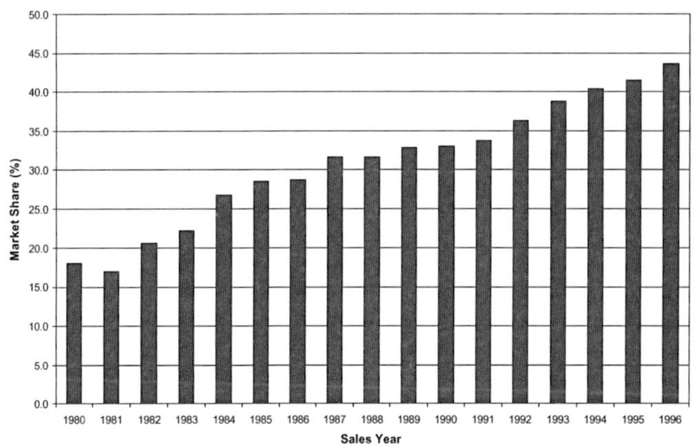

Figure 5. U.S. Sales of LTVs from 1980-1996 expressed as a fraction of light vehicle market share (Automotive News Market Data Book).

PROBLEM DEFINITION

The research program examined U.S. crash statistics to determine the characteristics and extent of the vehicle compatibility problem. One obstacle to quantifying the compatibility of a vehicle is the lack of an accepted measure of compatibility. A primary objective of our research effort was to develop a clearly defined a metric for measurement of vehicle aggressivity. To date, the NHTSA aggressivity research program has developed two potential aggressivity metrics.

Option 1:

$$\text{Aggressivity} = \frac{\text{fatalities in collision partners}}{\text{Registrations of subject vehicle}}$$

Option 2:

$$\text{Aggressivity} = \frac{\text{Driver Fatalities in collision partners}}{\text{Number of Crashes of subject vehicle}}$$

The first metric was used in our early Aggressivity research as reported at the 15th ESV conference [2]. For each vehicle make / model, this metric determines the number of fatalities in the collision partner resulting from collisions with the subject vehicle. Only two-vehicle crashes in which both vehicles were either a car or an LTV are considered. The fatality count is normalized by the total number of registrations of the subject vehicle so that vehicles with large populations are not unfairly penalized. Using this metric, the U.S. fleet was rank ordered by aggressivity as presented at the 15th ESV conference. This initial study indicated that LTVs as a group were twice as aggressive in crashes as passenger cars -- i.e., per vehicle, LTVs caused more than twice as many fatalities in their collision partners as do cars.

The second, more recent, metric represents a refinement to the earlier definition of aggressivity. The second metric defines aggressivity to be the number of driver fatalities in the collision partner normalized by the number of vehicle-to-vehicle crash involvements of the subject vehicle. Only two-vehicle crashes in which both vehicles were either a car or an LTV are considered in computing the fatality count and the crash involvement count. One of the confounding factors in determining aggressive vehicle designs is aggressive driver behavior. Because aggressive drivers are involved in more crashes than less aggressive drivers, normalizing by the number of crashes rather than vehicle registrations focuses the metric more on vehicle performance and less on driver behavior. Note also that the second metric keys on driver fatalities rather than all fatalities in the struck vehicle. Because all vehicles have only one driver, this refinement avoids any biases accruing from differences in vehicle occupancy rate between, for example, pickup trucks and minivans.

APPROACH – The analysis for the second metric used statistics from the 1991-94 Fatality Analysis Reporting System (FARS) to determine the number of fatalities, and statistics from the 1991-94 General Estimates System (GES) to determine the number of crash involvements. FARS provides a comprehensive census of all U.S. traffic related fatalities. The GES is a large sample of over

60,000 police reported crashes collected annually. The scope of our analysis was constrained to cars, light trucks, and vans under 10,000 pounds in Gross Vehicle Weight Rating. The focus was further narrowed to two vehicle collisions in which the vehicles were either cars or LTVs. The fatality counts in the struck vehicle were limited to driver fatalities.

Initial analysis of GES revealed that approximately half of the make and model codes in this database were listed as unknown. For those GES cases with valid Vehicle Identification Numbers (VINs), the make and model was obtained by decoding the VIN using a combination of the VINDICATOR code, developed by the Highway Loss Data Institute, and the VINA code, developed by the R.L. Polk Company [3]. However, even after decoding the VINs, approximately 20 percent of all vehicle make and models remained unknown. The number of crash involvements for all vehicles was weighted accordingly in order to preserve total number of crashes. Although this strategy maintains the total count of crash involvements, this approach has the disadvantage of preserving any reporting biases. An improved approach would be to explore the missing data as a function of vehicle body type and model year, and prorate unknown make-models within these categories if biases exist.

Overall Fleet Aggressivity Ranking – The second metric, hereafter referred to as the aggressivity metric (AM), was used to rank order all passenger vehicles, cars and LTVs, by their relative aggressivity. Only current production vehicles with at least 10,000 police-reported crashes over the period of 1991-94 were included in the ranking. Figure 6 presents the top twenty vehicles from that ranking in descending order by their relative aggressivity. The most striking feature of Figure 6 is that every vehicle in the rank ordering of the top twenty aggressors is either a light truck or van.

Figure 6 suggests that LTVs as a group are more aggressive than passenger cars. To further pursue this observation, the aggressivity ranking was regrouped by vehicle family into five categories of LTVs – sports utility vehicles, full-sized pickups, small pickups, minivans, and full-sized vans – and four categories of passenger cars – large, midsize, compact, and subcompact. The categories assigned to each vehicle are as tabulated in the Automotive News Market Data Book [1]. This study groups luxury, near luxury, and large cars into a single large car category.

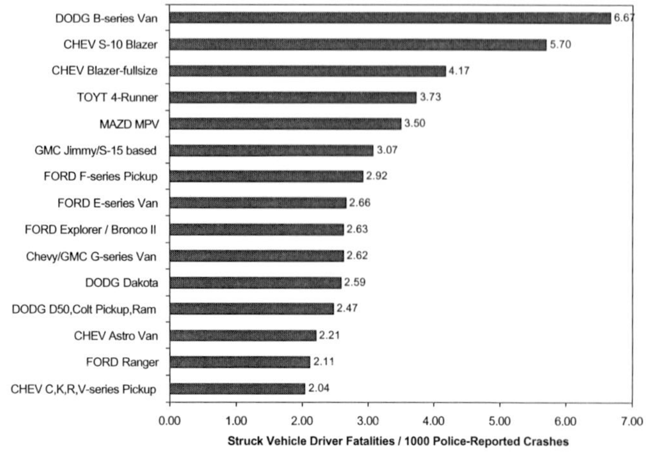

Figure 6. Aggressivity Ranking: Top 15 Vehicles (FARS/GES 1991-94).

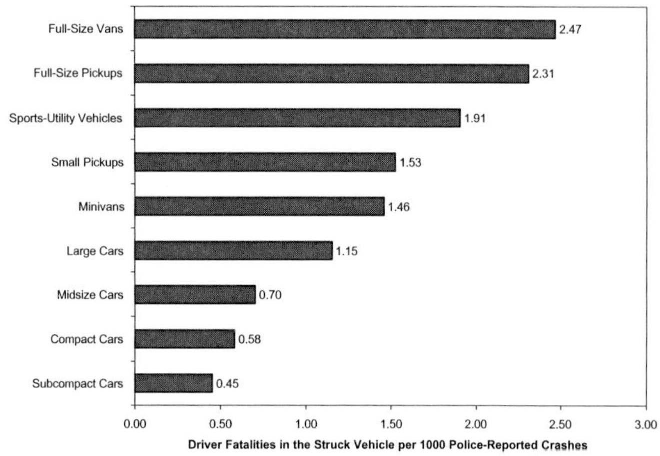

Figure 7. Vehicle Aggressivity by Vehicle Category (FARS/GES 1991-94).

As shown in Figure 7, full-sized vans were found to be the most aggressive vehicle category with an AM = 2.47. This category was closely followed by Full-Size Pickups (AM=2.31), Sports-Utility Vehicles (AM = 1.91), and small pickups (AM = 1.53). Minivans were the least aggressive of all LTV groups with an average AM = 1.46. The AM of passenger cars was significantly lower and ranged from AM= 0.45 for subcompacts to AM=1.15 for large cars.

Vehicle weight is not always the overriding factor dictating aggressivity as clearly demonstrated by Figure 7. Mid-sized cars, e.g., the Ford Taurus, and the small pickups, e.g., the Ford Ranger, both have approximately the same curb weight of 3,000 pounds. However, small pickups (AM = 1.51) are over twice as aggressive as mid-sized cars (AM = 0.70). The higher aggressivity of the small pickup class may be due to its greater structural stiffness and its higher ride height.

Among cars, the Aggressivity Metric is a strong function of vehicle weight. AM for the large car category, e.g., the Ford Crown Victoria, is 1.15. This is two to three times higher than the AM for the subcompact car category, e.g., Geo Metro, which is 0.45. The conservation of momentum in a collision places smaller cars at a fundamental disadvantage when the collision partner is a heavier vehicle. The importance of car size in providing occupant protection has been demonstrated in several studies of the U.S. crash statistics [4,5].

Aggressivity by Impact Mode – Having established that LTVs are incompatible with cars in traffic crashes, the next requirement was to determine the relationship between aggressivity and impact direction. Joksch, et al. [6], in a NHTSA-sponsored study, computed the ratio of driver fatalities in the striking vehicle vs. driver fatalities in the struck vehicle for cars versus each of three LTV categories. In this study, minivans and full-sized vans were aggregated into a single 'van' category, and small and full-sized pickups were grouped into a single 'pickup' category. The counts of fatalities were obtained from 1991-94 FARS. To eliminate the complication of determining occupant restraint effectiveness, the study excluded airbag-equipped vehicles. It should be noted in the discussion which follows that this analysis was based on small numbers of fatal crashes (on the order of a hundred for each case), and the results should be regarded as preliminary.

The ratio of striking-to-struck driver fatalities resulting from frontal-frontal impacts is presented in Figure 8. In collisions between sports-utility vehicles (SUVs) and cars, 5.6 drivers died in the car for every driver who was killed in the striking SUV. In collisions between vans and cars, 5.4 drivers died in the car for every driver who was killed in the striking van. In collisions between pickup trucks and cars, 3 drivers died in the car for every driver who was killed in the striking pickup. Clearly, the fatality toll in car-LTV frontal crashes is disproportionately shouldered by the drivers of passenger cars.

The ratio of striking-to-struck driver fatalities resulting from side impacts to the left (driver) side of the struck vehicle are presented in Figure 9. As a control configuration, note first that in car-to-car impacts approximately 6 side-struck drivers are fatally injured for every fatally-injured driver in the bullet car. This imbalance is not unexpected as the side structure of passenger vehicles provides little protection for the side-struck occupant when compared with the significantly greater protection afforded by the front structure to the bullet vehicle driver.

The analysis is even more startling for LTVs striking cars in side impact. As shown in Figure 9, 30 side-struck car drivers are fatally injured for every driver who dies in a bullet SUV. For every driver who dies in a striking van, 13 side-struck car drivers are fatally injuried. For every fatally-injured driver of a striking pickup truck, 25 side-struck car drivers are killed. By contrast, when cars strike LTVs in side impact, the ratio of striking car to struck LTV driver fatalities is roughly 1:1.

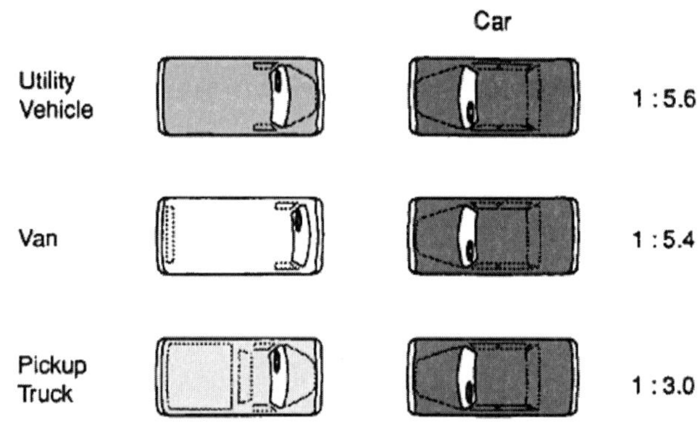

Figure 8. Ratio of Fatally-Injured Drivers in LTV-to-Car Frontal Collisions. FARS 1991-94. (Joksch, et al., 1997)

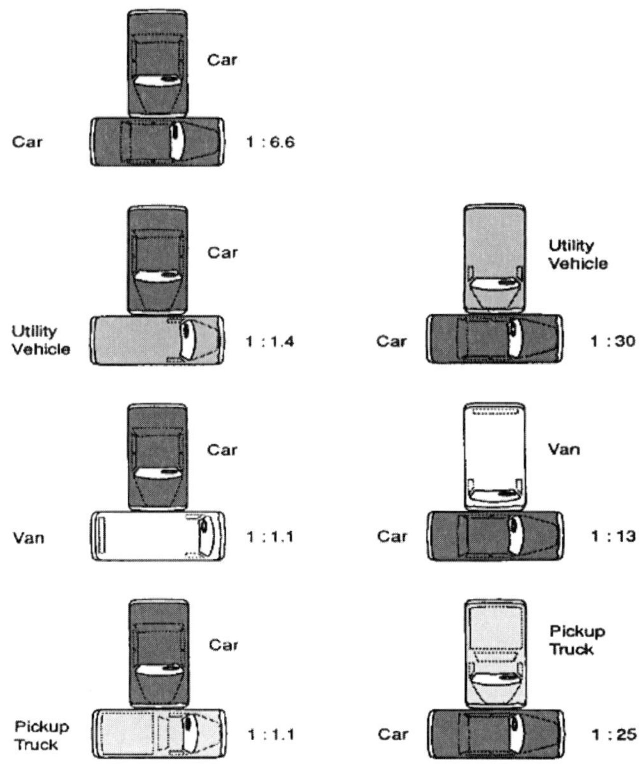

Figure 9. Ratio of Fatally-Injured Drivers in LTV-to-Car Left Side Impacts. FARS 1991-94. (Joksch, et al., 1997)

WHY ARE LTVS MORE AGGRESSIVE?

The preceding analysis of crash statistics has clearly demonstrated the incompatibility between cars and LTVs in highway crashes. Still remaining to be determined however are the design characteristics of LTVs which lead to their incompatibility with cars. In general, crash incompatibility arises due to the three factors:

- Mass Incompatibility.
- Stiffness Incompatibility
- Geometric Incompatibility.

The following section will examine the relationship between LTV-car compatibility and these sources of incompatibility.

MASS INCOMPATIBILITY – LTVs are 900 pounds heavier than cars on average [5]. The conservation of momentum in a collision places smaller vehicles at a fundamental disadvantage when the collision partner is a heavier vehicle. As shown in Figure 10, LTVs, as a group, tend to be heavier than passenger cars [7]. Figure 10 crossplots AM as a function of vehicle weight, and demonstrates the strong relationship between mass and aggressivity. The mass incompatibility between cars and LTVs appears to be growing. As shown in Figure 11, during the time frame of 1985-1993, average mass of both cars and LTVs has steadily increased, and the weight mismatch between the two classes of vehicles has slowly grown as well.

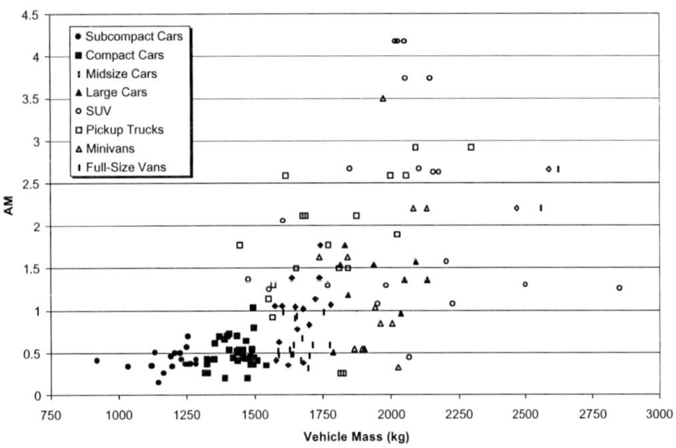

Figure 10. Aggressivity as a function of Vehicle Mass.

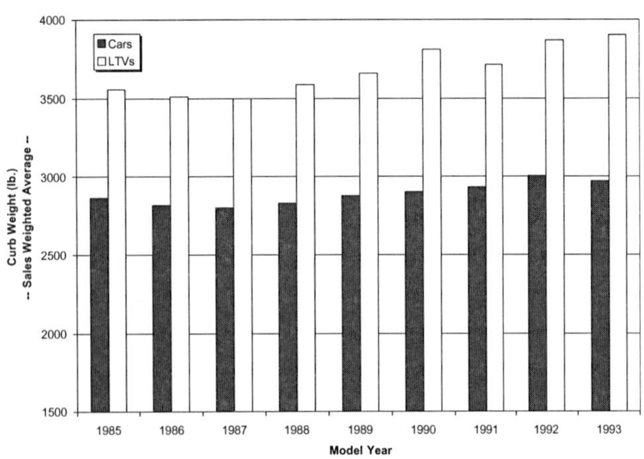

Figure 11. Sales-averaged Mass of LTVs and Passenger Cars from 1985-93 (Kahane, 1997).

STIFFNESS INCOMPATIBILITY – As a group, LTV frontal structures are more stiff than passenger cars. LTVs frequently use a stiff frame-rail design as opposed to the softer unibody design favored for cars. Drawing on NHTSA New Car Assessment Program crash test results, the linear stiffness of a selection of LTVs and cars was estimated using the following relationship:

$$k = (mv^2) / x^2 \qquad \text{(Eq. 1)}$$

where m is the mass of the vehicle, v is the initial velocity of the vehicle, and x is the maximum dynamic crush of the vehicle. The relationship between linear stiffness and AM is shown in Figure 12. Figure 12 indicates that stiffness is a contributing factor to the aggressivity of a vehicle. Because the stiffness of a vehicle is also somewhat related to its mass, as shown in Figure 13, stiffness may not prove to be as dominant an aggressivity factor as mass. Although stiffness and mass are related in many cases, stiffness is not totally driven by the mass of the vehicle. Figure 13 shows that for any given mass, there is a wide distribution of stiffness values. For example for 1750 kg vehicles, the least stiff vehicles are passenger cars while the most stiff vehicles are LTVs.

Figure 14 compares the frontal stiffness of a Ford Taurus and a Ford Ranger pickup. Both vehicles have approximately the same mass, but note that the Ranger pickup is significantly stiffer than the Taurus. In a frontal collision between the two, the bulk of the crash energy would be absorbed by Taurus and the Taurus occupants. Far less energy would be absorbed by the Ranger. From a compatibility perspective, a more ideal scenario would be for the Taurus and Ranger structures to each share the crash energy rather than forcing one of the collision partners to absorb the bulk of the crash.

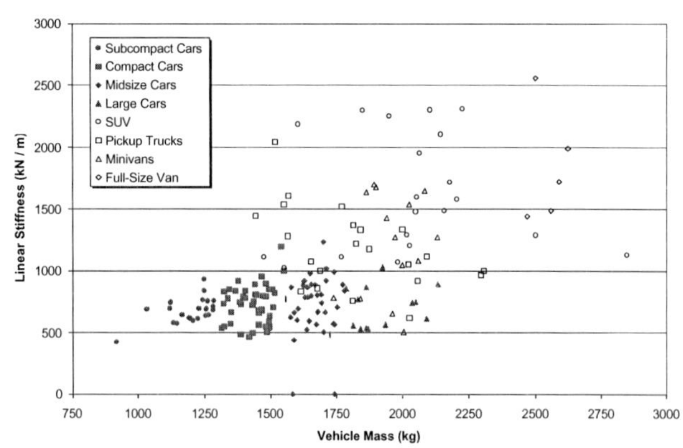

Figure 12. Aggressivity as a Function of Linear Stiffness as computed from NCAP crash test results.

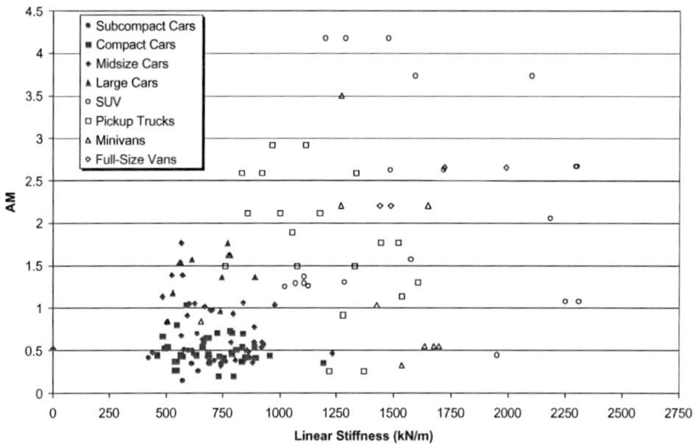

Figure 13. Relationship between Frontal Stiffness and Vehicle Mass as determined from NCAP Crash Tests.

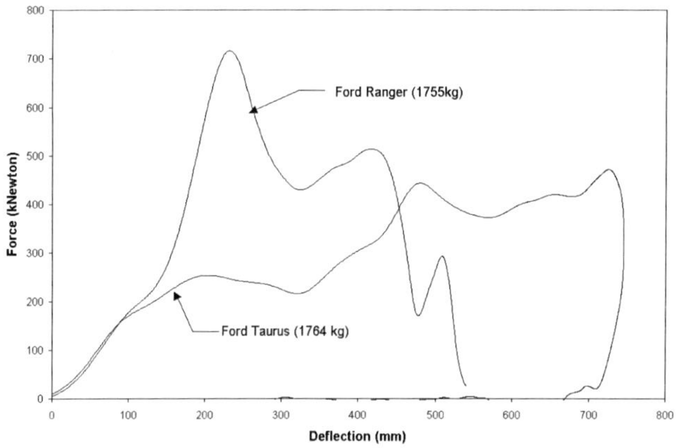

Figure 14. Frontal Stiffness: Small Pickup (Ford Ranger) vs. Midsize Car (Ford Taurus)

GEOMETRIC INCOMPATIBILITY – LTVs, especially four-wheel drive sport utility vehicles, ride higher than cars. This creates a mismatch in the structural load paths in frontal impacts, and may prevent proper interaction of the two vehicle structures in a collision. In a side impact, this imbalance in ride height allows the LTV structure to override the car door sill, and contributes to the intrusion of the side-impacted vehicle.

Ideally, the ride height used in an analysis of this type would be the height of the forward-most load bearing structural member of the vehicle. The location of this forward-most structural element however has no precise definition, and must be estimated from other measurements. Some analyses have used bumper height as the height of this load bearing member. However, because in the U.S., the bumper must only meet a 2-½ mile/hour bumper impact standard, and LTVs have no bumper standard, our belief is that, with respect to occupant protection, bumpers are largely ornamental, and their location provides little evidence of the location of load bearing members. The rocker panel, on the other hand, is a much more substantial structural member, and because the rocker panel is typically lower than the forward-most structure, serves as a superior lower bound on the location of the frame structure.

Figure 15 shows that ride height is related somewhat with vehicle mass. For this analysis, ride height is defined to be the rocker panel height as measured at the trailing edge of the front wheel well [7]. However note that the rocker panel height across all masses of passenger cars is relatively consistent – perhaps due to the bumper standard with which all passenger cars must comply. On the other hand, LTVs, which have no bumper standard, exhibit a wide variation in ride height and are in general much higher than passenger cars.

Figure 16 presents average ride height by vehicle category. Sport utility vehicles have the highest ride height with an average rocker panel height of 390 mm. Subcompact cars have the lowest-riding height with an average rocker panel height of 175 mm. SUVs ride almost 200 mm higher than mid-sized cars – a geometric incompatibility that would readily permit the SUV to override any side structure in a car and directly strike the car occupant.

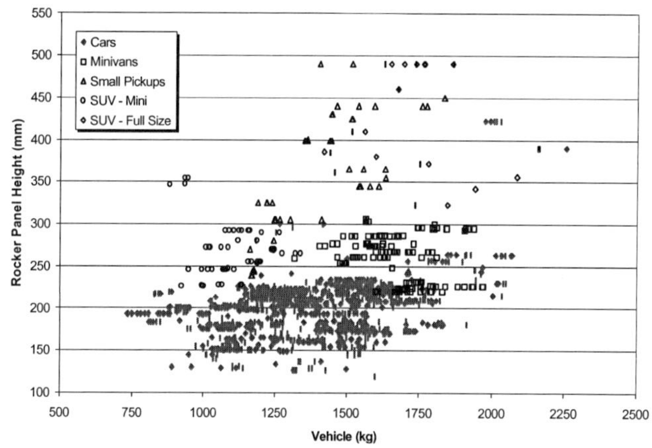

Figure 15. Relationship between Vehicle Mass and Ride Height as estimated by Rocker Panel Height. Reference: AAMA Vehicle Specification Sheets (1990-94).

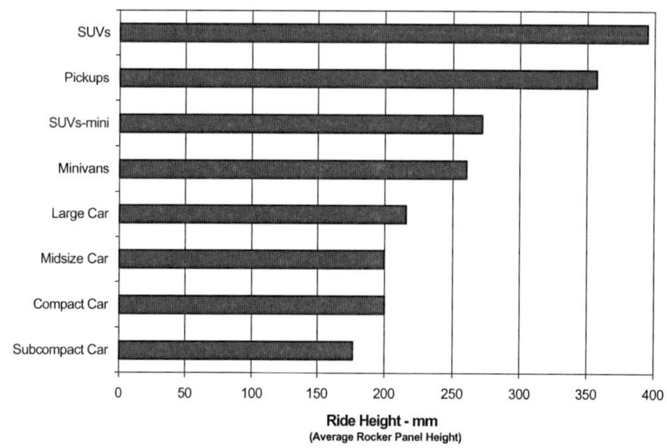

Figure 16. Geometric Compatibility: Average Ride Height vs. Vehicle Category as determined from AAMA Vehicle Specification Sheets (1990-94).

It should be noted that the data for the preceding analysis was drawn from AAMA Vehicle Specification Sheets supplied by vehicle manufacturers, and collected in the NHTSA Vehicle Attributes Database [7]. While geometric data was available for most passenger car models, the Vehicle Specification sheets for LTVs was much more limited. The LTV data presented here was primarily obtained from foreign manufacturers, and contains no data on full-sized pickups or vans.

DISCUSSION

The study presented in this paper based its measure of aggressivity upon fatalities per 1000 police reported crashes. No effort was made to control for the severity of the crashes as this information is not available in the GES files. Some make-model vehicles, such as high performance sports cars, may have more severe crashes more because of the driver than because of the vehicle structure. Normalizing fatalities by number of crash involvements removes much of this driver aggressivity effect but does not completely eliminate this effect. Future work will explore refinements to the aggressivity metric which account of crash severity in addition to crash frequency.

The aggressivity metric used in this study assumes that all make-models strikes the same cross-section of the vehicle population, i.e., the same proportion of small cars, large cars, minivans, pickups, and so forth. The influence of this assumption upon the aggressivity ranking will be explored in future work. Joksch (1997) has noted that the age distribution of struck drivers varies somewhat from striking make-model to make-model. As injury tolerance is a strong function of age, his analysis suggests an additional refinement to the aggressivity metric which accounts for the average age of struck drivers for a given striking make-model.

The crash statistics presented in this paper clearly demonstrate a clear incompatibility between cars and LTVs. A comparison of mass distribution, stiffness distribution, and ride height geometry confirm that these two categories of vehicles are incompatible from a design point-of-view. However, this study has not attempted to assign what proportion of the aggressivity of LTVs is a function of each of these three separate sources of incompatibility. To determine the relationship between LTV design features and crash aggressivity, NHTSA plans to conduct a series of LTV-to-car crash tests in conjunction with a series of finite element simulations of LTV-to-car crash events.

FUTURE WORK

NHTSA research to date on LTV aggressivity has focused on problem definition based upon U.S. crash statistics and their correlation with vehicle design features. Follow-on work will expand upon these initial analyses as a precursor to potential rulemaking in this area. Specific future tasks include:

- **Crash Testing.** To demonstrate and better understand the nature of the incompatibility between cars and LTVs, NHTSA plans to conduct a series of crash tests during the next year in which LTVs will be used to impact mid-sized passenger cars. Our current plans are to conduct both frontal-frontal crash tests as well as frontal-side crash tests.
    - ◊ Striking Vehicles
        - — SUV
        - — Minivan
        - — Pickup
    - ◊ Struck Vehicle
        - — Midsize passenger car

  Additional tests will be conducted based on results of the first test series.
- **Simulation and Systems Modeling**. This task will develop a large scale systems model which will evaluate vehicle crashworthiness based on the safety performance of the vehicle when exposed to the entire traffic accident environment, i.e., across the full spectrum of expected collision partners, collision speeds, occupant heights, occupant ages, and occupant injury tolerance levels. The foundation for the Systems model will be a comprehensive suite of finite element models and articulated mass models constructed to represent nine light vehicle categories -- five LTV and four passenger car – and their occupants.
- **Test Procedure Development**. Development of test procedures and test devices for a standardized evaluation of vehicle aggressivity/compatibility.
- **International Harmonization Efforts**. Collaboration with international regulatory bodies and research organizations in vehicle compatibility research. This committee was organized at the 15th ESV and is led by representatives of the EC/EEVC. This will be a challenging effort due to differences in U.S. and international fleet composition (i.e., the U.S. has a large LTV fleet constituent which is not present in other continents/countries).

CONCLUSIONS

This paper has examined the compatibility of LTVs and cars in vehicle-to-vehicle collisions. Using struck driver fataltíes per crash involvment of the subject vehicle as an aggressivity metric, examination of U.S. crash statistics has clearly shown a striking incompatibility between cars and all categories of LTVs. LTVs now account for over one-third of light vehicles on U.S. highways, but collisions between cars and LTVs lead to over 50% of all fatalities in light vehicle-to-vehicle collisions. Furthermore, a disproportionate number of the fatalities in LTV-car crashes are incurred by the car occupants. A comparison of LTVs and cars reveals that the LTVs are more aggressive than cars for a number of reasons including their greater weight, stiffer structure, and higher ride height. This mismatch in design has serious consequences for crash safety as approximately one-half of all passenger vehicles sold in the U.S. are LTVs, and presents a growing source of incompatibility within the fleet.

ACKNOWLEDGEMENTS

The authors would like to acknowledge the contributions of the Volpe Transportation Systems Center to this study. In particular, we wish to acknowledge Hsi-sheng Hsia and Roger Flanders for their expert and insightful analyses of the FARS and GES databases which made this study possible. The authors also gratefully acknowledge Hans Joksch of UMTRI for his many important contributions to the analysis presented in this paper and for the insights he has shared with us during the preparation of this work.

REFERENCES

1. ____, *Automotive News Market Data Book*, Crain Communications, Inc., Detroit, MI., 1980-96.
2. Hollowell, W.T., and Gabler, H.C., ANHTSA=s Vehicle Aggressivity and Compatibility Research Program", *Proceedings of the Fifteenth International Enhanced Safety Vehicle Conference*, Paper No. 96-S4-O-01. Melbourne, Australia (1996).
3. ____, "VINA: Vehicle Identification Number Analysis Software", The R.L. Polk Company, June 20, 1997.
4. Evans, L., "Driver and Fatality Risk in Two-Car Crashes Versus Mass Ratio Inferred", *Accident Analysis and Preventation*, Vol. 26, 1994, pp. 609-616.
5. Kahane, C. J., "Relationships between Vehicle Size and Fatality Risk in Model Year 1985-93 Passenger Cars and Light Trucks", DOT HS 808 570, January 1997.
6. Joksch, H., Massie, D., and Pichler, R., "Vehicle Aggressivity: Fleet Characterization Using Accident Data", University of Michigan Transportation Research Institute, Report No. UMTRI-97-16, September 1997
7. McCullough, C.A., Hollowell, W.T., and Battisti, P. "Applications of NHTSA's Vehicle Parameter Database", Society of Automotive Engineers Paper No. 950360.

2003-06-0172

NUMERICAL FLEET OPTIMIZATION STUDIES FOR IMPROVED COMPATIBILITY

Paul Lemmen
Cor van der Zweep
Floris Leneman
TNO Automotive
The Netherlands
Paul Altamore
TNO-MADYMO North America
United States
Paper No. 445

ABSTRACT

On behalf of NHTSA and the Dutch Ministry of Traffic and Transport the Safety department of TNO Automotive is performing numerical fleet studies using multi-body models. Aim is to develop strategies for optimization of front-end structures minimizing the total harm in car-to-car crashes on a fleet-wide basis. For these studies multi-body models are being constructed from existing finite element models. Front-end structure and passenger cell are modeled in detail to provide realistic deformation modes. Furthermore dummies, airbags, belts and main interior parts like dashboard and steering wheel are included. Currently four models are available, each of a different vehicle class.

To indicate the performance of the multi-body vehicle models for crashworthiness optimization of a fleet a study on offset frontal impacts is performed. Using the multi-body models a series of parameter sweeps over relevant accident and design parameters were performed. The accident parameters included vehicle type, belt usage and occupant size. The design parameters relate to the front-end geometry of the two smaller vehicles and the front-end stiffness of all vehicles. A total set of 2500 scenarios was simulated.

INTRODUCTION

Compatibility is an important subject in road traffic safety research, because in many accidents more than one road user is involved. In that case the passive safety of the different road users is often in unbalance. This leads to an incompatible situation in which one of the parties suffers from the relative aggressiveness of the other. A solution to this problem may be found in improved vehicle compatibility which combines self and partner protection characteristics. During the last decades extensive research was done on the statistics of car-to-car crashes giving a/o interesting rates of aggressiveness [1,2]. Although the occupant safety has improved significantly car-to-car crashes form an increasingly important class of accidents to be examined making it one of the most important safety issues for the car industry and governmental bodies [1,3, 4, 5, 9].

There are two main injury-causing aspects to car collisions in general but also with respect to compatibility: excessive deceleration and intrusions [8]. The deceleration aspects relate to the phenomenon that lighter cars undergo larger decelerations than heavier cars in a collision between both. So, in the lighter car the occupants can get injured more easily, due to these large decelerations and contacts directly resulting from these decelerations. On the other hand, intrusions relate to entering of structural car parts into the passenger cabin that should be avoided as much as possible. A first important step to avoid intrusions is the avoidance of geometrical mismatch. Shearlaw and Thomas [6] show that it is very difficult to tackle the question whether or not cars are compatible with respect to these intrusion effects.

Furthermore, the passenger compartment integrity should be preserved as much as possible: collapse of the compartment should be avoided. For this purpose, the global strength of the passenger compartment should be larger than the strength of the front and of course large enough to withstand the forces during the whole crash. This also means that the strength of two cars in a crash should be optimized such, that the collision energy is dissipated without compartment collapse of any of the cars [3]. Of course, the strengths of the cars are closely related to the deceleration of the cars.

When assessing compatibility the overall safety of a fleet should be considered. For the evaluation of the overall safety of an automotive fleet systems modeling approaches have been developed. In the nineteen-seventies Ford Motor Company developed a method for maximizing a single vehicle's safety performance in frontal crashes [10]. This program was updated by the University of Virginia to include new biomechanical transforms and updated accident data as well as multivariable analysis capability [11]. Other car manufacturers also developed programs, mainly for optimizing single vehicle design. On behalf of NHTSA Volpe developed a model that predicts the total harm over a range of vehicle types rather than a single subject vehicle [12]. The model estimates injuries over a given set of crashes considering air bags, seat belts and occupants of varying size. It incorporates updated accident data for the statistical accident environment model and injury risk functions that convert injury measures into the AIS scale [14]. Injury values are obtained from MADYMO occupant kinematic models of a car and LTV loaded by crash pulses obtained from one-dimensional lumped parameter models. Figure 1 summarizes the methodology.

The method does not include an optimization strategy to minimize the overall Harm in a fleet environment.

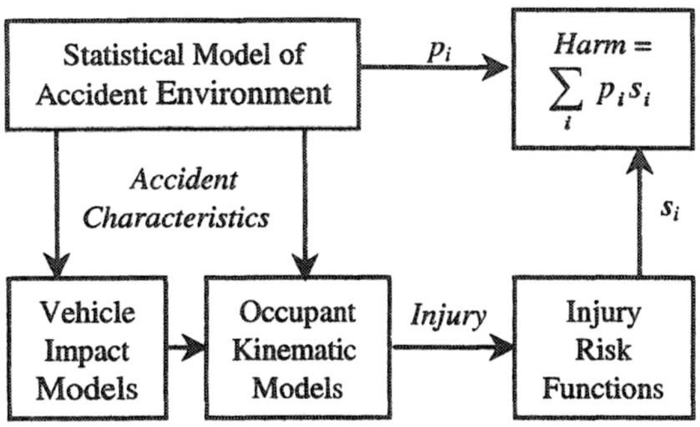

Figure 1. Fleet Systems Model Methodology [12]

Ideally injury values are to be computed using detailed vehicle FE models. However, while potentially very accurate these are computationally too expensive to execute for fleet systems models that a substantial amount of simulations [15]. Alternatively simpler but faster running lumped mass models in MADYMO may be used. Although less accurate these models require substantially less computer time making it easier to conduct the necessary simulations. The models were successfully applied in numerical optimization studies bringing all injury levels in a small fleet (limited number of scenarios) below critical values by adjustment of the frontal stiffness [9, 16]. The result was obtained by optimizing the stiffnesses of main load carrying members in the front-ends. Geometrical interaction was not considered, as it could not be translated into a continuous parameter for numerical optimization using direct methods. However, while the front-end stiffness does affect compatibility, geometrical interaction is regarded as the prime factor for good compatibility. Therefore structural variants should be considered in the optimization process which may be stated as [15]

$$\text{Minimize Inj }(x,u) = p_I\, s_I(x,u)$$

subject to:

$$Wgt(x) < Wgtmax$$
$$Cost(x,w(x)) < Costmax$$
$$x_{min} < x < x_{max}$$

where:

- x = Vector of design variables
- u = Belt usage rate
- Inj(x,u) = Total injuries
- Wgt(x) = Incremental weight for with design 'x'
- Cost = Incremental cost for x and Wgt(x)
- Wgtmax = Upper constraint on incremental weight
- Costmax = Upper constraint on incremental costs
- p_i = Probability of event *i*
- s_i = Injuries resulting from event *i*

The restraints on the design variables x are included to limit weight as well as costs of the proposed design modifications and to ensure that modifications remain within realistic ranges. Each crash event *i* may be characterized by six accident variables namely vehicle types, impact mode, impact speed, seat position, occupant size and belt usage [12, 15].

In view of the large number of scenarios to be considered in fleet studies optimization by considering structural variants can only be achieved using design of experiments (DOE) or identical methods that scan the design space by variation of relevant design parameters. Correct interpretation of results requires an adequate formulation of the object or target function.

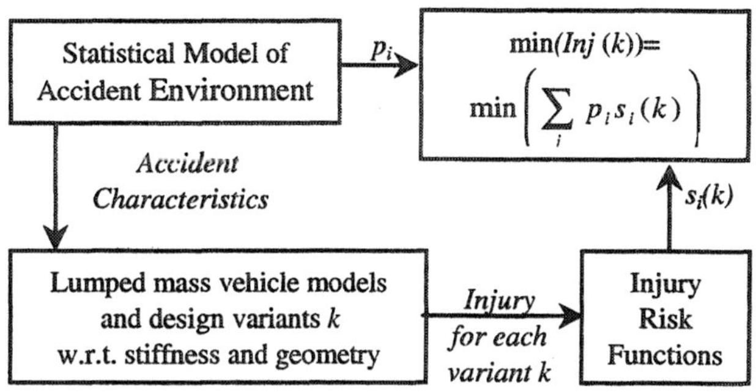

Figure 2. Fleet Systems Model Methodology using multi-body vehicle models with occupants

Figure 2 depicts the fleet systems model using multi-body vehicle models to predict injuries. In this paper the vehicle models, the injury risk functions that form the basis of the object function and results of a fleet study considering design variants will be discussed.

VEHICLE MODEL DEVELOPMENT AND VALIDATION

Based on existing FE models four lumped mass models representative for the US fleet were developed. The vehicles are listed in table 1.

Table 1. Available vehicle models

Model	Class	Mass [kg]	Test Mass [kg]
Geo Metro	Subcompact	800	1191
Chrysler Neon	Compact pass.	1085	1371
Ford Taurus	Midsize pass.	1488	1728
Ford Explorer	SUV	1971	2205

The front-ends and the side structures were modeled in detail to describe the actual interaction for frontal and side impacts. The rigid bodies are connected by non-linear spring and damper elements, which represent the stiffness behavior. Characteristics of these elements were derived using FE models and component test data. Attention was focussed on the main load carrying components like longitudinals and shotguns. The occupant compartment intrusion is described using contact surfaces. The interior of the car is modeled including a dashboard, steering wheel, belts, airbag and Hybrid III dummy at the driver side. Figure 3 shows the models of the Ford Taurus and the Chrysler Neon in the undeformed configuration.

Interaction between the vehicles is realized with contact facets at the front and the side.

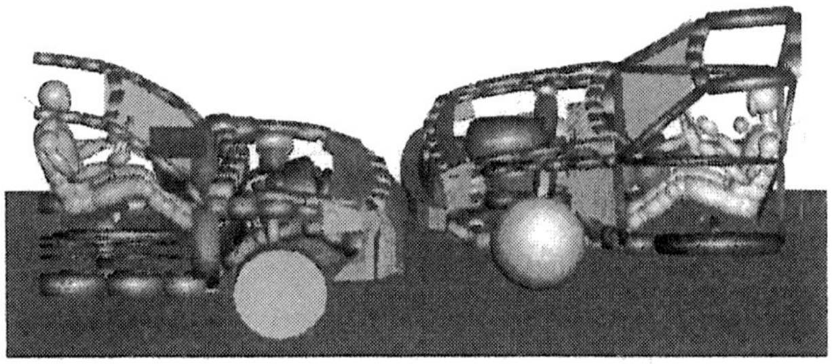

Figure 3. Undeformed frame model of Neon Chrysler (left) and Ford Taurus (Right)

The frame models were validated against Full Width barrier NCAP test data. In addition, the vehicle signals are validated against FE simulations under different angels and different crash scenarios. Figure 4 shows typical results for vehicle signals (left) and dummy response (right). Results correlate well.

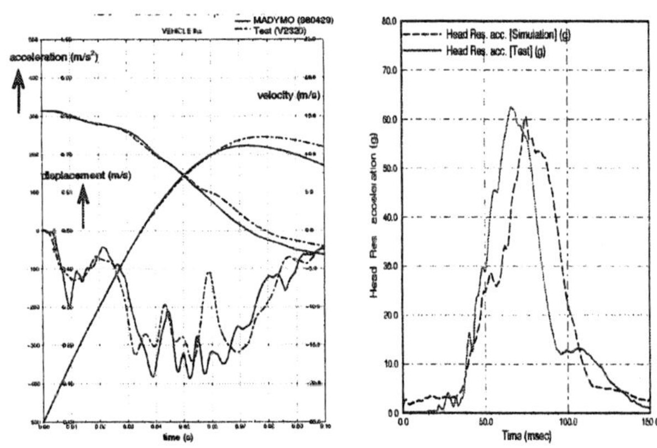

Figure 4. Validation frontal Chrysler Neon model: frontal car structure (left) and resultant head acceleration of a 50th percentile Hybrid-III (right)

In addition to the vehicle and dummy signals the proposed compatibility measurables were compared. Table 2 compares the Average Height of Barrier Force (AHoBF) which is calculated as follows [17]

$$AHoF(t) = \frac{\sum_i \left(F_i(t) h_i\right)}{\sum_i F_i(t)}$$

$$AHoBF = \frac{\sum_t \left(AHoF(t) \sum_i F_i(t) \right)}{\sum_t \sum_i F_i(t)} \qquad (1)$$

where:
- F_i = Force on cell i
- h_i = height of cell i

Except for the Chrysler Neon all vehicle models show a good correlation with test results. Based on NCAP test data NHTSA has identified AHoBF ranges for different vehicle classes [17]. Results for all vehicle models except the Neon are within the specified class range. Findings for the Neon are currently being investigated in more detail using FE models.

Table 2. Comparison of Average Height Of Barrier Force.

Car	AHoBF [m]	AHoBF (NCAP)	Class range [m]
Geo Metro	0.44	0.42 1)	0.41 – 0.47
Chrysler Neon	0.51	0.45	0.43 – 0.48
Ford Taurus	0.49	0.50 1)	0.43 – 0.50
Ford Explorer	0.58	0.63	0.50 – 0.62

1) Data somewhat different than values indicated by NHTSA in ref. [20]

The Average Height of Barrier Force is known as a relevant measure, however, for compatibility the force distribution on the barrier is even more important. Figure 5 therefore compares load cell data at two time frames for the Ford Taurus.

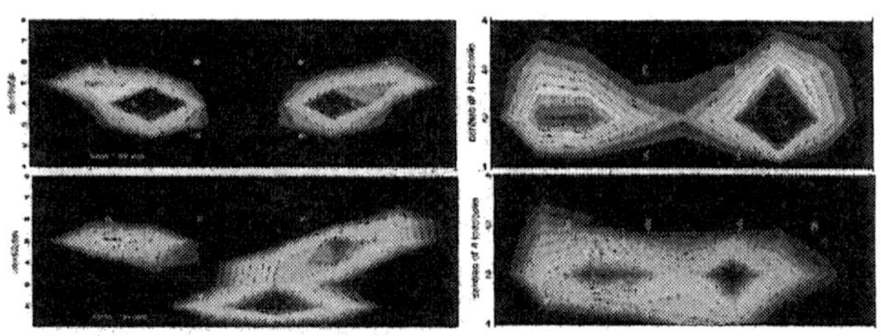

Figure 5. Comparison of simulated (left) and experimental (right) load cell wall data for the Ford Taurus at t = 25 ms (top) and t = 60 ms (bottom). Note that the dimensions of the grid sizes in the simulation are somewhat different from the test (simulation: 8*8; test: 4*9).

At 25 ms high loads are located at longitudinal locations. Note that the dimensions of the cells and therefore the grid sizes are somewhat different, which may affect the results to some extend. At 60 ms the sub-frame and engine of the simulation model partake in the load transfer which is not the case in the actual test. Despite this discrepancy the simulated results generally correlate very well with experimental data and the models may be regarded adequate for usage in fleet studies.

INJURY RISK FUNCTIONS

For frontal impacts the most commonly used injury measures include Head Injury Criterion (HIC), Viscous Criterion (VC), 3 millisecond acceleration (3 MS), Combined Thoractic Index (CTI), Femur Force Compression (FFC), Nij, FNIC, Tibea Index (TI) and TCFC. See also figure 6. All of these injury measures used as regulatory criteria except for the CTI. CTI though is recommended by NHTSA for research use [13].

Prediction of the lower leg injuries requires accurate representation of intrusions, which can only be achieved by use of detailed finite element models with correct geometry and material modeling. The multi-body models have insufficient detail to represent intrusions correctly and therefore TTI and TCFC are not considered here.

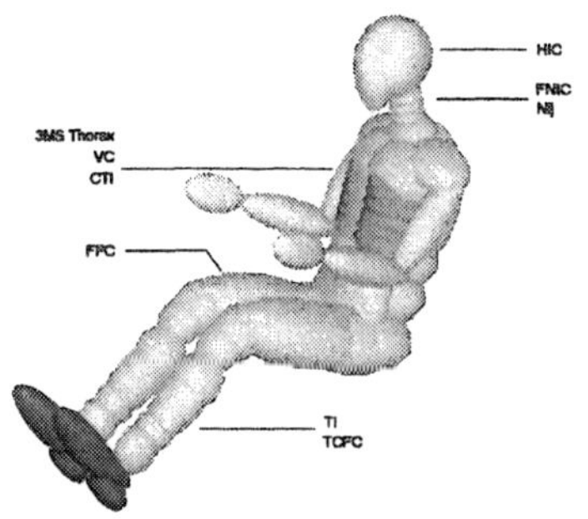

Figure 6. Hybrid III 50th percentile frontal impact dummy with injury criteria. Except for the lower leg injuries all indicated mechanisms are considered in this study.

Each MADYMO simulation results in a set of injury values for the drivers in both vehicles. To compare risks in the different scenarios results need to be converted into a measure that gives an indication for the overall injury risk (AIR). In previous studies into the optimization of the front-end stiffness this was achieved by summing squared normalized injury values for head, upper leg and chest [16]

$$obj = \sum_i \left(\left(\frac{HIC(i)}{HIC_{crit}(i)} \right)^2 + \left(\frac{FFC(i)}{FFC_{crit}(i)} \right)^2 + \left(\frac{3MS(i)}{3MS_{crit}(i)} \right)^2 \right) \quad (2)$$

The injury values used in eq. (2) were selected based on results of parametric studies [16]. This function was found to be quite effective as it is very discriminative for critical or near critical injuries. Main disadvantage tough is that all injuries have identical weights, which is not realistic when considering the harm. Therefore injury significance ratings should be introduced:

$$obj = \sum_i \left(\alpha \left(\frac{HIC(i)}{HIC_{crit}(i)} \right)^2 + \beta \left(\frac{FFC(i)}{FFC_{crit}(i)} \right)^2 + \gamma \left(\frac{3MS(i)}{3MS_{crit}(i)} \right)^2 \right)$$

(3)

where α, β and γ are the weight factors for the respective injury types. Estimates for the weight factors are provided in table 3. These numbers, based on field studies, were derived in the early nineties to evaluate the performance of restraint systems [18].

Table 3. Injury significance factors [18]

Body region	Significance	Weights
Head	60%	$\alpha = 0,60$
Chest	35%	$\beta = 0,35$
Extremities	5%	$\gamma = 0,05$

Alternatively an approach based on the Abbreviated Injury Scale (AIS) may be used, see e.g. [19]. Injury risk functions are used to convert injury values into AIS levels, which subsequently may be transformed into an overall injury risk using the Injury Severity Scale (ISS).

Figure 7 and figure 8 show mathematical models to transfer CTI and HIC values into the AIS probabilities. Identical models have been derived for 3ms, CD, FFC, and Nij. The models, generally known as the injury risk functions, have been proposed by NHTSA on the basis of experimental data and previous research [14]. The experiments were performed within the regulatory range of interest up to critical values. For higher injury values the plotted approximations are therefore more heuristic.

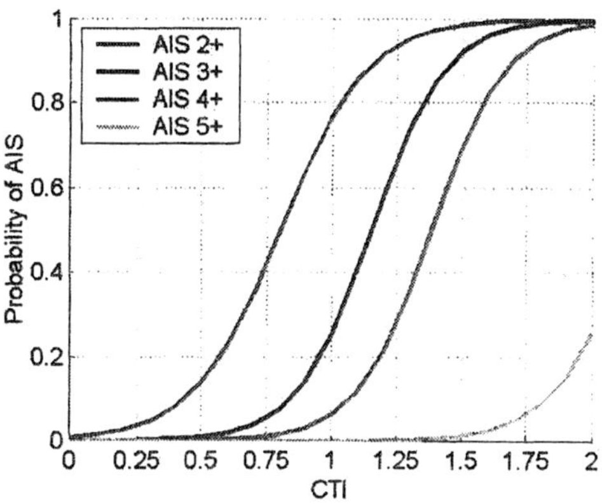

Figure 7. Injury risk function for CTI [14]

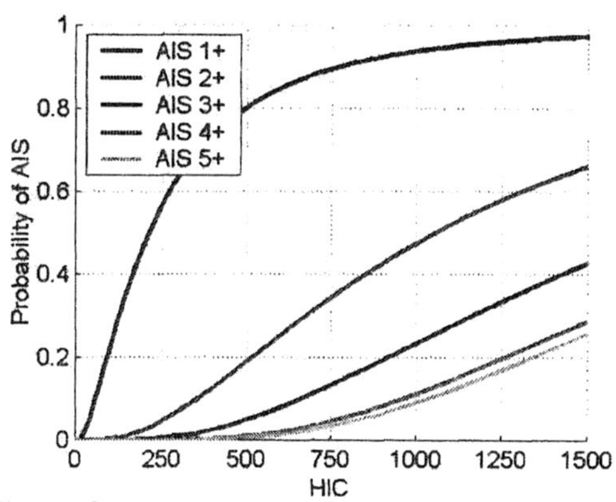

Figure 8. Injury risk function for HIC [14]

Mathematical expressions for the injury risk functions can be found on the NHTSA website (http://www.nhtsa.dot.gov/cars/rules/rulings/AAirBagSNPRM/PEA/pea-III.n.html). Using these cumulative functions a vector of AIS probabilities (AIS=0,1,2,3,4,5,6) is obtained by subtracting each AIS probability at the computed injury level from the next AIS probability. For each injury type a vector of AIS probabilities is computed which is converted into an expected AIS value according to

$$E(X) = \sum_{i=1}^{6} w_i P_{wi}(X) \quad (4)$$

where

X = Injury value

E = Expected AIS value

P_{wi} = Probability of AIS level w_i

w_i = AIS level 0,1,2,3,4,5,6

The expected AIS values for each injury mechanism may be converted into an overall body criterion using normalized cost functions to obtain communal costs (HARM) or using the Injury Severity Scale (ISS) [19]. In this paper the ISS will be used.

FLEET SETUP AND ACCIDENTS SCENARIOS

To explore the potential of the multi-body models for usage in optimizing the crashworthiness behavior of a fleet a study with four vehicles was performed. Frontal offset impacts at a closure speed of 50 km/h were analyzed in a fleet consisting of the Geo Metro, Neon Chrysler, Ford Taurus and Ford Explorer. Figure 8 shows the accident scenarios. Crashes between the Explorer and the Geo Metro are not considered, as this scenario is strongly incompatible, even when applying the design modifications suggested below. The accident variables occupant size (5th percentile female, 50th percentile male and 95th percentile male) and seat belt usage (belted an unbelted) were varied.

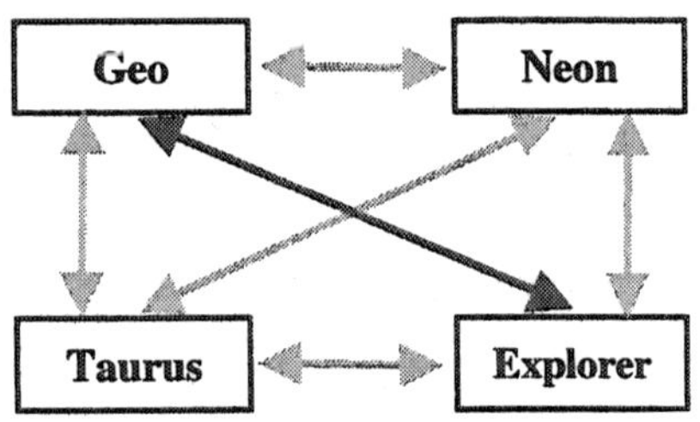

Figure 8. Crash scenarios between vehicles. The Explorer Geo scenario is not considered, as it is highly incompatible.

For the optimization two design variables were introduced namely front-end stiffness and front-end geometry. For the stiffness structural components relevant for the crash behavior were identified in each vehicle and related bodies grouped such that their connecting springs and damper characteristics can be changed simultaneously. The scaling of the characteristics corresponds to overall changes in elastic and plastic stiffness of the component. The allowable range of the stiffness was set between 75 and 150% of the original values to be within physically realistic bounds. Weight and cost restraints were not considered here but are largely covered by the above-mentioned restraint. For the front-end geometry design variants of the Geo Metro and the Neon Chrysler were created, see figures 9 and 10. The depicted modifications are easily implemented in the multi-body models.

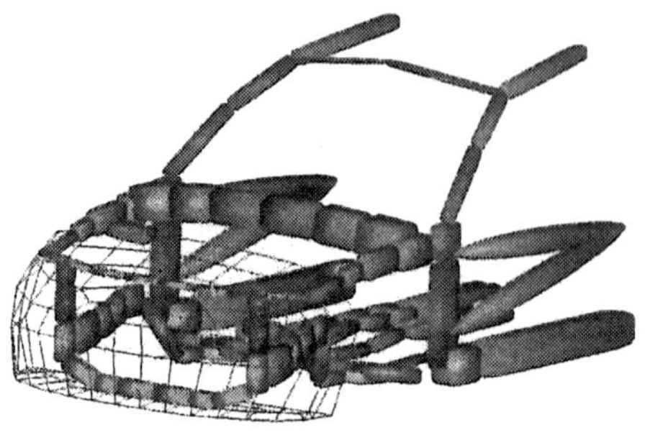

Figure 9. Modified front-end Geo Metro. To improve the interaction with other vehicles cross members have been reinforced (Green bodies) and two vertical members linking shotguns and longitudinals (red bodies) have been added.

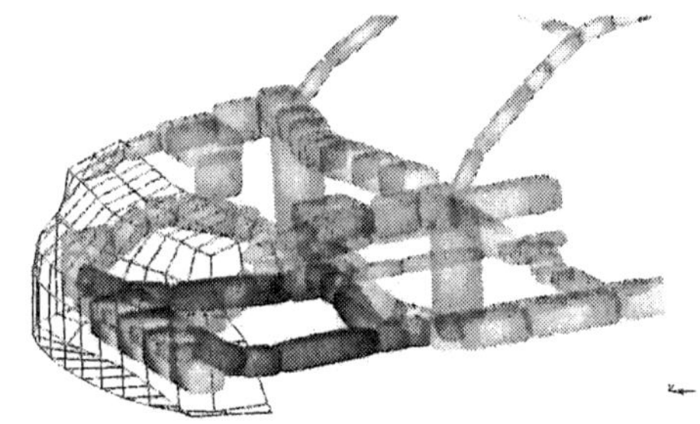

Figure 10. Modified front-end Chrysler Neon. To improve the geometrical interaction a sub-frame was added (red bodies) and the lower cross-beam was reinforced (yellow bodies)

SIMULATION RESULTS

Parametric simulation of all scenarios for relevant design variables yields a total of 2500 individual cases. Here the stiffness for each vehicle is varied in five discrete steps ranging between 75% and 150% of the original stiffness. Each case requires approximately 20 minutes of CPU time on a PC server system. The total required CPU was about 830 hours. For each case a vector of injuries is obtained which is processed into an overall body value using eq. (2) or the alternative method based on AIS and ISS.

Results of the simulations are analyzed with the SPSS statistical program [21]. The computed distributions for the entire subset are shown in figures 10 (ISS) and 11 (weighted squared injuries using eq. (2)). Both figures show identical trends but results based on eq. 2 appear to have larger relative differences which is mainly due to use of squared relative values rather then relative values, see also [16]. This finding may be important for optimization studies as it focuses the search towards critical or near critical cases. However, as it is based on a limited set of accident variants care should be taken when generalizing this finding. Therefore ISS will still be used in the sequel of this paper.

To obtain an indication for the contribution of different injuries to these results 3MS, HIC and FFC are plotted in Figure 12 through 14 for the entire subset. FFC values for the smaller vehicles are near or over critical (FFC_{crit} = 10000). Chest values are relatively high especially for the smaller cars ($3MS_{crit}$ = 60 g). HIC values are generally low at values up to 400 (HIC_{crit} = 1000).

Figure 15 shows ISS values sorted by belt usage. Only limited influence of the belt usage is observed, which is unrealistic. Figure 16 and 17 show FFC and 3MS values. From these figures it is observed that the upper leg load levels for belted drivers are significantly lower as to be as expected. However, 3MS values for the belted drivers are higher than for the unbelted. This is explained by the fact that the belt systems in the vehicle models do not have a load limiter resulting in high chest loads. For a more realistic representation of the fleet behavior a load limiter should be included.

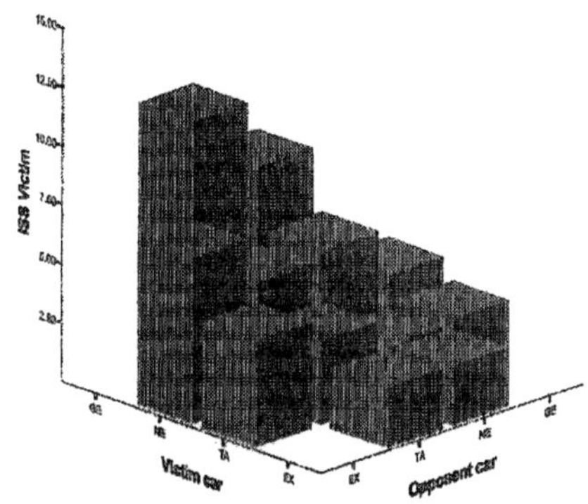

Figure 11. ISS distribution (mean values) for entire subset plotted as function of victim and opponent car.

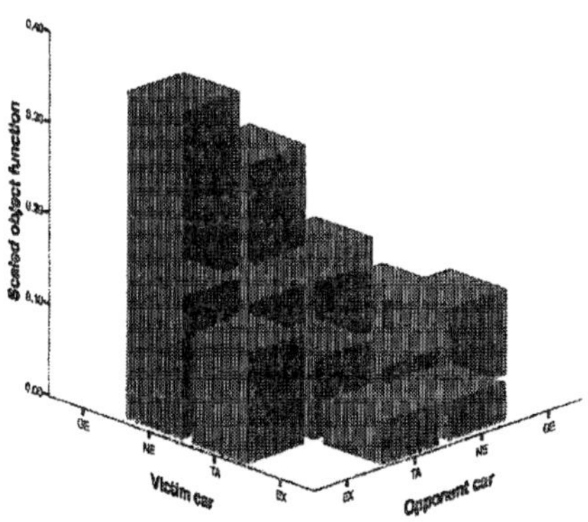

Figure 12. Weighted squared injury distribution (mean values) for entire subset plotted as function of victim and opponent car.

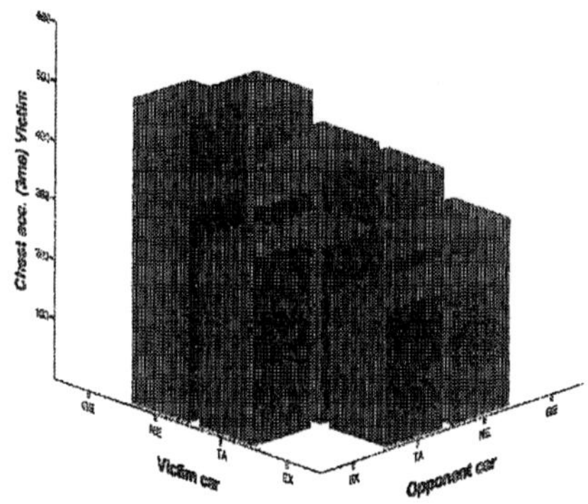

Figure 13. 3MS distribution (mean values) of entire subset plotted as function for victim and opponent car.

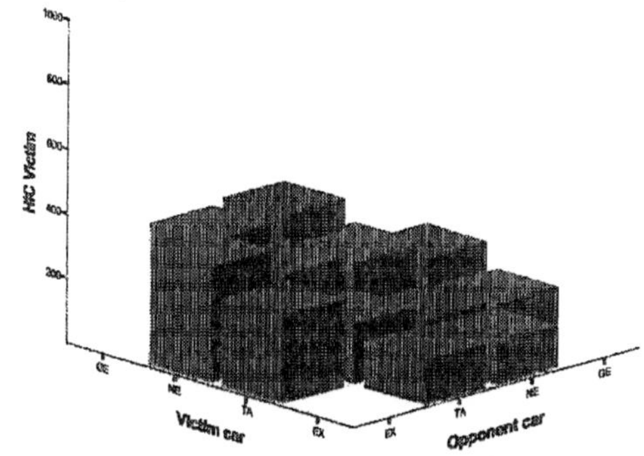

Figure 13. HIC distribution (mean values) of entire subset plotted as function of victim and opponent car.

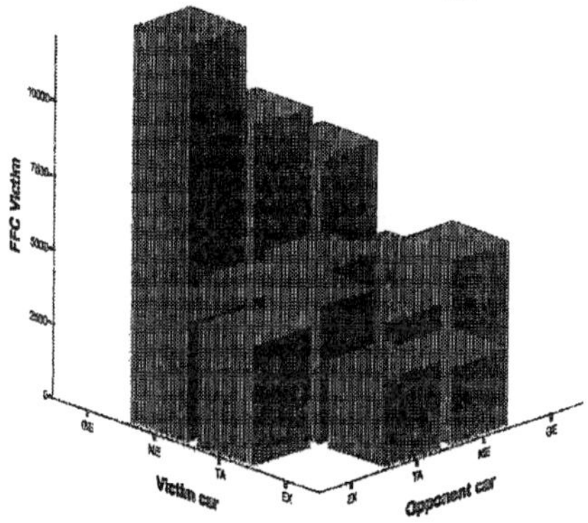

Figure 14. FFC distribution (mean values) of entire subset plotted as function of victim and opponent car.

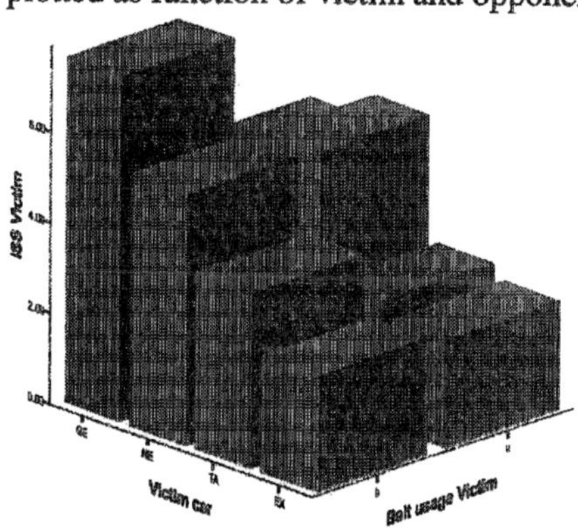

Figure 15. ISS distribution (mean values) of entire subset plotted as function of victim car and belt usage.

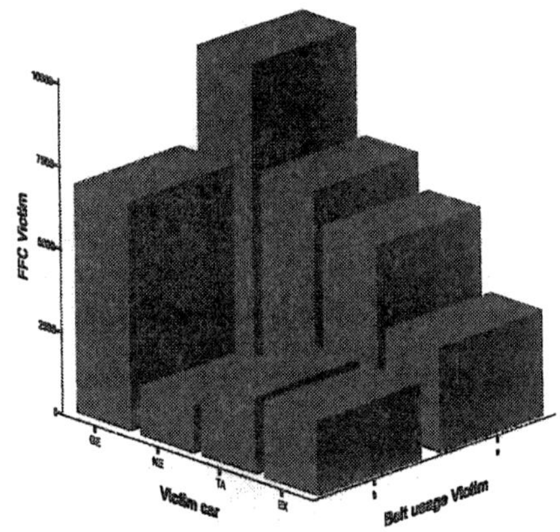

Figure 16. FFC distribution (mean values) of entire subset plotted as function of victim car and belt usage.

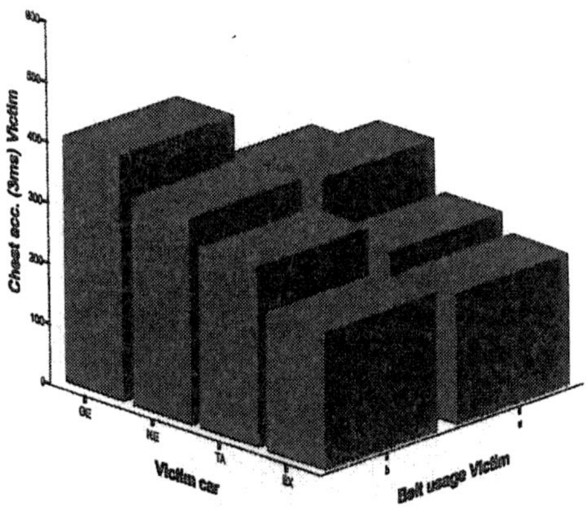

Figure 17. 3MS distribution (mean values) of entire subset plotted as function of victim car and belt usage.

Figure 18 shows ISS values of the Geo driver for stiffness variations in the front-end of the Geo. The columns indicate the 25% to 75% range of samples. The vertical lines related to each column indicate ultimate and meridian values. Minimum ISS values occur at the original stiffness. However, when considering separate injuries, figures 19 through 21, different trends are observed. HIC in figure 19 shows an identical behavior as the overall measure. 3MS in figure 20 is fairly insensitive to the front-end stiffness. FFC-left in figure 21 shows a clear reduction with front-end stiffness bringing the 95% range below critical. This trend is not observed in the ISS result due to the relatively low weight factor for FFC. The finding is in agreement with previous studies using direct optimization [9, 16].

Figure 22 and 23 show results of the Geo driver for stiffness variation in the Taurus front-end. The influence on ISS is fairly low but the FFC shows a trend with reduced injury for reduced stiffness of the Taurus front-end, as to be expected.

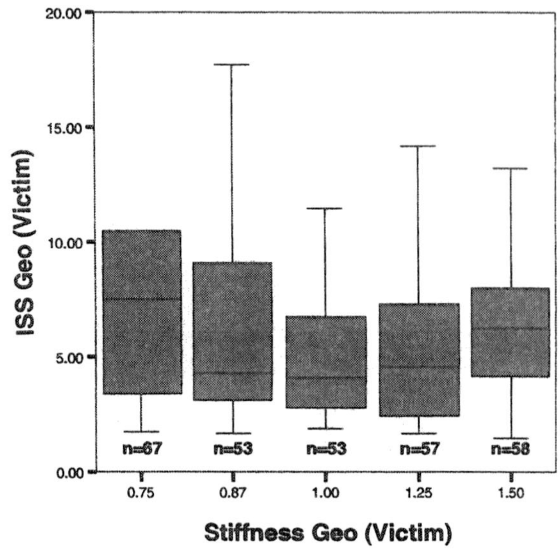

Figure 18. ISS of Geo driver as function of frontal stiffness

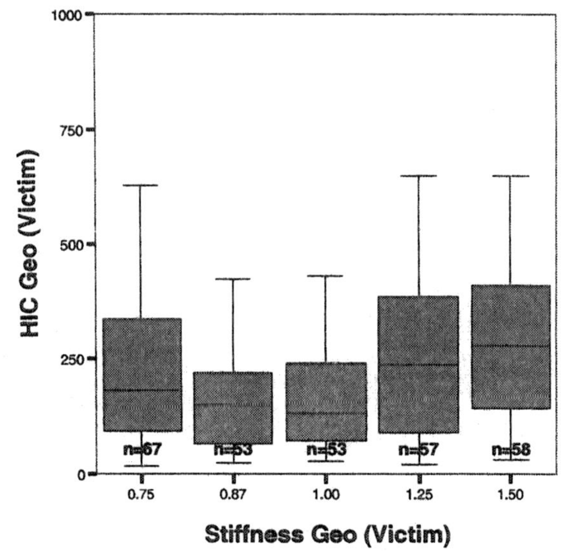

Figure 19. HIC of Geo driver as function of frontal stiffness

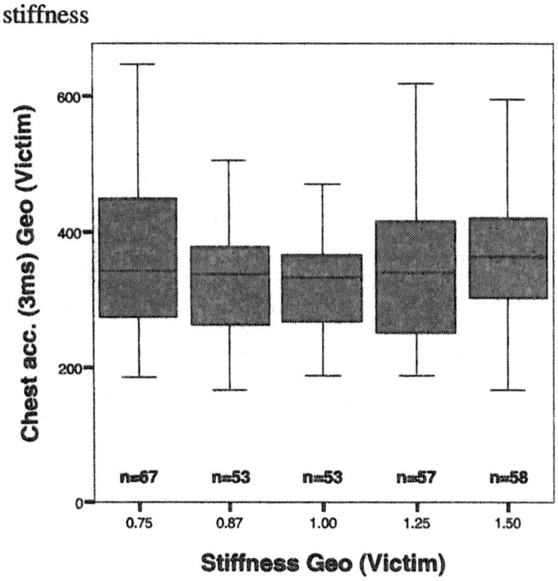

Figure 20. 3MS of Geo driver as function of frontal stiffness

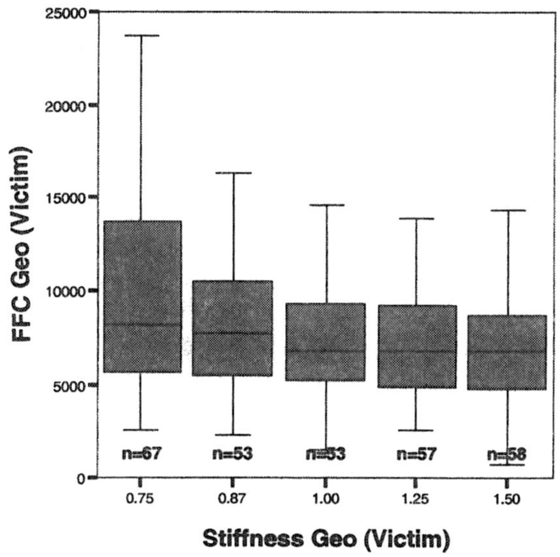

Figure 21. 3MS of Geo driver as function of frontal stiffness

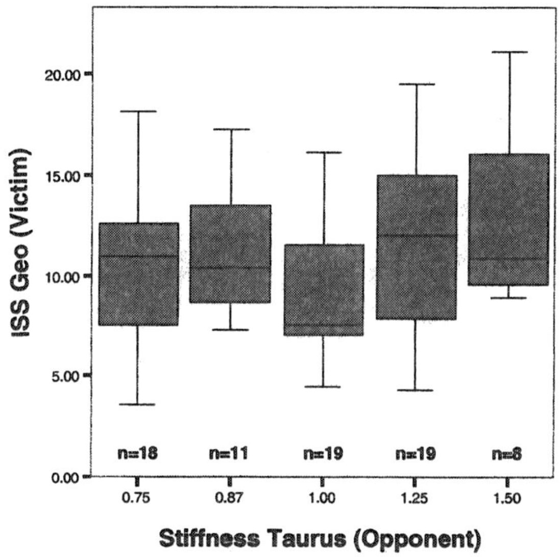

Figure 22. ISS of Geo driver as function of frontal stiffness Ford Taurus

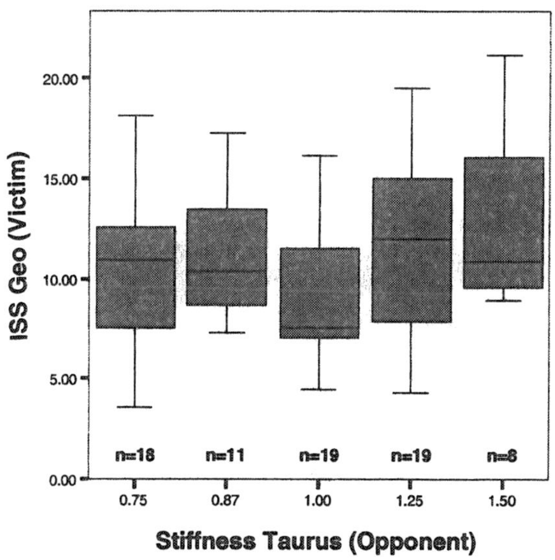

Figure 23. FFC of Geo driver as function of frontal stiffness Ford Taurus

Figures 24 through 26 compare injury values between the updated and the original Chrysler Neon. The figures show results for the reference configurations (original stiffness and a belted 50-percentile dummy) only as simulations for the complete statistical study were still in progress at the time of writing. The modification of the front-end depicted in figure 10 was meant to improve the structural interaction and as such reduce intrusions. Knowing the limitations for the lower extremities, results related to upper legs, chest and head are plotted. The results indicate that acceleration related injuries for head and chest remain nearly unaffected whereas intrusion related injuries for the upper legs reduce significantly. The reduction of intrusion becomes clear from figure 27 that shows deformed configurations for the Neon-Geo scenarios. Identical results where found for the adjusted Geo.

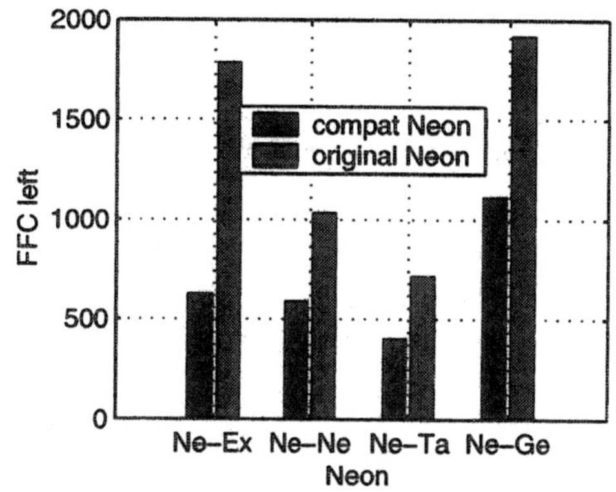

Figure 24. FFC values for original and updated Neon for crashes against other cars.

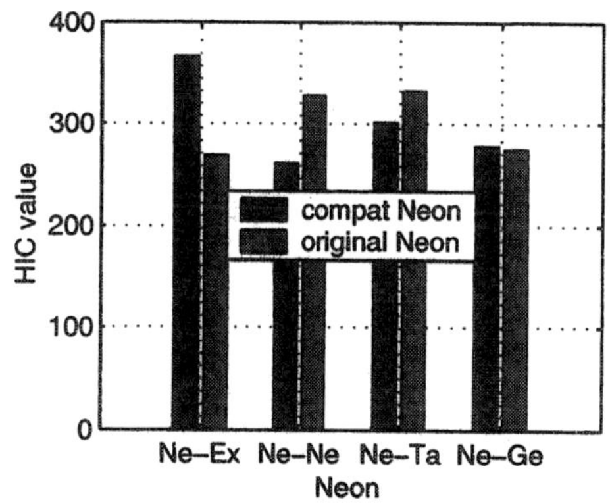

Figure 25. HIC values for original and updated Neon for crashes against other cars.

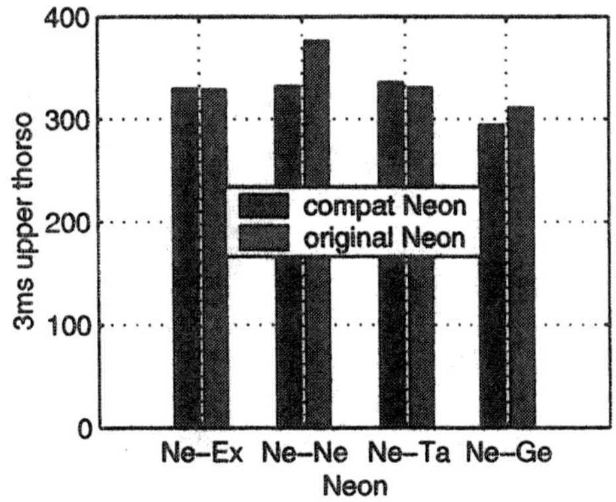

Figure 26. 3MS values for original and updated Neon for crashes against other cars.

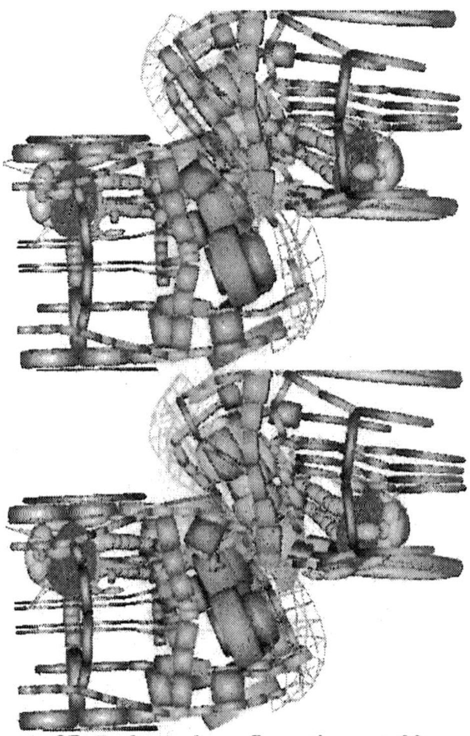

Figure 27. Deformed configurations at 80 ms for Neon-Geo scenarios: original Neon (top) and updated Neon (bottom). Neon is blue, Geo is green.

CONCLUSIONS AND RECOMMENDATIONS

To indicate the performance of the multi-body vehicle models for crashworthiness optimization of a fleet a study on offset frontal impacts was presented. Using models of four different vehicles, that represent the US car fleet, parameter sweeps over relevant accident and design variants were performed.

The vehicle models are 3-D rigid mass models derived from FE models. The rigid bodies are connected by non-linear spring and damper elements representing the stiffness behavior. Main interior parts and dummies are included. Comparison with test results shows that the models provide realistic crash and occupant behavior. The models integrate vehicle and occupant models that were separated in previous fleet studies. Front-end stiffness and geometry can be adjusted easily by scaling the stiffness of main members and adding new bodies and joints. Although less accurate than finite element models the multi-body models require substantially less CPU making them suitable for the large amount of simulations required in fleet studies.

In the fleet study frontal offset impacts between the four vehicles (Geo Metro, Chrysler Neon, Ford Taurus and Ford Explorer) were considered. Crashes between the different vehicles were simulated with belted an unbelted drivers of different size (5th percentile female, 50th percentile male and 95th percentile male dummies). The front-end stiffness of each vehicle was varied between 75 and 150% of their original value. For the two smallest vehicles design variants that provide improved structural interaction were considered. A full factorial parameter sweep over these accident and design variables resulted in 2500 scenarios where the stiffness of each vehicle was varied in five steps. Simulating these scenarios required an acceptable 830 hours of CPU on a PC server system. Statistical analysis of results showed that injury levels for these considered accident scenario can be reduced below critical values.

Evaluation of results for belt usage showed that the modeling of restraints needs improvements. Despite this the study showed that the models have high potential for this type of fleet studies since they provide realistic vehicle behavior at limited CPU costs. Also structural modifications are easily introduced.

In future work the modeling of the restraint systems should be improved. Comparison of belted and unbelted results showed that the belt models should include load limiters to provide more realistic chest loads. Also the triggering of the airbags should be made dependent on the accident scenario in terms of impact speed and other relevant factors. This allows the simulation of scenarios at different impact speeds. With these improvements fleet studies using scenario weight factors from a statistical accident environment model can be made. Resulting injury distributions can be compared with real world data for validation purposes. In these studies the accident scenarios can be extended, e.g. with side impacts for which validated models are available.

Apart from the fleet modeling work the multi-body models and their improved variants will be employed to improve proposals for compatibility test procedures. Car to barrier simulations with original and improved vehicles will be performed to evaluate proposed barrier design and assessment criteria. This activity will be performed in the European 5th framework project VC Compat.

ACKNOWLEDGEMENTS

The authors wish to thank Jos Huibers from TNO Automotive as well as Gijs Kellendonk and Francois Bronkers from the University of Eindhoven for support in the development of use of the multi-body frame models. They also thank Stephen Summers and Thomas Hollowell of NHTSA as well as Leo Schlosser and Jan Busstra from the Dutch Ministry of Transport for their support of the compatibility research at TNO.

REFERENCES

[1] Hollowell, W.T., and Gabler, H.C., 'NHTSA's Vehicle Aggressivity and Compatibility Research program', 15th Int. Technical Conf. on the Enhanced Safety of Vehicles, Melbourne, Australia, May 13-17, 1996.

[2] Evans, L., and Frick, M.C., 'Driver fatality risk in two-car crashes -- dependence on masses of driven and striking cars', in Proc. of the 13th ESV Conf., paper no. 91-S1-O-10, Paris, France, 1991.

[3] Zobel, R., 1997, 'Barrier impact tests and demands for compatibility of passenger vehicles', in: VDI Tagungsbericht 1354, Innovativer Insassenschutz im PKW, pp. 209-225, Düsseldorf, Germany, 1997.

[4] Faerber, E., 'Improvement of crash compatibility between cars', (project summary), Conference on road infrastructure and safety research in Europe, Risc '97, Brussels, 1997.

[5] Wykes, N.J., et al., 1998, Accident analyses for the review of the frontal and side impact directives, Report to the European Commission DGIII, TRL, UK.

[6] Shearlaw, A., and Thomas, P., ' Vehicle to vehicle compatibility in real world accidents', Proc. of the 15th ESV conference, paper no. 96-S4-O-04, 1996.

[7] H. G. Mooi, J. H. A. M. Huibers (1998). Simple and effective lumped mass model for determining kinetics and dynamics of car-to-car crashes. in:Proc. of Int. Crashworthiness Conf. IJCrash '98, 9-11 September 1998, Delft, The Netherlands.

[8] H. G. Mooi, T. Nastic., J. H. A. M. Huibers (1999). Modelling and Optimization of Car-to-Car compatibility. VDI berichte NR 1471 (ISBN 3-18-091471-8), pages 239-255, delft, The Netherlands.

[9] A.W.J. Gielen, H.G. Mooi, J.H.A.M. Huibers (2000). 'An optimization methodology for improving car-to-car compatibility', ImechE transactions C567/047/2000, Delft, The Netherlands

[10] Ford Motor Co. (1978). 'Safety Systems Optimization Model, Final Report', DOT Report Number DOT HS 6-01446, USA

[11] K.P. White, H.C. Gabler, W.D. Pilkey and W.T. Hollowell (1985). 'Simulation Optimization of the Crash Worthiness of a Passinger Vehicle in Frontal Collisions using Response Surface Methodology', SAE Paper 850512.

[12] A.C. Kuchar, R. Greif, G.W. Neat (2001). 'A Systems Methodology For Evaluation of Vehicle Aggressivity In The Automotive Accident Environment', SAE Paper 2001-01-1172

[13] M. Kleinberger, E. Sun, R. Eppinger, et al. (1998). 'Defelopment of Improved Injury Criteria for the Assessment of Advanced Automotive Restraint system', NHTSA Docket No. 1998-4405-9

[14] R. Eppinger, E. Sun, et al. (1985). 'Development of Improved Injury Criteria for the Assessment of Advanced Automotive Restraint Systems – II', Supplement to NHTSA Docket No. 1998-4405-9

[15] S.M. Summers, W.T. Hollowell (2001). 'NHTSA's Crashworthiness Modeling Activities', ESV Paper No. 251

[16] C.D. van der Zweep, P.L. Lemmen (2002). 'Improvement of Crash Compatibility between Passenger Cars by Numerical Optimisation of Front-end Stiffness', JSAE Paper No. 2002-5353

[17] S.M. Summers, A. Prasad, W.T. Hollowell (2001). 'NHTSA's Compatibility Research Program Update', SAE Paper No. 2001-01-1167

[18] D.C. Viano, S. Arepally (1990). 'Assessing the Safety Performance of Occupant Restraint Systems', SAE Paper No. 902328

[19] J.S.H.M. Wismans, et al. (1994), 'Injury Biomechanics', Course Notes Eindhoven University of Technology No. PRO 1411 – TUE

[20] http://www-nrd.nhtsa.dot.gov/ NHTSA's web-page

[21] SPSS Base 10.0 Applications Guide, SPSS Inc., Chicago, IL, USA, 1999

1999-01-0065

Accident Analysis and Measures to Establish Compatibility

Robert Zobel
Volkswagen AG

ABSTRACT

The vehicle fleet differs in mass, geometry, stiffness and many other parameters. These differences are consequences of different design objectives for these vehicles and result from consumer demand, environmental and safety considerations etc. Accident research shows that the injury outcome differs in some cases, when two vehicles collide. Scientists often discuss a list of features that are assumed to be relevant for compatibility of vehicles. The relevance of these potentially important compatibility features and expected compatibility measures is examined from the perspective of accident analysis. An overview of this accident research is given and crash tests and measures are discussed that correspond with these findings.

INTRODUCTION

Compatibility is expected to be the field of vehicle safety that promises the next step for an increase of vehicle safety with world-wide implications. Many countries perform research on compatibility with the goal of establishing standards for the structural interaction of vehicles. These expectations are based on observations that are well-known not only in the safety community, but also by many automobile users: Two vehicles collide. One vehicle is deformed , the other vehicle is virtually undeformed. These findings frequently consider the undeformed vehicle to be the safest. A more obvious observation is that different bumper heights and different vehicle geometry are considered to lead to incompatibilities It is the task of accident research to clarify this situation and to identify, what really is incompatibility.

But unfortunately accident researchers usually deal only with mass ratios and fail to study other effects in depth such as bumper height, geometry etc.. Workshops conducted at the beginning of 1998 showed a lack of accident data that can be used to evaluate various observations under public discussion regarding compatibility. The situation in the U.S. is probably somewhat different than in Europe, because the range of mass ratios encountered in the U.S. is probably greater than in Europe. In Germany e.g. 7.1% of all newly registered vehicles are MPVs or SUVs. The ratio between light trucks and passenger vehicles is 1:22.4, in the U.S. this ratio is 1:2.4. Even if MPVs, SUVs, and light trucks up to a laden weight of 7000 lbs are combined to a group, in Germany, the ratio to passenger vehicles changes to 1:8.7. Including all trucks, all MPVs, and all SUVs in Germany, the ratio is still 1:7.9. This calculation is based on information from the Kraftfahrtbundesamt 1997 for Germany and the number of registered vehicles in the U.S. for 1994. Compatibility of light trucks has a quite different importance in Germany than in the U.S. Although the author does not have the data, it may be expected that the situation in other European states to be similar to that in Germany.

Three major projects were undertaken to study compatibility in depth: NHTSA has initiated a project. The national safety organizations in Europe have started a project under the overall supervision of the Bundesanstalt für Straßenwesen, in Germany. The European car manufacturers are carrying out a similar project. The European project teams are co-operating with each other. IHRA, the International Harmonized Research Agenda, is offering a forum for a world-wide exchange of information. The European projects are structured in a similar manner. Accident analysis is performed to identify the areas of interest with regard to compatibility. Testing and simulation is performed to develop test procedures that allow assessment of vehicle compatibility in terms of self- and partner-protection.

The significance of testing must be discussed with regard to compatibility. All institutions conducted tests, before commencing compatibility research. These tests were a starting point for the interpretation of vehicle compatibility. E.g. different deformation stroke may be used as a characteristic to characterize a vehicle as being too stiff or too soft, depending what vehicle is sought to be the subject of criticism. Testing, however, is not able to tell us how relevant an observed behavior is. How often do these configurations occur? How often does a phenomenon, observed in testing, occur in real world accidents? Is there an increase or decrease of injury risk, when such a situation occurs? These questions can be answered only by careful analysis of accident data.

Thus, it is an acceptable method in the preliminary stages of a project to conduct tests and to interpret them using accepted engineering principles. But it must also be taken into account that testing is only capable of documenting a specific situation with an exact velocity, a given degree of overlap, a given impact angle etc. Even the prediction of occupant injury in this very particular situation is possible only up to a certain extent through analysis of dummy response, because dummy biofidelity is limited. Accident analysis reflects directly the field experience and can thus be directly correlated to occupant risk, testing detects phenomena that might be of interest, but the relevance of these phenomena is unknown as long as no accident data on this phenomenon are available.

Therefore the success of all compatibility research depends on the willingness of all researchers, to put aside all of their more or less justified prejudices and to perform accident analysis and to follow the results of this research.

There is an important aspect of accident analysis that should be kept in mind to avoid the potential for misinterpretation: Accident analysis reflects the current vehicle fleet on our streets today. Thus we can only study accidents with current vehicles. If compatibility research changes vehicles significantly (e.g. with regard to stiffness or other structural parameters), we create new collision modes that cannot be observed in current accident scene: The collision "new vehicle against new vehicle" and the collision "new vehicle against old vehicle" must be studied carefully utilizing accepted and proven testing and simulation tools, because there is no other accident experience available for study. The transition phase, when new and old vehicles are on the roads together must be predicted and evaluated with proven techniques.

We are now experiencing such a transition phase in Europe. By implementing Euro-NCAP, vehicle design is being influenced. These 3- and 4-star EURO-NCAP vehicles collide with one-another and with older vehicles.

ACCIDENT ANALYSIS

Accident analysis in the past always pointed out that there is a difference in fatalities when small vehicles and large vehicles collide. This was often used to criticize the smaller vehicle. A previous paper by this author showed that the main reason for this finding is the sympathetic characteristic of the small vehicle to be nearly incapable of causing fatalities in the large vehicle [Dreyer, 1981, Richter, 1982]. The lack of aggressiveness of the small vehicle shows that partner protection is possible. The question is whether there are possibilities to ensure partner protection within the population of large vehicles as well so as to also help prevent injuries in the small vehicles.

In the compatibility project being conducted by European car manufacturers, a list of possible compatibility features was defined which had already been published at the ESV-conference [Zobel, 1998]. The following characteristics were taken into account:

- vehicle mass
- vehicle stiffness
- lateral fork effect
- vertical fork effect
- low/high front end of the vehicle regarding frontal impact
- high front end of the striking vehicle regarding side impact
- longitudinal engine
- transverse engine
- additionally, for side impact: a well-balanced distribution of the force over the front of the vehicle.

This list was evaluated from the data sets that are available to Rover, Volkswagen, and Volvo, and to the University of Oulu. It was changed based on experience the group gained during the initial phases:

COMPULSORY COMPATIBILITY FEATURES – Compulsory compatibility features are parameters that inevitably have to be considered with respect to compatibility.

Mass is the dominant parameter and must be taken into account primarily. All other features have to be cross-checked against mass, because parameters that have a correlation to mass appear to influence self- and partner-protection.

Longitudinal and transverse engines are a good example. Studies, performed by the author, appeared to indicate that vehicles with longitudinal engines are more aggressive than vehicles with transverse engines. But repeating the studies, but now mass adjusted, that means within mass groups, this effect was no longer observed. So when a compatibility-feature is evaluated from accident analysis, a mass adjustment has to be made. Otherwise all features that are more often implemented in larger vehicles (this is e.g. true for longitudinal engines), seem to be aggressive due to the dominant influence of mass.

Compartment stiffness is the second item that was evaluated as a compulsory compatibility feature. Although compartment stiffness cannot be studied satisfactorily due to the lack of data, it was understood to be evident that a stiff compartment forms the basis for survival space and is thus necessary. The degree of compartment stiffness and its relative importance will be discussed later.

Override of sills is obviously an aspect which must also be considered. When the longitudinal frae members of a striking vehicle override the sill on the side of another vehicle, a negative effect on the near side occupant of the struck vehicle has been noted. This effect is detected, although it has a lower relevance than the mass effect.

POSSIBLE COMPATIBILITY FEATURES – Other parameter are not detected clearly enough or are not seen in accident statistics as relevant, but the result is not so clear that additional research can be considered superfluous. These items may be taken into account, when strong arguments about their influence on self- and partner-protection can be derived from testing or simulation.

Front end stiffness is one of these parameters. This is surprising, because from a theoretical point of view, front-end stiffness is one of the main parameters of compatibility. Thus, this parameter will be examined further. A possible reason for this negative result from accident analysis is that current vehicles have very similar behavior in frontal impact as a consequence of existing self protection requirements. The new vehicles that are designed with regard to EURO-NCAP, are not available today for accident analysis in sufficient numbers.

Vertical mismatch of longitudinals in frontal impact is examined with a rather peculiar result. Results show that having higher longitudinals than an opponent is an advantage with regard to self-protection, consequently disadvantages arise when the longitudinals are of lower than those of an opponent. Although frequently thought to be beneficial, more injury occurs, when longitudinals are of similar height. Therefore the often mentioned demand for similar height of longitudinals is of no relevance and might even be detrimental. A possible explanation is that overriding in frontal impact leads to lower accelerations in the vehicle, as long as no compartment intrusion occurs. If this should be a disadvantage in some cases, then these cases are rarely detected thorough accident analysis.

Engine orientation is either longitudinal or transverse. Only one study by the University of Oulu indicates with low statistical significance (appr. 50%) that there might be greater injury in an opposing vehicle, if it is struck by a vehicle with a transverse engine. Other studies show no effect after mass adjustment. This finding will remain open and it must be ascertained whether there are tests that explain this observation.

Homogeneous front ends generate a equal distribution of load into the side structure of the struck vehicle. These front-end designs will also improve load paths in offset impacts. However, in accident data a parameter "homogeneous front end" does not exist. It is difficult to define the group of vehicles having such a design. Thus, there are only indications which must be cross-checked through testing.

EXCLUDED ITEMS – Excluded were the following items because there was no indication that they are relevant with regard to compatibility:

- Lateral mismatch of longitudinals
- Lateral fork effect
- Vertical fork effect

These comments conclude the current state of our knowledge on compatibility features within the compatibility project being conducted by European car manufacturers. It is not a final statement of this project. The project will continue with additional accident analysis, especially regarding other collision modes that were not so thoroughly studied in the beginning, when the focus was mainly on the frontal and side impact.

SOME REMARKS ON DEFORMATION

Detailed accident analysis was undertaken to detect an influence of stiffness, because all theoretical calculations indicate this as a major problem. One approach was presented to the ESV-conference in Windsor [X]. This approach attempted to verify that stiff cars are those that have lower deformation than an opponent.

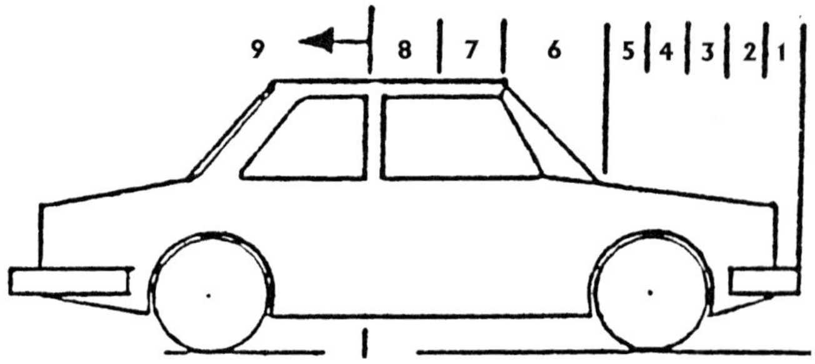

Figure 1. Vehicle deformation index VDI for deformation extend guide (VDI6). (Source: Manual for Coding. Variables in the in depth investigation work. Accident Research Unit, Hanover, 1990)

Another attempt tried to examine the energy required to deform a vehicle up to VDI6 as an measure of its stiffness.

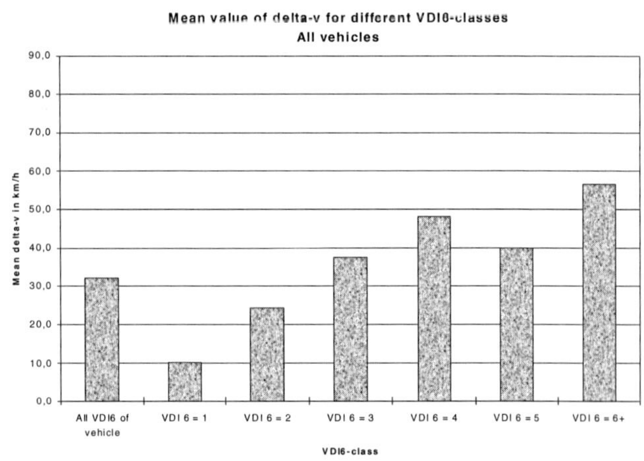

Figure 2. Comparison of VDI6 and average delta v for all vehicles. In higher VDI6 classes, delta v no longer increases linearly between VDI6=1 and VDI6=4. The lower value for VDI6=5 is a consequence of the fact that VDI5 has a higher proportion of accidents with lower overlap.

The decrease in delta v for the higher VDI6 is a consequence of the fact that VDI6=5 has a of 50% share of accidents with lower overlap up to 30%. This share is only 40% for VDI6=4. This explains the irregularity. This chart is used to define an average delta-v required to achieve a certain deformation in terms of VDI6. The assumption is that a vehicle is stiffer when it requires a higher delta v, it is less stiff, when this amount of deformation is already reached at a lower level. Delta v is used as an estimate for the amount of energy needed for this deformation. Since energy is not available in an accident database, this approach was taken.

To ascertain the consequences, a model was chosen and three model generations were taken to detect the influence of different stiffness of these vehicles. The results are shown in figure 3. Only in very high deformation categories does the delta-v of this vehicle differ from the average. Only in these classes does the opponent have a significantly lower delta v. What does this mean? It is submitted that this vehicle behaves like the average, as long, as the mass ratio is not too high. The VDI6=6+ class gathers the cases of high mass ratio. It is only here that the only significant difference can be detected. Regarding lower VDI6 classes, the stiffness average seems to have no significant influence.

These figures show that the focus of compatibility is not on the stiffness itself, the discrepancy between masses with regard to deformation becomes relevant only in such cases, were vehicles are over-crushed. In this case, with a high probability, mass differences are the reason for the over-crushing (fig.3).

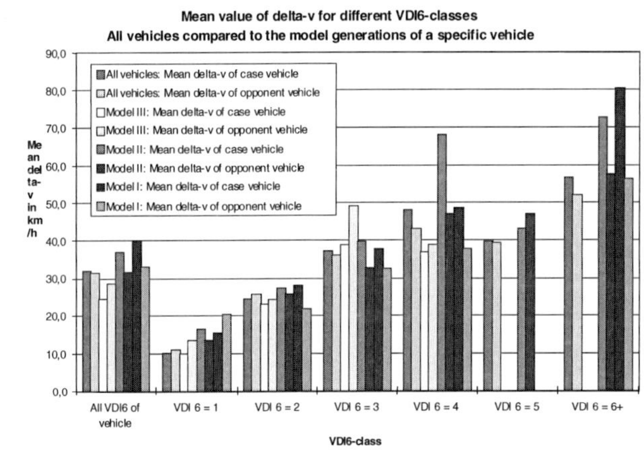

Figure 3. Comparison of VDI6 and average delta-v for different model generations of a specific vehicle. In higher VDI6 classes the delta v of the opponent is lower, this indicates lower mass of the opponent. No significant differences between model generations are evident.

DEFORMATION AND INJURY

The influence of deformation on injury supports the findings above. With increasing deformation, the risk of injury increases. The risk of fatal injury starts with VDI6=4 and becomes relevant with VDI6=6 and higher, that means, in case of over-crushing. Figure 4 provides an overview of Vehicle-MAIS that is the maximum MAIS that occurred in the vehicle to belted occupants. Cases with unbelted drivers were excluded. Figure 5 is similar to figure 4, it shows the percentage range between 80% and 100%, the higher MAIS classes.

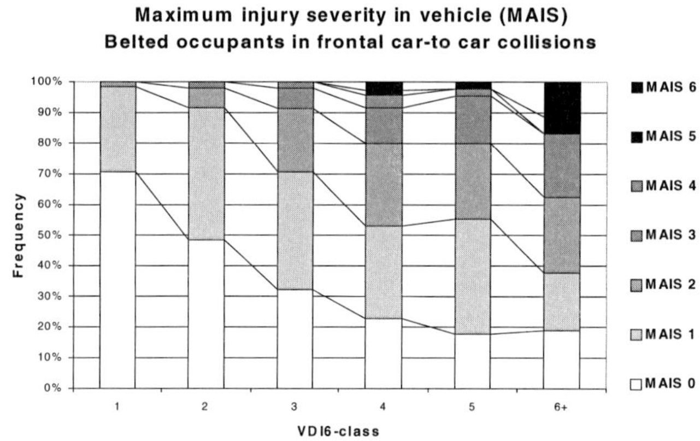

Figure 4. Deformation and occupant injury for belted occupants.

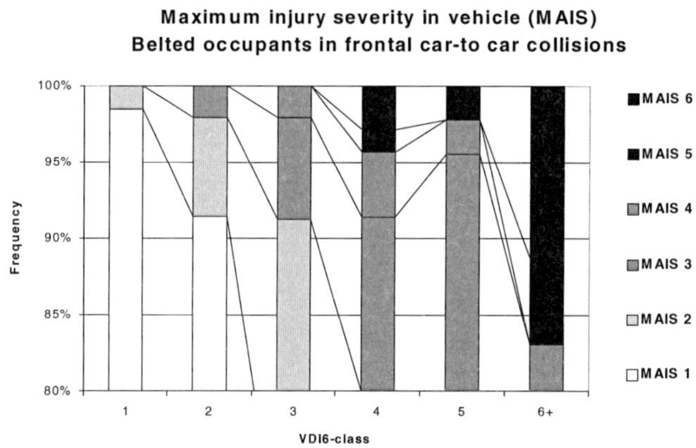

Figure 5. Sector of 80% up to 100% of figure 4.

The risk of fatalities increases with over-crush. This finding is true for head-injury related fatalities, which are understood to be acceleration related injuries. It is also true for thorax injury related fatalities, which are understood to be more deformation related injuries. Thus, this finding is not a contribution to the question of intrusion and acceleration related injuries. (Fig. 6 and fig. 7) Leg injuries react in a relatively moderate manner to high intrusion. Leg is more inluenced by the location of the intrusion, and not that much by the intrusion itself. (Fig. 8)

These observations are presented to show the dominance of over-crushing for severe injuries. They also show that even in case of over-crushing, not every accident is catastrophic in terms of its consequences to the occupants. The observations nevertheless prove that compatibility research must deal with this phenomenon. They are the reason for the following discussions on the question of differential masses in vehicle-to-vehicle collisions.

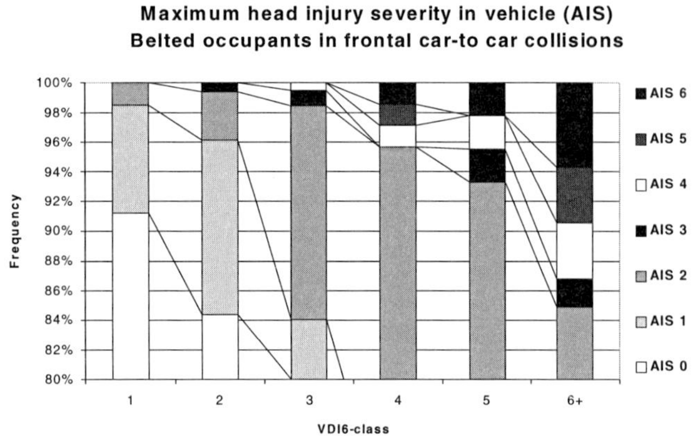

Figure 6. Sector of 80% up to 100% for head injuries.

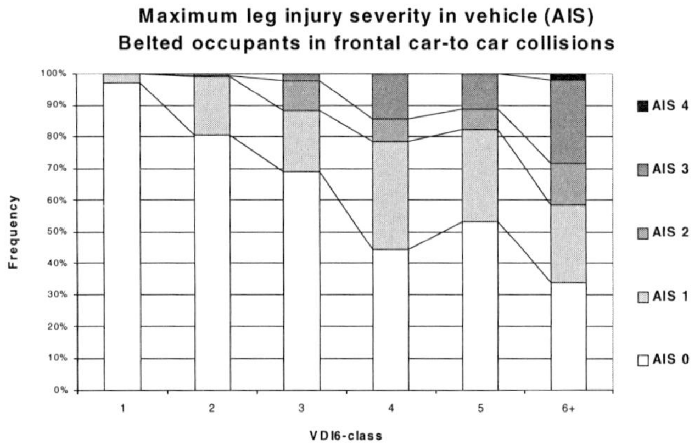

Figure 7. Sector of 80% up to 100% for thorax injuries.

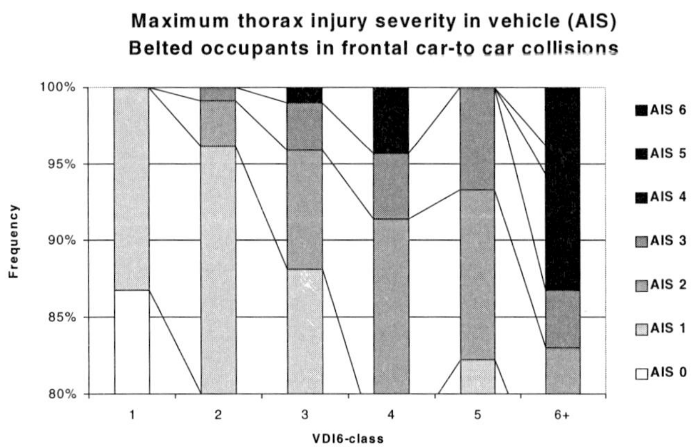

Figure 8. Leg injuries and deformation.

THE COMPATIBILITY RELATIONSHIP

The test against a fixed barrier forces manufacturers to provide sufficient deformation energy. If the barrier impact must be conducted with an impact speed v_B for a vehicle with mass m, then the required deformation energy is approximately equivalent to the kinetic energy of the barrier impact:

$$E = \frac{1}{2} * m * v_B^2$$

A frontal vehicle-to-vehicle collision involves vehicle 1 and vehicle 2 with masses m_1 and m_2 and velocities v_1 and v_2. The kinetic energy before the crash is

$$E_{before} = \frac{1}{2} * m_1 * v_1^2 + \frac{1}{2} * m_2 * v_2^2$$

The velocity after the crash is assumed to be the same for both vehicles v_{after}. Rebound is neglected in this case. Thus, v_{after} is computed from the theorem of momentum by

$$(m_1 + m_2) * v_{after} = m_1 * v_1 + m_2 * v_2$$

This results in a kinetic energy after the crash:

$$E_{after} = \frac{1}{2} * (m_1 + m_2) * v_{after}^2$$

The difference between E_{before} and E_{after}, the energy absorbed, is the deformation energy D in this accident

$$D = E_{before} - E_{after} = \frac{1}{2} * \frac{m_1 * m_2}{m_1 + m_2} * (v_1 - v_2)^2$$

Thus, this well-known computation shows that deformation energy depends on the two masses and the closing velocity $v_c = v_1 - v_2$ of the two colliding vehicles.

The following relationship is remarkable, because it shows the chances of compatibility.

If there is a collision between two vehicles of different mass and a closing velocity of v_c £ 2 * v_B and if each vehicle is designed for a barrier impact with v_B, then the deformation energy of both vehicles is sufficient to sustain the collision without intrusion into occupant compartments.

In this case the available deformation energy of the two vehicles is

$$E = \frac{1}{2} * m_1 * v_B^2 + \frac{1}{2} * m_2 * v_B^2 = \frac{1}{2} * (m_1 + m_2) * v_B^2$$

For this collision the energy absorption will be

$$D = \frac{1}{2} * \frac{m_1 * m_2}{m_1 + m_2} * (v_1 - v_2)^2 = \frac{1}{2} * \frac{m_1 * m_2}{m_1 + m_2} * v_c^2 \le \frac{1}{2} * \frac{m_1 * m_2}{m_1 + m_2} * (2 * v_B)^2$$

It has to be proven that D£E. This was done in the Windsor ESV-paper [X].

The result of this is a basic finding on compatibility:

$$D = \frac{1}{2} * \frac{m_1 * m_2}{m_1 + m_2} * v_c^2 \leq \frac{1}{2} * \frac{m_1 * m_2}{m_1 + m_2} * (2 * v_B)^2$$

$$\leq \frac{1}{2} * (m_1 + m_2) * v_B^2 = E$$

as long as $v_1 - v_2 = v_c \leq 2 * v_B$.

The computation looks rather theoretical, but the result has practical consequences. It is a fundamental relationship between the design speed of two vehicles and the deformation energy that is needed, when two vehicles collide:

When two vehicles collide, and if their closing velocity is less than their double design speed, then there is sufficient deformation energy available for this particular crash. This observation holds whatever the mass-ratio of the two vehicles is.

This means that over-crush is not due to physical laws, to a certain extent it is avoidable. This can be shown by the following bulkhead concept.

THE BULKHEAD CONCEPT

The bulkhead concept attempts to address the question of over-crush. Over-crush occurs when the stiffness of a compartment is too low to sustain the load of a striking vehicle. This means that the force level of the striking vehicle is too high for the compartment of the struck vehicle. Obviously there are limits to the application of this concept and there are accidents which are so severe that reasonable and practical engineering means are not available to adequately mitigate the severe consequences expected and observed.

To help avoid this, we define a force level that is the maximum force-level for front-end of a car to deform. A bulkhead must then be designed and built which is able to sustain this maximum force level. This bulkhead would avoid compartment collapse, as long as one of the vehicles is still deforming. This design concept is called the "bulkhead concept."

The bulkhead concept would force both vehicles to deform. It takes advantage of the fact that sufficient deformation energy is available. It will decide the question of too stiff and too soft. Forces up to the pre-defined maximum force-level are acceptable. Everything else is too stiff. If we had a bulkhead concept already, all collisions up to a closing velocity of 2^*v_B would occur without excessively deforming one of the collision partners, regardless of the mass ratio. This is possible, because the fundamental relationship of compatibility, derived above, holds.

DEFINITION OF THE FORCE LEVEL

A consequence for barrier impact speed will be shown here:

When a (small) car with mass m_S collides with another vehicle with deformation force F, then for the deceleration a_S, the following equation holds:

$$m_s * a_s = F$$

For the large vehicle with mass m_l, which is designed for a barrier impact speed v_B, we need deformation energy D, which is sufficient to compensate for the kinetic energy at barrier impact test speed:

$$D = \frac{1}{2} * m_l * v_B^2$$

On the other hand, deformation energy is computed by

$$D = \int F(s) ds = F * s_l$$

with deformation travel s_l of the large vehicle.

The equation only holds for the average force level F of the force deflection curve of the large vehicle's front structure.

$$\frac{1}{2} * m_l * v_B^2 = D = F * s_l$$

$$\left(\frac{1}{2} * m_l * v_B^2\right) \Big/ s_l = F$$

When this large vehicle and the small vehicle mentioned above collide, then pursuant to the principle of action and reaction, the forces are equal. This can be assumed by neglecting dynamic effects that more or less produce oscillation for which the static approach generates the average behavior. Thus, the static computation is more or less a lower limit of the dynamic computation. It leads to

$$\left(\frac{1}{2} * m_l * v_B^2\right) \Big/ s_l = F = m_s * a_s$$

or

$$\frac{1}{2 * s_l} * \frac{m_l}{m_s} * v_B^2 = a_s$$

This shows a relationship between barrier impact speed, mass ratio, and the available deformation stroke of the large vehicle. It is clear that there exist limits of feasibility for the deformation stroke. The length of vehicles must be restricted for several reasons, including those of an environmental nature. Vehicle deceleration of the must be restricted, because restraint systems are only able to load the occupant in an acceptable manner if vehicle deceleration remains within certain limits.

Therefore, there is a relationship between these principal vehicle parameters, as long as we want to keep the question of compatibility in mind. One question is the average deceleration that can be permitted for a vehicle, especially for a small vehicle. We should bear in mind that we are speaking about the average deceleration of the compartment. Today, with a compartment deceleration of 20 – 25 g we already need optimized restraint systems to achieve acceptable occupant loadings. A compartment deceleration of 30 g is already an upper limit in terms of acceptable dummy loads. Furthermore, this level of 30 g is probably not acceptable with respect to older vehicle occupants.

Now, it is easy to compute, what v_B is acceptable, as long as compatibility is possible. If we use a mass ratio of up to $\mu = \frac{m_l}{m_s} = 1{,}6$, it does not describe all vehicle-to-vehicle mass combinations, but covers approximately 90% of all frontal collisions in real-world accidents (in Germany). A deformation distance of $s_l = 0{,}7\ m$ is higher than available deformation stroke in the current fleet. If we accept a deceleration of 30 g that is higher than current restraint systems permit, then we find

$$30\ g = \frac{1}{2} \cdot 1{,}6 \cdot \frac{v_B^2}{0{,}7}$$

$$v_B = 16\ m/s = 57{,}8\ km/h$$

When we take into account that the 30 g level is an upper limit and that a mass ratio of 1,6 does not cover the whole fleet, than we must accept the fact that the current barrier test speed is at the upper limit. Thirty-five mph (56 km/h), the NCAP test speed for FMVSS 208 is already equivalent to the EES which is achieved when 64 km/h test is performed against a deformable barrier. Therefore, an increased barrier impact speed represents an additional decrease of the possibilities for compatibility. If the test speed were to be increased to 60 km/h, the mass ratio for which compatibility measures are possible will decrease by the formula

$$30\ g = \frac{1}{2}\mu \cdot \frac{(60\ km/h)^2}{0{,}7}$$

$$\mu = 1{,}48$$

If test speed were to be increased to 64 km/h, then the analogous computation leads to a mass ratio of $\mu = 1{,}30$. This means that compatibility is only possible in the range of 1000 kg to 1300 kg vehicles, e.g. that is not sufficient. If environmental considerations require, we can expect that small and fuel efficient vehicles will have an increasing share of the market and, consequently, we must allow for a larger compatibility range.

THE 90% RANGE OF MASS RATIOS

Using the 30g rule, established above, we can compute the acceptable force level. From the viewpoint of a 1000kg vehicle it is 1000kg*30g=300kN. From the viewpoint of a 500kg vehicle it is 150kN, and from the viewpoint of a 2000kg vehicle it is 600kN. If 600kN is allowed, then a 2000kg vehicle needs more than 40cm of deformation travel at this force level to establish 242kJ, the deformation energy, which it needs for a 56km/h barrier impact. 40cm is a theoretical lower limit, which is valid only for constant deformation forces, which cannot be designed in real cars. Thus, actual deformation travel is longer; for triangular (linearly increasing force) force-deflection curves, it is already 80cm. If only 300kN are allowed, this theoretical value is already 80cm, and on the basis of a triangular curve, it is already 160cm. From the viewpoint of a 500kg vehicle, 300kN are still too high, because they will produce 60g in this 500kg compartment. If 150kN are chosen, an appropriate figure for a small vehicle, the 2000kg vehicle will require at least 160cm deformation travel on the theoretical assumption of constant force. It requires 320cm on the basis of a triangular force-deflection curve. 320cm of deformation travel is, of course, only a computation. Realistic deformation travel is less than 80cm.

Thus, the question remains whether compatibility is actually impossible.

On the basis of a high level of self-protection, it is very hard to achieve this goal. This is clearly shown by the computations above. The same problem, however, occurs when we define a barrier impact speed. There are accidents with an impact speed of 100mph. When defining the barrier impact speed, we do not recommend to chose the maximum speed ever observed or theoretically possible, but to choose a speed that is relevant for 90% of frontal impacts. The same procedure is possible with mass ratios.

In Europe approximately 90% of all vehicle-to-vehicle accidents have a mass-ratio less than 1.6. This means that a 1600kg vehicle deals only with vehicles down to 1000kg. The design force for this vehicle can be derived from the demands of a 1000kg vehicle. This was already computed to be 300kN. This necessitates a deformation travel of 65cm on the theoretical approach of constant force. It requires 130cm, when triangular force-deflection curves are assumed.

Therefore, even the 90% approach is very ambitious, when the high level of self protection, which is reflected by a barrier impact speed of 56km/h, is considered. But this approach proves that there is a partial solution for this problem. Some possible approaches are shown by the following figures.

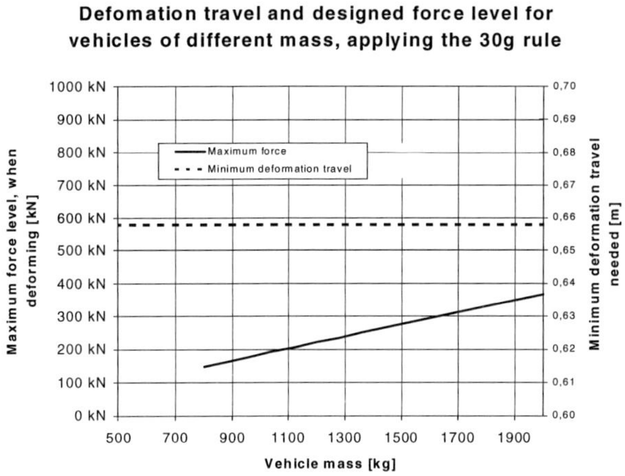

Figure 9. Design parameters, if compatibility is applied up to a mass ratio of 1.6 and a self-protection level of 56 km/h barrier impact speed. Compartment deceleration is limited to 30g.

This situation will change in a somewhat more realistic manner, when the barrier impact speed is reduced to 50km/h. Then the minimum deformation travel also decreases. This means that the possibility of applying this to real cars, increases. If we repeat the calculation on the basis of the reduced barrier impact speed, the force level which depends on the 30g rule, does not change, but the minimum deformation travel changes from 0,65m to 0,52m. (fig. 10)

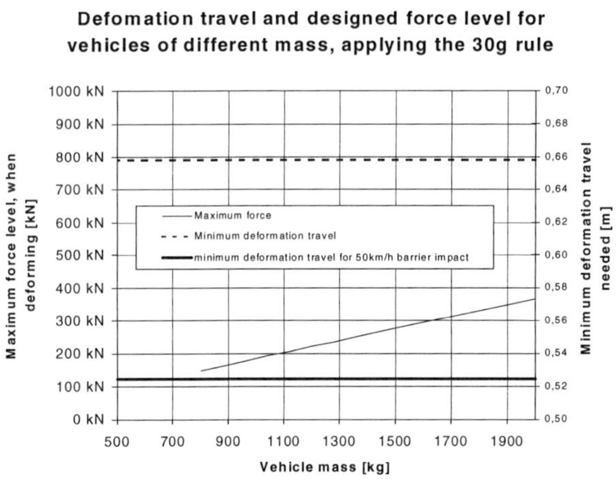

Figure 10. Same as fig. 9, but compared to deformation travel, resulting from 50 km/h barrier impact speed.

When we assume 50km/h, then there is more freedom to increase the range of mass ratios, for which this approach may apply. In fig. 11, the situation is shown for a mass ratio of 1.6 and 1.8. In the case of higher mass ratios, the force level must be lowered, by reason of the 30g rule. The demand for deformation travel increases, due to the lower force level. But with 59cm it is still less than the 66cm in the case of the 56km/h barrier impact speed and a mass ratio 1.6. Figure 11 shows these relationships.

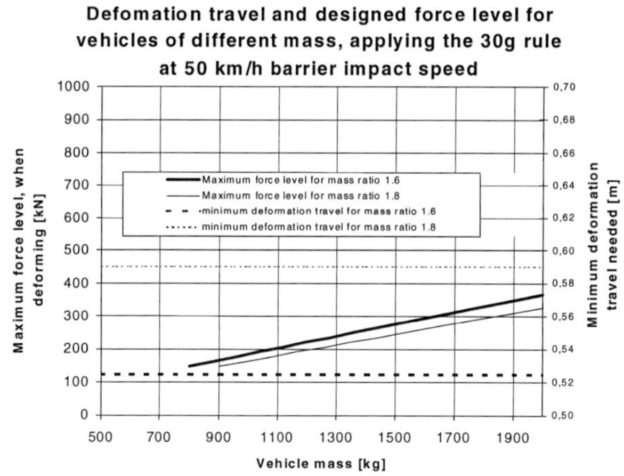

Figure 11. Different deformation travel at 50km/h barrier impact speed for mass ratio 1.6 and 1.8.

THE SMALL SOLUTION OF THE COMPATIBILITY PROBLEM

The previous chapter shows that taking into account a 90% range of mass ratios is already an acceptable solution of the compatibility problem in Europe, because the 90% range leads to a mass ratio of up to 1.6. In the U.S.A., this approach might not be applicable, because the 90% range of mass ratios is significantly wider than 1.6. So some scientists in the U.S.A. discuss a compatibility-solution which only solves the geometrical problem without offering adequate force levels. When we discuss the car-to-truck compatibility, it will be obvious that the force level cannot be defined adequately. So in this case it is obvious that only a geometrical solution is applicable.

Solving the geometrical problem means that the geometry of two colliding vehicles is designed in such a way that deformation structures in one vehicle find support in the other vehicle, so that they are able to deform and to ensure a designed deceleration of the vehicle. It does not mean that the longitudinals of the two vehicles directly contact. An extreme counterexample is a car underriding a truck, where the longitudinals of the car do not touch the truck and the first contact between truck and car occurs, when the A-pillar of the car touches the frame of the truck.

Solving the force level problem means solving the questions mentioned in the chapter before. This means a car design that a car is deforming at a force level which is still within the range, acceptable for the opponent with regard to its restraint system.

The following computations will show that we gain additional partner protection, even if we are not able to solve the problem of an adequate force level completely. We discuss this question by introducing a factor GAMMA, γ, which defines the portion of the deformation energy of the large vehicle that is available to the small vehicle. The range of g is between 0 and 1: $0 \leq \gamma \leq 1$.

In this case the available deformation energy of the two vehicles is if vehicle 1 is the large vehicle:

$$E = \frac{1}{2} * m_1 * v_B^2 * \gamma + \frac{1}{2} * m_2 * v_B^2$$
$$= \frac{1}{2} * (m_1 * \gamma + m_2) * v_B^2$$
$$= \frac{1}{2} * m_2 * (\mu * \gamma + 1) * v_B^2$$

with the mass ratio $\mu = \frac{m_1}{m_2}$, mass of large vehicle by mass of small vehicle.

For a collision with a closing velocity of v_c, a deformation energy *D* is needed:

$$D = \frac{1}{2} * \frac{m_1 * m_2}{m_1 + m_2} * v_c^{\ 2}$$
$$= \frac{1}{2} * m_1 * \frac{1}{1+\mu} * v_c^{\ 2}$$
$$= \frac{1}{2} * m_2 * \frac{\mu}{1+\mu} * v_c^{\ 2}$$

This offers an approach to show the energetic relationships (fig. 12). This figure shows the energy, which both vehicles provide depending on m_1=2000kg and the mass ratio of 1 up to 3, this means m_2=2000kg down to m_2=667kg. This available deformation energy is described in bold lines. It is compared to the deformation energy that is needed for a vehicle-to-vehicle collision of the two vehicles at a collision velocity of 112km/h (35mph) and on lower collision velocities. This is marked in standard lines. The energy needed at the smallest collision velocity is provided in dotted lines.

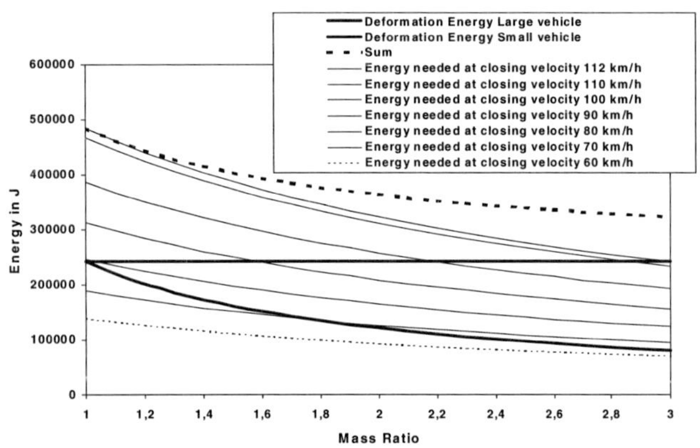

Figure 12. Available deformation energy in a 2000kg vehicle and its opponent versus mass ratio, compared to the requested deformation energy at different closing velocity levels.

It can be seen that at high ratio of 3, the deformation energy of the larger vehicle would be sufficient for a crash at a closing velocity of 112km/h. At a closing velocity of 60km/h, even the deformation energy of the smaller vehicle is sufficient.

So solving the geometrical problem without solving the force level problem would mean that at a mass ratio of 2 e.g., we will deal with all closing velocities up to 70km/h. Solving the force level problem would deal with closing velocities of more than 112km/h. So for mass ratio of 2, the benefit of solving the geometrical problem is gained up to a closing velocity of 70km/h, if we solve the force level problem as well, we gain benefit up to a closing velocity of more than 112km/h.

Taking the factor GAMMA into account, the following computation can be made. The highest closing velocity which is still within the car design range is the closing velocity which fulfills the equation

$$E =$$
$$= \frac{1}{2} * m_2 * (\mu * \gamma + 1) * v_B^2$$
$$= D$$
$$= \frac{1}{2} * m_2 * \frac{\mu}{1+\mu} * v_c^{\ 2}$$

or directly

$$\frac{1}{2} * m_2 * (\mu * \gamma + 1) * v_B^2 = \frac{1}{2} * m_2 * \frac{\mu}{1+\mu} * v_c^{\ 2}$$

$$(\mu * \gamma + 1) * v_B^2 = \frac{\mu}{1+\mu} * v_c^{\ 2}$$

$$v_c^{\ 2} = (1 + \frac{1}{\mu}) * (1 + \mu * \gamma) * v_B^2$$

$$v_c = v_B * \sqrt{(1 + \frac{1}{\mu}) * (1 + \mu * \gamma)}$$

This computation is the basis for figure 13.

Even if GAMMA=0, there is still a benefit in solving the geometrical problem. Depending on the mass ratio, it is in the range of 60km/h to 80km/h. The higher the value of GAMMA will be, the higher the benefit will be. If the force level problem would completely be solved, the benefit for mass ratio 3 would even be higher than for mass ratio 1. It would be up to a speed of 130km/h. If we take a closing velocity of 112km/h as a desirable goal, for mass ratio 3 we could allow a GAMMA=70%.

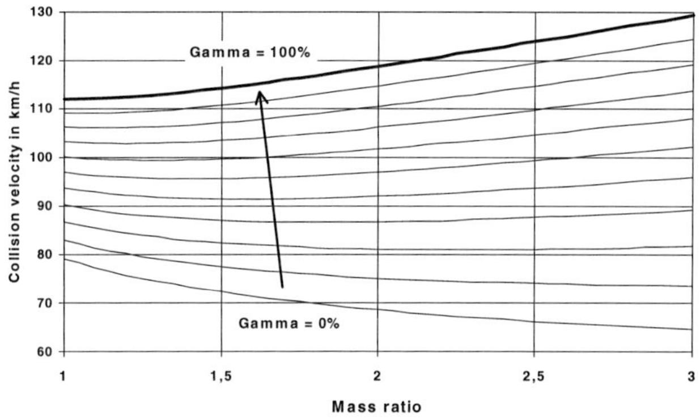

Figure 13. Attributed collision velocity related to mass ratio and percentage Gamma of available deformation energy of large vehicle.
Basis: Barrier impact speed of 56 km/h

This raises the question, how figure 13 will change, if we proceed to very large vehicles, to trucks. In this case we have mass ratios of up to 30. Much more than GAMMA=0% or GAMMA=10% will not be achievable in this case. In this case, the situation of a barrier impact against a rigid barrier more or less occurs. This is also shown by figure 14 which shows the asymptotic behavior of the formulas with a limit of 56km/h closing velocity.

It is clear that there exists no realistic background for GAMMAS≥10%. But these curves were not eliminated, to allow even serious safety researchers to smile a little bit and to dream about trucks with 10m or 20m of deformation travel, which make car-to-truck accidents at a closing speed of more than 300km/h to a normal even of minor accidents.

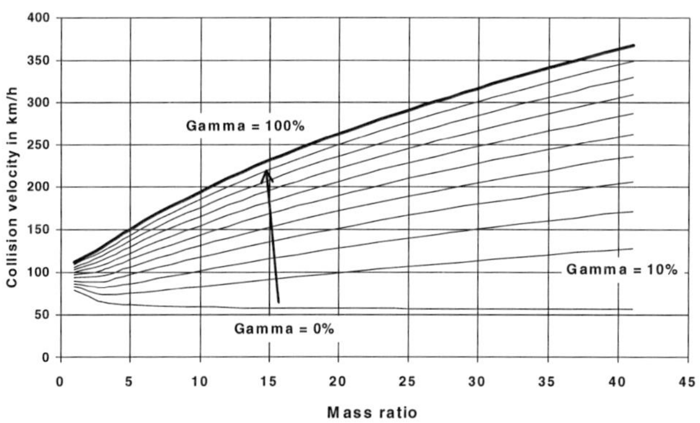

Figure 14. Same computation as figure 13, but with higher mass ratios, thus showing the benefit, if for vehicle-to-truck accidents the geometrical problem would be solved.
Basis: Barrier impact speed of 56 km/h

CONCLUSION

Accident analysis within the European project on compatibility clearly proves that mass has the predominant effect on compatibility. A compatibility program must therefore deal with this issue. A restraint system is able to make crashes survivable as long as compartment deceleration is not too high. This means that the deceleration of the small vehicle must be restricted to a certain level. The consequence is a restriction of the force decelerating these vehicles in the event of a collision. As long as the impact velocity of two vehicles is not too high (less than twice the barrier impact speed, for which the vehicles were designed), the amount of available deformation energy of the two vehicles is sufficient, regardless of the mass ratio. The deformation of the larger vehicle is available in case of a collision, when the small vehicle is stiff enough to force this deformation before its own compartment collapses. It is therefore necessary to design the compartment stiffness sufficiently high so that the deformation force of the large vehicle is lower. But the first restriction has to be taken into account, too. Both ideas form the basis of the bulkhead concept:

Restrict force levels of the front-end of the vehicles in such a manner that a certain (e.g. 30g) compartment deceleration in the small vehicle is not exceeded. (Definition of a F_{max})

Design the compartment of a vehicle in such a way that does not permit excessive intrusion as long as the deformation force is less than this maximum force F_{max}.

This concept is capable of managing a vehicle-to-vehicle collision. When every vehicle is equipped with a restraint system that is able to sustain 30g without exerting excessive loads on the occupant, a high number of vehicle-to-vehicle collisions will become survivable because no overcrushing and no excessive acceleration occurs.

This concept has restrictions. Because deformation travel of a large vehicles is also restricted, , it is probably not applicable for the entire range of mass ratios, but probably for 80% or 90% of them.

This concept is in clear conflict with self-protection. The higher the degree of self-protection (i.e. the higher the barrier impact speed of vehicles), the smaller the range of mass ratios, to which this concept applies.

ACKNOWLEDGEMENT

This paper partially reports on results which were obtained received the project "Development of Criteria and Standards for Compatibility". This project is being funded in part by the European Union (European Commission, Directorate General XII, Nr. BE 97-4049). It is run by Volkswagen AG, Ford-Werke AG, Daimler-Chrysler AG, Adam-Opel AG, GIE PSA Peugeot Citroën, GIE Renault Recherche et Innovation, Rover Group Limited, SEAT SA, TÜV Rheinland Luxemburg, AB Volvo, Gesamtverband der deutschen Versicherungswirtschaft e.V., Technische Universität Berlin, University of Oulu. Volkswagen AG is co-ordinating this project.

The results presented are discussed within this project and published with the goal of encouraging public discussion on this very relevant safety issue. The conclusions presented are those of the author and not those of individual consortium companies. Conclusions drawn by consortium companies will be presented at the end of 2000 in the final report for this project.

The author thanks all participants in the project consortium for the open and fruitful discussions and contributions to the challenging task of establishing higher levels of compatibility. This appreciation also extends to the partners within the EEVC-project on compatibility and the NHTSA-project on the same subject.

REFERENCES

1. Appel, H. Compatibility, guideline for passive safety and even for the total transportation system. FISITA-Congress, Prague, 1996
2. Dreyer, W., Richter, B., Zobel, R. Handling, Braking, and Crash Compatibility of Small Front Wheel Drive Vehicles. SAE-Paper 810792, Dearborn, MI, U.S.A., 1981
3. Richter, B., Zobel, R. Aspects of the Passive Safety of Motor Vehicles. 9th international conference on the enhanced safety of vehicles. Kyoto, Japan, 1982
4. Faerber, E. Improvements of crash compatibility between cars. Conference on road infrastructure and safety research in Europe (RISC '97. Brüssel, 1997
5. Gabler, H.C., Hollowell, W.T. The Aggressivity of light trucks and vans in traffic crashes. SAE-paper 980908, Detroit, 1998
6. Hobbs, C.A., Williams, D.A., Coleman, D.J. Compatibility of cars in frontal and side impact. 15th international conference on the enhanced safety of vehicles. Paper 96-S4-O-05. Melbourne, 1996
7. Hollowell, W.T., Gabler, H.C., NHTSA's vehicle aggressivity and compatibility research program. 15th international conference on the enhanced safety of vehicles. Paper 96-S4-O-01. Melbourne, 1996
8. IIHS. Crash compatibility. How vehicle type, wheight affect outcomes. Insurance Institute for Highway Safety. Statusreport Vol.33 Arlington, 1998
9. Klanner, W., West, F.van, Felsch, B. Evaluation of occupant protection and compatibility out of frontal crash tests against the deformable barrier. 16th international conference on the enhanced safety of vehicles. Paper 98-S3-O-06. Windsor, 1998
10. Schoeneburg, R., Zakmak, M., Zobel, R., Busch, D. Evaluation of crash compatibility of vehicles with the aid of finite element analysis. 15th international conference on the enhanced safety of vehicles. Paper 96-S4-W-25. Melbourne, 1996
11. Schoeneburg, R. Implementation and assessment of measures for compatible crash behavior using the aluminium vehicle as example. 16th international conference on the enhanced safety of vehicles. Paper 98-S3-O-07. Windsor, 1998
12. Zobel, R., Barrier impact tests and demands for compatibility of passenger vehicles. VDI-Tagung: Innovativer Insassenschutz im PKW. Berlin, 1997
13. Zobel, R. Principles for the development of a passenger car safety information system for consumers, based on real-life accident evaluation. Crash-Tech special '98, München, 1998
14. Zobel, R., Demands for compatibility of passenger vehicles. 16th international conference on the enhanced safety of vehicles. Paper 98-S3-O-07. Windsor, 1998

CHAPTER THREE – SELF PROTECTION

2001-06-0086

THE ESSENTIAL REQUIREMENTS FOR COMPATIBLE CARS IN FRONTAL COLLISIONS

Mervyn Edwards, Julian Happian-Smith, Huw Davies, Nigel Byard and Adrian Hobbs.
TRL Ltd
United Kingdom
Document Number 158

ABSTRACT

The work reported here forms part of a research project that is being undertaken to further the understanding of compatibility in car to car collisions and develop crash evaluation procedures that are suitable for consumer and legislative testing. For frontal impact, full scale crash testing, accident analysis case studies and supportive finite element modelling studies have been used to identify the major factors that influence compatibility. One result is that the geometrical interaction of car structures has a large effect and it is now believed that obtaining good structural interaction is an essential prerequisite for frontal impact compatibility. Having achieved this, the next step is to control the global stiffness of the cars to ensure that they are able to absorb the collision energy, with minimal occupant compartment intrusion, without compromising the vehicle's deceleration pulse profile. Frontal impact evaluation procedures are being developed, which use load cell wall measurements to assess a car's compatibility. The current state of development of possible procedures is described, with an emphasis on results from full width deformable barrier tests. Procedures to assess side impact compatibility may be added following further research. This reported research is being used to support the European Enhanced Vehicle-safety Committee (EEVC) and the International Harmonisation of Research Activities (IHRA) Compatibility Working Group activities, and is funded by the Department of the Environment, Transport and the Regions (DETR).

INTRODUCTION

Following the introduction of the European Frontal and Side Impact Directives in October 1998, compatibility offers the next greatest potential benefit for improving car occupant safety and reducing road casualties. For the UK, this can be illustrated by examining the STATS19 accident database. Approximately 60 percent of all car occupant casualties occur in accidents where the car collides with another vehicle, the type of accident that compatibility primarily addresses (Table 1). A Renault study has suggested that improved compatibility could reduce the number of fatalities and serious injuries by as much as a third in accidents where a car collides with one other vehicle (1). If this was realised in GB it would have equated to a saving of 278 fatalities and 3511 serious injuries in 1999 in accidents where the car collided with one other vehicle. In addition, benefits should also occur in multi-vehicle and many single vehicle accidents.

Table 1.
Distribution of car occupant casualties by collision type in GB

Collision Type	Fatalities		Serious Injuries	
	1998	1999	1998	1999
Single car	491 (29%)	495 (30%)	5132 (26%)	4825 (26%)
Car to one other vehicle	821 (49%)	835 (50%)	11179 (57%)	10533 (57%)
Car to more than one other vehicle	364 (22%)	348 (20%)	3467 (18%)	3144 (17%)

In 1995, on behalf of the DETR, TRL commenced a project to research compatibility. This ongoing work is being used to support the European Enhanced Vehicle-safety Committee (EEVC) and the International Harmonisation of Research Activities (IHRA) compatibility working groups. This project aims to identify how vehicle safety may be improved by developments to their structures and subsequently implement these changes in the vehicle fleet. This requires an understanding of the factors that influence compatibility and the development of new or modified evaluation procedures to bring about greater compatibility. The project also aims to identify the potential benefits that could be obtained from improved compatibility. In a complex area, the approach taken for this work was to focus on car to car frontal impact and car to car side impact separately, in order to understand the influencing factors more clearly. This paper reports on the frontal impact studies only.

FRONTAL IMPACT COMPATIBILITY

Continuing the drive of the European Frontal Impact Directive and EuroNCAP, the work performed to date has focused on the structural performance of the vehicles, with the aim of providing a safe environment in which the restraint system can operate. This approach is supported by the results from a recent European accident analysis study which shows that for the UK over 70 percent of the AIS 3+[1] injuries received by belted occupants were contact induced as opposed to restraint system induced (2). Once the structure provides a safe environment within which the restraint system can operate, the next step for further improvement will be to control the compartment deceleration pulse. Following this, intelligent restraint systems could offer a way to cope with higher compartment decelerations, and give the occupant an optimised ride-down for a variety of impact severities. Previously reported work conducted for this project has investigated the effect of the shape of the deceleration pulse on the performance of current restraint systems (3).

The results from this project have overturned the original views about compatibility, which thought that mass and mass ratio were the dominant factors. Now it is clear that although mass has an effect on stiffness, which affects intrusion, the most important factor is structural interaction. Good structural interaction is required to ensure that cars interact predictably to absorb the impact energy, in the designed manner with minimal occupant compartment intrusion. Without this essential prerequisite, the energy absorbing capability of the frontal structure is, generally, dramatically reduced. This leads to excessive compartment intrusion. Once good structural interaction has been achieved, some form of stiffness matching between vehicles is necessary to ensure that the impact energy is absorbed without exceeding the strength of the occupant compartment. The evidence leading to these conclusions is presented below, together with proposals for test procedures to assess and control compatibility.

[1] An Abbreviated Injury Scale 3+ (AIS3+) injury severity level relates to a severe or greater injury.

Structural Interaction

Real road accident configurations are very varied; impact angle, overlap, impact point and speed are just a few of the parameters describing an accident. The concentration of structural stiffness in elements such as the frontal lower rails can adversely affect safety performance in accidents. Misalignment of these stiff lower rails is normal and can result in high passenger intrusion levels due to inadequate energy absorption by these stiff elements. This can manifest itself in a number of different ways, such as override, where one vehicle tends to ride up over the other, or the penetrating fork effect where the stiff members of one vehicle penetrate the soft areas of the other vehicle due to lateral misalignment.

The override effect has been seen in accidents even with recent car designs that perform well in the EuroNCAP assessment. An example of this effect is described. The accident occurred between a Volvo S40 and a VW Polo (mass ratio 1.26). The collision was head on with approximately 30 percent overlap. The young healthy female driver of the Polo was killed but the driver of the S40 survived with serious injuries. There was poor structural interaction between the frontal structures of the two vehicles resulting in the Volvo overriding the Polo. Comparing the relative deformation of the upper and lower rails for the two cars shows this. The S40 upper rail showed little deformation whereas the Polo upper rail was significantly deformed (Figures 1 and 2). In contrast, the S40 lower rail was lifted up whereas the Polo lower rail was bent down and out. The result of this non-ideal structural performance was that the energy absorption efficiency of both cars was reduced, which lead to additional compartment intrusion.

Figure 1. Volvo S40, note the small deformation of the upper rail and the lifting of the lower rail indicating that this car has overriden the VW Polo.

Figure 2. VW Polo, note the large deformation of the upper rail and the downward deformation of the lower rail indicating that this car has been overriden by the Volvo S40.

In summary, the overriding was a major contributory factor to the large intrusion seen in the Polo at fascia rail level, in comparison to the small intrusion in the S40. Even so, the mass and stiffness difference between the cars was likely to have contributed as well.

The lack of good structural interaction between current vehicle designs is further illustrated by comparing the relative performances of the Peugeot 806 and VW Sharan in EuroNCAP tests with those in a car to car impact test. The EuroNCAP test results show that there was less door aperture deformation for the 806 compared to the Sharan (Figure 3).

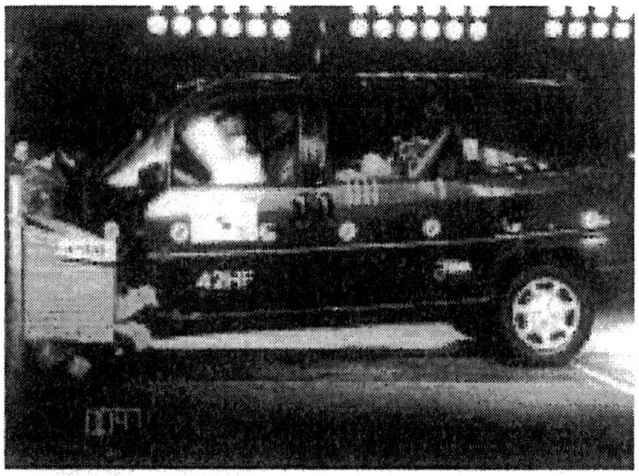

Figure 3. The Peugeot 806 (above) showed less door aperture deformation than the VW Sharan (below) in EuroNCAP tests[2].

However, in the car to car test the Sharan (Galaxy)[3] overrode the 806 with the result that the 806 occupant compartment experienced far greater intrusion than in the equivalent EuroNCAP test (Figure 4).

The sensitivity of structural interaction with current cars has been illustrated in previous work (3). This showed how a 100 mm difference in ride height for an impact between two identical cars resulted in a significantly different structural performance for each car. The raised car overrode the lowered car resulting in greater intrusion in the lowered car at fascia rail level. In contrast there was greater intrusion in the raised car in the footwell area.

---

[2] Photographs courtesy of EuroNCAP.

[3] Please note that the MPV in the car to car test was a Ford Galaxy. However, the VW Sharan and Ford Galaxy are structurally identical.

Figure 4. Peugeot 806 (above) is overriden by VW Sharan (Galaxy) (below) in car to car impact resulting in increased occupant intrusion in 806[4].

These results show that in order to achieve a compatible fleet an essential prerequisite is good structural interaction. Until vehicle designs enable structures to interact better in car to car impacts, any compatibility improvements in stiffness matching are unlikely to be fully realised. To achieve good structural interaction, the implications for car design are that they will require better vertical, lateral and shear connections. These connections will increase the number of active load paths into the main energy absorbing structures. This will help to ensure that predictable behaviour occurs over a wider range of impacts, hence improving crashworthiness performance.

The introduction of such connections will most likely result in cars having a more uniform frontal stiffness distribution. Following this logic, a test procedure has been proposed. This test uses load cell wall measurements, from a full width frontal impact test with a deformable element, to assess and control a car's frontal stiffness distribution. It is recognised that one failing of this proposed test concept is that it would not exercise the shear connectivity between load paths such as the upper and lower rails. However, it is unlikely that a car could achieve a homogeneous stiffness distribution without good shear connections.

Full Width Deformable Barrier Test

A series of full width tests with a deformable element have been performed with current cars varying in size from small family to Sports Utility Vehicle (SUV) using an impact velocity of 56 km/h. High resolution load cell wall measurements were recorded using a wall which consisted of 128 load cells of size 125 mm by 125 mm arranged in an 16 by 8 matrix. The deformable element used, consisted of a 150 mm deep aluminium honeycomb face with a longitudinal crush strength 0.34 MPa.

The depth and stiffness of the barrier were chosen primarily for two reasons. The first was so that, compared to a rigid wall test, the initial high decelerations at the front of the car were attenuated to make the test more representative of a vehicle to vehicle impact. The second was to ensure that the element had little effect on the occupant compartment deceleration pulse so that the test could also be used as a frontal impact test similar to US FMVSS 208. Previous work has found that this deformable face also reduces the magnitude of the engine loading on the wall, which may be more realistic as in accidents the engine can rotate or move horizontally or vertically. However, the results of the latest series of tests have shown that local hard points on the car can dramatically reduce loading from adjacent structures indicating that the barrier depth and / or stiffness may be need to be altered. FE modelling studies are currently addressing this issue.

To demonstrate the potential of this test procedure, the load cell wall results for a family car exhibiting features likely to benefit compatibility such as an engine subframe load path have been compared to those for a less compatible vehicle, a SUV. A subjective visual inspection of contour plots of the peak force measured on each load cell shows a greater spread of the higher loads for the family car caused by the presence of the engine subframe indicating the potential of this test procedure (Figure 5). Please note that the load cell array was raised 125 mm for the SUV.

[4] Photographs courtesy of ADAC.

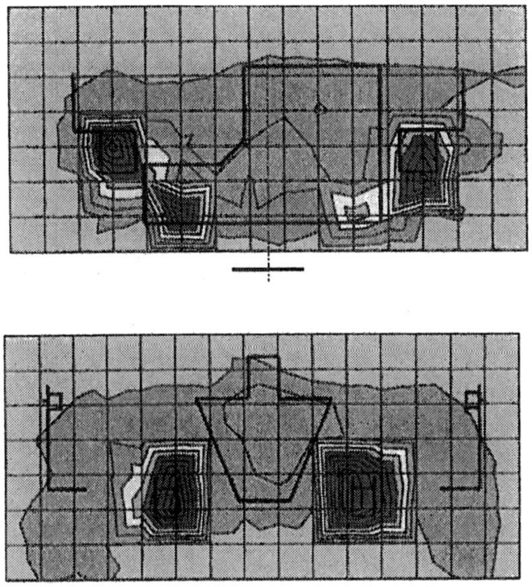

Figure 5. Comparison of load cell peak forces for family car (above) and SUV (below). Note, vehicle main structure is superimposed on plots to show load paths.

In order to assess the stiffness homogeneity in a more objective manner, further analyses of the results tried various statistical techniques to differentiate the stiffness homogeneity of these vehicles. These included the application of coefficient of variance (CV), skewness and kurtosis techniques to the load cell wall data for ten cars. The coefficient of variance (CV), which is defined below, was found to be the most promising parameter to date.

$$CV = \text{Standard Deviation/ Mean}$$

Inclusion of all load cell wall data in an analysis would have distorted any potential homogeneity measure by including many unloaded cells. To solve this problem, a rectangular block of load cells was chosen to represent the vehicle footprint, using the criterion that at least one load cell in the outside row or column must experience a peak load greater than 5 kN.

Two problems were found in applying the CV technique to the load cell wall data set. Firstly, the CV was too variable when the mean value of the load was low at the beginning and end of the impact, because small changes in the loading caused large changes in the standard deviation. Secondly, the bridging of load cells by structural members of the vehicle sometimes caused changes in the CV, although these were smaller than expected. Bridging is the term used to describe the loading of two or more load cells by one structural member of a vehicle, such as the longitudinal. The low load variability problem was solved by defining that the CV measure was only applied at times when the total load on the wall was greater than 50 kN. The bridging problem was solved by averaging the outputs from four adjacent load cells and allocating this value to one of the cells in a stepwise manner for each of the load cells. Calculating the CV, following the above process, proved to be stable as the implementation of an artificial 5 percent noise on the load cell force input only created a 1 percent difference to the output values.

Using the above methodology, the CV was calculated for the family car and SUV and shows substantially higher values for the SUV (Figure 6). The dark shaded area defines the time for which the CV is considered valid, i.e. when the total wall loads are over 50 kN.

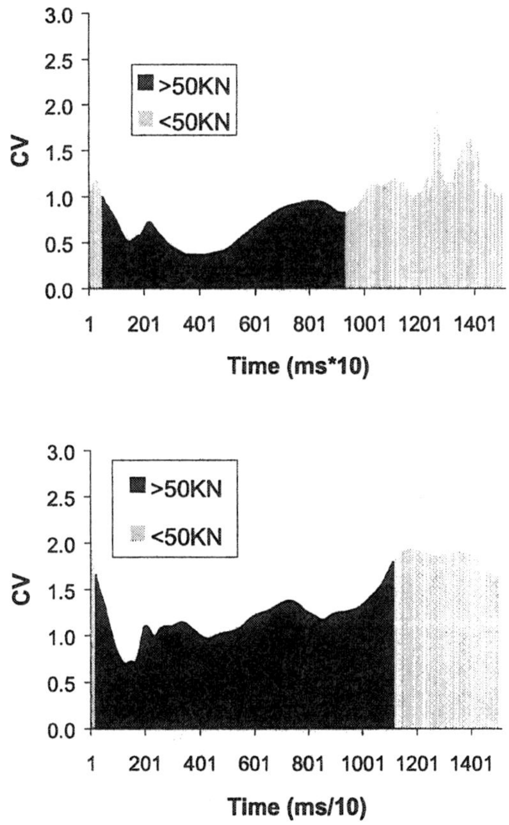

Figure 6. Coefficient of variance (CV) for family car (above) and SUV (below) showing higher values for the SUV.

To enable easy direct comparison of the results the CV was integrated over the defined impact time and normalised to a time of 100 ms to give a single value. For the more compatible family car this was 0.756 whereas for the less compatible SUV it was 1.174. The difference in these values demonstrates the potential of this parameter as a tool to evaluate the stiffness homogeneity of car frontal structures.

The above proposal could be used to control a vehicle's stiffness homogeneity over its frontal crash footprint, but to ensure good structural interaction these footprints need to overlap. One way to achieve this would be to control the height of the centre of force measured on the load cell wall as proposed by NHTSA (4). The centre of force height may also need to be controlled throughout the duration of the impact. In addition, a limit controlling a vehicle's minimum footprint area will be required.

The centre of force height measured from the ground for the family car and SUV show a substantial difference as might be expected (Figure 7). The average height weighted by the load on each cell was 438 mm and 587 mm, respectively. To ensure good structural interaction between these vehicles this difference needs to be reduced. However, for the cars tested, not including SUVs, the force centre height range was 31 mm indicating a large overlap of the car's footprints, which shows the potential for good structural interaction.

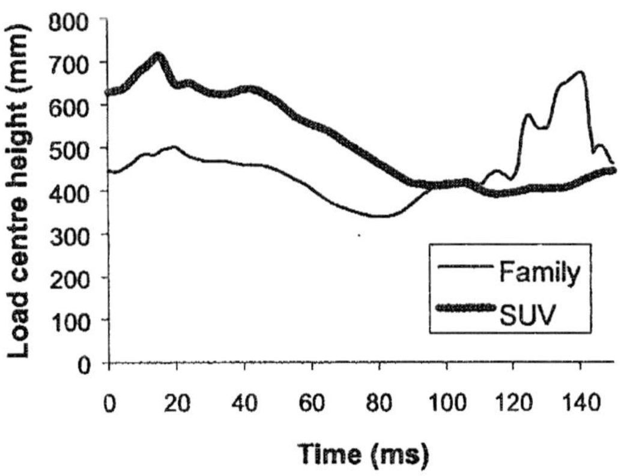

Figure 7. Variation of centre of force height showing substantial difference between family car and SUV.

Structural Stiffness

The primary aim of compatibility is to achieve minimal occupant compartment intrusion levels in both vehicles, in car to car collisions. If we consider a car, of mass m, impacting a rigid barrier with an initial velocity of v and assume that its occupant compartment undergoes minimal intrusion, then that car's front structure has an energy absorption capability of at least its initial kinetic energy, $\frac{1}{2}mv^2$. Simple mathematics shows that if two cars of differing masses crash together with a closing velocity of less than 2v, then there is sufficient energy absorption capability available to absorb the impact energy with no more occupant compartment intrusion than in the rigid barrier test (5). In order to ensure that the energy absorption capability is available for use, then the stiffness of the two cars needs to be matched. However, in general, heavier vehicles have a higher global stiffness than lighter vehicles, as heavier vehicles have to absorb their larger kinetic energy in approximately the same deformation length. In a car to car collision, this leads to greater intrusion in the less stiff car as it absorbs more energy relative to an equivalent barrier test (1,6). Previous work has shown that occupants of lower mass vehicles have higher injury risks due to both lower vehicle mass and stiffness (7). There is, however, no certainty that the stiffness differential between small and large cars will continue to exist at the current ratio.

The 'interface force' is defined as the force between the car and the barrier in a car to barrier test or the force between the car and the opposing car in a car to car test. It can be obtained using a load cell wall or by the measurement of the deceleration of the constituent parts of the vehicle (1,3). The interface force for a small family car and an MPV in a 64 km/h Offset Deformable Barrier (ODB) test is shown as a function of B-pillar displacement (Figure 8). The end of crash force is the interface force at the end of crash and is shown (Figure 8).

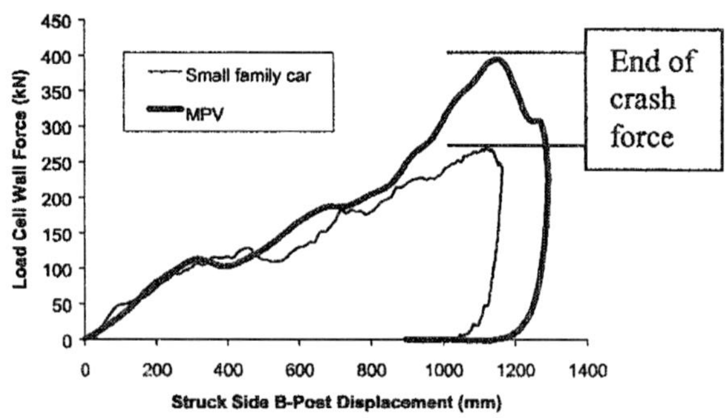

Figure 8. Interface force plotted as function of B-pillar displacement for small family car and MPV showing end of crash force.

To control a vehicle's frontal stiffness and occupant compartment strength, two test procedures are proposed. To control the frontal stiffness, the end of crash force would be measured using a load cell wall in a 64 km/h ODB test and limited to a maximum value, yet to be determined. To ensure that the occupant compartment strength is sufficient to withstand the loads imposed by an opposing vehicle, a second ODB test is proposed at an increased impact speed in which the end of crash force would be measured and controlled to exceed a minimum value yet to be determined. To assess the feasibility of these proposals and determine suitable end of crash force values, much further work is required. The inertial contribution of the power train to the interface force should be taken into account in this work.

As part of this continuing investigation, load cell wall end of crash force measurements have been taken for a number of recent EuroNCAP tests (Figure 9). These results indicate that, in general, lighter cars have smaller end of crash forces. However, it was noticeable that each mass range of cars has a wide spread of 'end of crash' force levels, which would tend to indicate that some uniformity across the mass range should be possible to create a more compatible car fleet.

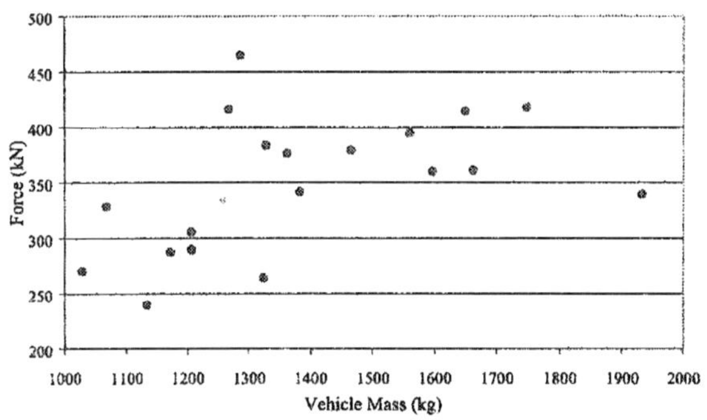

Figure 9. End of crash force measurements for recent Euro NCAP tested cars.

In order to test the validity of using the end of crash force concept to control a car's stiffness, two car to car crash tests were conducted with a 50 percent overlap and a closing speed of 112 km/h. The height of the main load carrying structures of these vehicles were checked to help ensure that good structural interaction would occur in the tests. The first test was between two small cars with a mass ratio of 1.01 with different end of crash force values, 270 kN and 210 kN. These values were measured in a 64 km/h ODB test. The result was that the car with the higher end of crash force, indicating a higher stiffness, absorbed less than its share of the impact energy. As a result, it suffered less intrusion relative to that measured in the ODB test, compared to the car with the lower end of crash force (Figure 10). The intrusion values shown are an average of four measurements, footwell intrusion, fascia displacement and A-pillar waist and sill displacement. This result supports the validity of the end of crash force concept.

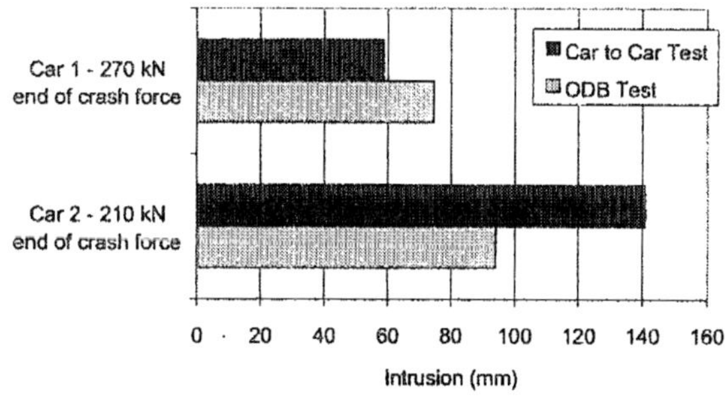

Figure 10. Intrusions measured in ODB and car to car test showing car with higher end of crash force suffered less intrusion relative to its performance in an ODB test.

In contrast, the second test was between a small car and a family car with a mass ratio of 1.3 with similar end of crash force values, 270 kN and 290 kN, respectively, indicating similar stiffnesses. For this test the expected result was that the intrusion measured in each vehicle would be similar relative to that measured in the ODB test. Unfortunately, the family car overrode the small car, which caused greater intrusion in the small car than expected. This shows that without good structural interaction any stiffness matching will not be effective.

As mentioned above, to ensure that the occupant compartment strength is sufficient to withstand the loads imposed by an opposing vehicle, a second ODB test is proposed at an increased impact speed in which the end of crash force would be measured and controlled to exceed a minimum level. The test would not require instrumented dummies.

A number of trials have been conducted to investigate the possibility of using an already crashed vehicle for this test to save the use of an expensive prototype. However, this 'double testing' approach proved to be infeasible so a 'new' car will have to be used for this test. Reusing a car from a full width barrier test was shown to be unrepresentative of a single impact test because of the different structural behaviour. For example, the engine to firewall interaction was much greater for the double impacted car compared to the single impacted one. To reuse a car from an ODB test was deemed infeasible because it would not always be possible to latch the driver's door following the first test.

To investigate what impact speed should be used for this test other researchers have performed ODB tests at 80 km/h, (50 percent greater KE than 64 km/h) (8). They concluded that if the passenger compartment became unstable, repeatability was poor. They also found that if the occupant compartment intrusion was large the driver's femurs could support the car's structure. To attempt to overcome these problems future tests will be conducted at lower impact speeds, possibly 72 km/h, (25 percent greater KE than 64 km/h) and require a stable occupant compartment.

SUMMARY OF PROPOSED TEST PROCEDURES FOR COMPATIBILITY AND FRONTAL IMPACT

In order to address the issues above and improve frontal impact compatibility, it is proposed that the use of three tests for both compatibility and frontal impact should be considered.

1. A full width deformable barrier test at 56 km/h with a high resolution load cell array behind the deformable element. From the load cell measurements the centre of force height over time could be controlled. The force distribution could be controlled using a technique such as the Coefficient of Variance (CV). This should encourage the development of frontal structures that behave in a more homogeneous manner, which should lead to an improvement in structural interaction in car to car impact. Damage to the deformable face might also be used as a check for load concentrations that are not revealed by the load cell wall.

2. An Offset Deformable Barrier (ODB) test at 64 km/h with a load cell array behind the deformable element. From the load cells, the car's global stiffness characteristic could be assessed using the end of crash force concept. In the future, control of the pulse shape could be used to limit the occupant compartment deceleration pulse and restraint loading. Control of the centre of force height could also be included for this test as well as the full width test.

3. A second ODB test with a load cell wall at an elevated speed to assess the strength of the passenger compartment. This test would not require instrumented dummies.

For frontal impact, one of the advantages of these proposed tests is that they have quite different deceleration pulses and hence would not encourage optimisation of the performance of restraint systems to one pulse. The full width would generate a 'hard' deceleration pulse on the vehicle and restraint system, whereas the 64 km/h ODB test would generate a 'soft' pulse.

At this stage, it is not expected that the introduction of evaluation procedures detailed above would have any detrimental effect on car to car side impacts or impacts with pedestrians, HGV's, or other obstacles, indeed they are expected to be beneficial.

CONCLUSIONS

The results from this project indicate that in order to improve structural performance for better compatibility in frontal impact, cars need to interact in a predictable manner to absorb the impact energy with minimal occupant compartment intrusion over a broad range of collision types. To achieve this, an essential prerequisite is good structural interaction. Following this, some form of stiffness matching between vehicles will be necessary to ensure that the impact energy is absorbed without exceeding the strength of the occupant compartment.

In order to address these issues and improve compatibility, outlines for three possible test procedures to assess and control both compatibility and frontal impact have been proposed. This is an evolving area and further work is required to develop these outlines to a level suitable for consumer or legislative test procedures.

Implementation of these three test procedures should be sufficient to control intrusion and provide a safe environment within which the restraint system can operate. This should address contact induced injuries but not restraint induced injuries. A next step to reduce injuries caused by the restraint system could be to control the shape of the interface force profile or the car's deceleration pulse. This would provide a deceleration pulse, which would enable optimum performance of the restraint system.

ACKNOWLEDGEMENTS

The work described in this paper forms part of a Department of the Environment, Transport and the Regions, Vehicle Standards and Engineering Division funded research project, being carried out by TRL Ltd.

REFERENCES

1. STEYER C, DELHOMMEAU M, AND DELANNOY P, (1998). Proposal to Improve Compatibility in Head On Collisions, Sixteenth International Technical Conference on the Enhanced Safety of Vehicles, Windsor 1998.

2. WYKES N, (1998). Accident Analyses for the Review of the Frontal and Side Impact Directives. Report to the EC DGIII, Contract Reference, SUB/97/501215/TRL.

3. WYKES N, EDWARDS M AND HOBBS C, (1998). Compatibility Requirements for Cars in Frontal and Side Impact, Sixteenth International Technical Conference on the Enhanced Safety of Vehicles, Windsor 1998.

4. DIGGES K AND EIGEN A, (2000). Load Cell Barrier Measurements of Geometric Compatibility, Vehicle Safety 2000 Conference, London, June 7-9 2000.

5. ZOBEL R (1998). Demands for Compatibility of Passenger Vehicles. Sixteenth International Technical Conference on the Enhanced Safety of Vehicles, Windsor 1998.

6. EDWARDS M, HAPPIAN-SMITH J, BYARD N, DAVIES H, AND HOBBS C, (2000) Compatibility – the Essential Requirements for Cars in Frontal Impact, Vehicle Safety 2000 Conference, London, June 7-9 2000.

7. GRIME G AND HUTCHINSON T, (1982). The Influence of Vehicle Weight on the Risk of Injury to Drivers. Ninth International Technical Conference on the Experimental Safety Vehicles, Kyoto 1982.

8. ACEA / EUCAR Vehicle Compatibility Workshop (2001). Wolfsberg, Germany, Feb 13-14th 2001.

2003-06-0074

INNOVATIVE BODY STRUCTURE FOR THE SELF-PROTECTION OF A SMALL CAR IN A FRONTAL VEHICLE-TO-VEHICLE CRASH

Masuhiro Saito
Tetsuya Gomi
Yoshinori Taguchi
Takeshi Yoshimoto
Tomiji Sugimoto
Honda R&D Co., Ltd. Tochigi R&D Center
Japan
Paper Number 239

ABSTRACT

Preservation of passenger compartment space during a frontal vehicle-to-vehicle collision is extremely significant for the self-protection of small cars.

It is well known that crash speed, mass, stiffness and geometric interaction all have an influence on the intrusion of the passenger compartment in a frontal impact between vehicles. This paper reports on a new enhanced body structure to reduce passenger compartment intrusion in a crash between large and small cars. The test discussed in this report set the crash speed of both cars at 50kph, the mass of the large car at almost twice that of the small car, and the small car over lap at 50%. The proposed innovative body structure for the front end of small cars achieved a previously unavailable level of efficiency of energy absorption and was able to maintain cabin integrity.

INTRODUCTION

In recent years the use of stationary barrier crash tests as a method of evaluation of crash safety performance has increased internationally. This has been very effective in improving vehicle crash safety performance and reducing the number of casualties in traffic accidents.

However, in the case of frontal collisions between small cars and large cars in the real world accident, it is said that the risk of injury to the small car's occupants is higher than that to occupants of the large car. This is caused by incompatibility between 'mass' 'stiffness' and 'geometry' in vehicle-to-vehicle collisions. A collision in which the mass and stiffness ratios of the vehicles are large is equivalent to an extremely high speed stationary barrier crash for a small car.

Small cars which receive good evaluations in full lap and offset frontal crash barrier tests are therefore not always sufficiently safe in a small car to large car collision in the real world accident. And it begins to be pointed out the necessity to have an another manner to evaluate it in the collisions with relatively different sized vehicles.

In particular in the case of narrow offset collisions in which overlap distance is relatively small in the direction of the vehicle' s width and collisions with differing bumper height, it is very difficult for conventional body structures to maintain crash safety performance. It is therefore necessary to propose innovative body structures based on new design concepts. Of course it goes without saying that it is extremely significant for compatibility not only to consider the progress of self-protection but also partner-protection (for opposite vehicles). This research reports on the possibility to improve the self and partner protection which are discussed in the society by modifying the body structure such as one of the unique technique.

THE SAFETY IMPROVEMENT FOR SMALL CARS

In frontal vehicle-to-vehicle collisions it has been discussed that mainly 'mass' 'stiffness' and 'geometry' are the factors for incompatibility. In this part we report on some subjects and solutions in the view of the self-protection for small cars.

The portion of the velocity change before and after collisions is influenced by mass ratio as follows.

$$Vr = 1/Mr \qquad (1.)$$

$Vr = \Delta V_2/\Delta V_1$ ΔV_i: velocity change of Vehicle i $Mr = M_2/M_1$ M_i: mass of Vehicle i

The lighter vehicles are forced to a higher deceleration level than the heavier vehicles. As a result the risk of injuries in small cars is higher than in large cars. We will be able to solve such mass incompatibility developing superior restraint systems for small cars and to reduce mass in large cars.

The portion of energy absorption in vehicle deformation depends on stiffness ratio of two vehicles as follows.

$$Er = 1/Kr \qquad (2.)$$

$Er = E_2/E_1$ E_i: energy absorption of Vehicle i

$Kr = K_2/K_1$ K_i: stiffness of vehicle i

The problem is how to balance several vehicles' stiffness. To be realistic we need to improve the stiffness for small cars.

Certainly as stated above the difference between mass and stiffness is a problem for the compatibility, however we should first improve geometrical compatibility.

In recently years the level of self-protection for small cars has advanced greatly because of the adoption of offset deformable barrier testing. However the deformable barrier is very uniform in stiffness. But in reality, the front of vehicle in vehicle-to-vehicle collisions lack uniform stiffness, which causes severe damage to small cars. After this we focus on the improvement of interaction in frontal structures.

NEW DESIGN CONCEPT

To improve the offset and full lap frontal crash performance, conventional body structures are generally designed with two main frames located on each side of the engine compartment to absorb vehicle energy and to control vehicle deceleration.

However, in the case of vehicle-to-vehicle collisions such as narrow offset collisions in which there is less overlap distance in the direction of vehicle width and collisions between a passenger cars and a Sports Utility Vehicle (SUV) in which the bumper beam heights of the vehicles differ (as shown in Figure 1), this body type allows structural penetration into the engine compartment.

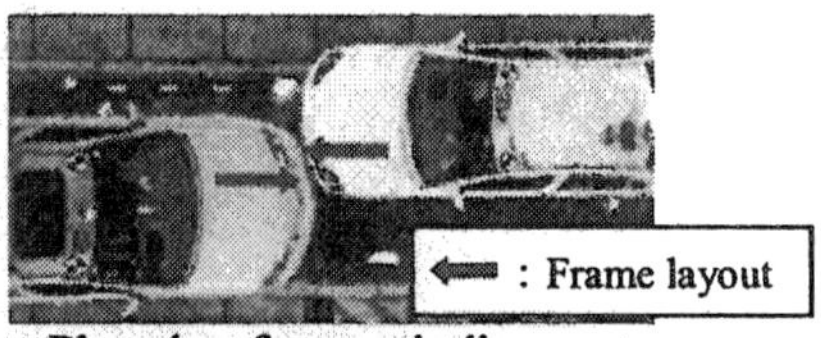

Plan view frame misalignment

Side view frame misalignment

Figure 1. Misalignment of stiffness between vehicles in vehicle-to-vehicle collision.

In particular when there is a significant difference in vehicle weight, the bumper beam and main frame of the large car passes into the frame of the small car without sufficient energy absorption and deceleration, penetrating the weaker part of the small car. As a result, deformation extends to the small car's passenger compartment, increasing injuries to occupants from secondary collisions in the passenger compartment, as shown in Figure 2.

It is important for the improvement of the level of protection offered by small cars to prevent structural penetration of major frontal components, increase the homogeneity of strength distribution and improve energy absorption in the engine compartment in the event of vehicle-to-vehicle collisions between vehicles with misalignment of stiffness. The design concept will be expected not only to achieve progress for self-protection but also the effect of partner-protection for compatibility.

Figure 2. Large deformation of the passenger compartment after narrow offset vehicle-to-vehicle collision.

STRUCTURAL OUTLINE AND CRASH PERFORMANCE

The proposed structure consists of three components, as shown in Figure 3.

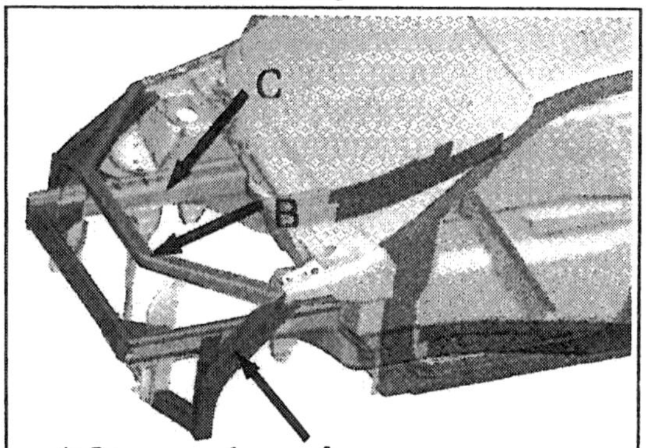

A; Lower member

B; Closed bulk head upper cross member

C; Polygonal main frame

Figure 3. Structural outline.

These components are A; A lower member to prevent penetration. B; A closed bulk head upper cross member to assist energy absorption in the upper part of the engine compartment. and C; A polygonal main frame enabling high efficiency energy absorption. This paper will offer a structural outline of the lower member system and discuss the predicted effectiveness of this system as determined by computer simulations.

The new 'lower member' was positioned in front of the tires extending from the wheel house upper member, and was connected to the main frame and bulk head cross member. This prevents the penetration of the frames of the respective vehicles in narrow offset collisions and collisions between vehicles with differing bumper heights. (See Figure 4). On impact, the lower member makes contact with the front structure of the other vehicle and deforms, thus achieving a high level of energy absorption (as shown in Figure 5).

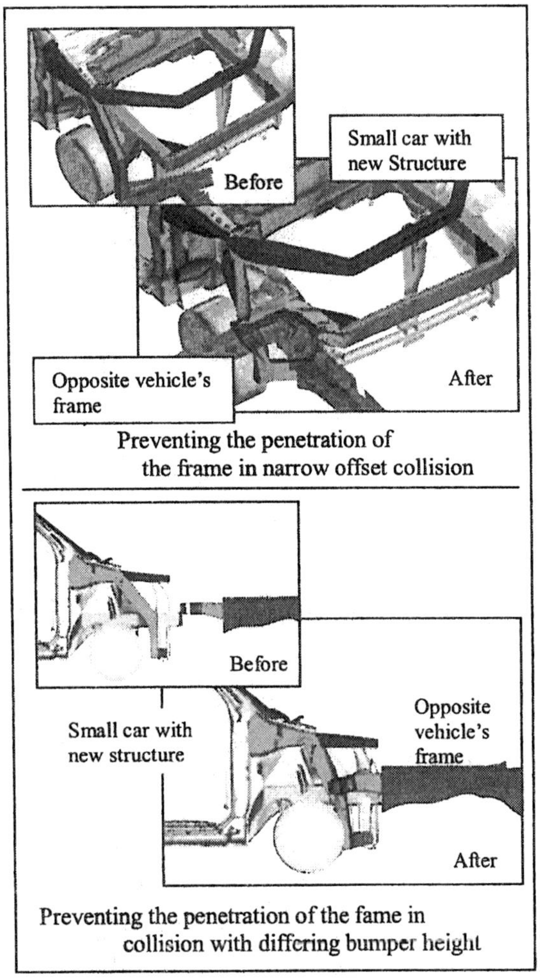

Figure 4. Preventing the penetration of the Frame.

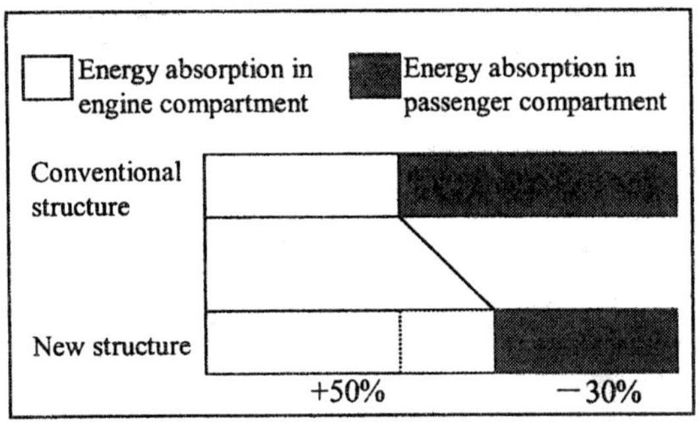

Figure 5. Increasing energy absorption in the engine compartment from vehicle-to-vehicle collision.

This leads to a significant reduction in the proportion of energy absorption in the cabin and of the degree of passenger compartment intrusion. The effectiveness of the new structure in reducing passenger compartment intrusion is shown in Figure 6.

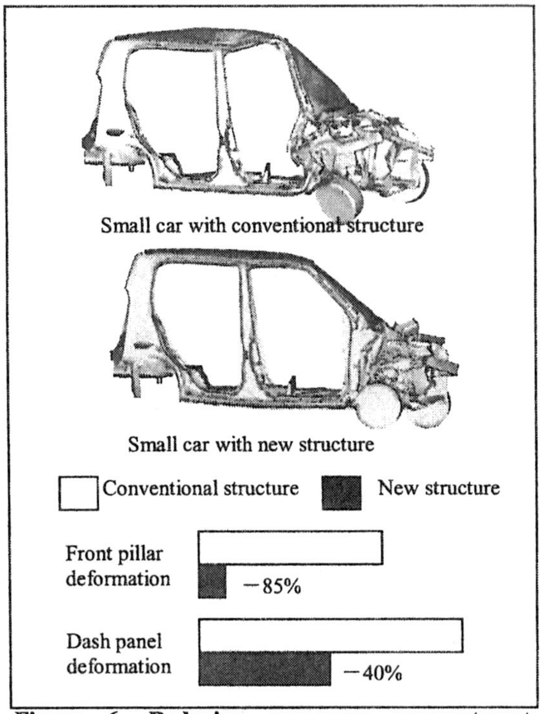

Figure 6. Reducing passenger compartment intrusion in vehicle-to-vehicle collision.

The new structure also enables reduction of the strength and weight of the main frame, allowing a more homogeneous distribution of strength in the front end structure.

It is predicted that this structure will be effective in reducing passenger compartment intrusion for various overlap distances in the direction of vehicle width (as shown in Figure 7), for the difference of bumper height (as shown in Figure 8) and for angle of approach (as shown in Figure 9).

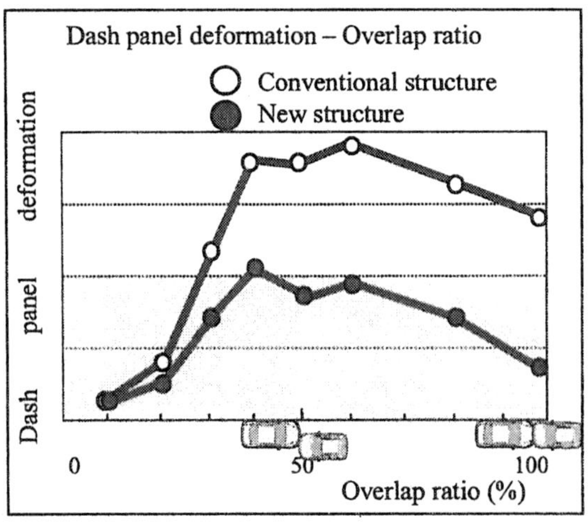

Figure 7. Reducing passenger compartment intrusion in various offset vehicle-to-vehicle collisions.

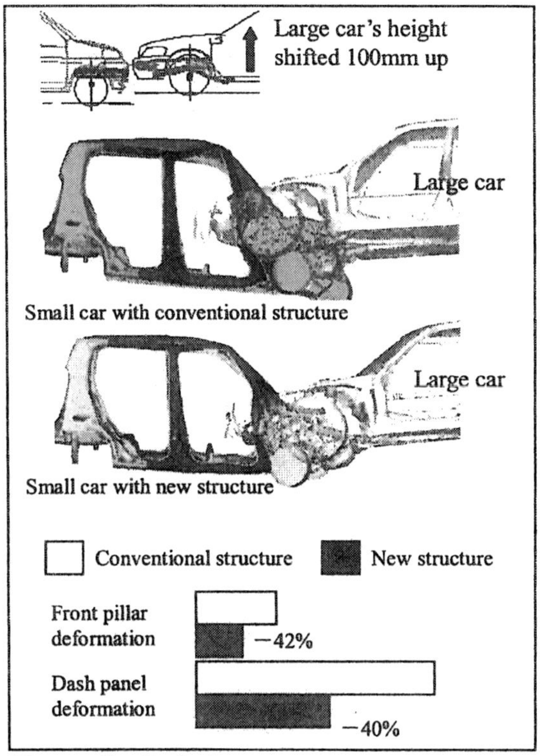

Figure 8. Reducing passenger compartment intrusion for vehicle-to-vehicle collisions with differing bumper height.

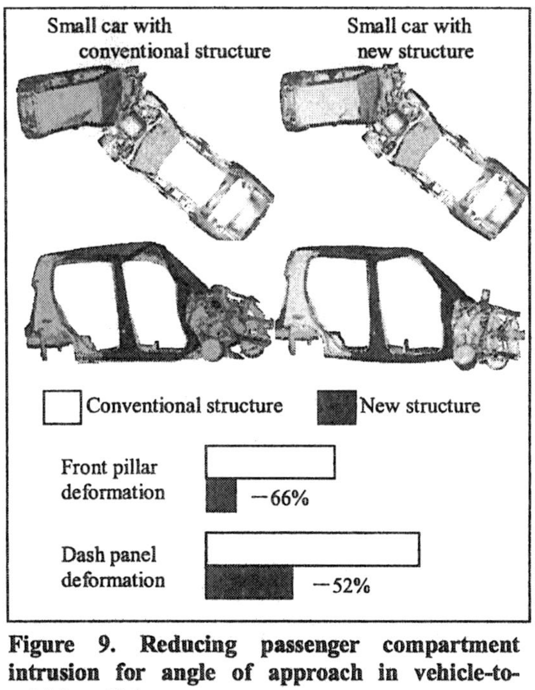

Figure 9. Reducing passenger compartment intrusion for angle of approach in vehicle-to-vehicle collision.

AGGRESSIVENESS

Secondly we give consideration to aggressiveness. As a result of above-mentioned vehicle-to-vehicle simulations, energy absorption in the engine compartment has increased slightly and the energy absorption in the cabin is decreased in large cars with conventional body structures by means of the effect from the new structure. It can therefore be predicted that the new body structure will not increase aggressiveness towards the partner vehicle in a collision (as shown in Figure 10).

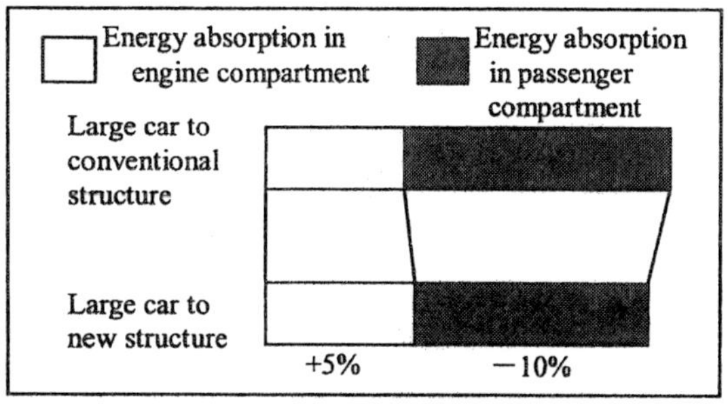

Figure 10. Improving of energy absorption.

Further we analyzed aggressiveness towards small cars as follows. We describe the simulation results in a frontal collision between similar small cars A and B with conventional structure (case 1), and between a small car A with conventional structure and a small car C with the new proposed structure (case 2)in Figure 11. Opposite car's intrusion in the passenger compartment was approximately similar in both cases. We could assess that our proposed new structure greatly improved self-protection, and doesn't increase the aggressiveness towards the small car, and we could find the possibility for compatibility.

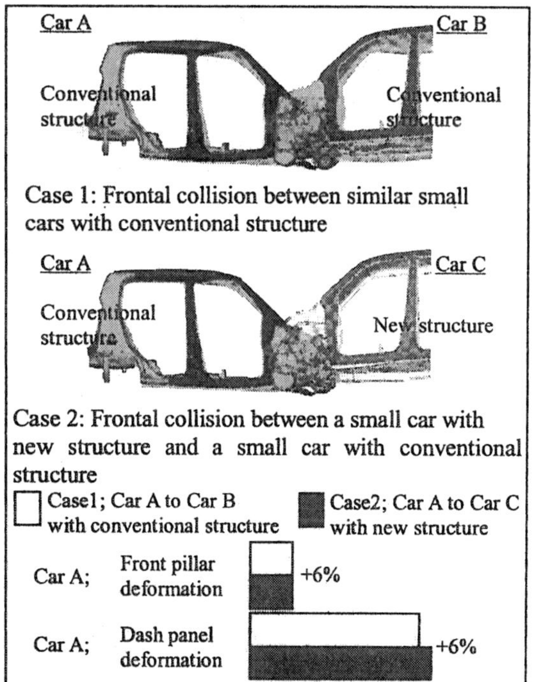

Figure 11. Not increasing aggressiveness to small cars.

ACCIDENT ANALYSIS

This time the condition in a vehicle-to-vehicle crash test was based on the accident analysis in Japan. Some accident data were provided from the Institute for Traffic Accident Research and Data Analysis (ITARDA) established in 1992.

Almost half of the number of occupant deaths is in the case of frontal collisions as shown in Figure 12[1]. About 90% of the frontal collision deaths were at speeds lower than 50kph as shown in Figure 13 [1]. And when the opposite vehicles are heavier than the subject vehicles, the driver deaths in frontal collisions are about 75% as shown in Figure 14 [1]. With relation to overlap in frontal collisions, 30% and 50% overlap cases are the primary overlap conditions for offset collisions (as shown in Figure 15) [1].

For reduction casualties, we would get effective test results from test condition based on accidents analysis.

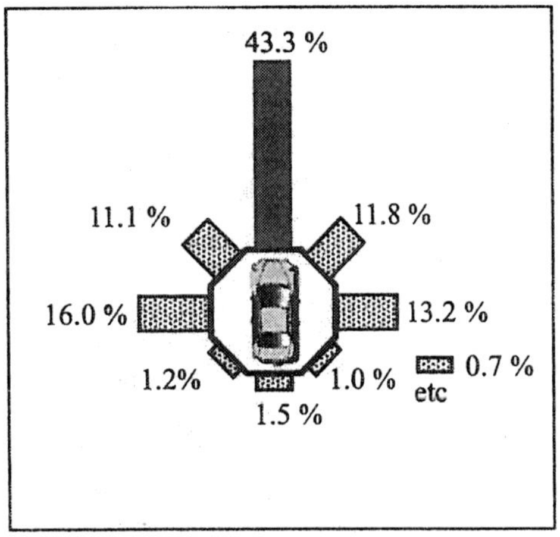

Figure 12. The collision direction in the fatal accidents for vehicle-to-vehicle collisions.

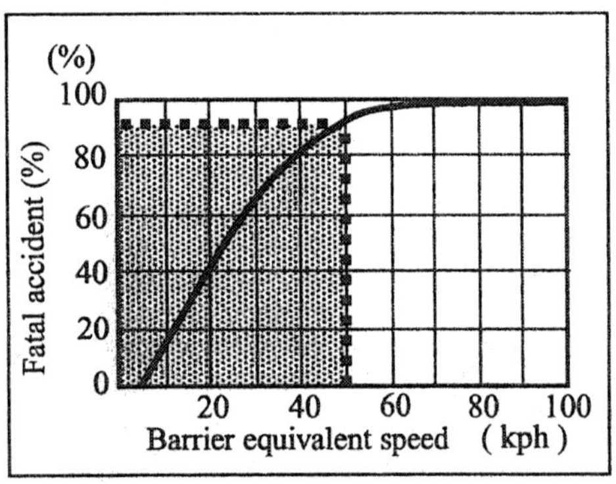

Figure 13. Fatal accident speed and percentage in frontal collisions.

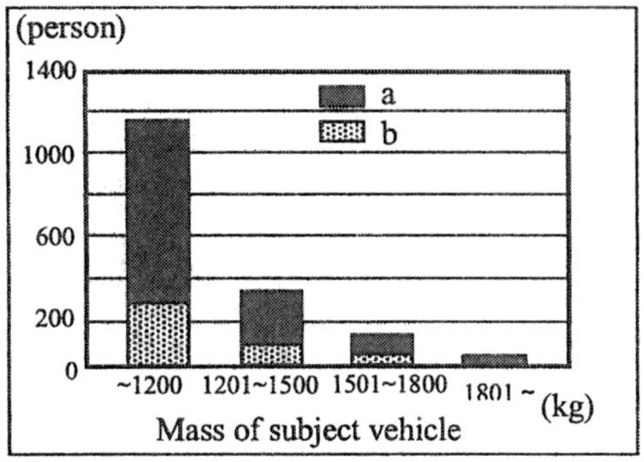

Figure 14. Vehicle mass and the driver fatalities of subject vehicle in frontal collisions.
; a) The opposite vehicle is heavier than the subject vehicle.
; b) The opposite vehicle is lighter than the subject vehicle.

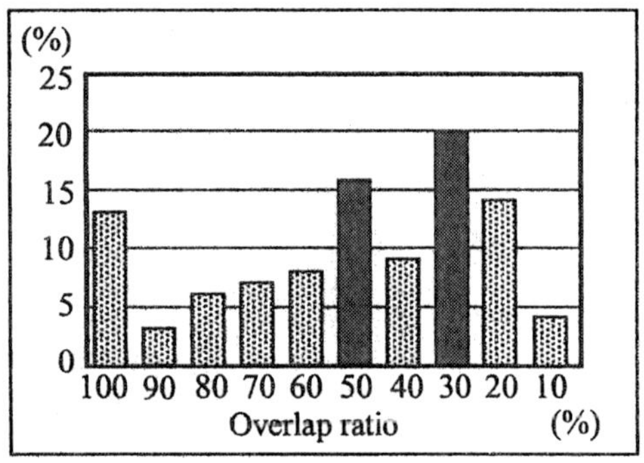

Figure 15. Overlap ratio in frontal collisions.

TEST CONDITIONS

The target in this test is the verification for a small car' s self-protection in the conditions based on real world accidents, especially the improvement of geometrical interaction in the frontal structure.

Small car: Prototype model
(This time only the 'Lower member system' was added on.)
Large car: Conventional model
Speed: 50kph per car
Mass : Small car 985kg Large car 1855kg
Mass ratio: 1.9 (Large car／Small car)
Overlap ratio: 50% of small car

The 50% overlap case is the reason that the risk of injury is higher than 30%. (as shown in figure 7) Impact angle is 0°to the car's longitudinal axis.

Figure 16 shows the two cars before the crash test. The main structural layouts of the cars are shown in Figures 17.

Figure 16. Two cars before the crash test.

Bumper heights are approximately the same.

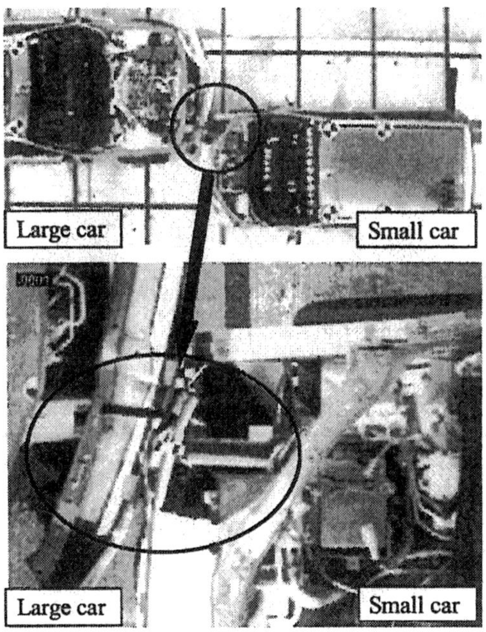

The main frame of the vehicles did not overlap in the direction of width.

Figure 17. Main structural layout of two cars.

In this test the main frames of the cars did not overlap in the direction of the cars' width, and the bumper heights were approximately the same.

Dummy: HybridIII

Restraint system: airbag and seat belt pretensioner with load limiter.

The restraint specification is similar to a small car without suitable modification for the change of the vehicle's deceleration characteristic.

TEST RESULTS

Speed: Small car 50.0kph

Large car 49.9kph

Overlap ratio: 48% of Small car

Figure 18 shows two cars after the crash test.

The structural deformation of the small car and the large car is shown in Figures 19 and 20 respectively.

The mode of structural deformation in the engine compartment of the small car during the crash is shown in Figure 21. Figure 22 shows deformation in the engine compartment of the small car after the crash.

Each part of deformation in both cars is listed in Table 1. Figure 23 shows the velocity change of both cars.

Figure 18. Two cars after the crash.

Figure 19. Deformation of small car with the new structure after the crash.

Figure 20. Deformation of large car after the crash.

0msec 10msec 38msec

Figure 21. Deformation mode of small car's new structure during the crash.

Figure 22. Deformation of small car in the engine compartment after the crash.

Table 1. Deformation

		Small car with new structure	Large car
Deformation	Front pillar(mm)	13	9
	Dash panel(mm)	114	162

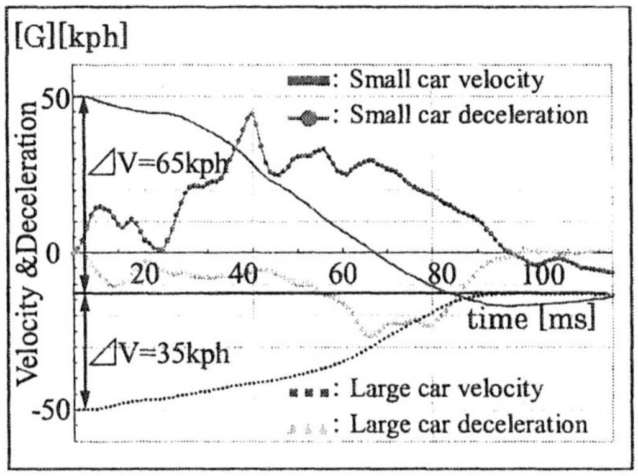

Figure 23. Velocity & Deceleration of both cars.

The vehicle-to-vehicle crash test confirmed the effectiveness of the new design in increasing the level of self-protection of small cars. The mode of deformation on impact confirmed that the bumper beam and main frame of the large car collided with the new lower member fitted in the small car.

The lower member restrained intrusion into the small car by making contact with the tire and wheel. Intrusion of the passenger compartment was therefore significantly reduced and the integrity of the cabin was maintained for occupants. The large car deformations in each part were approximately similar with the small car deformations.

The risk of injury to the small car's occupants were generally low by the prevention of secondary collisions in the passenger compartment.

The test results show the small car driver's injury risk is higher than the large car driver's injury risk. The reason is that the small car's Δ V[65kph], (namely the velocity change of before and after impact) is higher than the large car's Δ V[35kph], due to the influence by mass ratio.

CONCLUSION

A small car to large car crash test confirmed that the new body structure is one of the ways in increasing the level of self-protection of small cars.

It is very difficult for small vehicles with conventional body designs to maintain cabin integrity in narrow offset collisions and collisions with SUV because of the intrusion of the frame of the other vehicle

The innovative body structure proposed in this research reduces passenger compartment intrusion and occupant injury by restraining frame intrusion and enabling a high level of energy absorption in the engine compartment.

The structure improves the level of self-protection in small cars, it is also expected to improve the level of partner-protection offered by large car. As a further step we are going to research the aggressiveness for large cars based on the proposed new structure in this report.

Therefore, new design concept in making vehicles isn't an individualistic one and doesn't aim for only superior self-protection. Rather, the concept is harmony with the society of automobiles.

Finally it is hoped that the proposal of this new structure will trigger further research on body structures enabling reduction of traffic accident casualties in the future.

REFERENCES

[1] The Annual Reports of Traffic Accident Research and Data Analysis.1996.ITARDA.

CHAPTER FOUR – FRONTAL IMPACT

2001-06-0031

A COMPARATIVE ANALYSIS OF VEHICLE-TO-VEHICLE AND VEHICLE -TO-RIGID FIXED BARRIER FRONTAL IMPACTS

Saeed Barbat, Xiaowei Li, and Priya Prasad
Ford Motor Company, USA

ABSTRACT

The relationship between designing for both rigid fixed barrier (RFB) and vehicle-to-vehicle tests is a topical area of research. Specifically, vehicle-to-vehicle compatibility has been a topic of keen interest to many researchers, and the interplay between the two aspects of design is presently addressed.

In this paper, the studied vehicles for potential vehicle-to-vehicle impacts included: sport utility vehicles (SUVs), Pickups (PUs), and passenger cars. The SUV/PU-to-Car frontal impact tests were compared to those obtained from vehicle-to-rigid fixed barrier frontal impacts. Acceleration pulses at the B-pillar/rocker as well as dash and cabin intrusions were monitored and compared. Additionally, the energy distributions in SUV/PU-to-Car crash tests were compared to those of single vehicle-to-RFB tests.

It was concluded from the analysis that vehicle weight and front-end stiffness were not always the overriding factors dictating performance. Design alternatives that have positive impact on the distribution of energy on both vehicles involved in a crash were shown to provide improvement in vehicle compatibility. In the present work, it was also shown that good geometrical interaction in SUV/PU-to-Car impact was fundamental in providing self and partner risk-reducing potential. Moreover, the effect of geometry was shown to possibly mask the effects of mass and stiffness.

1. INTRODUCTION

Improvement in occupant risk-reducing potential and vehicle crashworthiness has traditionally been studied and addressed through a well-defined laboratory crash test procedures. Examples of such procedures include regulated full-frontal vehicle-to-rigid fixed barrier testing conducted by the National Highway Traffic Safety Administration (NHTSA) in the United States, Transport Canada in Canada, the Federal Office of Road Safety (FORS) in Australia, and the Ministry of Transport in Japan. Similar crash testing, but at a higher speed, is used in the New Car Assessment Program (NCAP) which serves to provide consumer safety information via a star ratings system (which represents various levels of attendant life-threatening combined head-chest injury potential).

Full-frontal rigid fixed barrier crash tests are used to evaluate the performance of restraint systems and safety devices. These devices may or may not provide the same level of risk-reducing potential to the occupants in SUV/PU-to-Car impacts as observed in real-world accidents [1]. This is mainly due to cabin intrusions caused by incompatibility between vehicles on the road, especially in the case of a SUV impact with a passenger vehicle. Serious injury and fatalities have been reported in conjunction with cabin intrusions [2]. Cabin and windshield intrusions observed in SUV/PU-to-Car crashes are generally not observed in controlled vehicle-to-rigid fixed barrier impacts. Therefore, a different structural design philosophy needs to be introduced in order to reduce injuries related to vehicle intrusions subject to the development of a front-end structure capable of dissipating crash energy in both vehicle-to-rigid fixed barrier and SUV/PU-to-Car impacts. Clearly, the distribution of stiffness and the force levels by which a vehicle's front-end components can absorb kinetic energy are important factors.

In most cases, vehicle development for front impact is concentrated on vehicle occupant protection without considering partner protection in SUV/PU-to-Car impact. Moreover, developments in frontal and side impact research are taking place with little interaction. For instance, improved occupant performance in frontal impact through higher stiffness of the front-end structure may have a trade off effect in side impact [3]. Therefore, it is very important to consider interactions between different crash configurations to ensure both self as well as partner protections.

The work described in this paper provides a comparative analysis of the vehicle structural performance obtained from vehicle-to-rigid fixed barrier and SUV/PU-to-Car impacts. A series of full-frontal SUV/PU-to-Car and vehicle-to-rigid fixed barrier crash tests, in conjunction with a series of full-scale finite element simulations were carried out in this study to support the comparative analysis. The comparative analysis in this paper was intended to provide a basis for possible future improvements, e.g., an upgrade to current test procedures, development of a complementary test to that specified in FMVSS 208, or a new design approach.

2. OBJECTIVES

The main objective of this work was to provide a comparative summary of results obtained from the frontal rigid fixed barrier and SUV/PU-to-Car crash tests. Results for a reference (Target) vehicle, in terms of crash deceleration pulse at B-pillar/rocker, dash and cabin intrusions, lower tibia index, and crash energy measurements were used in the comparison. The secondary objective was to identify the effect of vehicle design imbalances in SUV/PU-to-Car impact on the toe-board intrusions and energy management in the reference vehicle. The final objective was to recommend design actions to manage the crash energy in SUV/PU-to-Car impacts without performance degradation in related single-vehicle rigid fixed barrier tests.

3. FULL-SCALE CRASH TEST MATRIX

Table 1 shows the selected vehicles and their associated weight used in both vehicle-to-rigid fixed barrier and SUV/PU-to-Car impacts. The reference vehicle used for rigid fixed barrier impact was selected to be the struck vehicle in SUV/PU-to-Car impact for structural performance evaluation. The reference vehicle was a four-door sedan representing an average mid-size passenger vehicle in the fleet.

Table 1.

Vehicle Description and Weight

Frontal Impact Test	Striking Vehicle	Bullet Vehicle Test Weight	Target Vehicle Test Weight
Unit		kg	kg
50 Kph Rigid Fixed Barrie	-	-	1775
57 Kph Rigid Fixed Barrie	-	-	1776
Vehicle-to-Vehicle	Sport Utility (SUV I)	2132	1863
Vehicle-to-Vehicle	Pickup (PU)	2127	1778
Vehicle-to-Vehicle	Sport Utility (SUV II)	2694	1777

Two sport utility vehicles and a Pickup, viz., SUV I, SUV II, and PU, with different mass, stiffness and geometry were selected for the striking vehicles in SUV/PU-to-Car impacts. For the remainder of the text, the striking vehicle is referred to as the bullet vehicle (B).

3.1 Vehicle-to-Rigid Fixed Barrier Impact Tests

Two RFB crash tests that presently exist in the vehicle design process include: (1) FMVSS 208 with belted mid-sized male, instrumented test dummies, and (2) The USA NCAP test involving belted mid-sized male, instrumented test dummies. The FMVSS 208 test is conducted at 48 kph while the USA NCAP is conducted a 57 kph.

Typical velocity changes, ΔVs, that the occupants undergo in FMVSS 208 and NCAP crash tests, are 48 kph and 57 kph, respectively (neglecting the rebound velocities). The restraint system is designed to handle such changes in velocity. Two rigid fixed barrier crash tests were conducted with the target vehicle at 50 kph and 57 kph. The kinetic energy of the vehicle impacting the rigid fixed barrier was mainly absorbed by the vehicle structural deformation and the restraint system. It should be noted that the height from the ground of the bumper and front rail has negligible effect on the vehicle structural performance and energy management during a rigid fixed barrier impact. However, it may have a significant effect in SUV/PU-to-Car impacts.

3.2 SUV/PU-to-Car Impact Tests

US cars and light trucks are designed to 57 kph rigid fixed barrier impact (NCAP) for belted occupants. Therefore, relatively low occupant response rates are expected in real-world collisions up to 57 kph fixed barrier equivalent velocities. In the test configuration shown in Figure 1, the target vehicle was initially at rest in all of the SUV/PU-to-Car tests carried out in this study. The bullet vehicle's velocity was selected based on the relative masses involved, i.e., the bullet vehicle impact velocity was mass-adjusted to result in a 57 kph barrier-equivalent velocity for the target vehicle. Lighter vehicles generally experience higher velocity changes, ΔVs, while heavier vehicles experience lower ΔVs in SUV/PU-to-Car impacts. The initial speed of the bullet vehicle ranged from 92.3 kph to 95 kph. The induced velocity changes in the target vehicle ranged from 50 kph - 57 kph. The velocity changes in both the bullet and target vehicles could be calculated either from the basic momentum conservation or from velocity time histories obtained from tests.

Figure 1. SUV/PU-to-Car test configuration.

Table 2 shows the initial impact velocity, mass ratios, and velocity changes for all of the rigid fixed barrier and SUV/PU-to-Car crash tests conducted in this study.

Table 2.
Impact Speeds, Mass Ratios, and Velocity Changes

Frontal Impact Test	Initial Velocity		Mass Ratio	Velocity Change	
	Target	Bullet	Bullet/Target	Target	Bullet
Unit	kph	kph		kph	kph
Car-to-Rigid Fixed Barrier	50	-	-	49.6	-
Car-to-Rigid Fixed Barrier	56.7	-	-	56.7	-
SUV I-to-Car	0	94.9	1.14	50.6	44.2
PU-to-Car	0	101.6	1.20	55.3	46.3
SUV II-to-Car	0	92.2	1.52	55.5	36.6

Selection of bullet vehicles was based on providing different levels of mass, front-end stiffness, and bumper/rail heights. In SUV/PU-to-Car impacts, the effect of geometry can mask the effect of mass and stiffness, due to override or underride. Figure 2 shows the geometrical alignment, in terms of bumper heights, of the three bullet vehicles compared to the target vehicle. Table 3 provides bumper heights in (mm) for the bullet vehicles. These heights were measured from the center of the bumper of each bullet vehicle relative to the center of the bumper for the target vehicle. In other words, the reference point was the bumper center height of the target vehicle.

Table 3.

Relative Bumper Heights

Frontal Impact Test	Relative Bumper Height (mm)
Car-to-Rigid Fixed Barrier	-
Car-to-Rigid Fixed Barrier	-
SUV I-to-Car	94.0
PU-to-Car	116.8
SUV II-to-Car	12.7

SUV I-to-Car PU-to-Car SUV II-to-Car

Figure 2. Geometrical differences between bullet and target vehicles.

In all the SUV/PU-to-Car tests, both the bullet and target vehicles used a Hybrid III 50th percentile, male dummy in the driver mid position and a Hybrid III 5th percentile, female dummy in the passenger full forward position. All the dummies were belted and the airbags were active.

4. TEST RESULTS AND DISCUSSIONS

The safety performance of the restrained occupants as measured in rigid fixed barrier impact was dependent on several parameters, which included:

- The crash pulse and the design of the restrained system,
- The intrusions and the structural integrity of the vehicle cabin, and
- The energy management performance of the energy absorbing elements such as the front rail or frame.

It should be acknowledged that a passenger car whose structural design was specifically tailored to achieve high performance in a rigid fixed barrier impact might not necessarily perform as well in a SUV/PU-to-Car impact. That same passenger car might be subjected to both increased levels of intrusion and energy management in a SUV/PU-to-Car impact. The following sections provide the discussion and comparison of the target vehicle performance only, in terms of crash pulse, intrusions and crash energy management, between the rigid fixed barrier and SUV/PU-to-Car impacts.

4.1 Vehicle Deceleration Pulse Comparison

In vehicle crashworthiness, vehicle deceleration pulse, measured at B-Pillar/rocker, during a crash is generally considered to be an important characteristic of vehicle safety performance. However, the characteristics of the time history of the crash pulse for the target vehicle at a certain ΔV may vary with the crash modes. In a target vehicle-to-rigid fixed barrier impact, the crash pulse depends only on the inherent front-end stiffness and structural design. While in SUV/PU-to-Car impact, it also depends on the structural stiffness, mass, and geometry of the bullet vehicle. Front-end design of a vehicle, which achieves better performance during a full-frontal crash into a rigid fixed barrier, can be less effective during SUV/PU-to-Car impact.

Figures 3 and 4 show the comparisons of the deceleration pulse of the target vehicle obtained from various rigid fixed barrier tests and SUV/PU-to-Car tests as listed in Table 1.

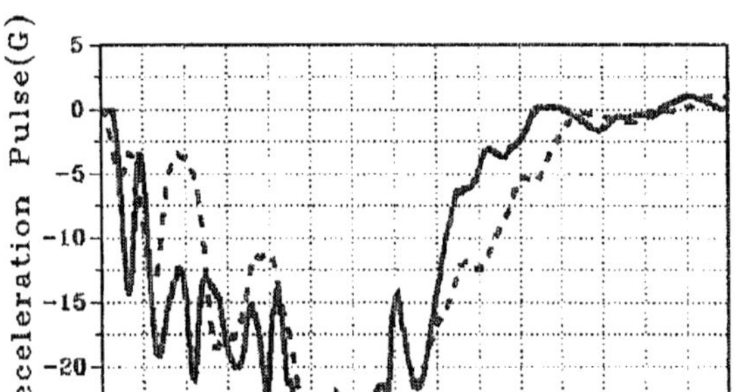

Figure 3. Comparison of crash pulse in a passenger car obtained from rigid fixed barrier and SUV I-to-Car impacts (50 kph velocity change).

In order to eliminate the effect of the velocity change, ΔV, on the deceleration pulse characteristics, the crash pulse of the target vehicle obtained from 50 kph rigid fixed barrier test was examined against that obtained from SUV I-to-target impact test which induced a 50 kph velocity change in the target vehicle. The crash pulse comparison of these two tests is shown in Figure 3. Similarly, the 56.7 kph crash pulse from the NCAP test was examined against those obtained from Pickup-to-target vehicle and SUV II-to-target vehicle impact tests that generated a 56.7 kph barrier equivalent velocity in the target vehicle. This comparison is shown in Figure 4.

As shown in Figures 3 and 4, and in the case of SUV/PU-to-Passenger car impacts, the target vehicle experienced a lower deceleration level early in the collision (up to 40 ms) and a higher deceleration later in the collision compared to that observed in rigid fixed barrier impact. This was due to the fact that in full-frontal rigid fixed barrier impact, at either 50 kph or 56.7 kph, the energy absorbing elements of the target vehicle's front-end fully engaged with the rigid fixed barrier at 40 ms. This leads to a progressive dissipation of the crash energy (kinetic energy) through a progressive and controlled deformation of the front-rails exhibited by the front-end stiffness of the target vehicle only. However, in SUV/PU-to-target vehicle impacts, the energy dissipation within the first 40 ms shared by the target vehicle was much less compared to that in rigid fixed barrier impact. The target vehicle front-end stiffness had little effect on the pulse outcome due to partial or minimum engagement with the bullet vehicle at 40 ms.

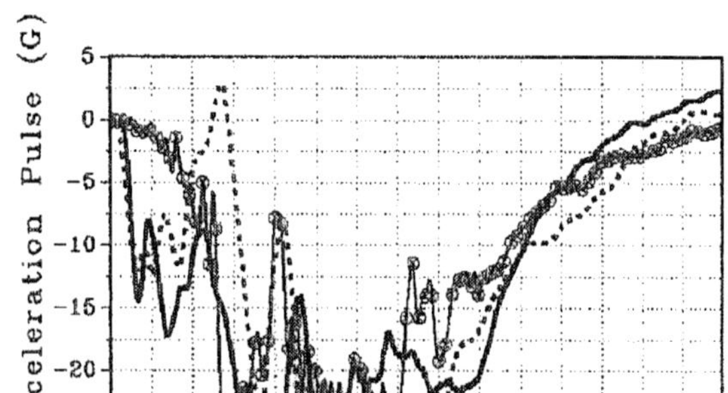

Figure 4. Comparison of crash pulse in a passenger car obtained from rigid fixed barrier and SUV II/PU-to-Car impacts (57 kph velocity changes).

In the case of the PU-to-target vehicle impact, as shown in Figure 4, there was a clear evidence that the Pickup overrode the target vehicle due to incompatible geometries (Table 3). This led to a very low level of the deceleration pulse early in the collision. The front-end stiffness effect of the target vehicle was almost negligible within the first 20–25 ms since there was no engagement of the front-end.

After 40 ms, the bullet vehicle was close to having full engagement with the target vehicle in all the three SUV/PU-to-Car impact tests. The deceleration pulses of the target vehicle appeared to be similar, but higher, compared to that in rigid fixed barrier impact. This was due to the lighter (target) vehicle experiencing a higher level of both velocity change and deceleration compared to the heavier (bullet) vehicle. In conclusion, the deceleration pulse of the target vehicle in SUV/PU-to-Car impacts appeared to be similar to that in rigid fixed barrier impact in terms of peak amplitude and time to peak.

4.2 Overall Deformation and Intrusion Comparison

For a full-frontal rigid fixed barrier impact, intrusions are of secondary importance compared to the crash pulse, which is generally considered to be an important factor in the occupant response outcome. However, both the crash pulse and intrusions are of comparable importance in occupant response outcome in SUV/PU-to-Car impact. A similar level of occupant risk-reducing potential in a target vehicle can be achieved in rigid fixed barrier and SUV/PU-to-target vehicle impacts, if the intrusions in the target vehicle are assumed the same. Post-crash static intrusion measurements in SUV/PU-to-target vehicle impacts indicated that cabin intrusions in the target vehicle were considerably greater than those measured in a target vehicle-to-rigid fixed barrier impact. Therefore, a SUV/PU-to-Car or similar impact test [4] might be a suitable test protocol to complement the full-frontal rigid fixed barrier impact test. This additional test will evaluate vehicle design to provide occupant protection and cabin structural integrity in single and multi-vehicle crashes.

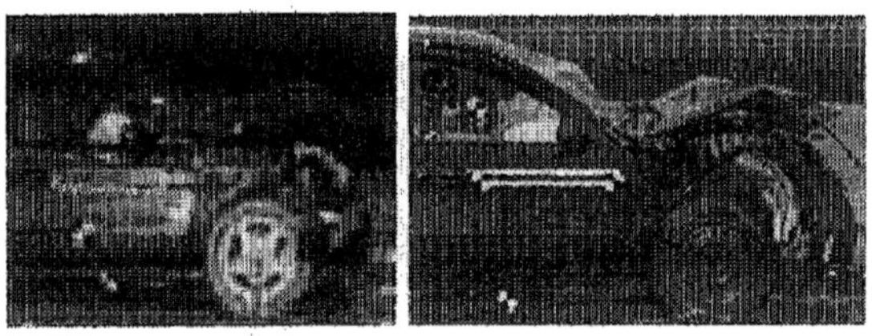

(a) Rigid Fixed Barrier (b) SUV I-to-Car

(c) PU-to-Car (d) SUV II-to-Car

Figure 5. Overall, deformations in target vehicle in different crash configurations.

The resulting deformation of target vehicle's front structure for all frontal impact tests listed in Table 1 is shown in Figure 5. It was apparent that the collision type is an important factor governing front-end structural performance since it determined how the vehicle's front-end structures interacted. Figure 5 (a) shows a progressive deformation of the front-end structure when engaged with a rigid fixed barrier. However, Figure 5 (b, c and d) show that the front rail deformations were ineffectively used, and had noticeable deformation at the shotgun and hood areas. As a result, higher occupant compartment intrusions were observed due to the effect of mass, and possibly stiffness and geometrical differences.

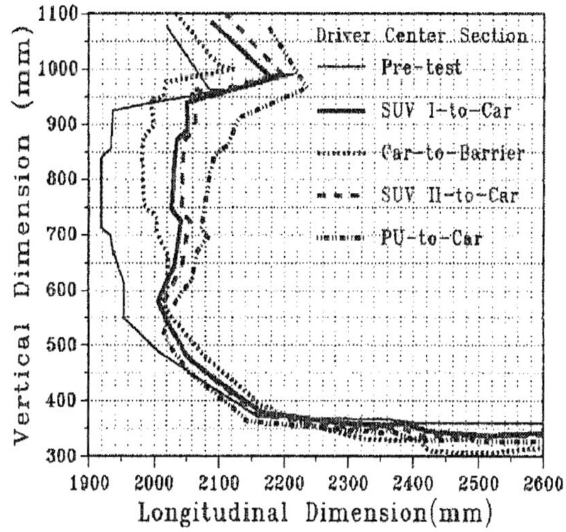

Figure 6. Comparison of dash intrusion profiles between rigid fixed barrier test and SUV/PU-to-Car tests.

Pre- and post-crash dimensional analyses on target vehicles were carried out to obtain intrusion profiles at the driver center section. The results are shown in Figure 6, where the intrusion profiles from the cowl top to the floor panel at this section are illustrated. As shown in Figure 2, there was considerable difference in the vertical heights of the significant front-end structures in the case of the PU-to-target vehicle impact. Table 3 also indicates a maximum of 117 mm difference in the vertical heights between the bumper center points of the Pickup and the target vehicle. The Pickup caused significantly more intrusion in the occupant compartment of the target vehicle compared to those caused by SUV I and SUV II in SUV/PU-to-Car frontal impact (see Figure 6). It was clear that there was little structural interaction due to geometrical differences that caused the Pickup to override on to the target vehicle. There was very little difference between the dash intrusion profiles caused by SUV I and SUV II impacts. However, this dash intrusion was significantly greater than that seen in related full-frontal rigid fixed barrier impacts.

Figures 7 and 8 show the effect of bumper height differences and mass ratio of the bullet to target vehicles on the maximum dash intrusions in the target vehicle, respectively. These figures, as well as the results demonstrated in Figure 6, suggested that the effect of geometry in SUV/PU-to-Car impacts can reduce the effect of mass and stiffness.

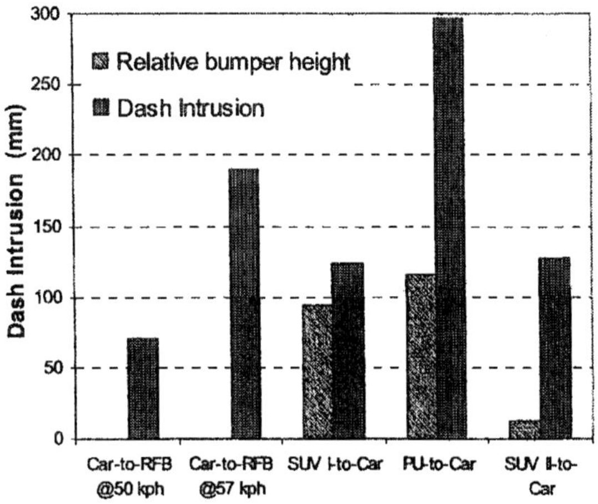

Figure 7. Effect of relative bumper height on dash intrusion in various crash tests.

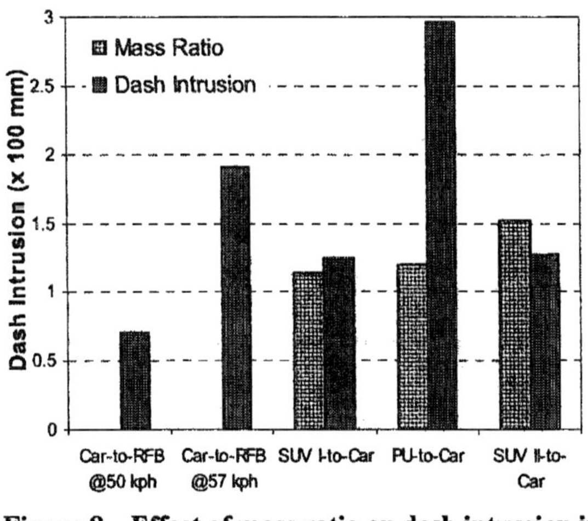

Figure 8. Effect of mass ratio on dash intrusion in various crash tests.

4.3 Lower Leg Tibia Index Comparison

Tibia Index is a function of axial force and bending moment measured by the lower Tibia load cell during crash tests. With a good, crashworthy front-end structural design for a target vehicle, occupants in full rigid fixed barrier impact have a much lower risk of potential lower leg injury than those in SUV/PU-to-target vehicle impact. The relatively high Tibia Index expected in SUV/PU-to-Car impacts was caused by higher intrusions.

Figure 9 shows the normalized values of the lower leg Tibia Index as a function of crash test configurations. It is shown that target vehicles with low intrusions, such as in vehicle-to-rigid fixed barrier, have lower Tibia Index. Since the maximum intrusions in SUV/PU-to-Car impact are associated with the Pickup overriding the target vehicle, this resulted in a very high Tibia Index of 1.8 exceeding the Injury Assessment Reference Values (IARV) of 1.0 [5]

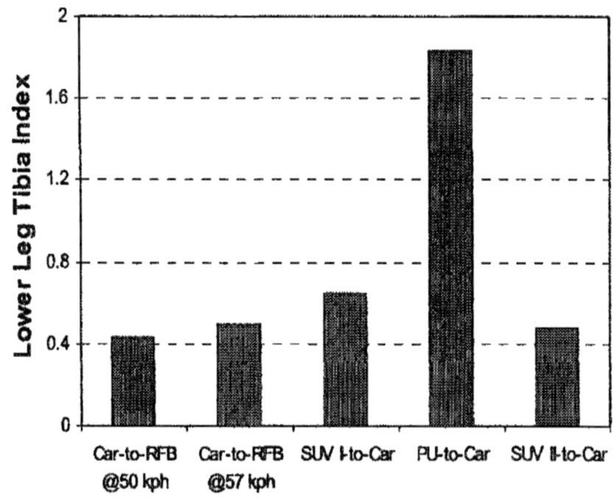

Figure 9. Lower leg Tibia Index comparison.

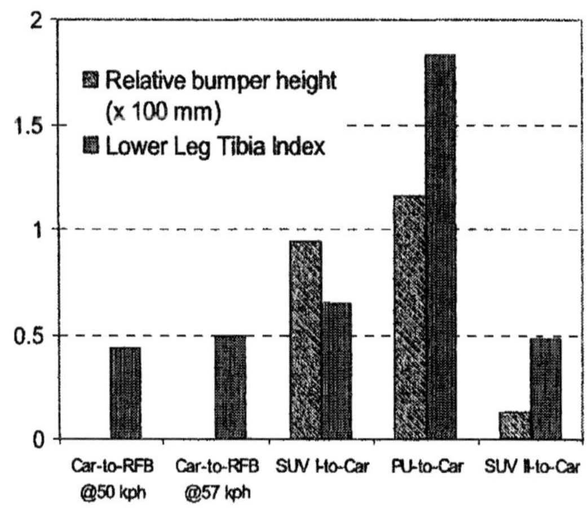

Figure 10. Relative bumper height effect on lower leg Tibia index.

4.4 Energy Absorption Comparison

Comparisons of crash energy management of a target vehicle involved in a single fixed barrier and multi-vehicle impacts can be an appropriate measure for examining good safety vehicle performance. In this section, energy absorbed by the target vehicle in a rigid fixed barrier test was compared to that from a SUV/PU-to-Car impact.

All vehicles consume the same amount of energy per unit weight when crashed against a rigid fixed barrier at any given speed (assuming negligible rebound speeds). However, uneven distribution of energy absorption is noticeable between the colliding vehicles in SUV/PU-to-Car crashes. The lighter and less stiff vehicle absorbs more of the impact energy while the heavier and stiffer vehicle deforms less and therefore absorbs less energy.

Various methods for calculating the energy absorbed by a vehicle when impacting a fixed barrier or another vehicle have been reported in previous studies [4]. Momentum and energy conservation equations do not provide details of how energy is shared between the two colliding vehicles, or how energy is absorbed by main components of the structure. Computer simulations were used to estimate the energy absorbed by each vehicle. Detailed finite element models of the target and bullet vehicles used in this study were developed. The models were fully validated at various speeds against a rigid fixed barrier and SUV/PU-to-Car tests (see Figure 11). Details of the models are beyond the scope of this paper.

Figure 11. Comparison of SUV-to-Car crash test (bottom) and related simulation (top).

Figure 12 shows the comparison between the energy absorption resulting from rigid fixed barrier and SUV/PU-to-Car impacts. The energy presented in this figure represented the internal energy absorbed by each vehicle as calculated in the finite element analyses. This figure clearly shows that the lighter (target) vehicle absorbed more energy than the heavier (bullet) vehicle in SUV/PU-to-Car impacts. It also indicated that the target vehicle absorbed more energy in SUV/PU-to-Car impacts (compared to that in rigid fixed barrier impact when similar levels of velocity changes were experienced). This was due to the deformation in the target vehicle resulting from the SUV/PU-to-Car impact.

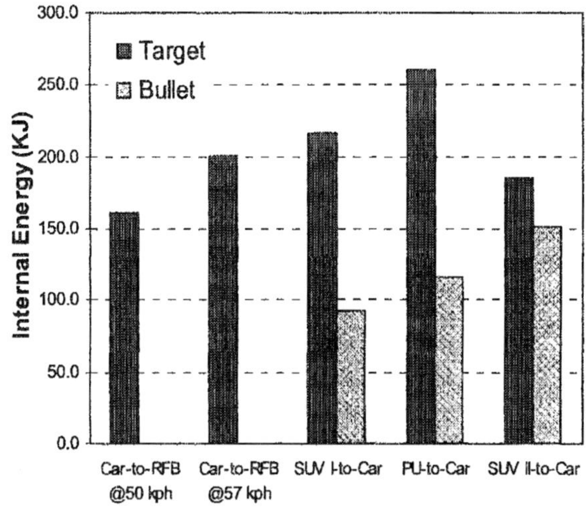

Figure 12. The comparison between the energy absorption.

To achieve good compatibility in SUV/PU-to-Car impacts, it is necessary for both vehicles to share the impact energy and to absorb a quantity of energy proportional to their own masses. In the following discussion, the energy absorbed by a target vehicle in a rigid fixed barrier impact at a certain speed was assumed to be equivalent to that absorbed by each of two identical target vehicles in a full overlap frontal SUV/PU-to-Car impact. It was also assumed to be true only when the closing speed between the two target vehicles was double that of the same target vehicle speed against a rigid fixed barrier. If the term " Energy Ratio" is defined as the ratio of the energy absorbed by the target vehicle to that of the bullet vehicle in SUV/PU-to-Car impact, then that Energy Ratio for the rigid fixed barrier impact will always be equal to one at any given speed.

Figures 13 and 14 present the effect of geometry and mass on the energy management of the target vehicle in SUV/PU-to-Car impact. It was evident that the geometry effect dominated the effect of mass. The relatively high Energy Ratio resulted from the SUV and PU impact to the target vehicle was mainly due to high difference in bumper height compared to that of the target vehicle. This was due to the SUV/PU override that caused deformation and higher energy absorption by the target vehicle. Conversely, the Energy Ratio resulting from the heaviest SUV II impacting the target vehicle was close to one which also confirmed that the alignment of energy absorbing elements and bumper heights in SUV/PU-to-Car impact was a very important design element that helped ensure the aforementioned characteristics of good compatibility were achieved. Also, it was observable that the effect of geometry can reduce the effect of mass in a SUV/PU-to-Car impact.

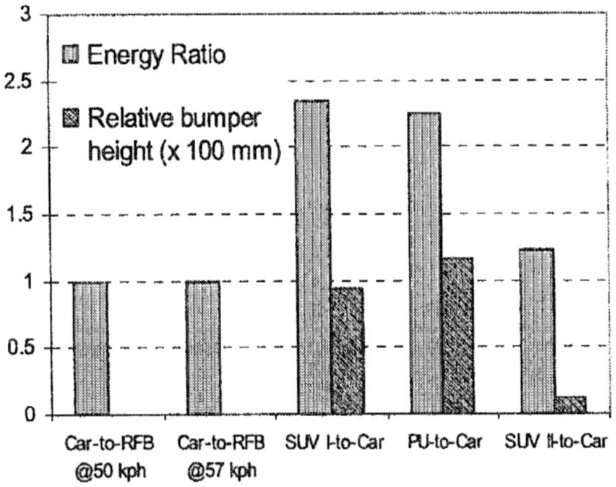

Figure 13. Variation of internal energy ratio with relative bumper height.

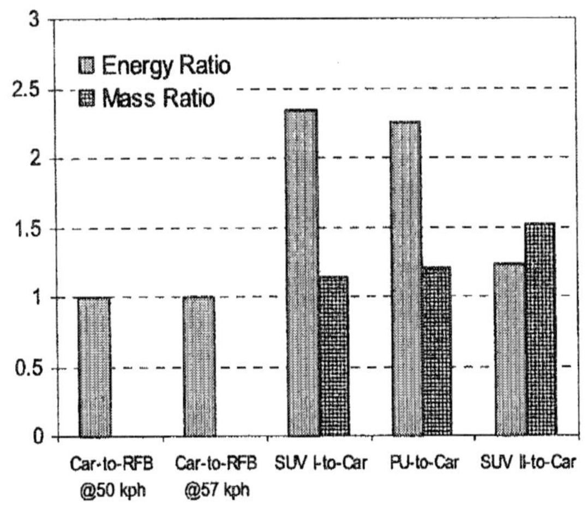

Figure 14. Variation of internal energy ratio with vehicle mass ratio.

6. CONCLUSIONS

- Vehicles that achieve good safety performance in rigid fixed barrier impacts may not necessarily achieve the same level of performance in SUV/PU-to-Car impacts.
- Crash pulse is generally considered to be an important factor for the occupant response outcome in a rigid fixed barrier impact. It was observed that similar crash pulses existed (after 40 ms into the crash events) for both the rigid fixed barrier tests and target passenger car in the studied SUV/PU-to-Car collisions.
- Significant deformation and dash intrusions were observed in the target vehicles in SUV/PU-to-Car impacts compared to that in rigid fixed barrier impacts.
- Higher intrusion-related Tibia Indices were observed in the studied SUV/PU-to-Car impacts
- The energy absorbed by target vehicles in the studied SUV/PU-to-Car impacts was higher than that resulting from rigid fixed barrier impacts.
- A SUV/PU-to-Car or an equivalent test to complement the current rigid fixed barrier test was proposed to help further assess vehicle structural performance.
- In the studied frontal SUV/PU-to-Car impacts, it was shown that good geometrical interaction and alignment of the energy absorbing elements were fundamental to crash compatibility.

7. ACKNOWLEDGEMENTS

The authors wish to acknowledge Mr. Kris Warmann, a technical specialist at Ford Motor Company, for conducting the PU-to-Car test and providing the authors with the test data. The authors also wish to thank Mr. Tony Laituri, a technical specialist at Safety R & D department of Ford Motor Company, for editorial comments.

8. REFERENCE

[1] Zobel, R., " Evaluation of the safety performance of passenger vehicles", 11th ESV Conference, May 1987.

[2]. Foret-Bruno, J. Y., Le Coz, J.Y., Thomas, C. Brutel, G., " In depth analysis of frontal collisions as regards the influence of overlaps and intrusion on occupant severe and fatal injuries", 14th ESV Conference, Munich, May 1994.

[3] Hobbs, A. C., "The need for improved structural integrity in frontal car impacts", 13th ESV Conference, Paper No. S9-0-12, November 1991.

[4] Ragland, C.," Offset test procedure development and comparison", 16th ESV Conference, Paper No. 98-S-0-03, Windsor, Canada, May 1998.

[5] Mertz, H., "Anthropomorphic Test Devices", in Accidental Injury: Biomechanics and Prevention, by A. Nahum and J. Melvin, eds., Spring-Verlag, New York, 1993.

2001-06-0097

EVALUATION OF VEHICLE COMPATIBILITY IN VARIOUS FRONTAL IMPACT CONFIGURATIONS

Saeed Barbat, Xiaowei Li, and Priya Prasad
Ford Motor Company, USA

ABSTRACT

Light truck vehicles (LTVs), sport utility vehicles (SUVs), and vans collectively make up a growing segment of the total automotive fleet sales, particularly in the United States. The National Highway Traffic Safety Administration (NHTSA) has identified this trend and has increased the extent of its research in vehicle-to-vehicle compatibility. Additionally, vehicle compatibility concerns have also been emphasized by International Harmonization Research Activity (IHRA). Accordingly, with intention to further enhance road safety, research in the area of crash compatibility between cars and LTVs in different crash configurations is of significant importance.

This paper describes a part of ongoing research at Ford Motor Company to further investigate the effect of compatibility in SUV/LTV-to-Car crashes. Test results of SUV/LTV-to-Car crashes involving various frontal impact configurations were analyzed in order to develop test procedures and requirements to help assess vehicle compatibility. Specifically, three SUV-to-Car frontal impact configurations were assessed in the present study: full overlap collinear impact, 50% offset collinear impact, and 30-degree oblique impact. In each of the tests, both the target and bullet vehicles contained a Hybrid III 50th percentile instrumented test dummy (HIII50) for the driver and a Hybrid III 5th percentile instrumented test dummy (HIII05) for the passenger. Analysis of the tests yielded the following results: (1) Structural and occupant responses were used to help quantify the effect of mass, stiffness, and geometry, (2) A robust and repeatable vehicle-to-vehicle test procedure was proposed, and (3) Preliminary results indicated that geometric incompatibility was the dominating factor in the studied vehicle design characteristics.

1. INTRODUCTION

Compatibility concerns in vehicle-to-vehicle crashes in the United States have received considerable attention due to recent studies and publications by NHTSA and other agencies worldwide [1,2,3,4]. These concerns have been noted for the entire fleet of vehicles on the road, although government agencies and the media have concentrated on the SUVs and LTVs. The SUV and LTV market share has risen steadily during the last decade and has become a growing component of the U.S. fleet. As a group, these vehicles are generally heavier, stiffer, and have higher ground clearance compared to those of the passenger cars.

In LTV/SUV-to-passenger car frontal crashes, the potential occupant risk to the occupants in the passenger car is generally higher. This was due to the larger velocity change (ΔV) and the higher intrusion experienced by the generally lighter, smaller, and less stiff, passenger cars. However, to achieve enhanced real-world safety, reducing the aggregate potential risk of the involved occupants needs to be considered. This presents vehicle design challenges to provide self risk-reducing potential as well as partner risk-reducing potential. The partner protection issue led to several studies and investigations to help better understand the effect of vehicle design parameters such as mass, geometry, and stiffness on crash compatibility.

The present vehicle fleet differs in mass, geometry, stiffness, and many other design parameters. Vehicle compatibility has been investigated in many studies using different approaches such as real-world crash statistics, crash testing and computer modeling. NHTSA used U.S. crash statistics from the Fatality Analysis Reporting System (FARS) to determine the number of fatalities in vehicle-to-vehicle collisions [1]. Moreover, statistics from the General Estimates System (GES) were used to determine the number of vehicle-to-vehicle collisions [2]. One objective of these studies was to identify and demonstrate the extent of the problem of vehicle compatibility. Another objective was to demonstrate, through statistical analysis, the aggressivity metric as a function of vehicle mass, stiffness, and geometry. Other statistical analysis such as that conducted by Evans [5,6] indicated that mass is one of the most significant factors affecting potential risk of occupant injury in vehicle-to-vehicle collisions. His study indicated that the ratio of the injury rate in a lighter vehicle to that in a heavier one can be expressed as the power function of the mass ratio (of the heavier to the lighter vehicle). Accident data in Japan and computer modeling techniques were also used by Mizuno and Kajzer [7] to investigate the compatibility of mini cars in traffic accidents.

Most recently, Summers et al. [3] from NHTSA conducted a series of full-scale vehicle-to-vehicle crash tests to evaluate potential crash compatibility issues. Tests were conducted with four different striking vehicles representing various vehicle classes impacting a Mid-size sedan struck vehicle in both side and oblique frontal impacts. For the oblique offset crash test series, a good correlation was obtained between the driver injury measures of the struck vehicle with vehicle mass. However, NHTSA concluded that more research needed to be conducted to further understand the vehicle design parameters that effect vehicle compatibility.

In the present study, NHTSA's test procedures for the frontal oblique crash tests, as there were issues related to the repeatability and reproducibility of those crash tests, were further evaluated. Specifically, a series of vehicle-to-vehicle crash tests were conducted in various frontal impact configurations. A series of 30-degree offset oblique frontal impacts, similar to NHTSA's test configurations, were carried out to address issues associated with this test procedure. Two additional SUV-to-Car collinear frontal impact configurations, e.g., full overlap and 50% offset, were also investigated and presented in this study. Structural and occupant performances to quantify the effect of vehicle design parameters on vehicle compatibility were analyzed and reported. A robust test procedure for vehicle-to-vehicle frontal impact was also proposed to help assess vehicle compatibility.

2. VEHICLE-TO-VEHICLE CRASH TESTING

Because of the limitations associated with the statistical approach, crash testing was deemed necessary to examine the influence of vehicle mass, stiffness, structural interaction, and geometry on vehicle compatibility. Impact configuration and collision type in frontal impact were determined to be the most important factors governing vehicle performance since they directly affected how the two vehicle's structures interacted. For example, the effect of vehicle stiffness and geometry on the deformation and occupant response outcome in the struck vehicle changed significantly from a full frontal and offset collinear impact to a frontal oblique impact (in two-vehicle crashes).

Figure 1. 30-degree oblique offset test configuration with the struck vehicle in horizontal and the striking vehicle in oblique.

NHTSA [3] conducted a series of frontal oblique vehicle-to-vehicle impact tests using a test procedure developed under NHTSA's Advanced Frontal Offset Research Program [8]. The test configuration is shown in Figure 1. Test results showed good correlations between the occupant response parameters (measured on occupants in the struck vehicle) and the striking vehicle mass and aggressivity metric (AM) defined by NHTSA as the ratio of the driver fatalities in collision partners to the number of crashes of the subject vehicle. However, it was also concluded that there were additional factors other than the mass difference that affected vehicle aggressivity. The present authors considered it difficult to extract the effects of stiffness and geometry in the oblique frontal vehicle-to-vehicle impact tests. Therefore, crash tests in various frontal impact configurations, including NHTSA's configurations, were examined in the present work.

2.1 30-degree Oblique Offset Test

In this test series, the target vehicle (T) was a four-door sedan representing an average Mid-size passenger vehicle in the fleet. This target vehicle was impacted by a series of four bullet vehicles (B): a SUV, a Small Pickup, a Mini-Van, and another Mid-size car (i.e., the same as the target). The bullet vehicles were selected to provide various potential aggressivity characteristics associated with mass, stiffness, and geometry. This selection of the vehicles was similar to that reported by NHTSA [3].

The frontal oblique tests were conducted according to the test configuration shown in

Figure 1. Both the target and bullet vehicles were moving at approximately 56 kph at the time of contact. At impact, the left side of the bullet vehicle aligned with the front center of the target vehicle, which was struck on the left front side. In all the tests presented in this paper, both the target and bullet vehicles used a Hybrid III 50th percentile male dummy for the driver and a Hybrid III 5th percentile female dummy for the passenger. The driver seat was at the mid-position while the passenger seat was full-forward. All the dummies were belted and the airbags were active. Table 1 shows vehicle test weights, relative bumper heights, initial speeds, and vehicle velocity changes. The relative bumper height was represented by the vertical distance between reference points, i.e., the center of the target vehicle's bumper and the center of the bumper of the bullet vehicle. It should be noted that a positive number indicated that the bumper of the bullet vehicle was initially above the reference point on the target vehicle.

Table 1.

Test Condition of 30-degree Oblique Vehicle-to-Vehicle Tests.

Bullet Vehicle		Small SUV	Mini-Van	Pickup	Mid-size Car
Target Vehicle		Mid-size Car			
B\Vehicle Mass	kg	2084	2040	1727	1843
T\Vehicle Mass	kg	1845	1845	1847	1849
Mass Ratio		1.13	1.11	0.94	1.00
T\Initial Velocity	kph	57.2	57.2	54.4	55.7
B\Initial Velocity	kph	56.4	56.7	56.4	56.2
T\Velocity Change	kph	56.1	55.6	51.1	52.8
B\Velocity Change	kph	49.7	50.2	54.7	53.0
Relative Bumper Height	mm	94.0	25.4	116.8	10.2

2.2 Full Frontal and 50% Offset Tests

In this test series, the target vehicle was the same Mid-size sedan used in the 30-degree oblique offset tests. The target vehicle was impacted by various bullet vehicles (i.e., two different SUVs) in two different frontal impact configurations (i.e., full frontal and 50% offset). The test setup is illustrated in Figure 2. The first bullet sport utility vehicle, SUV I, was selected to be the same as that used in the oblique offset test configuration listed in Table 1 (designated as "Small SUV"). This provided the opportunity to evaluate the performance of the target vehicle and its occupants when struck by the same bullet vehicle in three different frontal impact configurations. The second bullet sport utility vehicle, SUV II, was selected to be heavier and stiffer than SUV I. However, its rail height from the ground was roughly equal (therefore, roughly geometrically compatible) to that of the target vehicle. This selection was made to isolate the effects of geometry. Figure 3 shows the geometrical alignment of the bumpers and frame rails of the two bullet vehicles, SUV I and SUV II, with that of the target vehicle. Table 2 shows vehicle test weights, initial speeds, and velocity changes for both full frontal and 50% offset impact tests.

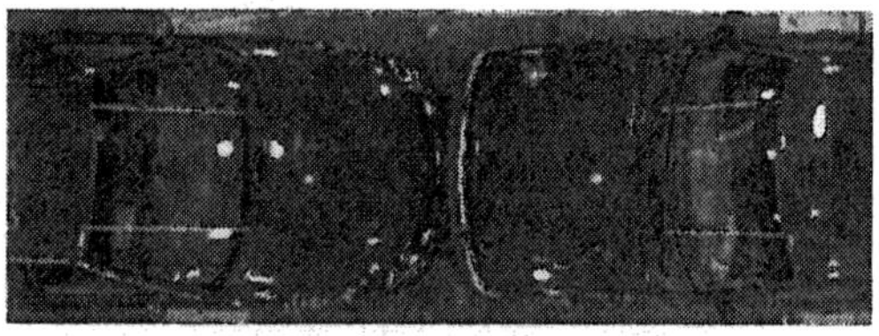

Full frontal collinear

50 % offset collinear

Figure 2. SUV-to-Car test configurations.

The target vehicle was initially at rest in both the full frontal and 50% offset impact tests. The bullet vehicle velocity was selected based on relative masses involved, i.e., the velocity was mass-adjusted to result in approximately 56 kph barrier-equivalent velocities for the target vehicle. The same dummies, restraint systems, and seating positions used in the oblique offset tests were also used for this series of tests.

SUV I-to-Car SUV II-to-Car

Figure 3. Geometrical differences between bullet and target vehicles.

Table 2.

Test Condition for Full Frontal and 50 % Offset SUV-to-Car Impacts

Bullet Vehicle		SUV I	SUV II	SUV I	SUV II
Target Vehicle		Mid-size Car			
Impact Configuration		**Full frontal**		**50 % Offset**	
Bullet/Target Speed	kph	96.2/0	93.3/0	96.1/0	92.8/0
B\Vehicle Mass	kg	2132	2694	2131	2680.2
T\Vehicle Mass	kg	1867	1777	1859	1781.3
Mass Ratio		1.14	1.52	1.15	1.50
B\Velocity Change	kph	44.2	35.3	44.3	31.9
T\Velocity Change	kph	52.0	58.0	51.7	60.9

3. TEST RESULTS AND DISCUSSIONS

When considering lighter vehicle-to-heavier vehicle impacts, the lighter vehicles experience higher velocity changes while heavier vehicles experience lower velocity changes. The velocity changes experienced by both vehicles during impact were calculated and presented in Tables 1 and 2 for the oblique offset and collinear frontal impact tests, respectively. Since both the target and bullet vehicles were subjected to the same impact durations, the occupants in the lighter vehicle experienced a higher level of acceleration. Occupant responses such as resultant head acceleration, chest acceleration, chest deflection, and femur compression loads were measured for both the driver and passenger dummies in both vehicles during impact. The aforementioned occupant responses were normalized by their corresponding occupant response injury assessment reference values (IARVs) reported by Mertz [9].

3.1 Dummy Response in 30-degree Oblique Offset Tests

Table 3 summarizes the driver (HIII50) responses for both the target and bullet vehicles along with related IARVs. Similarly, Table 4 summarizes the passenger (HIII05) responses for both the target and bullet vehicles. The normalized, derived occupant responses, such as the 36ms-based Head Injury Criteria (HIC36), 3ms-based accumulative chest acceleration, chest deflection, and the maximum femur compression forces were plotted for all four oblique offset tests. Test results, in terms of occupant responses, were plotted on the x-axis in a descending order of the aggressivity metric values associated with the bullet vehicle used in each test. The largest AM was associated with the Small SUV, and the smallest AM was associated with the Mid-size car (as reported in Reference 3). The aggressivity metrics were quantified as the ratio of the driver fatality in collision partners to the total number of crashes of the subject vehicle for various subject vehicle categories.

Figure 4 shows the comparison of the driver dummy responses in the target vehicle for the four oblique offset crash tests. Similar comparisons for the passenger dummy responses in the target vehicle are shown in Figure 6. The comparisons of the driver and passenger dummy responses for the bullet vehicle in these tests are shown in Figures 5 and 7, respectively. The occupant responses were normalized by the IARVs.

For the driver in the target vehicle as shown in Figure 4, there was a general trend of increasing normalized occupant responses with increasing aggressivity metric. With the exception of the Small Pickup, a very similar trend is observed with increasing weight. The weight of the Small Pickup was the lowest of the bullet vehicles used in the tests. It was also observed for the driver of the target vehicle, that the Small SUV produced significantly higher occupant responses than those associated with the Mini-Van (although the difference between their weights was only 44 kg c.f., Table 1). These two observations indicated that there might be factors other than the weight that influenced the aggressivity metric of these bullet vehicles. In such a complicated test configuration, it was difficult to extract the effect of the individual design parameters such as mass, stiffness, and geometry. The Small SUV produced occupant responses for the driver in the target vehicle that exceeded the cited IARVs.

All of the occupant responses for the occupants of the bullet vehicles were below the IARVs. The vehicle design parameters of the bullet vehicle were considered to have little effect on the related occupant responses (see Figures 5 and 7). However, they had a major influence on response outcome of occupants of the struck target vehicle, in particular the driver. The driver occupant responses in the Mid-size bullet car were slightly higher compared to the responses produced by the other three striking vehicles. Since the target and bullet vehicles were identical in this case and fully compatible, the velocity change experienced by the occupants of the bullet vehicle was the same as that of the target vehicle (see Table 1). This level of velocity change was considered to have caused the slightly higher occupant responses compared to the related driver's responses of the other bullet vehicles. Figure 8 shows selected results of Figure 4 plotted with the descending order of the relative bumper height (representing potential geometrical incompatibility). The occupant responses produced by the Mid-size vehicle striking the same Mid-size target vehicle were the lowest among the four tests. Presenting the results in this manner confirmed that aligning the geometrical height of the rails or bumpers of both vehicles, in vehicle-to-vehicle impact, was an important design feature to achieve compatibility. The significantly higher occupant responses experienced by the driver of the target vehicle struck by the Small SUV was attributed to a combination of mass and geometry effects.

Table 3.

Driver Dummy Responses in Target and Bullet Vehicles in the Oblique Offset Tests

			Occupant Responses for Driver in Target Vehicle				Occupant Responses for Driver in Bullet Vehicle			
Bullet Vehicle		IARV	Small SUV	Small Pickup	Mini-Van	Mid-Size Car	Small SUV	Small Pickup	Mini-Van	Mid-Size Car
Target Vehilce		Ref[9]	Mid-Size Car				Mid-Size Car			
Occupant Responses			*Driver 50th Percentile Male*				*Driver 50th Percentile Male*			
HIC		1000	931	377	693	318	234	286	221	549
Chest Acceleration	G	60	119	32	37	30	36	35	30	48
Chest Deflection	mm	63	56	16	21	15	25	22	16	30
L/Femur Compression	N	10,000	22347	2970	2713	2730	2608	6653	1738	3652
R/Femur Compression	N	10,000	8153	4784	4519	4561	5241	6027	2635	777

Table 4.

Passenger Dummy Responses in Target and Bullet Vehicles in the Oblique Offset Tests

			Occupant Responses for Passenger in Target Vehicle				Occupant Responses for Passenger in Bullet Vehicle			
		IARV	Small SUV	Small Pickup	Mini-Van	Mid-Size Car	Small SUV	Small Pickup	Mini-Van	Mid-Size Car
Target Vehicle		Ref[9]	Mid-Size Car				Mid-Size Car			
Occupant Responses			*Passenger 5th Percentile Female*				*Passenger 5th Percentile Female*			
HIC		1000	355	89	385	388	76	146	24	83
Chest Acceleration	G	60	36	24	35	28	31	35	21	30
Chest Deflection	mm	52	16	7	10	9	20	23	12	19
L/Femur Compression	N	6,800	3734	3750	3381	2824	2138	2790	1730	1910
R/Femur Compression	N	6,800	1511	1414	850	608	1601	1697	1399	1753

3.2 Dummy Response in Full Frontal and 50% Offset Tests.

For both full and 50% overlap frontal impacts, the measured occupant responses for the drivers and the passengers of tested vehicles are presented in Tables 5 and 6, respectively. Again, the occupant responses were normalized by the IARVs and presented graphically in Figures 9 – 12. As indicated in Table 2, the bullet vehicle, SUV II, was heavier than the SUV I, but its rail height was geometrically compatible with that of the target vehicle (see Figure 3). Figure 9 indicates that the driver dummy responses in the target vehicle struck by the SUV II were substantially less than those associated with the strike by the SUV I and were attributed to the better geometrical matching of the SUV II rails with those of the target vehicle. As observed from Table 2, the tests conducted with SUV II in both the full frontal and 50% offset induced velocity changes of 58 kph and 61 kph compared

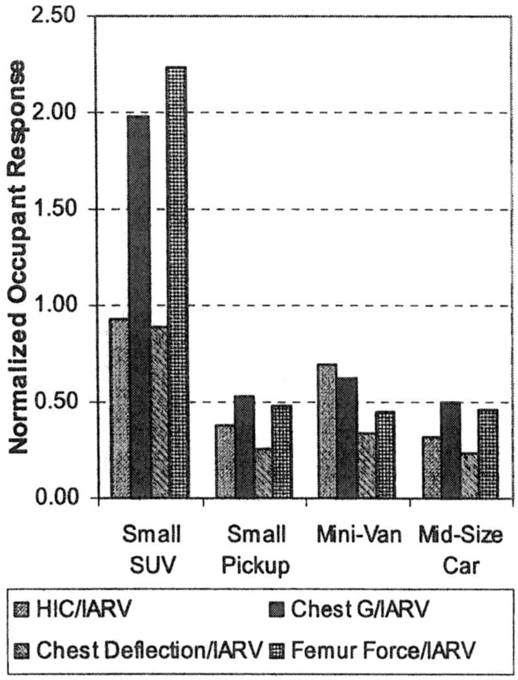

Figure 4. Occupant responses for driver of the target vehicle.

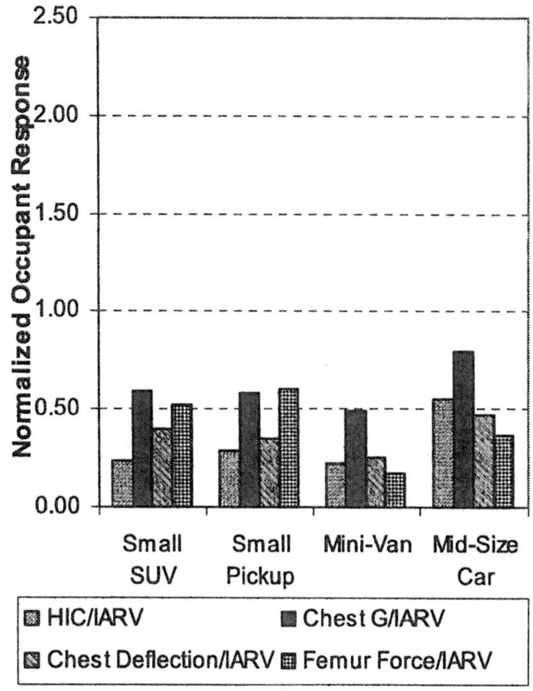

Figure 5. Occupant responses for driver of the bullet vehicle.

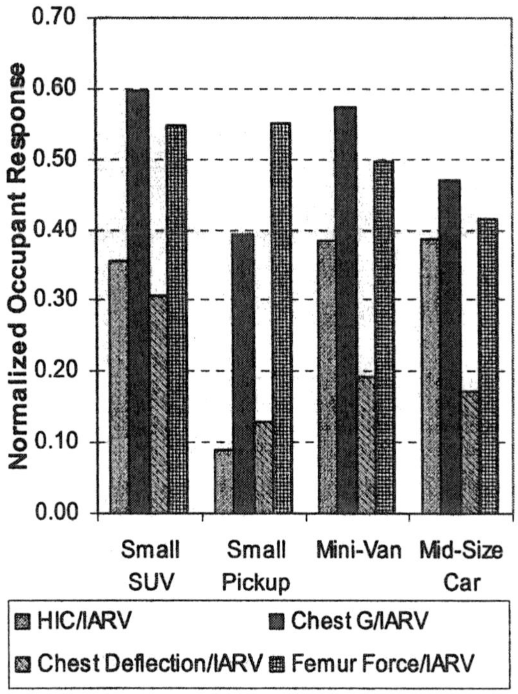

Figure 6. Occupant responses for passenger of the target vehicle.

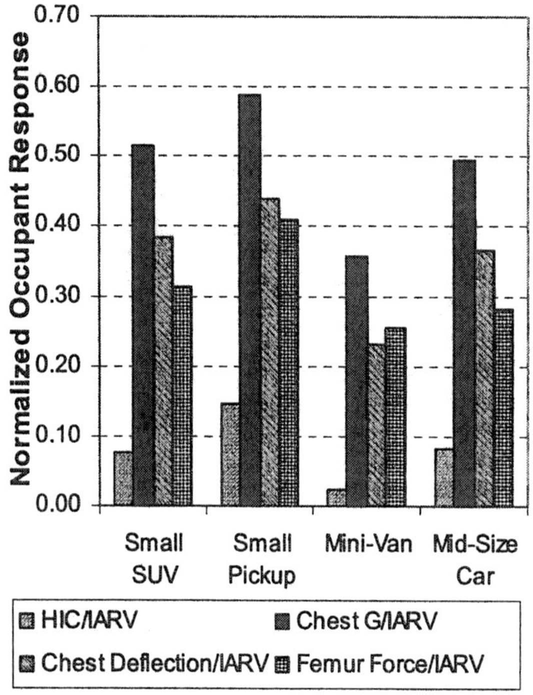

Figure 7. Occupant responses for passenger of the bullet vehicle.

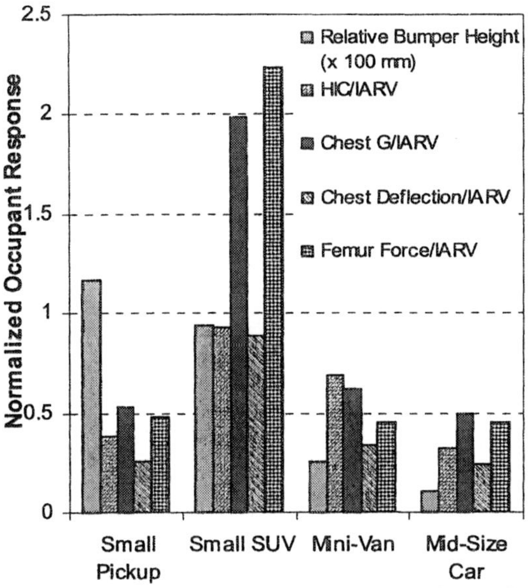

Figure 8. Occupant responses for driver of the target vehicle.

to those of 52 kph and 51.7 kph in the SUV I test, respectively. Due to a higher velocity change, it was expected that the occupant response outcome of the target vehicle occupants struck by SUV II to be higher than those of SUV I. However, the responses shown in Figures 9 and 11 for the driver and passenger in the SUV II tests were still generally better. Similar trends were also observed for the driver and passenger of the bullet vehicles in both the full frontal and 50% offset tests (see Figures 10 and 12).

Pre- and post-crash dimensional analyses on the driver center section for cabin intrusion profile were conducted to help assess the effect of mass, stiffness, and geometry on the structural performance of the target vehicle. It was expected that intrusions of the target vehicle struck by heavier and stiffer vehicles would be greater than if struck by lighter vehicles. The comparison shown in Figure 13 indicated that intrusions, from the cowl top to the floor panel at the driver center section caused by SUV II were significantly lower than those caused by SUV I. This observation held true for both full frontal and 50% offset impact tests. It was clear that the reduced structural interaction (due to the geometrical incompatibility) caused SUV I to override the target vehicle; this caused significant cowl intrusions at the windshield level. These findings served as further evidence that matching the front rail heights for the target and bullet vehicles was a very important design feature for compatibility.

Table 5.

Driver Dummy Responses in Target and Bullet Vehicle in the Full Frontal and 50 % Offset Tests

			Occupant Responses for Driver in Target Vehicle				Occupant Responses for Driver in Bullet Vehicle			
Impact Configuration			Full Frontal		50 % Offset		Full Frontal		50 % Offset	
Bullet Vehicle			SUV I	SUV II	SUV I	SUV II	SUV I	SUV II	SUV I	SUV II
Target Vehilce		IARV	Mid-Size Car				Mid-Size Car			
Occupant Responses		Ref[9]	*Driver 50th Percentile Male*				*Driver 50th Percentile Male*			
HIC		1000	531	757	386	443	327	166	148	174
Chest Acceleration	G	60	52	43	49	39	47	29	30	27
Chest Deflection	mm	63	45	32	43	34	34	28	19	25
L/Femur Compression	N	10,000	3576	2904	5716	4038	6783	1819	3376	3193
R/Femur Compression	N	10,000	5235	3309	6837	3193	3821	765	2793	3162

Table 6.

Passenger Dummy Responses in Target and Bullet Vehicle in the Full Frontal and 50 % Offset Tests

			Occupant Responses for Passenger in Target Vehicle				Occupant Responses for Passenger in Bullet Vehicle			
Impact Configuration			Full Frontal		50 % Offset		Full Frontal		50 % Offset	
Bullet Vehicle			SUV I	SUV II	SUV I	SUV II	SUV I	SUV II	SUV I	SUV II
Target Vehicle		IARV	Mid-Size Car				Mid-Size Car			
Occupant Responses		Ref[9]	*Passenger 5th Percentile Female*				*Passenger 5th Percentile Female*			
HIC		1000	349	649	125	536	307	201	72	157
Chest Acceleration	G	60	40	33	29	31	40	30	26	28
Chest Deflection	mm	52	24	25	17	28	21	30	13	30
L/Femur Compression	N	6,800	4977	5688	3109	6030	2366	1983	2126	2224
R/Femur Compression	N	6,800	4323	4710	1939	2486	2695	1147	2059	98

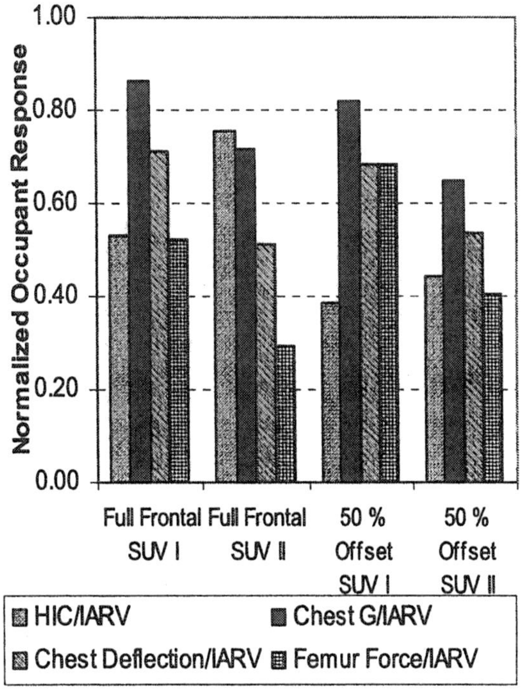

Figure 9. Occupant responses for driver of the target vehicle.

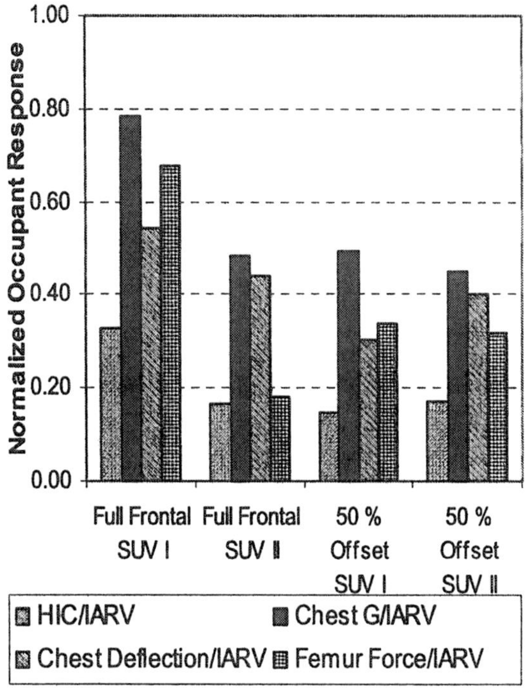

Figure 10. Occupant responses for driver of the bullet vehicle.

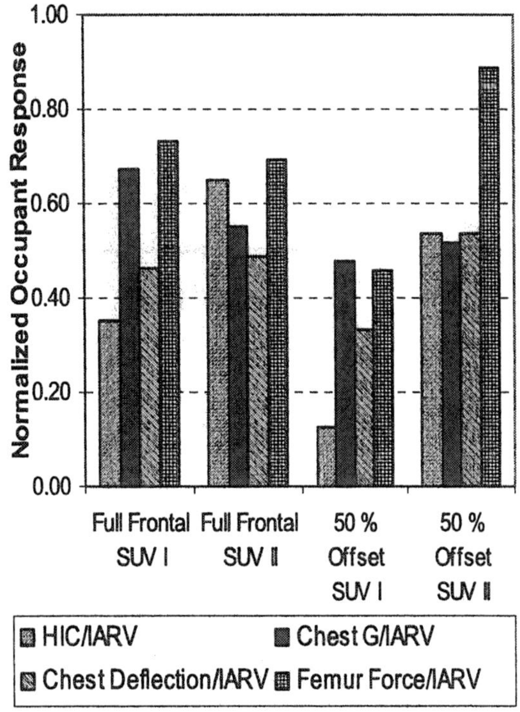

Figure 11. Occupant responses for passenger of the target vehicle.

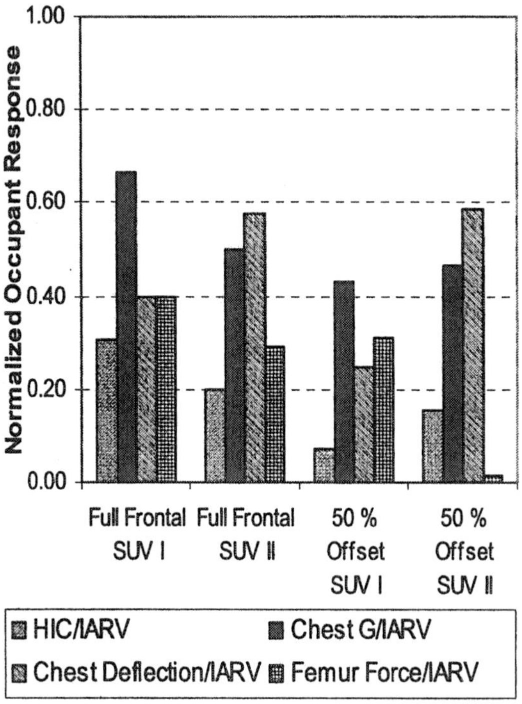

Figure 12. Occupant responses for passenger of the bullet vehicle.

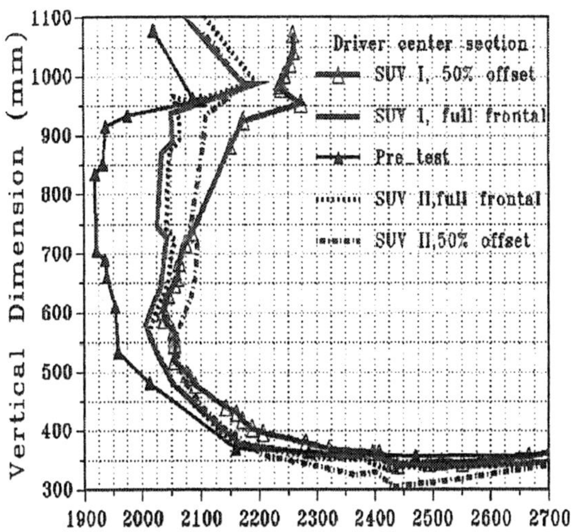

Figure 13. Comparison of dash intrusions in the target vehicle struck by SUV I and SUV II.

4. EFFECT OF COLLISION TYPE IN VEHICLE-TO-VEHICLE IMPACT

The same target vehicle, Mid-size car, was used in all the tests involved in this study. The same bullet vehicle, referenced by the "Small SUV" or "SUV I," was used to strike the target vehicle in (1) full frontal collinear, (2) 50% offset collinear, and (3) 30-degree oblique offset as shown in Figure 14. This selection provided the opportunity to investigate the effect of collision types on vehicle compatibility. Table 7 shows the mass ratio, initial impact speed, and velocity changes associated with the three different test configurations.

Table 7.

Mass Ratios, Impact Speeds and Velocity Changes in Each Test

Impact Configuration		Full Frontal Collinear	50 % Offset Collinear	30 dg. Oblique Offset
Bullet Vehicle Mass	kg	2132	2131	2084
Target Vehicle Mass	kg	1867	1859	1845
Mass Ratio		1.14	1.15	1.13
Bullet/Target Initial Speed	kph	95 / 0	94.9 / 0	56.4 / 57.2
T\Velocity Change	kph	51	51.8	57
B\Velocity Change	kph	47	44.6	58.2

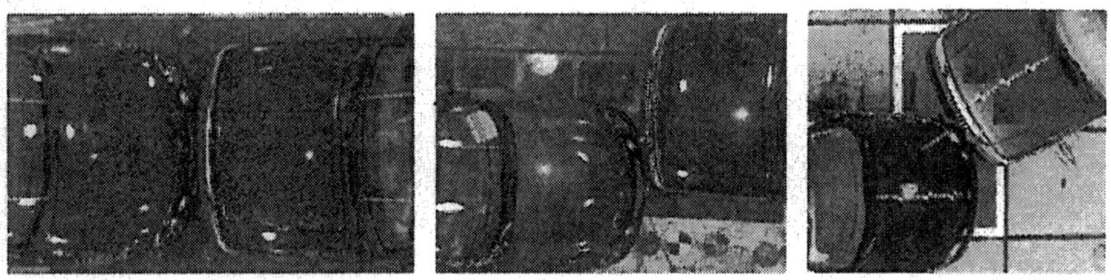

Figure 14. Vehicle-to-vehicle test configurations.

4.1 Occupant Response Comparison

The occupant response results for the occupants of the target vehicle in full frontal collinear, 50% offset collinear and 30-degree oblique offset crash tests are presented in Table 8 and Figures 15 and 17. Similar responses associated with occupants of the bullet vehicle are also presented in Table 9 and Figures 16 and 18. The performance of the occupants in the target vehicle varied significantly with the type of impact configuration. As shown in Figure 15, none of the IARVs were exceeded in the collinear impact tests while two out of four IARVs were exceeded and the other two were marginal in the oblique offset test. The overall deformation mode shown in Figure 19 was a clear explanation for these results. In the oblique test, only the driver-side front corner engaged with the bullet vehicle. The energy absorbing structure of the target vehicle completely missed structural interaction with the bullet front end. In addition, the geometrical incompatibility between both vehicles resulted in the bullet vehicle overriding the target vehicle towards the driver A-Pillar of the target vehicle. This override resulted in considerably higher intrusions in the occupant compartment. This intrusion caused significant rearward movements of the base of the A-pillar at the windshield and the center of the steering wheel. This resulted in significantly higher chest deflections, chest accelerations, and femur loads.

The occupant responses of the passenger of the target vehicle appeared to be higher in the full frontal compared to those in 50% offset and 30-degree oblique (see Figure 17). This was not surprising since the passenger of the target vehicle engaged less with the impact in the 50% offset and oblique tests. However, all the occupant responses for the occupants of the bullet vehicle were below the IARVs.

Table 8.

Driver Dummy Responses in Target and Bullet Vehicle in Various Impact Configurations

			Occupant Responses for Driver in Target Vehicle			Occupant Responses for Driver in Bullet Vehicle		
Collision Type		IARV	Full Frontal	50 % Offset	30-dg. Oblique	Full Frontal	50 % Offset	30-dg. Oblique
Occupant Responses		Ref[9]	*Driver 50th Percentile Male*			*Driver 50th Percentile Male*		
HIC		1000	531	386	931	327	148	234
Chest Acceleration	G	60	52	49	119	47	30	36
Chest Deflection	mm	63	45	43	56	34	19	25
L/Femur Compression	N	10,000	3576	5716	22347	6783	3376	2608
R/Femur Compression	N	10,000	5235	6837	8153	3821	2793	5241
R/Tibia Index, Driver		1	0.65	0.81	1.16	0.37	0.43	0.56
L/Tibia Index, Driver		1	0.55	1.00	0.90	0.26	0.36	0.68

Table 9.

Passenger Dummy Responses in Target and Bullet Vehicle in Various Impact Configurations

			Occupant Responses for Passenger in Target Vehicle			Occupant Responses for Passenger in Bullet Vehicle		
Collision Type		IARV	Full Frontal	50 % Offset	30-dg. Oblique	Full Frontal	50 % Offset	30-dg. Oblique
Occupant Responses		Ref[9]	*Passenger 5th Percentile Female*			*Passenger 5th Percentile Female*		
HIC		1000	349	125	355	307	72	76
Chest Acceleration	G	60	40	29	36	40	26	31
Chest Deflection	mm	52	24	17	16	21	13	20
L/Femur Compression	N	6,800	4977	3109	3734	2366	2126	2138
R/Femur Compression	N	6,800	4323	1939	1511	2695	2059	1601
R/Tibia Index, Driver		1	-	0.14	0.72	-	0.10	0.13
L/Tibia Index, Driver		1	-	0.15	0.12	-	0.16	0.13

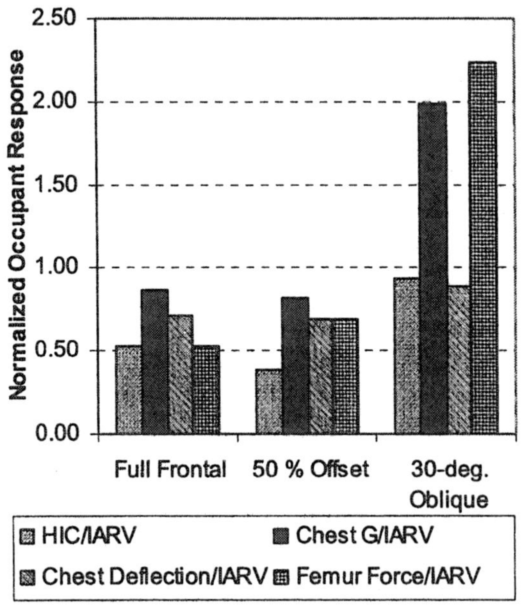

Figure 15. Occupant responses for driver of the target vehicle.

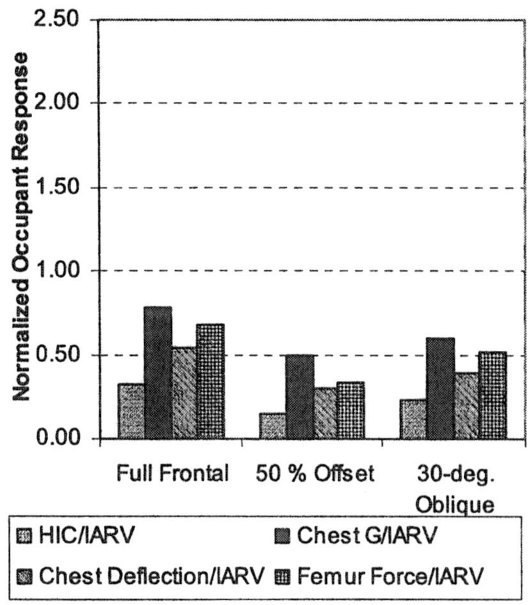

Figure 16. Occupant responses for driver of the bullet vehicle.

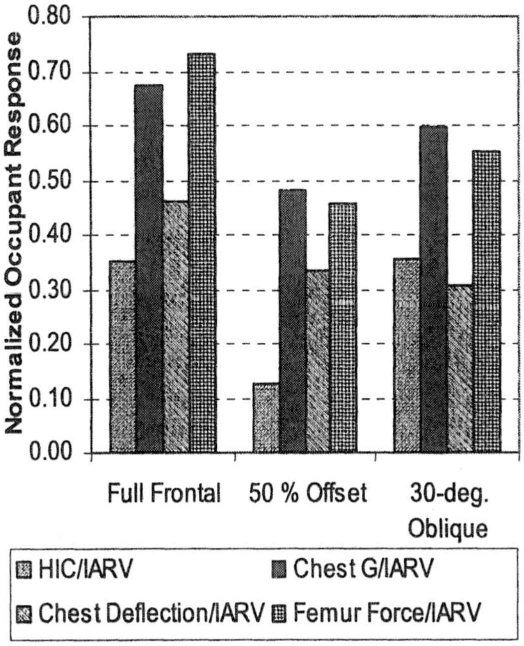

Figure 17. Occupant responses for passenger of the target vehicle

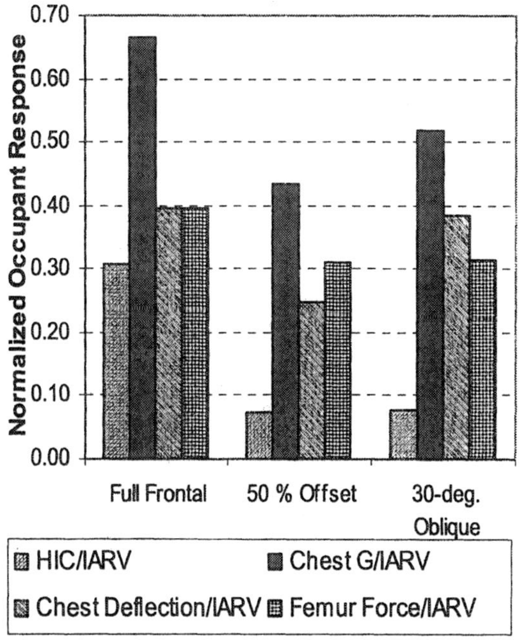

Figure 18. Occupant responses for passenger of the bullet vehicle

Full frontal collinear **50 % offset collinear** **30-degree oblique offset**

Figure 19. The overall deformations in the three types of impact tests.

5. CONCLUSIONS

- In the 30-degree oblique offset tests, it was difficult to extract the effect of vehicle design parameters such as mass, stiffness, and geometry, on vehicle compatibility.
- It was also concluded that this type of test configuration resulted in unacceptable test-to-test repeatability and reproducibility, and vehicle trajectories were sensitive to initial contact points.
- In the 30-degree oblique tests, the occupant responses in the target vehicle correlated well with both the vehicle mass and aggressivity metric of the bullet vehicle.
- An alternative compatibility test procedure consisting full frontal and 50% offset collinear impact was deemed more practical in terms of intrusion limits, occupant responses, and repeatability.
- Results indicated that geometrical compatibility was a necessary design feature in order to achieve compatibility in the fleet. Moreover, the effect of geometry was shown to possibly reduce the effects of mass and stiffness.

6. ACKNOWLEDGEMENT

The authors wish to acknowledge Mr. Tony Laituri, a Technical Specialist at the Safety R & D Department of Ford Motor Company, for editorial comments.

7. REFERENCES

[1] Hollowell, W.T. and Gabler, H.C. "NHTSA's Vehicle Aggressivity and Compatibility Research Program", 15th International Conference on the Enhanced Safety of Vehicles, Paper No. 96-S4-O-01, Melbourne, Australia, 1990.

[2] Gabler, H.C. and Hollowell, W.T., " The Aggressivity of Light Trucks and Vans in Traffic Crashes", SAE Paper No. 980908, Detroit, 1998.

[3] Summers, S., Prasad, A. and Hollowell, W.T., " NHTSA's Vehicle Compatibility Research Program", SAE Paper No. 1999-01-0071, Detroit, 1999.

[4] Hobbs, C.A., Williams, D.A. and Coleman, D.J., "Compatibility of Cars in Frontal and Side Impact", 15th International Conference on the Enhanced Safety of Vehicles, Paper No. 96-S4-O-05, Melbourne, 1996.

[5] Evans, L. and Frick, M.C., "Mass Ratio and Relative Driver Fatality Risk in Two-Vehicle Crashes", Accident Analysis and Prevention, Volume 25, pp 213-214, 1993.

[6] Evans, L., "Driver Injury and Fatality Risk in Two-Car Crashes versus Mass Ratio Inferred Using Newtonian Mechanics", Accident Analysis and Prevention, Volume 26, No.5, pp 609-616, 1994.

[7] Mizuno, K. and Kajzer, J., " The Compatibility of Mini Cars in Traffic Accidents", 16th International Technical Conference of the Enhanced Safety of Vehicles, Paper No. 98-S3-O-08, Windsor, Canada, 1998.

[8] Stucki, S.L., Hollowell, W.T., and Fessahaie, O., " Determination of Frontal Offset Test Conditions Based on Crash Data", 16th International Technical Conference on the Enhanced Safety of Vehicles, Paper No. 98-S1-O-02, Windsor, Canada, 1998.

[9] Mertz, H., "Anthropomorphic Test Devices", in Accidental Injury: Biomechanics and Prevention, by A. Nahum and J. Melvin, eds., Spring-Verlag, New York, 1993.

[10] Barbat, Saeed., Li, Xiaowei and Prasad, Priya., " A Comparative Analysis of Vehicle-to-Vehicle and Vehicle to Rigid fixed Barrier Frontal Impacts", 17th International Conference on the Enhanced Safety of Vehicles, The Netherlands, 2001.

2004-01-1163

Aggressivity-Reducing Structure for Large Vehicles in Frontal Car-to-Car Crash

Shinji Fujii, Takayuki Sunakawa, Akiko Abe, Masanobu Fukushima, Kenji Kawaguchi, Shigeru Ogawa
Mazda Motor Corporation

ABSTRACT

This paper clarifies aggressivity reduction approach for MPV, Multi-Purpose Vehicles, derived from large passenger vehicles toward small passenger vehicles. The effects of aggressivity-reducing approach were measured through full-frontal rigid barrier crash simulations with TRL aluminum honeycomb by Finite Element Method. The front-end structures of large vehicles studied in this paper based on this aggressivity reduction approach show good front-end homogeneity and low average height of force. The structures were also found to effectively reduce aggressivity toward small vehicles by car-to-car simulation. However, there are some cases where the effect was influenced by overlap ratios. From this result, overlap ratio is considered to be one of the important factors to improve compatibility performance.

INTRODUCTION

Crash safety performance has drastically improved for the last several years as one can see from the regulation and NCAP crash test results worldwide (1). The fact is supported by the recent Japanese field data where fatality is on the decline while accident number itself is on the increase (2). However, there still exists an issue of compatibility where fatality rate of smaller vehicles exceeds that of larger vehicles when they are collided with each other. Therefore, there is much room for improvement in the area of "vehicle compatibility" (3)(4).

At present, compatibility evaluation and test procedures are studied by various institutions and auto-manufacturers (5)(6)(7). In reference to front-to-front compatibility evaluation, aggressivity of larger passenger vehicles is to be evaluated by full-frontal rigid barrier tests to see front-end homogeneity and average height of force (AHOF), and offset tests with PDB to see front-end homogeneity. Whether or not to use TRL aluminum honeycomb in full-frontal rigid barrier test is also a matter of discussion. As an evaluation method for small vehicle self-protection performance, high-speed ODB test has been proposed to see cabin strength.

Researchers have discussed in various papers the mechanisms and countermeasure approach concerning aggressivity reduction of large vehicles (hereinafter called "large vehicle") and self-protection improvement of small passenger vehicles (hereinafter called "small vehicles"). However, few papers have discussed specific and realistic structures. This paper describes specific aggressivity-reducing structures of large vehicles and the effect measured by full-frontal car-to-rigid barrier simulation with TRL aluminum honeycomb and car-to-car simulation in various overlap ratios by Finite Element Method.

AGGRESSIVITY OCCURRENCE MECHANISM AND COUNTERMEASURE STRUCTURE

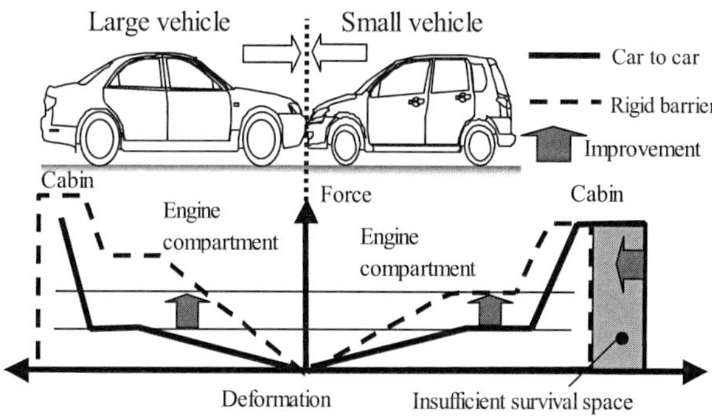

Figure 1 Aggressivity Occurrence Mechanism of Large Vehicles

In general, when a small vehicle is collided with a large vehicle, passenger compartment of the small vehicle deforms more than that of the large vehicle, leaving less occupant survival space. The reason for this lies in different vehicle strength when deformation occurs between rigid barrier impact and car-to-car impact as shown in Figure 1. In a car-to-car crash, strength and energy absorption amounts in the engine compartments of both vehicles are reduced influenced by geometry difference. Because of this, the crash energy is left without being absorbed in the engine compartments of both vehicles. The residual crash energy is used in the cabins of both vehicles. However, much of the residual crash energy is absorbed by the less stiff cabin of the small vehicle, resulting in less survival space for small vehicle occupants. Influence of geometry here refers to the phenomena such as override and fork effects caused by front side member height difference, etc. Such phenomena spoil energy absorption ability of structural parts in the engine compartment that are originally designed to absorb crash energy efficiently.

The concept to improve compatibility is to eliminate the influence of geometry to increase the crash energy absorption amount in the engine compartments of both small and large vehicles. One of the measures to realize this concept is to make the front-end of both vehicles similar to the rigid barrier, i.e. to homogenize the stiffness of the front-end of both vehicles. With such a structure, we consider vehicle strength and energy absorption in engine compartments of both small and large vehicles increase when they are collided with each other, resulting in cabin deformation reduction of the small vehicle as shown in Figure 1. This means that, for the large vehicle, the aggressivity against small vehicle is reduced, and for the small vehicle, the self-protection of its own is increased. In this paper, focus is on the aggressivity reduction of large vehicle, and four front-end countermeasure structures of large vehicles are designed and studied by CAE. The designed structures are as follows.

The base vehicle in Figure 2 is an MPV type prototype vehicle weighing 2,000kg with sub-frame type suspension.

Case 1 has a forward-extended sub-frame as shown in Figure 3. Crash force is dispersed to this sub-frame and side members.

Case 2 has forward-extended sub-frame and enlarged front bumper beam in the longitudinal direction as depicted in Figure 4. By preventing front bumper from breaking, stiffness of the front end is considered to remain homogeneous.

Case 3, as shown in Figure 5, has a forward-extended sub-frame and reinforced radiator frame. The reinforced radiator frame is considered to work as a wide rigid area that absorbs energy homogeneously.

Case 4 in Figure 6 has a forward-extended sub-frame and laterally extended bumper beam that is connected with forward-extended shotguns on both sides. With this structure, stiffness area is extended to the outboard area of side members.

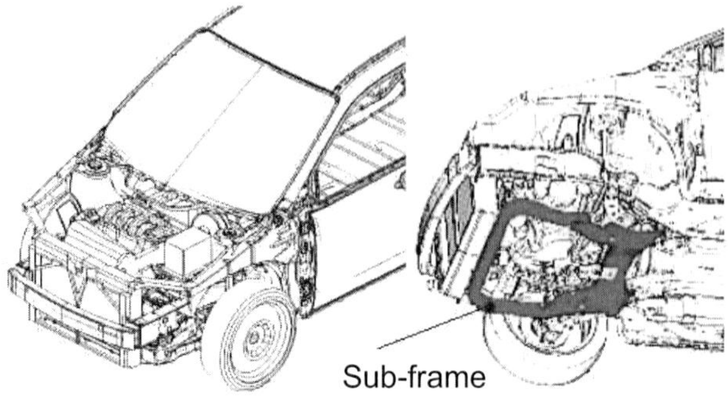

Figure 2 Large Vehicle Structure (Base Model)

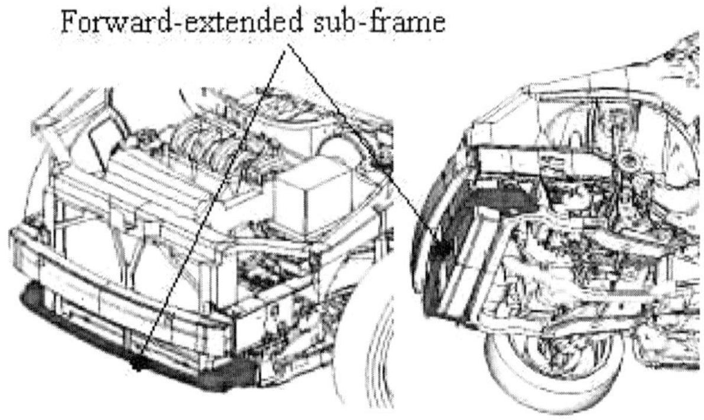

Figure 3 Large Vehicle Structure (Case 1)

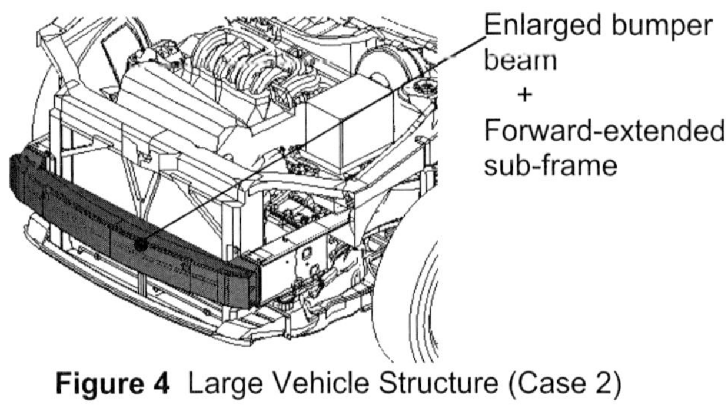

Figure 4 Large Vehicle Structure (Case 2)

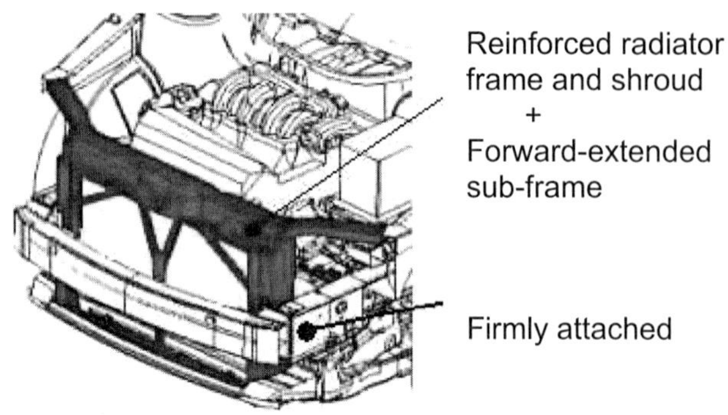

Figure 5 Large Vehicle Structure (Case 3)

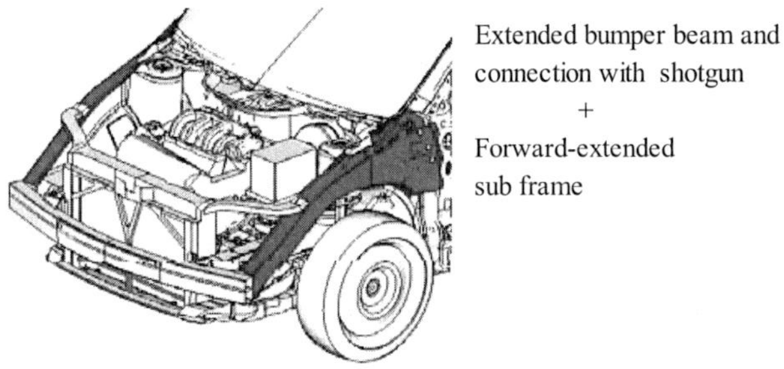

Figure 6 Large Vehicle Structure (Case 4)

AGGRESSIVITY REDUCTION EFFECT

To verify if our designed structures of large vehicles reduce aggressivity, CAE simulations were conducted as follows.

1. Confirmation of homogeneity

   To verify if the designed front-end structures of the large vehicles have homogenous front-end, car-to-rigid barrier full frontal crash simulations were conducted.

2. Confirmation of homogeneity effect on force and energy absorption

   To see if homogeneous front-end actually increases force and energy absorption in engine compartment, car-to-rigid barrier full frontal crash simulations were conducted.

3. Confirmation of aggressivity reduction effect

   To see if increase in force and energy absorption in engine compartment of large vehicles leads to aggressivity reduction toward small vehicles, car-to-car crash simulations were conducted

The results are described by crash mode.

CAR-TO-BARRIER CRASH SIMULATION

Homogeneity of the four countermeasure structures of large vehicles and the effect of homogeneity were studied by full-frontal rigid barrier crash simulation with TRL aluminum honeycomb discussed within IHRA compatibility working group. The study was done by comparison of the base vehicle results with modified vehicles results (cases 1 through 4).

Parameters used for the homogeneity evaluation are homogeneity evaluation indices calculated from loadcell wall data. The homogeneity evaluation indices include coefficient of variance (hereinafter "CV") and Homogeneity Assessment Measure ("HAM"). We also evaluated homogeneity effect on Average Height of Force ("AHOF"). The outlines of these evaluation indices are described below:

- CV: Standard deviation of loads divided by mean, showing stiffness homogeneity. Outer rows or columns with <5kN peak force are ignored.
- HAM: Load variance by all (Vc), rows (Vr), and columns (Vcl), showing force homogeneity. Load measured during the entire crash event is used for the calculation. Each value is weighed by coefficient. HAM= a x Vc + b x Vr + c x Vcl (8)
- AHOF: The height of the center of force of a vehicle weighed by the loadcell load and divided by mean

Parameters used for the evaluation of homogeneity effect on force and energy absorption include engine compartment stiffness and absorbed energy amount.

<u>CAE Conditions</u>

Full-frontal car-to-rigid barrier crash simulations with loadcell wall and TRL aluminum honeycomb were conducted by CAE at 56km/h (see Figure 7). Those simulations were done on the base model and large vehicle models with modifications (cases 1 through 4). Full-car FEM model was used for the simulation.

In order to eliminate the weight effect, weight of all models, base model and models for cases 1 through 4, was equalized. Software adopted was RADIOSS. FEM model specifications for large vehicles and honeycomb are summarized in Table 1. A physical test using the base vehicle was conducted to prove enough accuracy of the simulation. Several honeycomb component tests were also conducted for simulation accuracy.

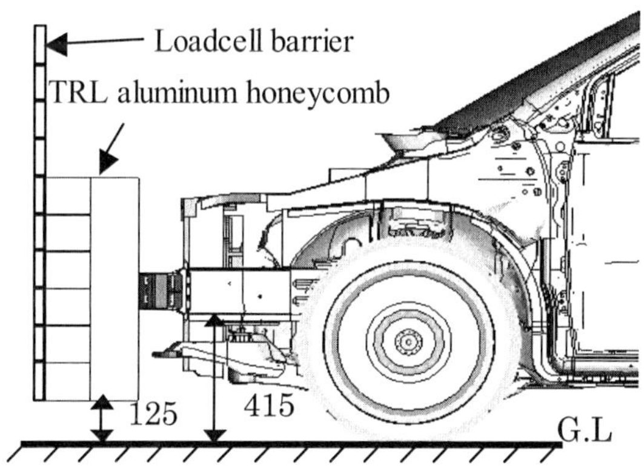

Figure 7 Full-Frontal Car-to-Rigid Barrier Crash Simulation Mode

Table 1 FEM Specification

	Large vehicle					Honey-comb
	base	case1	case2	case3	case4	
Nodes	418,506	415,856	420,149	418,506	422,483	50,796
Length	4,810mm					
Width	1,820mm					
Mass	2,000kg					

CAE Results and Discussion

The values of homogeneity evaluation parameters, i.e., CV and HAM described below proved that the designed four countermeasure structures have more homogeneous front-end compared to the base model.

As for CV, as illustrated in Figure 8, every modified model showed homogeneity compared to the base. Figures 9 through 13 show load distribution when vehicle deformed by 270mm where difference in homogeneity by models appeared, judging from the CV result. Figure 9 is for base model, and figures 10 through 13 are for modified models (cases 1 through 4). Compared to the load distribution of the base model, modified models have wider distribution with homogeneous stiffness.

Figure 14 illustrates the assessment area of HAM. Note that as the bottom row of the loadcell wall we use for this study is 125mm from the ground, it is impossible to calculate the value in the same manner as is discussed within the IHRA working group. HAM calculated when the overall and vertical and lateral coefficient is set as 1 shows improved homogeneity for those modified cases compared to the base model. However, there is no conspicuous improvement for case 3. See Figure 15.

In addition, the effect of homogeneous structure on AHOF was investigated. The results are plotted in Figure 16. In comparison with the AHOF of the base model, AHOF of modified models were lowered. This indicates that modification with the purpose to make the front end homogeneous may contribute to lowering AHOF.

Load-vehicle displacement curve and energy absorption amounts in the engine compartment of the large vehicles obtained by these car-to-rigid barrier simulations show that homogeneous front end increases force and energy absorption amount in the engine compartment.

Figure 17 compares the load-displacement curve among each model. The results show that modified models have stiffer engine compartment and smaller vehicle displacement compared to the base model.

Energy absorption amount comparison in Figure 18 tells that modified models absorbed more energy in the engine compartment compared to the base model.

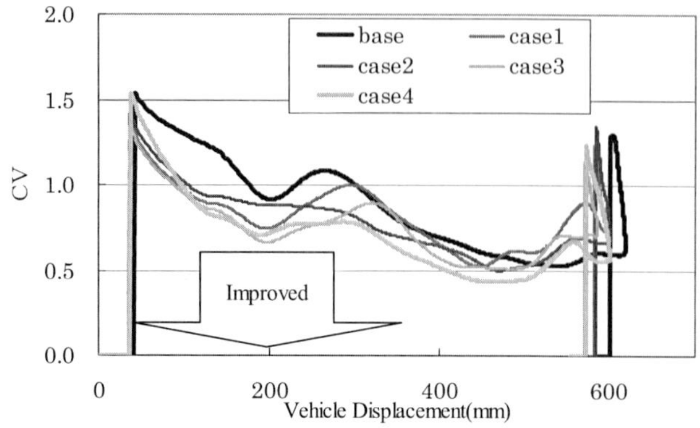

Figure 8 CV Comparison

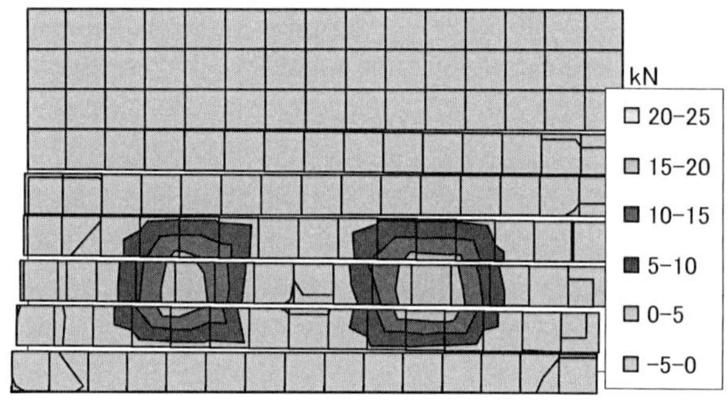

Figure 9 Load Distribution at 270mm Displacement (Base Model)

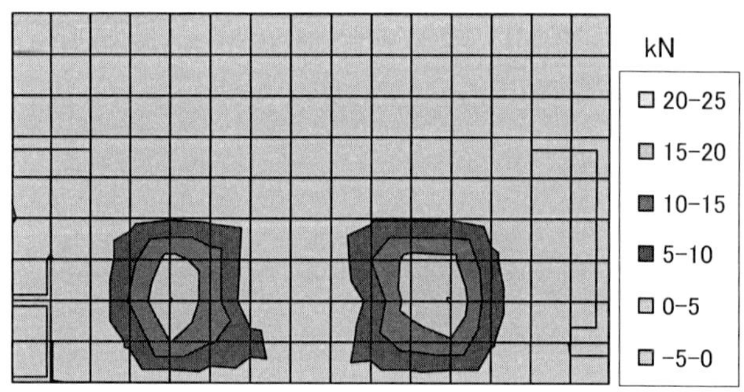

Figure 10 Load Distribution at 270mm Displacement (Case 1)

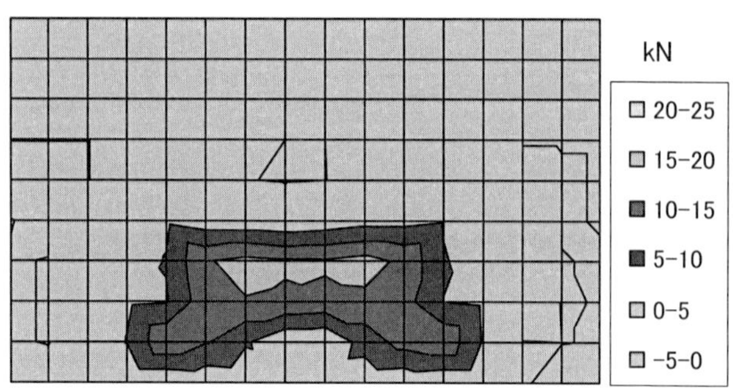

Figure 11 Load Distribution at 270mm Displacement (Case 2)

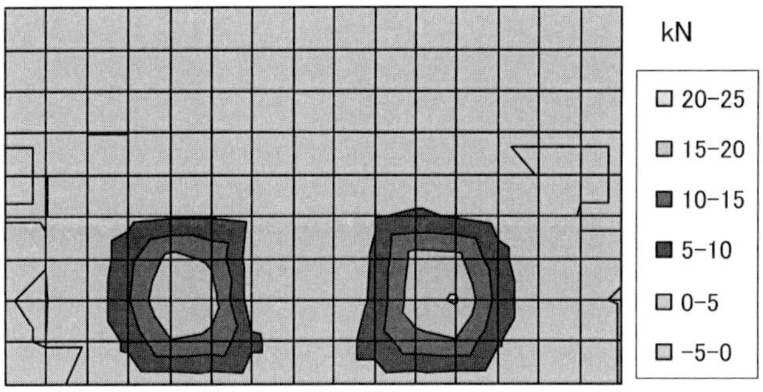

Figure 12 Load Distribution at 270mm Displacement (Case 3)

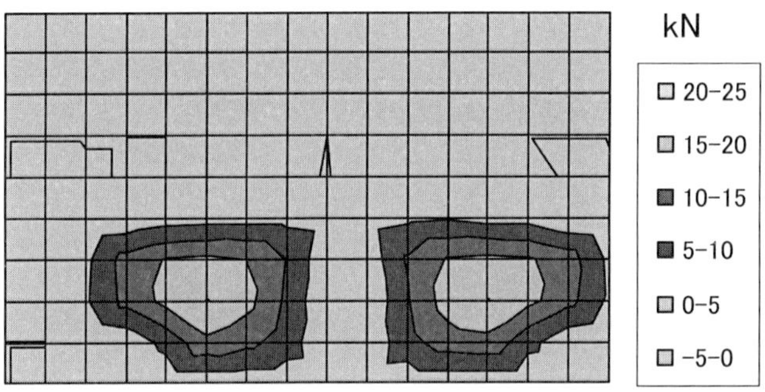

Figure 13 Load Distribution at 270mm Displacement (Case 4)

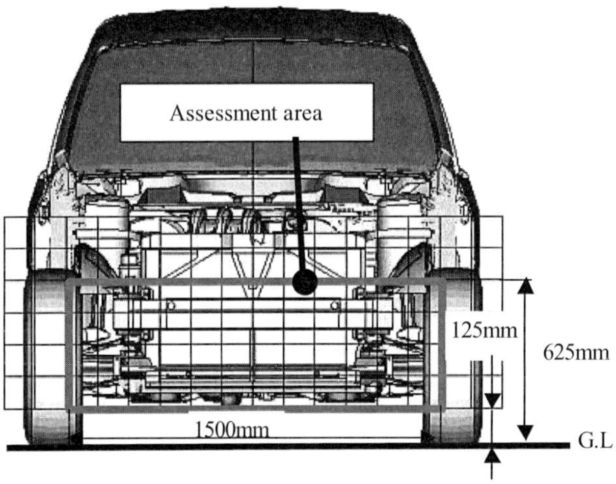

Figure 14 HAM Assessment Area

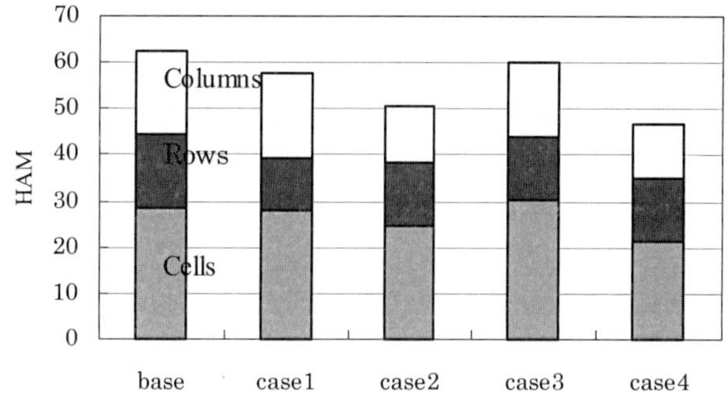

Figure 15 HAM Comparison

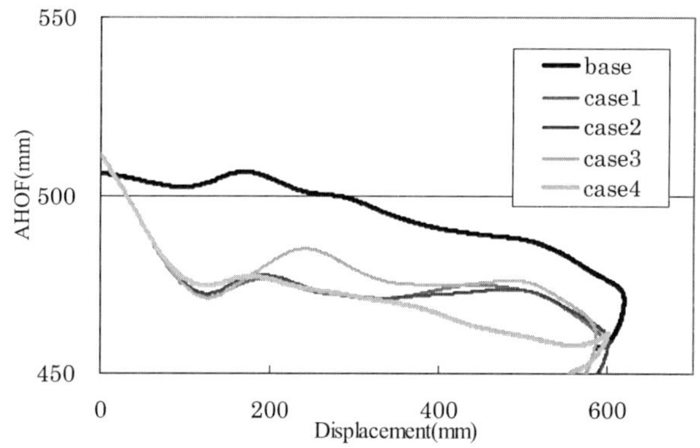

Figure 16 AHOF Comparison

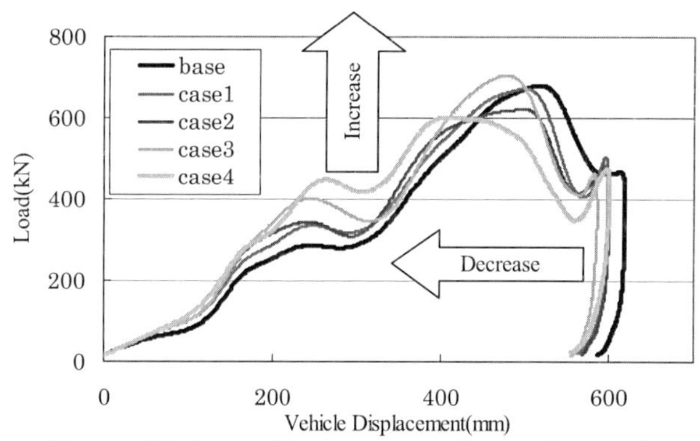

Figure 17 Load-Displacement Curve Comparison

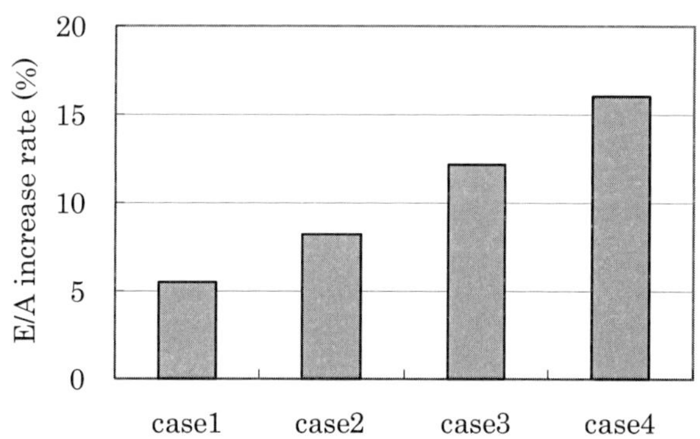

Figure 18 Increase Rate of Energy Absorbed in Engine Compartment

CAR-TO-CAR CRASH SIMULATION

Car-to-rigid barrier crash simulation proved that the designed structures are homogeneous and homogenous structure increases force and energy absorption in the engine compartment of the large vehicles. According to our theory, this suggests a strong possibility of aggressivity reduction against small vehicles. Therefore, to see if the homogeneous front-end of a large vehicle actually reduces aggressivity toward a small vehicle, car-to-car crash simulation was conducted by CAE. Software adopted in this simulation was RADIOSS. Full car FEM models were used for both large and small vehicles. The full car FEM models for large vehicles were the same ones as were used in the car-to-rigid barrier crash simulation mentioned earlier.

CAE Simulation Conditions

Figure 19 shows overlap ratio of frontal crash accidents in Japan from 1969 to 1992 reported by Japan Automobile Research Institute. 30% and 50% overlap ratios on the right side and 100% overlap ratio are the highest in both total accident number and fatal/serious injury accident number. Figure 20 indicates barrier equivalent speed (BES) in passenger car-to-car injured and fatal frontal crash accidents in Japan in 1995. 90% of those accidents occurred at less than 50km/h of BES. Based on these accident data, we decided to conduct CAE simulation at 50km/h under 30%, 50%, and 100% overlap ratio of small vehicles. 75% overlap was also studied to see the trend.

Small vehicle model is a prototype vehicle which weighs 1305kg with a suspension cross member commonly used for small vehicles. The structure is illustrated in Figure 21 and model specification is summarized in Table 2. As depicted in Figure 22, the ground height of the side member of the large vehicle locates at 42mm higher than that of the small vehicle. CAE model accuracy was proved by a physical 50% overlap car-to-car test using the large vehicle base model vehicle and the small vehicle. Car-to-car simulation matrix is shown in Figure 23.

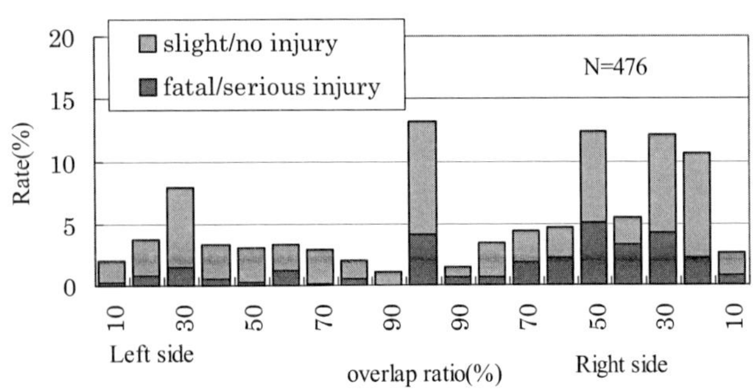

Figure 19 Overlap Ratio of Frontal Crash Accidents in Japan (1969-1992)

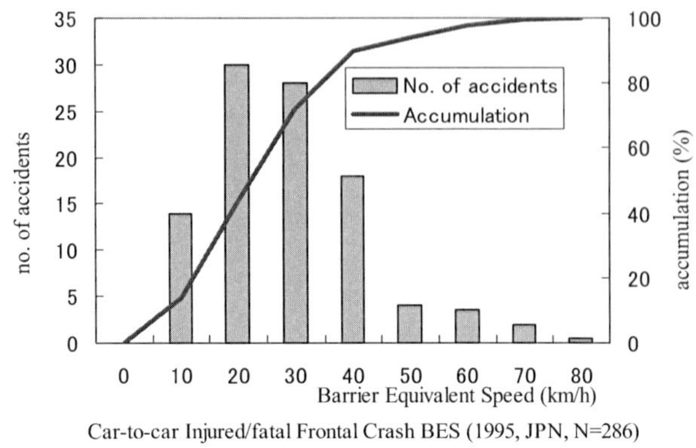

Figure 20 Car-to-Car Injured/Fatal Frontal Crash Barrier Equivalent Speed (1995)

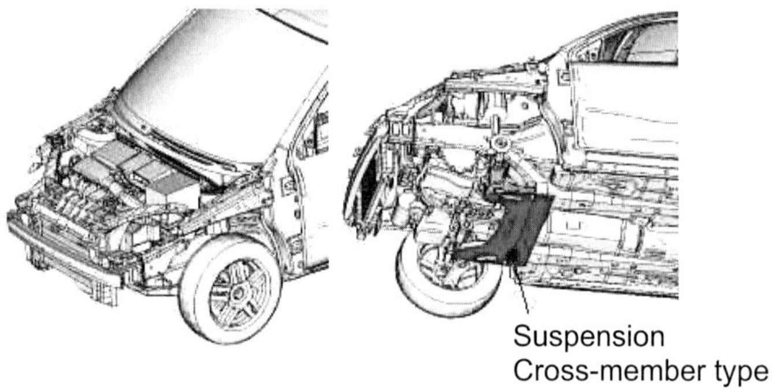

Figure 21 Small Vehicle Structure

Table 2 Small Vehicle FEM Specification

	Small vehicle
Nodes	265,000
Length	3,925mm
Width	1,680mm
Mass	1,305kg

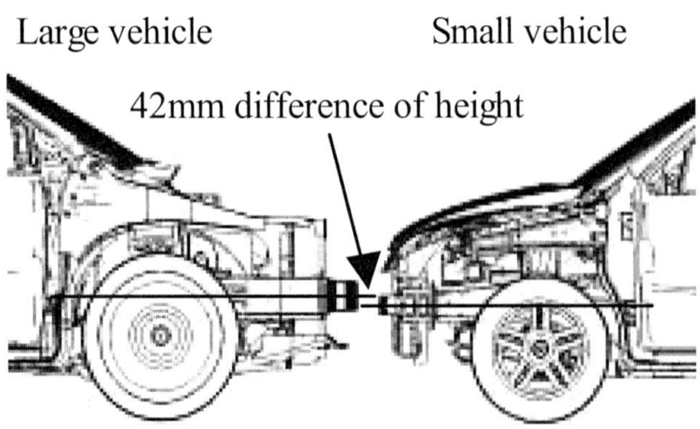

Figure 22 Front Side Member Height Difference

The type of lap		base	case1	case2	case3	case4
Large Small				+ case1	+ case1	+ case1
30%		×	No study	No study	No study	×
50%		×	×	×	×	×
75%		×	×	×	×	×
100%		×	×	×	×	No study

Figure 23 Car-to-Car Simulation Matrix

CAE Results and Discussion

The aggressivity reduction effect was evaluated by driver-side toeboard and A-pillar intrusion of small vehicles. These intrusion amounts are considered to be the parameters to measure survival space.

Figures 24 and 25 illustrate small vehicles' toeboard and A-pillar intrusion reduction ratios against the intrusions of the small vehicle collided with the base large vehicle. The intrusion amount is reduced for almost all cases regardless of overlap ratio, meaning that aggressivity reduction effect was proved. However, as shown in Figure 25, A-pillar intrusion of case 1 under 75% overlap and case 2 under 75% overlap didn't show any effect.

Displacement at the maximum intrusion of the small vehicle collided with the large vehicle of case 1 in 75% overlap ratio is shown in Figure 26. The deformation of the small vehicle is illustrated on the left side of this Figure. The damage on the hinge pillar and the door of the small vehicle is enormous and the A-pillar inclined greatly. Force from the forward-extended sub-frame of the large vehicle (case 1) was excessively transmitted to the hinge pillar of the opponent small vehicle via tire, increasing A-pillar intrusion. The same can be said to the large vehicle of case 2 that has the same structure. On the contrary, for case 4, A-pillar intrusion of the small vehicle is decreased as shown in Figure 27. This is because the upper structural parts of the large vehicle of case 4 dispersed the force from the extended sub-frame. The same is true for case 3.

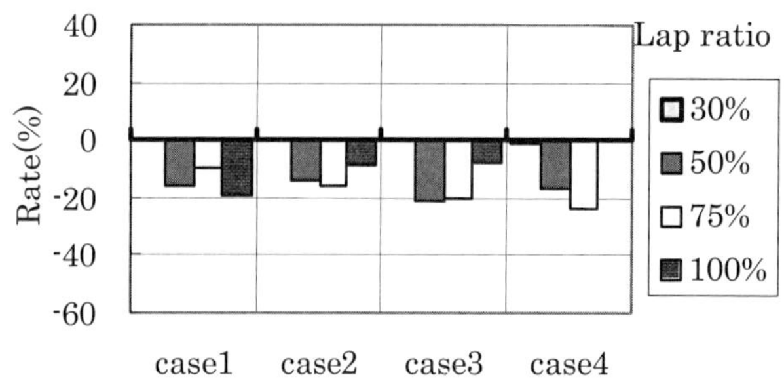

Figure 24 Toeboard Intrusion Comparison on Driver Side of Small Vehicle

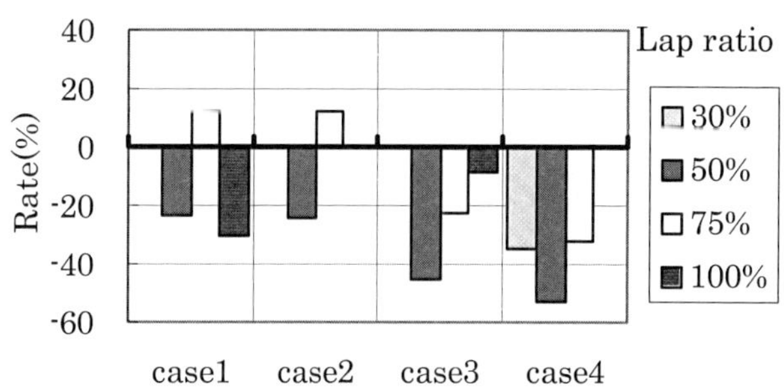

Figure 25 A-pillar Intrusion Comparison on Driver Side of Small Vehicle

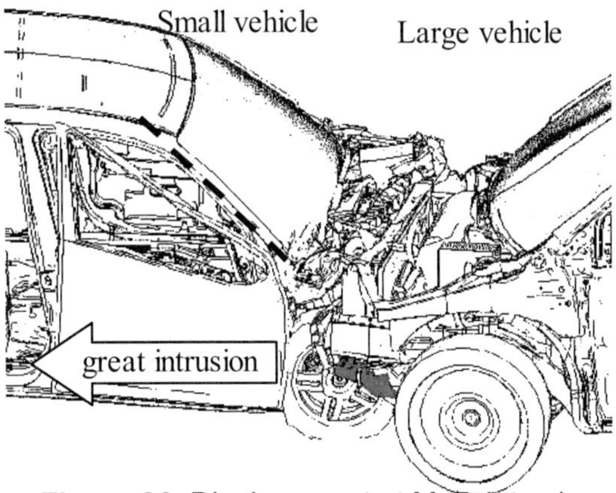

Figure 26 Displacement at Max. Intrusion of Small Vehicle, Case 1, 75% Overlap

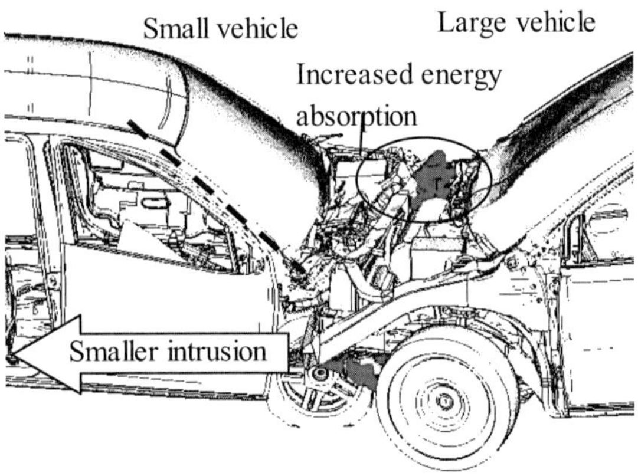

Figure 27 Displacement at Max. Intrusion of Small Vehicle, Case 4, 75% Overlap

Figures 28 and 29 depict large vehicles' toeboard and A-pillar intrusion reduction ratio against the base model. Except for 75% overlap, the intrusion amount is decreased compared to the base, meaning that aggressivity was reduced without sacrificing self-protection of large vehicles. As for 75% overlap, toeboard intrusion of all large vehicles increased. However, when the absolute toeboard intrusion amount of these large vehicles is compared to that of the small vehicle, it is only 16% against small vehicle toeboard intrusion. Accordingly, influence on self-protection of large vehicles is rather small compared to the small vehicles.

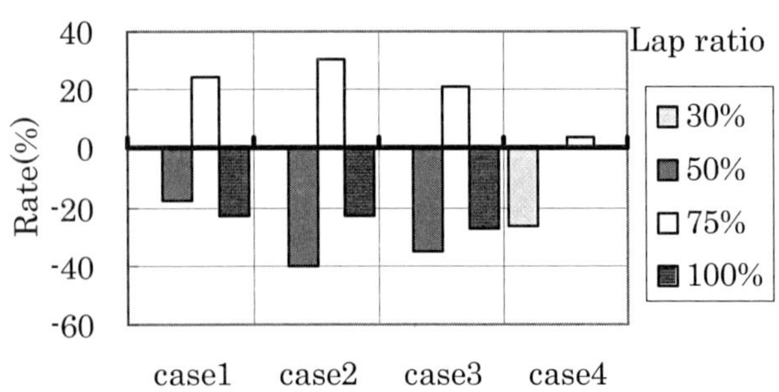

Figure 28 Toeboard Intrusion Comparison on Driver Side of Large Vehicle

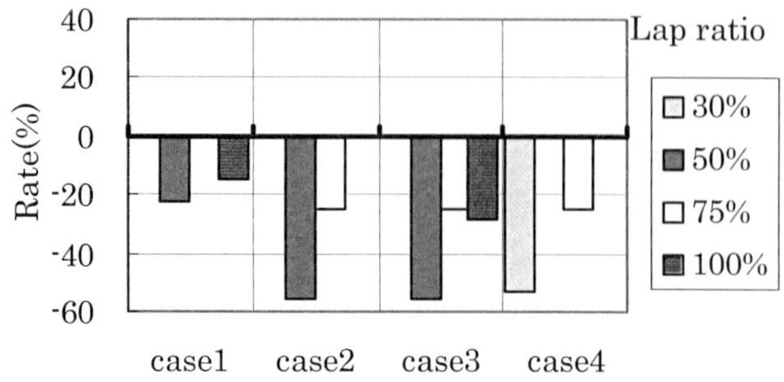

Figure 29 A-Pillar Intrusion Comparison on Driver Side of Large Vehicle

Increase in energy absorption amount in the engine compartment is considered to reduce toeboard and A-pillar intrusion of small vehicles. To prove this, energy absorption in engine compartments of small vehicles was measured and is shown in Figure 30. Data in Figure 30 are energy absorption increase rate against the energy absorbed by the small vehicle collided with the base large vehicle. Energy absorption in engine compartment by every case was confirmed increased. This result supports our theory that increase in energy absorption amount by engine compartment reduces toeboard and A-pillar intrusion, i.e., cabin deformation.

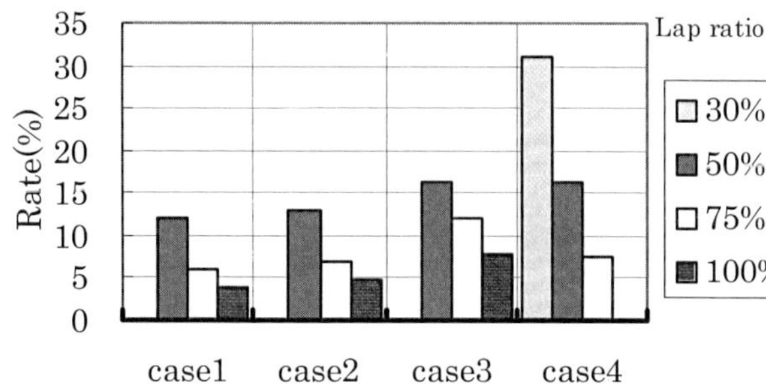

Figure 30 Energy Absorption in Engine Compartment of Small Vehicle

Aggressivity-reduction countermeasure effect was further studied within the same overlap ratio. From this study, we intended to find the effective countermeasure for any overlap ratio. The data used for this study were 50% and 75% overlap ratio results for cases 1 through 4. 50% and 75% overlap ratios were chosen for this study because simulations were done on all cases (1 through 4) as shown in Figure 23. The study result is summarized in Table 3. It is clear that countermeasures implemented for cases 3 and 4 are effective for any overlap ratio. As for case 4, modifications on the outer side of side members seem to be effective in any overlap ratio (50% and 75%).

Table 3 Aggressivity Reduction Effects by Overlap Rate

	Cabin deformation				Energy absorption of engine room	
	Toeboard		A-pillar			
Lap ratio	50%	75%	50%	75%	50%	75%
case1	Good	Good	Good	Poor	Good	Good
case2	Good	Good	Good	Poor	Good	Good
case3	Excellent	Excellent	Excellent	Good	Excellent	Excellent
case4	Good	Excellent	Excellent	Excellent	Excellent	Good

Poor: in case the value is worsened compared to base
Good: in case the value is improved compared to base
Excellent: in case the value is above average among those excluded the values of "poor"

CONCLUSION

1. Following front-end homogeneous structures are found to be effective for aggressivity reduction. CAE full-frontal car-to-rigid barrier impact and car-to-car impact simulations verified the effect.

- Forward-extended sub frame
- Enlargement of front bumper beam
- Stiffness increase in radiator frame and shroud
- Laterally-extended front bumper beam and connection of the extended beam with shotgun

2. The following effects of homogeneous front-end structures were confirmed by the CAE full-frontal car-to-rigid barrier impact simulations:

- Increase in engine compartment force and energy absorption amount
- Improvement of CV and HAM that show homogeneity
- Lowered AHOF

3. From car-to-car impact simulation results, it was found that homogeneous front-end structure of a large vehicle increases energy absorption amount in the engine compartment of a small vehicle thus reducing small vehicle cabin deformation. At the same time, it has a tendency to increase energy absorption amount in the engine compartment of the large vehicle itself, thus reducing the cabin deformation of its own.

4. Structural parts added on the outboard area of the side members reduces aggressivity in any overlap ratio such as, 1) slight overlap ratio where large vehicle's side members do not interfere with small vehicle's side members, and 2) significant overlap ratio.

5. The degree of cabin deformation reduction effect varies by overlap ratio by aggressivity-reduction structure. Accordingly, we consider overlap ratio is one of the important factors to improve comprehensive compatibility performance.

REFERENCES

1. Lund, A., Nolan, J. M., "Changes in Vehicle Designs From Frontal Offset and Side Impact Crash Testing", SAE Paper No.2003-01-0902
2. Miyake, T., "Analysis of Traffic Accident Data and In-depth Investigation Data in Japan, Institute for Traffic Accident Research and Data Analysis", ESV2003 Paper No. 496
3. Kato, H., "An Approach for Compatibility Improvement Based on US Traffic Accident Data", SAE Paper No.2003-01-0906
4. Summers, S. M., Hollwell, W. T., Prasad, Al, "NHTSA's Research Program for Vehicle Compatibility", ESV2003, Paper No. 307
5. Mizuno, K., Tateishi, K., Yonezawa, H., Minda, H., "Evaluation of Passenger Compartment Strength in Car-To- Car Frontal Crashes", SAE Paper No.2003-01-0909
6. Kitagawa, Y., "Evaluation and Research of Vehicle Body Stiffness and Strength for Car-To-Car Compatibility", SAE Paper No.2003-01-0908
7. O'Reilly, P., "Status Report of IHRA Compatibility and Frontal Impact Working Group", ESV2003 Paper No. 402
8. Edwards, M., Hobbs, A., "Development of Test Procedures and Performance Criteria to Improve Compatibility in Car Frontal Collisions", ESV2003 Paper No. 86

CONTACT

Shinji Fujii

Senior Staff Engineer

Mazda Motor Corporation

Advanced Safety Technology Development Group

Crash Safety Development Department

3-1, Shinchi, Fuchu-cho, Aki-gun,

Hiroshima 730-8670, Japan

TEL +81-82-287-5423

FAX +81-82-287-5127

fujii.s@mazda.co.jp

2003-01-0909

Evaluation of Passenger Compartment Strength in Car-to-Car Frontal Crashes

Koji Mizuno
Nagoya University

Kazumasa Tateishi
Japan Automobile Research Institute

Hideki Yonezawa and Hiroko Minda
National Traffic Safety and Environment Laboratory

ABSTRACT

The strength of the passenger compartment is crucial for occupant safety in severe car-to-car frontal offset collisions. Car-to-car crash tests including minicars were carried out, and a low end of crash force was observed in a final stage of impact for cars with large intrusion into the passenger compartment. From overload tests, the strength could be evaluated from collapsing the passenger compartment. Based on the test, the end of crash force as well as the maximum forces might be important criteria to determine the passenger compartment strength, which in turn could predict the large intrusion into the passenger compartment in car-to-car crashes. A 64 km/h ODB test was insufficient to evaluate the potential strength of the passenger compartment because the maximum forces could not be determined in this test.

INTRODUCTION

Accident data show that the intrusion into the passenger compartment is the main cause of death and serious injuries [1]. The structural integrity of the passenger compartment secures the survival space in collisions and is important for the effective operation of restraint devices, which is fundamental to self-protection. To ensure the passenger compartment integrity in car-to-car collisions, Zobel proposed the bulk-head concept [2], according to which the passenger compartment must not collapse until the front structure of the other car in a collision has completely absorbed the possible deformation energy. To evaluate the compatibility of cars, Steyer proposed two impact tests against a fixed barrier [3]. One is a crash test at a low velocity to assess the partner-protection based on the front stiffness of the car. The second is a high velocity test to assess the self-protection based on the passenger compartment strength. Newland proposed a constant energy PDB (progressive deformable barrier) test which can predict the collapse of the passenger compartment of a small car [4]. The MDB (moving deformable barrier) test can also evaluate the strength of the passenger compartment of a small car theoretically because the impact severity depends on the vehicle mass in this test [5].

As a high velocity impact test into a fixed barrier, an overload test has been proposed which is an extremely high impact ODB (Offset Deformable Barrier) test by making the passenger compartment fully collapse to evaluate the strength of the passenger compartment. This test has been included as one of the candidate test procedures for compatibility proposed by the

Table 1 Car-to-car crash test matrix

Car model	Model year	Velocity (km/h)	Curb mass (kg)	Test mass (kg)	Overlap ratio	Remarks
Daihatsu Move (minicar)	2000	55.9	845	929	50%	with driver dummy
Toyota Crown (large car)	1999	55.9	1510	1595	–	with driver dummy
Suzuki Wagon R (minicar)	2002	56.2	822	907	50%	with driver dummy
Toyota Crown (large car)	2002	56.2	1553	1638	–	with driver dummy
Honda Civic (small car)	1997	55.9	1095	1180	40%	with driver dummy
Toyota Crown (large car)	1999	55.9	1510	1595	–	with driver dummy
Daihatsu Move (minicar)	2001	56.0	845	929	40%	with driver dummy
Toyota Corolla (Small car)	2001	56.0	1082	1167	–	with driver dummy

International Harmonized Research Activities (IHRA) Compatibility Working Group [6]. The criteria of the overload test are not determined yet. Edwards suggested that the final crash force measured from barrier load cells are important criteria [7]. To evaluate the strength of the passenger compartment, these criteria should be validated with car-to-car crashes.

The present research mainly focused on the evaluation of the strength of the passenger compartment by comparing car-to-car, overload, and ODB tests (64 km/h). The acceleration of the passenger compartment and load cell forces were examined as the criteria in these tests. Test series included four crash tests between cars of different sizes. Three overload tests were also performed, and the effectiveness of the criteria for determining the passenger compartment strength were examined. The criteria were also compared for 64 km/h ODB tests to show the capabilities of the ODB test to assess the passenger compartment strength.

CAR-TO-CAR CRASH TESTS

In order to evaluate the compatibility of cars in car-to-car crashes, four tests were conducted. The test matrix of car-to-car crashes is shown Table 1. In Japan, one of the important issues in vehicle compatibility is related to minicars and small cars crashing into large cars. Thus, the test matrix involves two minicar-to-large car crashes, one small car-to-large car crash and one minicar-to-small car crash. The Daihatsu Move (GH-L900S-GPSF, engine displacement 660 cc) and Suzuki Wagon-R (LA-MC22S, 660 cc) are the minicars, and the Honda Civic (E-EK3, 1500 cc) and Toyota Corolla (TA-NZE 121-AEPNK, 1500 cc) is the small cars with the transverse engines, and the Toyota Crown (TA-JZS171-AEPSH, 2500 cc) is the large car with longitudinal engine. In tested cars, only the Daihatsu Move has a lower frame, which can distribute impact energy in lower structures. The impact velocity of each car is 56 km/h, and the overlap of the car is 50% for the smaller car. In all tests, only a driver dummy is used.

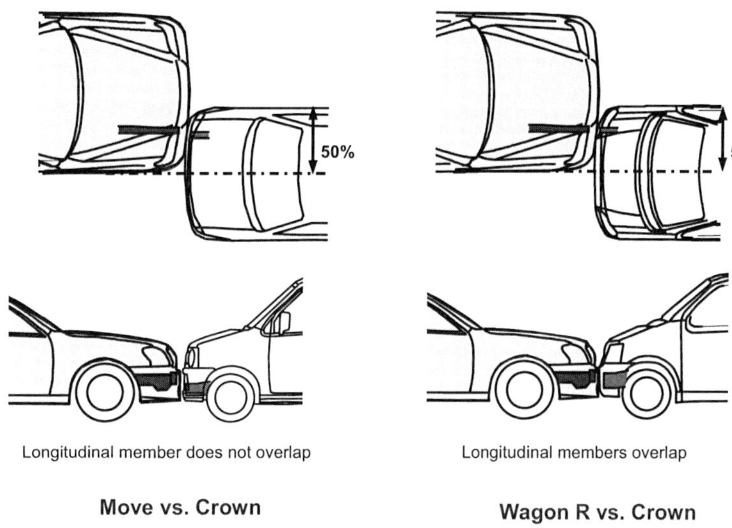

Fig. 1 Longitudinal members in crashes involving minicars.

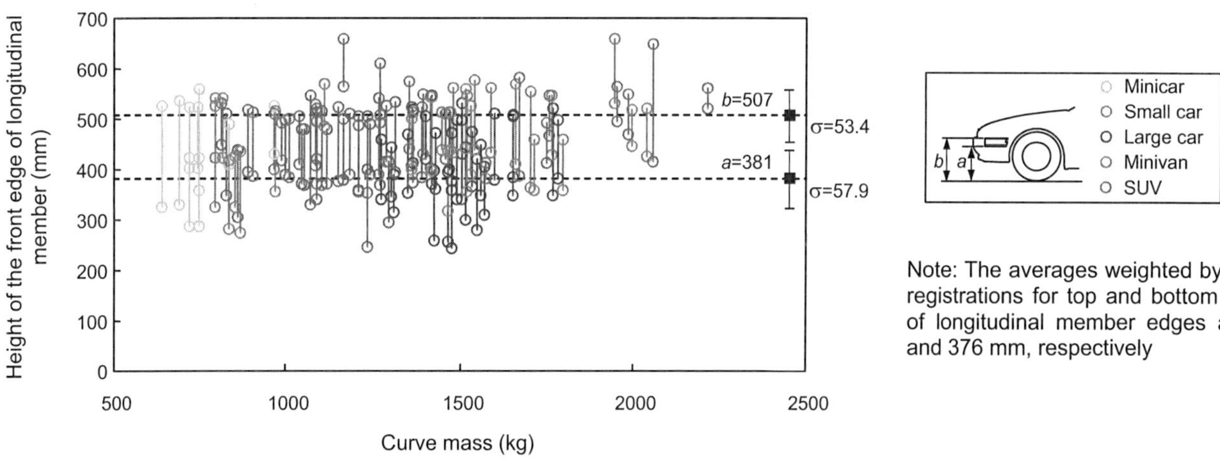

Note: The averages weighted by vehicle registrations for top and bottom heights of longitudinal member edges are 504 and 376 mm, respectively

Fig. 2 Height distributions of front edges of longitudinal members.

GEOMETRICAL AND STRUCTURAL INTERACTION

Before crash tests, a structural interaction was examined for the Move and Wagon R with respect to the Crown from the front end geometry. The relation of longitudinal member geometry in 50% overlap of minicars is illustrated in Fig. 1. The top and bottom heights of the longitudinal member were 280, 420 mm for the Move, 330, 530 mm for the Wagon R, and 390, 500 mm, respectively, for the Crown. In the crash between the Move and Crown, the longitudinal members would make only partial contact in the lateral direction, and insufficient contact in the vertical direction. In the crash between the Wagon-R and Crown, the longitudinal members would make contact in the lateral as well as the vertical direction. Fig. 2 shows the distributions of the height of longitudinal members. The average top-edge height is 507 mm and the average bottom-edge height 381 mm. It is advisable to bring the height of the longitudinal member close to the mean height.

The Move and Wagon R deformations after the crash are presented in Fig. 3. The structural interaction in the vertical direction was poor for the Move as predicted. Since the longitudinal member of the Move impacted below that of the Crown, the longitudinal member of the Move did not deform so much, and underrode the Crown. As a result, the lower frame of the Move did not deform enough for absorbing impact energy, but the deformation of fender and A-pillar beltline became large. If the Crown has the lower frame, the structural interaction between the Crown and the Move may be favorable since the lower frame contacts the longitudinal member of the Move. The lower frame of the Crown also makes for more uniform stiffness, which would be more effective in various impact conditions. For the Wagon R, the structural interaction was acceptable, and a longitudinal member and shotgun absorbed the internal energy efficiently. The front structure of the Wagon R shifted upward due to large pitching of the minicars in the final stage of impact in collisions.

The intrusion into the passenger compartment was smaller for the Move than for the Wagon R, irrespective of poor structural interaction. Accordingly, the passenger compartment strength of the Move was considered to be greater than that of the Wagon R.

The deformation of cars from the bottom is presented in Fig. 4 to show the structural interaction as viewed from the bottom. In the crash between the Move and the Crown, a lateral mismatch occurred so that the longitudinal members did not make lateral contact with one another. Move's lower frame did not deform enough as well. Consequently, its right side deformed. In the Wagon R vs. Crown and Move vs. Corolla crashes, the minicars' longitudinal members contacted one another, resulting in relatively uniform deformation of both vehicles.

In crashes involving minicars, the tire of a minicar impacted the bumper beam of a large car, which prevented the fork effects of longitudinal members. Therefore, the bumper beam of a large car plays a key role in generating a load path from the tire to the side sill of the minicar. A mismatch may occur in the lateral direction between a minicar and a large car, increasing the load on the side of the minicar. Thus, this load path must be duly considered in maintaining the integrity of the minicar passenger compartment.

FORCE-DEFORMATION CHARACTERISTICS

Force-time histories

In a simple crash model, the impact force was assumed to be the sum of the mechanical force and structural force [7]. The mechanical force was the product of the engine mass and acceleration. The structural force was also obtained from the product of the affiliated mass and the average acceleration of both side sills. In the present research, the test mass including dummies was used to estimate the affiliated mass.

The force-time histories are shown in Fig. 5. Generally, if the same cars with a transverse engine crash into each other, the engine acceleration shows high peaks. Similarly, when the Move crashed into the Corolla, the mechanical force had a spike at 27 ms. After it neared zero, the structural force had the maximum value. When the Move or Wagon-R impacted the Crown, they made contact with the soft part of the Crown. Thus, the engine acceleration was generally low.

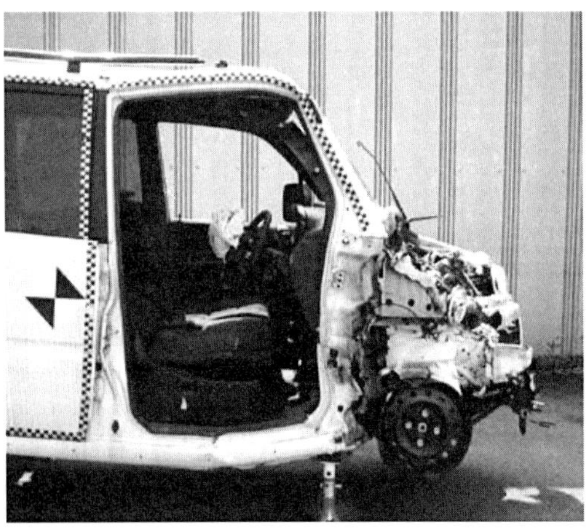

Move (vs. Crown)

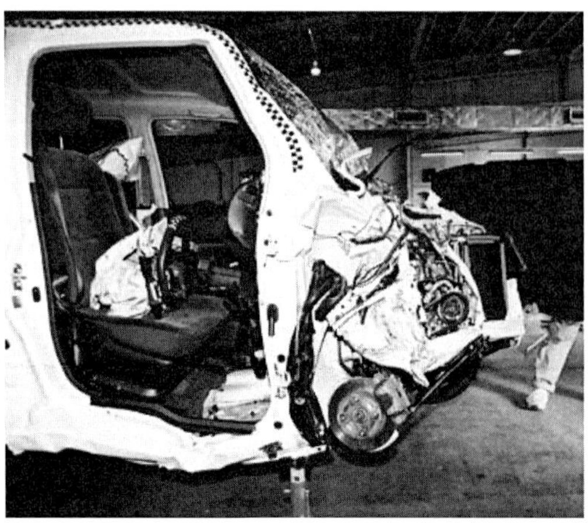

Move (vs. Crown)

Fig. 3 Deformation of Move and Wagon R in car-to-car crash with a Crown.

(a) Move vs. Crown

(b) Wagon R vs. Crown

(c) Move vs. Corolla

Fig. 4 Car deformation viewed from bottom.

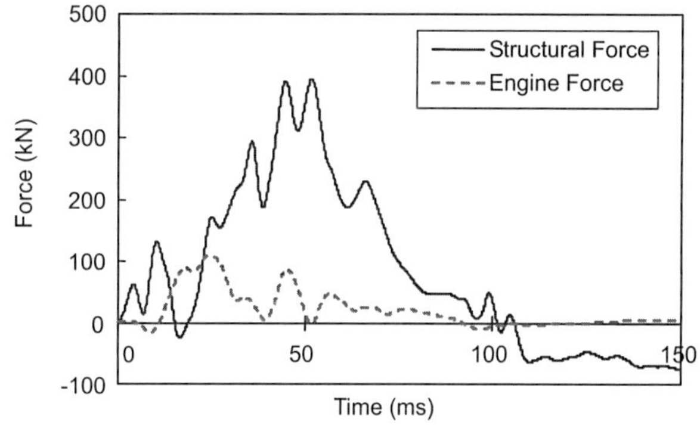

(a) Move (vs. Crown)

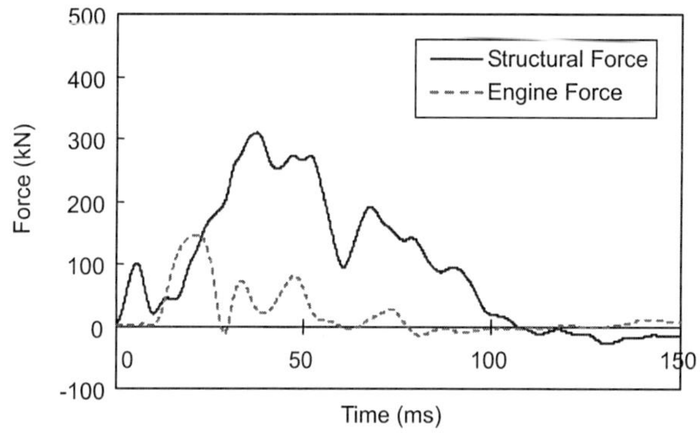

(b) Wagon R (vs. Crown)

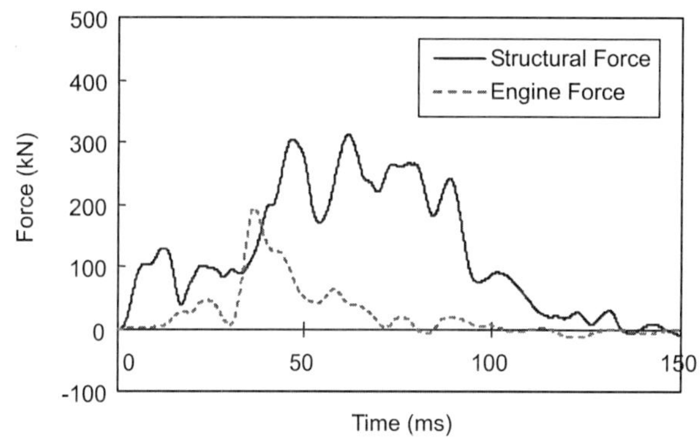

(c) Civic (vs. Crown)

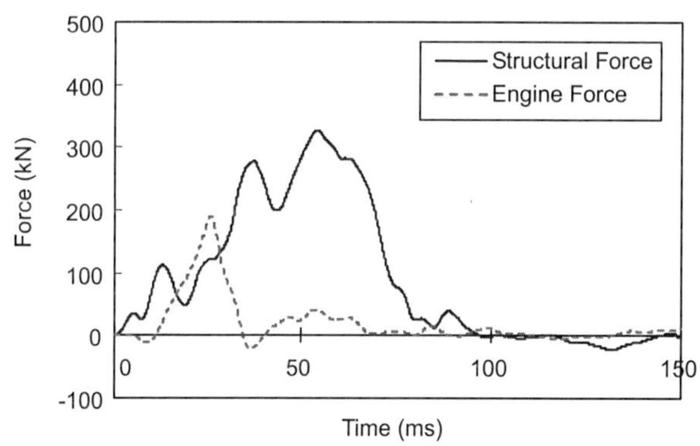

(d) Move (vs. Corolla)

Fig. 5 Force-time histories for smaller car in car-to-car crashes.

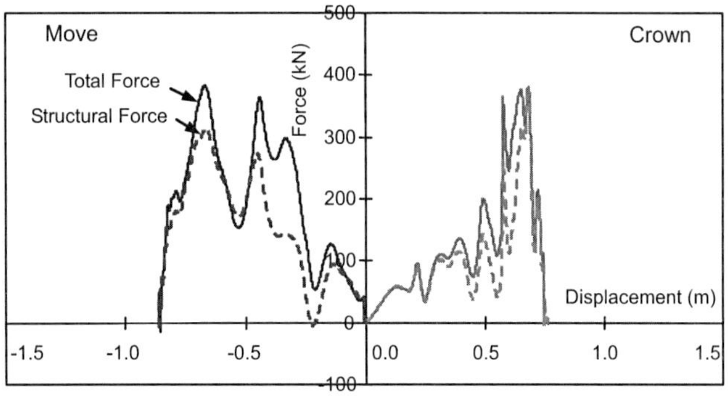

(a) Move vs. Crown

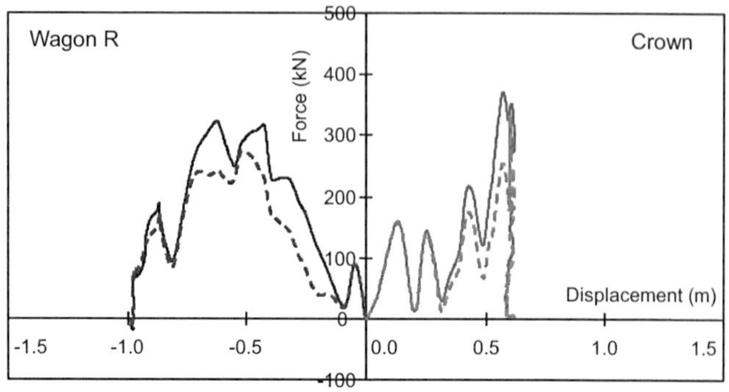

(b) Wagon R vs. Crown

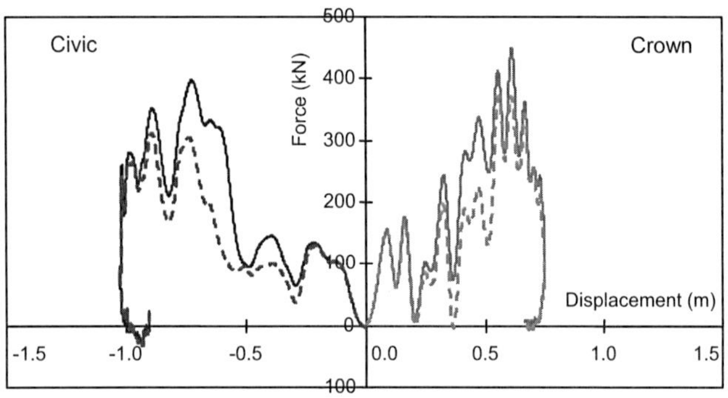

(c) Civic vs. Crown

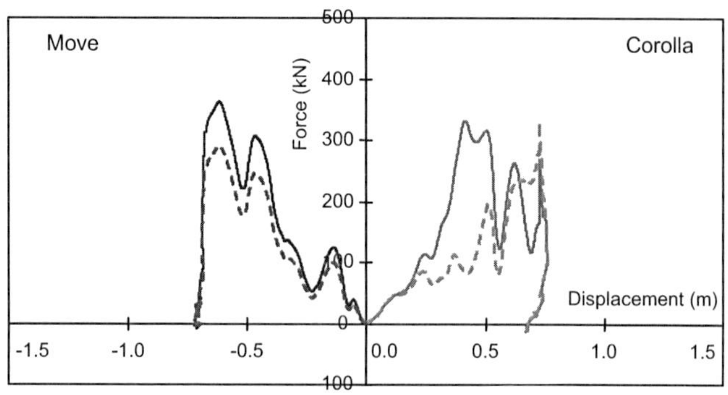

(b) Move vs. Corolla

Fig. 6 Force-deformation curves for car-to-car crash tests. Solid lines: total force; dotted lines: structural force.

When the intrusion into the passenger compartment was large (e.g., the Wagon-R or Civic collision with Crown), the maximum structural force level was less than 300 kN. In these crashes, the low structural force levels in the final stage of impact after engine acceleration became small, while the compartment intrusion became large without high resistance force.

Force-deformation characteristics

Usually, it is difficult to identify the crash interface as well as the car deformation with time in car-to-car crashes from analyzing measurements from accelerometers attached to the vehicles. In this analysis, the total displacement between the center of gravity of the two cars was obtained from double integral of the mean acceleration of the right and left side sills. The crash interface was decided each time from high-speed videos, and the ratio of the distances between the crash interface and the center of gravity of each car was calculated. This ratio was applied to the total displacement, and the deformation of each car was determined.

The force-displacement characteristics for each car are shown in Fig. 6 including the total force, which is the sum of the mechanical and structural force. In a car-to-car crash, the smaller car absorbed more internal energy than the larger car. The absorbed internal energy was 169 and 97 kJ for the Move and the Crown, 183 kJ and 85 kJ for the Wagon R and the Crown, 184 and 147 kJ for the Civic and the Crown, and 131kJ and 107 kJ for the Move and the Corolla. If more energy were applied to the Civic, the passenger compartment would collapse because the structural force was low and the intrusion was already large in a crash with a Crown.

OVERLOAD TESTS

An overload test is a high-velocity ODB test causing a significant deformation leaving a large intrusion into the passenger compartment. In the present study, overload tests were conducted using Move, Wagon R and Civic models. An impact velocity of 80 km/h with 40% overlap on the ECE R94 barrier was used to collapse the passenger compartment completely. The test matrix for the overload test is summarized in Table . A barrier force and an average acceleration of both B pillars were used to measure the passenger compartment strength.

Table 2 Overload test matrix

Car model	Model year	Velocity (km/h)	Curb mass (kg)	Test mass (kg)	Overlap ratio	Remarks
Daihatsu Move (minicar)	2000	80.0	845	845	40%	Without dummy
Suzuki Wagon R (minicar)	2002	80.3	822	822	40%	Without dummy
Honda Civic (Small car)	1998	80.0	1094	1095	40%	Without dummy

Figure 7 shows force-time histories of the barrier force, mechanical force and structural force. The tested cars are also presented over time in Figs. 8, 9 and 10. The barrier force reached its maximum when the mechanical force peaked due to engine impact against the barrier. The maximum barrier force was almost the same as the total force. At the time, the engine bottomed out the barrier, and at the same time contacted the fire wall, presumably beginning the intrusion into the passenger compartment. After the engine stopped, the floor and side sill began to deform and buckle, which resulted in the collapse of the passenger compartment. At this final stage of the passenger compartment collapse, the deformation became large, especially when the end of crash force obtained from the barrier was low, as in the case of Civic. The Civic indicated clearly the area of low end of crash force, though the Move kept high level. As seen in the car-to-car crash with the Crown, the passenger compartment intrusion became large for the Civic and for the Wagon R which had the low end of crash force in the overload test.

The acceleration curve of the Civic shows a vibration. With the passenger compartment collapse and consequent destabilization of the Civic, the acceleration curves measured at the right and left sills and rear cross member showed an obvious deviation. The Move accelerations at these locations are similar, which indicates the stability of the passenger compartment of this car.

Based on the force-time histories, two criteria can be considered to evaluate the strength of the passenger compartment. The first is 'the maximum force level' which basically makes for comparable deformations of both cars, and also reduces the intrusion of the bulkhead into the passenger compartment. The second criterion is the 'end of crash force' [7], which represents the resistance force that prevents the collapse of the passenger compartment. This criterion will have a relation with life threatening-injuries with sufficient survival space in the final stage where an excessive load is applied to the car structures. In the present study, the end of crash force was determined as the maximum barrier force after the engine stopped. For the Move, Wagon R and Civic, the maximum force level was 313, 299 and 295 kN, and the end of crash force was 234, 152 and 115 kN, respectively.

The barrier force distribution was also measured in the Wagon R overload test using high-resolution load cells (125 x 125 mm). Fig. 11 shows the force distribution at 55 ms at which the force was maximum. There was a strong force concentration at the barrier edge due to contact with the engine. This force concentration continued because the car rotated around the engine. Force was distributed also out of the car since the impact force was transmitted and dispersed behind the thick honeycomb. Thus, it may be difficult to evaluate the force distribution from overload tests.

Taking the honeycomb energy absorption into account, the equivalent barrier speed of the Move, Wagon R and Civic was 71.1, 73.1 and 73.4 km/h, respectively, which were higher than those in a car-to-car crash with a Crown. To keep the internal energy absorption in the overload test larger than for car-to-car crash with a large car, an impact velocity of 80 km/h may be appropriate at least for cars with low mass.

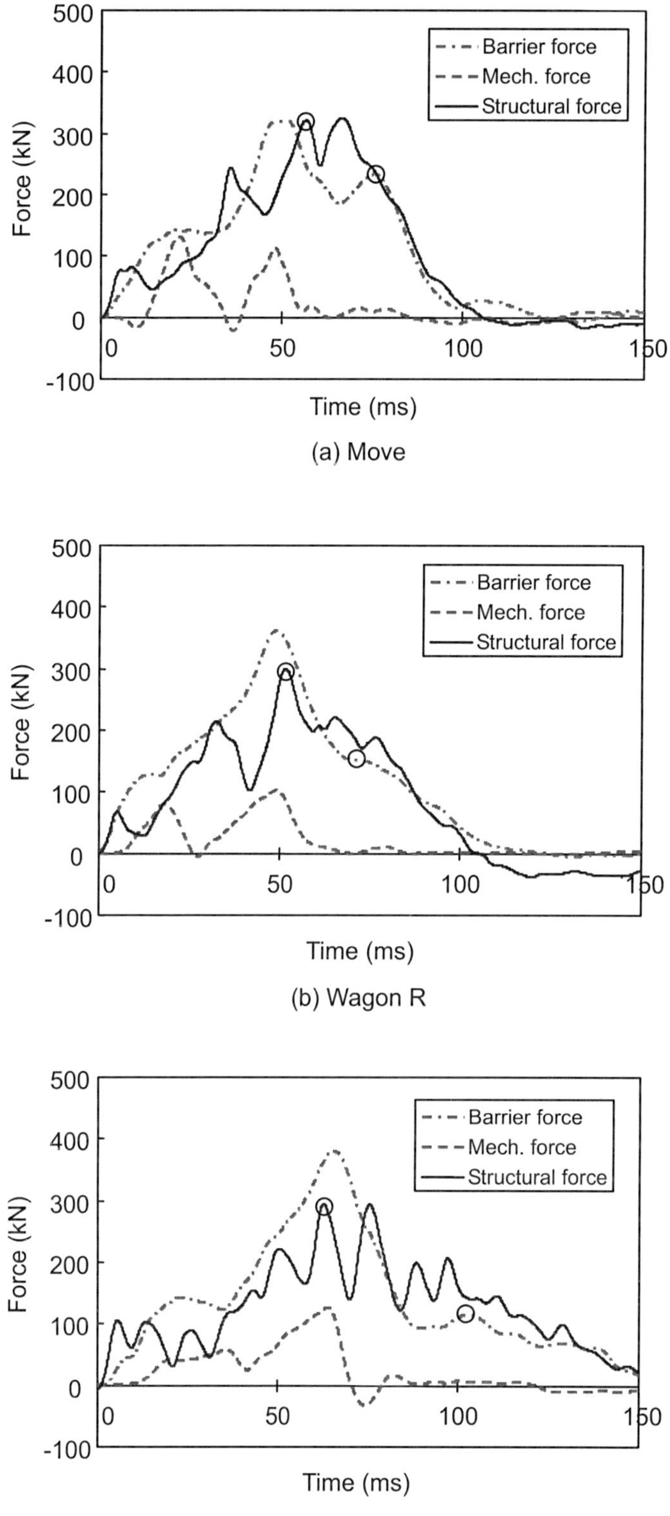

(a) Move

(b) Wagon R

(c) Civic

Fig. 7 Force-time histories in overload tests.

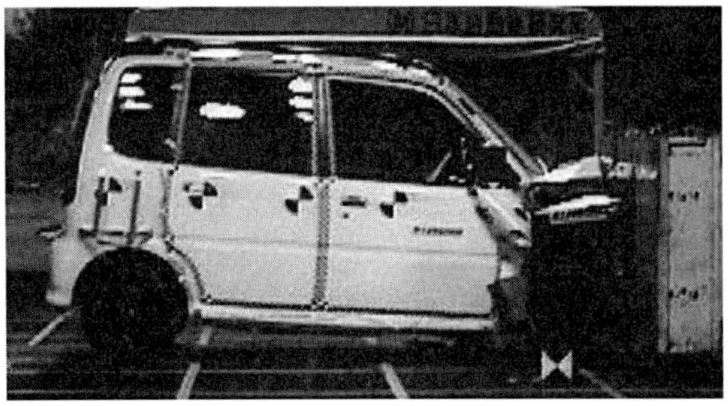

Barrier force maximum (50 ms)

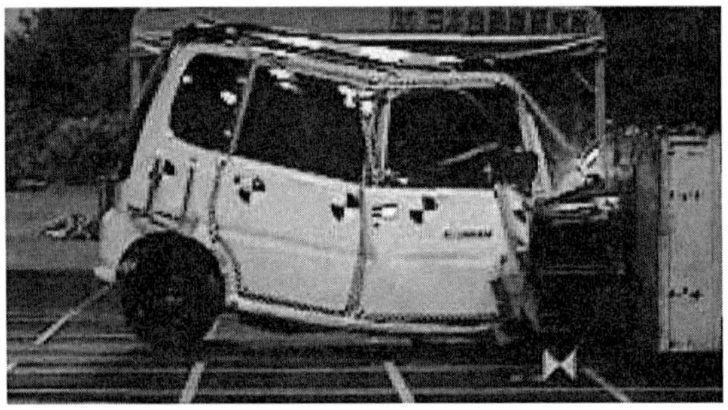

Deformation maximum (80 ms)

Fig. 8 Deformation of Move in overload test.

Barrier force maximum (66 ms)

Deformation maximum (100 ms)

Fig. 10 Deformation of Civic in overload tests.

Barrier force maximum (66 ms)

Deformation maximum (80 ms)

Fig. 9 Deformation of Wagon R in overload tests.

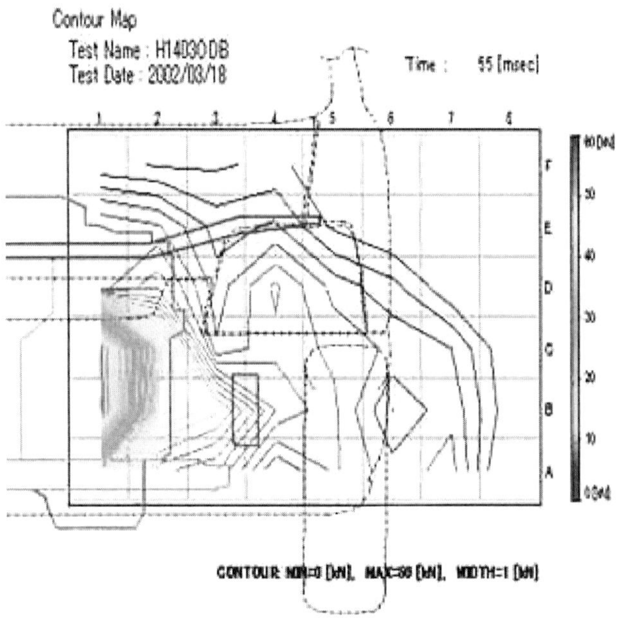

Fig. 11 Load cell force distribution of Wagon R in overload test at 55 ms.

ODB 64 KM/H TESTS

The data from 64 km/h ODB tests of the Move and the Wagon R from the JNCAP (Japan New Car Assessment Program) were used to estimate the capacity for prediction of structural interaction and compartment collapse in a car-to-car crash with a large car. In the JNCAP, six load cells are attached behind the ECE R94 honeycomb. Therefore, only the inclination of the force distribution can be discussed.

STRUCTURAL INTERACTION

Figures 12 and 13 show the load cell force, and the deformation of the car and the honeycomb for Move and Wagon R. For the Move, the load cell force was higher for the lower part. The lower area of the honeycomb bottomed out whereas the upper area still remained without absorbing energy. The stiff structure of the Move was distributed in the lower parts, which could lead in underriding in a car-to-car crash with the Crown. The Wagon R showed more balanced upper and lower load cell force distribution, and the honeycomb deformation was more uniform than that of the Move. Local holes of the honeycomb were also observed for the Move, which reflects the high local stiffness of this car.

By using high-resolution load cells, the force distribution can be analyzed more in detail. However, as described in the overload test, the force could be spread behind the thick honeycomb, making it difficult to evaluate the local force concentration even using a high-resolution load cell force.

The Progressive Deformable Barrier (PDB) proposed by Renault [8] allows the honeycomb deformation to be analyzed more precisely since this barrier represents the front stiffness of the average small car in a realistic way, and the deformation distribution is easy to measure due to the simple geometry of the barrier. Using the criteria proposed in the PDB tests, we now clarify the overall and local stiffness of the tested cars [8].

FORCE-DISPLACEMENT CHARACTERISTICS BY IMPACT VELOCITIES

The structural force curves with displacement are shown in Fig. 14 for the Move, Wagon R and Civic, according to impact velocity. Since the structural force-displacement characteristics described a similar curve, regardless of velocity, the car deformation mode at 80 km/h may be considered to match those at lower velocities. For the Move, the maximum force was 330 kN in the overload test, which was higher than that in the ODB test (220 kN).

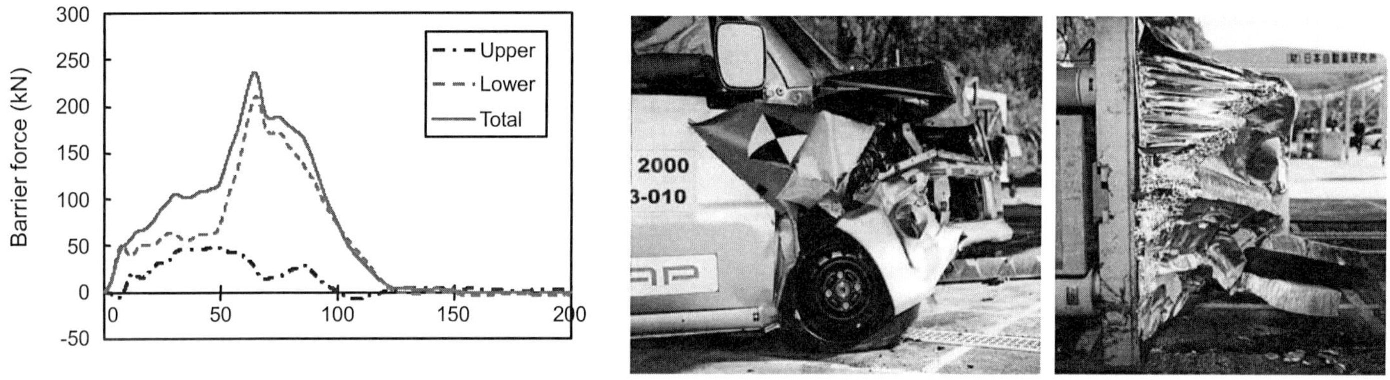

(a) Barrier force (b) Deformation (c) Honeycomb

Fig. 12 Barrier load cell force and deformation in ODB 64 km/h test (Move).

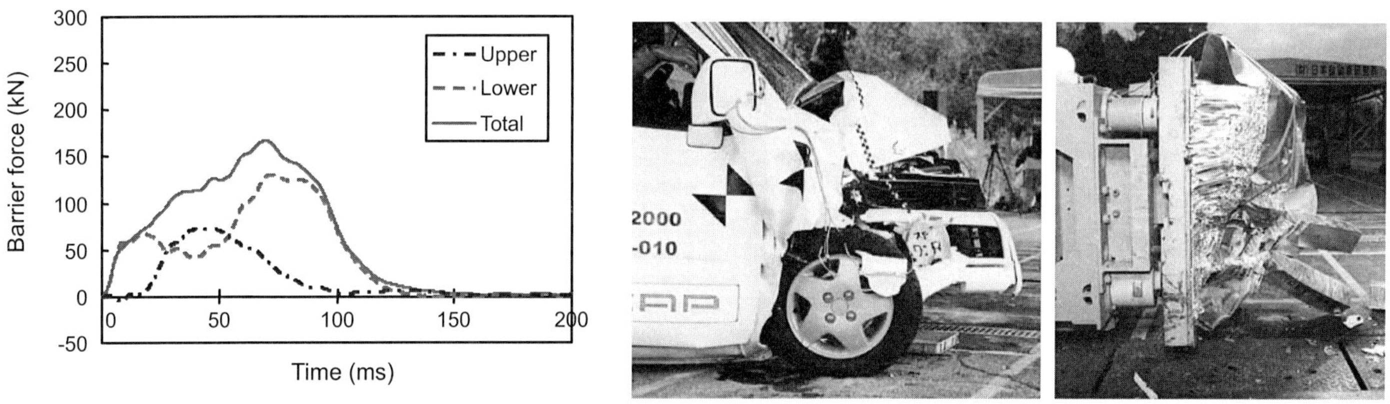

(a) Barrier force (b) Deformation (c) Honeycomb

Fig. 13 Barrier load cell force and deformation in ODB 64 km/h test (Wagon-R).

Therefore, whereas the force of Move and Wagon R did not reach its passenger compartment strength at 64 km/h ODB test, the overload test could determine the actual passenger compartment strength. The maximum forces of the Civic in the overload test and in the 64 km/h ODB test were very similar at around 250-280 kN. However, the low end of crash force could be detected from the overload test. Although the large intrusion into the passenger compartment was found in the ODB 64 km/h tests of the Wagon R and Civic, the 64 km/h ODB test might be insufficient to evaluate the threshold of the criteria for passenger compartment strength.

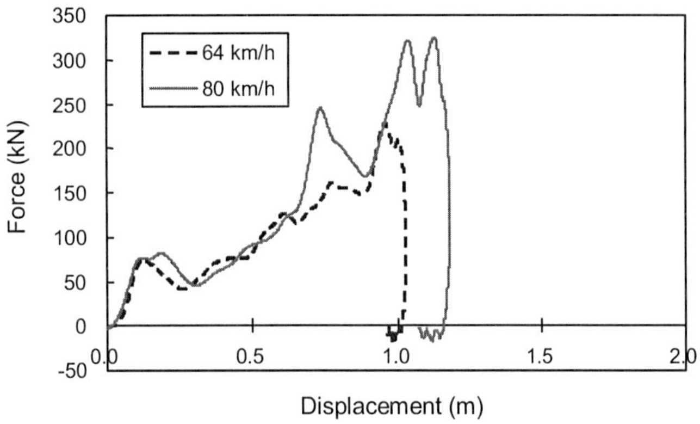

(a) Move

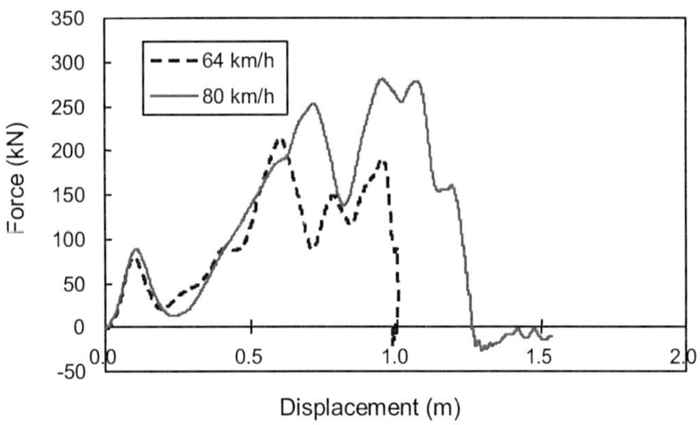

(b) Wagon R

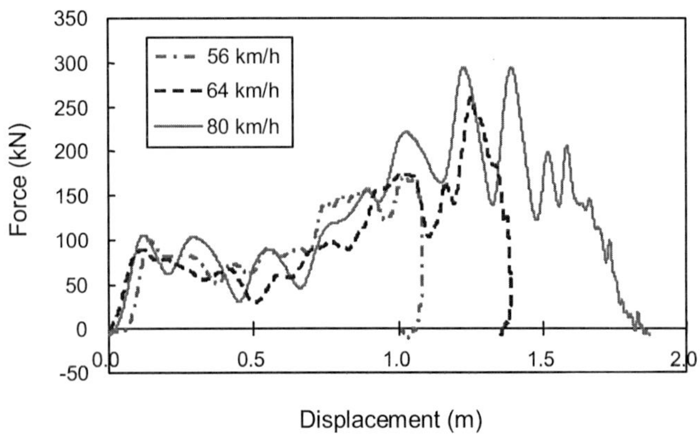

(c) Civic

Fig. 14 Force-displacement curves in ODB tests according to impact velocities.

INJURY PARAMETERS

A 64 km/h ODB test is adopted in EuroNCAP, JNCAP, ANCAP and IIHS, and is most widely accepted to assess and compare injury risks to occupants by car model. In the present research, injury parameters of the driver dummies in the Move and Wagon R were compared for ODB 64 km/h test and car-to-car crash with a Crown (Fig. 15). In general, the injury parameters were lower for the Wagon R, and the JNCAP score of the Wagon R in the offset frontal test was also better than that of the Move. However, in the car-to-car crash with the Crown, the injury parameters of Wagon R were elevated, especially intrusion-based parameters such as femur force and tibia index. From the results, it is demonstrated that although a less-stiff car may score well in the ODB 64 km/h tests, this score can not be used to predict the injury parameters of the driver in car-to-car crash. In a car-to-car crash with a large car, cars with a high passenger compartment strength may have an advantage that can reduce the intrusion-base injury parameters when the impact loading applied to the car is greater than in a 64 km/h ODB test.

EVALUATION OF PASSENGER COMPARTMENT STRENGTH

The intrusion into the passenger compartment is shown in Fig. 16. The deformation and intrusion were most severe in an overload test. This kind of test may therefore be used to evaluate the passenger compartment strength of the kind when a large crash transmits the massive load on the compartment in a severe crash where compartment can collapse.

The maximum force level and end of crash force determined for car-to-car and overload tests are shown in Fig. 17. In the car-to-car test, the end of crash force was determined from the structural force since barrier forces could not be determined. Since the acceleration of impact side B-pillar of the CIVIC showed vibrations, the structural force obtained from the non-struck side B-pillar was used. The maximum force level is likely to show a correlation between the overload test and the car-to-car crash with the Crown. Consequently, overload tests can predict the maximum force levels in car-to-car crashes. In Fig. 17, the end of crash force of the Civic is seen to be higher in the car-to-car test than the overload test. The main reason is that the force applied to the passenger compartment crashing into the Crown did not reach the severity of the end of crash force determined from the overload test. However, with the Move and Wagon R, there also seemed to be some correlations of the end of crash force between the car-to-car and overload tests. Based on the test results, the overload tests will be effective to predict the compartment integrity based on the maximum force level and the end of crash force in a car-to-car crash with a large car. The threshold of maximum force level might be 300 kN and that of the end of crash force might be 250 kN, at least given the performance of the Move when crashing into a large car. In order to determine the criteria or their threshold, more research will be necessary to confirm the effectiveness of the overload test of the compartment integrity in car-to-car tests.

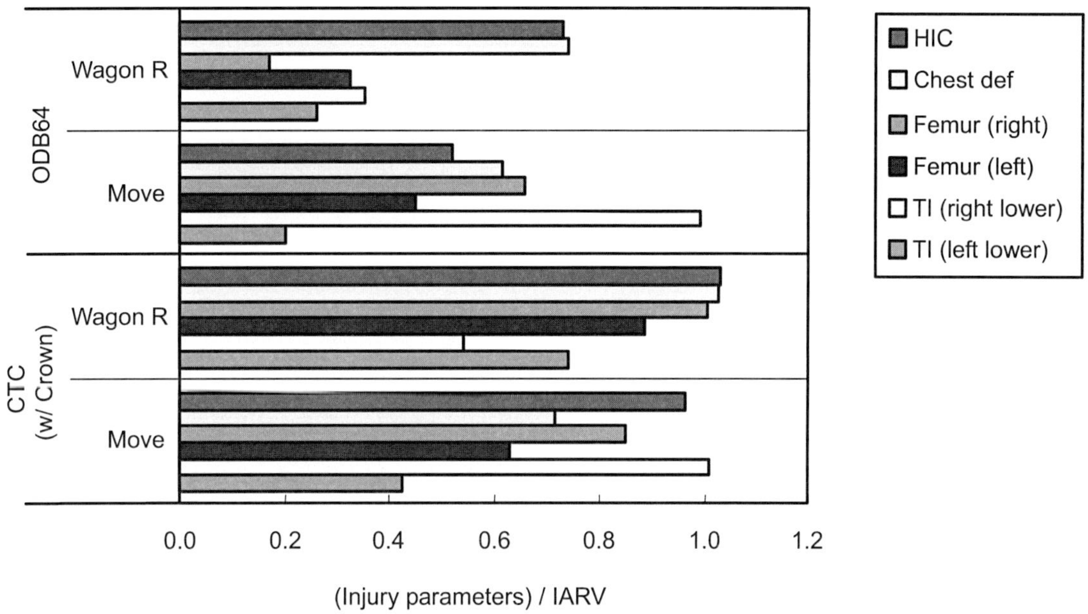

Fig. 15 The ratio of injury parameters to injury assessment reference values (IARV) for the driver dummy in minicars used for 64 km/h ODB crash tests and car-to-car crash with a Crown.

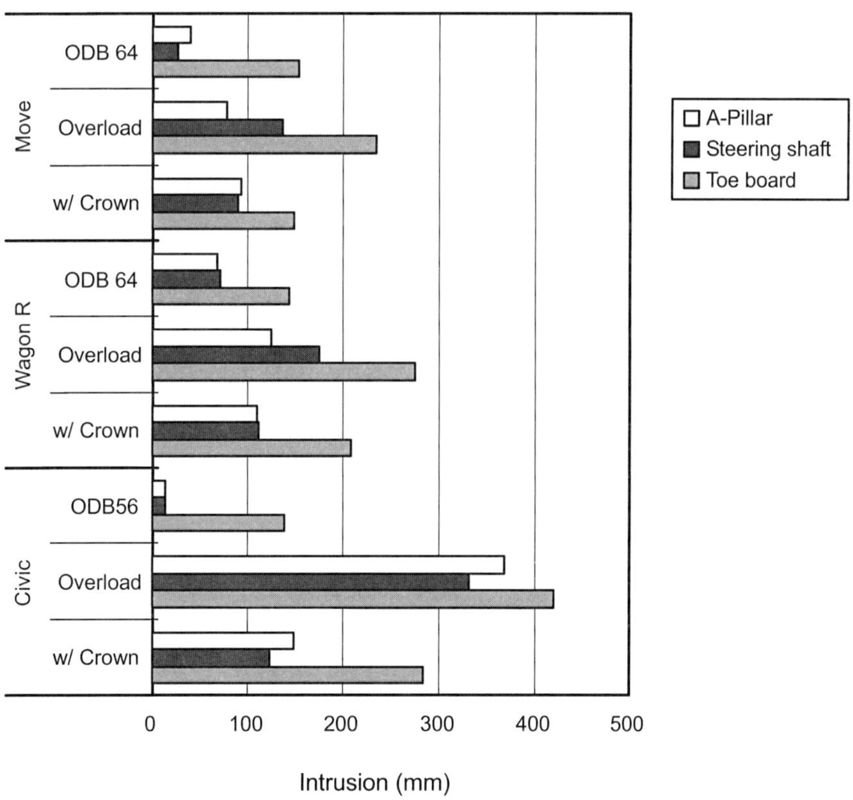

Fig. 16 Intrusion into the passenger compartment in ODB test, overload test and car-to-car crash with a Crown.

The force path or deformation mode can be different between overload test and car-to-car test because the structural interaction varies in both tests or the bottoming-out of the honeycomb occurs in the ODB tests which can make for a different force path around the engine. For example, lateral and vertical mismatches can occur in car-to-car tests, which cannot be predicted in ODB tests. Irrespective of these differences, the maximum force level and end of crash force will be evaluated from the overload tests, which may predict the performance of the passenger compartment integrity in a severe car-to-car crash with a large car.

To improve the compartment strength of minicars or small cars, the first priority will be to maintain the end of crash force at a high level. In this stage of the end of the crash, the floor of the car can buckle which may lead to life-threatening injuries to the occupants. The high level of the end of crash force can prevent the passenger compartment from collapsing. The second priority will be increasing the maximum level of the structural force, which ensures balanced deformation between cars in a crash.

The passenger compartment strength seems to be most important because it involves the survival space. Furthermore, the compartment strength is most effective when the structural interaction is good, where the occupants can be exposed to the optimized deceleration in ODB tests.

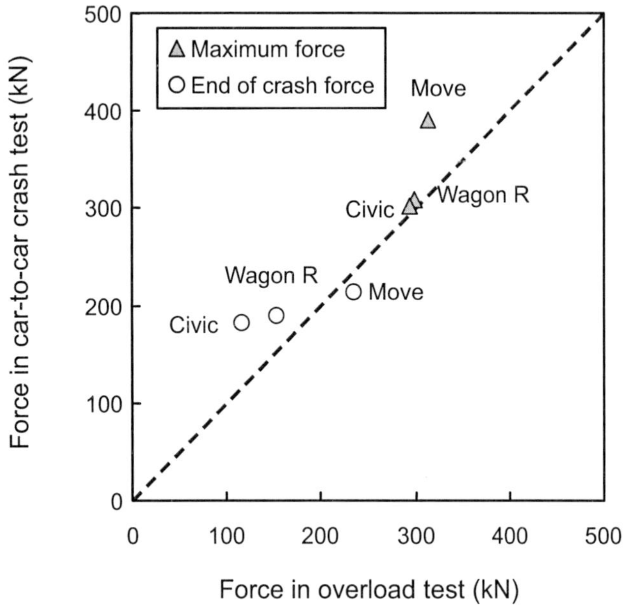

Fig. 17 Maximum force level and end of crash force in overload tests and in crashes with Crown.

CONCLUSIONS

The strength of the passenger compartment was examined using the test results including car-to-car, 64 km/h ODB, and overload tests. The conclusions are summarized as follows:

In car-to-car crashes, there were cars with low end of crash force, which could lead to a large intrusion into the passenger compartment. In overload tests, this low end of crash force could be also identified for cars with large intrusion. Criteria such as the maximum force and end of crash forces had correlations with those in overload tests involving a great load on the passenger compartment or in severe car-to-car offset crashes.

From 64 km/h ODB tests, it was difficult to evaluate the strength of passenger compartment because the structural force did not reach its maximum. Comparison of the injury parameters between two cars in the 64 km/h ODB test could not be directly applied to predict the injury parameters in a car-to-car crash with a large car.

ACKNOWLEDGMENTS

This research was conducted as one of the research projects of the Japanese Ministry of Land, Infrastructure and Transport (JMLIT). The authors also wish to thank the National Organization for Automotive Safety & Victims' Aid (OSA) for providing JNCAP data.

REFERENCES

1. Hobbs, A., “The Rationale and Development of the Offset Deformable Frontal Impact Test Procedure”, SAE Paper 950501.
2. Steyer, C., Diaboine, A., Delhommeau, M., Delannoy, P., “Taking Compatibility Aspects into Account When Designing Vehicles”, VDA Technical Congress, Frankfurt, 1999.
3. Zobel, R., “Accident Analysis and Measures to Establish Compatibility”, SAE Paper 1999-01-0065, SAE Congress, 1999.
4. Newland C., “Australian Research to Develop a Vehicle Compatibility Test”, IMechE, Conference Transactions, Vehicle Safety 2002.
5. Hollowell, T., “NHTSA’s Research Program for Vehicle Aggressivity and Fleet Compatibility”, IMechE, Conference Transactions, Vehicle Safety 2002.
6. O’Reilly, P., “International Harmonized Research Activity (IHRA) – Vehicle Compatibility, IMechE, Conference Transactions, Vehicle Safety 2002.
7. Edwards, J., Happian-Smith, J., Byard, N., Davies, C., Hobbs, A., “Compatibility–the Essential Requirements for Cars in Frontal Impact”, IMechE, Conference Transactions, Vehicle Safety 2000, pp.3-17, 2000.
8. Diaboine, A., Delannoy, P., “Improvements in Car to Car Compatibility: Physics, Design Constraints and Assessment Test Methodology Criteria, IMechE, Conference Transactions, Vehicle Safety 2002.

2004-01-1169

NHTSA's Frontal Offset Research Program

James W. Saunders III, Shashi Kuppa and Aloke Prasad
National Highway Traffic Safety Administration

ABSTRACT

The National Highway Traffic Safety Administration (NHTSA) is conducting a research program to investigate the use of the 40 percent offset deformable barrier (ODB) crash test procedure to reduce death and injury, in particular debilitating lower extremity injuries in frontal offset collisions. This paper presents the results of 22 ODB crash tests conducted with 50th percentile male and 5th percentile female Hybrid III (HIII) dummies fitted with advanced lower legs, Thor-Lx/HIIIr and Thor-FLx/HIIIr, to assess the potential for debilitating and costly lower limb injuries. This paper also begins to investigate the implications that the ODB test procedure may have for fleet compatibility by evaluating the results from vehicle-to-vehicle crash tests.

INTRODUCTION

In the United States, driver and right front passenger air bags are required in all passenger cars and light trucks under Federal Motor vehicle Safety Standard (FMVSS) No. 208, Occupant Crash Protection. However, NHTSA estimates that over 8,000 fatalities and 120,000 AIS 2+ injuries will continue to occur in frontal crashes even after all passenger cars and light trucks have frontal air bags. Therefore, NHTSA has focused on the development of performance tests not currently addressed by FMVSS No. 208, particularly the high severity frontal offset crashes. These tests result in a large amount of occupant compartment intrusion that could compromise occupant survival space and may increase the potential for lower leg injury.

Since the EU directive 96/79 for frontal crash protection became effective in 1998, other countries and consumer rating programs have adopted the use of a fixed ODB crash test procedure. The Australian and European regulation conduct the ODB crash test at 56 km/h while the consumer rating programs, Euro NCAP (European New Car Assessment Program), Australian NCAP and IIHS (Insurance Institute for Highway Safety) conduct the ODB crash test at 64 km/h. IIHS has evaluated over 150 vehicle models using their 64 km/h ODB crash test procedure. IIHS indicated in a status report (September, 2001) that various vehicle models which had received a "poor" ratings were redesigned and are now obtaining "good" or "marginal" rating in their ODB crash tests. This improved performance in the redesigned vehicles was attributed to vehicle structure design that prevents major collapse of the occupant compartment. Nolan and Lund (2001) found that the majority of vehicles (mainly passenger cars) whose structural performance improved in the IIHS frontal offset crash tests after being redesigned did so without significant alteration to the stiffness of the vehicle for the first half meter of deformation. However, General Motors Corp. (Verma et al., 2003) indicated that the use of fixed barriers as a test device may lead to higher force levels for front ends of larger vehicles and could act to increase the incompatibility between large and small vehicles.

Kahane (2003) analyzed fatality rates in two vehicle crashes based on the mass and stiffness of each vehicle and the height mismatch between vehicles. The result of this analysis presented the individual effect of stiffness, geometry, and mass on vehicle aggressivity. Kahane found that in a front-to-front collision between an LTV and a passenger car, the fatality risk of the driver in the car was directly correlated (statistically significant) to the weight of the LTV, the front end stiffness of the LTV and the height mismatch between the LTV and the car. Kahane noted that crash testing with existing and, eventually, modified vehicles is another essential step in learning about what makes LTVs aggressive.

Preliminary ODB crash tests conducted by NHTSA (Park et al. 2002) using the HIII dummy indicated that a high speed ODB crash test would yield few benefits over that of FMVSS No. 208 for upper body injuries. However, if lower leg instrumentation was incorporated into the FMVSS No. 208 and ODB test procedure, Park et al. (2002) concluded that this action would yield considerable benefits in addressing lower extremity injuries. FMVSS No. 208, through its current test protocols and dummy instrumentation, does not fully address hip and lower limb injuries that are disabling and associated with high societal costs (Kuppa et al., 2001a, 2003). However, recent advances in a lower extremity device (Thor-Lx/HIIIr) and associated injury criteria allow for better assessment of lower leg and foot/ankle injuries. NHTSA published an Advanced Notice of Proposed Rulemaking on lower leg instrumentation for use in full frontal and offset frontal vehicle crashes in May, 2002. The Thor-Lx/HIIIr (Kuppa et al., 2001b) is a test device that fits on the distal femur of the Hybrid III dummy family. Its biofidelity and injury assessment capability has been reported to be superior to the standard Hybrid III legs (Rudd et al., 1999, Petit et al., 1999).

This paper presents an update of real world frontal crash data and an evaluation of offset deformable barrier crash tests conducted by NHTSA. Vehicle-to-vehicle crash tests are also examined to study the effect of the ODB procedure on vehicle fleet compatibility.

REAL WORLD CRASH DATA

The NASS/CDS data files for the years 1993-2002 were examined for drivers and front outboard passengers involved in frontal crashes in vehicles equipped with air bags, in a similar manner as outlined by Kuppa et al. (2001). Crashes involving rollovers and ejections were excluded in the analysis. All data presented in the paper are weighted according to NASS/CDS recommendations.
Among all frontal crashes, 25% were full frontal, 37% were left offset and 38% were right offset frontal crashes. The risk of AIS 2+ injuries to various body regions in different crash modes for belted occupants is presented in Figure 1.

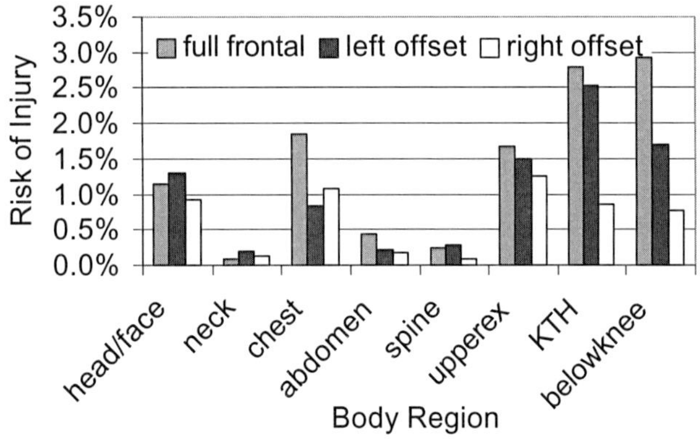

Figure 1 - Risk of AIS 2+ injuries to different body regions for belted front outboard occupants in air bag equipped vehicles involved in different frontal crash modes.

The lower extremities (knee-thigh-hip (KTH) and below knee) are the most frequent AIS 2+ injured body region among front seat occupants in air bag equipped vehicles. Though the risk of lower limb injuries is higher in full frontal crashes than in offset crashes, the annual number of occupants who sustain lower limb injuries is higher in left offset crashes than full frontal crashes since the number of left offset crashes is greater (Kuppa et al. 2003). The risk of injury in right offset crashes is lower than that in left offset crashes because 80 percent of front outboard occupants were drivers who were not in proximity to the impact location.

The distribution of AIS 2+ injuries below the knee sustained by belted front seat occupants in air bag equipped vehicles involved in different frontal crash modes is presented in Figure 2. Leg shaft injuries include tibial plateau and tibia and fibula shaft fractures. Ankle injuries include fracture of the calcaneus, talus, tibial plafond, and malleoli, and ankle ligament injuries. Foot injuries are fractures of the midfoot and forefoot bones. Fifty percent of below knee injuries are to the ankle, while leg shaft injuries account for only 25 percent of below the knee injuries.

The risk of AIS 2+ injuries to the knee-thigh-hip complex is higher in SUVs and minivans than in the passenger cars (Figure 3). However, the risk of AIS 2+ injuries below the knee is higher in passenger cars. The risk of lower limb injuries in SUVs is significantly higher than that in any other vehicle body type.

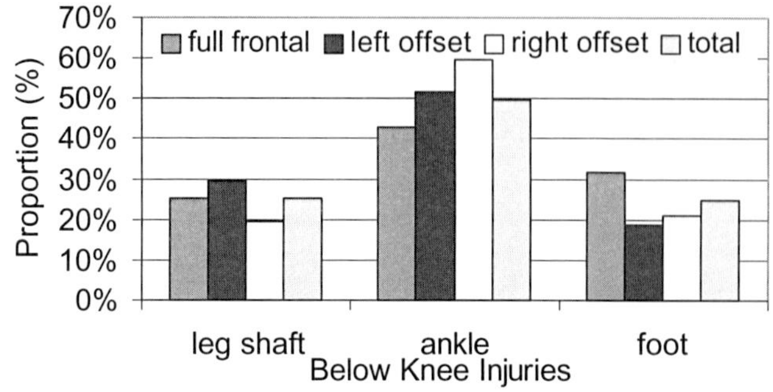

Figure 2 - Distribution of AIS 2+ below knee injuries sustained by belted front outboard occupants in air bag equipped vehicles involved in frontal crashes.

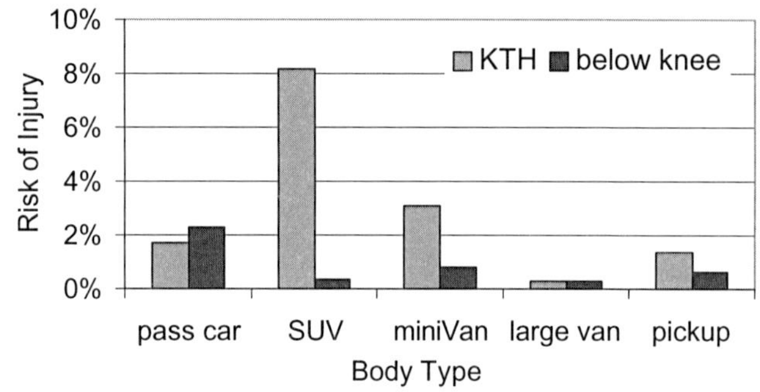

Figure 3 - Risk of AIS 2+ lower limb injuries sustained by belted front outboard occupants in frontal crashes for different vehicle body types with air bags

The risk of knee-thigh-hip injuries in air bag equipped vehicles involved in frontal crashes is increasing in newer vehicle models (Figure 4) while foot and ankle injuries are about the same for new and old vehicle models.

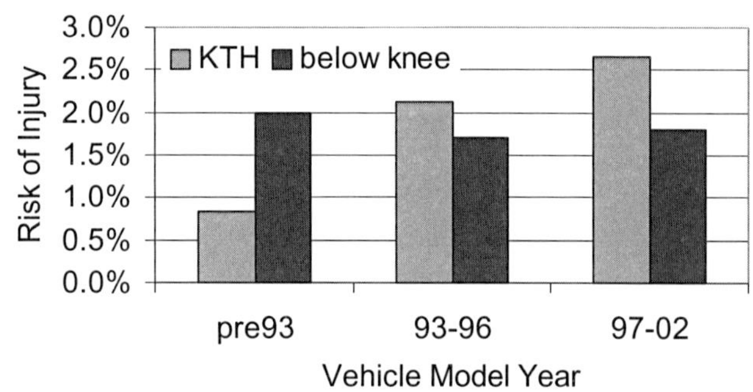

Figure 4 - Risk of AIS 2+ lower limb injuries sustained by front outboard occupants in air bag equipped vehicles of different vehicle model years involved in frontal crashes

In order to compare the risk of injury in real world crashes with that obtained from ODB crash tests, the risk of injury for front seat occupants in air bag equipped vehicles involved in left offset frontal crashes with known DeltaV>48 km/h was determined. The risk of AIS 2+ injuries below the knee in left offset frontal crashes is 48 percent while the risk of AIS 2+ chest and head injuries is 17 and 7 percent, respectively (Figure 5). The risk of AIS 3+ head injuries in left offset frontal crashes with DeltaV>48 km/h is 4 percent while that for neck and chest injuries is 1 percent and 13 percent, respectively.

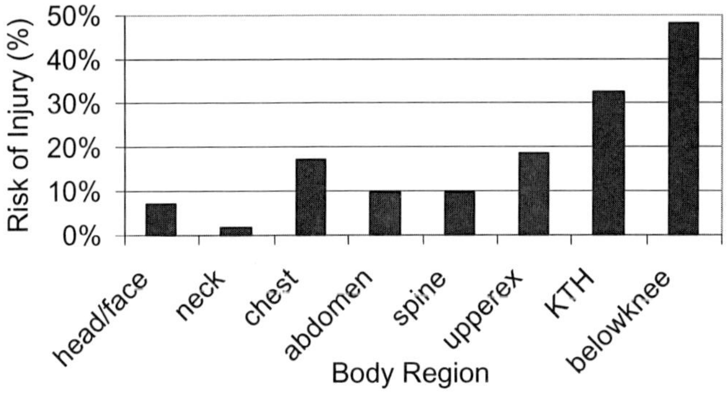

Figure 5 - Risk of AIS 2+ injuries to different body regions for belted front seat occupants in air bag equipped vehicles involved in left offset frontal crashes with DeltaV>48 km/h.

40 PERCENT OFFSET DEFORMABLE BARRIER CRASH TEST

TEST PROCEDURE – The 40 percent frontal offset deformable barrier (ODB) crash test series was conducted using the procedure defined in FMVSS No. 208, Occupant Crash Protection (S18). The test configuration is shown in Figure 6. The vehicles were moving at 56, 60 or 64 km/h, such that 40 percent of the vehicle engaged the barrier (Tables A.1 and A.2). In all tests the driver and front seat passenger were two Hybrid III 50th percentile males (HIII 50M) with the Thor-Lx/HIII retrofit lower leg or two Hybrid III 5th percentile females (HIII 5F) with the Thor-FLx/HIII retrofit lower leg. The HIII 50M and HIII 5F dummy positioning was done in accordance with FMVSS No. 208.

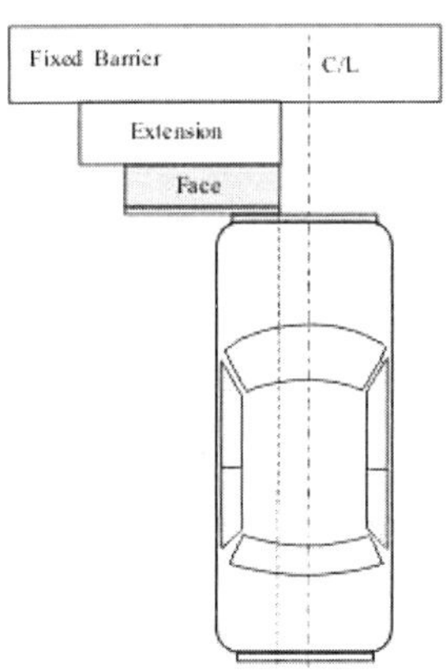

Figure 6 - ODB Test Configuration

VEHICLE MEASUREMENTS – Figure 7 and Figure 8 show the vehicle measurements taken pre and post test for the ODB crash tests. Points 1 through 9 (Figure 8) are located on the toe pan, point 10 is the brake pedal and point 11 is the footrest. The dummy's left foot was placed approximately at the footrest (Point 11) and the right foot was placed at the accelerator (near point 6). The upper instrument panel measurements pre and post test where taken approximately at the left and right knees of the dummy. The steering column was also measured pre and post-test. All measurements were made with respect to the rear of the vehicle as the reference frame and according to SAE J1733.

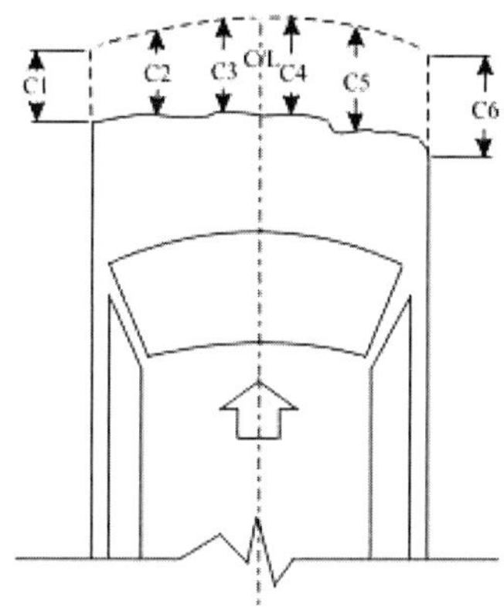

Figure 7 - Front Crush Measurments

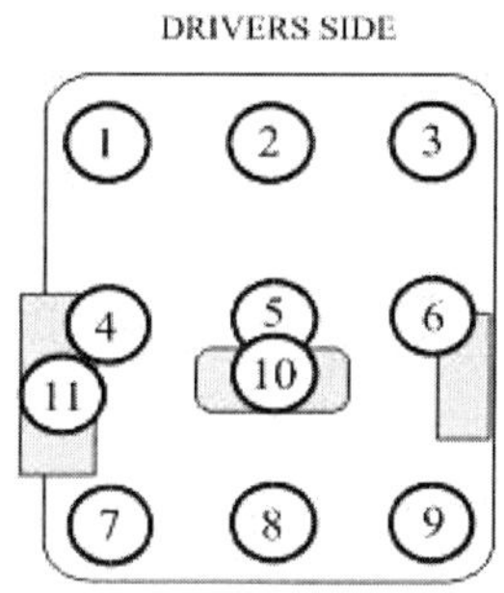

Figure 8 - Toeboard Measurements

TEST RESULTS–The Injury Assessment Reference Values (IARV) for the HIII 50M and 5F dummies that were developed for the FMVSS No. 208 Advanced Air Bag Final Rule were used. The injury assessment for the lower leg was conducted according to Kuppa et al., (2001b). The IARVs used to assess injuries below the knee are presented in Table 1.

Table 1. Injury Assessment Reference Values for injuries below the knee (Kuppa et al., 2001a, 2001b)

Injury Criteria	IARV for HIII 50M	IARV for HIII 5F
knee shear	15 mm	13 mm
Upper tibia axial force	5600 N	4000 N
Lower tibia axial force	5200 N	3750 N
Upper tibia index *	F/12000+M/240<0.91	F/8640+M/146<0.91
Lower tibia index *	F/12000+M/240<0.91	F/8640+M/146<0.91
Dorsiflexion	35 deg	35 deg
Inversion/eversion	35 deg	35 deg

* F= axial force in N, M is resultant moment in Nm.

The injury assessment values for the head, neck, and chest in the ODB tests with the HIII 50M and HIII 5F dummies in the driver's seat as a function of resultant steering wheel displacement are presented in Figure 9 through Figure 12. The HIC15 injury threshold limit of 700 was exceeded in one test with HIII 50M dummies and one test with HIII 5F dummies (Figure 9) because the air bag did not deploy in one case and the air bag deployment was delayed in the other. In each case where the HIC limit of 700 was exceeded, the dummy's head contacted the steering wheel.

Figure 10 shows that over 50 percent of the tests performed with HIII 5F dummies in the driver's seat exceeded the Nij injury threshold. None of the tests performed exceeded the injury threshold for chest Gs or chest displacement (Figure 11 and Figure 12).

The measured axial femur force and knee shear for the HIII 50M and 5F dummies in the driver's seat with respect to the resultant instrument panel displacement are presented in Figure 13 and Figure 14. These figures show that the femur force and knee shear did not exceed the injury threshold level in any of the tests even though there was knee contact with the knee-bolster. The injury assessment values for the legs, foot, and ankle of the HIII 50M and 5F dummies with respect to the maximum resultant toe pan intrusion are presented in Figure 15 through Figure 20. There is seemingly no correlation between any of the injury measures and the post test intrusion measurements (Figure 9 through Figure 20).

The percentage of tests that exceeded each of the IARVs are presented in Table 2. The IARVs of the lower tibia index, lower tibia axial force and dorsiflexion for the HIII 50M and 5F dummies were exceeded in 30 to 56 percent of the tests. The Nij value for the HIII 5F dummy exceeded the IARV of 1 in 56 percent of the tests due to delayed air bag deployment and dummy interaction with the air bag.

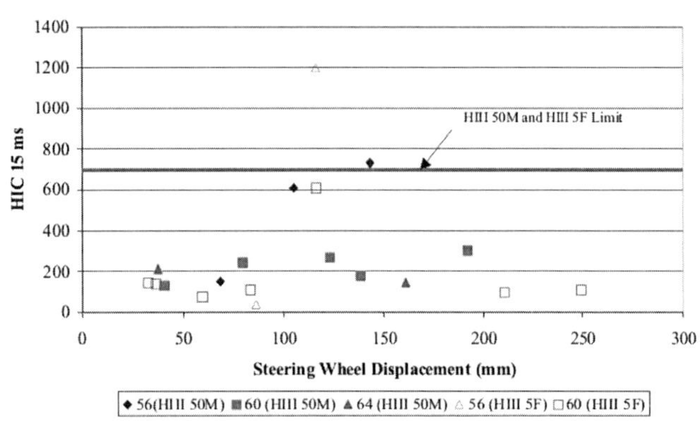

Figure 9 - HIC15 for HIII 50M and HIII 5F dummies in the driver's seat (in ODB crash tests at 56 km/h, 60 km/h and 64 km/h impact speed) versus resultant steering wheel displacement.

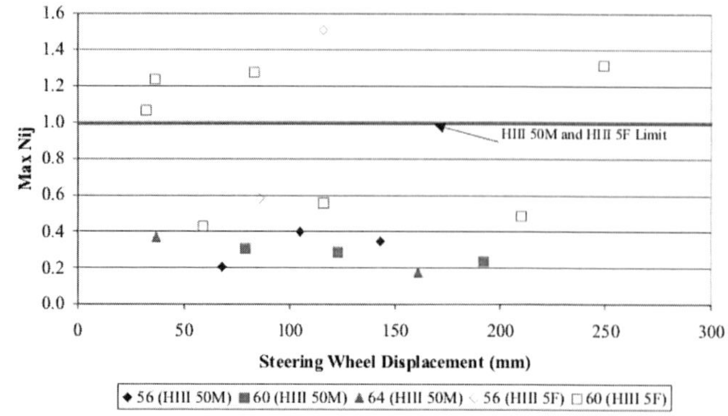

Figure 10 – Maximum Nij for HIII 50M and HIII 5F dummies in the driver's seat (in ODB crash tests at 56 km/h, 60 km/h and 64 km/h impact speed) versus resultant steering wheel displacement.

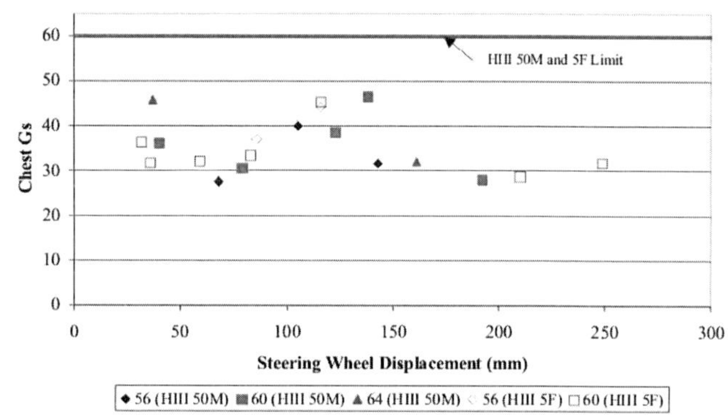

Figure 11 – Chest Gs for HIII 50M and HIII 5F dummies in the driver's seat (in ODB crash tests at 56 km/h, 60 km/h and 64 km/h impact speed) versus resultant steering wheel displacement.

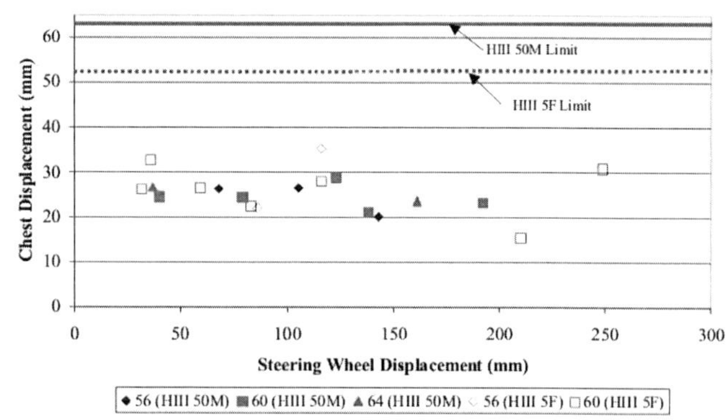

Figure 12 – Maximum chest displacement for HIII 50M and HIII 5F dummies in the driver's seat (in ODB crash tests at 56 km/h, 60 km/h and 64 km/h impact speed) versus resultant steering wheel displacement

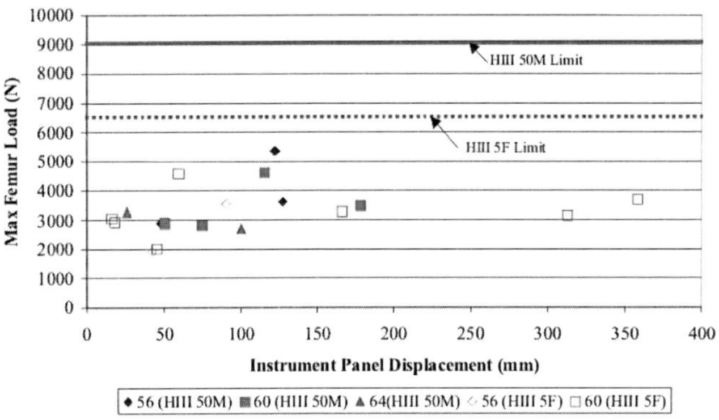

Figure 13 – Max femur load for HIII 50M and HIII 5F dummies in the driver's seat (in ODB crash tests at 56 km/h, 60 km/h and 64 km/h impact speed) versus resultant instrument panel displacement.

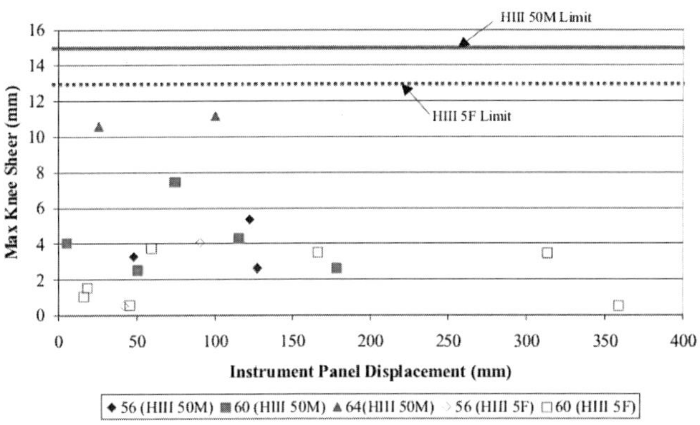

Figure 14 - Maximum knee shear for the HIII 50M and HIII 5F in the driver's seat in ODB crash tests at 56 km/h, 60 km/h and 64 km/h impact speed versus resultant instrument panel displacement.

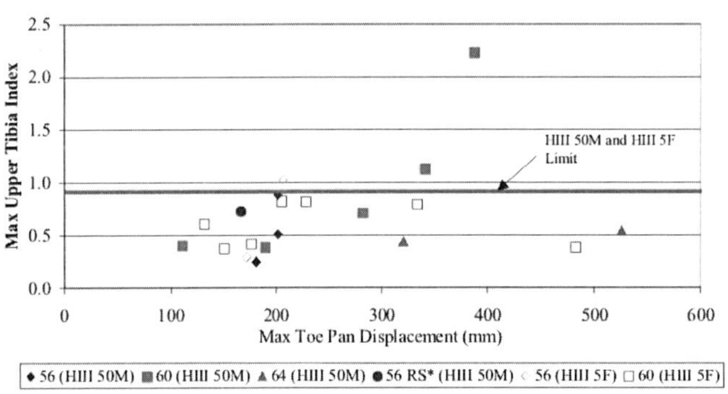

* RS = Impact on the right side of the vehicle, using passenger data

Figure 15 - Maximum upper tibia index for HIII 50M and HIII 5F dummies in the driver's seat (in ODB crash tests at 56 km/h, 60 km/h and 64 km/h impact speed) versus maximum resultant toe pan displacement.

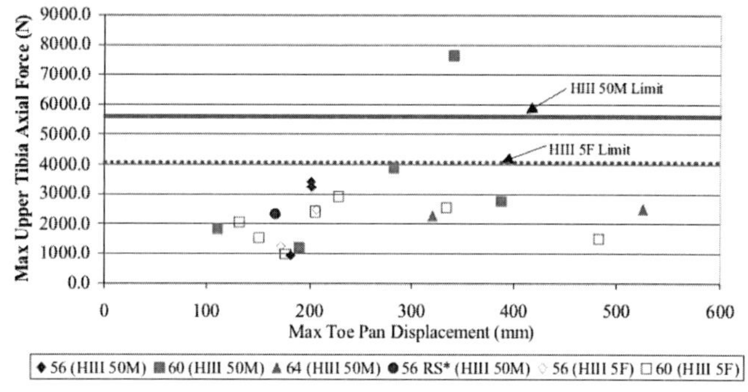

* RS = Impact on the right side of the vehicle, using passenger data

Figure 16 - Max upper tibia axial force for HIII 50M and HIII 5F dummies in the driver's seat (in ODB crash tests at 56 km/h, 60 km/h and 64 km/h impact speed) versus max resultant toe pan displacement.

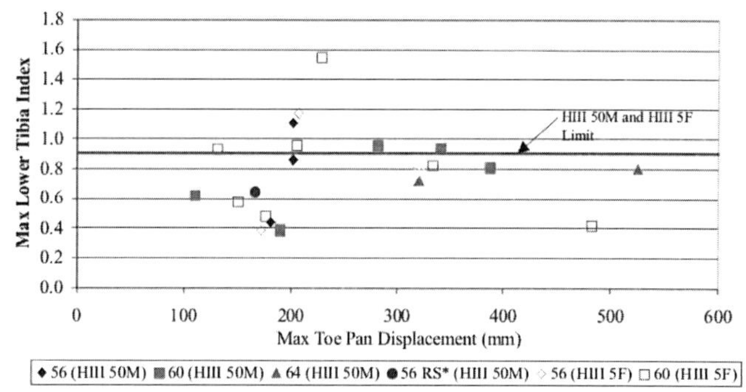

* RS = Impact on the right side of the vehicle, using passenger data

Figure 17 - Maximum lower tibia index for the HIII 50M and HIII 5F in the driver's seat versus maximum resultant toe pan displacement.

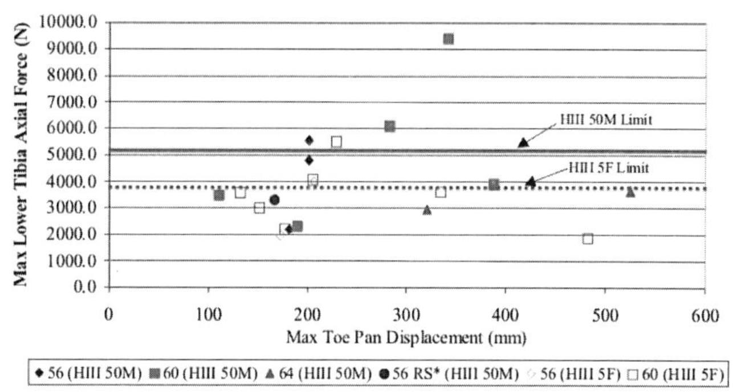

* RS = Impact on the right side of the vehicle, using passenger data

Figure 18 - Maximum lower tibia axial force for HIII 50M and HIII 5F dummies in the driver's seat (in ODB crash tests at 56 km/h, 60 km/h and 64 km/h impact speed) versus maximum resultant toe pan displacement.

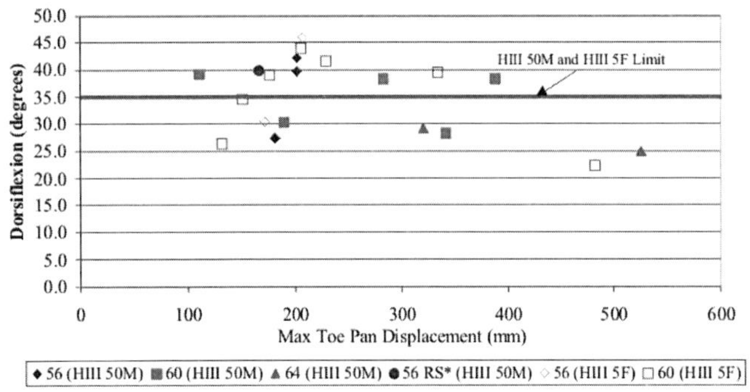

* RS = Impact on the right side of the vehicle, using passenger data

Figure 19 – Ankle dorsiflexion for HIII 50M and HIII 5F dummies in the driver's seat (in ODB crash tests at 56 km/h, 60 km/h and 64 km/h impact speed) versus maximum resultant toe pan displacement.

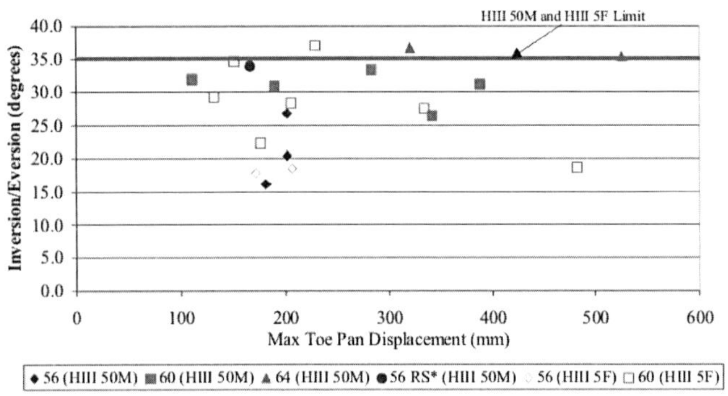

* RS = Impact on the right side of the vehicle, using passenger data

Figure 20 – Ankle inversion/eversion for HIII 50M and HIII 5F dummies in the driver's seat (in ODB crash tests at 56 km/h, 60 km/h and 64 km/h impact speed) versus maximum resultant toe pan displacement.

Table 2 - Percent of tests that exceeded the IARVs for the HIII 50M and HIII 5F dummies in the drivers position in ODB crash tests at 56 km/h, 60 km/h and 64 km/h impact speed.

	HIII 50M	HIII 5F
Chest G	0.0%	0.0%
Chest Displacement	0.0%	0.0%
HIC 15	10.0%	11.1%
Nij ver. 10	20.0%	55.6%
Neck Tension	0.0%	11.1%
Neck Comp.	0.0%	0.0%
Max Femur	0.0%	0.0%
Max Knee Shear *	0.0%	0.0%
Max Up TI*	20.0%	11.1%
Max Low TI*	30.0%	44.4%
Max Upper Tibia Axial Force*	10.0%	0.0%
Max Lower Tibia Axial Force*	30.0%	33.3%
Max Dorsiflexion Angle*	50.0%	55.6%
Max Inversion / Eversion Angle*	20.0%	11.1%

* IARVs are not part of FMVSS No. 208 Advanced Air Bag Final Rule. They are from Table 1 and are used for injury assessment with the Thor-Lx/HIIIr and Thor-FLx/HIIIr (Kuppa et al., 2001a, 2001b).

VEHICLE-TO-VEHICLE CRASH TESTS

Since some researchers have reported that the use of fixed barriers as a test device would lead to an increase in the front-end stiffness of larger vehicles and thereby result in an increase in the incompatibility between large and small vehicles (Verma et al., 2003), NHTSA began to investigate the effect that the ODB test procedure may have on fleet compatibility. For this purpose, three paired bullet vehicles were run into a moving 1996 Honda Accord as the target vehicle (Table 3). The bullet vehicles in each pair were one large pickup or two sport utility vehicles of the same vehicle make and model with one being an older model and rated "poor" in the IIHS ODB test while the other was a newer redesigned model and rated "marginal" or "good" in the IIHS ODB test[1].

TEST PROCEDURE – The frontal oblique vehicle-to-vehicle crash test series was conducted using a test procedure developed under NHTSA's Advanced Frontal Offset Research Program (Stucki, et al., 1998). Both vehicles were moving at approximately 56 km/h with the 1996 Honda Accord (target vehicle) struck on the left side (Figure 21). At impact, the left side of the striking vehicle aligns with the center of the front of the Accord. In all tests, both vehicles used a Hybrid III 50th percentile male dummy in the driver position. The passenger position for both vehicles was occupied by a Hybrid III 5th percentile female dummy.

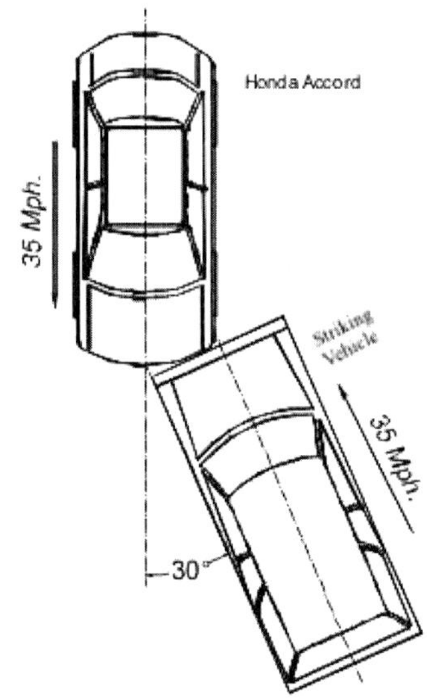

Figure 21 - Oblique Offset Test Configuration.

VEHICLE SETUP - The upper and lower bumper height and the hood leading edge were measured before each test and used to create a general profile of the front of the vehicle (FP). Also, the upper and lower height of the longitudinal members (LM) was measured according to an INSIA report (INSIA, 1999). The newer, redesigned vehicles had a lower bumper

[1] http://www.hwysafety.org/vehicle_ratings/ratings.htm

height and the longitudinal members were slightly higher (Figure 22). Figure 23 through Figure 25 show the initial position of the old and new models of the Blazer/TrailBlazer, Montero Sport and the RAM 1500 tests with respect to the Accord, respectively.

Table 3 – Striking vehicle-to-Accord test matrix

Before Redesign	After Redesign
1997 Chevrolet Blazer IIHS Rating = Poor Test Weight = 2130 kg Vehicle Width = 1700 mm NHTSA Test # = V4363	**2002 Chevrolet TrailBlazer** IIHS Rating = Marginal Test Weight = 2355 kg Vehicle Width = 1847 mm NHTSA Test # = V4364
1999 Mitsubishi Montero Sport IIHS Rating = Poor Test Weight = 2112 kg Vehicle Width = 1705 mm NHTSA Test # = V4474	**2001 Mitsubishi Montero Sport** IIHS Rating = Good Test Weight = 2143 kg Vehicle Width = 1750 mm NHTSA Test # = V4438
2001 Dodge Ram 1500 IIHS Rating = Poor Test Weight = 2531 kg Vehicle Width = 2035 mm NHTSA Test # = 4581	**2002 Dodge Ram 1500** IIHS Rating = Good Test Weight = 2572 kg Vehicle Width = 2024 mm NHTSA Test # = 4617

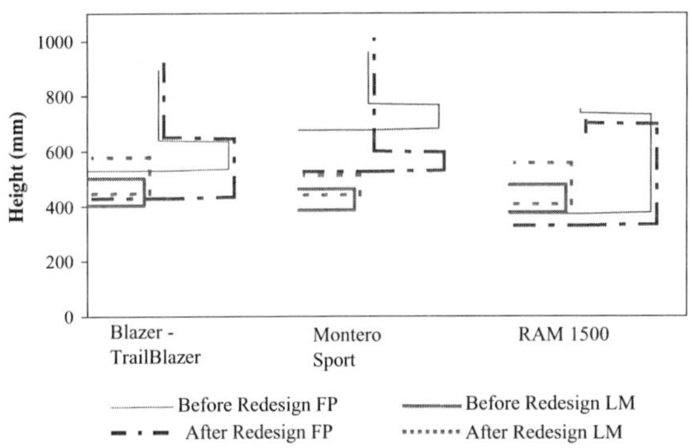

Figure 22 - Frontal profile (FP) and longitudinal member (LM) height of the before and after redesign.

Before Redesign

After Redesign

Figure 23 - Blazer - TrailBlazer before and after redesign.

Before Redesign

After Redesign

Figure 24 - Montero Sport before and after redesign.

Before Redesign

After Redesign

Figure 25 - RAM 1500 before and after redesign.

VEHICLE FORCE-DEFLECTION- NHTSA's New Car Assessment Program (NCAP) measures the total force applied to the rigid wall in a full frontal rigid barrier crash test. Figure 26 and Figure 27 show the force-deflection profiles obtained from the NCAP tests of the pre and redesigned RAM 1500 and the Blazer/TrailBlazer models. Similar force-deflection profiles are not available for the Montero Sport. The general trend of the force-deflection profiles is that the redesigned vehicles have a higher onset rate of force and a peak force that is higher and occurring earlier in the event as compared to the pre-redesign vehicles. In addition, the deflection of the redesigned vehicle is lower than the corresponding pre-redesigned vehicles.

The initial stiffness of the vehicles was computed according to the method outlined by Swanson et al. (2003) and are presented in Table 4. The initial stiffness is higher for the redesigned vehicles than for the pre-redesigned vehicles.

Table 4 - Initial Stiffness of the Ram 1500 and the Trailblazer

Vehicle Model	Pre-redesigned	Redesigned
Ram 1500	1985 N/mm	2732 N/mm
Blazer/Trailblazer	1528 N/mm	2479 N/mm

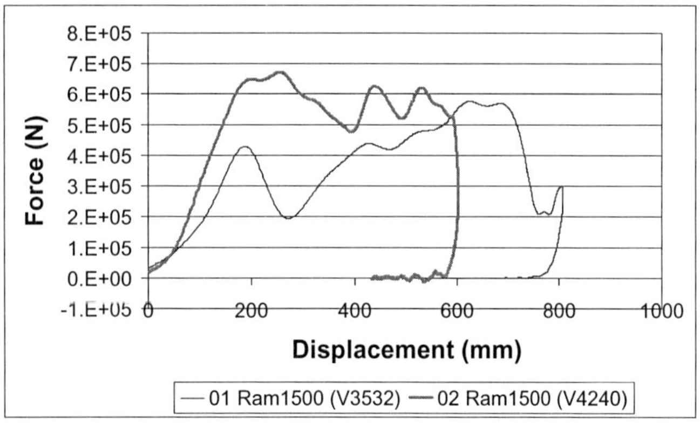

Figure 26 - RAM 1500 Force deflection profile from NCAP test.

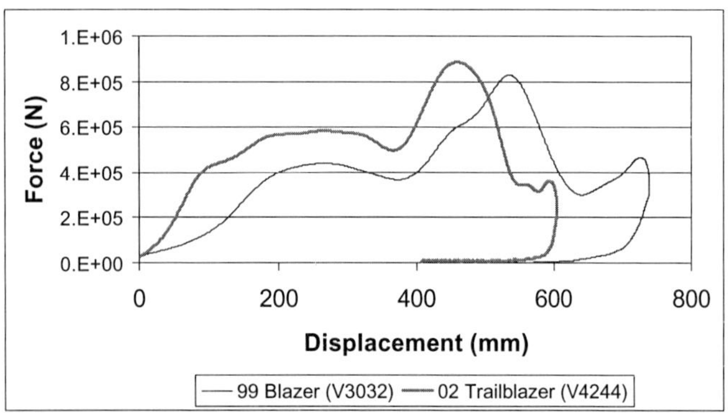

Figure 27 - Blazer/Trailblazer Force deflection profile from NCAP test.

VEHICLE-TO-VEHICLE CRASH TESTS RESULTS-

Vehicle Crush and Intrusion – Figure 28 through Figure 30 show the crush of the Accord for each test series. For these tests, the interior occupant compartment measurements were taken pre test as described in the section VEHICLE MEASUREMENTS for the ODB crash test procedure. However, due to extreme intrusion of the occupant compartment, shattering of the toeboard, dislodging of the brake pedal and instrument panel, post-test measurements were not possible. These pictures are presented to provide a qualitative assessment of the amount of vehicle crush. The Accords struck by the redesigned Trailblazer and the Montero Sport appear to have more occupant compartment deformation, as indicated by the A-pillar deformation, than the Accords struck by the corresponding older models. The redesigned Ram 1500 and its older counterpart appear to cause about the same amount of occupant compartment deformation to the Accord.

The pre-redesigned RAM 1500 rolled over while the redesigned RAM 1500 did not rollover in the crash with the Accord. The rollover of the pre-redesigned RAM 1500 initiated after 180 msec, which was after fully engaging the front end of the Accord. However, it may override the Accord before the rollover event.

Accord crush from 1997 Blazer

Accord crsuh from 2002 TrailBlazer

Figure 28 - Honda Accords' crush after being impacted by the before and after redesign Blazer / TrailBlazer.

Accord crush from 1999 Montero Sport

Accord crush from 2001 Montero Sport

Figure 29 - Honda Accords' crush after being impacted by the before and after redesign Montero Sport.

Accord crush from 2001 RAM 1500

Accord crush from 2002 RAM 1500

Figure 30 - Honda Accords' crush after being impacted by the before and after redesign RAM 1500.

Dummy Performance – Figure 31 through Figure 34 present the injury measures for the HIII 50M drivers of the Accord and the striking vehicle in each of the paired crash tests along with the IARVS specified in the FMVSS No. 208 Advanced Air Bag Final Rule.

All the IARVs for the driver in the Accord were higher in the crash test with the redesigned vehicle than in the crash test with the corresponding older model. The only exception was the chest Gs of the driver in the Accord in the crash test with the redesigned RAM 1500, which were slightly lower than the crash test with the corresponding older model. A possible explanation of this reversal in trend may be associated with the overriding of the Accord by the pre-redesigned RAM 1500 and its eventual rollover, which did not occur in the crash test with the re-designed RAM 1500. Though the rollover event occurred after the occurrence of peak injury measures, the overriding of the Accord by the pre-redesigned RAM 1500 may have occurred earlier.

The high HIC for the driver in the Accord in the crash test with the redesigned 2002 TrailBlazer was due to head contact with the hood of the TrailBlazer. There was also head contact for the driver in the Accord with the hood of the corresponding older model (Blazer), but it was not as severe.

In all three crash tests with the redesigned vehicles, the injury measures for the driver in the Accord exceeded their respective IARVs except for HIC in the crash with the redesigned Montero Sport. In all three vehicle-to-vehicle crashes of the Honda Accord with the redesigned vehicles, the injury measures of the driver in the Honda Accord were all significantly higher than those for the driver in the redesigned striking vehicle.

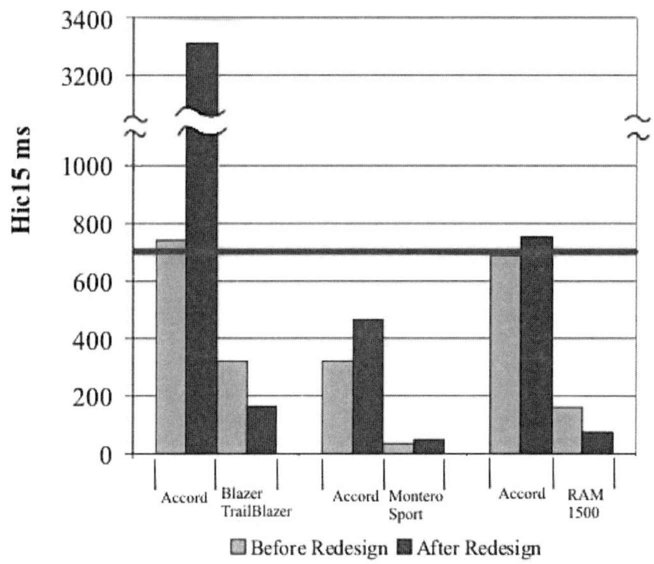

Figure 31 – HIC 15 for the HIII 50M driver in the Accord and the striking vehicle in the three paired crash tests.

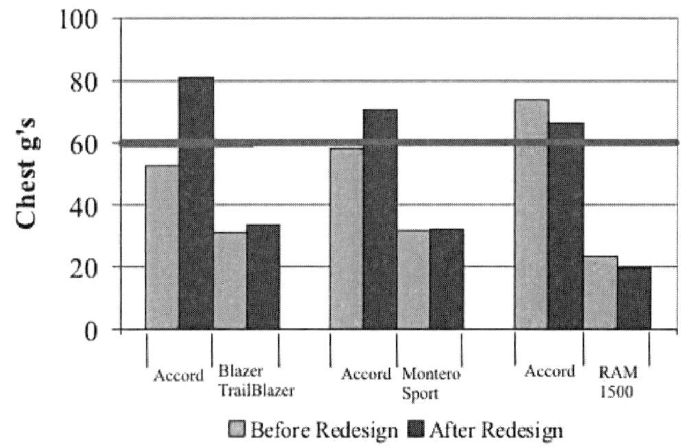

Figure 32 –Chest Gs for the HIII 50M driver in the Accord and the striking vehicle in the three paired crash tests.

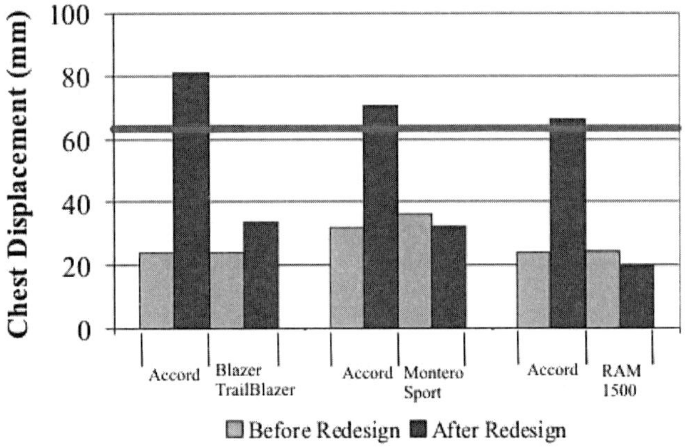

Figure 33 –Chest displacement the HIII 50M driver in the Accord and the striking vehicle in the three paired crash tests.

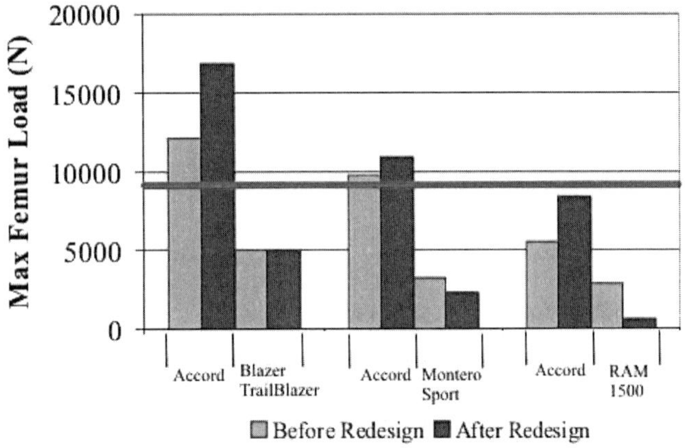

Figure 34 –Maximum femur load for the HIII 50M driver in the Accord and the striking vehicle in the three paired crash tests.

DISCUSSION

COMPARISON OF REAL WORLD CRASHES AND OFFSET DEFORMABLE TEST RESULTS: The HIC limit was exceeded in two crash tests due to delayed deployment or no deployment of the air bag. The vehicles tested were not designed to meet the requirements of the FMVSS No. 208 Advanced Air Bag Final Rule (May 2000) and so the air bag deployment was not optimized for the ODB crash tests or the associated injury criteria. It is expected that, if the vehicles met the requirements of the Advanced Air Bag Final Rule, there would be more consistency in the timing of the air bag deployment in offset crashes and the HIC limits would not be exceeded in the high speed ODB tests. While the risk of AIS 3+ chest injuries in severe real world left offset crashes is 13 percent, all vehicles tested passed the chest injury criteria. This suggests that the ODB test procedure may under-represent the risk of thoracic injury observed in real world crashes.

The Nij threshold levels were exceeded in more than half the tests with the HIII 5F dummy (Figure 10) while the risk of AIS 3+ neck injuries is relatively low (1 percent) in real world left offset crashes. The 5th percentile female occupant represents a very small percentage of the driving population. Therefore, the average injury risks obtained from real world crash data (Figure 5) do not accurately reflect the injury risks associated with a 5th percentile female. More real world crash data of 5th percentile females in offset frontal crashes is needed to better understand their injury risks.

The risk of AIS 2+ injuries below the knee in high severity offset crashes is as much as 50 percent (Figure 5), which is similar to that observed in the ODB vehicle crash tests (Table 2). In addition, the proportion of ankle injuries is 50 percent of all injuries below the knee while that of leg shaft injuries is 30 percent. This proportion of ankle to tibia shaft injuries observed in real world crashes is also reflected in the ODB test results (Table 2). This suggests that the ODB test procedure correctly reflects the risk and proportion of injuries below the knee observed severe real world offset frontal crashes. While the risk of knee-thigh-hip injuries in severe left offset real world crashes is about 33 percent, the femur force limit was not exceeded in any of the crash tests. Therefore, the ODB crash tests under represent knee-thigh-hip injuries observed in real world crashes.

The lower extremity injury assessment values in these offset tests showed no relationship to the maximum toeboard intrusion (Figure 15 through Figure 20). The analysis of real world crashes indicates that the risk of AIS 2+ lower extremity injuries is higher in a full frontal crash than an offset crash (Figure 1). This implies that maximum intrusion is not the only factor related to the incidence of lower extremity injuries.

VEHICLE-TO-VEHICLE CRASH TESTS – The paired vehicle-to-Accord crash test series indicated that the large pickup and SUV models that were redesigned and achieved a higher rating from IIHS, generally increased the potential for head, chest, and femur injuries in the driver of the Accord as compared to the corresponding older models. This suggests that the redesigned vehicles were more aggressive than their corresponding older models.

The redesigned TrailBlazer had differences in both geometry and mass. The width and the mass of the redesigned Trailblazer is 147 mm wider and 225 kg heavier then the Blazer (Table 3) and the bumper was lowered (Figure 22). Therefore, the greater injury potential for the driver of the Accord in the crash with redesigned TrailBlazer than in the crash with the older Blazer model suggest that it is related to the combined effect of changes in the vehicle mass, bumper and longitudinal member geometry, and the front-end stiffness. The individual effect of these three parameters that influence the aggressivity of the vehicle cannot be separated for the test with the TrailBlazer.

The Montero Sport and RAM 1500 had similar widths and mass for both the old design and the redesigned vehicles, and the results showed an increase in injury potential for head, chest and femur injuries on the driver of the Accord (target vehicle) as compared to the old design. However the redesigned Montero Sport had a lower bumper and a higher longitudinal member compared to the older model (Figure 22). The redesigned RAM 1500's bumper and longitudinal members were only changed slightly from the older model (Figure 22). Therefore, while the greater aggressivity of the redesigned Montero Sport compared to the older model may be related to the combined effect of changes in geometry and the front-end stiffness, the greater aggressivity of the redesigned RAM 1500 may be related to changes in its front-end stiffness alone. As indicated in Figure 26 and Table 4, the initial stiffness of the redesigned Ram 1500 computed from an NCAP full frontal barrier crash test was greater than its corresponding older model. However, the relative ranking of initial stiffness computed from a full frontal barrier crash may not be the same as that computed from an offset barrier crash. The difference in initial stiffness between a full frontal and an offset barrier could be due to the additional loads being transferred into the longitudinal beam through the transverse members, since both longitudinal members are not engaged during an offset crash.

In general, various parameters are altered while redesigning a vehicle. Therefore, it is difficult to determine the individual effect of stiffness, mass, and geometry on the aggressivity of vehicles by conducting point comparisons as presented in this paper. Among the three paired vehicle-to-vehicle crash tests presented here, only the RAM 1500 demonstrated the most direct association of increased front-end stiffness of the redesigned vehicle to its increased aggresivity. For the other two vehicles tested, the effect of stiffness on aggressivity was confounded by geometry and/or vehicle mass.

CONCLUSION

Lower extremities are the most frequent AIS 2+ injured body region in frontal crashes of air bag equipped vehicles and 50% of these injuries are below the knee. Foot and ankle injuries account for 50 percent of below knee injuries in frontal crashes. The risk of knee-thigh-hip injuries is higher in newer vehicle models than in older vehicle models while the risk of injuries below the knee has remained relatively steady. The risk of knee-thigh-hip injuries is higher in SUVs than in passenger cars and minivans.

The ODB test procedure appears to predict the risk and proportion of injuries below the knee in severe real world offset crashes. However, it under represents thoracic and knee-thigh-hip injuries that are observed in real world crashes.

The magnitude of toe board intrusion did not correlate to the injury risk measures from the Thor-Lx/HIIIr for injuries below the knee.

Three redesigned vehicles used in this study that obtained a better rating in the IIHS ODB tests than their respective older models, were found to be more aggressive in vehicle-to-vehicle crash tests than their older counterparts. In two of the vehicle models tested, the effect of increased stiffness on their aggressivity was confounded by the effect of increased mass and a change in geometry of the vehicle. In one of the paired vehicle-to-vehicle crash tests, the greater aggressivity of the redesigned vehicle may be related to its greater front-end stiffness as was noted from the NCAP barrier crash test data. However, greater relative front-end stiffness in a full frontal barrier crash may not necessarily imply a greater relative front-end stiffness in an offset barrier crash test. Further study is needed on this issue.

REFERENCES

INSIA report, "Structural Survey of Cars, Methodology of the Main Resistant Elements in the Car Body", March 1999

C. Kahane, "Vehicle Weight, Fatality Risk and Crash Compatibility of Model Year 1991-99 Passenger Cars and Light Trucks," NHTSA Technical Report, DOT HS 809 662, October 2003.

S. Kuppa, J. Wang, M. Haffner, Rolf Eppinger, "Lower Extremity Injuries and Associated Injury Criteria," 17th International Technical Conference on the Enhanced Safety of Vehicles, Amsterdam, The Netherlands, June 4-7, 2001a.

S. Kuppa, M. Haffner, R. Eppinger, J. Saunders, "Lower Extremity Response and Trauma Assessment using the Thor-Lx/HIIIr and the Denton Leg in Frontal Offset Vehicle Crashes," 17th International Technical Conference on the Enhanced Safety of Vehicles, Amsterdam, The Netherlands, June 4-7, 2001b.

S. Kuppa, O. Fessahaie, "An Overview of Knee-Thigh-Hip Injuries in Frontal Crashes in the United States," 18th International Technical Conference on the Enhanced Safety of Vehicles, Nagoya, Japan, May 19-22, 2003.

J. Nolan, Lund, A., "Frontal Offset Deformable Barrier Crash Testing and its Effect on Vehicle Stiffness," 17th International Technical Conference on the Enhanced Safety of Vehicles, Amsterdam, The Netherlands, June 4-7, 2001

B. Park, S. Partyka, R. Morgan, J. Hackney, J. lee, L. Summers, J. Lowrie, N. Beuse, "Comparison of Vehicle Structural Integrity and Occupant Injury Potential in Full-frontal and Offset-frontal Crash tests," SAE Paper 2000-01-0879.

P. Petit, X. Trosseille, "Comparison of the Thor, Hybrid III, and Cadaver Lower Leg Dynamic Responses in Dorsiflexion," Forty-Third Stapp Car Crash Conference, Paper No. 99SC10, 1999.

R. Rudd, J. Crandall, J. Butcher, "Biofidelity Evaluation of Dynamic and Static Response Characteristics of the Thor Lx Dummy Lower Extremity," International IRCOBI Conference on the Biomechanics of Impact, Spain, 1999.

Stucki, Sheldon L., Hollowell, William T., and Fessahaie, Osvaldo, "Determination of Frontal Test Conditions Based on Crash Data," Sixteenth International Technical

Conference on Enhanced Safety of Vehicles, Windsor, Canada, June, 1998.

Swanson, J., Rockwell, T., Beuse, N., Summers, L., Summers, S., Park, B., "Evaluation of Stiffness Measures from the U.S. New Car Assessment Program," 18th International Technical Conference on the Enhanced Safety of Vehicles, Nagoya, Japan, May 19-22, 2003.

Verma, M., Lange R., Levelle J., "Relationships of Crash Test Procedures to Vehicle Compatibility," Paper 2003-01-0900, Society of Automotive Engineers, March 2003

APPENDIX

Table A. 1 – Injury assessment values for the HIII 50 th percentile male with Thor-Lx/HIIIr legs driver

Make	Model	Year	NHTSA Test #	Closing Speed (kmph)	Chest G	Chest Disp (mm)	HIC 15	Nij ver. 10	Neck Tension (N)	Neck Comp. (N)	Max Femur (N)	Max Knee Shear (mm)	Max Up TI	Max Low TI	Max Upper Tibia Axial Force (N)	Max Lower Tibia Axial Force (N)	Max Dorsi-flexion Angle (degrees)	Max Inversion / Eversion Angle (degrees)	Max Resultant Toepan/ Footrest (mm)
			IARV		60.0	63.0	700.0	1.00	4170.0	4000.0	9040.0	15.00	0.91	0.91	5600.0	5200.0	35.0	35.0	
NISSAN	QUEST	2002	4439	56.0	27.6	26.2	148.8	0.21	602.2	133.7	2893.6	3.27	0.51	1.11	3242.1	5554.6	42.2	20.5	201
LINCOLN	NAVIGATOR	2003	4441	56.0	31.7	20.1	731.5	0.35	1758.4	234.4	3607.7	2.62	0.25	0.44	930.8	2191.6	27.4	16.2	181
DODGE	NEON	2002	4428	56.1	40.0	26.5	610.5	0.40	1499.3	50.7	5351.2	5.33	0.88	0.86	3418.2	4802.8	39.7	26.8	201
TOYOTA	CAMRY	1996	3459	60.7	30.7	24.5	245.6	ND	944.2	693.4	2893.3	2.56	0.38	0.39	1205.7	2344.1	30.4	31.0	189
DODGE	NEON	1998	3466	60.8	38.6	28.9	271.4	ND	1707.8	442.4	4611.1	4.32	0.71	0.96	3915.8	6099.3	38.4	33.4	282
NISSAN	QUEST	2000	3857	59.5	28.1	23.3	304.8	0.31	1723.8	207.5	3491.4	2.64	2.23	0.81	2778.2	3920.1	38.4	31.3	387
CHEVROLET	TAHOE	2000	3855	60.4	46.5	21.2	180.5	0.29	1565.1	170.6	6303.8	7.50	1.13	0.94	7648.8	9404.1	28.4	26.6	341
NISSAN	ALTIMA	2002	4461	60.2	36.1	24.6	132.6	0.24	704.9	402.2	2810.1	4.06	0.40	0.62	1857.2	3505.7	39.2	32.0	110
NISSAN	ALTIMA	2000	3460	63.6	32.1	23.6	146.5	0.18	1022.0	54.4	2701.9	11.19	0.54	0.80	2485.8	3653.0	25.0	35.4	526
SUBARU	LEGACY	2000	3467	64.5	45.8	26.7	213.0	0.37	1956.9	574.5	3264.2	10.60	0.44	0.72	2268.9	2942.8	29.2	36.7	321
DODGE *	NEON	2002	4442	56.1	33.9	20.1	525.0	0.29	835.5	56.7	2949.3	3.41	0.72	0.65	2343.7	3323.2	40.0	34.1	166

*Data for passenger dummy in a right side impact

Table A. 2 - Injury assessment values for the HIII 5 th percentile female with Thor-FLx/HIIIr legs driver

Make	Model	Year	NHTSA Test #	Closing Speed (kmph)	Chest G	Chest Disp (mm)	HIC 15	Nij ver. 10	Neck Tension (N)	Neck Comp. (N)	Max Femur (N)	Max Knee Shear (mm)	Max Up TI	Max Low TI	Max Upper Tibia Axial Force (N)	Max Lower Tibia Axial Force (N)	Max Dorsi-flexion Angle (degrees)	Max Inversion / Eversion Angle (degrees)	Max Resultant Toepan/ Footrest (mm)
			IARV		60.0	52.0	700.0	1.00	2620.0	2520.0	6510.0	13.00	0.91	0.91	4000.0	3750.0	35.0	35.0	
DODGE	NEON	2002	4377	56.0	44.1	35.3	1201.8	1.51	3532.7	183.4	3550.1	4.06	1.02	1.17	2468.7	4005.3	46.0	18.5	206
NISSAN	ALTIMA	2002	4431	56.1	37.1	22.0	39.4	0.58	1009.3	307.7	1911.6	0.55	0.29	0.39	1230.6	1986.0	30.5	17.8	172
TOYOTA	CAMRY	1996	3664	59.9	31.8	32.9	141.7	1.24	1600.6	158.8	3050.8	1.07	0.42	0.48	995.4	2213.5	39.1	22.4	176
SUBARU	LEGACY	2000	3665	59.9	36.4	26.2	146.0	1.07	1702.0	77.4	2935.6	1.55	0.38	0.58	1545.9	3006.4	34.7	34.7	150
NISSAN	ALTIMA	2000	3666	60.0	33.5	22.6	110.9	1.28	1629.8	349.7	4582.9	3.78	0.82	1.55	2930.0	5519.3	41.7	37.1	228
DODGE	NEON	1998	3667	60.4	45.4	28.1	611.0	0.56	2080.7	842.9	3152.2	3.48	0.83	0.96	2425.6	4086.6	44.0	28.5	205
NISSAN	QUEST	2000	3856	60.0	28.8	15.6	96.7	0.49	1080.5	119.6	3291.0	3.54	0.79	0.83	2555.2	3628.3	39.7	27.6	334
FORD	EXPLORER	2000	3850	59.9	31.8	31.0	111.3	1.32	1741.8	444.9	3700.8	0.56	0.38	0.42	1519.4	1908.3	22.4	18.7	482
NISSAN	ALTIMA	2002	4440	59.8	32.2	26.6	76.8	0.43	907.9	337.3	2019.3	0.58	0.61	0.94	2058.9	3586.0	26.6	29.3	131

986059

COMPATIBILITY REQUIREMENTS FOR CARS IN FRONTAL AND SIDE IMPACT

Nial J. Wykes
Mervyn J. Edwards
C. Adrian Hobbs
Transport Research Laboratory
United Kingdom
Paper Number 98-S3-O-04

ABSTRACT

In support of the European Enhanced Vehicle-safety Committee (EEVC) research programme and through it, the International Harmonisation of Research Activities work on compatibility, TRL is investigating the compatibility of cars in frontal and side impact scenarios. Initial research has focused on identifying the major factors which influence compatibility and determining the extent to which they might influence injury outcome. Experimental crash test research is backed with Finite Element simulation modelling. For frontal impacts, full scale testing has been used to examine the influence of vehicle mass, stiffness, structural interaction and geometry. The modelling work has studied how non contact, deceleration related injuries might be minimised by optimising the deceleration pulse. For side impact, full car finite element models have been used for parametric studies to aid our understanding of the effects of the bullet vehicle mass, geometry and stiffness and to help predict more compatible designs. This has been backed by full scale crash testing, aimed at determining the ideal characteristics of interacting car front and side structures. All of this work is aimed at developing crash test requirements that are capable of assessing a car's compatibility.

INTRODUCTION

It is now generally accepted by the international vehicle safety community that the 'compatibility' of vehicles needs to be addressed. The development of vehicles which simply 'self protect' will no longer be acceptable and growing emphasis must be placed upon the protection of all road users.

Each subsystem within the global fleet may currently possess its own unique 'incompatibilities'. A large number of U.S. publications are written about side impact and the problem associated with 'aggressive' sport utility vehicles and their mismatch with the midsize family saloon. This problem is exacerbated by a road system that gives rise to a large number of side impacts. In European publications the emphasis is on frontal impact. However, it is well established that a compatible fleet can not be achieved through concentrating on a single impact scenario.

Research, guided by EEVC WG 15, which has been set up to study compatibility, will concentrate on both frontal and side car to car impacts. In addition it will also consider the requirements for car impacts with pedestrians, large vehicles and roadside obstacles. This should ensure that developments to improve car to car impacts do not reduce protection in these other types of impact.

In 1995 TRL commenced compatibility research stimulated by their findings from frontal and side impact testing. The programme is funded by the UK Department of Environment, Transport and the Regions and the European Commission. The work contributes to EEVC WG 15 and the International Harmonisation of Research Activities Compatibility Working Group.

The programme aims to identify how vehicle safety may be improved by developments to vehicle structures which are designed to interact better in an impact, and subsequently implement these changes in the vehicle fleet. This requires an understanding of the factors which influence compatibility and the development of new or modified legislative procedures to bring about greater compatibility. The project also aims to identify the potential benefits that could be obtained from improved compatibility.

This paper discusses those factors that are currently seen as having the greatest influence on compatibility and reports on the work carried out so far.

CAR TO CAR IMPACTS

Current activities are focusing on car to car frontal impact and car to car side impact separately, in order to understand more clearly the controlling factors. It is envisaged that conflicting requirements will exist and that a compromise will have to be established. It is possible that the current fleet differences between the U.S., Europe and other continents, will encourage researchers to reach differing compromises. It is for this reason that great importance must be placed upon international harmonisation in an attempt to reach a single, internationally appropriate, assessment procedure. One step towards compatibility may be the worldwide harmonisation of legislative and consumer testing.

CAR TO CAR FRONTAL IMPACT

This section discusses those parameters which are currently seen as having the greatest influence on compatibility in frontal impact and presents the results from some of the work carried out so far.

Collision Type

Collision type is the most important factor governing vehicle performance since it determines how the two car's structures will interact. For this reason, collision type will have a modifying influence on the effect of other parameters such as geometry and stiffness. For example, the effective global stiffness of a vehicle changes significantly from a full frontal impact to frontal offset impact, highlighting the importance of selecting the appropriate testing procedure (1). In this example, the differences in effective stiffness between the two collision types are attributed to the specific structures used for energy absorption and the modes by which they fail.

Geometry

Car structures are designed to perform a multitude of functions as well as good crashworthiness. As a consequence, car fronts often have local areas of stiff structure within a much larger area of weaker structure. Due to this lack of stiffness uniformity it is usual in a car to car frontal impact, for the stiff parts of one vehicle to penetrate the weaker parts of the other. This may result in penetration fork effect or over-ride. Since in these situations the vehicle's energy absorbing structure is ineffectively used, higher occupant compartment intrusions are often observed. In one reported offset frontal test between a small and medium sized car, the poor structural interaction masked the effect of mass and stiffness and dominated the outcome (2). In this test the structure of the small car over-rode that of the medium sized car.

More recently this test has been repeated. The aim was to select a vehicle which would remove the previously reported geometrical incompatibilities and permit the assessment of mass and stiffness influence. The chosen substitute was a car that performed structurally well in the European Offset Deformable Barrier (ODB) test, whilst being of a similar mass to the car it replaced. In this repeat test the medium sized car, possessing better frontal structure tie-up, virtually eliminated over-riding (Figure 1.).

Figure 1. Medium size passenger car with well-connected front structures minimised over-ride by the small car.

Unfortunately, geometrical compatibility is not straight forward to assess. This was highlighted in an offset frontal car to car impact carried out between an off road vehicle and the medium size passenger car. It can be seen from the pre test photographs that there was a considerable difference in the vertical height of the significant structures of the two vehicles (Figure 2.).

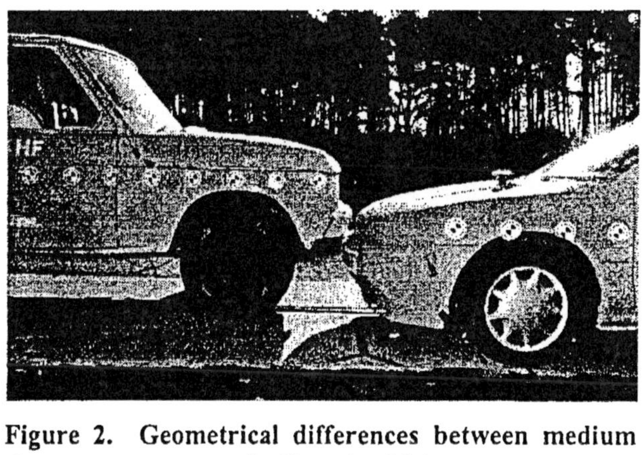

Figure 2. Geometrical differences between medium size passenger car and off road vehicle.

The results were contrary to expectations. The geometry of the two vehicles interacted well and there was no significant over-riding. In this particular test, the chassis rail of the off road vehicle engaged the suspension turret of the passenger car becoming retained by the engine mount attachment. A significant proportion of the loading was then transmitted through the engine and onto the firewall (Figure 3.).

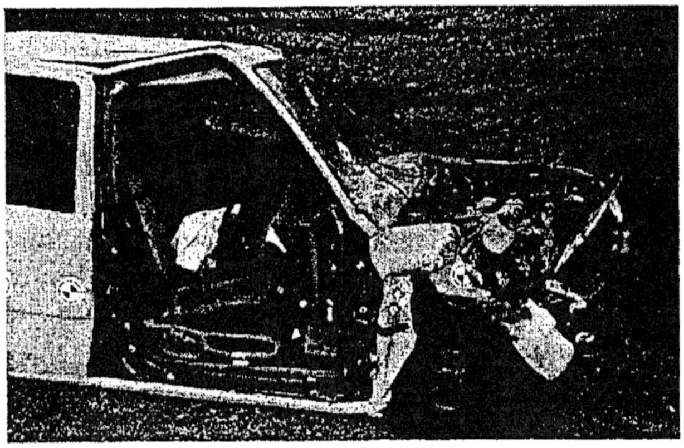

Figure 3. Good geometrical interaction resulted in no significant over-riding by impacting 1900 kg off road vehicle, note intrusion at facia level.

The resulting deformation of the car's front structure and door aperture were not dissimilar to the results obtained from a car to car impact between two of the same medium size vehicles (Figure 4.)

Figure 4. Deformation of medium sized car after offset frontal impact with identical vehicle.

One significant difference, however, was the intrusion at facia level. In the car impacted by the off road vehicle, the additional loading directed onto the fire wall resulted in high occupant compartment intrusion, and as a consequence caused a higher chest deflection on the dummy. An important observation to note from this test is that the structures of two apparently incompatible vehicles interacted in such a way as to virtually eliminate over-riding. Even though the good structural performance was not reflected in satisfactory dummy injury criteria (high chest deflection), this test goes part way to demonstrate that compatibility between dissimilar vehicles may be achievable.

In all of the tests commented on, it is a fair observation to make that geometry is not the only variable. Hence, in order to study the importance of structural compatibility, tests need to be conducted which have the sole variable of geometry.

A medium size, car to car offset test was performed using identical vehicles, at the same test weight, but having undergone modifications to their ride heights. The vertical difference in height was 100mm, well within the fleet variations for this segment of car (3).

In the lowered car, a large proportion of the loading was carried by the upper load path (Figure 5.). This resulted in excessive intrusion at facia rail level and to a higher degree than that observed in the car impacted by the off road vehicle.

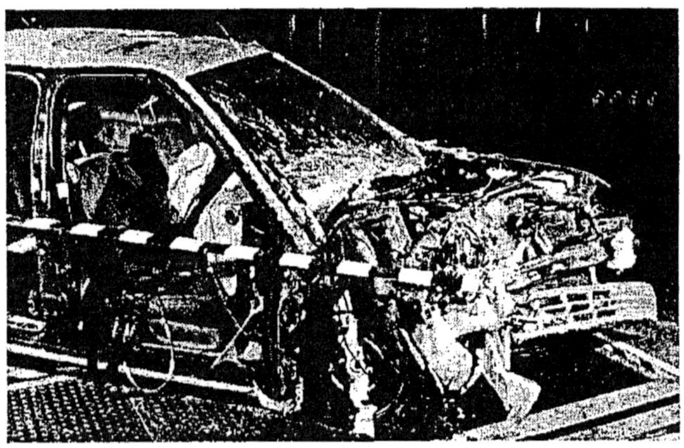

Figure 5. Lowered car, note 'A' pillar pushed down and high deformation at facia rail level.

In the lowered car the front profile was noticeably sloped back, and the 'A' pillar had been pushed downwards. The excessive deformation of the top load path meant the suspension turret had been displaced 480mm rearwards. By contrast, the suspension turret of the higher car moved rearwards by only 295mm. In this vehicle the 'A' pillar was pushed upwards increasing the deformation seen on the cant rail. More use was made of the lower load path and the frontal profile can be seen to be flatter (Figure 6.).

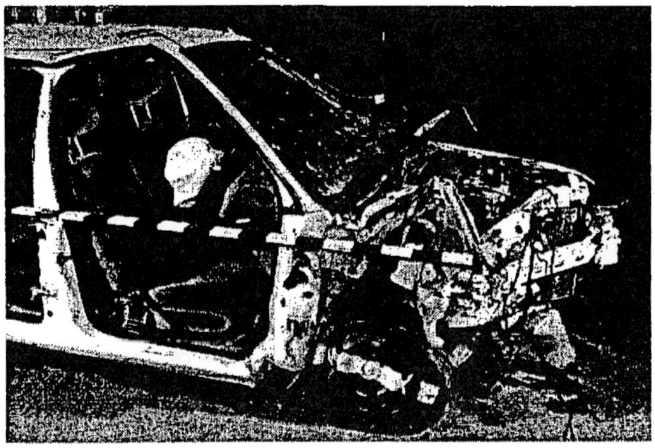

Figure 6. Raised car - note 'A' pillar pushed up and little deformation at facia rail level.

In this test the vehicle's mass, stiffness and occupant compartment strength were matched. However, the 100mm vertical height difference was sufficient to noticeably change the vehicles structural response. When considering that the vertical height difference of the two vehicles was less than the height of the bumper beam, these results become significant. Until vehicle designs enable structures to interact better in car to car impacts, any compatibility improvements in mass ratios, or stiffness matching, are unlikely to be fully realised. The authors suggest that in order to achieve a compatible fleet it will be necessary to firstly establish good geometrical compatibility.

Mass and Structural Stiffness

The effect of mass and mass ratio on injury risk have been studied for many years and dominate the literature on compatibility. There is plenty of evidence that in car to car impacts the risk of injury in the heavier car is lower than that in the lighter car (4). However, whilst the authors do not dispute that mass has a significant effect on injury, it must be remembered that mass is a surrogate measure for other factors such as vehicle size, length of front structure and presence and quality of safety features. These factors could help to exaggerate the mass effect. It is also recognised that mass is a common parameter to record during accident investigation and as a consequence is available for data analysis.

In frontal accidents, occupants of lower mass vehicles often have higher injury risks due to both lower vehicle mass and stiffness (5,6). Ignoring crashworthiness, heavier cars are stiffer for other reasons, with structures often being stronger to take the higher engine and suspension loads (2). There is, however, no certainty that the stiffness differential of small and large cars will continue to exist at the current ratio. There is clear evidence that the stiffness of smaller cars is increasing as a consequence of new crash test requirements.

There is virtually universal agreement amongst independent accident investigators that passenger compartment intrusion is a major cause of fatal and serious injuries to restrained car occupants. This view is also supported by many accident investigators employed by car manufacturers (5). In order to limit and manage occupant compartment intrusion in car to car frontal impacts, the crush zones of the cars involved must be capable of absorbing the full energy of the impact. A number of simple concepts have been proposed to control the global stiffness of the vehicle and hence improve compatibility.

The first proposal to note is that of a 'semi-rigid' passenger compartment (7). This suggestion involves designing the vehicle in such a manner that the occupant compartment is sufficiently stiff so that it can resist the deformation force put on it by any colliding car. This ensures that the impact energy is absorbed by the front structures of both cars.

Another suggestion has be made in which a limit is placed on the maximum crush force that must not be exceeded in a given impact (8). This concept has been extended with suggestions being made that both maximum and minimum force requirements are needed or that force corridors should be defined. These proposals are effectively equivalent to controlling the vehicle's stiffness.

Two fundamental questions need to be answered before pursuing either of these logical proposals:

1. What test could be performed to ensure that the occupant compartment could withstand the maximum crush force level?
2. What should the crush force level(s) be?

Currently there is no internationally recognised or legislative test which is capable of addressing the first question. It would be necessary to introduce an additional test to make such an assessment. Two possibilities may be a quasi-static crush, or a frontal test at an elevated speed. Neither of these solutions are however without drawbacks.

In order to be in a position to address the second of these questions it is important to have an understanding of current interface force levels between the impacting cars, and to be able to measure them in both barrier and car to car impacts. The proposals which have been commented on in this paper are theoretical and the first step must be to establish if such requirements could be achieved.

Interface force can be obtained through the use of a load cell wall or the measurement of the deceleration of the constituent parts of the vehicle. Both of these methods have been used at TRL and for barrier tests have been shown to be in good agreement (Figure 7.).

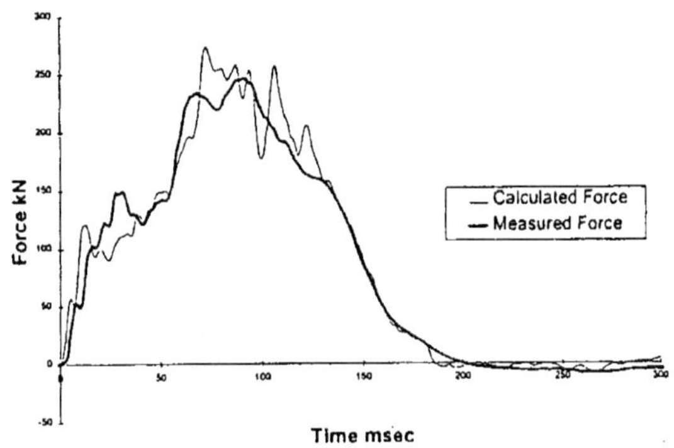

Figure 7. Interface force from frontal ODB test, note agreement between load cell wall data and interface force calculated from accelerometers.

In order to calculate the interface force from the car's deceleration, it is necessary to select and instrument discrete components and areas of structure. The selection is made based upon how it is believed these discrete areas of the vehicle will move and deform during the impact. It is assumed that the selected area will act as a lumped mass and the deceleration force for each lumped mass can be calculated from its associated deceleration. The summation of these forces gives the overall interface force (Figure 8.).

Differing views exist on how the total interface force could be limited or controlled. However, it should be possible to exercise control over vehicle local stiffness by limiting the interface force distribution. This type of control could be one way to improve geometrical compatibility.

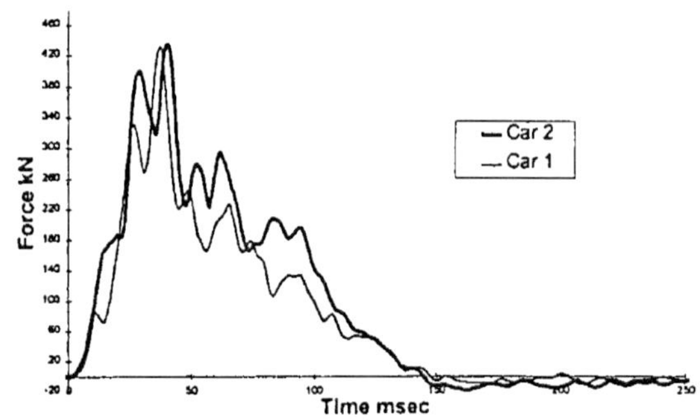

Figure 8. Interface force calculated in car to car offset impact.

With the semi rigid passenger compartment and the crush force limit theoretical concepts (7,8), it would be possible to achieve the requirements without controlling the shape of the deceleration pulse. It is the form of deceleration seen by the occupant compartment that drives the performance of the restraint system, and influences any restraint system induced injury. If we assume that our compatible vehicle has minimised intrusion due to the 'rigid' occupant cell, this type of injury will be dominant.

For this reason a study was initiated into the influence of the deceleration pulse on restraint system related injuries for frontal impact, with the aim of modifying the deceleration pulse to minimise this type of injury. Having identified the most desirable shape we will, in later work, assess if it is possible to achieve such a pulse in real cars.

<u>Computer Simulation</u> was selected as the most appropriate method by which to address this question. The MADYMO software package was used to simulate the deceleration of the occupant compartment. (Figure 9.).

Various simple shaped, analytical and experimental deceleration pulses were applied to the model in the fore aft direction. The Fourier Equivalent Wave (FEW) method (9) was used to generate a wide range of analytical deceleration pulses with predetermined ride down distances and times. This method sums three sine waves to produce a deceleration pulse :

$$a(t) = \sum_{N=1}^{3} a_n \sin(n\omega t) \qquad (1.)$$

Figure 9. Occupant compartment model consisting of a HYBRID III dummy held by a typical restraint system.

Chest injury criteria were monitored as the chest is directly loaded by the seat belt. Peak chest compression was used as the measure of chest injury in the current study in preference to Viscous Criteria. Previous researchers who have studied the mechanism of impact induced soft tissue injury show that for a properly designed restraint system, the best indicator of injury is the peak chest compression (10).

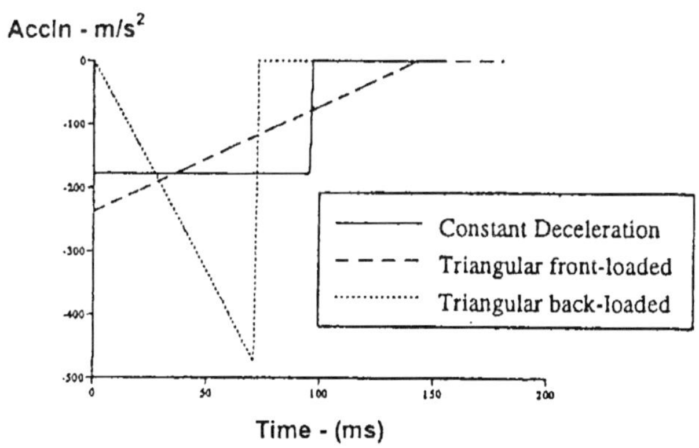

Figure 10. Comparison of simple shaped deceleration pulses for a ride down distance 0.8m.

Table 1.
Chest Injury; Simple Deceleration Pulse Shapes, Ride down Distance 0.8m, Initial Velocity 61km/hr.

Shape	Ride Down Time (msec)	Chest Injury	
		Peak Comp (mm)	Peak Accln (ms^{-2})
Constant	95	41 @62ms	359 @59ms
Triangular front loaded	142	44 @57ms	408 @55ms
Triangular back loaded	71	56 @78ms	618 @76ms

Investigations of the relationship between Hybrid III sternal deflection and thoracic injury severity on occupants have established that there is a 5 percent risk of injury greater or equal to AIS3 for a chest compression of 22mm. For a chest compression of 50mm the risk of injury greater or equal to AIS3 increases to 50 percent (11). Chest acceleration is also shown as a supplementary measure of injury.

The first parameter sweep applied three simple shaped deceleration pulses; constant, front loaded triangular and back loaded triangular, to the model to produce a ride down distance of 0.8 m (Figure 10.). This is a typical ride down for a 50 percent offset car to car impact. The results show that the constant pulse gives the lowest chest compression and the triangular back loaded the highest (Table 1.). An explanation for this emerges upon examination of the seat belt loads (Figure 11.).

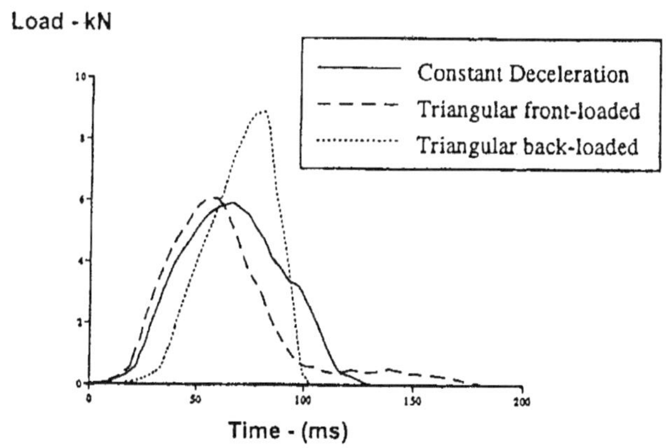

Figure 11. Comparison of seat belt loads for simple shaped deceleration pulses, ride down distance 0.8m.

The high deceleration at the beginning of the constant pulse, loads the occupant into the restraint system early. The constant deceleration, applied throughout the duration of the long ride down time, keeps the load applied over as long a time as possible to give a well rounded chest compression profile with a low peak (Figure 12.).

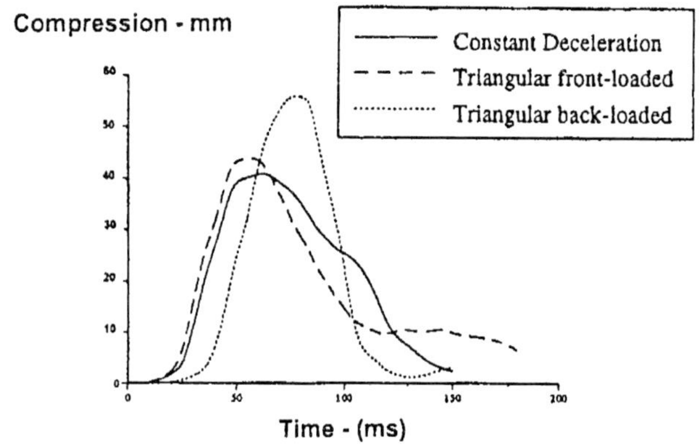

Figure 12. Comparison of chest compression for simple shaped deceleration pulses, ride down 0.8m.

In contrast, the low deceleration at the beginning of the back loaded triangular pulse loads the occupant into the restraint system late. As a consequence, a high restraining load is applied to the occupant for a short time, which results in a high peak chest compression. This also causes the peak chest compression to be at a much later time compared to the constant pulse.

A similar result is seen for a ride down distance of 1.2m which is a typical ride down for a car to ODB impact (Table 2.).

Table 2.
Comparison of Chest Injury; Simple Deceleration Pulse Shapes, Ride down Distance 1.2m, Initial Velocity 61km/hr.

Shape	Ride down Time / msec	Chest Injury	
		Peak Comp (mm)	Accln (ms^{-2})
constant	173	31 @67ms	226 @63ms
triangular front loaded	214	35 @63ms	266 @60ms
triangular back loaded	107	40 @104ms	404 @104ms

It is interesting to note that decreasing the ride down distance from 1.2 m to 0.8 m for a constant pulse has approximately the same effect on chest compression as keeping the ride down distance constant at 1.2 m and changing the shape of the pulse from a constant to a triangular back loaded pulse. These results are supported by previous work carried out at TRL (12).

Parameter sweeps were conducted using the analytically derived FEW deceleration pulses for various ride down distances. For this part of the study the airbag and steering wheel assembly was removed from the model. This eliminated the sensitivity of results to airbag trigger time. The double humped FEW pulse that gave the lowest chest compression for a ride down of 0.8m, is shown (Figure 13.), together with a graph of chest compression (Figure 14.).

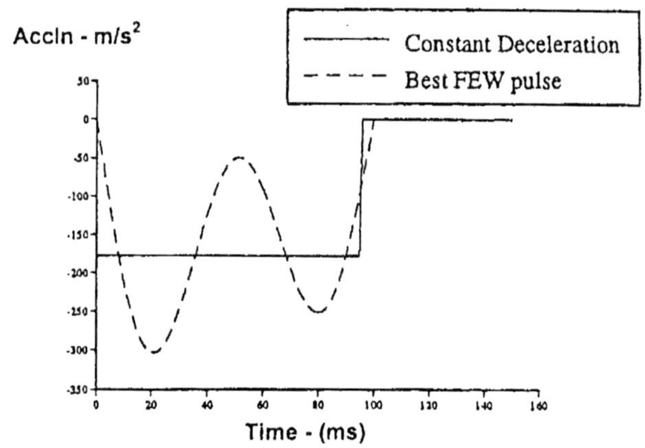

Figure 13. Comparison of doubled humped FEW and constant deceleration pulses.

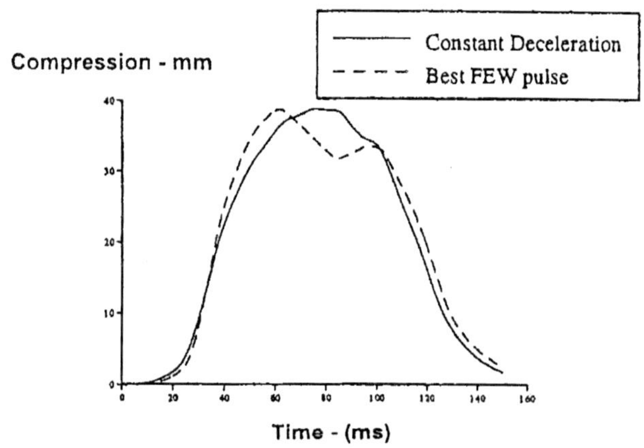

Figure 14. Comparison of chest compression for double humped FEW and constant deceleration pulses.

If these graphs are examined in detail it is seen that the double humped FEW pulse gives a low chest compression because of a similar behaviour to the constant pulse, i.e. the high deceleration at the beginning of the pulse loads the occupant into the restraint system early and the high deceleration at the end keeps the load applied for as long as possible.

Summarising, the results from this study indicate that :

1. Peak chest compression is minimised by maximising ride down distance.
2. Peak chest compression is minimised by having a passenger compartment deceleration pulse profile that is constant in shape as opposed to triangular back loaded.

The next part of the study abstracted the fore aft occupant compartment deceleration pulses from the following crash tests:

1. Car to car frontal impact with 50 percent overlap at an initial velocity of 56 km/hr.
2. European ODB test with 40 percent overlap but at an initial velocity of 61 km/hr.
3. Light car impacted by heavy car (mass ratio 1.4) with 50 percent overlap at an initial velocity of 56 km/hr.
4. Heavy car impacted by light car (mass ratio 1.4) with 50 percent overlap at an initial velocity of 56 km/hr.
5. Car to rigid barrier test with 100 percent overlap at an initial velocity of 56 km/hr.

It should be noted that in all of these crash tests the same model of mid sized family car was used. These pulses were applied to the model, in order to determine occupant response without any passenger compartment intrusion. Two restraint systems were modelled; the first was a seatbelt only system, and the second a seatbelt and airbag system. The results of applying these pulses to the model for both restraint systems are shown below (Table 3.).

Firstly, we will discuss the differences between the model response for the car to car 50 percent overlap and the ODB deceleration pulses. A comparison of the deceleration pulses is shown (Figure 15.).

For the seatbelt only system, it is seen that the chest compression is higher for the ODB deceleration pulse (Figure 16.). This shows that the advantage of having a longer ride down distance is out-weighed by the increase in velocity and the change in the shape of the pulse, i.e. to a more triangular back loaded form.

Table 3.
Chest injury model results obtained by applying deceleration pulses abstracted from crash tests.

Deceleration Pulse Description	ΔV (km/hr)	Ride-down Distance (m)	Seatbelt restraint system		Seatbelt and airbag restraint system	
			Chest Comp (mm)	Chest Accln (ms^{-2})	Chest Comp (mm)	Chest Accln (ms^{-2})
1. Car to car 50% overlap	56	1.0	37	305	38	327
2. Offset Deformable Barrier 40% overlap	61	1.2	41	335	35	341
3. Light car impacted by heavy car 50% overlap	65	0.8	42	370	44	422
4. Heavy car impacted by light car 50% overlap	47	1.2	33	252	33	257
5. Full frontal rigid barrier 100% overlap	56	0.7	43	365	43	405

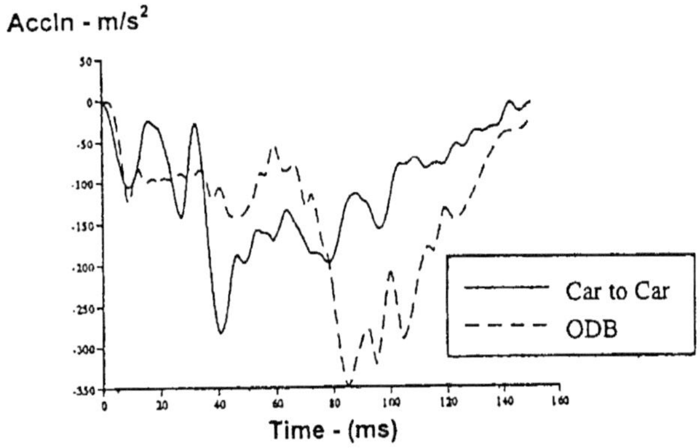

Figure 15. Comparison of deceleration pulse shape for ODB and 50 percent car to car.

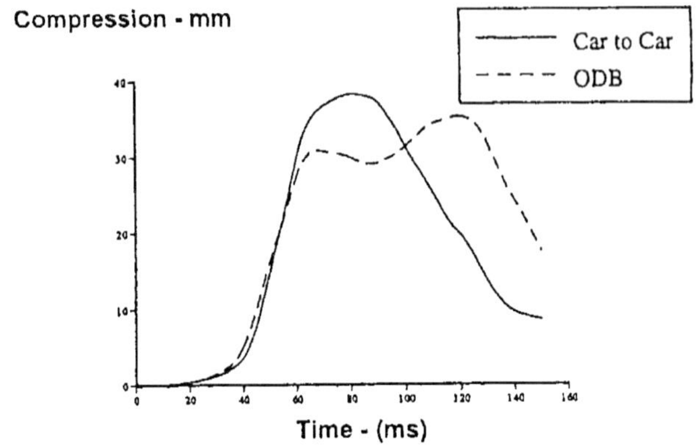

Figure 17. Comparison of chest compression for car to car and ODB impacts with seatbelt and airbag restraint system.

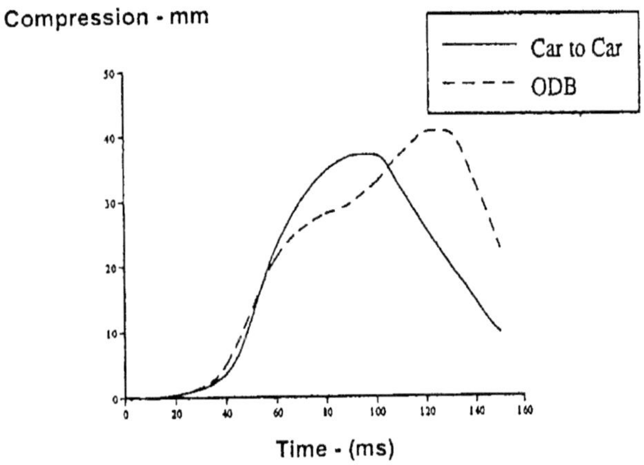

Figure 16. Comparison of chest compression for car to car and ODB impact with seatbelt restraint system.

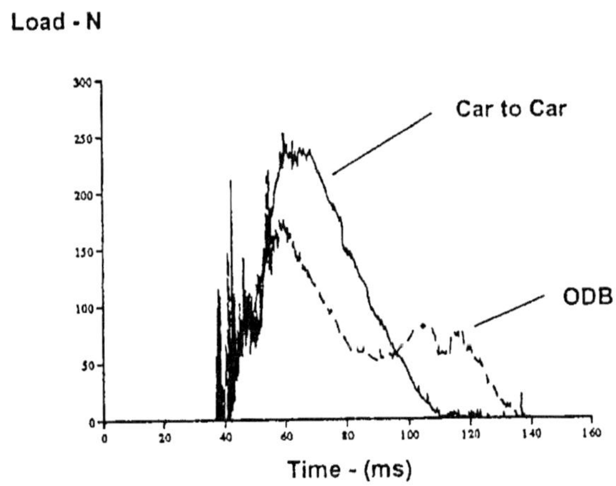

Figure 18. Comparison of sternum load from airbag for car to car and ODB impacts.

For the seatbelt and airbag system it is seen that the chest compression is higher for the car to car deceleration pulse (Figure 17.). The reason for this is that the higher deceleration at the beginning of the car to car pulse causes the occupant to contact the airbag with a higher velocity giving a high sternum to airbag load (Figure 18.), so causing higher chest compression. One possible explanation for this is that the restraint system is 'tuned' to perform well in the ODB test. Manufacturers should be aware that the additional deceleration at the beginning of a car to car pulse compared to an ODB pulse can cause the occupant to contact the airbag with a greater velocity hence causing higher chest loads.

This would be difficult to allow for by adjusting the airbag trigger time, as the beginning of the pulses are very similar.

Secondly, we will compare the differences in the model response by applying the deceleration pulse from the car to car test with the pulses from the cars in the light to heavy car test (Figure 19.). The deceleration pulse from the light car has a more triangular back loaded shape, whilst the pulse from the heavy car has a more constant form. In addition the ride down is less and the change in velocity greater for the lighter car (Table 3.). It should be noted again that these cars are the same model and hence have the same stiffness

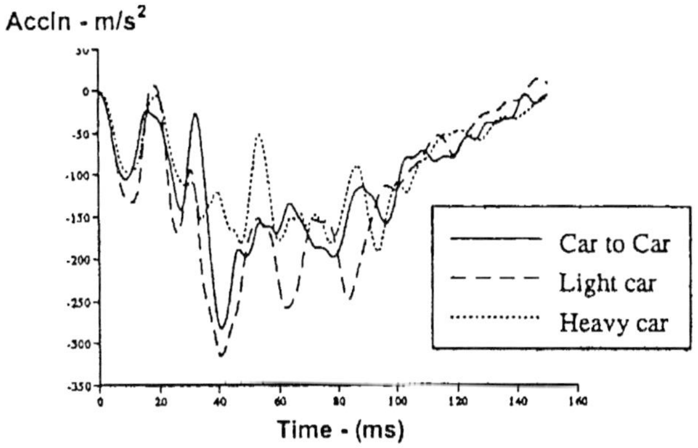

Figure 19. Comparison of deceleration pulse shapes for car to car and light to heavy car impacts.

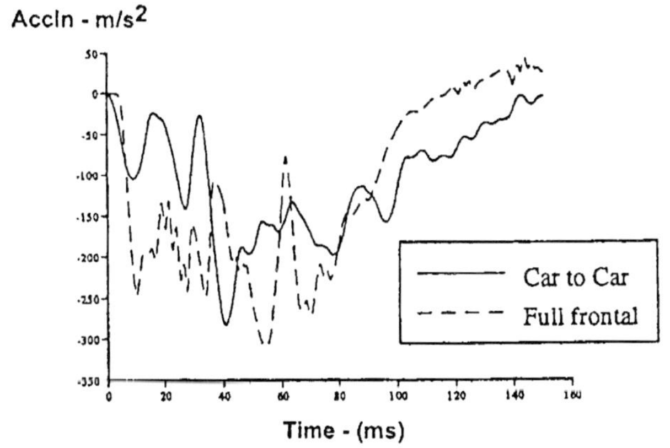

Figure 21. Comparison of deceleration pulse shapes for car to car and full frontal rigid wall impacts.

In summary, in a car to car impact with cars of a similar stiffness, the occupant in the heavier car is subjected to a deceleration pulse that causes least injury in terms of ride down distance and shape. The converse is true for the occupant of the lighter car (Figure 20.).

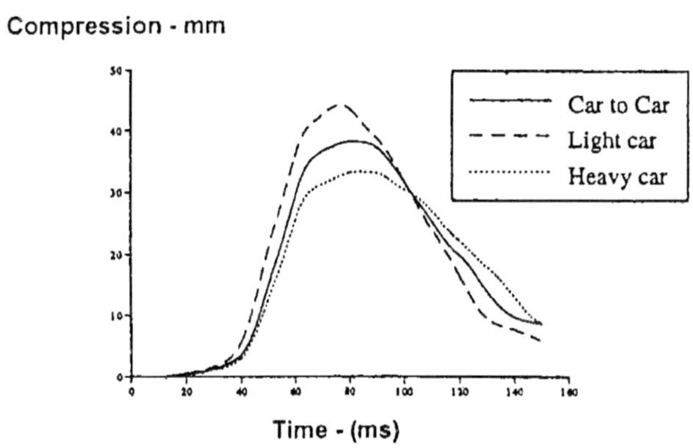

Figure 20. Comparison of chest compression for car to car and light to heavy car impacts.

Finally, we will consider the results obtained by applying the deceleration pulse from the full frontal rigid barrier test to the model (Figure 21.). The ride down distance for this deceleration pulse is the lowest (0.7 m) and hence considering this factor alone would be expected to give a high peak chest compression. In fact, the chest compression is not particularly high (Figure 22.). The low chest compression is a consequence of the near constant deceleration profile. This profile is most likely the result of manufacturers designing cars to meet and perform well in full frontal rigid wall legislative and consumer tests.

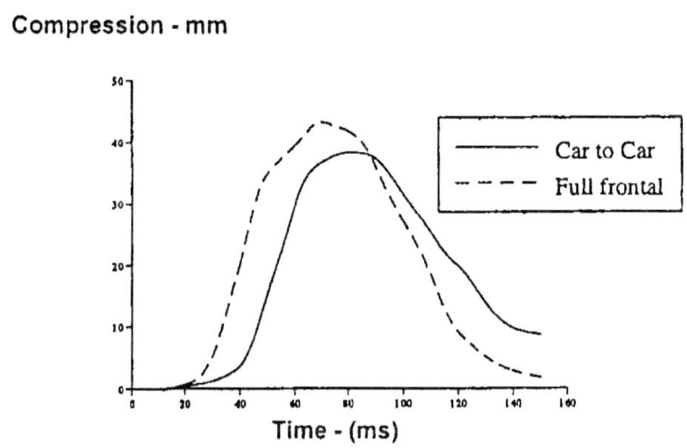

Figure 22. Comparison of chest compression for car to car and full frontal rigid wall impacts.

This study has indicated that in order to minimise chest injury the passenger compartment deceleration pulse should have a constant profile as opposed to a triangular back loaded profile, and ride down distance should be maximised. This has also been demonstrated with pulses from experimental car crashes.

The authors recognise that this study has used some theoretical deceleration pulses which may be impractical. However, a comprehensive range of pulses was investigated in order to quantify possible benefits before the study was restricted with practical considerations. In the future this study will be extended to help quantify the potential benefits of controlling the stiffness and deceleration pulse shape of vehicles in a 'compatible fleet'.

CAR TO CAR SIDE IMPACT

For side impact, parametric studies have been carried out using full car finite element models. The aim of this work was to aid our understanding of the effects of the bullet vehicle's mass, geometry and stiffness on the impacted car's structure and occupants response. This will help us to identify car structure characteristics which will improve compatibility. The modelling has been supported by full scale crash testing.

Finite Element Modelling

The purpose of this study was to understand the effect of changing bullet vehicle parameters on the impacted car's structure and dummy response. In order to undertake such a parametric study, the European Mobile Deformable Barrier (MDB) was chosen as the bullet vehicle. It was assumed that changing the MDB characteristics would indicate trends similar to those from changing the characteristics of an impacting car.

The FE model of the small four door car, EUROSID and MDB, used for the study is shown below (Figure 23.).

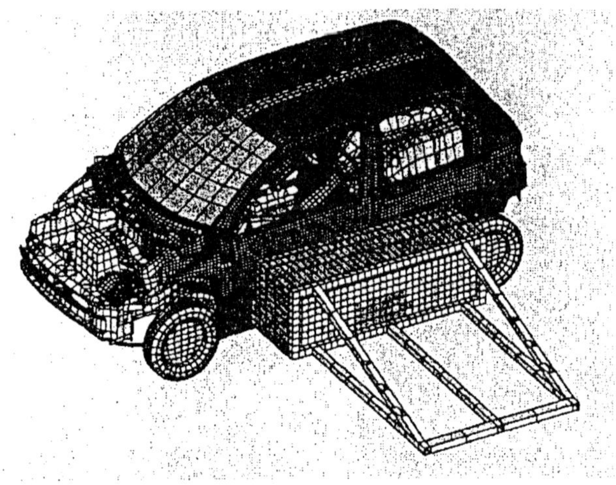

Figure 23. Small car FE model.

The model was validated for an European side impact test and shown to give reasonable agreement. The resulting vertical intrusion profile from this test is an indicator of a good structural response expected from a well designed modern car (13). In order to understand the results from the study it is important to understand the structural interaction between the barrier, the car and EUROSID. For this purpose the position of the barrier relative to EUROSID and the main side structure of the car is shown. It is seen that the bottom of the barrier just interacts with the sill and that the bottom stiffer half of the barrier is just in line with lower part of the EUROSID pelvis. There is limited contact with the 'A' and 'C' pillars (Figure 24.).

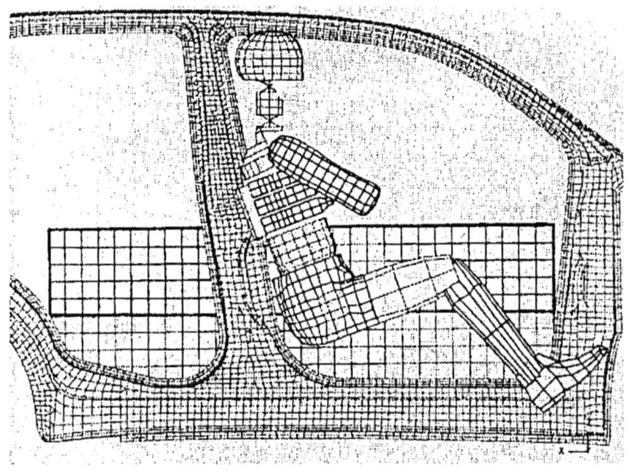

Figure 24. Relative positions of MDB, EUROSID and car structure in European side impact test.

A number of parameter sweeps were performed changing the following barrier characteristics:

1. Barrier centre impact point.
2. Barrier mass.
3. Barrier front face geometry.
4. Barrier stiffness.

Firstly, we will consider the effect of changing the point of impact of the barrier centre. For a standard European side impact test this is 550 mm above ground level in line with the R-point. This gives a barrier ground clearance of 300 mm (Table 4).

The form of these results can be explained in terms of the load paths into the car and its subsequent structural response. The two paths that we are considering are the load path through the door into the occupant and through the car's structure to its distributed mass. Ideally, we would like to reduce the load through the door into the occupant by putting more load directly into the car's structure. Whilst achieving this, the vertical intrusion profile should be maintained and any unnecessary delay in the occupant's acceleration should be avoided.

Table 4.
Injury Parameters with Varying Barrier Impact Point - European Side Impact Test as Reference.

Barrier centre impact point	Reference (Euro)	Lowered 100 mm	Raised 100 mm	Fore 200 mm	Aft 200 mm
Chest Compression (mm)					
Top rib	35	24	47	31	37
Middle rib	35	21	49	32	37
Bottom rib	32	20	44	30	32
Viscous Criteria (m/s)					
Top rib	0.58	0.22	0.74	0.50	0.59
Middle rib	0.63	0.26	0.93	0.55	0.63
Bottom rib	0.53	0.20	1.0	0.48	0.53
Abdomen Total					
Force (kN)	1.9	1.6	2.2	2.1	2.0
Pubic Symphysis					
Force (kN)	3.7	3.2	4.1	3.6	4.4

Lowering the barrier achieves this by giving better structural engagement with the sill. Raising the barrier has the opposite effect and changes the intrusion profile to put additional load on the chest. Moving the barrier fore and aft does not have such a large effect on the injury criteria as moving the barrier up and down. This is an expected result as the amount of structural engagement does not change greatly. Any additional interaction with the 'A' and 'C' pillars is with the weak edges of the barrier. These parameter sweeps indicate that more structural engagement with the sill can result in lower injury. However, a test conducted at TRL in which good structural engagement with the sill was achieved caused significant amounts of roll on the target car (Figure 25.). This may not be desirable as excessive roll may lead to head impacts on the cant rail.

Figure 25. Good Structural Engagement Inducing Significant Role in Target Vehicle.

Secondly, we will consider the effect of changing the barrier mass (Table 5.). The mass of the car modelled was about 800 kg. The effects on the dummy injury criteria are not as great as one might expect, considering the large changes in mass. The reason for this is that the most of the injury criteria peak before 40 ms whereas the momentum transfer is not complete until 80 - 100 msec. However, this effect is larger than that observed in a previous test conducted at TRL where no significant change in injury criteria was seen when changing the barrier mass from 950 kg to 1350 kg, with a car mass of 1080 kg (14). A possible explanation for this is that the stiffness of cars in side impact has increased, hence the momentum transfer is earlier, so mass can have a greater effect on injury criteria.

Table 5.
Injury Parameters with Varying Barrier Mass - European Side Impact Barrier as Reference.

Barrier mass	950 kg (Euro)	500 kg	1500 kg
Chest Compression (mm)			
Top rib	35	32	37
Middle rib	35	32	36
Bottom rib	32	27	33
Viscous Criteria (m/s)			
Top rib	0.58	0.51	0.79
Middle rib	0.63	0.50	0.73
Bottom rib	0.53	0.37	0.58
Abdomen Total			
Force (kN)	1.9	1.3	2.4
Pubic Symphysis			
Force (kN)	3.7	2.9	5.1

Thirdly, we will consider two simple geometry changes. The European side impact barrier has a bottom half which is 60 mm deeper than the top half (Figure 26.).

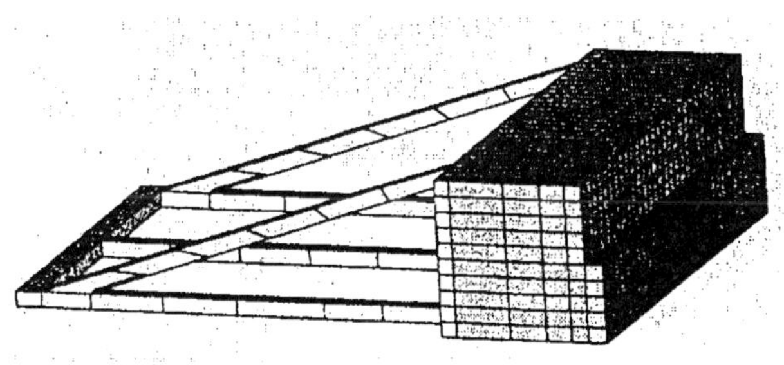

Figure 26. FE Model of European side impact barrier.

The first change was to extend the top half of the barrier 60 mm forward to give the barrier a planar front. The second change was to shorten the top half of the barrier by 60 mm so that the barrier bottom half was 120 mm deeper than the top half. The certification test results for the standard barrier model are near the centre of the specified corridor. Even with these changes in geometry, the barrier model still gives a total force deflection characteristic within the corridor. This shows that the effect of the geometry change on overall barrier stiffness was minimal. The effect of the geometry changes on the EUROSID injury criteria, were also small (Table 6.). Some slight improvement is seen from shortening the top half of the barrier. The most likely cause of this is the transfer of load from the load path into the occupant to the load path directly into the car.

Finally, we will consider the results of the stiffness parameter sweeps (Table 7). It is seen that stiffening the whole of the barrier increases all of the injury parameters, stiffening just the top of the barrier causes an even larger increase, but stiffening just the bottom of the barrier reduces the chest injury.

Table 6.
Injury Parameters with Varying Barrier Geometry-European Side Impact Barrier as Reference.

Barrier Geometry	Reference (Euro)	Planar Front	Bottom half 120mm deeper than top
Chest Compression (mm)			
Top rib	35	33	29
Middle rib	35	34	32
Bottom rib	32	31	30
Viscous Criteria (m/s)			
Top rib	0.58	0.57	0.52
Middle rib	0.63	0.65	0.56
Bottom rib	0.53	0.57	0.57
Abdomen Total			
Force (kN)	1.9	2.1	2.0
Pubic Symphysis			
Force (kN)	3.7	3.5	3.2

Table 7
Injury Parameters with Varying Barrier Stiffness.

Barrier Stiffness	Reference (EU)	Stiffness 2X	Stiffness 4X	Stiffness Top 2X	Stiffness Top 4X	Stiffness Bot 2X	Stiffness Bot 4X	Stiffness 2X	Stiffness Bot 2X
Geometry Changes:	None							Bottom half 120mm deeper than top	
Chest Compression (mm)									
Top rib	35	36	37	40	41	30	30	26	22
Middle rib	35	38	39	41	44	32	29	28	25
Bottom rib	32	36	37	36	42	30	27	28	26
Viscous Criteria (m/s)									
Top rib	0.58	0.54	0.51	0.60	0.64	0.56	0.44	0.25	0.23
Middle rib	0.63	0.65	0.64	0.60	0.68	0.64	0.54	0.40	0.38
Bottom rib	0.53	0.54	0.58	0.60	0.67	0.67	0.56	0.49	0.46
Abdomen Total									
Force (kN)	1.9	2.9	2.9	2.2	3.0	1.9	2.0	1.9	1.8
Pubic Symphysis									
Force (kN)	3.7	4.0	5.1	4.3	4.7	3.7	4.2	3.1	3.0
Barrier Peak Loads (kN)									
Barrier Bottom	85	101	115	79	71	106	120	111	111
Barrier Top	31	37	41	47	64	23	18	28	19
Barrier Total	112	128	149	124	133	126	136	127	119

If we combine this with a geometry change, i.e. shorten the top half of the barrier by 60 mm, a further reduction in the chest injury is achieved.

These results can be explained in terms of the load share between the load path into the occupant and the load path directly into the car. Stiffening the top half of the barrier increases the load through the door into the occupant and hence increases injury. In contrast, stiffening the bottom half decreases the load through the door into the occupant, by transferring load directly into the car, so reducing chest injury. For this particular car, as the middle of the barrier is approximately in line with the bottom of the pelvis, the ratio of the barrier top and bottom loads can be used as an indicator of the load share between the two major load paths (Table 7). It should be noted that in the case of stiffening the top of the barrier, there is an additional factor which could increase chest injury. This is the change in the intrusion profile from vertical to one that preferentially loads the chest.

In summary, the results of this modelling study indicate that in order to improve compatibility for side impact, the bullet vehicle should be designed such that it engages the structure of the target vehicle more effectively, through improved geometrical interaction. However, this should be achieved without compromising the intrusion profile or causing excessive roll in the target car. Stiffening of the bullet car's upper load path, without stiffening the lower path should be avoided as this will lead to increased occupant injury. It should be noted that these conclusions have yet to be validated by experimental test.

DISCUSSION

In both frontal and side impact, it has been shown that good geometrical interaction is fundamental to compatibility. In frontal impact the effect of geometry can mask the effect of mass and stiffness and dominate the outcome. In side impact, it is seen that better structural interaction can result in lower injury criteria by transferring loads from the occupant load path directly to one directly into the car's structure.

For frontal impact it is envisaged that in order to achieve compatibility between vehicles, the frontal structure will need to have a more uniform stiffness, with better structural tie-up. This may mean that the upper load path stiffness will increase. At a first glance this would appear to be in conflict with the requirements for side impact, but this does not have to be the case. In side impact the crush distance is small compared to frontal impact. Hence, in order to resolve the possible conflict, the upper load path of the car could have a stiffness profile that is soft for the initial crush and stiff for the remainder, satisfying both side and frontal requirements. This is not in conflict with current design, since the initial crush of cars is generally soft so that stiffer members are not deformed in low speed impacts, to meet insurance company repair ratings.

ACKNOWLEDGMENTS

The work described in this paper forms part of a Vehicle Standards and Engineering Division, Department of the Environment, Transport and Regions funded research programme supported by DGVII of the European Commission, conducted by the Transport Research Laboratory.

The authors gratefully acknowledge the contribution of the Ford Motor Company Ltd. and the Daewoo Motor Company Ltd.

REFERENCES

1 Lowne, R W, (On behalf of EEVC Working Group 11), *The Validation of the EEVC Frontal Impact Test Procedure* Fifteenth International Technical Conference on the Enhanced Safety of Vehicles, Melbourne 1996.

2 Hobbs, C A, D A Williams and D J Coleman, *Compatibility of Cars in Frontal and Side Impact,* Fifteenth International Technical Conference on the Enhanced Safety of Vehicles, Melbourne 1996.

3 INSIA, *Structural Survey of Cars. Definition of the Main Resistant Elements in the Car Body,* EEVC WG 15 Document, Doc No 20, 1997.

4 Grime, G, and T P Hutchinson, *The Influence of Vehicle Weight on the Risk of Injury to Drivers,* Ninth international Technical Conference on Experimental Safety Vehicles, Kyoto 1982.

5 Lowne, R W (On behalf of EEVC Working Group 11), *EEVC Working Group 11 Report on the Development of a Front Impact test Procedure*, Fourteenth International Technical Conference on the Enhanced Safety of Vehicles, Munich 1994.

6 Tarriere, C, Morvan, Y, Steyer, C and Bellot, D *Accident research and experimental data useful for an understanding of the influence of car structural incompatibility on the risk of accident injury,* Fourteenth International Technical Conference on the Enhanced Safety of Vehicles, Munich 1994.

7 Zobel, R, *Barrier impact Tests and Demands for Compatibility of Passenger Vehicles,* EEVC WG 15 Document, Doc No 43, 1998.

8 Steyer, C and A Dubos, *Renault's Proposal to Improve Compatibility*, EEVC WG 15 Document, Doc No 31, 1997.

9 Huang, M, G P Lawson, B K Powell and J H Walker, *Characterization of Vehicle Deceleration Time Histories in the Analysis of Impact Dynamics*, SAE paper 770013, Detroit Michigan, 1977.

10 Lau, I V, and D C Viano, *The Viscous Criterion - Bases and Applications of an Injury Severity Index for Soft Tissues*, 30th Stapp, 1986.

11 Mertz, H, G Horsch, G Horn and R Lowne, *Hybrid III Sternal Deflection Associated with Thoracic Injury Severities on Occupants restrained with Force-Limiting Shoulder Belts*, SAE paper 910812, 1991.

12 Neilson, I D *The Dynamics of Safety Belts in Motor Car Head-on Impacts. Ergonomics and Safety in Motor Car Design*, 1966.

13 Hobbs, C A, *Dispelling the Misconceptions about Side Impact protection,* SAE paper 950879, Detroit Michigan, 1977.

14 Hobbs, C A; *The Influence of Car Structures and Padding on Side Impact Injuries.* Twelfth International Technical Conference on Experimental Safety Vehicles, Goteborg 1989.

CHAPTER FIVE – SIDE IMPACT

Factors Contributing to Front-Side Compatibility: A Comparison of Crash Test Results

Joseph M. Nolan, Michael R. Powell, Charles A. Preuss, and Adrian K. Lund
Insurance Institute for Highway Safety

ABSTRACT

The occupants of passenger vehicles struck in the side by another vehicle are more likely to be fatally injured than are occupants of the striking vehicle. The risk of fatality in a side struck car is higher still when the striking vehicle is a pickup or utility vehicle rather than a passenger car of the same mass. This suggests there are other factors inherent to pickup and utility vehicle design in addition to mass that contribute to this increased risk. In this paper, results are presented from a series of six 90-degree front-to-side crash tests conducted with both vehicles moving. The side-struck vehicle, a Mercury Grand Marquis with a BioSID (biofidelic side impact dummy) in the driver position, was moving at 24 km/h (15 mi/h) in all tests. The striking vehicles, moving at 48 km/h (30 mi/h) in all tests, included a passenger car, a baseline pickup with greater front-end stiffness than the car but of comparable mass and ride height, a pickup with increased mass, a pickup with increased ride height, a pickup with increased mass and ride height, and a pickup with increased mass, ride height, and stiffness. The effects of these variations in striking vehicle characteristics were evaluated in terms of the struck and striking vehicles' deformation patterns and accelerations, as well as dummy injury measures and kinematics. Results from these six tests show that increasing the mass and ride height of the striking vehicles produced deeper deformation of the struck vehicle and increased the likelihood of injury, as measured by the BioSID. In addition, the increased hood height of the striking pickups, compared with the striking car, resulted in driver head contacts with the hood as an additional potential source of injury. Increasing the front-end stiffness of the striking vehicles, which was varied by selecting vehicles with different stiffness characteristics, also resulted in deeper deformation of the struck vehicle, but effects on dummy injury measures were inconsistent. These findings suggest that variations in vehicle front-end geometry and associated localized stiffness differences, which could not be separated from overall stiffness in this study, critically affect the outcomes for occupants of the struck vehicle.

INTRODUCTION

As of July 1998, utility vehicles and pickups accounted for 25 percent of all registered passenger vehicles in the United States, and their popularity is growing. As shown in Figure 1, registrations of these vehicles accounted for 16 percent of all new passenger vehicle registrations in model year 1975, rising to 35 percent of all new passenger vehicle registrations in model year 1997 [1].*

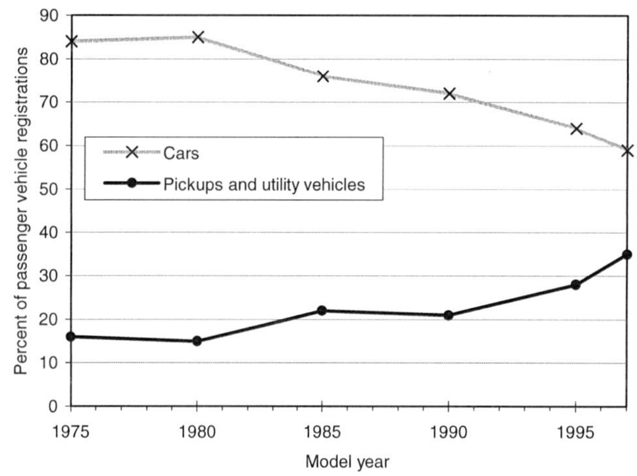

Figure 1 – Percent of Passenger Vehicle Registrations by Vehicle Type and Model Year

As the population of utility vehicles and pickups increases, so does their crash exposure. This has implications not only for the occupants of these vehicles but also for the occupants of other vehicles with which they may collide. Utility vehicles and pickups, compared with cars of equivalent mass, are more likely to cause fatalities to occupants of other vehicles; the risk is doubled, compared with cars, when they strike other vehicles in the side [2]. Recent studies by the National Highway Traffic Safety Administration (NHTSA) reach similar conclusions [3,4]. These results suggest there are inherent design differences between utility vehicles and pickups, on the one hand, and cars, on the other, that influence the severity of collisions with other vehicles in side impact crashes.

*Numbers in brackets designate references at the end of the paper.

Pickups and utility vehicles generally have higher front bumper heights and greater mass compared with cars. In addition, most pickups and many utility vehicles are body-on-frame construction versus the unibody construction of virtually all passenger cars. The frames of these vehicles tend to be stiffer than cars in order to facilitate load-hauling, towing, and increased engine torque, but this stiffness also may contribute to more aggressive front ends in crashes. These three variables — mass, stiffness, and front geometry — are the focus of most research evaluating crash compatibility among passenger vehicles.

Research into the compatibility of vehicles is not a new subject. Early in the 1970s, organizations concerned about the effects of newly proposed compliance test procedures and consumer information crash testing cited vehicle compatibility as a source of concern. A systematic series of crash tests in the late 1970s was conducted by Renault to study the effects of striking vehicle mass, ride height, and front-end stiffness in side impact crashes [5]. This study reported that increased striking vehicle mass had little effect on struck vehicle driver injuries, but increased striking vehicle front-end stiffness increased both thoracic and pelvic impact severities. The authors also reported significant reductions in pelvic impact severity and lesser thoracic impact severity when the striking vehicle's front end aligned more closely with the side structure of the struck vehicle.

Researchers at TRL conducted parametric mathematical modeling of the European moving deformable barrier (MDB) in side impacts with a small car [6]. Some dummy injury measures increased with striking vehicle mass, but the increases were not as much, proportionally, as the large increases in mass (42 percent). The authors noted that momentum is not fully transferred by the time peak injury measures occur. These results were consistent with earlier vehicle crash tests conducted by Hobbs [7], which reported no mass effect on dummy injury values with increased MDB mass. This work reported significant changes in dummy injury measures with changes in MDB ground clearance (normal, +10 cm, –10 cm). However, results for the stiffness variations, like those for mass, were mixed. Doubling and quadrupling barrier stiffness yielded slightly higher rib deflections but largely unchanged, and in some cases lower, thoracic viscous criteria. Increasing stiffness on the upper portion of the barrier produced slightly higher injury values, whereas increasing stiffness of the lower portion produced slightly lower injury measures. The most favorable dummy results were calculated when the barrier model's bottom stiffness was increased, and the bottom also protruded 12 cm more than the top.

Other work by Hobbs et al. [8] suggests that homogeneity of the front structure rather than simple stiffness dominates the injury risk in side impact crashes. As an example, the authors found that the European side impact MDB is more aggressive than the stiffer U.S. side impact MDB because the European MDB is softer on the edges and its narrower width does not bridge the pillars of struck vehicles. The authors cite real car-to-car testing, showing that stiffer cars with more homogeneous front-end stiffness are less aggressive than softer cars with less homogenous front-end stiffness.

NHTSA has conducted a series of side impact crash tests of a 1997 Honda Accord using different vehicle types (car, pickup, utility vehicle, passenger van) as striking vehicles [9]. These tests showed a substantial range in accelerations experienced by the dummy's chest and pelvis, but the accelerations did not correlate well with either vehicle mass or with an "aggressivity metric" derived from real-world crash injury rates for these striking vehicles. However, because the tests used real vehicles that inherently differ in terms of front-end stiffness and geometry as well as mass, the individual contributions of these factors to side impact compatibility could not be assessed in the study. General Motors also has conducted side impact tests with the 1997 Honda Accord using a 1996 Ford Taurus, a 1996 Honda Accord, and a U.S. side impact MDB [10]. These tests, like NHTSA's, could not completely separate the effects of mass, geometry, and stiffness, but major differences were observed in the injury measures for the dummy in the side struck vehicle when mass was held constant. The authors concluded that other factors, including front-end stiffness and geometry of the striking vehicles, are important contributors to aggressiveness.

The research conducted to this point indicates that front-end geometry has been the most consistent predictor of vehicle aggressiveness, whereas the effects of vehicle mass and front-end stiffness have been mixed. However, the test data that support this conclusion are limited, and most studies have test matrices that do not enable assessment of the individual contributions of these factors. The test matrix reported here facilitates an assessment of the separate contributions of mass, stiffness, and geometry with a new set of vehicles.

METHOD

STRIKING VEHICLES – A series of six crash tests was conducted with a common vehicle impacted on the driver side by vehicles that varied in mass, front-end stiffness, and front-end geometry. This variation was achieved by using three different vehicles: a 1995 Lincoln Town Car, a 1997 Ford F-150 4x2 pickup, and a 1997 Ford F-150 4x4 pickup. The Town Car and F-150 4x2 are similar in mass and height of lower structural elements, but the F-150 4x2 is stiffer than the Town Car. The F-150 4x4 is heavier, stiffer, and higher than the F-150 4x2. The first three tests with each of these as the striking vehicle provided baseline measures of the aggressiveness of typical large cars, pickups, and utility vehicles in the U.S. market today. To separately assess the contributions of mass, stiffness, and geometry to issues of compatibility, three additional tests were conducted with modified F-150 4x2s: one to match the height of the F-150 4x4, one to match the mass of the F-150 4x4, and one to match the mass and height of the F-150 4x4. This resulted in a test matrix containing three stiffness levels, two ride heights, and two mass levels, with tests in six of the cells (Table 1).

Table 1 – Test Matrix of Striking Vehicle Characteristics

Bumper Height	Front-End Stiffness	Vehicle Mass	
		Low	High
Low	Low	1995 Lincoln Town Car	no test
	High	Normal 1997 Ford F-150 4x2	Heavy 1997 Ford F-150 4x2
High	High	Raised 1997 Ford F-150 4x2	Raised-Heavy 1997 Ford F-150 4x2
	Highest	no test	1997 Ford F-150 4x4

Front-end stiffness – Vehicle stiffness was assessed from force deflection curves derived from NHTSA's New Car Assessment Program (NCAP) tests of these vehicles. In these flat, rigid barrier frontal crash tests, the barrier is fitted with load cells that measure the force exerted against the barrier in 6 (in some cases 36) zones over time. The forces measured in the different zones were summed to obtain the total force at regular intervals during the crash. In addition, the vehicle's longitudinal acceleration measured at the B-pillars was integrated twice to estimate forward displacement of the vehicle's center of gravity at regular intervals after the crash began. Figure 2 gives a comparison of the force deflection curves for the Town Car and F-150 4x2, showing that the F-150 4x2 has greater stiffness from the beginning of the crash. There are no comparable public data on the stiffness of the F-150 4x4; however, its frame rails have a larger cross section, and there is more substantial interconnection between the front longitudinal rails. Thus the front structure of the F-150 4x4 is most likely stiffer than that of the F-150 4x2, though it is not possible to quantify the difference.

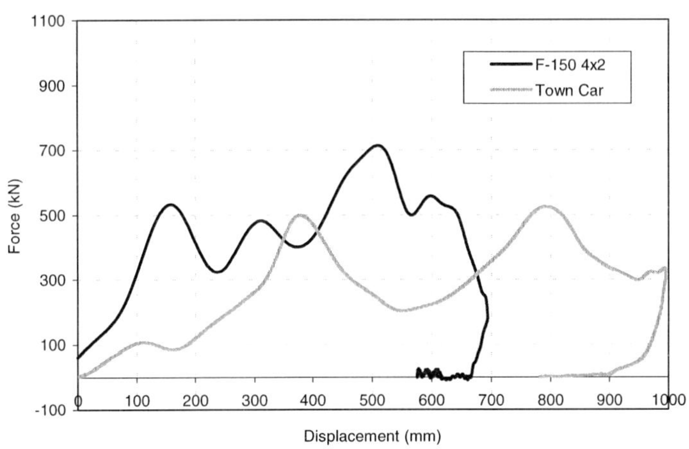

Figure 2 – Force Deflection Curves from NCAP Tests (56 km/h), 1995 Lincoln Town Car and Normal 1997 Ford F-150 4x2

Front-end geometry – Three aspects of front-end geometry are potentially important: height of the lower structural elements, height of the highest structure that could impact the struck vehicle or its occupants, and homogeneity of stiffness across the vehicle front. The Town Car and F-150 4x2 have similar heights of their lower structural elements but considerably different hood heights, making the possibility of a head strike on the intruding vehicle much more likely with the F-150 4x2 (Figure 3). In addition, different placement and size of the bumpers, as well as differences in the lateral location of the lower structural elements (not shown), almost certainly introduce differences in localization of stiffness.

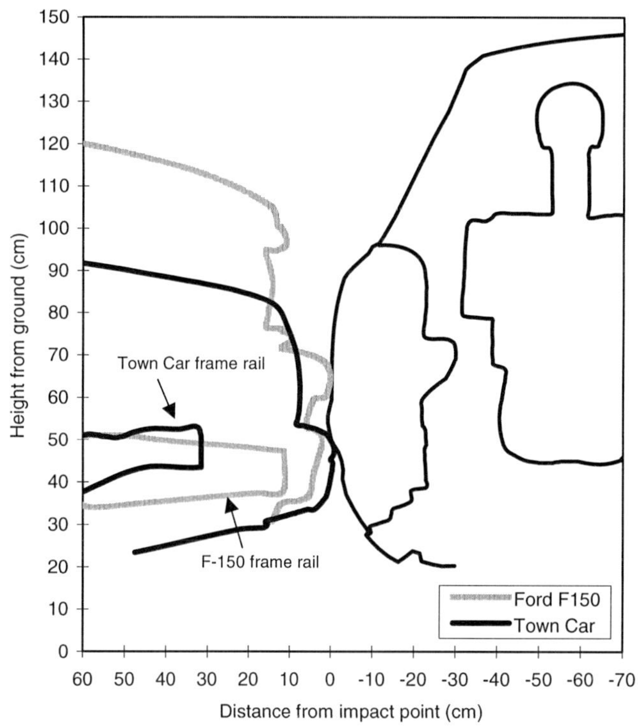

Figure 3 – Striking Vehicle Side Profiles, 1995 Lincoln Town Car and Normal 1997 Ford F-150 4x2 Compared with 1992-93 Mercury Grand Marquis

Comparison of the profiles of the F-150 4x2 and F-150 4x4 (Figure 4) shows that the lower structural elements of the F-150 4x4 are about 10 cm higher, as is the top of the hood. This variation provided a measure of the effect of ride/hood height, but as noted earlier, it is confounded by the F-150 4x4's increased stiffness, as well as its heavier mass. In order to assess the effect of vehicle height without these confounding factors, an additional F-150 4x2 was tested after raising it to match the height of a normal F-150 4x4. The F-150 4x2's ride height was adjusted by installing spacers between the coils of the front springs and spacers between the rear leaf springs and axle. Further ride height adjustment was made, if necessary, by selecting different tire sizes.

Vehicle mass – Mass was the third parameter varied in the striking vehicles. The normal F-150 4x2 and Town Car each weighed approximately 2,080 kg, whereas the F-150 4x4 weighed 2,400 kg, or about 15 percent more. One F-150 4x2 of normal height and one of raised height were tested after bolting weights into the truck beds to

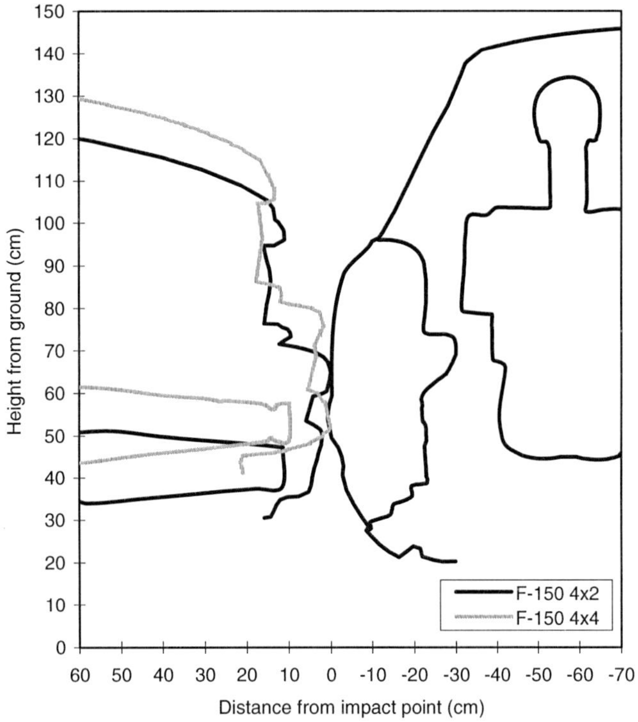

Figure 4 – Striking Vehicle Side Profiles, 1997 Ford F-150 4x2 and 1997 Ford F-150 4x4 Compared with 1992-93 Mercury Grand Marquis

increase their mass to match that of the F-150 4x4. The weights were placed close to the longitudinal center of the wheelbase in order to minimize changes to the front-rear mass distribution.

STRUCK VEHICLES – The struck vehicles were all 1992 or 1993 model year Mercury Grand Marquis four-door cars. This vehicle is approximately the same mass as the Town Car and F-150 4x2, and its corporate sibling, the Ford Crown Victoria, had performed well in NHTSA's Lateral Impact New Car Assessment Program (LINCAP) [11]. Therefore, it was expected that its structure would not suffer catastrophic deformation in the test series, even when struck by heavier vehicles. Avoidance of catastrophic structural failure was important in order to improve the reliability of observed results and maximize the ability to differentiate the effects of variations in the striking vehicles. The impact was coordinated so that the longitudinal centerline of the striking vehicle initially contacted the Grand Marquis 10 cm forward of the driver dummy's H-point. In two of the tests (normal F-150 4x2 and raised F-150 4x2), a slight misalignment caused the striking vehicle to impact the struck vehicle 7 cm rearward of the intended impact location. This misalignment did not result in any discernible increase of interaction with the A- and C-pillars on the struck vehicle compared with those tests with correct alignment. But the misalignment did cause some minor changes in the striking vehicles' interaction with the struck vehicle driver. The misalignment caused the F-150 to strike the region of the Grand Marquis driver slightly later (less than 1 ms) in both misaligned tests due to the curvature of the F-150's front structure, where there is less than a 1 cm difference between the normal front curvature and the shifted curvature in the region of the Grand Marquis driver. The struck vehicle was instrumented with accelerometers at the vehicle center of gravity, above the rear axle, and on the front and rear of each door sill.

TEST PROCEDURES

The impact angle was 90 degrees, and both vehicles were moving, the striking vehicle at 48 km/h and the struck vehicle at 24 km/h. These speeds were selected to nominally match the crash energy of the LINCAP test of the Ford Crown Victoria.

A driver BioSID (biofidelic side impact dummy) was positioned in the struck vehicle, arm down. The BioSID was instrumented to record head accelerations (triaxial); neck axial and shear forces, as well as biaxial bending moments; shoulder forces; thoracic and abdominal rib deflections and accelerations; T1, T4, and T12 spine accelerations; pelvic accelerations; lumbar forces; iliac wing force; pubic symphysis force; and sacrum force.

The recorded injury measures were processed in accordance with standard Society of Automotive Engineers J211 procedures and guidelines. Head acceleration data for one test (raised-heavy F-150 4x2) were lost due to a technical problem. Some lumbar load cell data were lost in each test, and it was decided not to use the lumbar load cell in the last two tests (raised-heavy F-150 4x2 and F-150 4x4). In addition, many of the dummy injury measures varied little among the six tests and never exceeded injury assessment reference values (IARV). These measures are not discussed in the paper, but a summary of all recorded dummy injury measures is shown in Appendix 3.

RESULTS

Table 2 includes those dummy injury measures that appeared to be most sensitive in these tests. Before addressing the compatibility issues of vehicle mass, front-end stiffness, and front-end geometry, several general aspects of the results are noteworthy. First, in the Town Car test, none of the dummy measures indicated a high risk of serious injury; thus, the Grand Marquis appears to provide a high level of occupant protection in side impacts when the striking vehicle is of similar size and mass. Second, abdominal and pelvic loadings were uniformly higher in the tests with the pickups than with the Town Car, and the abdominal loadings exceeded IARVs; however, variations in the mass, height, and stiffness of the pickups did not seem to affect these measures in a consistent manner. Finally, in each of the pickup tests, the dummy's head was at risk of being struck by the hood, although it narrowly missed the hood of the normal F-150 4x2 (possibly due to the impact point misalignment) and did not experience exceptionally high accelerations except in the F-150 4x4 test (and perhaps the raised-heavy F-150 4x2; the head acceleration data were lost, but the head strike appeared hard in the high-speed film).

Table 2 – Struck Vehicle Driver Dummy Test Results

Injury Measure	IARV	1995 Lincoln Town Car	Normal 1997 Ford F-150 4x2	Raised 1997 Ford F-150 4x2	Heavy 1997 Ford F-150 4x2	Raised-Heavy 1997 Ford F-150 4x2	1997 Ford F-150 4x4
Head injury criterion (HIC)-15	1000	75	182	227	249	no data	717
Resultant head acceleration (g)	80	34	47	54	59	no data	105
Thoracic rib acceleration (g)		58	74	104	116	143	112
Thoracic rib deflection (mm)	42	41	37	**60**	32	**62**	**48**
Thoracic viscous criterion (m/s)	1.00	0.40	0.40	**1.51**	0.37	**1.54**	0.44
T12 lateral acceleration (g)		41	63	69	87	79	66
Thoracic trauma index (TTI)	85	50	68	**86**	**101**	**111**	**87**
Abdominal deflection (mm)	39	55	**75**	**76**	**71**	**72**	**77**
Abdominal viscous criterion (m/s)	1.00	0.70	**1.10**	**1.46**	**1.80**	**1.65**	**1.42**
Pelvic acceleration (g)	130	48	105	105	103	114	109
Struck vehicle test weight (kg)		2,007	2,008	2,008	2,007	2,007	2,007
Striking vehicle test weight (kg)		2,067	2,084	2,085	2,402	2,400	2,401

STIFFNESS COMPARISONS – Two test comparisons address the effects of increased stiffness. The first comparison is between the Town Car and the normal (low height and mass) F-150 4x2. Figure 5 shows a top view of the bumper deformation of the Town Car and the F-150 4x2. Immediately apparent is the greater bending of the front structure in the Town Car and the deformation caused by the B-pillar of the Grand Marquis. The Town Car exhibited both axial crush and lateral bending of the front longitudinal rails. The F-150 4x2's frame rails did not crush; instead they bent laterally at the intersection of the longitudinal rails and front suspension (Figure 6). In general, this pattern of deformation was seen in all the F-150 4x2 tests, and the variation in striking pickup deformation was solely in terms of the extent of lateral bending.

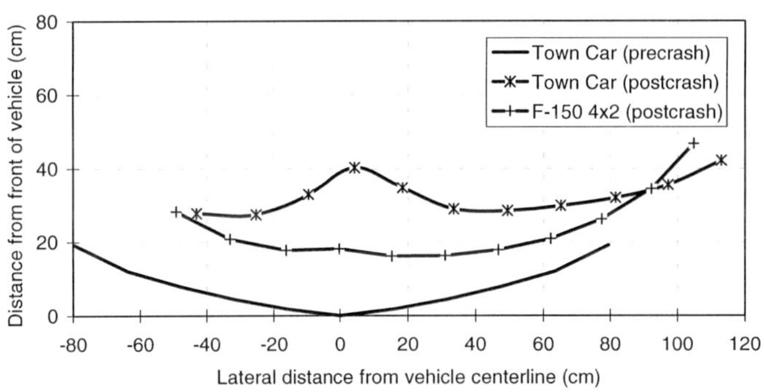

Figure 5 – Top View of Bumper Deformation, 1995 Lincoln Town Car and Normal 1997 Ford F-150 4x2

1995 Lincoln Town Car 1997 Ford F-150 4x2

Figure 6 – Postcrash Underbody View of Front-End Structure, 1995 Lincoln Town Car and 1997 Ford F-150 4x2

The Grand Marquis crush profiles reflect these differences in deformation of the striking vehicles (Figure 7). The normal F-150 4x2 intruded into the Grand Marquis, occupying space where the dummy's shoulder had been located. The Town Car's frame rails punched into the front and rear doors of the Grand Marquis, but the bumper collapsed around the B-pillar, and the extent of intrusion was less. A closer look at the vertical deformation pattern at the longitudinal location of the dummy (Figure 8) revealed that, although the magnitude of deformation was greater for the F-150 4x2, the vertical deformation pattern was similar.

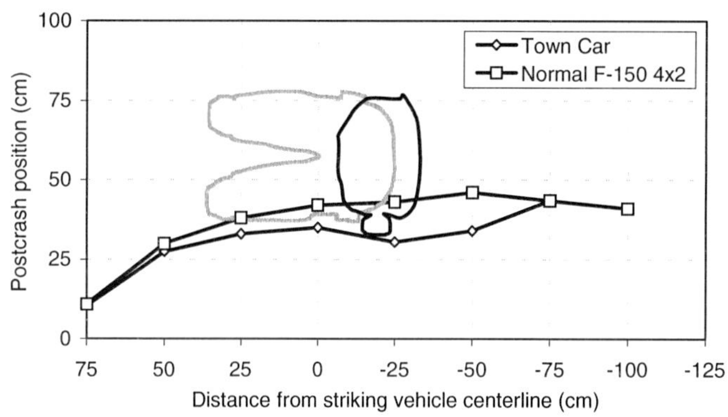

Figure 7 – Top View Crush at Mid-Door Level of 1992-93 Mercury Grand Marquis, Striking Vehicles 1995 Lincoln Town Car and Normal 1997 Ford F-150 4x2

These vehicle deformation patterns were consistent with the greater stiffness of the pickup than the car. However, the dummy injury measures did not match these patterns. Pelvic acceleration, thoracic rib and spine accelerations, and abdominal deflections were substantially greater in the F-150 4x2 test, but thoracic viscous criteria were similar, and thoracic rib deflections were lower in the F-150 4x2 test (Table 2). No thoracic injury measures for this test condition exceeded the IARVs.

The timing of the dummy responses may help explain the small increase in injury risk relative to the stiffness differences (Figures 9 and 10). Most peak injury measures were recorded later in the Town Car test than in the F-150 4x2 test, suggesting the intruding door contacted

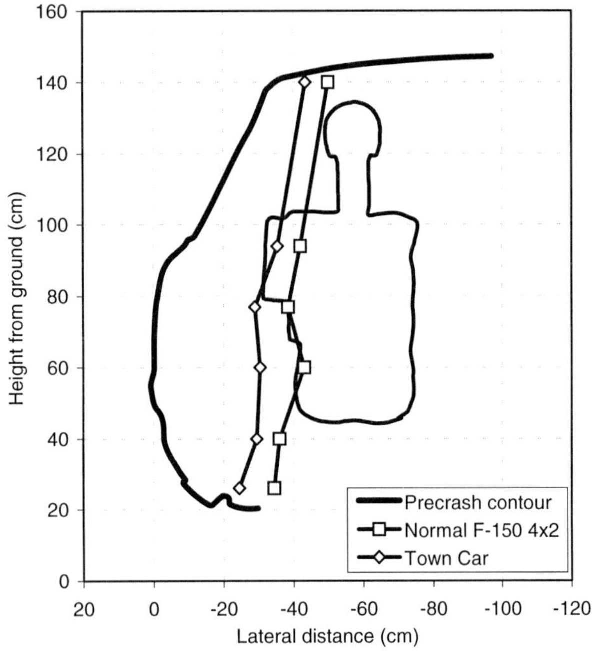

Figure 8 – Vertical Crush Profile of 1992-93 Mercury Grand Marquis at Dummy Location, 1995 Lincoln Town Car and Normal 1997 Ford F-150 4x2

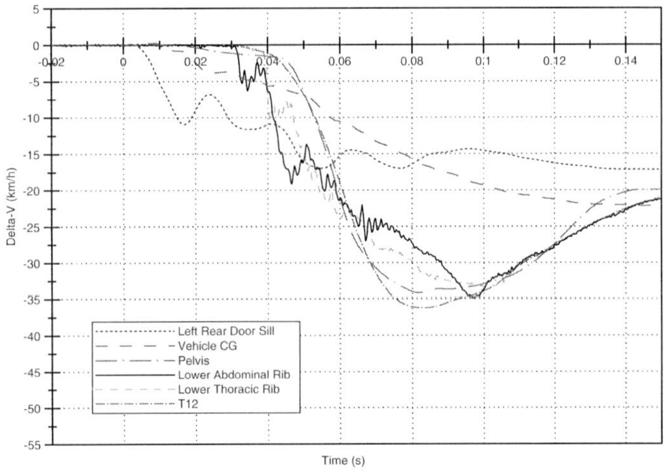

Figure 9 – Side Impact Loading Sequence: Vehicle and Dummy Velocity Changes, 1995 Lincoln Town Car

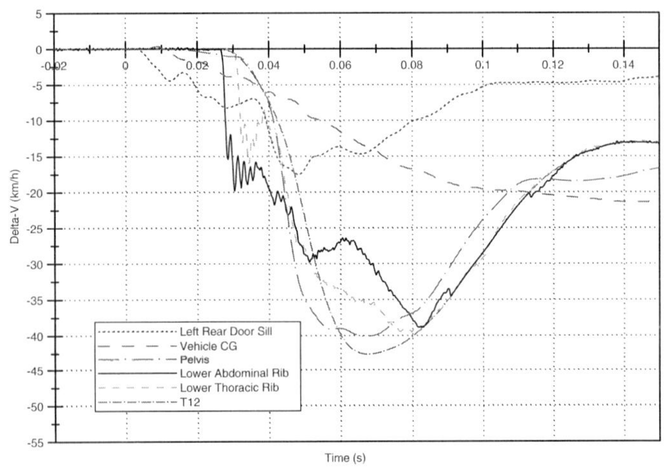

Figure 10 – Side Impact Loading Sequence: Vehicle and Dummy Velocity Changes, Normal 1997 Ford F-150 4x2

the dummy later. The pelvis and abdominal rib, in particular, began to move much earlier in the F-150 4x2 test. These body parts were directly in front of the pickup's larger, heavier bumper (Figure 3). In contrast, the Town Car's smaller bumper contacted the Grand Marquis' door lower down, and a comparison of the door sill accelerations indicated the Town Car loaded the sill more than the F-150 4x2. This difference of local stiffness geometry seems to explain the higher abdominal and pelvic loads in the F-150 4x2 test. In addition, the early loading of the lower torso and pelvis may have limited the direct forces experienced at the thorax by indirectly accelerating the upper torso and shoulder away from the intruding door.

A second stiffness comparison was available, based on the tests involving the raised-heavy F-150 4x2 and the F-150 4x4. The greater stiffness of the F-150 4x4 was confirmed by the fact that its frontal deformation was much less than that of the raised-heavy F-150 4x2. In addition, the crush of the Grand Marquis was greater for the F-150 4x4 (Figure 11). However, higher dummy injury measures were not recorded in the car with more deformation. The stiffer F-150 4x4 produced lower thoracic deflections, viscous criterion, thoracic trauma index (TTI), and pelvic acceleration (Table 2) than the raised-heavy F-150 4x2, despite more residual crushing of the Grand Marquis.

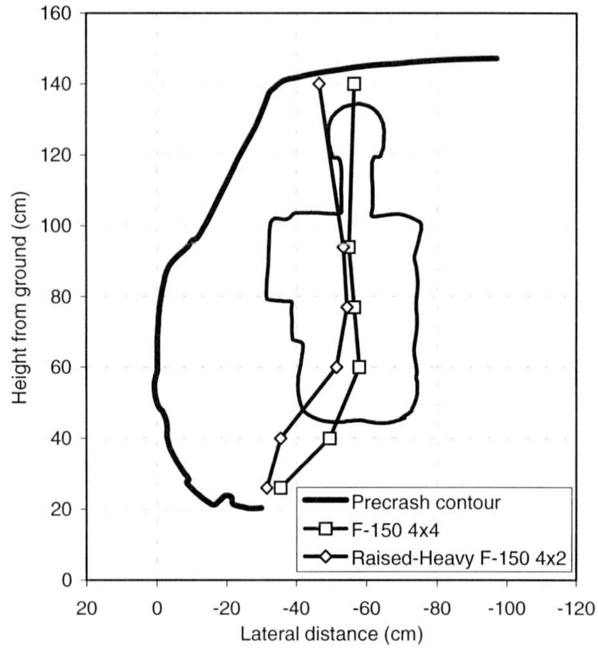

Figure 11 – Vertical Crush Profile of 1992-93 Mercury Grand Marquis at Dummy Location, Raised-Heavy 1997 Ford F-150 4x2 and 1997 Ford F-150 4x4

Again, the timing and vertical profile of the intrusion may have been more important than the amount of intrusion in determining occupant outcome (Figure 12). In the F-150 4x4 test the pelvis and spine were loaded together, whereas in the raised-heavy F-150 4x2 test the significant loading of the spine (through the thorax) occurred prior to the significant loading of the pelvis. Previous research by Hobbs [12] and Dalmotas et al. [13]

shows a strong link between the timing of dummy loading and injury severity. According to their research, it is beneficial for the pelvis to be loaded before the thorax.

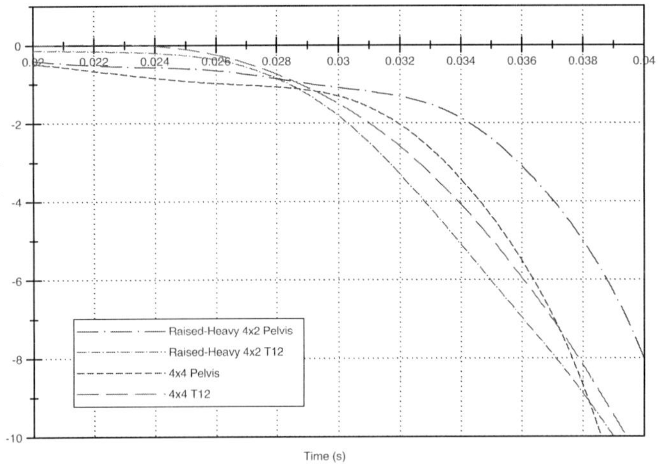

Figure 12 – Side Impact Loading Sequence: Spine (T12) and Pelvis Y-Axis Velocity Changes, Raised-Heavy 1997 Ford F-150 4x2 and 1997 Ford F-150 4x4

MASS COMPARISONS – Two pairs of tests can be compared to assess the effects of striking vehicle mass. The first comparison is between the normal F-150 4x2 and the heavy F-150 4x2. In this test pair, both vehicles have the same front-end profile, structure, and ride height; the only difference is mass. As expected, the additional mass of the striking vehicle increased the deformation of the Grand Marquis (Figure 13); however, the heavy F-150 4x2 did not appear to absorb any of the additional energy associated with its increased mass. As Figure 14 shows, there was little change in frontal deformation between the normal F-150 4x2 and the heavy F-150 4x2.

Despite the greater struck vehicle deformation associated with the heavier striking vehicle mass, not all injury measures increased. TTI and abdominal viscous criterion were substantially higher in the heavy F-150 4x2 test, but chest deflection and viscous criterion were lower compared with the normal F-150 4x2 test. The heavy F-150 4x2 transmitted loads through the door sill sooner (Figure 15) than the normal F-150 4x2 (Figure 10), but timing of dummy loads appeared similar between the two tests.

Another mass comparison is available between the raised F-150 4x2 and the raised-heavy F-150 4x2. Again, the additional mass of the striking vehicle increased the side deformation of the struck vehicle (Figure 16) but had minimal effect on the frontal deformation of the striking vehicle. Also, TTI and abdominal viscous criterion increased between the two tests, but other injury criteria seemed little affected (Table 2). These results are essentially similar to those of the first mass comparison at the low ride height. The timing of the pelvic and thoracic (T12) loading was very similar between the two tests (Figure 17). In both tests, the spine was loaded through the thorax prior to loading through the pelvis.

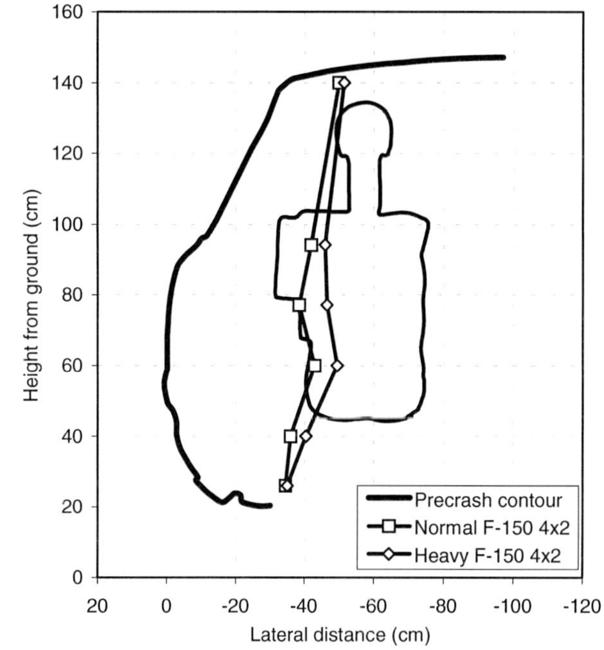

Figure 13 – Vertical Crush Profile of 1992-93 Mercury Grand Marquis at Dummy Location, Normal 1997 Ford F-150 4x2 and Heavy 1997 Ford F-150 4x2

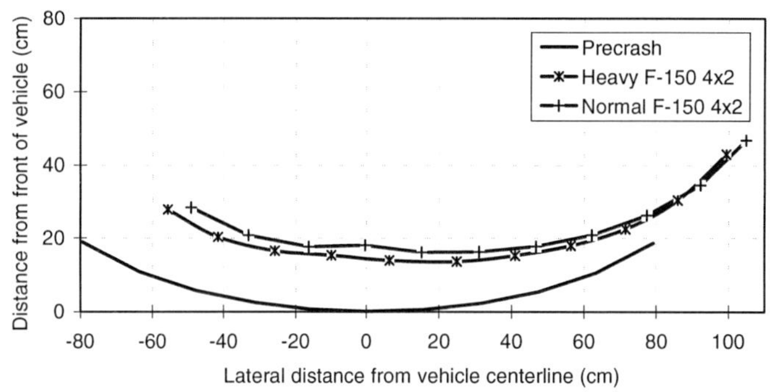

Figure 14 – Top View of Bumper Deformation, Normal 1997 Ford F-150 4x2 and Heavy 1997 Ford F-150 4x2

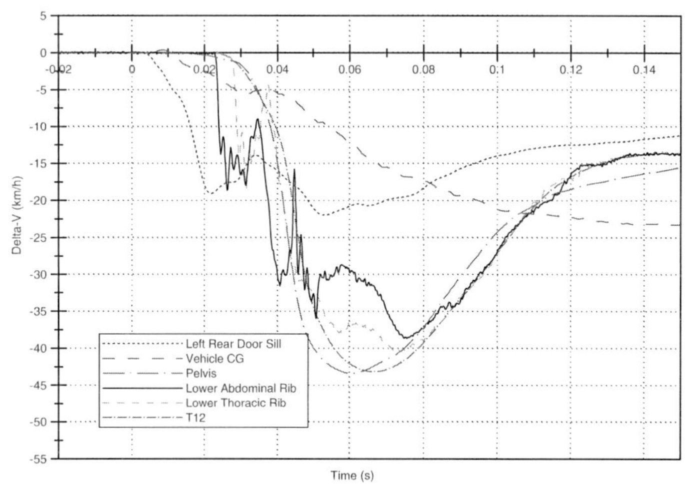

Figure 15 – Side Impact Loading Sequence: Vehicle and Dummy Velocity Changes, Heavy 1997 Ford F-150 4x2

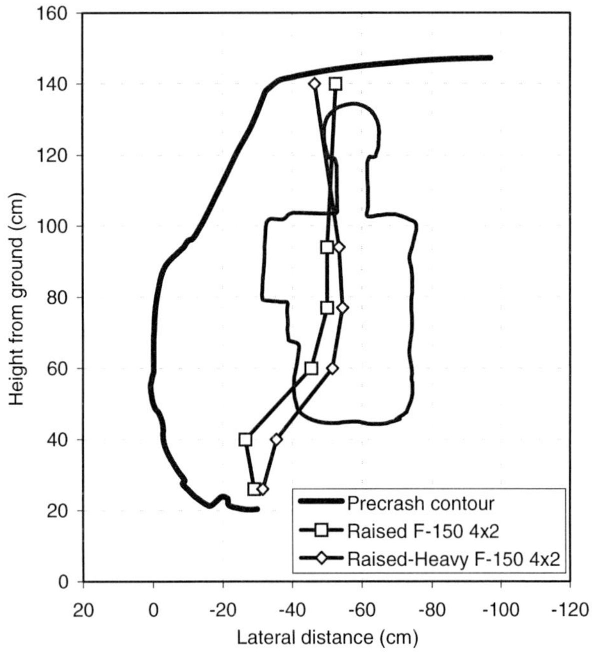

Figure 16 – Vertical Crush Profile of 1992-93 Mercury Grand Marquis at Dummy Location, Raised 1997 Ford F-150 4x2 and Raised-Heavy 1997 Ford F-150 4x2

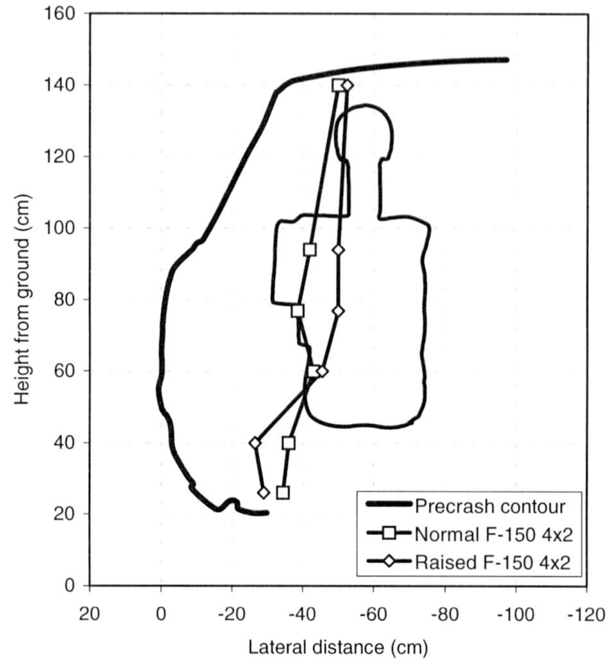

Figure 18 – Vertical Crush Profile of 1992-93 Mercury Grand Marquis at Dummy Location, Normal 1997 Ford F-150 4x2 and Raised 1997 Ford F-150 4x2

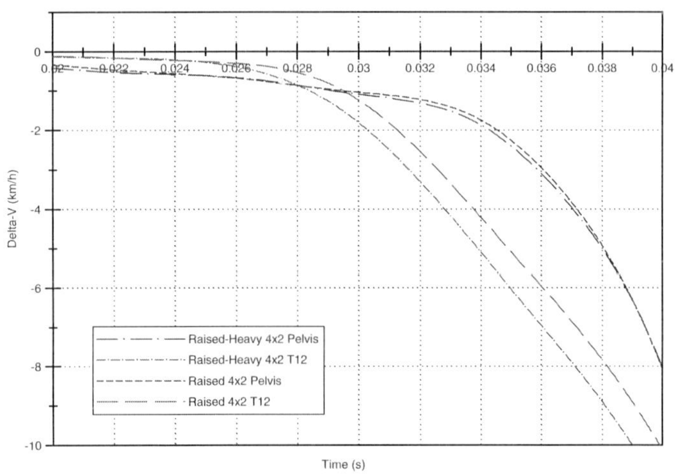

Figure 17 – Side Impact Loading Sequence: Spine (T12) and Pelvis Y-Axis Velocity Changes, Raised 1997 Ford F-150 4x2 and Raised-Heavy 1997 Ford F-150 4x2

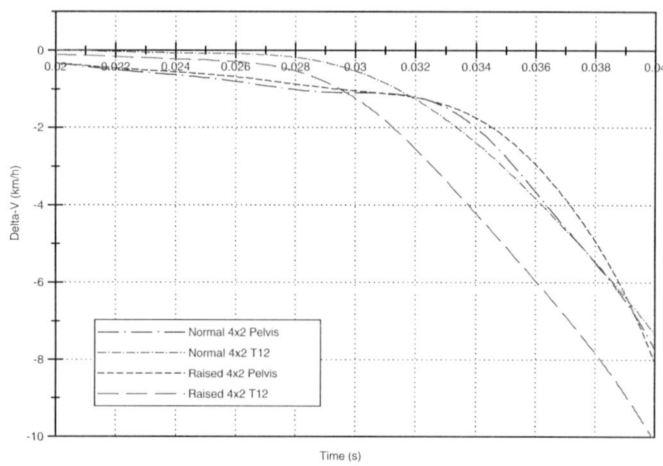

Figure 19 – Side Impact Loading Sequence: Spine (T12) and Pelvis Y-Axis Velocity Changes, Normal 1997 Ford F-150 4x2 and Raised 1997 Ford F-150 4x2

GEOMETRY COMPARISON – Two pairs of tests assessed the effects of striking vehicle ride height. In the first comparison, the normal F-150 4x2 and the raised F-150 4x2 have the same front-end profile, structure, and mass; the only difference is ride height, a 26 percent (10 cm) increase. The raised F-150 4x2 caused less crush of the lower door and door sill of the Grand Marquis and more crush at the higher levels (Figure 18). At the same time, there was less lateral bending and shifting of the front structure of the raised F-150 4x2 (not shown).

Consistent with the altered intrusion pattern, thoracic deflection, viscous criterion, and to a lesser extent TTI were higher in the raised F-150 4x2 test (Table 2). Timing of the pelvic and thoracic (T12) responses for these two tests indicated a more rapid increase in thoracic loading in the raised F-150 4x2 test, whereas in the normal F-150 4x2 test the significant loading of the thorax and pelvis occurred at approximately the same time, with some slight loading of the pelvis occurring before any significant loading of the thorax (Figure 19).

The second assessment of the effect of ride height compared tests involving the heavy F-150 4x2 and the raised-heavy F-150 4x2. Again, the only difference between the two vehicles is ride height (10 cm). An examination of the vertical crush profiles for the Grand Marquis in the two tests shows the same pattern as in the previous comparison (Figure 20). The deformation on the lower portion of the door was unchanged, or slightly reduced, and the middle and upper door deformation increased in the raised-heavy F-150 4x2 test. And, as in the tests at the lower mass levels, lateral bending and shifting of the front structure of the striking vehicle was reduced with the higher ride height (not shown).

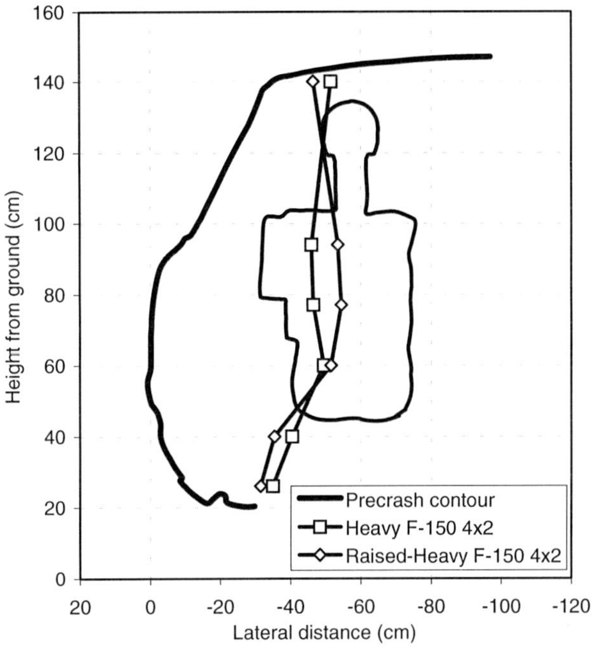

Figure 20 – Vertical Crush Profile of 1992-93 Mercury Grand Marquis at Dummy Location, Heavy 1997 Ford F-150 4x2 and Raised-Heavy 1997 Ford F-150 4x2

As with the first comparison of ride height at the lower mass levels, the increased intrusion at the shoulder and thorax level in the raised-heavy F-150 4x2 test was associated with greater thoracic deflection, viscous criterion, and to a lesser extent TTI (Table 2). Again, the significant thoracic loading occurred earlier than the significant pelvic loading in the raised-heavy F-150 4x2 test (Figure 21).

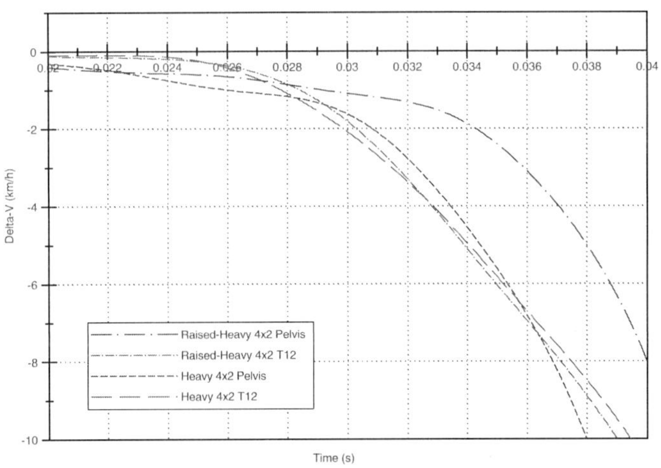

Figure 21 – Side Impact Loading Sequence: Spine (T12) and Pelvis Y-Axis Velocity Changes, Heavy 1997 Ford F-150 4x2 and Raised-Heavy 1997 Ford F-150 4x2

DISCUSSION

The tests in this series confirm the general expectation that, among the three striking vehicle types in this study, the least aggressive vehicle was the car. Despite the fact that the car was of similar mass and body-on-frame construction, the standard F-150 4x2 was somewhat more aggressive in terms of both struck vehicle intrusion and dummy abdominal injury measures. The F-150 4x4, with stiffer front-end structure, higher ride height, and higher mass, caused the most damage to the struck vehicle and the highest injury values among the three unmodified vehicles.

An obvious change in risk to occupants of the struck vehicle occurred for the head, where total height of the front structure of the strlklng vehicle was critical. When struck by the Town Car, the BioSID driver's head missed the intruding vehicle by a wide margin. In contrast, the dummy's head was struck by the higher hood in all of the pickup tests except the normal F-150 4x2 test; in that case, the encroaching hood missed the head by only a narrow margin (possibly as a result of the impact point misalignment). In the case of the F-150 4x4 (head injury criterion – 15 ms interval of 717) and probably the raised-heavy F-150 4x2 (head acceleration data lost), the head contact with the striking truck hood was fairly hard. A HIC-15 of 717 corresponds to about a 5 percent risk of severe head injury in frontal impacts [14] and a 40 percent risk of serious injury in lateral impacts [15]. Although HICs (and corresponding head injury risks) were very low (less than 1 percent risk of severe injury) in the other tests in which the dummy's head was struck by the truck hood, the fact that the head contact occurred is indicative of a potential injury mechanism. These results confirm the risk posed by increased mass, ride height, and front-end stiffness and clearly account for some of the increased fatality risk in cars struck in the side by pickups or utility vehicles compared with cars; they also highlight the importance of deployable head protection (airbags) and side airbags, which could mitigate head contacts with striking vehicles in side impact crashes [16,17].

Changes in injury risk to other body regions also reflected the increased aggressiveness of pickups and utility vehicles in side impacts, but the results indicated a complex interaction of mass, front-end stiffness, and front-end geometry, as some injury measures did not always respond to changes in these factors. For example, pelvic accelerations, which were higher in tests with the pickups than with the car (though less than IARVs in all tests), varied little among the pickups, despite large differences in mass, stiffness, and ride height. In addition, in all the pickup tests regardless of configuration, the abdominal deflections and viscous criteria exceeded published IARVs by a wide margin, and differences in the abdominal deflections among the pickup tests were small and inconsistent.

Changes in risk of thoracic injury demonstrated the most complex interactions. Comparisons among the Town Car and F-150 4x2 tests demonstrated reasonably consistent effects of increased mass, stiffness, and geometry on thoracic measures. However, results for the F-150 4x4 test, the vehicle that combined high ride height and mass with the stiffest front-end structure, indicated the effects of these variables are sensitive to other vehicle design factors. In this test, some thoracic injury measures were much lower than those observed in tests with the less stiff but equally heavy and equally high F-150 4x2. These counterintuitive results may have been due to the presence of tow hooks mounted to the frame rails of the F-150 4x4. These hooks, absent on the F-150 4x2, extend through the lower portion of the bumper. The tow hooks contacted the side structure of the struck vehicle lower and sooner than the bumper on the F-150 4x2, resulting in a more uniform vertical crush profile of the struck vehicle. The raised and raised-heavy F-150 4x2, without tow hooks, caused much less crush in the lower door of the struck vehicle than did the F-150 4x4. Also, the pelvic and thoracic loading occurred nearly simultaneously in the F-150 4x4 test, whereas the significant pelvic loading lagged the thoracic loading in both the raised and raised-heavy F-150 4x2 tests.

This pattern of results suggests that homogeneity of stiffness may be more important than overall stiffness in improving vehicle compatibility [8]. That is, the aggressive effects in side impacts of increased front-end stiffness, and to some extent ride height and mass, may be mitigated by front-end designs that spread the load of the striking vehicle both horizontally and, as demonstrated in these tests, vertically. Assuring distribution of the impacting force to lower, stiffer parts of the struck vehicle may not reduce the total intrusion (in this series, the F-150 4x4 caused the deepest penetration of the struck car), but it has the benefit of creating a more uniform intrusion profile and assuring that the occupant's pelvis is accelerated with or before the more vulnerable thorax. Other researchers also have remarked on the protective effects of this intrusion profile [8,13].

CONCLUSIONS

Increasing striking vehicle stiffness does not necessarily result in higher dummy injury measures, despite increased struck vehicle deformation. These results are consistent with the work of Hobbs et al. [8] who indicate that stiffness alone is not a sufficient measure of aggressiveness. Rather, the manner in which the striking vehicle deforms the struck vehicle has a much greater effect on the injury outcome than the simple aggregate front-end stiffness. A useful first step in achieving vehicle compatibility would address the geometry of the front structure to allow better engagement with the side structures of other vehicles.

Improving the geometric match between vehicle front ends and side structures would not address all issues of compatibility in side impacts, however. The simple change in height of front structure greatly increases the risk of head injury in some circumstances, which is not sufficiently mitigated by optimization of the vertical intrusion profile. For the head, the key issues are height of the intruding structure and the extent of intrusion, which is greatest for the heavier, stiffer, and more massive vehicles.

ACKNOWLEDGMENT

This work was supported by the Insurance Institute for Highway Safety.

REFERENCES

1. R.L. Polk and Company. 1997. National vehicle population profile 1997. Detroit, MI.
2. Lund, A.K. and Chapline, J.F. 1999. Potential strategies for improving crash compatibility in the U.S. vehicle fleet. SAE Technical Paper Series 1999-01-0066. Warrendale, PA: Society of Automotive Engineers.
3. Gabler, H.C. and Hollowell, W.T. 1998. The aggressivity of light trucks and vans in traffic crashes. SAE Technical Paper Series 980908. Warrendale, PA: Society of Automotive Engineers.
4. Gabler, H.C. and Hollowell, W.T. 1998. NHTSA's vehicle aggressivity and compatibility research program. *Proceedings of the 16th International Technical Conference on the Enhanced Safety of Vehicles*, 640-49. Washington, DC: National Highway Traffic Safety Administration.
5. Provensal, J. and Stcherbatcheff, G. 1981. Identification of compatibility factors in side collisions. *Proceedings of the Eighth International Technical Conference on Experimental Safety Vehicles*, 587-99. Washington, DC: National Highway Traffic Safety Administration.
6. Wykes, N.J.; Edwards, M.J.; and Hobbs, C.A. 1998. Compatibility requirements for cars in frontal and side impact. *Proceedings of the 16th International Technical Conference on the Enhanced Safety of Vehicles*, 667-81. Washington, DC: National Highway Traffic Safety Administration.
7. Hobbs, C.A. 1989. The influence of car structures and padding on side impact injuries. *Proceedings of the 12th International Technical Conference on Experimental Safety Vehicles*, 954-62. Washington, DC: National Highway Traffic Safety Administration.
8. Hobbs, C.A.; Williams, D.A.; and Coleman, D.J. 1996. Compatibility of cars in front and side impact. *Proceedings of the 15th International Technical Conference on the Enhanced Safety of Vehicles*, 617-24. Washington, DC: National Highway Traffic Safety Administration.
9. Summers, S.; Prasad, A.; and Hollowell, W.T. 1999. NHTSA's Vehicle Compatibility Research Program. SAE Technical Paper Series 1999-01-0071. Warrendale, PA: Society of Automotive Engineers.
10. Lugt, D.V; Connolly, T; and Bholsod, D. 1999. Vehicle compatibility – analysis of the factors influencing side impact occupant injury. SAE Technical Paper Series 1999-01-0067. Warrendale, PA: Society of Automotive Engineers.
11. Travale, D.J. and Levan, W.E. 1992. Side impact testing of a 1992 Ford Crown Victoria four-door sedan (final report). Report No. 214-CAL-92-2. Washington, DC: U.S. Department of Transportation.
12. Hobbs, C.A. 1995. Dispelling the misconceptions about side impact protection. SAE Technical Paper Series 950879. Warrendale, PA: Society of Automotive Engineers.

13. Dalmotas, D.; German, A.; Zygmunt G.M.; Green, R.N.; and Nowak, E.S. 1991. Prospects for improving side impact protection based on Canadian field accident data and crash testing. SAE Technical Paper Series 910321. Warrendale, PA: Society of Automotive Engineers.
14. Mertz, H.; Prasad, P.; and Irwin, A.1997. Injury risk curves for children and adults in frontal and rear collisions. SAE Technical Paper Series 973318. Warrendale, PA: Society of Automotive Engineers.
15. McIntosh, A.; Kallieris, D.; Mattern, R.; and Miltner, E. 1993. Head and neck injury resulting from low velocity direct impact. SAE Technical Paper Series 933112. Warrendale, PA: Society of Automotive Engineers.
16. Insurance Institute for Highway Safety. 1997. Federal Motor Vehicle Safety Standard 201, head impact protection; notice of proposed rulemaking, supplementary side impact into pole crash test results. Comment submitted to the National Highway Traffic Safety Administration (Docket No. NHTSA-92-028, Notice 8). Arlington, VA.
17. Insurance Institute for Highway Safety. 1999. Impressive crash test performance for Ford with side airbags that also protect the head. *Status Report* 34(4):4-5. Arlington, VA.

APPENDICES

Appendix 1 – Dummy Clearance Measurements

		Striking Vehicle					
	Test ID:	CS98001	CS98002	CS98003	CS99001	CS99004	CS99002
Measurement	Code	1995 Lincoln Town Car	Normal 1997 Ford F-150 4x2	Raised 1997 Ford F-150 4x2	Heavy 1997 Ford F-150 4x2	Raised-Heavy 1997 Ford F-150 4x2	1997 Ford F-150 4x4
Head to header	HH	323	315	330	325	310	315
Head to windshield	HW	555	546	560	525	525	531
Nose to rim	NR	337	332	353	350	330	340
Chest to dash	CD	563	554	588	563	586	564
Rim to abdomen	RA	131	125	130	152	135	126
Knee to dash, left	KDL	256	260	260	228	279	273
Knee to dash, right	KDR	227	235	252	215	225	219
Steering wheel to chest, horizontal	SCH	241	236	257	256	234	255
Steering wheel to chest, perpendicular	SCP	391	378	348	392	375	333
Steering wheel to chest, reference	SCR	322	321	320	326	307	321
Hub to chest, minimum	HCM	187	177	186	213	180	128
Pelvis angle	PA	22.3°	22.5°	24.2°	25.0°	24.6°	24.4°
Seat back angle	SA	21.6°	21.6°	21.6°	21.6°	21.6°	21.6°
Neck angle, torso 90	NAT90	n/a	n/a	n/a	n/a	n/a	n/a
Neck angle, seated*	NAS	2.1°	4.1°	3.2°	5.3°	4.0°	4.3°
Torso angle (NAT90 – NAS)	TA	n/a	n/a	n/a	n/a	n/a	n/a
Striker to knee*	SK	588	585	570	567	578	573
Striker to knee angle*	SKA	0.7°	1.0°	0.0°	2.9°	0.8°	0.5°
Striker to H-point, horizontal	SHH	201	193	195	183	194	194
Striker to H-point, vertical	SHV	102	100	105	98	104	109
Ankle to ankle	AA	315	312	313	300	305	316
Knee to knee	KK	310	310	315	330	305	323
Arm to door	AD	98	133	147	113	110	122
H-point to door	HD	178	194	207	178	176	187
Head to A-pillar	HA	485	490	500	460	480	499
Head to roof	HR	185	190	195	183	190	186
Head to side window	HS	270	275	282	268	283	245

*These measurements were made in a vertical plane containing the striker and parallel to the driver door sill.
All distance measurements are in millimeters (mm).
In all tests, the seat back angle was positioned to 21.6 degrees. The upper belt anchorage was in the third notch, and the steering column tilt was adjusted to the third notch.

Appendix 2 – Test Information

	Striking Vehicle					
Test ID:	CS98001	CS98002	CS98003	CS99001	CS99004	CS99002
	1995 Lincoln Town Car	Normal 1997 Ford F-150 4x2	Raised 1997 Ford F-150 4x2	Heavy 1997 Ford F-150 4x2	Raised-Heavy 1997 Ford F-150 4x2	1997 Ford F-150 4x4
Striking vehicle mass (kg)	2,067	2,084	2,085	2,402	2,400	2,401
Struck vehicle mass (kg)	2,007	2,008	2,008	2,007	2,007	2,007
Impact speed (striking) (km/h)	48.7	48.9	48.8	48.7	49.2	48.9
Impact speed (struck) (km/h)	24.3	24.4	24.4	24.3	23.9	24.4

Appendix 3 – BioSID Results from Struck Vehicles (All 1992-93 Mercury Grand Marquis)

	Striking Vehicle											
Test ID:	CS98001		CS98002		CS98003		CS99001		CS99004		CS99002	
	1995 Lincoln Town Car		Normal 1997 Ford F-150 4x2		Raised 1997 Ford F-150 4x2		Heavy 1997 Ford F-150 4x2		Raised-Heavy 1997 Ford F-150 4x2		1997 Ford F-150 4x4	
Measure	Value	Time	Value	Time (ms)	Value	Time (ms)	Value	Time (ms)	Value	Time (ms)	Value	Time (ms)
Head resultant acceleration (g)	34	74	47	67	54	60	59	64	—	—	105	69
Head 3 ms clip acceleration (g)	32	74-77	46	64-67	50	59-62	53	62-65	—	—	83	67-70
HIC (36 ms)	120	60-96	261	58-94	326	47-83	381	55-91	—	—	1082	115-151
HIC (15 ms)	75	65-80	182	59-74	214	54-69	249	56-71	—	—	717	122-137
Neck X-force (+) (kN)	0.3	75	0.4	80	0.4	64	0.4	66	9.4	62	0.7	126
Neck X-force (–) (kN)	–0.1	350	–0.1	192	–0.2	169	–0.1	194	–0.9	62	–0.2	344
Neck Y-force (+)(kN)	0.8	81	1.1	86	1.2	75	1.3	85	1.4	69	1.1	82
Neck Y-force (–) (kN)	–0.2	195	–0.5	177	–0.7	162	–0.6	181	–1.1	155	–1.1	154
Neck tension (kN)	–1.1	75	–1.8	64	–2.1	60	–2.1	64	–2.6	129	–3.4	129
Neck compression (kN)	0.1	39	0.1	35	0.2	36	0.1	33	0.4	34	0.2	35
Neck X-moment (+) (Nm)	23	57	43	175	57	158	47	178	94	156	99	156
Neck X-moment (–) (Nm)	–64	91	–83	82	–85	82	–95	82	–96	80	–78	97
Neck Y-moment (+) (Nm)	7	140	12	87	7	91	12	92	13	157	41	158
Neck Y-moment (–) (Nm)	–10	350	–17	194	–21	175	–19	185	–24	123	–44	122
Neck Z-moment (+) (Nm)	20	84	25	83	29	72	31	82	34	72	41	79
Neck Z-moment (–) (Nm)	–5	134	–5	118	–11	155	–6	173	–23	153	–36	155
T1 Y-acceleration (+) (g)	42	56	47	54	53	46	8	126	16	113	21	117
T4 Y-acceleration (–) (g)	–39	59	–48	51	–55	47	–60	46	–62	49	–51	53
T12 Y-acceleration (–) (g)	–42	61	–64	50	–70	49	–87	45	–79	49	–66	49
Rib 1 deflection (mm)	30	62	30	55	43	52	24	50	51	51	41	57
Rib 2 deflection (mm)	36	63	33	55	53	52	27	50	57	51	44	56
Rib 3 deflection (mm)	41	63	37	54	60	48	32	48	62	47	48	55
Rib 4 deflection (mm)	38	62	59	53	57	51	62	48	72	50	63	53
Rib 5 deflection (mm)	55	63	75	52	76	48	71	44	57	49	77	52
Rib 1 V*C (m/s)	0.30	56	0.27	48	0.53	43	0.17	45	0.77	42	0.44	53
Rib 2 V*C (m/s)	0.35	55	0.33	47	0.98	42	0.24	45	1.15	42	0.44	53
Rib 3 V*C (m/s)	0.41	54	0.40	48	1.51	42	0.37	44	1.54	42	0.42	50
Rib 4 V*C (m/s)	0.39	47	0.69	42	1.02	43	1.57	41	1.65	43	0.89	38
Rib 5 V*C (m/s)	0.68	48	1.10	41	1.47	42	1.84	41	1.81	49	15.28	155
Rib 1 Y-acceleration (–) (g)	–54	52	–70	32	–87	29	–76	41	–96	28	–82	29
Rib 2 Y-acceleration (–) (g)	–59	39	–85	32	–111	29	–98	29	–124	51	–103	29
Rib 3 Y-acceleration (–) (g)	–69	39	–79	44	–127	39	–116	41	–143	38	–111	29
Rib 4 Y-acceleration (–) (g)	–70	43	–114	29	–105	39	–151	38	–102	39	–126	46
Rib 5 Y-acceleration (–) (g)	–82	42	–49	48	–68	36	–148	38	–75	25	–108	46
Lumbar X-force (+) (kN)	—	—	—	—	—	—	—	—	—	—	—	—
Lumbar X-force (–) (kN)	—	—	—	—	—	—	—	—	—	—	—	—
Lumbar Y-force (+) (kN)	0.3	61	2.5	50	—	—	2.2	45	—	—	—	—
Lumbar Y-force (–) (kN)	–1.5	80	–2.9	98	—	—	–2.8	102	—	—	—	—
Lumbar Z-force (+) (kN)	0.7	156	0.6	111	—	—	0.3	234	—	—	—	—
Lumbar Z-force (–) (kN)	–1.7	64	–2.9	63	—	—	–3.7	57	—	—	—	—
Lumbar X-moment (+) (Nm)	92	77	141	103	180	87	141	107	—	—	—	—
Lumbar X-moment (–) (Nm)	–15	61	–150	50	–56	50	–146	45	—	—	—	—
Lumbar Y-moment (+) (Nm)	13	52	21	49	9	52	53	48	—	—	—	—
Lumbar Y-moment (–) (Nm)	–43	156	–72	123	–87	95	–92	93	—	—	—	—
Pelvis resultant acceleration (g)	48	59	113	44	112	44	103	43	114	44	109	42
Pelvis iliac crest force (kN)	0.7	61	1.6	45	1.9	45	2.9	44	3.0	48	3.1	43
Pelvis sacrum force (+) (kN)	2.2	52	4.8	44	4.0	46	4.3	42	3.7	45	3.7	43
Pelvis sacrum force (–) (kN)	–1.6	85	–3.6	93	–3.5	84	–3.4	83	–3.4	81	–3.4	88
Pelvis pubic symphysis force (kN)	2.4	66	4.6	56	4.0	52	4.4	54	4.5	52	5.9	54
Shoulder X-force (+) (kN)	0.2	107	0.5	256	0.3	257	0.4	273	0.6	169	0.4	212
Shoulder X-force (–) (kN)	–0.3	97	–0.4	93	–0.2	84	–0.3	44	–0.4	85	–0.4	89
Shoulder Y-force (–) (kN)	–2.8	59	–3.4	54	–3.5	50	–2.5	58	–4.0	49	–4.0	57
Shoulder Y-deflection (mm)	25	62	32	58	32	51	21	57	43	51	42	58
Shoulder Z-force (+) (kN)	0.5	114	0.6	99	0.7	89	0.3	105	0.6	89	1.1	93
Shoulder Z-force (–) (kN)	–0.6	40	–0.5	76	–0.5	71	–0.8	44	–0.7	121	–0.7	123

— no data

2001-06-0045

THE DEPENDENCE OF SIDE IMPACT INJURY RISK ON MDB CONFIGURATION

Mark Terrell, Craig Newland, Keith Seyer

Department of Transport and Regional Services, Australia

Dainius Dalmotas

Transport Canada

399

ABSTRACT

This paper reports on a parametric study of side impact crash tests. Relative changes in injury risk are assessed for both front and rear struck side occupants in tests with variation of mass, stiffness, geometry and speed of the impacting mobile deformable barrier.

The study concludes that the ground clearance of the MDB face and impact velocity have a significantly greater effect on injury risk than the other parameters.

The paper also includes consideration of tests to further investigate the effects of mass ratio between the struck and striking vehicle.

This cooperative project between the Australian Department of Transport and Regional Services and Transport Canada includes analysis of intruding door behaviour and consequent effects on injury risk.

INTRODUCTION

In the development of vehicle standards to provide improved protection in side impact crashes there has been considerable debate surrounding the relative importance of the various parameters of impact configuration. The two existing crash regulations, ECE Regulation 95 and US FMVSS 214 employ similar principles with a moving deformable barrier striking a stationary vehicle, however the barrier and trolley configuration differ considerably.

The intent of this study is to provide a quantitative assessment of the effect of variations in trolley mass, element stiffness and geometry, and speed. While this study concentrates on one target vehicle type, a large Australian passenger car, a second vehicle type was included to support consideration of mass variation.

This parametric study varies characteristics of the impacting trolley and assesses the effect on vehicle side structure and dummy responses.

The results of this series of tests are being used in the development of a harmonised test procedure to improve protection in a wide range of side impact crashes.

METHOD

A single vehicle model was used, with dummy responses being measured with variation in parameters of the striking trolley:

- Mass
- Deformable element stiffness
- Deformable element stiffness distribution
- Ground clearance of deformable element
- Angle of impact (crabbed or perpendicular)
- Impact Speed

Target Vehicle

The chosen target vehicle was the Ford Falcon EL model, a typical large Australian passenger sedan. This model is structurally the same as the previous EF model, which was used in previous tests conducted by DoTRS[1]. The vehicle was claimed by the manufacturers to comply with the requirements of US FMVSS 214, which was confirmed in the earlier DoTRS test series.

With the exception of test 9, where target vehicle mass was varied, all target vehicles in the main test series had a test mass of 1765kg. Identical dummies, instrumentation and camera set-up were used for all tests. The mass variation for test 9 was achieved by removing the engine and drivetrain, most engine bay components, non-struck side doors and other parts that were considered not to affect the test outcome.

The springs were modified to maintain the same ride height as the heavier target vehicle.

Vehicle Instrumentation

Table 1 lists the accelerations recorded on the target vehicles:

Table 1.
Vehicle Instrumentation

Lower B-Pillar	Right (struck)	A_X, A_Y, A_Z
	Left	A_X, A_Y, A_Z
Upper B-Pillar	Right	A_X, A_Y, A_Z
	Left	A_Y
Mid B-Pillar	Right	A_Y
A-Pillar	Right	A_Y
C-Pillar	Right	A_Y
Trans. Tunnel		A_Y
Driver Door	Front	A_Y
	Rear Upper	A_Y
	Rear Mid	A_Y
	Rear Lower	A_Y
Driver Seat	Lower Frame	A_Y

The accelerometers placed on the struck side of the vehicle were used to calculate the contact velocity of the side structure to occupants in the vehicle. Driver door accelerometers were mounted to the inner door structure, underneath the door trim panel. Limited mounting space required the use of uniaxial accelerometers only. Accelerometers at the rear of the driver door were approximately aligned with the dummy thorax and pelvis. Relative velocity of door to dummy was calculated by integration of the acceleration. Contact switch signals were used to determine contact time and hence contact velocity. It is recognised that there may be some error in the calculated velocity of the door relative to the dummy due to rotation of the door structure. However, analysis of the deformed vehicles showed that the accelerometer mounting points on the deformed door structure remain close to vertical, therefore the degree of error from rotation is expected to be fairly small. In addition the consistency of the static deformations from test to test is such that comparative assessments can be made.

Dummies

There are a number of side impact dummies currently in use or under development for compliance and research testing. The dummies used for this test series were considered to be the most biofidelic available [2]. A 50%ile BioSID dummy was used in the driver seating position (front outboard struck side) and a 5%ile female SID-IIs (β+) in the rear outboard struck side seating position. The SID-IIs was chosen as a surrogate for an adolescent child, though the results would also be applicable for a small-statured adult female.

Table 2.
Dummy Instrumentation

Head	C of G	A_X, A_Y, A_Z
Upper Neck		F_X, F_Y, F_Z M_X, M_Y, M_Z
Upper Spine (T1)		A_X, A_Y, A_Z
Shoulder		F_X, F_Y, F_Z A_Y,S_Y
Thorax	Upper Rib	A_Y,S_Y
	Middle Rib	A_Y,A_X**,S_Y
	Lower Rib	A_Y,S_Y
Abdomen	Upper Rib	A_Y,A_X**,S_Y
	Lower Rib	A_Y,S_Y
Opposite Thorax	Upper Rib	A_Y
Opposite Abdomen	Upper Rib (T12)	A_X,A_Y,A_Z
Pelvis		A_X,A_Y,A_Z
Iliac		F_Y
Pubic		F_Y
Sacrum		F_Y*

*BioSID only; ** SID IIs only

Contact Switches

Digital contact switches, constructed using horn-switch membranes were located at points on the occupants and in the occupant compartment to enable analysis of the timing of crash events. These were used in the determination of contact velocities. Switches were as described in **Table 3**:

Table 3.
Contact Switch Locations

Driver Dummy	Pelvis	Struck Side
	Upper Arm	Struck Side
Driver Seat	Bolster	Struck Side
		Non-struck Side
Rear Passenger Dummy	Femur	Struck Side
	Upper Arm	Struck Side
	Knee	Struck Side

Data Collection

All test data were collected at a 20kHz sampling rate. Tests were recorded by 8 high speed film cameras including three on-board the target car and one on the impacting trolley. The target vehicle was marked with a 100x100mm grid and points digitised in 3 dimensions in the vehicle's coordinate system before and after the test. Similar measurements on a 100x50mm grid were taken of the barrier element.

Barrier Elements

Four types of deformable aluminium honeycomb elements were used to vary the parameters of stiffness and stiffness distribution. Elements meeting the specification for ECE R95 and FMVSS 214 were used. The ECE R95 element was chosen as a reference and two custom elements based on ECE R95 were manufactured with modified stiffness and stiffness distribution. With the exception of the FMVSS 214 element, all deformable elements used in this test series were supplied by AFL Honeycomb Structures and were manufactured using a chemical etching process.

Baseline (ECE R95) – Element properties as specified in the ECE R95 test procedure.

Type 1 - A custom element design with each block having a nominal stiffness of twice the ECE R95 barrier face. Dimensions were the same as the ECE R95 face.

Type 2 - A second custom element, designed to represent a Sports Utility Vehicle, was also manufactured. Compared to the ECE R95 barrier face, this element had an increased stiffness and modified stiffness distribution with an enlarged centre lower (engine) block of high stiffness.

All elements were mounted using a perforated backing frame to allow venting of the honeycomb in accordance with the manufacturer's specifications.

Details including certification test results for the custom elements can be found in a previous paper documenting this test series [3].

Test Matrix

Table 4 provides a summary of the test matrix used in the parametric study.

The test matrix was designed to independently vary each parameter such that a comparison of two tests would yield information about the effect of a single parameter.

The baseline Test 1 was based on the Economic Commission for Europe (ECE) Regulation 95. A trolley conforming to ECE R95 was fitted with an AFL "Progress" etched aluminium honeycomb element.

Test 2 was the same as Test 1 except the barrier ground clearance was raised from 300 mm to 400 mm.

Test 3 was conducted as a repeat of Test 1 as a result of data acquisition failures for the rear dummy and also provided some repeatability assurance for the driver dummy.

Table 4.
Test Matrix

Test Number	1	2	3	4	5	6	7	8	9	10
Mass of target car										
1765 kg	●	●	●	●	●	●	●	●		●
1045 kg									●	
Impact configuration										
Perpendicular	●	●	●	●	●		●	●	●	
Crabbed						●				●
Mass of MDB										
950kg	●	●	●	●				●		
1365kg					●	●	●		●	●
Type of Element										
AFL Progress ECE R95	●	●	●		●	●		●	●	
FMVSS 214										●
Other				#1			#2			
Barrier ground clearance										
280mm										●
300mm	●		●	●	●	●		●	●	
400mm		●					●			
MDB Speed										
50 km/h	●	●	●	●	●		●		●	
56 km/h @27°						●				●
60 km/h								●		

Test 4 was the same as Test 1 except for the barrier element, which was a custom manufactured element (Type 1) with each block of the element having double the stiffness of the standard R95 element.

Test 5 was the same as Test 1 except that the trolley mass was increased from 950 kg to 1365 kg.

Test 6 was the same as Test 5 except that a crabbed trolley (as defined in FMVSS 214) was used. The barrier element and targeting were the same as for Test 5.

Test 7 was the same as Test 1 except for the following changes to the trolley that were aimed at mimicking a small Sports Utility Vehicle (SUV):

- Mass was 1365 kg
- Barrier face ground clearance was 400 mm
- Custom element (Type 2) with modified stiffness and stiffness distribution.

Test 8 was the same as Test 1 except the impact speed was raised from 50 km/h to 60 km/h.

Test 9 was the same as Test 5 except that the target vehicle mass was reduced to 1045 kg, ie the impacting trolley was heavier than the struck vehicle.

Test 10 was the same as Test 6 except that the barrier face and targeting were in accordance with US Federal Motor Vehicle Safety Standard (FMVSS) 214 ie:

- Trolley mass was 1365 kg
- Trolley crabbed at 27°
- Element ground clearance of 280mm
- FMVSS 214 element (Homogeneous stiffness)

Additional Test

In consideration of the effect of mass it was suggested that the MDB to vehicle mass ratio had a more significant effect on injury measures than impacting mass alone. Test 9, with a reduced target vehicle mass, introduced a number of uncertainties with the unknown effect of changes in mass distribution and rotational inertia of the struck vehicle.

An additional test was conducted using a small European passenger car with a test mass of 1009kg This allowed the use of an impacting trolley with a mass of 1365kg to achieve an MDB to vehicle mass ratio of 1.35. The results from this test were compared to manufacturer's results from an otherwise identical test using a 950kg MDB (MDB to vehicle mass ratio of 0.94).

In order to use the available baseline result it was necessary to test using a EuroSID 1 dummy (driver only), whereas the original test matrix used BioSID driver and SIDIIs struck side rear passenger. It is recognised that this introduces a further variation in test configuration, but this was considered reasonable for the purpose.

RESULTS

Vehicle Responses

Table 5 lists the responses recorded on the target vehicles. Figure 1 shows selected vehicle static intrusion profiles, measured on the outer body panels at the height of the driver H-point.

Table 5.
Vehicle Accelerations

Test Number	1	2	3	4	5	6	7	8	9	10
Peak Acceleration (CFC 60)										
Right B-Pillar (g)	107	115	141	152	85	83	131	125	144	184
@ time (ms)	13.3	30.8	23.4	9.5	14.5	30.0	23.2	11.9	26.7	11.1
Trans. Tunnel (g)	20.9	15.5	23.6	20.5	36.8	17.4	17.9	22.3	19.0	16.8
@ time (ms)	45.3	50.0	37.8	39.4	48.2	26.0	62.1	41.8	47.5	31.4
Left B-Pillar (g)	14.8	12.1	15.6	16.2	26.7	13.5	13.7	14.7	16.8	18.6
@time (ms)	26.6	11.3	26.1	24.9	23.9	74.2	10.9	26.7	46.1	26.2

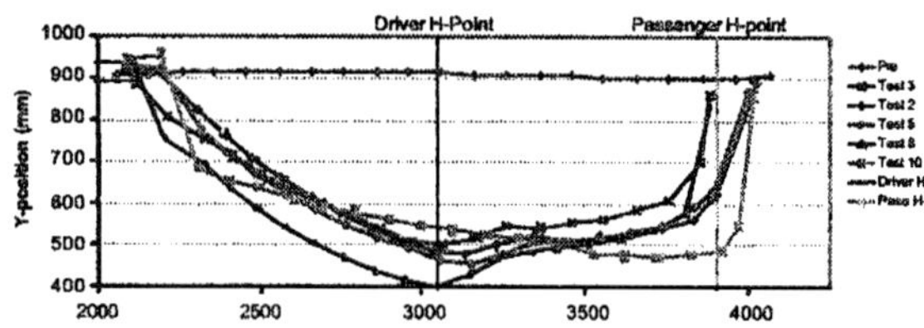

Figure 1 - Static Intrusion profiles

Dummy Responses

Table 6 shows the time and velocity of contact for the driver dummy and the contact times for the passenger dummy. No velocity information was recorded for the passenger door.

Table 6.
Contact Time and Velocity

Test Number	1	2	3	4	5	6	7	8	9	10
Driver Dummy										
Time of closure [ms]										
Shoulder	23.8	18.9	22.1	20.3	25.3	25.9	18.6	19.3	21.2	22.8
Pelvis	26.6	24.5	25.9	22.1	27.2	29.4	23.0	*	26.4	24.0
Door to Dummy Contact Velocity [m/s]										
Shoulder	8.4	10.3	9.2	9.4	8.8	5.7	12.1	9.3	9.5	10.4
Pelvis	7.2	9.2	7.1	6.8	6.1	5.1	8.2	*	7.6	10.7
Passenger Dummy										
Time of closure [ms]										
Knee	31.0	28.7	30.2	24.7	31.3	34.3	23.4	26.2	28.4	21.5
Pelvis	33.8	35.5	32.4	25.9	35.7	39.3	25.4	28.0	32.0	18.3
Shoulder	32.1	28.6	23.7	33.5	34.7	37.0	29.0	30.4	33.4	25.0

* Contact switch failure. Driver pelvis closure time and contact velocity unavailable for Test 8.

In test 1 a number of channels were not recorded for the passenger dummy. These included rib and pelvis accelerations. The baseline ECE R95 test was therefore repeated (test 3).

There were consistent measurement problems with the LVDT transducers measuring rib deflections in the SID-IIs. For a number of tests no values were obtained for either ribs 1 or 2. There were also 'spikes' in the data of several ribs in most tests. This did not generally affect the peak deflection, however the Viscous Criterion algorithm, which uses the differential of deflection to obtain rib velocity, produced unrealistic peaks (resulting from differentiation of the data 'spikes'). Therefore an alternate method of VC calculation was used for the SID-IIs, with rib velocity being calculated by integration of the acceleration of each rib relative to the spine. This produced results which were more realistic. This could not be done for test 1, therefore VC results are not included for this test.

Table 7 shows the measured dummy responses. The shaded cells show the relationship between dummy response and IARVs. It is not the intent of this paper to comment on biofidelity or reference values for the chosen dummies – however where accepted indicative values are available these do provide some guidance as to the important parametric effects. Green shaded cells are well below the IARV. Cells that are shaded yellow are within ± 20% of the IARV and cells shaded orange are well above the IARV.

Table 7.
Dummy injury responses

Driver (BioSID)										
	1	2	3	4	5	6	7	8	9	10
HIC15	137	266	134	278	252	19	416	491	232	46
3ms Head Accel [g]	44	55	42	58	55	18	60	94	52	25
Chest VC [m/s]	0.64	1.33	0.58	0.58	0.65	0.14	1.41	1.48	1.50	0.45
Chest defl [mm]	29.6	58.1	33.4	30.9	36.8	21.2	56.9	48.8	52.2	29.1
TTI [d]	93	127	104	113	97.9	43	122	150	128	81.4
Ab VC [m/s]	0.87	1.15	0.81	0.93	0.91	0.4	1.37	1.0	1.13	0.78
Ab defl [mm]	50	54.2	45.3	51.6	44.3	35.3	62.2	41.9	49.6	45.3
Pelvis Accel (g)	96	91	100	112	109	25.4	105	160	106	86
Pub. F [kN]	5	2.9	4.5	4.9	6.5	0.4	5.2	6.6	3.24	2.2
Passenger (SID-IIs)										
HIC15	49	28	58	80	114	138	157	169	391	221
3ms Head Accel [g]	20	21	29	34	40	48	48	49	80	52
Chest VC [m/s]	*	0.17	0.25	0.3	0.34	0.21	0.58	0.44	0.68	0.79
Chest defl [mm]	25.3	21.9	22.7	30.2	23.1	23.1	28.4	26.8	33.4	35.9
TTI [d]	*	48.9	49.1	46.4	56.1	46	73.6	63.9	74.7	90.2
Ab VC [m/s]	*	0.61	0.63	0.57	0.8	0.41	0.67	0.78	0.92	1.07
Ab defl [mm]	31.8	31.2	33.3	38.3	38.9	32.4	28.8	36.7	44.2	38.1
Pelvis Accel (g)	*	69.1	55.1	63.3	*	41	89	77.3	75.6	140
Pub. F [kN]	0.61	0.54	0.66	0.65	0.70	0.29	0.88	0.91	0.91	1.00

* Instrumentation failure

Further detailed results for the driver dummy can be found in the previous paper [3].

Additional Test

Table 8 shows the EuroSID-1 injury measures for the baseline test (950kg trolley) and the test with an increased mass trolley (1365kg).

Table 8
Dummy responses - additional test

	950 kg MDB (A1)	1365 kg MDB (A2)
Head Accel (3ms) (g)	48.4	78.6
HPC	222	404
Rib1 deflection	38.3	36.2
Rib2 deflection	38	36.7
Rib3 deflection	27.9	32.5
Rib1 VC	0.57	0.59
Rib2 VC	0.62	0.76
Rib3 VC	0.37	0.47
Peak Ab Force	2.01	1.96
Pubic Force	4.22	3.85

Figure 2 shows the relative change in injury responses for the two trolley mass configurations, normalised with respect to the ECE R95 injury limits.

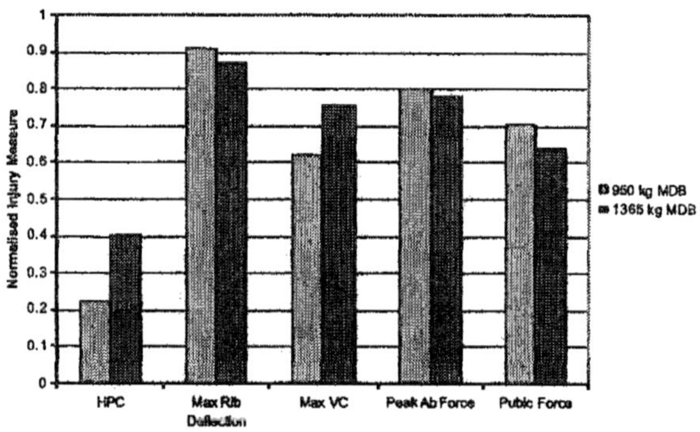

Figure 2 - EuroSID injury responses

DISCUSSION

In the following discussion, injury risk and other metrics are compared with those recorded in Test 1 & 3 which were based on the standard ECE R95 test procedure and used as the baseline. Where results for the passenger dummy were not available from Test 1 the Test 3 results only are used for comparison.

Tests 1 & 3 show a variation of approximately 10% when comparing injury outcomes for these two identical tests. Therefore it is assumed that any change to the injury outcome of less than 10% is unable to be regarded as significant as it is indistinguishable from test-to-test variation.

In Test 6, the etched ECE R95 deformable element was used on a crabbed trolley. During the test, the element failed in shear, producing negligible vehicle deformation and dummy readings that were significantly below the baseline tests for all measures on the driver dummy.

The values for the passenger in Test 6 were below the baseline tests for most measures other than head injury. However, the injury risk was not significant in either the baseline tests or Test 6.

Tests 5 and 6 were designed to assess the effect of crab angle, as this was the only trolley parameter varied between these two tests. Due to the dramatic change in behaviour of the deformable element under shear load, a comparison of these tests does not provide a true indication of the effect of crabbing, particularly with reference to the rear dummy. The results of Test 6 are therefore not discussed further. It is intended to further investigate the effect of crabbing by repeating this test with an element that is stable in shear and / or a test in the FMVSS 214 configuration with a non-crabbed trolley.

The figures in Appendix 1 show the relative change in a range of vehicle and injury measures, in comparison to the baseline tests (1 and 3). The values shown are calculated as:

$$\frac{\text{Test Value} - \text{Average Baseline Value}}{\text{Average Baseline Value}}$$

This provides a graphical representation of the relative results of each test configuration compared to the baseline tests.

Contact Velocity and Timing

An examination of the contact times and door to dummy impact velocities for the driver dummy shows that the shoulder was always struck before the pelvis (i.e. the pelvis lags the shoulder).

An increase in the pelvic lag coupled with an increase in contact velocity results in increased risk of thoracic injury as can be seen in Tests 2, 7 and 9.

By contrast, a decrease in pelvic lag coupled with an increase in contact velocity results in increased risk of injury in only the pelvic region as evidenced by Tests 4 and 10.

This relationship is not seen for the rear passenger dummy, where the kinematics are quite different, with the knee generally being contacted before the thorax or pelvis.

Increased Ground Clearance

Test 2 was the same as Test 1 except the barrier ground clearance was raised from 300 mm to 400 mm. This increase in height is such that the barrier no longer engages the sill of the target vehicle. This resulted in an increase in both the shoulder and pelvis contact velocities for the driver dummy, as well as an increase in the pelvic lag. The results indicate that the upper door area was contacting the dummy 5.6 ms earlier than the lower door contacting the pelvis. When compared to Tests 1 & 3, the higher intrusion profile is clearly seen to affect the thoracic loading of the driver dummy with the chest VC and deflection doubling. The injury measures for the pelvic area remained static. These results suggest that the high thoracic loading is caused by the high degree of pelvic lag.

The increased barrier ground clearance did not have a significant affect on the injury measures for the rear dummy when compared to the baseline tests.

Increased Stiffness

Test 4 used a Type 1 modified deformable barrier face (with a nominal stiffness twice that of the ECE R95 barrier face). Compared to Tests 1 and 3, Test 4 showed a slight increase in shoulder contact velocity for the driver dummy, but a slight reduction in pelvis contact velocity and a reduction in pelvic lag, resulting in increased injury risk for the head and pelvic region. Abdominal and thoracic injuries remained unchanged.

The increased stiffness produced a slight increase in most injury measures for the passenger dummy. The most significant increase was in thorax rib deflection, which was approaching the IARV of 34mm.

Effect of Mass

Test 5 was the same as Test 1 except that the trolley mass was increased from 950 kg to 1365 kg. The shoulder contact velocity was the same as for the baseline tests and the pelvis contact velocity decreased slightly. In addition, the pelvic lag was reduced. With the exception of pubic force, head and pelvic g's, this test caused no significant increase in injury risk.

There was a slight increase for most injury measures for the rear dummy, however all but the abdomen deflection were below the injury reference values.

The increased trolley mass in Test 5 did produce an increase in the peak acceleration of the vehicle, measured both at the left B-pillar and the transmission tunnel. This was not observed in other tests such as test 7, where the trolley mass was also 1365kg. This is likely because Test 7 also has a raised element with minimal sill engagement, therefore the additional energy may have been used in intrusion rather than acceleration of the target vehicle.

Test 5 used a 1365kg MDB to impact the 1765kg target vehicle, whereas the baseline tests used a 950kg MDB. In both cases the struck vehicle mass was significantly greater than the MDB mass resulting in an MDB to target vehicle mass ratio of les than 1 as shown in Table 9.

It was considered infeasible to conduct a test using an MDB with a mass greater than 1765kg. In an attempt to conduct a test with an MDB to target vehicle mass ratio greater than 1.0, an EL Falcon target vehicle was prepared with the test mass reduced to 1045kg. This was achieved by removal of driveline components and mass from the non-struck side of the vehicle. No mass was removed from the impacted structure of the vehicle.

The chest and abdominal injury measures for the driver dummy in Test 9 were similar to Test 8 and amongst the highest recorded. However, the pelvic responses were similar to the baseline. This is in contrast to Test 5, where the increased trolley mass produced thorax and abdomen injuries similar to the baseline, with some increase in pelvic load, particularly pubic force. It is noted that Test 9 had the 2nd highest amount of pelvic lag (5.2 ms).

Passenger dummy responses were also high for this test. A head contact was recorded and there was also a significant increase in thorax VC. Other measures showed a consistent increase over the baseline.

By removing the driveline components to reduce the mass of the target vehicle in Test 9 it is expected that the centre of gravity would be increased relative to the ground. It is unknown what effect this has on the kinematics of the target vehicle. Analysis of the high speed film indicates this vehicle rolls more than the other target vehicles.

The additional test (A2) was included in comparison with the manufacturer's data (A1) to provide further understanding of the effect of mass ratio change on dummy responses. This achieved an MDB to target vehicle mass ratio of 1.35, but without the removal of large amounts of mass from the struck vehicle. As shown in Figure 2, there was a slight increase in injury risk for some body regions (head and thorax VC) but these were not significant in magnitude and all measures remained below the regulatory limits. It is notable that a head contact to the door beltline was observed with the increased trolley mass that did not occur with the 950kg trolley.

The results from this test support the observations of Test 5 that increased trolley mass has only a slight effect on dummy responses.

Table 9
Vehicle / Bullet Mass Comparison

Test	1	5	9	A1	A2.
Target Mass [kg]	1765	1765	1045	1009	1009
Bullet Mass [kg]	950	1365	1365	950	1365
Mass Ratio Bullet / Target	0.54	0.77	1.31	0.94	1.35

Crab Angle

Test 6 was the same as Test 5, but used a crabbed (27°) trolley with a chemically etched ECE R95 barrier. Unfortunately the front (etched) section of this element failed in shear and therefore this test does not provide any further information for this parametric study.

Compound Variation

Test 7 was based on Test 1 but aimed at mimicking a small Sports Utility Vehicle (SUV) by using a trolley mass of 1365 kg, ground clearance of 400 mm and a custom deformable element (Type 2). For the driver dummy this configuration greatly increased the door to dummy shoulder contact velocity as well as increasing the pelvis contact velocity and pelvic lag. This resulted in greatly increased injury risk in both the thoracic and abdominal areas. There was also an increase in head injury risk.

Test 7 combined the heavy mass and increased ground clearance coupled with a modified stiffness and stiffness distribution. Compared to Tests 2, 4 and 5, Test 7 produced the highest driver dummy shoulder contact velocity, but Test 2 produced the highest pelvis contact velocity and largest pelvic lag. The driver thoracic injuries for Tests 2 and 7 were similar, and approximately twice that for the baseline tests.

This suggests that for the driver thorax, increasing the barrier face ground clearance has a very large effect on injury risk, whilst mass and stiffness have only a minor effect. In the abdomen, Test 7 showed the highest injury risk of all tests, with Test 2 showing the second highest. This further supports the notion that ground clearance is the predominant factor of all the compound variations in Test 7 and that mass, stiffness and stiffness distribution are less significant. In the pelvic region, Tests 4 and 5 showed a more significant influence on sacrum, iliac and pubic force than Tests 2 or 7. However, these increases in injury risk were all moderate and pelvic acceleration and TTI[d] were similar for Tests 2, 4, 5 and 7. This indicates that mass and stiffness may have a moderate influence on injury risk in the pelvic area.

For the passenger dummy the compound variation of Test 7 produced a significant increase in head acceleration (without contact) and chest VC was more than double the baseline value. There were also increases in TTI and pelvis acceleration that were greater than those measured for any of the parametric variations separately. Abdomen responses however remained essentially unchanged from the baseline.

Increase In Test Speed

Test 8 was the same as Tests 1 & 3 except the impact speed was raised from 50 km/h to 60 km/h. In Test 8, the driver pelvis contact switch did not indicate closure until after 70 ms, whereas the dummy's pelvic acceleration signal clearly showed an impact with a peak at 28.9 ms. In addition, the relative velocity of the door to dummy pelvis changes rapidly from 8 m/s to 4.8 m/s over the time interval from 21 to 23 ms. This further supports the notion that the door contacts the pelvis around that time.

Despite the fact that the closure time and contact velocity at the driver pelvis were unable to be determined, the pelvic acceleration is delayed in time relative to the thoracic acceleration, suggesting that a pelvic lag also occurs with this test. Abdominal rib deflection was the lowest recorded, however, abdominal VC was the fourth highest and almost 20% higher than the baseline test. The pelvic acceleration was the highest recorded while the chest VC was the 2nd highest. Iliac, sacrum and pubic force were also amongst the highest. These highly increased injury risks for the driver dummy were recorded without a correspondingly high increase in door to dummy contact velocity at the shoulder. As the pelvis contact velocity and pelvic lag are unknown for this test, it is not possible to come to a firm conclusion regarding the cause of the high dummy measures recorded. The rear of the driver head also recorded a glancing contact to the B-Pillar.

The passenger dummy also recorded a consistent increase in injury measures, though of slightly lesser order than the driver dummy.

It is useful to note that the kinetic energy of the trolley in Tests 5 and 8 was the same. Both these tests used a trolley with 132 kJ of impact energy (an increase of 44% over the baseline test which had 92 kJ). Test 5 used an increased trolley mass to raise the kinetic energy, whereas Test 8 used an increased impact velocity. The results clearly show that the increased impact velocity produces higher injury risk than the test with the same impact energy using a lower impact speed with higher mass.

Crab Angle and Stiffness Distribution

Test 10 used the same target vehicle conditions as Test 1 however, the trolley configuration, targeting and the deformable element were in accordance with US Federal Motor Vehicle Safety Standard (FMVSS) 214. All injury outcomes for the driver dummy decreased or were similar to the baseline except for sacrum force, which was about 60% higher. High contact velocities were measured at both the shoulder and pelvis, but this test resulted in the smallest pelvic lag of all tests (1.2 ms). This indicates that the shoulder and pelvis were impacted at almost the same time and with the same velocity, spreading the loads across the dummy. Compared to the baseline tests, this vertical intrusion profile produced in Test 10 leads to reduced or unchanged injury risk in all body regions. It is noted that the target vehicle was designed to meet FMVSS 214.

The responses for the passenger dummy however showed a different trend with increases over the baseline tests in all measures. Thorax VC was more than 3 times that of the baseline, with pelvis acceleration more than double. There were also significant increases in head acceleration, TTI and abdomen VC.

The static intrusion profiles shown in Figure 1 show that the FMVSS 214 barrier element has produced significantly greater intrusion at the point of the rear dummy than all other configurations. It is not possible to determine from this series whether this is a result of the element geometry or the crabbed configuration or both. This may be established by a further test, using the FMVSS 214 configuration, but without the crabbed MDB.

SUMMARY

While the results of this parametric study are specific to the subject target vehicle, some further support has been added with the inclusion of results A1 and A2. The results are also in strong agreement with other studies using variation of the impacting object [4],[5].

Driver Dummy

For the driver dummy the two most significant effects on injury risk were increase in element ground clearance and increase in test speed. This can be seen in Test 2, where the increased MDB element height produced a significant increase in mostly thoracic injury, and Test 8 where increased test speed produced increased injury to all body regions.

Increase in MDB mass had only a marginal effect on injury risk for the driver dummy (Test 5 and Test A2). A greater effect was observed when the target vehicle mass was reduced, however it is unclear to what extent this result was affected by changes to kinematics resulting from changes to the vehicle's moments of inertia and centre of gravity.

Doubling the MDB element stiffness produced an increased driver injury risk only in the pelvic area.

Test 7, using a MDB with increased mass, height and stiffness produced increases to both pelvic and thoracic loading.

Impact to the driver dummy pelvis consistently occurred after impact to the shoulder. The amount of 'pelvic lag' and the door contact velocity appears to influence the severity of thoracic loading.

The test using the FMVSS 214 configuration (Test 10) produced driver dummy responses reduced from those of the baseline tests (with the exception of sacrum force).

Passenger Dummy

The trends in results for the rear dummy with parametric variation are generally consistent with those observed for the driver dummy with the following exceptions:

- Pelvis contact does not consistently lag shoulder contact – pelvis contact does not appear to be a good predictor for injury to thorax or other body regions with the most severe tests showing pelvic lead.

- While head contact was observed in only 1 test (Test 9), with a moderate HIC (391) there was a high head acceleration (80g) resulting from contact with the C-pillar.
- Test 10, with the FMVSS 214 configuration, showed high injury measures, including the highest pelvis acceleration and iliac force. This is in contrast to the driver dummy, where this test produced injury measures that were consistently lower than the baseline tests. It is unclear whether this is a result of the crabbed impact configuration or the FMVSS214 barrier element, which is wide with homogeneous stiffness over the complete barrier face. Element aiming may also be a factor.

The measures for the passenger dummy were generally either well below the IARV or borderline. There were only two results that significantly exceeded the IARV. Both of these were for abdominal rib deflection (Tests 5 and 9).

REFERENCES

1. Seyer K.A., Terrell M.B., Fildes B., Dyte D. ,Digges K., "Development and Benefits of a Harmonised Dynamic Side Impact Standard", 16th ESV Conference, Windsor, June 1998, Paper 98-S8-O-04

2. Mertz H., Irwin A., ISO/TC22/SC12/WG5 Document N288.

3. Seyer K., Newland C., Terrell M., Dalmotas D. "The Effect of Mass, Stiffness and Geometry on Injury Outcome in Side Impacts – A Parametric Study", Stapp Car Crash Journal, November 2000, Paper 00S-01.

4. Nolan J.M., Powell M.R., Preuss C.A., Lund A.K., "Factors Contributing to Front-Side Compatibility: A Comparison of Crash Test Results", 1999 Stapp Car Crash Conference, SAE Paper 99SC02.

5. Wykes N.J., Edwards M.J., Hobbs C.A., "Compatibility Requirements for Cars in Frontal and Side Impact" 16th ESV Conference, Windsor, June 1998, Paper 98-S3-O-04.

APPENDIX 1

Driver Dummy

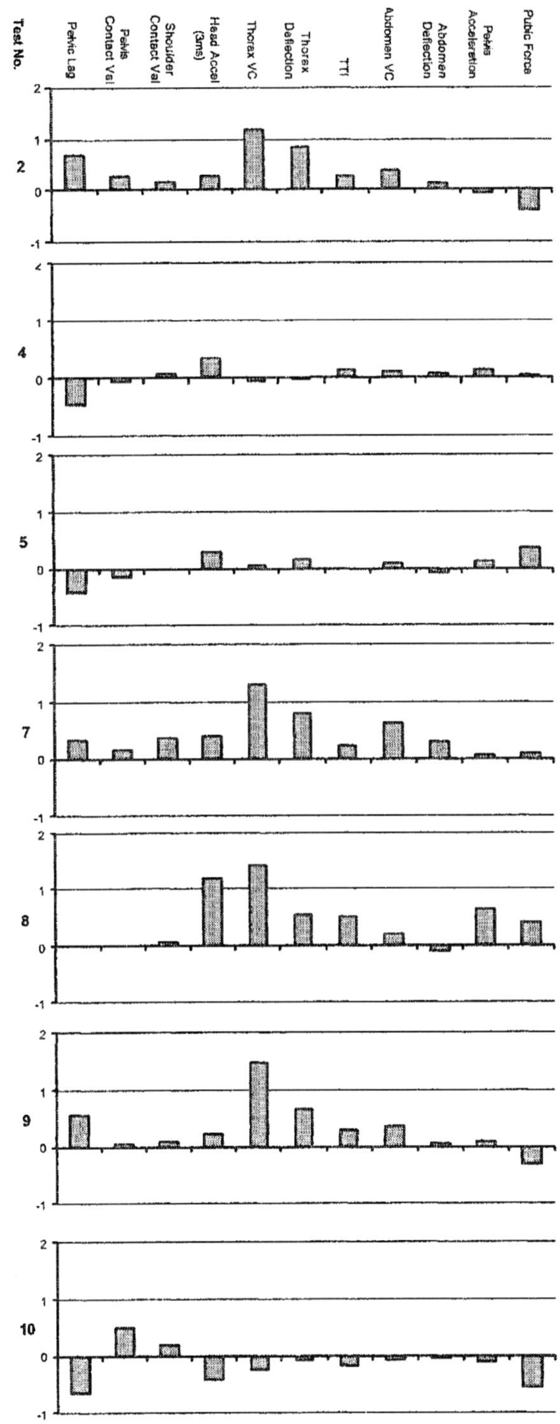

Passenger Dummy

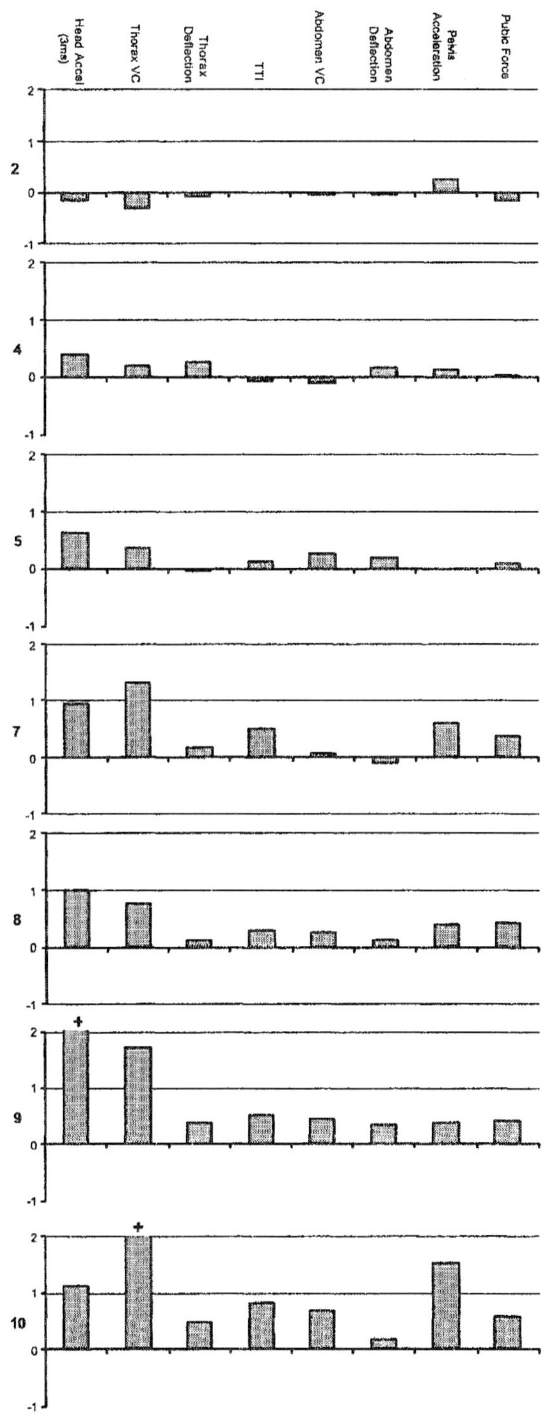

1999-01-0067

Vehicle Compatibility - Analysis of the Factors Influencing Side Impact Occupant Injury

Don Vander Lugt, Terry Connolly and Dilip Bhalsod
General Motors Corporation

ABSTRACT

This paper discusses a study conducted by GM to better understand the factors that influence injury potential in vehicle-to-vehicle side impacts. A number of other studies have been done which focus primarily on frontal vehicle-to-vehicle compatibility. GM focused on side impact compatibility in this study due to the risk of harm generally associated with this type of crash. Real world field performance was studied through an extensive six-state field analysis of recent model year ('94+) vehicles. Of particular interest in this study was an efficacy analysis of the MVSS 214 dynamic side impact standard, which was phased-in starting with some 1994 model year passenger cars. Physical side impact crash testing of a 1997 passenger car was used to investigate the relationship of impacting mass, speed, geometric profile and stiffness on side impact intrusion and occupant injury. Included in this test series was an assessment of the injury response differences between the SID and BioSID anthropomorphic test device. To further investigate the factors influencing side impact injury, computer simulation was used to investigate how changes in impacting vehicle speed, mass, height, angle, overlap, plan view curvature and width effect occupant injury and intrusion.

INTRODUCTION

Recently much attention has been paid to vehicle compatibility, especially SUV and Light Truck compatibility with passenger cars. Some in the media have tended to portray the subject of truck-to-car compatibility with significant emotion -- as a newly created danger caused by the increasing market share of trucks. In reality, compatibility concerns have existed since the first shared usage of roadways. A quick observation of the infrastructure of industrialized countries reveals that users come in all shapes and sizes -- from locomotives to pedestrians. Despite this wide variety of relatively incompatible users, traffic safety has improved significantly as a result of improvements in vehicle design, roadway design, emergency response and perhaps most importantly, operator (and passenger) behavior. Recently introduced safety features and performance requirements for frontal, side and head impacts offer additional opportunities for improved traffic safety. Traditionally, efforts to mitigate vehicle related injuries have focused mainly on areas which offered the greatest potential for societal harm reduction and primarily on self protection. As stated, this effort has proven to be very successful at improving traffic safety, but in order for the industry to continue to the next level of harm reduction, attention should be paid to vehicle-to-vehicle compatibility.

Crash compatibility between vehicles is not easily understood since performance can vary significantly with seemingly minor changes in the type of impact or specific vehicle characteristics. Considerations for crash compatibility vary substantially between different markets and road systems around the world. In Europe, the emphasis is on car-to-car frontal compatibility, but truck-to-car and side impact appears to take on additional significance in the U.S.

Fundamental physics and numerous field studies have clearly shown a higher fatality risk for occupants in smaller and lighter vehicles when colliding with heavier vehicles. Field data analysis also shows that side impacts, which are less frequent than frontals, can be severe, harm producing crashes. The harm potential associated with side impact makes it an important candidate for study and is particularly germane to understanding the issue of vehicle-to-vehicle compatibility. Vehicle designs have evolved such that front ends are often inherently incompatible with the side of another vehicle. Better understanding of the role of these design factors may present opportunities to diminish side impact harm by improving side impact protection and/or by improving front-end crash friendliness.

Most vehicle manufacturers agree that there are gaps in understanding the fundamental science in the area of compatibility. For example, little is known about the detailed field compatibility performance of recent model year vehicles. There is limited quantitative knowledge of the numerous design factors that influence the outcome of vehicle-to-vehicle crashes. In general, technical authors have speculated that three different factors are relevant to the compatibility issue; namely mass, stiffness and geometric characteristics. [1,2,3,4]

This paper discusses a study which was conducted to examine side impact compatibility and the factors influencing side impact occupant injury. Real world field side impact performance was studied through an extensive six state field data analysis of recent model year ('94+) vehicles. A particular focus of this study was the investigation of the effectiveness of MVSS 214. In addition, physical crash testing was performed to understand the factors influencing side impact compatibility. This test series was chosen to investigate the relationship of impacting vehicle mass, speed, profile, and stiffness on side impact intrusion and injury potential. The testing also examined injury response as a function of which anthropomorphic test device (ATD) was used, SID or BioSID. The results of these tests were analyzed and compared with test results generated from NHTSA's compatibility research and LINCAP program.

Computer simulation was utilized to understand the factors influencing side impact compatibility. Review of the physical test data (from this study and others) suggest an understanding of the influence of impacting speed, mass, height, stiffness, angle, overlap, plan view profile, and width in side impact crashes would require extensive analytical work supported by appropriate physical testing.

Interaction effects, the complexity of the number of relevant factors, and physical test variation make it premature to draw definitive conclusions on side impact compatibility factors based on this body of work. This should come as no surprise to crashworthiness engineers since they are very familiar with the effect subtle design changes can have on the measured injury outcome in tests (including the MVSS 214 protocol). This paper attempts to contribute to the base of physical understanding about side impact performance by sharing observations gained through field data analysis, physical crash testing and computer simulation.

FIELD DATA

It is instructive to first understand the involvement of side impact in motor vehicle crashes and the extent to which compatibility may affect the crash outcome. Many studies on this subject have been reported, including a paper by NHTSA/UMTRI [5]. These papers do not comprehend the effects of substantial vehicle safety modifications made in the late 1980's and early 1990's. In fact, some intentionally excluded vehicles with these upgrades. To fill this void, police reported collision data of recent model year vehicles ('94+) from six states (AR, FL, ID, MD, NC, PA) was combined and analyzed. States were selected based on availability, collision and injury code information and vehicle identification data. Results of this study can be projected to all states and provide insight on crash causation and outcome. The collision data for this study was taken from crashes in calendar years 1993-96.

Frontal crashes (two-vehicle and single-vehicle) are known to represent over 50% of crashes for the aggregate fleet. (Figure 1) This study categorized the 24% of crashes that are side impacts into "midship" (near + far) side impacts and off-side impacts, which are usually less harm-producing. Off-side impacts were classified as those where the impact point was in the fender or quarter of the vehicle. Side impacts are over-represented in fatalities, with near-side representing about 16% of combined single and multiple vehicle crashes, far-side representing an additional 6% of fatalities and off-side crashes another 10% (Figure 2).

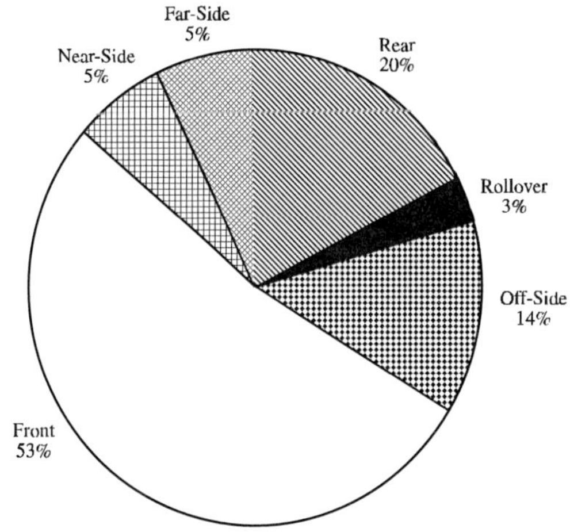

Figure 1. Percentage of Accident Involving Passenger Car Occupants by Type of Impact- Six States (Two Vehicle and Single Vehicle Accidents)

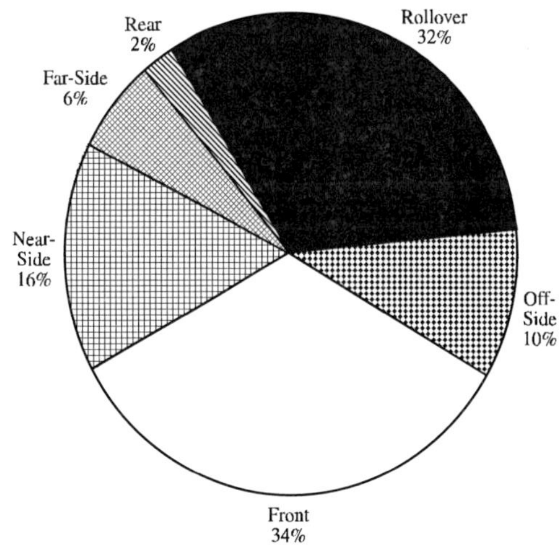

Figure 2. Percentage of Fatally Injured Occupants of Passenger Cars and Light Trucks by Type of Impact- Six States (Two Vehicle and Single Vehicle Accidents)

Near side and far-side crashes can be analyzed by broad categories of impacting vehicle. The breakdown of fatalities in both near and far side impacts is roughly one-third the result of single vehicle crashes, slightly less than one-third the result from being struck by a passenger car and slightly more than one-third the result of impact by a light truck, which for this analysis includes vans and SUVs

(Figure 3 and 4). Although single vehicle and smaller passenger car crashes are the largest individual contributors in near side impacts, other sources [5] have correctly pointed out that heavier truck striking vehicles are over-represented in fatality-producing collisions (in proportion to the frequency of such collisions).

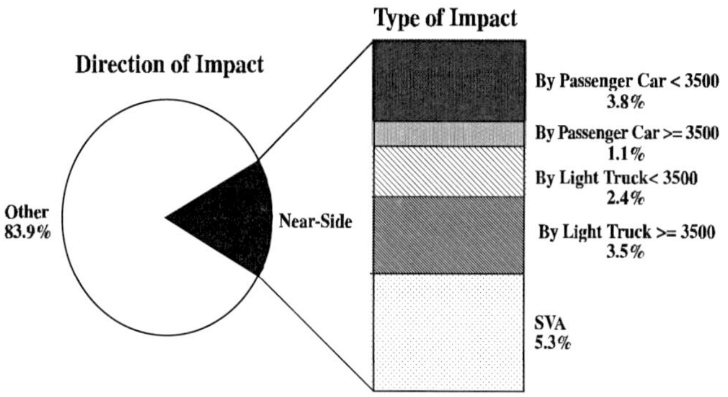

Figure 3. Percentage of Fatally Injured Occupants of Passenger Cars and Trucks for Near Side Impacts by Impact Type - Six States (Two Vehicle and Single Vehicle Accidents)

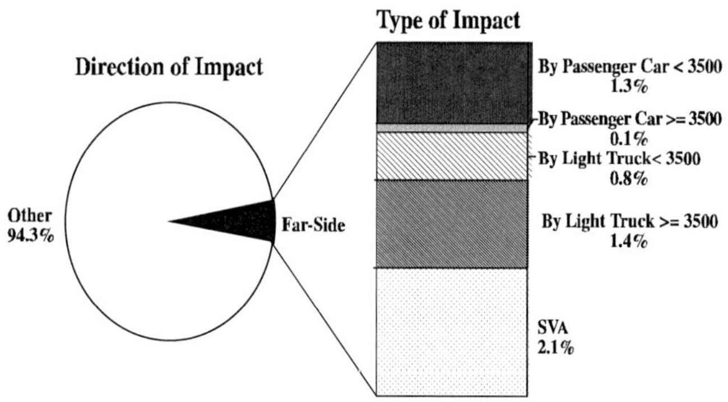

Figure 4. Percentage of Fatally Injured Occupants of Passenger Cars and Trucks for Far Side Impacts by Impact Type - Six States (Two Vehicle and Single Vehicle Accidents)

A study of 1996 FARS data is instructive as to the field effects of compatibility in our transportation system (Figure 5). Although these "trees" of fatalities in various types of crashes for cars and trucks are not normalized for exposure, a quick inspection suggests that "normal" high frequency car-to-car collisions produce the largest number of fatalities and infrequent events (car to train, or heavy truck to car) also result in substantial harm. Two important conclusions in guiding the emphasis of our research were derived from this analysis: 1) the breadth of compatibility crashes (pole vs. heavy truck) faced in the real world would offer very different guidance to the vehicle designer and 2) protecting occupants in the relative predominance of single vehicle crashes (whether car or light truck), cannot be compromised.

For frontal impacts, speculated compatibility "solutions" have often been vehicle design constraints that may increase the protection in one situation but decrease the protection in others, such as single vehicle crashes. The FARS analysis, discussed earlier, makes it evident that it would be inappropriate to undertake any measures that might reduce the protection of occupants in these single vehicle crashes. Viewed in aggregate, it is easily concluded that *side impacts constitute a compatibility mismatch for both cars and LTVs*, no matter what the configuration of the impacting vehicle. One reason for this is illustrated in Figure 6 which shows that the stiffness of a typical vehicle front end is more than 2 times that of the side.

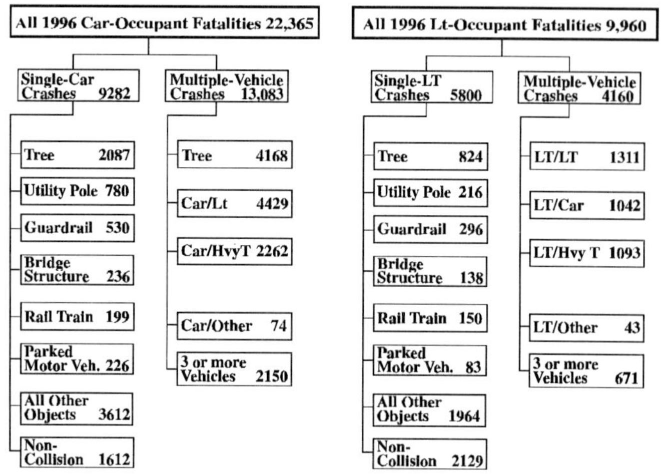

Figure 5. Distribution of Occupant Fatalities in Cars and Light Trucks in 1996 - FARS

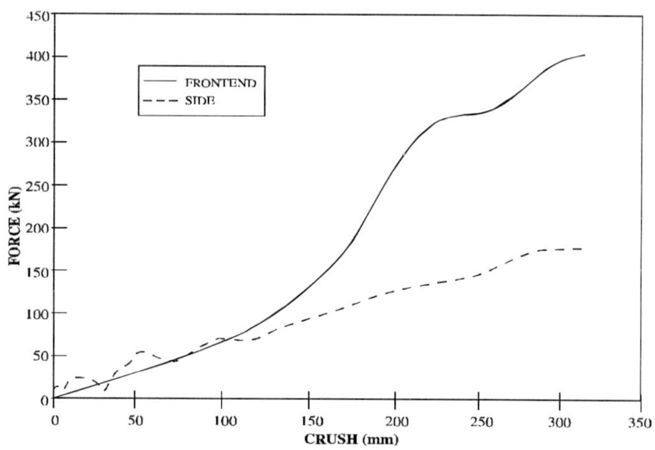

Figure 6. Force Deflection of Typical Frontend (35 mph NCAP Barrier Loads) and Passenger Car Side (Rigid MVSS 214 Barrier-Quasi-static Computer Simulation)

The effect of mass incompatibility in fleet crashes was qualitatively understood even before the first fatality ratios, summarized by Evans and Frick in 1993 [6], quantified the magnitude of such differences. Recent studies (particularly the NHTSA/UMTRI study previously mentioned) have attempted to relate other striking vehicle factors to crash outcome, but were constrained to use field data of older model vehicles and data which lacked the fidelity necessary to differentiate these factors. Unfortunately, the field data on the more recent vehicles is still not sufficiently mature to show accurately the relationship of vehicle characteristics other than by mass or car/truck classification.

Previous work by AAMA has put in perspective the growing SUV compatibility issue. SUV to passenger car collision represents only about 3% of vehicle occupant fatalities (Figure 7). Additional analysis, from our study, indicates approximately 3.75% of all reported crashes (regardless of injury) are LTV side impacts into cars. But clearly, both car and LTV occupants in <u>side impacts</u> are at higher risk than are occupants in frontal crashes regardless of the striking vehicle configuration. As a population, striking trucks result in higher injury risk than do cars of similar mass classification. One factor that might be contributing to this higher injury risk associated with trucks is that trucks in the field are more likely to be loaded in excess of their classification mass than cars. It is important to note, however, as work in this paper will demonstrate, injury producing characteristics of impacting vehicles can be present in either cars or trucks.

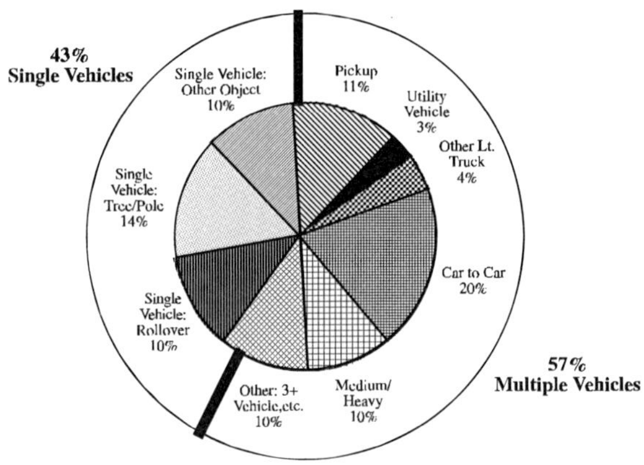

Figure 7. Percentage of Passenger Car Occupant Fatalities by Crash Type and Collision Partner

To better understand the trend of truck-to-car impacts, it is valuable to understand the incidence of crash involvement. A specific element of this study examined two-vehicle crashes in Maryland between 1977 and 1996 in more depth. As had been noted in other studies [1,2], truck market penetration is increasing and car market penetration is declining. As a natural result, crashes of cars and light trucks are on a long-term increase (as are truck to truck crashes) (Figure 8). However, the actual involvement of trucks in two vehicle crashes is less than that predicted by the registration database, even though trucks on annual average accrue more miles than cars (Figure 9)! Although the Maryland data is instructive, other state data may vary, depending upon truck fleet registrations and the predominance of urban vs. rural vehicle usage.

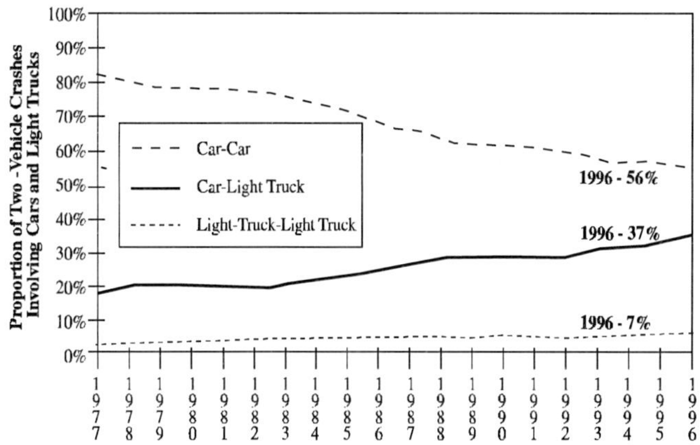

Figure 8. Percentage of Two-Vehicle Crashes in Maryland (Crashes Involving Only Cars and Light Trucks)

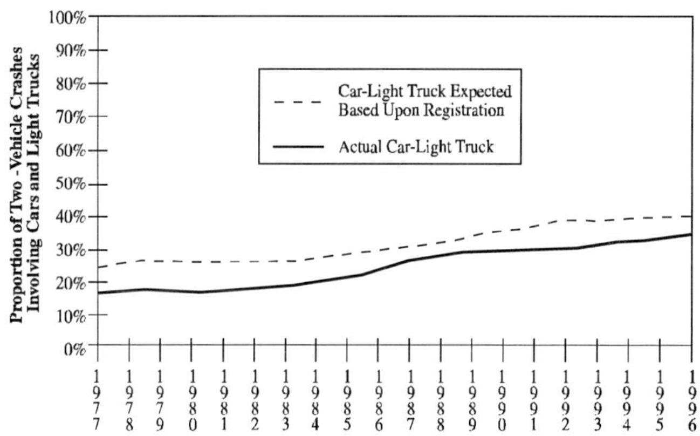

Figure 9. Percentage of Actual vs Expected Two Vehicle Crashes for Maryland (Crashes Involving Only Cars and Light Trucks)

MVSS 214 EFFECTIVENESS – Compliance to the dynamic side impact protocol, required for some passenger cars beginning in vehicle model year 1994 and required for all passenger cars in model year 1997, caused substantial re-engineering and improvement in side impact protection for many vehicles. Prior to examining further technical opportunity in side impact protection, it is appropriate to understand the effectiveness of this regulation. As part of the state data study, crash-involved populations of vehicles were normalized for exposure by registered vehicle years, then compared. For all types of crashes (frontal, side, etc.), the population of passenger cars reported to the NHTSA as MVSS 214 (dynamic) compliant showed substantial reduction in risk levels over vehicles not reported as meeting the dynamic MVSS 214 requirement (Figure 10). When impacted by either heavy trucks or light cars, these compliant vehicles showed statistically significant improvements for side impacts over vehicles not reported to the NHTSA as compliant (Figure 11). Similar trends were also noted for the population struck by large cars and small trucks, but these are not yet statistically significant. Typical reduction of harm (fatal or major injuries) for MVSS 214 compliant vehicles approached 20% overall for side impacts. Looking specifically at near-side impacts, MVSS 214 provides the most benefit for vehicles impacted by light passenger cars, heavy trucks and in single vehicle impacts (Figure 12).

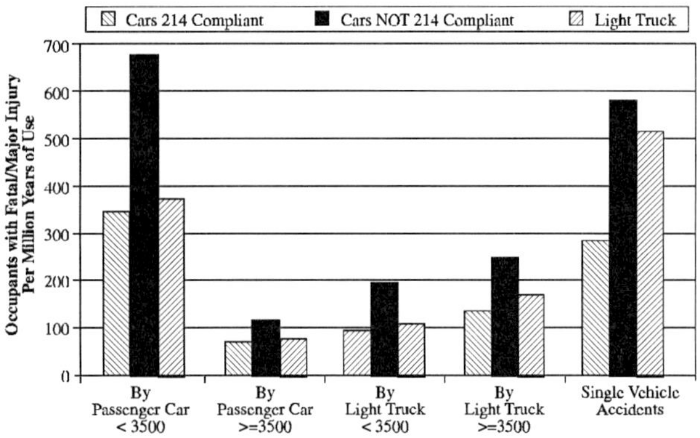

Figure 10. Risk of Fatal/Major Injury in All Impacts by Impact Type - Six States 1995-1996 (Two Vehicle and Single Vehicle Accidents)

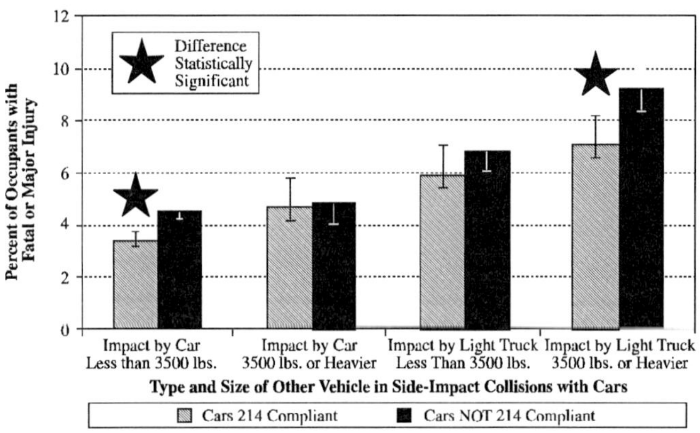

Figure 11. Percentage of Occupants with Fatal/Major Injuries for 214 Compliant and Non-Compliant Cars Involved in Side Impact with Lighter/ Heavier Cars and Trucks - Six State (Standardized Rates)

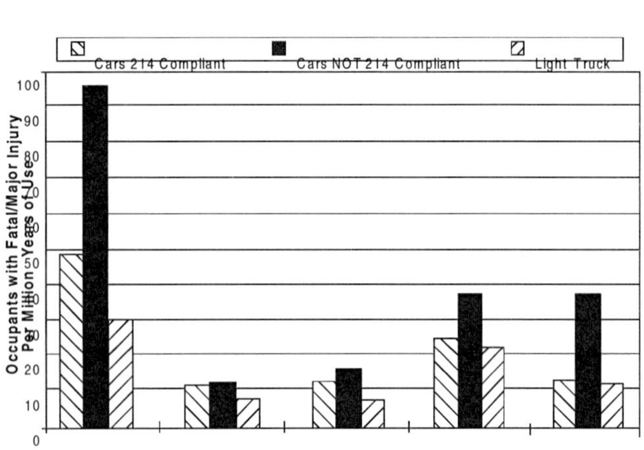

Figure 12. Risk of Fatal/Major Injury in Near-Side Impacts by Impact Type - Six States 1995-1996 (Two Vehicle and Single Vehicle Accidents)

The general occupant protection increases observed are likely the result of a number of changes being implemented across the industry in this time period, including: body side structure with positive benefits for frontal offset/ angle protection, the introduction of frontal airbags and the substantially deeper engineering capability offered by finite element and kinematics modeling. The study could be further refined as more data becomes available and if more information on the date of MVSS 214 dynamic compliance for vehicles was known. This study makes it clear that the positive improvements in both frontal and side impact protection of the last decade should not be dismissed in other compatibility studies. A comprehensive industry-cooperative study of side impacts as field data accrues would be beneficial.

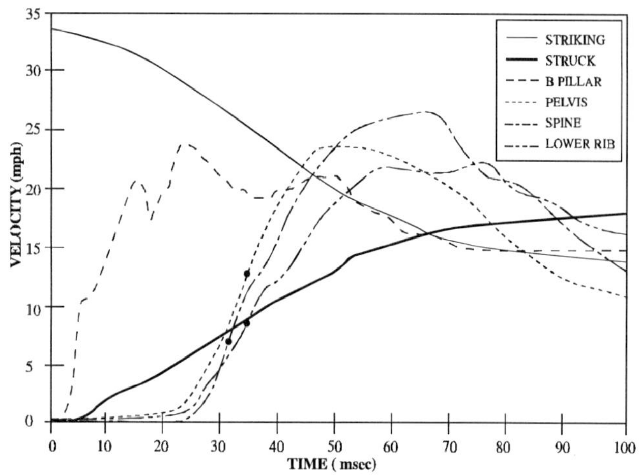

Figure 13. Velocity Time History of the Accord MVSS 214 Side Impact Test (Values Used to Determine Maximum TTI and Pelvis G Indicated by Solid Dots)

PHYSICAL CRASH TESTING

SIDE IMPACT CRASH EVENT – To provide insight into the results of the physical testing, it is helpful to consider the sequence and timing of events during a typical MVSS 214 side impact crash. To illustrate this sequence, the typical velocity time history of a number of components in the side impact is shown in Figure 13. The target vehicle contacts the bullet vehicle, resulting in the 'B' pillar/body-side being very quickly accelerated to match the velocity of the impacting vehicle. Simultaneously, the target vehicle slowly accelerates away from the impacting vehicle, but attains a velocity much lower than that of the incoming body- side. As the striking vehicle intrudes, the struck vehicle body-side velocity peaks and then begins to drop off. At some point, typically very early in the collision, the door contacts the occupant, causing the occupant to rapidly accelerate. In the SID, acceleration of the ATD's ribs, spine and pelvis is measured and used to evaluate injury risk. Typically, the velocity of contact can be viewed as a first order measure of impact severity. Higher contact velocities require more energy absorption, and therefore crush space in the side impact "restraint system" to manage the crash energy and limit occupant acceleration. Typical vehicle dimensions allow only limited crush space outboard of the occupant, thus limiting the energy absorbing capability of the side "restraint" system.

Contact velocity and thus injury risk can be altered by not only changes in the struck vehicle design but also by changes in the impacting characteristics. To better understand the effect impacting mass, stiffness, geometric profile and speed have on side impact injury potential, five vehicle side impact tests were conducted. The choice of ATD (SID vs BioSID) can also influence the indications of injury. To evaluate this, one test utilized BioSIDs rather than SID dummies. The struck vehicle in all impacts was a 1997 Honda Accord to allow comparison to NHTSA's 1998 compatibility research tests. The test configuration used was identical to NHTSA's compatibility tests. The SID test device used was also modified as in NHTSA's tests by replacing the standard SID neck and head with the Hybrid III neck and head.

In the first physical test group (three tests), a 1996 Taurus, 1996 Accord and heavy MDB, all with a mass of approximately 4200 lb., impacted the 1997 Accord. The striking Taurus and Accord were ballasted with additional rear ATDs, while the MDB was ballasted with additional weights. In the second group (two tests), a standard MVSS 214 test was conducted along with a LINCAP test using BioSIDs. Further test details and results are shown in Table 1 (NHTSA compatibility research and LINCAP results are shown in this table for ease of comparison).

RESULTS DISCUSSION – For clarity, only front seat ATD results are summarized below. They are more controlled and consistent due to the direction of loading, profile of the door deformation, etc., than are results from the rear seated ATD. Although numerous injury measurements were recorded, this discussion focuses primarily on TTI, HIC, and intrusion profile.

The first test series demonstrated the risk of injury, in the struck vehicle, was the highest for the MDB impact, followed by the Taurus and then the Accord. These striking vehicles were chosen based upon substantially different front stiffness and geometry as well as their high volume in the market. Post test exterior intrusion profile of the target vehicle, near the occupant "H" pt., showed the Taurus impact to have substantially deeper and more concentrated intrusion (Figure 14). As seen in numerous MVSS 214 tests, the MDB produced a flat and distributed intrusion pattern. Crush of the impacting vehicle varied significantly, with minimal crush for the MDB and significant crush for the Accord front-end. In this series, increased striking vehicle crush reflected lower injury measures in the struck vehicle, suggesting a relationship between striking stiffness (force level) and struck vehicle injury outcome. Geometric profile and contact area of each impact also affect injury outcome, thus these conclusions cannot be applied generally.

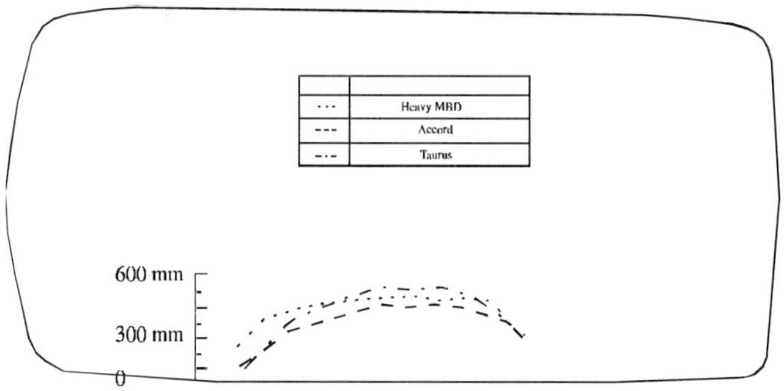

Figure 14. Post test intrusion profile of Taurus, Accord and MDB impacts.

It is important to understand the crash response due to different impacting types. Analyzing the rigid body velocity of the struck vehicle shows that the MDB impact resulted in a much higher rate of velocity change than either the Taurus or Accord impact, which interestingly were quite similar (Figure 15). This is probably due to the MDB contact area and stiffness being greater than the passenger vehicles'. Stiffness of the first 100 to 150 mm of the striking vehicle may materially affect the occupant contact velocity, which occurs during the first 20-30 ms of the side impact crash. These results support the injury response, indicating that the MDB is a more severe impact than a typical passenger vehicle impact of similar mass.

Table 1. Side impact test results

	GM Series 1			GM Series 2		NHTSA Compatibility Tests				LINCAP
Bullet Veh	96 Taurus	96 Accord	Heavy MDB	214 MDB	214 MDB	S10	Explorer	Caravan	Lumina	214 MDB
Mass (lb)	4234	4121	4229	3000	3000	3660	4680	4562	3983	3000
Speed (mph)	33.3	33	33.6	33.1	38.8	33	33	33	33	38.5
Target Veh	97 Accord	97 Accord	97 Accord	97 Accord	97 Accord	97 Accord	97 Accord	97 Accord	97 Accord	97 Accord
Mass (lb)	3341	3344	3384	3398	3407	3380				
Dummy	SID+	SID+	SID+	SID+	BioSID	SID+	SID+	SID+	SID+	SID
HIC	444	207	961	159	1023	314	395	193	242	n/a
TTI	56.6	42.8	78.2	65.2	115.5	88.1	114	89	91	96
Pelvis G	98.4	59.7	83.1	65.3	118.6	89.9	74	128	87	117
V*C					2.3					

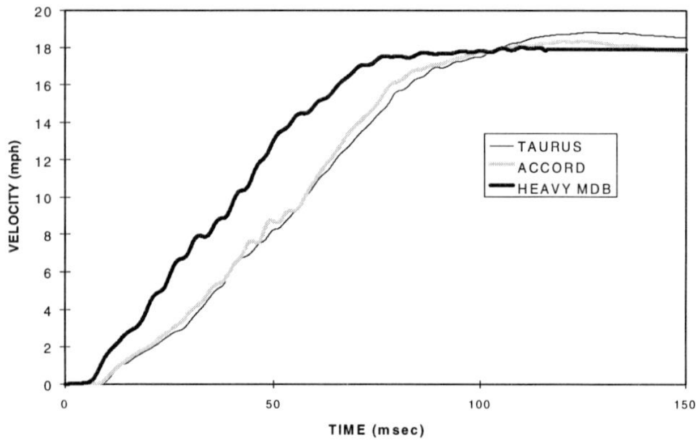

Figure 15. Struck Rigid Body Velocity Time History for Heavy MDB, Taurus and Accord Impacts

Mass effect conclusions are derived from these tests by comparing the heavy MDB, from the first group, with the standard MDB (MVSS 214) test of the second group. As expected, all injury values resulting from the heavy MDB were higher (front occupant); likely the result of higher contact velocity with the occupant. The heavy impact barrier vehicle decelerates at a much lower rate than the lighter barrier (Figure 16), translating directly to a higher door contact velocity and increased occupant loading.

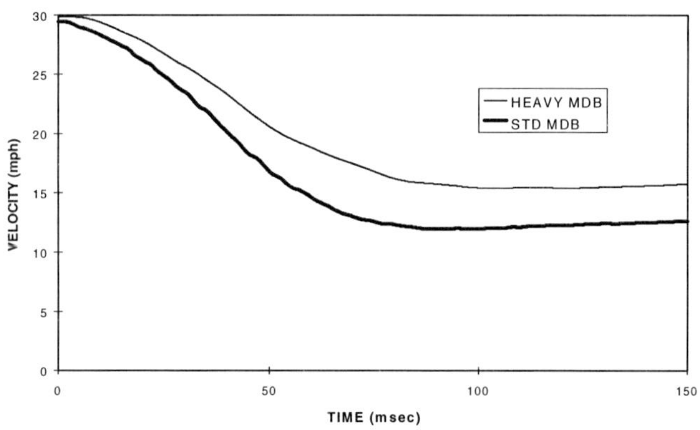

Figure 16. Striking vehicle velocity time history

Mass effects did not always dominate, which illustrates the confounding engineering difficulty of this issue. Despite the Taurus being loaded 15% heavier than the S10, tested by NHTSA, the S10 impact resulted in comparatively higher driver TTI. There are a number of differences between the two vehicles which could have contributed to this response, these are explored more in the analytical work below.

Comparison of the BioSID LINCAP test, performed in the second test group, and NHTSA's LINCAP (SID) test generates concern over the fidelity of the occupant injury metrics. Utilizing TTI, the BioSID showed a much higher response than the SID, resulting in nearly twice the risk of AIS 4+ injury (41% vs. 23%, respectively). The BioSID V*C metric indicated a risk of over 98% AIS 4+ injury [7]. Clearly, the BioSID identified higher injury risk in this crash configuration. This higher injury risk is collaborated by crash accident experts who indicate that substantial field injuries typically result from intrusions similar to those seen in these tests. Based on this and the limitations of SID, which have been well documented [8,9,10], the BioSID would be a more useful device in discerning the effect of various crash and design parameters. In total, the MDB may offer a more severe collision than similar vehicle-to-vehicle field crashes and the SID may under-represent occupant injury potential.

COMPUTER SIMULATION

MODEL DESCRIPTION AND APPROACH – Computer simulation of a midsize sedan MVSS 214 barrier test (Figure 17) was used to study the effects that impacting speed, mass, height, stiffness, angle, overlap, plan view curvature, and width have on occupant injury and intrusion. These factors were chosen based on engineering judgment from physical testing and field data analysis. For each run, only one parameter was changed, with the other seven factors being set to the MVSS 214 baseline value. Clearly, interaction effects also need further study. A computer model of a standard MVSS 214 SID was used in this analysis to measure the thorax and pelvis response to the parameter changes. Unlike the ATD used in the physical testing, the model did not include a Hybrid III head and neck because a correlated model was not available.

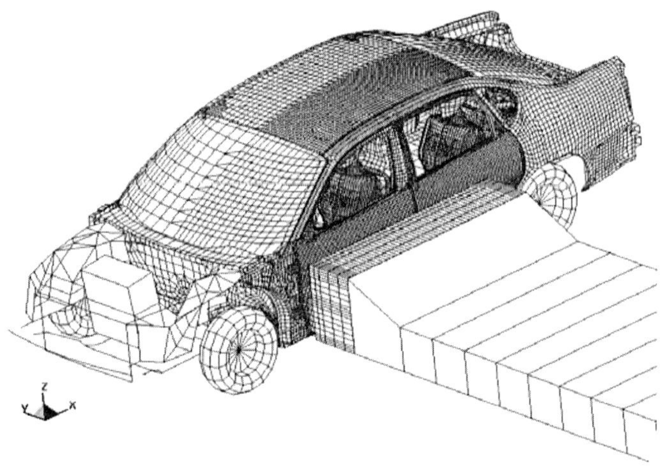

Figure 17. Finite Element Model of Mid Size Sedan and SID Dummies Used in Parameter Computer Simulation

PARAMETER VARIATION RESULTS – Again for simplicity, analysis of the SID response focused primarily around the TTI and Pelvic accelerations for the front ATD, and plan view intrusion profile of the vehicle at 60 msec. The reader is cautioned that generalizations beyond the specific architecture simulated may not be valid. Appendix 1 contains the detailed occupant and intrusion results of this study.

<u>Impact Velocity and Mass (Appendix 1a, 1b)</u> – Three impacting barrier speeds were evaluated: 28.5 mph, 33.5 mph and 38.5 mph. As expected, increasing the impact velocity increased the TTI and pelvis accelerations. The

intrusion differences between the three speeds are directly proportional to the level of energy involved in the impact.

The impacting barrier mass was evaluated at 3000 lb., 3500 lb. and 4500 lb. As the impacting mass is increased, the TTI and pelvis acceleration also increases. Again, the amount of intrusion is proportional to energy of the collision.

Stiffness (Appendix 1c) – The stiffness of the deformable barrier was changed to be 25% softer and 25% stiffer than the standard MVSS 214 barrier stiffness. Higher barrier stiffness caused higher injury potential as assessed by TTI and pelvis acceleration.

Height (Appendix 1d) – The barrier was raised 5 and 10 inches from the standard MVSS 214 height for this simulation. Raising the barrier 5 inches decreased both the TTI and pelvis value, showing a strong decrease in pelvis accelerations. However, raising the barrier another 5 inches (10 inches total) raised the TTI but further reduced the pelvic accelerations. Not surprisingly, intrusions near the pelvis changed little at 5 inches, but decreased once the barrier was raised to 10 inches. Intrusion near the thorax, however, shows continual increase as the barrier was raised (Appendix 1e). It should be noted that broad application of these results to other vehicles may be difficult since the response to impacting height changes is going to be highly dependent on the architecture of a vehicles' body-side.

Curvature and Width (Appendix 1f, 1g) – Two different plan view profiles were evaluated: the standard flat MVSS 214 barrier and a plan view curvature which matched a high volume midsize sedan (note, 214 barrier side view profile was not changed). Curving the plan view profile of the deformable barrier increased injury potential, with the TTI increasing significantly more than the pelvic accelerations. Interestingly, the amount of intrusion did not markedly change between the two profiles even though in physical testing, different curvatures clearly showed different post-test profiles.

The width of the barrier was evaluated at 10 inches narrower and 10 inches wider than the standard MVSS 214 barrier. Center point and targeting remained the same. Overall, the injury response was relatively insensitive to barrier width. Further study is needed to fully understand this result. Pelvic accelerations increased for the wider barrier and TTI decreased for both the wider and narrower barrier faces.

Intrusion profiles reflected deeper penetration of the narrowed barrier, as it loads a smaller area of the body side. By contrast, the widest barrier intruded the least but loaded more of the body side.

Fore/Aft Targeting and Angle (Appendix 1h, 1j) – The barrier was evaluated forward 12 inches and rearward 12 inches relative to the standard MVSS 214 location. As expected, the sensitivity to minor targeting changes is substantial. Moving the barrier forward strongly increased injury values, moving it rearward showed only minor effects. Deformation results matched the barrier impact location.

Plan view rotation of the barrier forward (-)20 degrees and rearward (+)20 degrees decreased injury measures in comparison to the standard 214 position. The intrusion profile was substantially less for the -20 degree impact compared to the standard impact, with the largest point of intrusion concentrated on the front door area. The +20 degree impact intruded nearly as much as the standard impact but was concentrated more on the rear door.

SUMMARY OF PARAMETER STUDY – Response gradients for the major parameters were compared to understand the relative significance of each of the changes described above. The percentage change of the injury response metrics (TTI or Pelvic Gs) were normalized by the percentage change of the individual parameter (stiffness, mass, etc.). Figure 18 exhibits this gradient for the eight parameters evaluated in this study. Gradient sensitivity outside the range of parameter change analyzed in this study have not been evaluated.

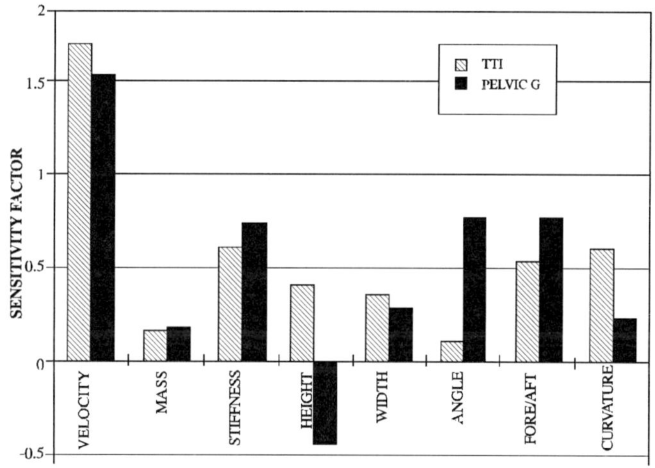

Figure 18. Response Sensitivities as a Function of Parameters Changed

Many factors can significantly influence side impact injury potential, some vehicle design-related (mass, profile, etc.) and some related to crash field circumstances (speed, angle of impact, etc.). As indicated by both the physical testing and the computer analysis, speed influences injury potential values more substantially than does any other factor. The importance of impacting vehicle stiffness on injury risk appears to overshadow that of mass. However, in considering the effects of mass and stiffness, we note the inherent vehicle engineering connections between these two physical characteristics. Larger, heavier vehicles nearly always are architected with greater structural stiffness to provide the customer with the necessary and expected utility and vehicle ride/

handling characteristics. Secondly (and perhaps more relevant to side impact injury), the rigid barrier test, utilized in NHTSA's MVSS 208 and NCAP protocols, creates a significant incentive for manufactures to make vehicle front-end initial stiffnesses (the front 250 mm) as steep as possible. Initially stiff front-ends improve occupant performance numbers in the U.S. consumer metric (NCAP) testing.

The influence of striking vehicle height changes is highly sensitive to architectural topology of the struck vehicle structure. It is clearly beneficial for pelvic accelerations not to have direct impact loading. However, the TTI gradient appears quick to reverse its response and magnitude depending on the initial height chosen for the striking vehicle.

In practice, vehicles have a nearly infinite set of combinations of the relevant parameters. Ultimately, the challenge to vehicle design is to make the impacted vehicle robust for a broad variety of real-world crash conditions - not just optimized for a single-point regulatory (or other) protocol. To accomplish this goal, it is critical that the protocols utilize robust ATDs (actual and analytical), not ones that are sensitive to point and angled loading.

CONCLUSION

As was stated in the introduction, tremendous improvement has been made in traffic safety and current vehicles have evolved to protect occupants at levels unattainable even a few years ago. There is little doubt that better understanding of crash compatibility will aid the industry in moving to the next frontier of occupant protection. Optimization of a vehicle's approach to crash compatibility will require a long range development effort. Although this and a few other studies offer analytical and physical insight, the depth of science supporting the understanding of relevant vehicle factors is limited.

Vehicle safety has improved significantly, most recently with the introduction of new safety features and North American performance requirements for side, frontal, and head protection. Truck-to-car incompatibilities resulting in major or fatal injuries are a relatively small piece of field harm. However, the U.S. data clearly shows that both car and LTV occupants impacted in the side are at higher risk than those in frontal crashes. Due to the nature of multi-vehicle side impacts (front of one vehicle impacting the side of another), a somewhat inherent incompatibility exists, regardless of striking vehicle characteristics. Vehicle size and utility differences, as well as distinct manufacturer safety engineering approaches, combine with the useful life of vehicles to make improvement of fleet crash compatibility a daunting task.

This study provides additional insight into field compatibility and the vehicle architecture factors influencing side impact injury. Significant injury reduction is demonstrated in vehicles that comply with the dynamic requirements of MVSS 214. Field studies which do not incorporate the effects of newer safety technologies cannot offer an accurate perspective on the magnitude of the field compatibility issue.

Compatibility issues are very complicated and not easily understood. Although the results of this study are not conclusive, injury risk increases as impacting vehicle mass, stiffness, plan view curvature, and height (of major structural stiffness elements) increases. The results of this work confirm other studies that side impact injury is influenced by factors other than just mass. The importance of impacting stiffness raises difficult questions concerning the influence that existing frontal regulatory and consumer metric test protocols have on vehicle engineering for frontal crash stiffness. As in other areas of crashworthiness, crash circumstance (impact velocity, impact angle, etc.) typically have a much more significant effect on occupant injury potential in a given collision than do vehicle design factors. Attention to the crashworthiness of vehicles must be accompanied by parallel efforts in side impact avoidance, not only on the vehicle through improvements in conspicuity and vision, but also through changes in transportation infrastructure and driver behavior. The benefit of these changes should not be taken lightly since they offer a broad and more immediate field harm reduction solution.

Although significant technical understanding is lacking before clear and practical action can be taken to address fleet compatibility in side impact collisions, guidance for further study is clear. Field performance, the factors influencing field harm and the mechanism of injury should be further studied, particularly as the number of MVSS 214 compliant vehicles grow in collision databases. Changes to objectives for vehicle crash performance need careful consideration; degrading frontal protection of a vehicle in any crash circumstance to aid crash-friendliness in side impact is likely not an optimal solution. Vehicle engineering studies are needed to understand the combined (interaction) effects of at least the vehicle design factors evaluated in this paper. Cooperative industry and regulatory work must encompass a thorough understanding of the regulatory and consumer metric protocols which now drive crash performance-related vehicle architecture decisions. In addition, a similar cooperative effort is needed to develop a meaningful and accurate metric of compatibility, including specifics of protocol and injury metrics. Since the BioSID is more biofidelic and capable of identifying injury, it (or possibly another more biofidelic ATD, like WorldSID) should be seriously considered for consensed use in future side impact compatibility studies. With this level of cooperative effort, a significant re-engineering of current crash performance protocols, which better reflect field relevance and balance self-protection with crash friendliness, is achievable.

ACKNOWLEDGMENTS

Pat Fyhrie, PhD, Michelle Du of GM and Rose Ray, PhD, Alan Donelson, PhD of Exponent/FaAA

REFERENCES

1. Gabler, H.C. and Hollowell, W.T., "The Aggressivity of Light Trucks and Vans in Traffic Crashes", SAE International Congress and Exposition, Paper No. 980908, Feb. 1998.
2. Gabler, H.C. and Hollowell, W.T., "NHTSA's Vehicle Aggressivity and Compatibility Research", 16th International Technical Conference on the Enhanced Safety of Vehicles, Paper No. 98 S3 O 01, June 1998.
3. Wykes, N.J., Edwards, M.J., Hobbs, C.A., "Compatibility Requirements in Frontal and Side Impacts", 16th International Technical Conference on the Enhanced Safety of Vehicles, Paper No. 98 S3 O 04, June 1998.
4. Sterer, C., Delhommear, M., Delannoy, P., "Proposal to Improve Compatibility in Head on Collisions", 16th International Technical Conference on the Enhanced Safety of Vehicles, Paper No. 98 S3 O 05, June 1998.
5. Joksch, H., Massie, D., Pichler, R., "Vehicle Aggressivity: Fleet Characteristics Using Traffic Collision Data", U.S. Department of Transpiration-NHTSA, DOT HS 808 679, Feb. 1998.
6. Evans, L., "Car size and safety: a review focused on identify causative factors", 14th Enhanced Safety of Vehicles Conference, Paper No. 94 S4 W 28, May 1994
7. Mertz, H. J., Priya, P., Irwin, A.L., "Injury Risk Curves for Children and Adults in Frontal and Rear Collisions", 41st STAPP Car Conference, Paper No. 973318, 1997.
8. Wasko, R. J., Campbell, K.L. and Henson, S.E., "Results of MVMA Full Vehicle Side Impact Tests on 1990 Model Year Pontiac 6000 Vehicles Using BioSID and SID", 13th Experimental Safety Vehicle Conference, Paper No. 91 S5 D 15, November 1991.
9. Irwin, A.L., Pricopio, L.A., Mertz, H.J., Balser, J.S. and Chkoreff, W.M., "Comparison of the EUROSID and SID Impact Responses to the Response Corridors of the International Standards Organization", SAE 890604, 1998.
10. Beebe, M.S., "What is BIOSID?", SAE 900377, February 1990.

APPENDIX

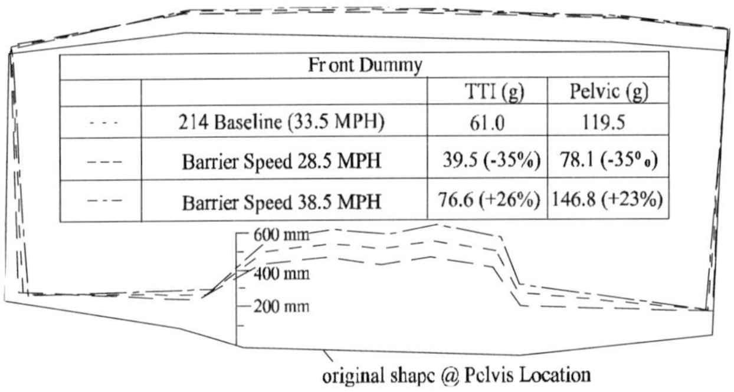

(a) Planview Intrusion Profile and SID Occupant Injury Response due to Change in Impact Velocity

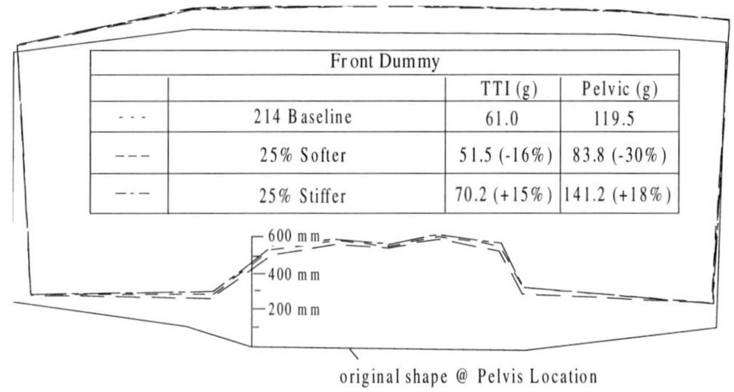

(c) Planview Intrusion Profile and SID Occupant Injury Response due to Change in Impact Stiffness

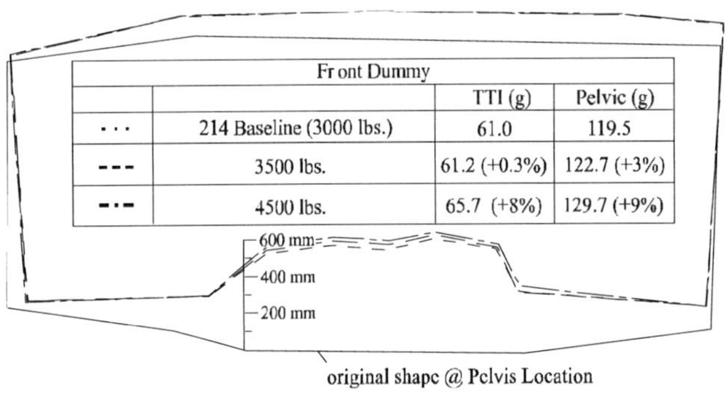

(b) Planview Intrusion Profile and SID Occupant Injury Response due to Change in Impact Mass

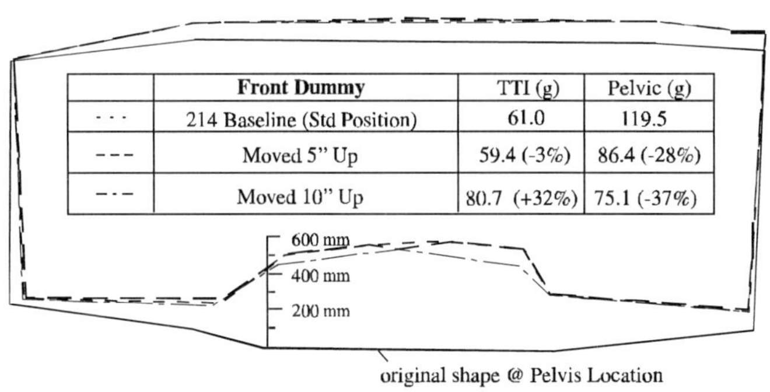

(d) Planview Intrusion Profile and SID Occupant Injury Response due to Change in Impact Height at Pelvis Location

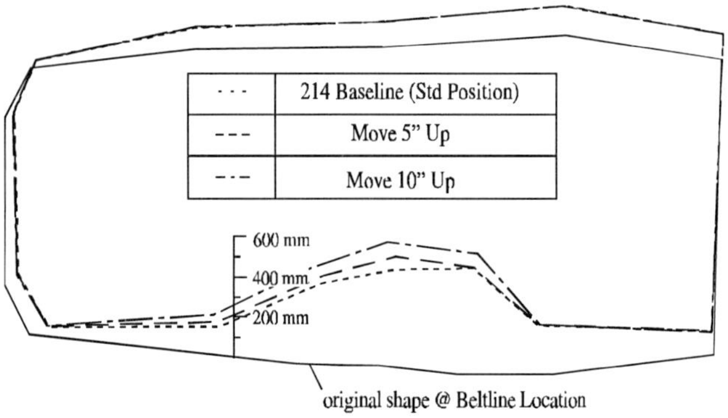

· · ·	214 Baseline (Std Position)
– – –	Move 5" Up
– · –	Move 10" Up

(e) Planview Intrusion Profile and SID Occupant Injury Response due to Change in Impact Height at Thorax Location

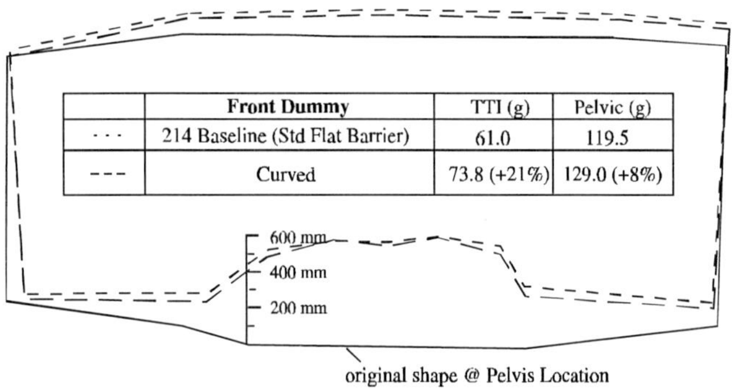

	Front Dummy	TTI (g)	Pelvic (g)
· · ·	214 Baseline (Std Flat Barrier)	61.0	119.5
– – –	Curved	73.8 (+21%)	129.0 (+8%)

(f) Planview Intrusion Profile and SID Occupant Injury Response due to Change in Impact Curvature

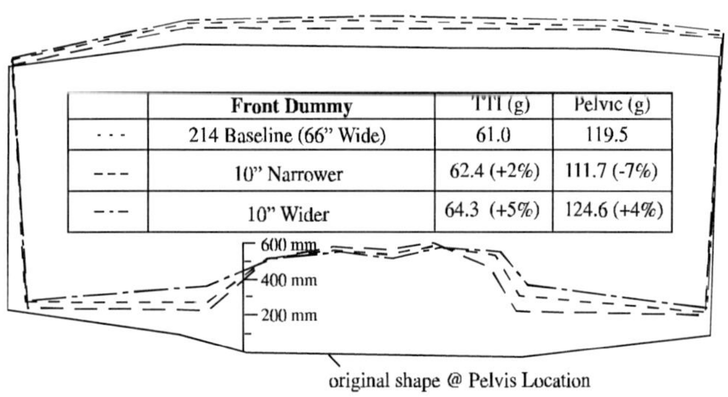

	Front Dummy	TTI (g)	Pelvic (g)
· · ·	214 Baseline (66" Wide)	61.0	119.5
– – –	10" Narrower	62.4 (+2%)	111.7 (-7%)
– · –	10" Wider	64.3 (+5%)	124.6 (+4%)

(g) Planview Intrusion Profile and SID Occupant Injury Response due to Change in Impact Width

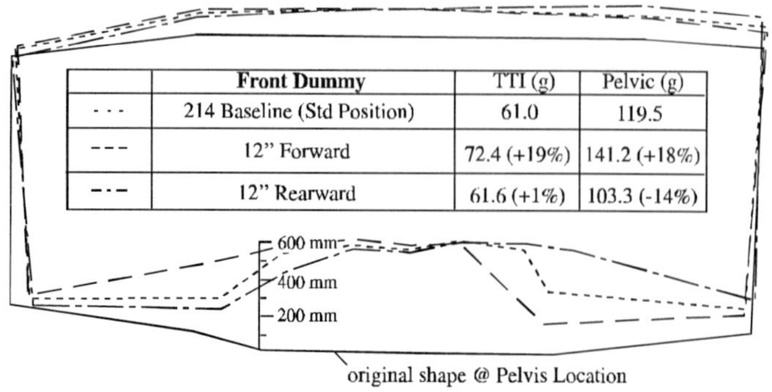

	Front Dummy	TTI (g)	Pelvic (g)
· · ·	214 Baseline (Std Position)	61.0	119.5
– – –	12" Forward	72.4 (+19%)	141.2 (+18%)
– · –	12" Rearward	61.6 (+1%)	103.3 (-14%)

(h) Planview Intrusion Profile and SID Occupant Injury Response due to Change in Impact Fore/Aft Location

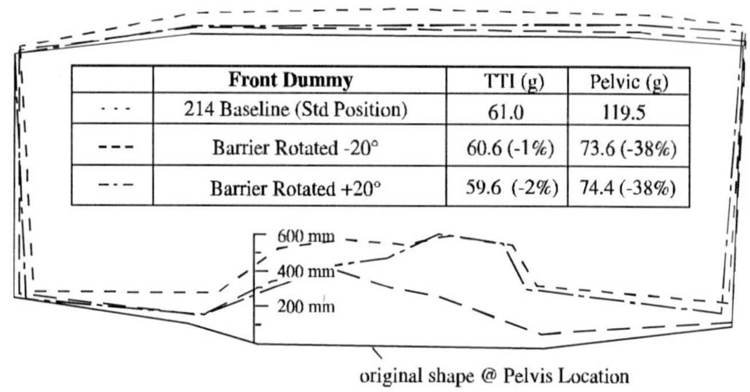

	Front Dummy	TTI (g)	Pelvic (g)
· · ·	214 Baseline (Std Position)	61.0	119.5
– – –	Barrier Rotated -20°	60.6 (-1%)	73.6 (-38%)
– · –	Barrier Rotated +20°	59.6 (-2%)	74.4 (-38%)

(i) Planview Intrusion Profile and SID Occupant Injury Response due to Change in Impact Angle

CHAPTER SIX – FACTORS INFLUENCING COMPATIBILITY

Measuring Compatibility

916010

Compatibility Problems of Small and Large Passenger Cars in Head on Collisions

G. Ernst; Dr. Brühning, E;
Glaeser, K.P.; Schmid, M.
Federal Highway Research Institute, BASt

Introduction

Road accidents may involve collisions between vehicles of different weights under a variety of circumstances. It is rare for vehicles of equal weight to collide. The range of vehicle curb weights (masses) extends from less than 700 kg (e.g. VW Polo) to over 2,000 kg (e.g. Daimler Benz, S-class). In accordance with the impact laws of mechanics, the consequences of collision involving smaller and larger vehicles are mostly more serious for the driver and passengers of the smaller vehicle. In the past, it has not always been possible to completely quantify the seriousness of accidents or the risk of being injured or killed in Germany because there is no direct link between vehicle mass and the seriousness of passenger injuries. All that is available at present is a study by an insurance association [5] based on single accident cases. This analysis covered front-seat passengers using seat belts as well, but not only the drivers.

An American study [7] estimate the risk of being killed in an accident in a small car (subcompact) very generally as being 8 times higher than the risk in a larger vehicle (fullsize car). Examining cars of different masses and of equal masses involved in accidents (comparison between vehicles weighing 900 kg and vehicles weighing 1,600 kg), in [4], on the basis of the FARS (Fatal Accident Reporting System), note a ratio of 2:1 for the risk of being fatally injured in a small car as against the risk of being fatally injured in a larger vehicle. However, the results of US studies cannot always be translated to European conditions, since America's car population is quite different from the European on account of the large number of extremely large vehicles on US roads.

A Swedish study by FOLKSAM Insurance—which also included single-car accidents—found that drivers of 800 kg vehicles have an injury frequency twice as high as that of 1,400 kg vehicles [6].

Following an evaluation of the accident data for North-Rhine/Westphalia now available for research purposes and extended to include automotive features, it is now possible to obtain results for the "compatibility" problem in Germany and in the car population involved in accidents here. First results were published in [8]. The study is concerned with the seriousness of injuries focused to car drivers involved in accidents with oncoming traffic.

Physical and Automotive Aspects of the "Compatibility" Problem

The seriousness of accidents can be described using the impact laws of mechanics; one measure is the "change of velocity" in the accident.

In the standard impact test against a rigid barrier, the deformation behaviour of the vehicle front is examined, and the protection afforded against serious injuries measured.

Still, although impact against the rigid barrier is a well established testing procedure, it does constitute a special case in the real accident world. Where a vehicle drives into a rigid barrier at e.g. 50 km/h, kinetic energy is transformed to deformation of the vehicle front. The vehicle undergoes a 50 km/h change of velocity.

To ensure optimal transformation of impact energy into work of deformation, the stiffness of the front is designed in such a way that, to meet the safety standard FMVSS 208, which provides for an impact at 30 mph (48 km/h) against a rigid barrier, advantage is taken of the max. deformation path of approx. 50 to 60 cm of the vehicle front, and also that the survival area in the vehicle cell is not destroyed. The vehicle structure, i.e. primarily the stiffness of the front side member, is optimated to this standard. This keeps the mean deceleration of the vehicle and its passengers (using seat belts) within limits. Where a large vehicle has, e.g., double the mass of a smaller vehicle, double the kinetic energy must be destroyed. This means—assuming the same front length (deformation distance)—double hardness for the front structure (stiffness), or—assuming the same stiffness of the front (force level)—double the length of front (or a combination of these two options).

Where two vehicles are involved in a head-on crash—for the sake of clarity, only head-on collisions are considered here—the change of velocity in the two vehicles results from the law of conservation of momentum applying to a plastic impact. This means—to remain with this example—in the case of an impact of 30 mph per vehicle and an assumed ratio between large vehicle mass and small vehicle mass of mL:ms=2:1, that the large vehicle continues driving in its direction at 10 mph, while the smaller car is reversed and is pushed back by the larger vehicle at a speed of 10 mph. The large vehicle undergoes a 20 mph change of velocity, the small vehicle one of 40 mph. We have a double change of speed in the smaller vehicle, whose normally softer structure must absorb double the energy.)[1]

Now, it is also possible to optimize vehicle fronts for a collision involving two moving vehicles, i.e. not a wall crash (self-protection) but a collision between vehicles of different size (partner protection). The collision tolerance involved here is called compatibility. The design for the fronts of "compatible" vehicles would have to look as follows:

- The small car must share the deformation of the larger vehicle. This can be done by choosing different stiffness characteristics (soft, medium, hard) or a linear increase in the stiffness characteristic for the front of the larger car, at the same time taking advantage of what will normally be greater front length and also a stiffer design in the front of the smaller car.
- If these two vehicles—now designed for compatibility—collide, first of all the stiff front of the smaller vehicle crushes the soft section of the front of the larger vehicle. In the process, the kinetic energy of the small vehicle is absorbed by the large vehicle. After that there is an even deformation of both vehicles.

Now, however, there are drawbacks in the 30 mph impact test against the rigid barrier described earlier for both vehicles. The small vehicle has been given a stiffer front design; as a result, the entire possible deformation path is no longer deformed; the mean deceleration and, hence, the strains on the occupants are higher. The large vehicle "gives away" opportunities for intaking energy. And although the choice of different degrees of stiffness can deal with the impact of 30 mph against the rigid wall thanks to the longer front, a constant choice of stiffness and optimation of stiffness to cope with the barrier impact might have allowed the occupants to survive an impact involving a speed of perhaps 35 mph without serious injuries. So far, vehicle stiffnesses have not yet been optimized for the protection of people. However, there are approaches to solutions, e.g. in [3].

Empirical Analyses

Data Base

In 1985, there were 77,009 accidents involving personal injuries between two cars in the Federal Republic of Germany. Out of these, 15,207 accidents with a total 28,757 casualties were accounted for by accident type "Collision with an oncoming vehicle." 686 of a total of 4,182 killed car occupants lost their lives in this accident type.

This analyses are based on data for accidents involving personal injuries and serious car damage for the State of North-Rhine/Westphalia, supplemented by automotive technology data available at the Federal Road Traffic Agency (Kraftfahrt-Bundesamt) [1]. Attention was paid in particular to the consequences of accidents for car drivers in accidents involving two passenger cars (station wagons excluded) with an car age of less than 10 years.

The risk of injury for car occupants depends very much on whether seat belts were used. The rate of seat belt use on front seats of cars averaged 60% until the introduction of fines in August 1984, since when, the rate has been over 90%. In order to rule out possible distortions owing to the different rates, this study is confined to accidents occurring between August 1984 and December 1988.

In order to target the problem of compatibility with the available data, the following remarks only concern accidents of type "Collision with an oncoming vehicle". By definition, therefore, the accidents to be considered here are collisions involving oncoming traffic without, e.g., one party intending to turn off to the left. In the period under review the accidents of this type in North-Rhine/Westphalia totalled 17,612.

Study Hypotheses

Compatibility problems always occur when large vehicles collide with smaller vehicles. The two main factors are:

- different vehicle structure optimized for self-protection
- different "mass aggressiveness" of vehicles of different weights.

This gives us the following two hypotheses:

Hypothesis 1: Differences in vehicle structure and "mass aggressiveness" mean that the consequences of an accident are less favourable for the occupants of the smaller and correspondingly more favourable for those in the larger vehicle.

Hypothesis 2: Where the vehicles have the same structure and "mass aggressiveness," this means that a head-on collision involving two small vehicles may be expected to have the same consequences as a head-on collision involving two large vehicles.

Characteristics for Assessing the Seriousness of Injuries

The central question concerns the extent to which the probability of serious consequences for the car occupants depends on vehicle size. Since considerable variations

[1]A double change of velocity normally means quadrupling the kinetic energy to be absorbed. However, this value is halved again, since the distribution of masses is 2:1 as indicated above.

are possible in the number and seating position of co-driver and passengers, our study is confined only to drivers and to the consequences for them.

For this study, the characteristic to indicate the mean seriousness of injuries to drivers involved in head-on collisions between two cars is the share of fatalities and serious casualties among all drivers involved in accidents with personal injuries and serious material damage is chosen:

$$AS = \frac{\text{number of fatalities + seriously injured drivers}}{\text{number of drivers in AI and AD}}$$

AI = accidents with personal injuries
AD = accidents with serious material damage

With regard to this specific characteristic the statistically significant variable of whether the driver felt "guilty" or "not guilty" does not have to be taken into account in the statistical analysis (cf. [2]).

Classification by Weight Classes

For the purposes of the present analysis, the vehicle population is subdivided by curb weight into four weight classes and the distribution is shown in Table 1.

Class 1: car with an curb weight of 600 to 799 kg,
Class 2: car with an curb weight of 800 to 999 kg,
Class 3: car with an curb weight of 1,000 to 1,199 kg,
Class 4: car with an curb weight of 1,200 to 1,599 kg.

Table 1. Cars involved in headon collision between two cars in North-Rhine/Westphalia in the period 1 August 1984 to 31 December 1988, classified by curb weight (accidents with personal injuries and serious material damage)

	Cars Involved	
Curb weight (kg)	Number	%
600 to 799	3,076	8.9
800 to 999	14,037	40.6
1000 to 1199	9,795	28.3
1200 to 1599	7,683	22.2
Total	34,591	100.0
up to 599/1600	615	
and more not known	18	

Classification by Design Features

As stated above, stiffness and/or vehicle design, which vary from type to type, have considerable implications for the relative collision tolerance of the vehicles.

Accordingly, two groups of vehicles were formed, taking account of empty weight and other design features:

Group 1: empty weight up to 999 kg;
"small" traverse front engine;
front-wheel drive

Group 2: empty weight up to 1,400 to 1,799 kg;
"large" longitudinally installed front engine;
standard drive

To form the above two groups, the empty weight classifications used deviate from those considered in Section 3.4. This was necessary in order to obtain sufficiently large groups for study when two further vehicle design features are taken into account.

The data material available for this study contains 13,554 vehicles in Group 1 ("small") and 3,750 in Group 2 ("large"). 8,742 of the 13,554 vehicles in Group 1 (64%) concern the car models Ford Fiesta, Opel Kadett, VW Rabbit and VW Polo, while 76% of the vehicles in Group 2 (2,863 of 3,750 vehicles) concern the car models Mercedes Benz W123 and W124 and BMW in the 5, 6 and 7 series.

Other Determinants

Alongside the vehicle features "vehicle structure" and "mass aggressiveness," there are other important factors—e.g. velocity, age of occupants, etc.—which may have a more or less favourable impact on the consequences of the accident for the occupants.

One factor of special importance for the consequences as they affect the occupants involved in a head-on collision is the speed at which the vehicles collide. The data material available for the present analysis does not explicitly include this feature, although we do know that the permissible max. speed and the mean speeds actually driven "inside built-up areas" are definitely below the values "outside built-up areas without autobahns" (rural roads). Alternatively, therefore, the factor "speed" is considered as explanatory variable via the feature "locality."

Also included in the analysis as a further explanatory variable is "driver age," since the seriousness of injuries under the same circumstance is, greater in the case of elderly people On the other hand, accidents involving younger drivers tend to correlate with excessive speeding.

In a consideration of collision compatibility, vehicle age is important for two reasons:

- old vehicles tend to be heavier and stiffer, which leads to special problems when a "light new" vehicle collides with an "old heavy" vehicle;
- older vehicles have, normally a higher degree of corrosion than newer vehicles, so that the accident may have different consequences for the occupants.

This was the reason to include only accidents with passenger cars of less than 10 years age.

Variables Used in the Study

In line with the remarks contained in the previous sections, the analysis uses the following features as independent variables (for dependent variables, see above):

Curb weight class of the car under consideration (CAR1) or the other car (CAR2)

(1)	600	to	799 kg
(2)	800	to	999 kg
(3)	1,000	to	1,199 kg
(4)	1,200	to	1,599 kg

Alternatively:

Size and design of the car under consideration (CAR_A) or the other car (CAR_B)

(1) small
(2) large

Locality (L)

(1) inside built-up area
(2) outside built-up area without autobahn (rural roads)

Age of driver in car under consideration (A)

(1) under 25 years
(2) 25 to 59 years
(3) 60 years and older

The Statistical Analysis Method

The counting of the fatalities and serious casualties for car drivers involved in head-on car crashes according to two or more variables produces a multi-dimensional matrix. The more variables are considered and the more categories the variables have, the more difficult it is—without adequate analytical procedures—to identify the relevant structures and to distinguish essential influences from the inessential. This is specially true where the occupation of the cells in the multi-dimensional matrix shows considerable variation, i.e. if the matrix has cells with a high as well as cells with a very low occupation frequency.

Statistical procedures based on logit models enable essential determinants to be identified using a multi-dimensional matrix; distinctions can be made between significant and nonsignificant determinants. Non-significant determinants can be excluded from the analysis, i.e. unlike a "conventional" evaluation involving a splitting up into multidimensional matrices, the use of logit models allows parameters to be excluded wherever they are not statistically significant on the basis of empirical data [2].

Results

Influence of "Locality"

The variable "locality" has, as expected, a very great influence on the mean seriousness of injuries to drivers involved in head-on collisions between two cars.

The percentage of fatalities and serious casualties among drivers (relative to all drivers involved in accidents with personal injuries and serious material damage) is about three times higher in the case of head-on collisions between two cars on rural roads than in built-up areas.

The inference is that the high mean seriousness of injuries on rural roads is basically statistically independent of the curb weight of one's own car and of the curb weight of the other car. Thus, the variable "locality" acts as a constant factor on the mean seriousness of injuries; the relative increase in the percentage of fatalities and serious casualties among drivers is equally high for all "curb weight classes."

In view of the high explanatory value of the variable "locality" accounting for the percentage of serious consequences, this variable is further considered again in the model.

Influence of Driver Age

The independent variable "driver age" has only a comparatively slight influence on the mean seriousness of injuries from head-on collisions. Of the three categories of the main effect, only the third category "drivers aged 60 years and more" is significantly different from 0 ($\alpha < 0.05$) and has a positive sign. This means that the percentage of fatalities or serious casualties among "old" drivers involved in head-on collisions is higher than the percentage of young drivers or middle-aged drivers. The absolute amount of this estimated value is small, however, relative to the values of the other parameters in the model, especially the "locality" variable.

Since no interactive effects significantly different from 0 occur between the three categories of the feature "driver age" in the car under consideration and the other parameters contained in the model, the inference is that a higher mean seriousness of injuries can only be established for older drivers, regardless of locality and the weight of one's own car and the weight of the other car.

An additional study of the connection between car driver age and car weight has shown that the percentage of older drivers increases markedly with the curb weight. In order to avoid any distortions to the results from a correlation between these two variables, the analyses in the following Sections ignore drivers aged 60 and older.

As was already established, the same accident consequences can be expected for young and middle-aged drivers involved in head-on collisions between two cars. Thus, if we ignore older drivers, the feature "age of driver of car considered" need no longer be considered in the analysis.

Seriousness of Head-on Crashes Involving Two Cars as a Function of Weight

Before dealing with the question of the collision compatibility of cars in the following Sections, the present Section examines the mean seriousness of injuries to drivers involved in head-on crashes between two cars as a function of the weight of one's "own" car; the weight of the "other" car is left out of account in this first analysis.

Figure 1 shows a curve for the mean seriousness of injuries to drivers involved in head-on crashes between two cars that falls with the curb weight of the car. The seriousness of the consequences fall substantially with the rise in the weight of the car under consideration. The percentage of fatalities and serious casualties among drivers of cars with 600 kg curb weight is 14.2%; and that for drivers in cars with an curb weight of 1,500 kg is 4.8%.

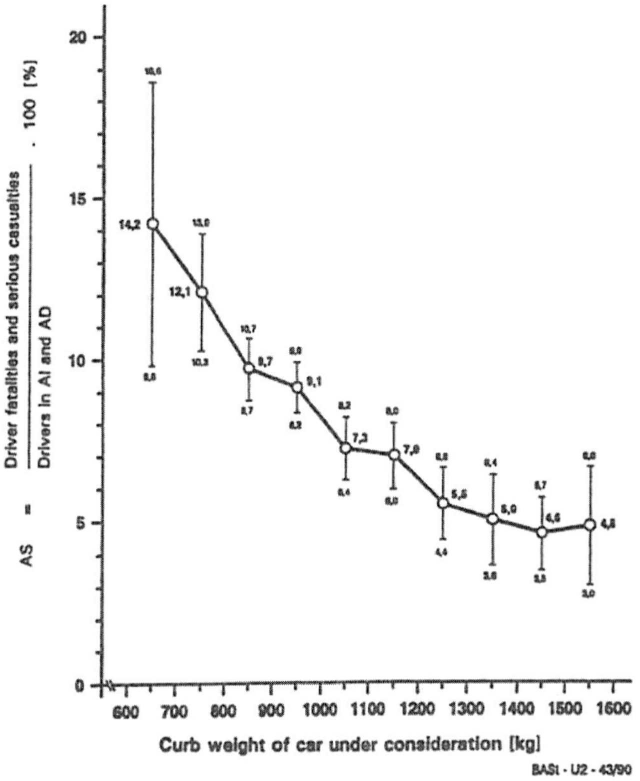

Figure 1. Seriousness of head-on crashes involving two cars on rural roads as function of the curb weight of the two cars (driver age up to 59 years)

Compatibility in a Consideration of Four Weight Classes

To describe the collision compatibility of cars, a model is estimated in what follows that takes as independent variables the feature "locality" and the car under consideration and/or the other car (each in four weight classes). The dependent variable considered is the percentage of fatalities and serious casualties among drivers relative to all drivers involved in accidents with personal injuries and serious material damage (AS).

Using this model, we can estimate expected values for the seriousness of accidents and the relevant confidence intervals for all covariable constellations. Fig. 2 shows the results for the rural road sector.

In the following the two hypotheses for the collision compatibility of cars involved in head-on collisions described above were tested (see Annex Table A1 and A2 and Figure 2):

Hypothesis 1 says that, in accidents involving cars of different weight, **different "mass aggressiveness"** can lead to more serious consequences of the accident for the occupants of the smaller and correspondingly less serious consequences for those in the larger vehicle. This hypothesis can only be maintained partly:

- Figure 2 makes it clear that seriousness of the consequences of the accident for the driver is related quite definitely to the curb weight of his own car. The consequences for drivers in cars with an curb weight of 600 to 799 kg are—regardless of the weight of the other car—some 2.5 times less favourable than for drivers of cars with an curb weight of 1,200 to 1,599 kg.

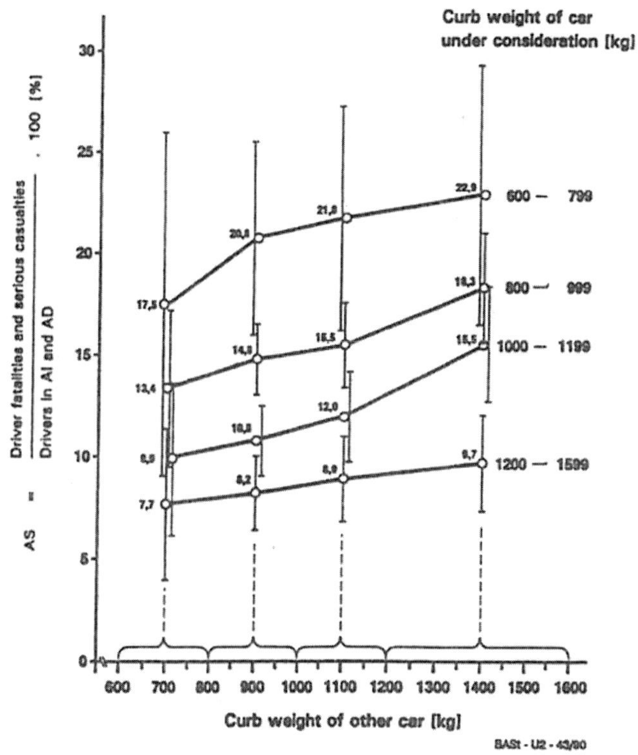

Figure 2. Compatibility—expected values for the seriousness of accidents in head-on crashes involving two cars on rural roads, as function of the curb weight of both cars (driver age up to 59 years)

On rural roads, for example, the expected value for the percentage of fatalities and serious casualties among drivers in cars with an curb weight of 600 to 799 kg in the case of head-on collisions with a car of 800 to 999 kg curb weight is 20.8%, with a car of 1,200 to 1,599 kg curb weight 22.9%; the corresponding expected values for drivers in cars with an curb weight of 1,200 to 1,599 kg, however, are only 8.2% (weight of other car 800 to 999 kg) and 9.7% (weight of other car 1,200 to 1,599 kg) resp.

- Within the various curb weight classes, however, hypothesis 1 can be confirmed in principle; it can be seen, e.g., that, in the case of drivers of cars with an curb weight of 800 to 999 kg involved in head-on collisions with a smaller car (600 to 799 kg weight), the consequences for the car with 800 to 999 kg curb weight are much less serious (13.4%) than in the case of a head-on collision with a heavier car (1,000 to 1,199 kg curb weight 15.5% and 1,200 to 1,599 kg curb weight 18.3%).

Hypothesis 2—equal "mass aggressiveness"—viz. that the consequences of a head-on collision between cars of equal weight classes are approx. the same for the occupants, cannot be confirmed. In fact, Fig. 2 shows that the seriousness of the consequences for the driver depends quite substantially on the curb weight of his own car. The values of the characteristic for the seriousness of injuries AS fall strikingly as the weight of the car under consideration increases. The expected value for the percentage of fatalities and serious casualties among drivers in head-on collisions involving two cars with an curb weight of 800 to 999 kg is 14.8%; in head-on collisions involving two cars with an curb weight of 1,200 to 1,599 kg, the figure is 9.7%.

Compatibility in a Consideration of Two Selected Vehicle Groups

To take account of these conditions, two vehicle groups ("small"/"large") were selected by size and design features, for the car under consideration and the other car—as independent variables and included in the multivariable analysis. The feature "locality" was considered as a further independent variable. The dependent variable is again the percentage of fatalities and serious casualties relative to all drivers involved in head-on crashes with personal injuries and serious material damage.

As stated above the consequences of head-on collisions are many times less favourable for drivers on country roads than they are inside builtup areas, **regardless** of the size and design of the vehicle. The accident severity (AS) for small and large cars is shown in Fig. 3. The hypotheses 1 and 2 for the collision compatibility of the cars can be checked again.

Overall, it can be said that the results obtained by classifying the vehicle populations into four empty weight classes for the car under consideration and the other car by postulating two vehicle groups based on masses and design features, were corroborated; in the case of head-on collisions involving a small and a large car, the mean seriousness of the injuries is some 2.5 times greater in the large car.

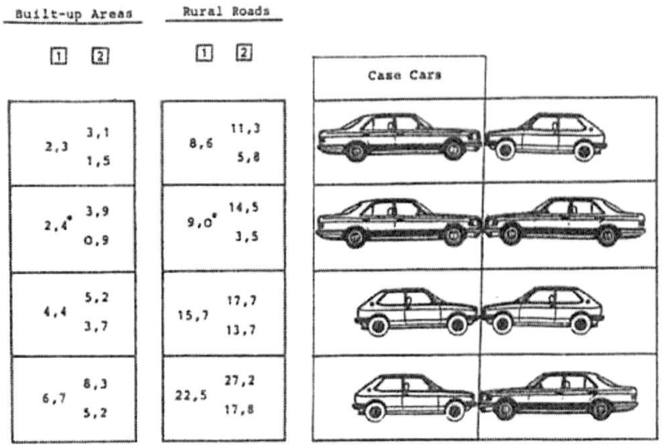

1 AS = Percentage of fatalities and serious casualties relative to all drivers involved in accidents with personal injuries and serious material damage
2 95 % confidence interval
* In this constellation, only 3 fatalities or seriously injured drivers were recorded

Figure 3. Expected value for seriousness of accidents involving car drivers in head-on crashes (accident type 4) between two cars *in built-up* and *on country roads* as a function of two selected vehicle groups (accident with personal injuries and material damage; driver age up to 59 years)

Summary, Discussion of the Results

Head-on crashes between cars are particularly serious. Where large and small vehicles collide, physical theory leads to expect compatibility problems. The main factors regarded as underlying the special collision intolerance of vehicles of different sizes are the different vehicle structures optimized for self-protection and the different "mass aggressiveness" of vehicles of different weights.

In 1985, there were 77,009 accidents involving personal injuries between two cars in the Federal Republic of Germany. Out of these, 15,207 accidents with a total 28,757 casualties were accounted for by accident type "Collision with an oncoming vehicle." 686 of a total of 4,182 killed car occupants lost their lives in this accident type.

The analysis undertaken in this report is based on the data for accidents involving personal injuries and serious material damage in the State of North-Rhine/Westphalia, supplemented by automotive data available at the Federal Road Traffic Agency. In order to rule out any distortions due to differences in rates of seat belt use, the study is confined to accidents occurring after August 1984. Since then, the rate of seat belt use in the Federal Republic has averaged well over 90%. The vehicle population was broken down by curb weight into four weight classes.

The assessment criterion in the statistical analysis was the mean seriousness of injuries to drivers involved in head-on collisions between two vehicles. The independent variables considered in addition to the features "curb weight class of the car under consideration" or "of the other car" were the features "locality" and "age of the driver of the car under consideration."

The statistical analysis was made using logit models, which enable the key determinants to be identified and their impact quantified within the scope of multi-dimensional analysis. As expected, the feature **"locality"** has a great impact on the mean seriousness of injuries to car drivers involved in collisions with oncoming cars; regardless of curb weight class of the vehicle, the consequences of accidents for drivers on rural roads are about three times less favourable than in the case of built-up areas.

The feature **"age of driver in the car under consideration"** has a comparatively slight impact on the mean seriousness of injuries to drivers involved in head-on collisions between two cars. In the case of older drivers, however, there is in fact a higher mean seriousness of injuries, regardless of locality and the curb weight of their own or the other car.

Study hypothesis 1, viz. that **different "mass aggressiveness"** in accidents involving cars of different weights produces more serious consequences for the driver of the smaller car—and correspondingly less serious consequences for the driver of the larger vehicle—were only confirmed with reservations.

Study hypothesis 2, viz. that **equal "mass aggressiveness"** produces approx. the same consequences for car occupants in head-on collisions involving cars of the same weight, was not confirmed.

In fact, the empirical analysis showed that the seriousness of the consequences for drivers was associated on average and quite significantly with the **curb weight** of the driver's own car. The mean seriousness of injuries is some 2.5 times as high for small cars or for cars with a weight between 600 and 799 kg—regardless of the weight of the other car—as it is for large cars or for cars with an curb weight between 1,200 and 1,599 kg.

The same results could be obtained by only looking to small cars (< 1000 kg) with traverse engine and large cars (> 1400 kg) with longitudinally installed engine.

Thus, there must be other factors operating to explain why safety is very much greater for the occupants of large vehicles than it is for small vehicles. One possible explanation is that the interior appointments of large—and hence more expensive—cars offer more scope or the use of safety-enhancing features. For example, the use of energy-absorbing materials in the interior (e.g. dashboard), the installation of safety steering wheels, etc. is easier to implement cost-wise for large cars than it is for small cars. Differences in car safety for different cars within each weight class cannot be made by means of the data base used for this study.

References

[1] Bock, O.; Brühning, E.; Dilling, J.; Ernst, G.; Miese, A.; Schmid, M.; 1989: Aufbereitung und Auswertung von Fahrzeug- und Unfalldaten zur Hebung der Verkehrssicherheit. Gemeinsamer Bericht der Bundesanstalt für StaBenwesen und des Kraftfahrt-Bundesamtes Herausgegeben im Auftrag des Bundesministers für Verkehr von der Bundesanstalt für StraBenwesen, Bereich Unfall-forschung, 1989, Heft 71.

[2] Bruhning, E.; Ernst, G.; 1988: Loglinear and Logit Modells in Accident Research. 9. IRCOBI Konferenz, Bergisch Gladbach. In: 1988 International IRCOBI Conference on the Biomechanics of Impacts, S. 35 - 46.

[3] Danner, M.; Appel, H.; Schimkat, H.; 1980: Entwicklung kompatibler Fahrzeuge. Abschluberícht des BMFI, 1980.

[4] Evans, L.; Wasielewski, P.; 1987: Serious or Fatal Driver Injury Rate vs. Car Mass in Head On Crashes of Similiar Mass. In: Accident Analysis and Prevention, Vol. 19, Nr. 2.

[5] Langwieder, K.; Schmelzing, W.; 1982: Kollisionen von Pkw unterschiedlicher Masse. BMFT Statusseminar. Münster 1982.

[6] Nygren, A.; Gustafsson, H.; Tingvall, C.; 1982: Injury Frequency and Injury Panorama among Drivers—A Comparison between Small and Big Cars and the Influence of Year Model and Production Start. 7. IRCOBI Konferenz, Koln 1982.

[7] William, Boehly, Lombardo (NHTSA); 1980: Safety Consequences of the Shift to Small Cars in the 1980' s. 8. ESV Konferenz, Wolfsburg 1980.

[8] Ernst,G. ,Bruhning,E., Glaeser,K.-P.,Schmid,M.; 1991 Safety in Small and Large Passenger Cars, IRCOBI Conf. Potsdam, Sept. 91.

Annex

Table A1: Expected value for seriousness of accidents involving car drivers in head-on crashes between two cars **inside built-up areas** as a function of four curb weight classes (Accidents with personal injuries and serious material damage;driver age up to 59 years)

Curb weight of considered CAR1	Curb weight of other CAR2							
	600 to 799 kg (1)		800 to 999 kg (2)		1000 to 1199 kg (3)		1200 to 1599 kg (4)	
	AS[1]	confidence-limits[2]	AS[1]	confidence-limits[2]	AS[1]	confidence-limits[2]	AS[1]	confidence-limits[2]
600 to 799 kg (1)	5.5*)	8.2 2.8	6.7	8.3 5.1	7.1	8.9 5.3	7.5	9.6 5.4
800 to 999 kg (2)	4.1	5.2 2.9	4.5	5.1 3.9	4.8	5.5 4.1	5.8	6.7 4.8
1000 to 1199 kg (3)	2.9	4.0 1.8	3.2	3.7 2.6	3.6	4.3 2.9	4.8	5.7 3.9
1200 to 1599 kg (4)	2.2*)	3.3 1.2	2.4	2.9 1.8	2.6	3.2 2.0	2.8	3.6 2.1

1) AS = percentage of fatalities and serious casualties relative to all drivers involved in accidents with personal injuries and serious physical damage

2) 95 % confidence interval

*) In the constellation, fewer than 10 fatalities or serious casualties were recorded

Table A2: Expected value for seriousness of accidents involving car drivers in head-on crashes between two cars **on rural roads** as a function of four curb weight (Accidents with personal injury and serious material damage; driver age up to 59 years)

Curb weight of considered CAR1	Curb weight of other CAR2							
	600 to 799 kg (1)		800 to 999 kg (2)		1000 to 1199 kg (3)		1200 to 1599 kg (4)	
	AS[1]	confidence-limits[2]	AS[1]	confidence-limits[2]	AS[1]	confidence-limits[2]	AS[1]	confidence-limits[2]
600 to 799 kg (1)	17.5	26.0 9.0	20.8	25.6 16.0	21.8	27.3 16.4	22.9	29.3 16.4
800 to 999 kg (2)	13.4	17.2 9.6	14.8	16.5 13.0	15.5	17.6 13.3	18.3	21.0 15.6
1000 to 1199 kg (3)	9.9	13.6 6.3	10.8	12.5 9.0	12.0	14.2 9.9	15.5	18.3 12.7
1200 to 1599 kg (4)	7.7	11.4 4.0	8.2	10.0 6.4	8.9	11.0 6.8	9.7	12.1 7.2

1) AS = percentage of fatalities and serious casualties relative to all drivers involved in accidents with personal injuries and serious physical damage

2) 95 % confidence interval

2003-06-0014

COMPATIBILITY ASSESSMENT PROPOSAL CLOSE FROM REAL LIFE ACCIDENT

Pascal Delannoy
Teuchos Group - Renault Safety Department -
Jacques Faure
Renault S.A.
France
Paper number 94

ABSTRACT

Accident studies show that incompatibility has become the main cause of fatal injury in car to car accidents. There is a general agreement today that improving compatibility is one of the most effective ways to reduce the number of road accident victims.
Therefore, structural car design must take into account other road users without decreasing self protection level supplied by all new passenger cars. In addition to these safety considerations, the front unit structural design has to account for an increasing number of constraints: improvement of real world performance in safety, fulfil current and future regulations like "CAFE" or pedestrian, reducing utilisation costs and so on.
Furthermore, European fleet is changing in mass and in size, as the world's ones, and new fashion vehicles appear different than the previous one.
This paper deals with the development of a more comprehensive approach in order to better take into account safety requirements coming from real life accidents and the work done over the past years on understanding the physic of compatibility. The aim of this paper is to propose a better assessment procedure and a new test methodology in a standard approach for improving compatibility.

INTRODUCTION

40 000 people die each year in Europe due to traffic accidents. In France there were 5 000 fatalities in a car and 2700 involved in car to car or car to vehicle accident. This figure represents more than half of people that die in a car collision. Our approach has been for many years to study real world accidents and try to understand what were and are the mechanisms of injury causation. Accident studies during the last twenty-five years clearly showed that car to car head-on collision is a major impact configuration to take into account in order to improve safety on the roads. With new self protection ratings all cars offer equivalent behaviour against a fixed obstacle. So, in the future, it is expected that the main progress will have to be made in car to car compatibility.

Compatibility between cars has for a long time been reduced to the simple image of heavy against light cars. Over the past ten years vehicle stiffness has been increased thanks to improved restraint systems. We also have a better understanding of the front end design energy absorption. Front end design is at the cross road of numerous contradictory constraints: self protection of occupants, protection of vulnerable users such as pedestrians, reparability, styling, aerodynamics, engine cooling and so on.
Therefore, each manufacturer has developed its own solution to solve this difficult equation which resulted in a wide variety of front end designs, structure and stiffness regardless of the overall mass of the vehicle.
Solutions however have been optimized against a rigid wall or soft obstacle but not in car to car configuration. The problem was first to understand the physics of car to car interaction then, to set design rules, to design for an improved compatibility. And finally, to find a test procedure and a set of relevant criteria in order to keep under control front end and passenger compartment design over the market production.
Whereas self protection can be achieved through a wide variety of structural design, compatibility (partner and self protection) can only be achieved through cooperative work and common design rules amongst manufacturers. This is what makes improvements in compatibility so difficult. To be effective in terms of lives saving, the procedure must absolutely take into account:

- accident research analysis
- heavy an light vehicles
- current self protection requirements
- current and future fleet
- current and future vehicle design.

CURRENT AND FUTURE EUROPEAN FLEET

In addition, research into compatibility must take into account the time taken to renew the fleet (figure1). Measures proposed for new vehicles must not create dangers for existing ones because the European market, as the other ones, is a mixture of old and recent cars.

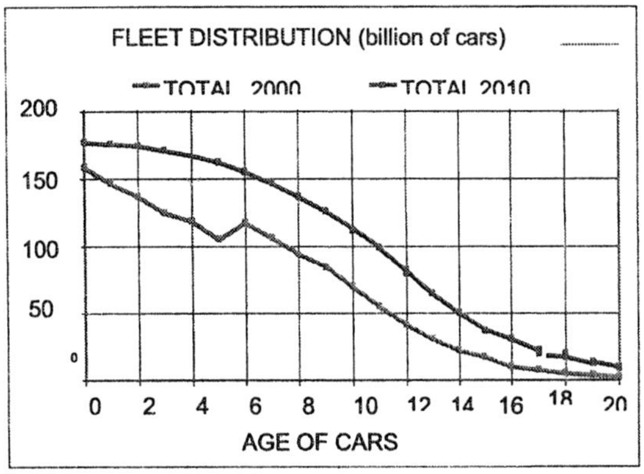

Figure 1: Age distribution of European fleet in 2000 and 2010 (estimate).

In marge of the global increase in weight, due to new fashion in car design, the vehicles segment distribution is changing into multi purpose or four wheel drive vehicles higher and heavier than the previous ones.

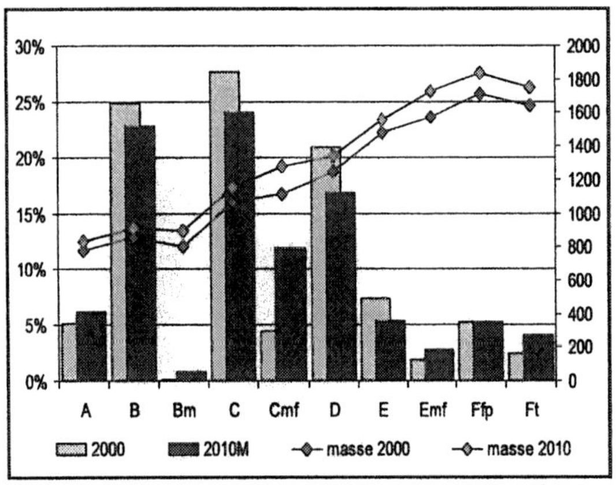

Figure 2: European mass and segment distribution change

It is necessary to point out that the procedure should last for a long time to avoid rupture effect in the vehicle fleet due to change in the rules. It is therefore better even if it takes some time, to opt directly for the best way of evaluating the compatibility, before adopting any standards.

So the future standards approach must be stable to avoid inhomogeneous fleet as we can see today. If not, the overall benefit of such measures may be severely compromised.

ACCIDENT RESEARCH FINDINGS

Accident configurations

Several generations of vehicles are now together on the same road network. By this term, we mean cars with different stiffness, in other words different crush force deformation. The introduction of ratings and reparability test over the past decade has led to stiffer cars and has increased the discrepancy between "old" and "new" generation of cars. There are now on the roads light cars which can be aggressive for heavy cars. There are also cars of the new generation which are aggressive for the same size of the old generation. The major difference between these cars is the stiffness of their compartment and front unit. This has clearly to be put under control through common assessment test.

Overlap and angle of collision

Thanks to the introduction of numerous actives safety systems helping the avoidance of crashes (ESP...), the overlap seems to decrease. However, the average is around fifty percent giving by current data. In the future we may need to consider decreasing this overlap.

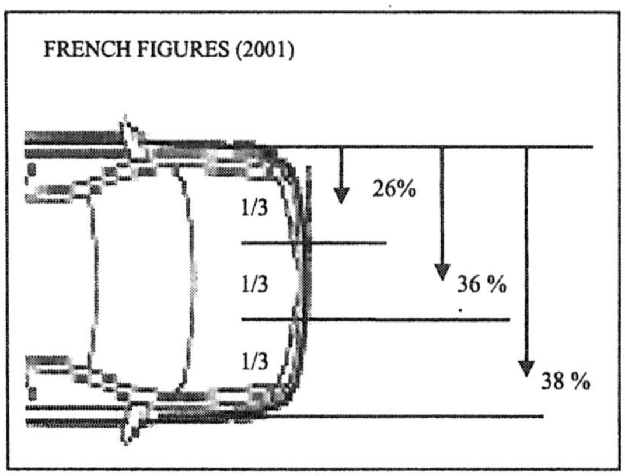

Figure 3: overlap distribution in head-on collision

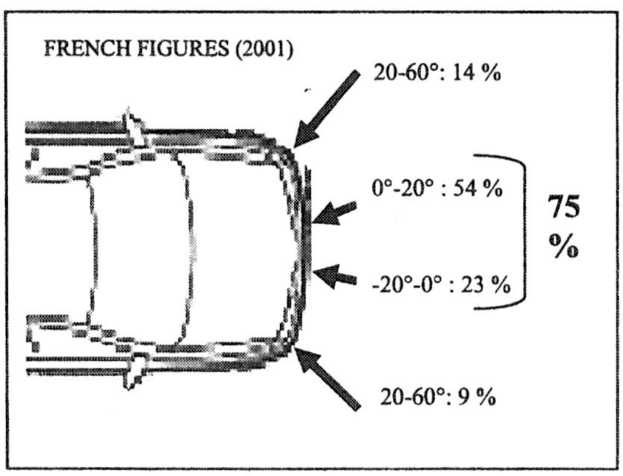

Figure 4: angle distribution in head on collision

This data (0° and 50% overlap) are largely accepted by the scientific community. Whether it be car to car ratings (ADAC, AMS), tests performed by different working groups (PREDIT, CEVE, EUCAR), by car manufacturers or states (Australia), all were carried out at fifty percent and zero degree angle. So, the future standard must take into account this point to be close from accident data and car to car test configuration.

Closing speed

The future frontal European regulation at 60 kph is linked to the fact that the improvement of the compatibility cannot be done to the detriment of self protection (especially heavy vehicle). It brings us to set the limit at 100 kph closing speed. Accident studies show us that 60% of people involved in the light car are covered (figure 5). It should be specified that these progress will be also applicable for higher closing speeds.

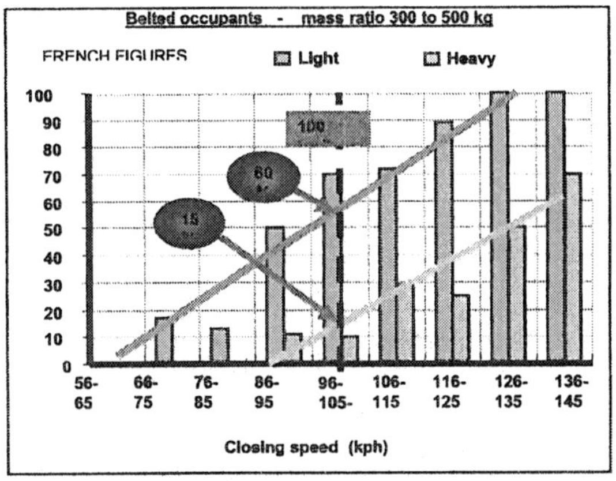

Figure 5: gravity rate covered vs closing speed

Breaking front unit clearance

In forty percent of cases, one of the two vehicles is in breaking (with the introduction of ABS, this information becomes more difficult to collect). The front structures are lowered (about 150mm) compared with the static position. Only a geometric treatment of frontal structures is therefore not sufficient to cover these cases.

STUCTURAL POINTS TO INVESTIGATE

The front end is collapsible whilst the occupant compartment must keep its integrity and play its role of survival cell. As a result, the intrusion into the occupant compartment has been greatly reduced thus reducing the risk of contact injury. Car design for front crash reflects also the ability to balance intrusion reduction which means stiff front structures and acceptable deceleration from the occupant point of view. Recent developments in restraint systems such as seat belt with load limiters, airbags and so on, have helped to make level of compartment deceleration acceptable, which would have killed people some years ago.
In order to reach the desirable intrusion level, the engine compartment has to absorb a certain amount of energy. Usually this is achieved through different load paths which absorb energy and transmit the load from the front to the occupant compartment. These load paths are designed and tuned against two types of obstacles: rigid barrier or deformable barrier. So far test s carried out on deformable barrier showed bottoming out phenomenon. This means that the front end design is not controlled by the barrier stiffness because the structure collapses with the help of the rigid wall behind the barrier. In each case the obstacle is far from representing a car front unit. That's why structural behaviour in car to car accidents are different due to bad connexion between load paths (figure 6), bad spreading surface in front of the load paths (figure 7 and 8) that cause fork effects, insufficient engagement and weak structural interactions.

Figure 6: Rupture of the connexion between subframe and longitudinal

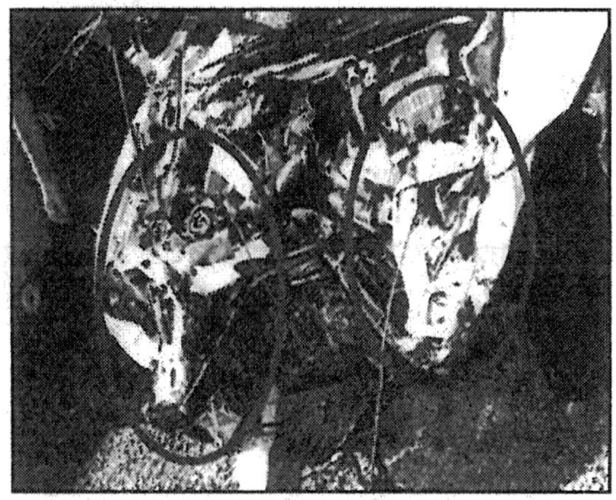

Figure7 : Undeformed longitudinals

Figure 8: severe intrusion in side impact due to local stiffness of the striking car.

Finally the amount of vehicle deformation should a least put under control. The total length of the frontal unit must be tested to detect structural behaviour and even the farthest away from the front end structures (figure 9). Sometimes, these structures help for a better engagement and structural interactions. The future procedure must be severe in order to check the front unit at least up to the compartment

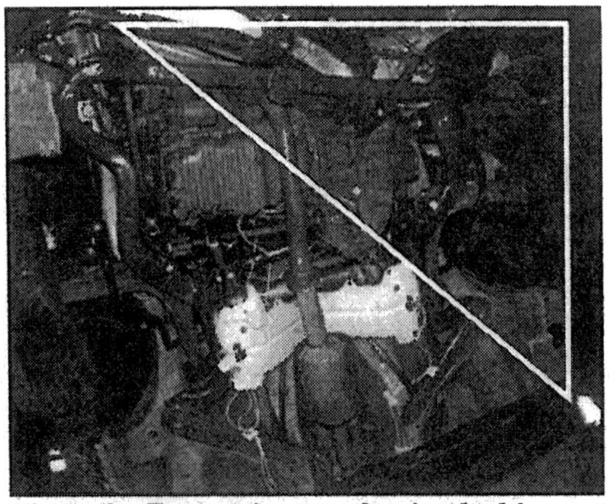

Figure 9 : Front unit area often involved in car to car accident

Energy absorption principles

Looking at real world car to car accidents, it is clear that the first cause of occupant death is due to intrusion of aggressive structural parts into the passenger compartment (figure 10). Unfortunately, structural behaviour and intrusion mechanism are completely different from those observed against in test with rigid wall or soft obstacle (EEVC barrier).

Figure 10: Overloaded compartment

This is where incompatibility arises. One definition of incompatibility between two cars is when one of the two cars is absorbing most of the crash energy whilst the other remains mostly undeformed.
The scientific community now agrees that mass does not play a direct role in compatibility. Its indirect role is through the stiffness or the deformation force.

In order to reach the same level of self-protection, the level of intrusion is roughly the same regardless of the vehicle mass. The increase in length of the car size does not usually compensate for the increase in energy to be absorbed. As a result the front end and the compartment deformation forces as to be increased to reach a higher stiffness. In addition, design against deformable barrier with bottoming out results directly in even stiffer heavy cars because this test is more severe for large car than for small ones. The fraction of energy absorbed in the barrier is roughly the same regardless of the car mass resulting in a higher fraction of energy to be absorbed by the large vehicle than by the small one.

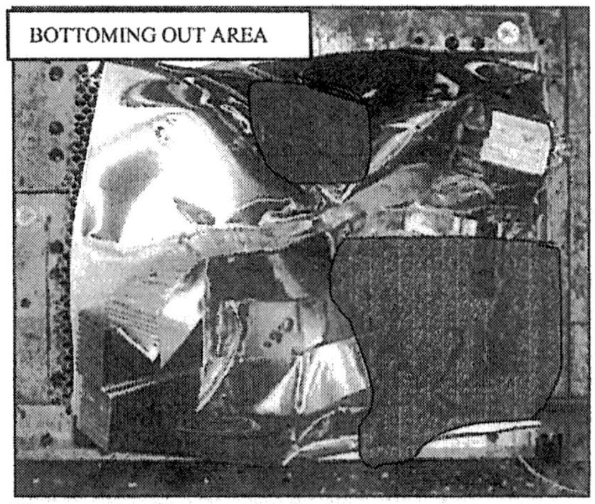

Figure 11: Current bottoming out phenomenon on EEVC barrier

FRONT END DESIGN AND COMPATIBILITY LIMITS

Front design limits

Front end design is at the cross road of a large number of conflicting requirements such as: aerodynamics, engine cooling, styling, pedestrian protection, accident protection or reparability. There are also some other constraints which at first sight does not appear related to crash performance but which could have an effect: overall length (cars have to fit into cities and allowed to be parked easily). In addition, the lower lip of the front bumper cannot be too low with respect to the front overhang in order to avoid difficult access to ramps.

Compatibility limits

Compatibility has two aspects: on one hand, self protection, which is controlled trough test against barrier. This can be achieved trough a variety of compartment force and design. On the other hand, partner protection is very seldom addressed. There is no agreed test methodology, neither regulation. This is typically a safety subject which has to be solved on an international cooperative basis as compatibility usually

involves two cars of different manufacturers. So there is only limited interest at designing its own range to be compatible.
Due to the requirement for self-protection and the wide range of vehicle's size, mass and stiffness, we have to define a limit for compatible design. For example, a car can only be made compatible up to a certain speed. Furthermore, this limit must be in line with other regulations. Then, if we take for instance an MPV of 1800kg at 60kph against the EEVC barrier (future ECE94 regulation), the corresponding EES is around 55kph. The intrusion level set by the regulation will make heavy cars much stiffer than small ones. One might think that we could always increase the length of the heavy cars but we have already seen that it is impossible before. The result is that heavy cars cannot be made compatible, in term of stiffness, with small ones at EES55 without decreasing their self protection. Thus, we have to choose a lower limit at EES50 which corresponds to a closing speed of 100kph. In such a crash configuration, two cars can be said compatible if the intrusions observed on both side are close to those observed in ODB test at 60kph.

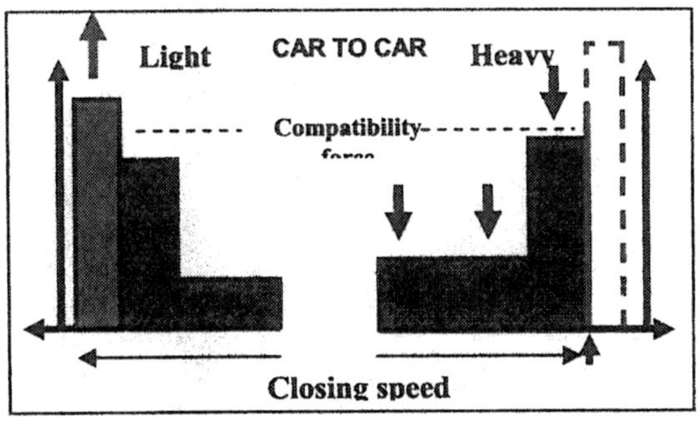

Figure 12: Car to car energy distribution

Structural improvements

As mentioned above, stiffness is one part of a compatible design. In order to take advantage of all the potential for energy absorption of both cars, their structure must interact correctly. This goes along with preventing over-ridding which is a specific aspect of energy deficiency. Improving energy deficiency is now something which is generally accepted.

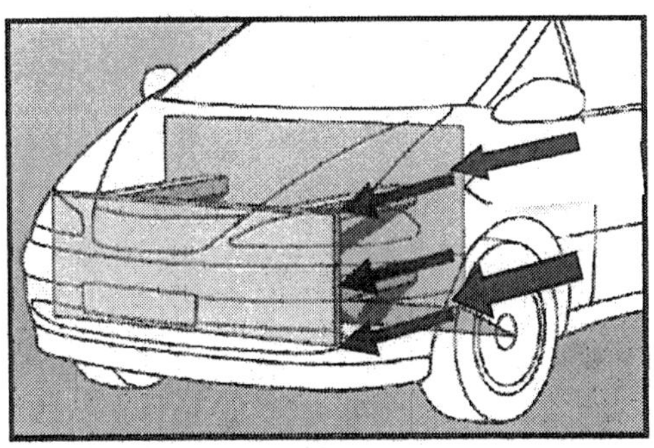

Figure 13 : Compatible front end

The problem now is to find common test procedure and criteria which will be representative of this phenomenon in order to put this item under control.
In term of design, one way to achieve structural interaction is to offer a front surface which is homogeneous in stiffness over a surface which is large enough. To illustrate this point, we have to imagine that we put a wall between both cars. The concept of the wall is to have a homogenous stiffness over a large surface. To achieve this result, the stiffness on the front block must be distributed along multiple load paths. Having this is not enough, as they cannot ensure that the stiffness is homogeneously spread over the front surface. The additional requirement, which is also confirmed by real world accident analysis, is that the different load paths must have strong connection both vertically and horizontally. To finish with, this result avoid to have localised load paths on small surfaces.
Equal stiffness also means corresponding deformation forces. Theses forces must be as close as possible for both cars in order to allow an equal deformation for both. As seen previously, this "compatibility" force is different from the self protection one.

PROPOSAL FOR A RELEVANT COMPATIBILITY ASSESSMENT

There are no effective proposed improvements unless they are applied by all manufacturers and for all passenger cars. The only way to reach that target is to define and apply a new standard project.
Several international task forces are working in that direction. Some orientations resulting from the preceding principles are proposed and can be discussed in that context. As we have seen before, the two main principles for a better compatibility are first to enforce a minimum resistance of the compartment, then check its stability and finally put under control the energy absorption and the force level of the front end of the car (called Bulkhead Principle by some compatibility experts);
We consider at the present time that one test is needed to cover these two issues. This procedure is a little bit different than the previous one proposed by Renault.

The procedure must check:

- load path positions and stiffness
- structural links among them
- global front unit and compartment force
- compartment stability

The procedure must generate:

- local and global longitudinal shears
- local and global horizontal shears
- compartment strength (overloading)
- large deformation depth

The procedure must be in line with:

- current and future regulations

- light and heavy vehicles

The assessment could be based on:

- barrier deformation
- force load measured behind the barrier

In other words, to cover the maximum of cases the procedure must reproduce a frontal car to car accident structural loading.

Several test procedures are proposed for compatibility assessment: against different rigid wall or soft obstacle (with bottoming out phenomenon) and different overlaps (between 40 to 100%). Unfortunately, as we have seen before against a rigid wall or soft barrier, the various load paths are not working the same way as they do in car to car interaction. The deformation process is at imposed displacement, whereas in car to car, the deformation is at imposed pressure or force.
A rigid wall might seem simpler; unfortunately it is not representative of a car front block and far from real world accident observations.

Test procedure

The objective is the ability to quantify:

- the capacity of the front unit to absorb energy corresponding to an 100 kph closing speed
- the capacity of car compartment to resist and be stable.

We have seen that heavy cars are stiffer than light ones due to their design only. Today we have to take into account the non aggression for the other car users (80% of cases). So, it is important to develop a testing procedure to put under control the energy absorbed by each car before reaching its self protection force. A car designed to be very stiff can reach the self protection force in an early phase of the crash and be very dangerous because it has not enough energy absorption capacity.
We are able to propose a new test procedure for all car range, after having analysed real world accidents, considered new and future generation of cars and current European regulation and all the work done by international experts.
If we focus on what we want to control in the test design the following configuration is proposed:

- offset :For controlling the strength of the lateral connections between the different load paths and generates global shears.
  Proposed offset: 50%

- obstacle : Progressive Deformable Barrier (PDB) is aiming at checking front unit in stiffness and geometry, generating shears in the vertical and the horizontal

- Mobile barrier (second step):
  The mass is supposed to be that of a medium car in Europe. The test progressively switches from a light car overload to a heavy car partner protection test.
  Proposed mass in Europe: 1300kg. Mass can be adapted with the continent fleet.

- Test severity : to be defined
  To be in line with 100 kph closing speed, current and future regulation.
  Fixed barrier: around 60 kph to 64 kph
  Mobile barrier: closing speed around 90 kph

In other words, the test procedure is fully representative of real world accident. Test dummies are employed – It could become a restraint-system dimensioning test also, but further investigation is required.

Barrier

The barrier is now well known (derived from ADAC barrier). The main difference is a progressive increase in stiffness in the depth, and two height dependant stiffnesses, which contribute to its name: PDB as Progressive Deformable Barrier (Figure 14). The barrier allows to checking the thrust surface of the vehicle, and the links between the transfer paths. Its dimensions and stiffness make the bottoming-out phenomenon very unlikely.
In car to car there is no bottoming out, thus the barrier must be designed in such a way as to prevent this phenomenon (figure 14).
Thus, the compression on the front block exhibits an increasing stiffness. Furthermore car force distribution in height should be represented; the lower front load path is usually stronger than the upper one.

Figure 14: PDB barrier version 7 - Vehicle collapse without the help of the rigid wall behind

Measurements

All criteria and investigation are based on the deformation of the barrier and the global force deflection measured behind. It looks like car to car accident or test analysis, except that in this case, the

barrier deformation is investigated instead of the car. An aggressive vehicle would be identified by large and non homogeneous deformation.
The PDB barrier is able to detect local stiffness but also transversal and horizontal links among load paths. The barrier represented on figure 15 clearly shows front cross member, lower cradle subframe and pendants that improve vehicles compatibility. That is the reason why the assessment is based on deformation in height and in depth.

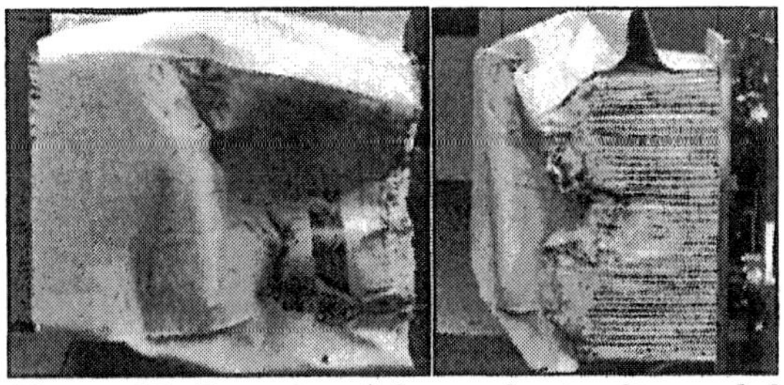

Figures 15: Footprint of the car front unit recorded by the PDB barrier- deformation is far from the wall

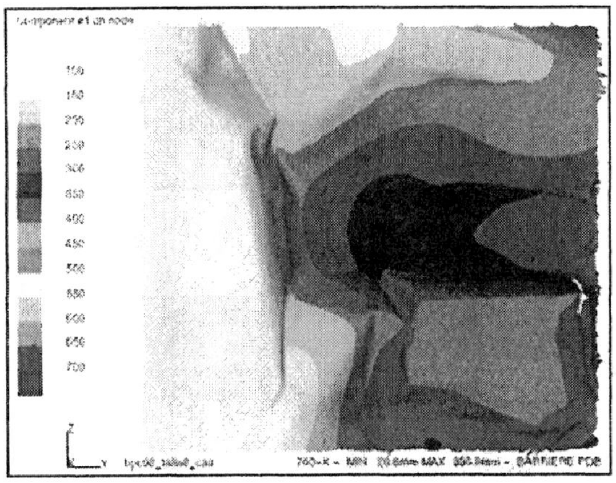

Figures 16: Contour plot representing front unit force distribution

Since the use of the last PDB version 7 with improved front sheet, localised tearing disappeared. Digitisation and interpretation is become easier.
Two limits are considered in the assessment: geometric in height and stiffness in depth. Both limits can be adapted with the fleet and what kind of cars we want to protect. The formula describes each colour surface weighted by its height (Z position) and its deformation (X position).

Load cell wall measurements

Before eventually using load cells wall behind the PDB, several tests have been performed to validate and control the accuracy of the measurement. Unfortunately, force measurement doesn't work well. In fact, the force distribution was not the expected one. The expected result could have been a homogeneous pressure because tests were performed with a rigid plan wall.
Unfortunately, the force was concentrated on one load cell (figure 17). The flatness of the multi load cell seems to be responsible for that result. Furthermore, the force measurement distribution seems to be better and more accurate without honeycomb. Further work is needed to clarify this point.

Figures 17: Force distribution corresponding to a plane rigid wall impact.

Tests already performed / Assessment

More than thirty tests have already been performed all around the world and the results confirm the previous ones. Vehicles have been selected for their front units in terms of design and stiffness, different size and mass, different generation and different driving position. These tests are often accompanied with a car-on-car test in order to validate the PDB assessment. The figure 18 shows the result vs kerb weight of the car.

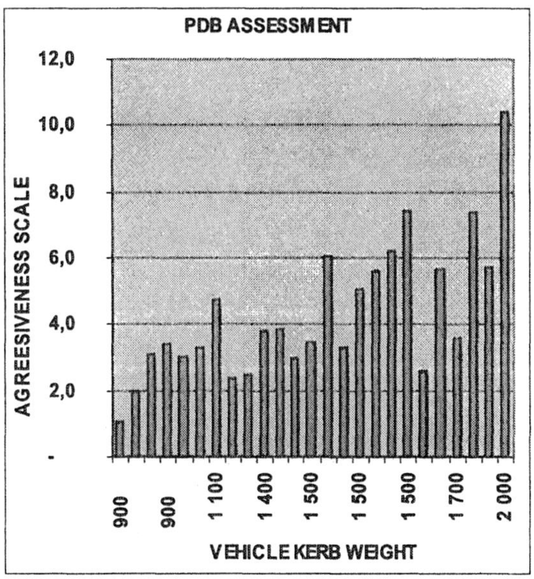

Figures 18: PDB assessment at 60 kph

Heavy vehicle performances

Some heavy cars have a better rank than small ones, because only the front unit behaviour is responsible for.

It is not a mass dependant test. Let's take for example a 1950 kg light commercial vehicle fitted with lower load paths (figure 19).

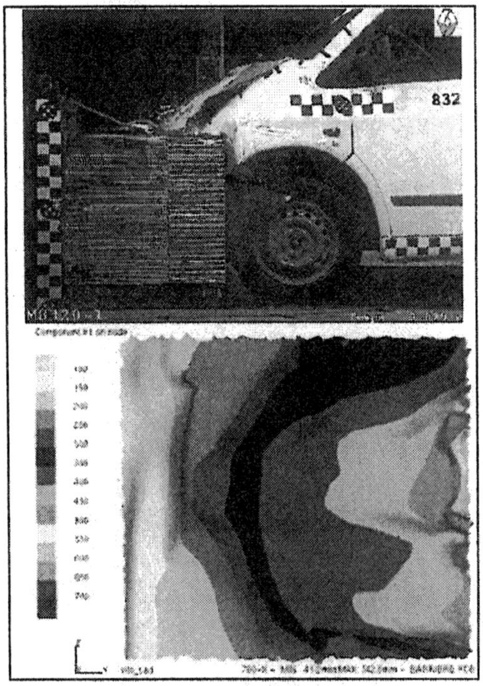

Figures 19: PDB assessment corresponding to a 1950 Kg Light Commercial Vehicle.

Multiple and moderate load paths force and good connexions among them allow to this vehicle getting acceptable compatibility assessment even if kerb weight is close from 2000 Kg.

FE model development

To complete the study, a numerical approach is needed. That's why, a Finite Element model of the PDB is under development. Actually, new test proposal must be easy to simulate. It will become a help to design the future compatible car.

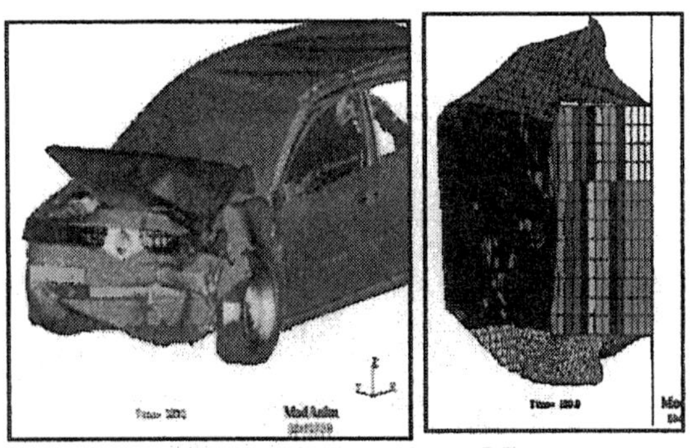

Figures 20: Family vehicle against PDB

The first approach is encouraging, however, further development are needed to tune the FE PDB. After that, the model will be marketed.

PROPOSAL TO ASSESS COMPATIBILITY

First step proposed by France – Mid term

The possibility to check both self and partner protection against PDB barrier was investigated. First results are encouraging to continue in this way. Up to now, two cars have been performed in this configuration.

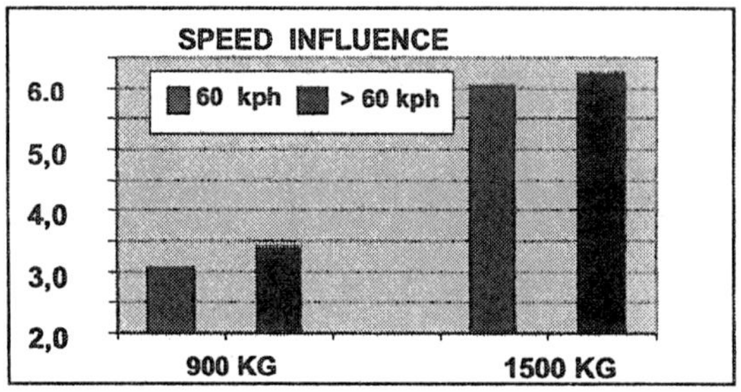

Figures 21: PDB assessment vs speed

The test speed was higher than 60 kph. We have seen that in both cases, the barrier was able to keep in mind the front unit footprint because the PDB assessment didn't change (figure 21). It means that the barrier seems not to be linked to the speed.

The French proposal (PSA Peugeot-Citroën, Renault and UTAC) is to replace the EEVC barrier by the PDB one (figure 22). The test speed could be eventually higher than the future regulation (60kph) and fixed for all cars. Furthermore, this proposal could generate higher deceleration pulse and could be able to combine both acceleration and intrusion. However, further researches are needed to set the optimal test speed.

Figures 22: New PDB test proposal with instrumented dummies.

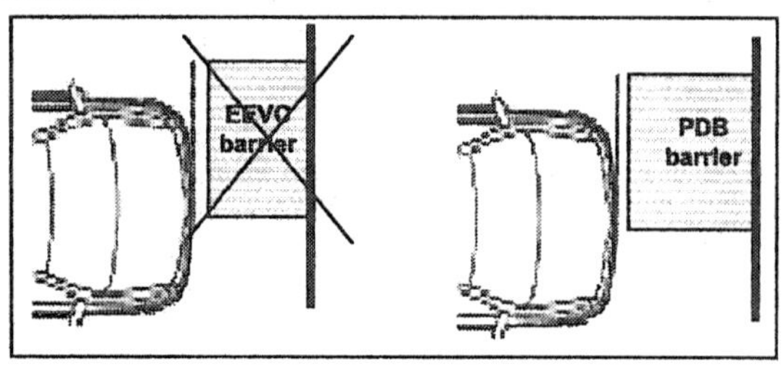

Figures 23: First step proposal to assess compatibility

Second step proposal – long term

To be closer to real life accident and be able to answer both partner protection especially for heavy cars and self protection especially for light cars with only one test. The PDB could be fixed on a mobile trolley as Australia investigated last year (figure 24). A quick energetically approach shows clearly than this test due to conservation of momentum associated to different energy absorbed in the barrier allows to progressively switch from a light car overload to a heavy car partner protection test (figure 25). We think that test answer all compatibility test configuration problems. However and before proposing this test as a standard, we have to investigate it.

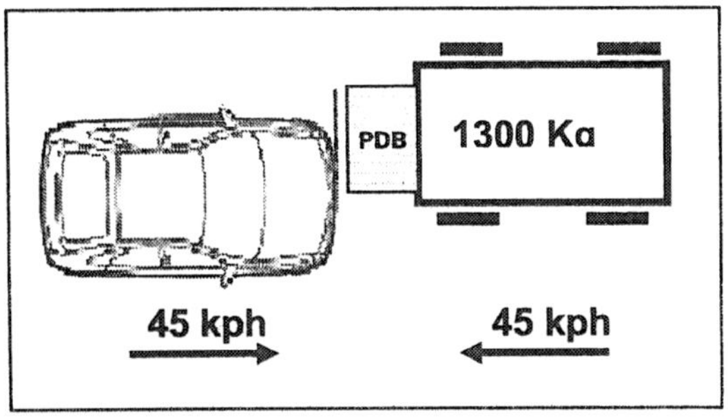

Figure 24: Second step proposal closer from real life accident

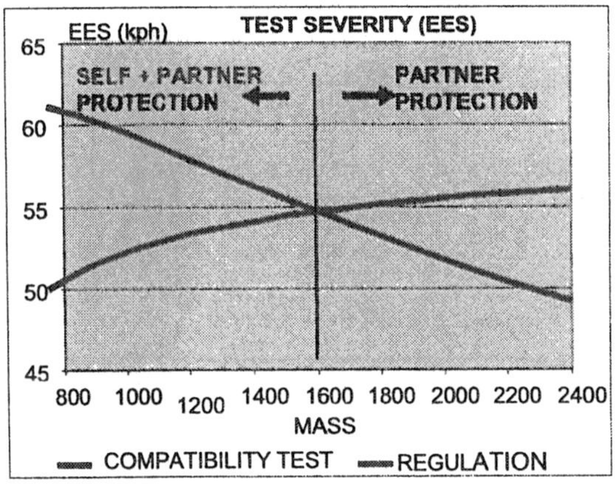

Figures 25: EES severity between future regulation at 60 kph and compatibility test vs mass

CONCLUSION

Improving compatibility is the most effective way to reduce the number of road accident victims. The study we made and the tests we conducted have showed that the front-end stiffness and compartment have an influence on compatibility.

A good structural interaction is a prerequisite to obtain a well balanced energy. It is not possible to separate stiffness and geometry of the front end. Furthermore, improvement that could be made on future vehicles for front impact will also produce benefit for side impact.

To answer these statements, the design constraints described (self protection, heterogeneous vertical distribution of stiffness and high local stiffness in front end) lead to define two compatibility targets:

- First of all, partner protection: vehicles should have an homogeneous front end and absorb a certain amount of energy before reaching self-protection force
- Then, self protection: vehicle should have a certain compartment crush force capacity and stability.

The development of future vehicles with respect to these targets would result in a compatible fleet of new vehicles. Moreover, considering the time taken to renew all the vehicles, it is necessary to propose measures that will limit dangers for existing vehicles and doesn't change every time to avoid rupture in the fleet.

The PDB test procedure already showed its capacity to verify the behaviour of new vehicles in regard to the partner protection targets. Aggressiveness assessment is allowed. The studies in progress confirm that statement. The concept, close from real life car to car accidents, clearly show the capacity of the front unit to be aggressive or not. Thanks to the propriety of the deformable element and the test configuration. Furthermore, interpretation is become easier with the current PDB version 7 that avoid rupture and tearing.

However, this test doesn't control the overload of compartment and nor answer the question of self protection. That is the reason why doing both self and partner protection with this test is investigating.
In a first step, that kind of evaluation could be introduced as a fixed barrier.

REFERENCE

1 W.Klanner, K Langwieder "The correlation between test and real-life accident for the car-to-car frontal crash", ESV 96

2. KA Seyer, CA Newland and MB Terrel "Australian Research to Develop a Vehicle Compatibility test" - Icrash 2002

3. LAB, JY Foret-Bruno, internal notes

4. PREDIT , French compatibility group: PSA, Peugeot Citroën, Renault S.A., INRETS – 2002

5. K. Mizuno "Vehicle compatibility – The new challenge for car crash safety" – 2002

6. EUCAR / WG15 Vehicle compatibility, workshop 2000 and 2001

7. IHRA / EEVC WG15 / ACEA OPC Vehicle compatibility, workshop 2002

Stiffness Effects and Geometry Effects

2002-01-1020

Stiffness and Geometric Compatibility in Front-to-Side Crashes

K. Digges, A. Eigen and G. Bahouth
George Washington Univ.

ABSTRACT

In this study, the comparative stiffness of vehicle side and frontal structures is determined by available static test and crash test data. NHTSA has conducted a series of staged crash tests where a Honda Accord is impacted by different bullet vehicles at a closing velocity of 32.5 mph. These staged front-to-side crash tests are examined to assess the extent of damage to both the bullet and struck vehicle. The load cell barrier data for the bullet vehicles used in NHTSA's vehicle-to-vehicle front-to-side crash tests are examined to determine the geometric and stiffness properties of the frontal structures as measured in the NCAP tests. The geometric and stiffness measurements during the early stages of frontal crush are most influential in front-to-side crashes. The barrier data provides useful stiffness information. However, the number of rows of load cells may be insufficient to provide geometric information. The bumper height and width appears to be a more accurate measurement of the load path that initially contacts the side of the struck vehicle. The bumper height for 20 representative vehicles was measured and summarized.

INTRODUCTION

Crash incompatibility between vehicles has been attributed to three vehicle factors: (1) mass incompatibility, (2) stiffness incompatibility, and (3) geometric incompatibility [Gabler, 1998]. The measurement of vehicle mass is relatively straightforward. However, the measurement of stiffness and geometric compatibility needs further definition.

For a stiffness metric, Gabler used the linear stiffness based on vehicle crush at the maximum barrier force in a 35 mph crash into a rigid barrier. In survivable front-to-side collisions, the level of frontal crush seen in the striking vehicle does not produce the maximum barrier forces seen during frontal tests. Consequently, the stiffness at lower values of crush is more applicable.

The Insurance Institute for Highway Safety reported a series of front-to-side crash tests to assess the influence of mass, stiffness, and vehicle ride height [Nolan, 1999]. The results were somewhat inconclusive, but suggested that the manner in which the striking vehicle deforms the struck vehicle has more influence than the vehicle stiffness alone. One significant test result indicated that the presence of a lower load path reduced dummy readings even for a heavy and stiff vehicle.

This finding is consistent with earlier observations made by Hobbs [1989] who reported that increasing the deformation of the lower door panel so that the door does not bow inward at the shoulder level was beneficial to dummy injury measures. These investigations emphasize the importance of the geometry of the impacting vehicle in addition to its stiffness and mass. In particular, the presence of a lower load path is important.

Past studies of compatibility by the authors have addressed primarily front-to-front compatibility [Digges 1999 and 2000]. In these studies, the height of the Center of Force was used as a geometric metric. This parameter was defined as the height at which a single force would balance the forces and moments exerted on the barrier by the vehicle [Digges 1999]. The Center of Force may vary across the front of the vehicle, particularly as the vehicle crushes more than 250 mm.

This analysis modifies the approach presented earlier to address front-to-side compatibility. In the past studies, the stiffness and geometric characteristics were examined at 250 mm and 375 mm of crush. In this study, we attempt to examine lower levels of frontal crush that are more appropriate for assessing front-to-side compatibility.

SELECTION OF DATA

In the NCAP tests, the frontal crush typically exceeds 400 mm. For survivable front-to-side impacts, the front of the striking vehicle crushes much less than

400 mm. To determine the extent of crush required, we examined the data from several sources. These sources include crush tests of the sides of vehicles to determine their stiffness, front to side vehicle-to-vehicle crash tests, and real world crashes.

There is some vehicle side crush data that has been reported in NHTSA research programs. In one research program, side crush tests were performed by mounting a vehicle in a static test fixture as described in the research report [DOT 1977]. Each test vehicle was positioned in a static test machine so a flat rigid platen could be forced into its side. The rigid platen had dimensions 1651 mm wide by 457 mm high and was intended to be representative of the frontal area of a midsize car. The side edge of the loading plate was 76 mm aft of the forward front door cut line. The bottom edge was positioned so that it loaded the side structure above the door sill. The loading ram was advanced in increments of 12.7 mm.

Tests of three production cars were reported. They were a 1976 Simca 1705, a 1975 Honda Civic, and a 1975 Plymouth Fury. The test results, shown in Figure 1, indicate an average initial stiffness of about 600 N/mm. At forces above about 50 kN, the stiffness was reduced. At 200 mm of crush, the applied force averaged about 80 kN.

In the reported research program the side of the Simca was reinforced and retested. The strengthened vehicle was called the Chrysler/Calspan RSV. The results of the strengthened Simca (RSV) are also included in Figure 1. An objective of the redesign was to increase the resistance to intrusion. The side structure was designed to limit the intrusion to 300 mm during a side impact severe enough to create a lateral acceleration of 20g's. The RSV achieved a TTI value of 55 when tested at 38.9 mph in the FMVSS 214 configuration using a Biosid driver dummy [Sankey 1991]. The side stiffness of the RSV represents the upper limit of strengthening that was practical at the time the research was conducted.

Another series of tests by NHTSA measured the door stiffness of seven different passenger cars, model year 1980. The results of the door tests found an average initial stiffness of 250 N/mm. However, the stiffness decreased after about 75 mm of crush. The results are summarizes in an earlier paper [Digges 1999].

NHTSA has not reported more recent static crush data on vehicle side strength. However, in 1998, NHTSA conducted four vehicle-to-vehicle front-to-side tests and reported the crush information on both vehicles (TRC, 1998). One test used a 1995 Chevrolet Lumina as the bullet vehicle.

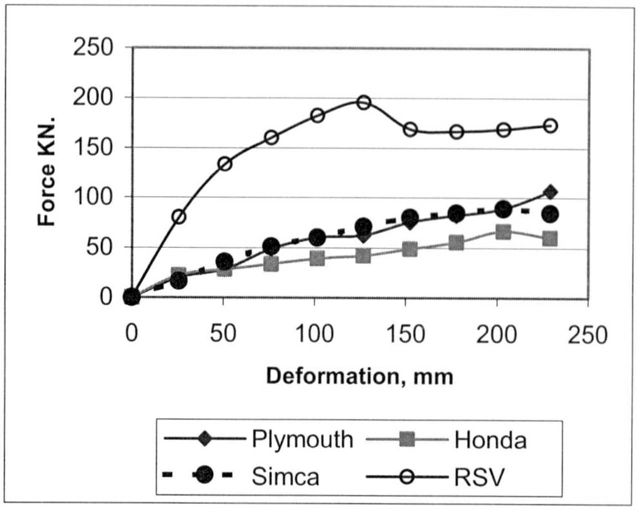

Figure1. Static Crush for Four Vehicles

The other bullet vehicles were a 1998 Chevrolet S-10, a 1997 Ford Explorer, and a 1997 Jeep Grand Cherokee. The impacted car was a 1997 Honda Accord in all crashes. The test configuration replicated the crash direction of the FMVSS 214 Side Impact Protection Standard with the impact speed of 53 kph.

In the FMVSS 214 test, the front of the barrier is parallel to the side of the car at the time of initial impact. However, the lateral velocity and impact location causes the struck vehicle to rotate about the edge of the barrier that engages the rear door. Maximum side crush frequently occurs at this location. The loading of the front door was assumed to be more comparative with the static crush tests reported earlier.

The crush profiles for the front of the four bullet vehicles are shown in Figure 2. The damage to the Lumina was measured at the bumper structure rather than at the fascia that covers the bumper. The higher amount of crush on the left side is associated with crushing of relatively soft frontal structure. In the central portion of the bullet vehicles, the crush ranged from 63mm for the Lumina bumper to 175 mm for the S-10.

The corresponding side crush profiles for the impacted Accord vehicles are shown in Figure 3. In this plot, the zero coordinate for the x-axis is the impact point for the left edge of the bullet vehicle. The plot shows both the direct and induced damage. The maximum damage occurs at a location on the rear door.

A comparison of Figures 2 and 3 indicates that the S-10 had more frontal damage and produced slightly less damage to the side of the Accord. However, in all cases, the maximum crush to the Accord ranged between 500 and 550mm. The small amount of crush

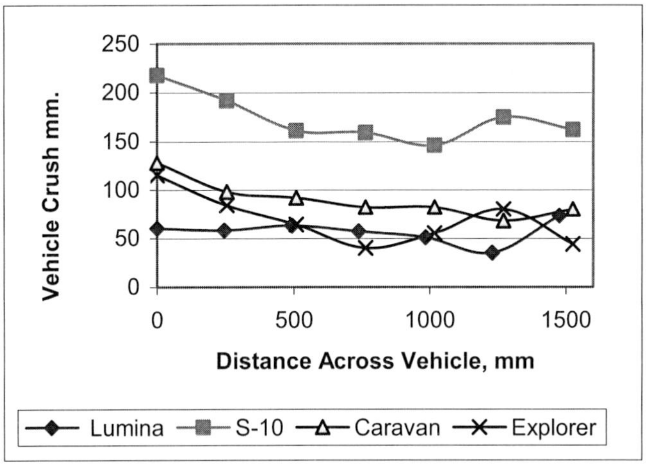

Figure 2. Frontal Crush for Bullet Vehicles in Side Impacts with Accord Vehicles

on the bullet vehicles suggests that the bumper is the primary loading element in front-to-side crashes. Consequently, the geometric and crush characteristics of the bumper and supporting structure determine front to side interactions.

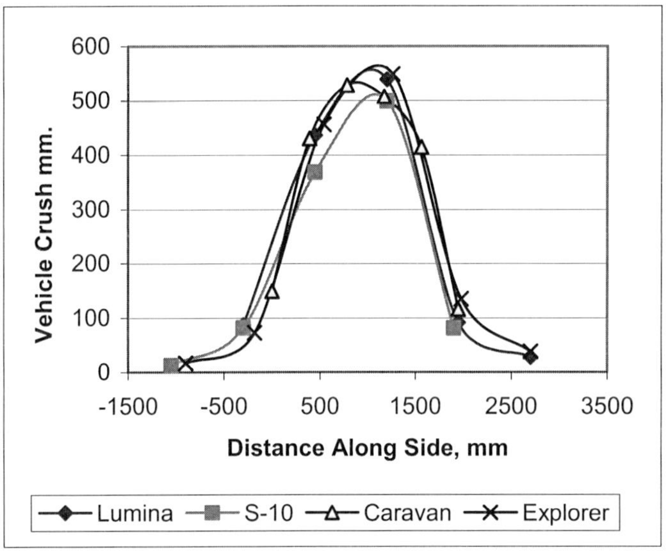

Figure 3. Side Crush for Target Accord Vehicles when Impacted by Four Different Bullet Vehicles

In the Lumina to Accord test, the maximum crush for the Lumina was 73 mm and the maximum front door crush for the Accord was 420 mm. The Lumina bumper is composed of a metal bar positioned 100 to 200 mm behind a fascia (see Figures 4 and 5). The 73 mm crush was the deformation of the metal bumper bar. The combined crush for the elastic fascia and plastic bumper was approximately 250 mm. It may be noted from Figure 4 that the vehicle frame to which the bumper is attached is approximately 175 mm behind the bumper. In Figure 4, the bumper measurement is taken at the centerline. As seen in the plot of the Chevrolet Lumina finite element model (Figure 5), the bumper actually curves across the front of the car, and is closer to the frame at the attachments. Curved brackets are also shown which separate the bumper beam from the vehicle frame. In the Lumina to Accord tests, the bumper damage did not exceed the bumper to frame clearance.

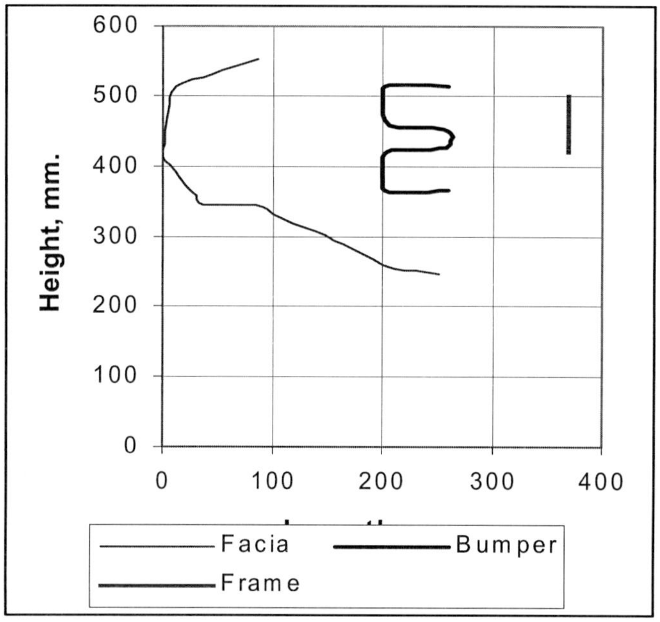

Figure 4. Profile of Lumina Bumper and Fascia.

Figure 5. Frontal Structure Of The 1996 Chevrolet Lumina Finite Element Model With Outer Fascia Removed.

Load cell data from the frontal NCAP test of the Lumina is useful in studying the crush characteristics of the frontal structure. The load cell matrix on the barrier consists of four rows and nine columns of load cells. The matrix is 984 mm high by 2108 mm wide. The bottom row of cells is positioned 66 mm above the ground. A plot of the Lumina load cell data is shown in Figure 6. This plot shows the total load on each column measured at five levels of vehicle crush. Only seven of the nine columns of load cells was loaded by the Lumina. Zero crush is determined by the beginning of bumper loading, rather than by a specific dimension of the elastic fascia.

Figure 6 shows how the force is distributed across the barrier face at five levels of vehicle crush. At 75 mm of crush, the total barrier force is 55 kN which is almost equally distributed across the three center columns of load cells. This loading is primarily from the bumper. At

125 mm of crush, the loading on columns C4 and C6 increases more than C5. This indicates that the stiff frame on each side of the vehicle is loading the barrier.

An examination of the barrier force characteristics of other passenger cars indicates a similar loading pattern to that in Figure 6. The load distribution during the bumper loading has a more uniform pattern than occurs after frame loading begins.

The NCAP frontal barrier crash tests show that passenger cars and light trucks have frontal stiffnesses ranging from 750 N/mm to 2070 N/mm [Digges, 2000]. This data is for average stiffnesses, measured at 375 mm of crush.

DISCUSSION

The NHTSA vehicle-to-Accord side crash test data suggests that the force generated by a vehicle's front structure during the initial period of frontal crush is most influential in controlling the side crush of the struck vehicle. The static side crush data indicates that force levels of 30 to 70 kN produce significant crush to the side structures of the on-the-road vehicles tested. Car models more recent than those tested may have higher stiffness. However, the vehicle-to-Accord tests indicate a very large difference remains between the frontal and side stiffness of vehicles.

Others have reported that more than 50% of the serious injuries in side impacts occur in crashes more severe than the tests specified by US and European Standards [Thomas, 1999]. In addition, real world vehicle-to-vehicle side crashes involve impacts in which the bullet vehicle impacts the struck vehicle at an angle other than 90 degrees. Consequently, the damage to the Lumina in the Lumina-to-Accord test conducted by NHTSA may not be representative of the bullet vehicle damage in real world side impacts that produce injury. The 1997 to 1999 data from National Automotive Sampling System/Crashworthiness Data System (NASS/CDS) was examined for cases that involved the Chevrolet Lumina impacting the driver's side of other passenger cars. A total of three Lumina cases were found in which the impact was at the occupant compartment of the struck car and the crash severity was sufficient to cause an AIS 1+ injury to the driver of the struck vehicle. The maximum AIS in the cases was 3. In all three cases, the measured deformation for the Lumina exceeded 125 mm. The loading on the Lumina was more complex than the car-to-car tests that produced 75 mm of bumper crush.

Earlier papers presented a geometric compatibility metric based on the height of the Center of Force [Digges 1999, 2000]. The values of the height of the Center of Force for 48 models of vehicles have been published previously [Digges 2000, 2001].

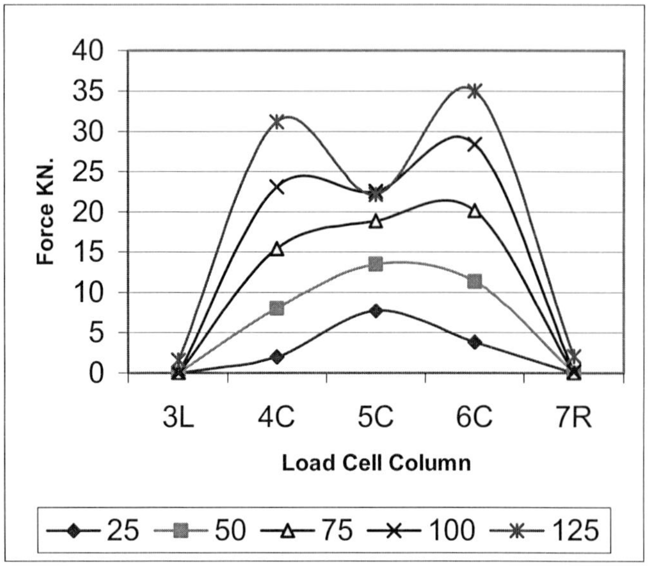

Figure 6. Force Distribution Across the Barrier for the Lumina NCAP Crash Test Measured at 25 to 125 mm of Crush

For the 48 vehicles tested, the height of the Center of Force, measured at 125 mm of crush, varied from 425 mm to 665 mm. For the Lumina, the value was 469 mm. The Center of Force shows a significant variation in the geometric height of the frontal loading for different vehicles. However, the height of the Center of Force does not give an indication of whether or not a load path exists near the level of the struck vehicle's lower door sill. For the Honda Accord, the top of the door sill was 320 mm above the ground. A striking vehicle would have to engage the lowest (A) row of load cells in order to directly load the sill of the Accord. The upper edge of the A row of load cells is 312 mm above the ground. Only one of the 48 vehicles tested produced significant loading of the A row. At crush levels that only engage the bumper, the lower height of the bumper structure may be a better indicator of the geometric compatibility than the height of the Center of Force, as measured by the NHTSA barrier. The reason for this observation is discussed in the paragraphs to follow.

For most of the passenger cars, the force was concentrated on only the second (B) row of load cells. Additional resolution would be desirable to improve the accuracy of the center of force calculation. It would be desirable to have better resolution of the B row level. The bottom of the B row is positioned 312 mm above the ground. The ground clearance of NHTSA barrier is 280 mm and the lower edge of the bumper on the barrier is at 330 mm. Tests reported in the literature indicate that increases in the barrier ground clearance of 50 to 100 mm result in large increases in struck vehicle deformation [Terrell, 2001]. Dummy injury measures also increase for these higher barrier conditions [Terrell, 2001]. The force distribution just above and below the intersection of the A and B rows of load cells appears to be a critical geometric compatibility parameter.

Consequently, additional resolution of the force distribution in this region is warranted.

Measurements of geometric and stiffness properties at low values of crush requires accuracy of the load cells at low load values. In addition, it requires precision in the determination of time zero for the acceleration and the barrier data. To date, the barrier data has been used primarily for assessing the maximum loads and the stiffness based on large amounts of crush. Consequently, accuracy at low loads has not been a major concern. If measurements at low loads become important, additional calibration may be required to assure the accuracy of the data.

To assess the variation in bumper height, representative vehicles were physically measured. The measurements were of the lower and upper height of the bumper bar. In many cases, the bumper is located behind a plastic fascia. As an example, the geometry for the bumper and fascia for the Lumina are shown in Figures 4 and 5. The region between the fascia and the bumper is partially filled with plastic foam.

The results of the bumper measurements are summarized in Table 1. These results show the height of the lower surface of the bumper, and the height of the upper surface of the bumper, both in millimeters. A plot of the lower edge of the bumper vs. vehicle weight is shown in Figure 7.

Table 1. Bumper Measurements for Twenty Vehicles

Year	Make	Model	Class	Lower Hgt. Of Bumper	Upper Hgt. Of Bumper
1995	Geo	Metro	Car	406	508
1999	Saturn	SL1	Car	381	508
1995	Chevrolet	Cavalier	Car	394	533
1999	Mazda	626	Car	381	527
1996	Pontiac	Grand AM	Car	387	508
1995	Chevrolet	Lumina	Car	425	546
1996	Ford	Taurus	Car	394	508
1999	Dodge	Intrepid	Car	419	540
1999	Oldsmobile	Intrigue	Car	381	546
1996	Jeep	Gr. Cherokee	SUV	508	654
1996	Toyota	4Runner	SUV	540	781
1997	Chevrolet	Blazer	SUV	438	610
1995	Ford	Explorer	SUV	565	686
1995	Isuzu	Trooper II	SUV	559	692
1998	Mercedes	M Class	SUV	406	540
1999	Ford	Expedition	SUV	578	673
1996	Ford	Ranger	PU	425	597
1999	Chevrolet	S-10	PU	406	584
1997	Dodge	Dakota	PU	660	737
1996	Chevrolet	C-1500	PU	495	610

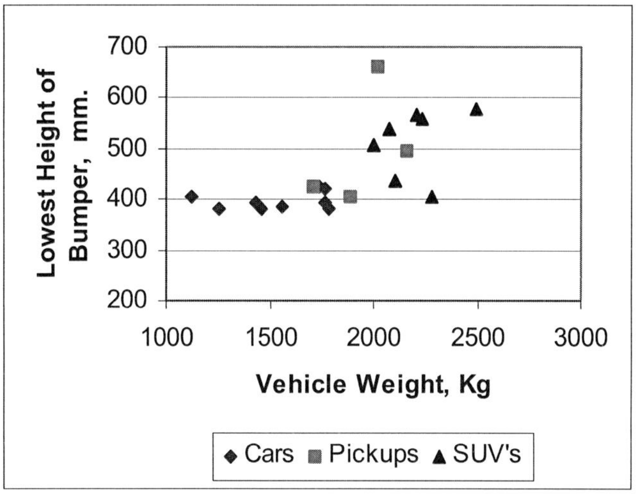

Figure 7. Height of Lowest Edge of Bumper vs Vehicle Weight for Cars, Pickups and SUV's

It is evident from Table 1 that the cars measured generally had bumper heights that were around 390 mm. Some pickups and SUV 's have bumper heights that are 100 to 200 mm higher. All the bumpers measured were higher than the 320 mm door sill on the Accord or the 330 mm bumper and the NHTSA barrier.

CONCLUSIONS

The load cell barrier data from NHTSA's NCAP tests provides a basis for measuring the forces exerted by the front of vehicles during the crash. For front-to-side crashes, the distribution of forces at low levels of frontal crush are most influential in the side impact compatibility. In NHTSA's Lumina-to-Accord crash test simulating the FMVSS 214 crash configuration, the crush of the bumper bar was less than 75 mm. This damage produced 539 mm of crush on the side of the Accord. For NASS/CDS real world injury producing crashes involving the Lumina as the bullet vehicle, the frontal crush was more than 100 mm. These observations suggest that vehicle stiffness and geometric measurements during the initial 125 mm of frontal crush are essential for evaluating front-to-side compatibility.

The accuracy of the barrier forces at low loads is critical to the measurement of front-to-side compatibility from NCAP load cell barrier data. Additional force resolution at the height of 300 to 350 mm would permit better assessment of the lower load path loading of the vehicle frontal structures. The lower load path loading appears to be a critical parameter in front-to-side compatibility.

In lieu of the more precise barrier measurements, the physical measurement of the bumper height provides a geometric measurement that may be suitable for characterizing front-to-side geometric compatibility.

REFERENCES

Digges, K., and Eigen, A., "Measurement of Stiffness and Geometric Compatibility in Front-to-Side Crashes", Paper Number 349, 17tth ESV Conference, May 2001

Digges, K., and Eigen, A., "Load Cell Barrier Measurements of Geometric Compatibility", Vehicle Safety 2000, I Mech E Conference Transactions, p.19-29, June 2000.

Digges, K., Eigen, A., and Harrison, J., "Application of Load Cell Barrier Data to Assess Vehicle Crash Performance and Compatibility", SAE, 1999-01-0720, March, 1999.

Gabler, H., and Hollowell, W., "The Aggressivity of Light Trucks and Vans in Traffic Crashes", SAE 980908, 1998.

Hobbs, A., "The Influence of Car Structure and Padding on Side Impact Injuries", SAE 896115, 1989.

Miller, P. and Green, J., "Design Development and Testing of the Calspan/Chrysler Research Safety Vehicle – Phase II Volume 1", DOT HS-802 250, 1977.

Nolan, J., Powell, M., Preuss, C., and Lund, A., "Factors Contributing to Front-Side Compatibility – A Comparison of Crash Test Results", 43rd Stapp Conference, pages 13-24, 1999.

Sankey, J.W., " Evaluation of the Biosid Dummy, MDB-to-Car Left Side Impact Test of a 27° Crabbed Moving Deformable Barrier into a Calspan RSV 5-Door Hatchback at 38.9 MPH" , DOT HS 807 758, June 1991.

Terrell, M., Newland, C., Seyer, K., and Dalmotas, D., "The Dependence of Side Impact Injury Risk on MDB Configuration", Paper 399, 17th ESV Conference, 2001.

Thomas, P., Frampton, R., "Injury Patterns In Side Collisions – A New Look With Reference to Current Test Methods and Injury Criteria", Paper 99SC01, 43rd Stapp Car Crash Conference Proceedings, p. 1-12, 1999.

TRC Test Report, "1997 Ford Explorer into Left Side of a 1997 Honda Accord", TRC Test Nr 980302, 1998.

TRC Test Report, "1998 Chevrolet S-10 into Left Side of a 1997 Honda Accord", TRC Test Nr 980220, 1998.

TRC Test Report, "1997 Dodge Caravan into Left Side of a 1997 Honda Accord", TRC Test Nr 980427, 1998.

TRC Test Report, "1995 Chevrolet Lumina into Left Side of a 1997 Honda Accord", TRC Test Nr 980512, 1998.

2003-01-0908

Evaluation and Research of Vehicle Body Stiffness and Strength for Car to Car Compatibility

Yuichi Kitagawa, Masashi Makita and Chinmoy Pal
NISSAN MOTOR CO., LTD.

ABSTRACT

In a CTC (car to car) crash, interaction between two vehicles is quite important. Interaction is primarily described by the contact area between two vehicles but interaction force (impact force) is also important for the entire crash phenomenon. In a frontal crash, impact force is resisted by the body structures, engine block, and tires. The resultant share of energy absorption, as well as the magnitude of body deformation, is greatly affected by the force profile. It is desired, therefore, to evaluate those factors of vehicle bodies in order to achieve CTC compatibility. There are some technical obstacles, however, in measuring those factors in testing. Impact force, for instance, cannot be measured directly in a CTC crash test unless load cells are installed in body frames. It is also difficult to analyze body deformation in a CTC crash test because both vehicles are moving. This study focused on methods to evaluate impact force and resultant deformation in a CTC frontal crash test, and proposes possible indicators to evaluate the interaction profile of the subject vehicle. First, a method has been developed to estimate the impact force and body deformation in a CTC test based on vehicle deceleration. The adequacy and consistency of the method were verified by using the principle of conservation of energy. By analyzing the estimated force-deformation curves, a possible mechanism in energy sharing between two vehicles was explained. Based on the mechanism, it was proposed to evaluate the interaction profile of the subject vehicle in a single vehicle test. The profile was well characterized by the stiffness of the front-end and the strength of the body. It was found that the interaction profile changes in CTC crashes. Barrier configurations and evaluation methods was also discussed to assess stability of the force profile.

INTRODUCTION

Compatibility among vehicles of different size is a big challenge in a crash safety research. Since frontal crashes are frequent in real world collisions, though its ratio varies depending on the market, frontal crash compatibility has been discussed from many aspects. One of the most severe cases is thought to be a collision between a small light vehicle and a large heavy vehicle. Zobel (1998) adopted a mathematical approach to find possible solutions to solve incompatibility. He suggested to define the maximum force level for the front-end among vehicles in a mass ratio of 1.6 based on his bulkhead concept. It was pointed out by Appel et al. (2000) that increasing stiffness and strength of small vehicles is relatively more important rather than controlling heavy vehicles. They also proposed possible test protocol with regard to force-deflection characteristics. Steyer et al. (1997) studied accident data collected for twenty-five years and concluded that the maximum force level should be controlled between 300 and 400 kN for passenger vehicles. It should be noted, however, compatibility issues can exist even in a crash between two vehicles of the same size, Edwards et al. (2000) conducted frontal crash tests between two identical vehicle models but the ride height was adjusted at different levels. Over-ride and under-ride occurred and resulted in different body deformation. They pointed out that geometrical interaction and shear stiffness can affect resultant body deformation. Based on these findings, Transport Research Laboratory (2000) is proposing three different test protocols to evaluate performance of a vehicle in occupant protection and compatibility. The first is an offset deformable barrier test at 64 kph and conducted primarily for measuring body deformation. The second is a full frontal test against a rigid barrier covered with a thin deformable element at 56 kph. This test includes two specific measurements. Deceleration pulse of the vehicle is used to evaluate performance of restraint system and damage to the deformable element is used to assess aggressivity. The last is another offset deformable barrier (ODB) test at an elevated crash speed (possibly at 80 kph) but without dummies. The purpose of this test is to determine the overload stiffness of the cabin. It is important especially for small vehicles because they are likely to collide with a relatively larger vehicle. For the test protocol to assess aggressivity, a progressive deformable barrier (PDB) is also proposed by Delannoy et al. (2001). They proposed that stroke, height and area on the deformed barrier face could be possible factors to assess aggressivity. Thus, many issues are currently discussed for compatibility assessment not only in terms of test protocol but also about indicators. Experimental and computational approaches are taken to better understand CTC crash phenomena and to discover best protocol and indicators for assessment.

METHOD TO ESTIMATE IMPACT FORCE AND BODY DEFORMATION

Because impact force cannot be directly measured in a CTC crash test, a method has been developed to estimate impact force based on deceleration measured on a vehicle. It was assumed that impact force of a vehicle was equivalent to a summation of inertial forces of mechanical parts and body structures. The mechanical parts includes the engine, transmission, and suspension parts in the front-end while the body structures mean the rest of the vehicle including the seats and rear suspension parts as well as the body frames and panels. In this paper, inertial force of mechanical parts is called 'mechanical force' while inertial force of body structures is called 'structural force' for convenience. In a frontal offset crash, only the mechanical parts on the struck side were taken into account when calculating mechanical force. The mechanical parts on the non-struck were added to the body group. Close observation on ODB tests conducted at NISSAN indicated that an engine and a tire-suspension system are two dominant components affecting load paths in the front-end. Mechanical force is, therefore, thought to be a combination of inertial forces calculated using decelerations at the engine block and at the suspension frame. It was assumed that inertial force in an offset crash would be generated by the parts on the struck side. Structural force was, therefore, considered separately for the struck side and the non-struck. A half body weight was used to calculate the structural force on the struck side while the other half and a half mechanical weight were used to calculate the structural force on the non-struck side. The calculating method is summarized as follows:

Mechanical Force:

$$Fm = 0.5 * Meng * Geng + 0.5 * Msusp * Gsusp \quad \ldots(1)$$

Structural Force:

$$Fs = 0.5 * Mbody * Gbody\text{-}r + (0.5 * Mbody + 0.5 * Msusp) * Gbody\text{-}l \quad \ldots(2)$$

Impact Force:

$$F = Fm + Fs \quad \ldots(3)$$

where Meng is the weight of the engine block, Msusp is the weight of the suspension parts, Mbody is the weight of the body structures, Geng is the deceleration at the right side of the engine block, Gsusp is the deceleration at the right arm of the suspension frame, Gbody-r and Gbody-l are the decelerations at the right and left sides on the cabin respectively. The concept of this method is similar to that proposed by Steyer et al. (1998) but the algorithm is simplified. Location of accelerometers was determined empirically.

In order to explain an energy sharing mechanism between two vehicles, it is necessary to analyze force-deformation curve for each vehicle. Body deformation is, however, not able to be measured directly in a CTC crash because both vehicles are moving. Displacement of the body, calculated by double-integrating the cabin deceleration, includes rigid body motion. In order to estimate body deformation on each vehicle, a method has been developed to remove rigid body motion. It was hypothesized that the center of gravity of the entire system including two vehicles can be regarded as an origin of gross deformation for both vehicles. Based on the hypothesis, rigid body motion can be known by calculating displacement of the center of gravity. The principle of conservation of momentum gives the final velocity of the system assuming no restitution:

$$V_1 = (M_1 * V_0 + M_2 * V_0) / (M_1 + M_2) \quad \ldots(4)$$

where V_1 is the velocity of the system, M_1 and M_2 are the weights of two vehicles, and V_0 is the initial velocity of two vehicles. Assuming that the principle is valid through the crash phenomenon, displacement of the system can be calculated by integrating the velocity. Then the body deformation on each vehicle was calculated by subtracting the system displacement from the body displacement.

VERIFICATION OF ESTIMATE METHOD

OFFSET DEFORMABLE BARRIER TEST

A barrier test was conducted to verify the validity of the developed method to estimate impact force. Load cells were placed behind the barrier to measure the actual crashing force to be compared with estimated impact force. A small-size passenger vehicle was crashed against an EEVC (European Enhanced Vehicle-safety Committee) deformable barrier with 40% overlap at 80 kph. It was expected that the cabin would be induced to crush in such a high speed crash. Figure 1 illustrates the test condition. Accelerometers were mounted on the necessary parts specified before. Deceleration at the body-side sill was taken as the cabin deceleration. Only longitudinal component of deceleration was considered in the test because body rotation was not significant up to the maximum body deformation. The analog data were collected by the data acquisition system with sampling rate of 10 kHz and pre-sampling filter of 4 kHz. Then the data were finally processed through a digital filter at 100 Hz (CFC60) for plotting. Data for weight were known for all mechanical parts and the body structure.

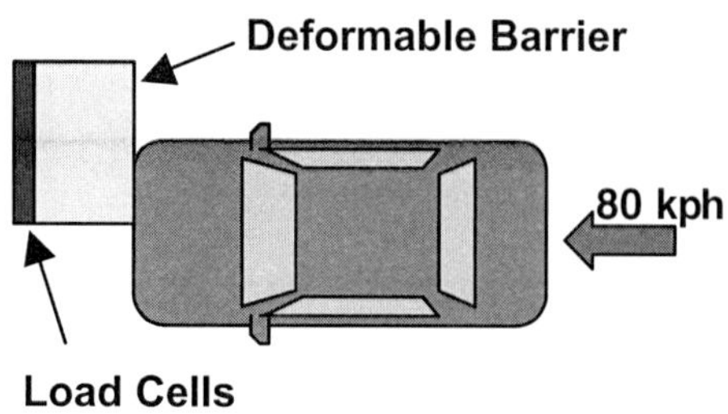

Figure 1. 80K-ODB Test Condition.

Figure 2 shows the estimated force curves calculated using the developed method. The mechanical force was dominant during the first half of the crash phenomenon but it dropped off at around 60 ms. It means that the mechanical parts stopped moving forwards and no more inertial force was working after that, though they worked as load paths throughout the crash. The structural force was the greatest part of the impact force in the later half. This means that the rest of the vehicle kinetic energy was absorbed by body deformation. Figure 3 compares the estimated impact force and the measured barrier force. The estimated force shows slight oscillation possibly because of body vibration during the crash. The entire force profile, including the gradient of force rising and the maximum force level, was almost identical between the two curves. The results suggested that the developed method was valid to estimate impact force of a vehicle.

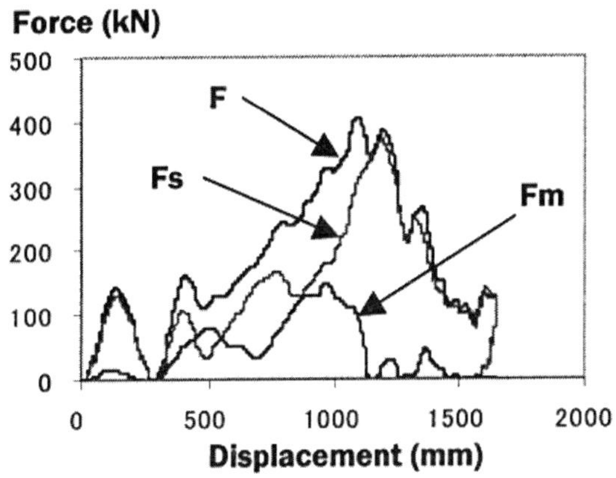

Figure 2. Estimated Impact Force.

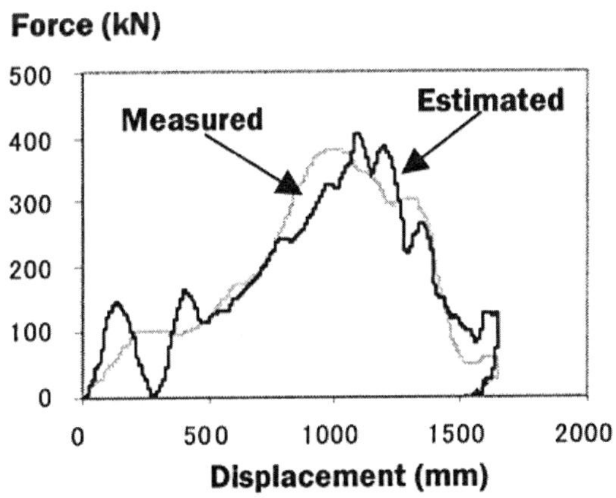

Figure 3. Estimated and Measured Forces.

CAR TO CAR FRONTAL CRASH TEST

A CTC offset frontal crash test was conducted between a small light vehicle and a large heavy vehicle. The small vehicle had a 1.5 liter 4 cylinder engine with lateral layout and front drive system while the large vehicle had a 3.0 liter V6 engine with longitudinal layout and rear drive system. The vehicle weights were 1,300 kg and 2,000 kg respectively including dummies and measuring devices. Each vehicle was accelerated by a cable and allowed a few meters of inertial motion before contact. The target speed was 56 kph for both vehicles with an overlap ratio of 50% for the small vehicle. Figure 4 illustrates the test condition. Actual impact speed was measured with phototubes that were placed at a few meters away from the contact point. The actual overlap ratio was verified by checking marks painted on the bumper surfaces. Accelerometers were mounted on the right side of the engine and the suspension frame, and both side of the body-side sill for each vehicle. The data acquisition system was the same with that used in the previous test. High speed digital video cameras were used to record crash behavior of both vehicles.

The resultant impact speed was 56.7 kph for both vehicles and the overlap ratio was 52% for the small vehicle. Post impact, the small vehicle was pushed back a few meters from the contact point and was rotated around 135 deg. to its final rest position while the large vehicle rotated 70 deg. Review of the high speed video indicated that most of the vehicle rotation occurred after maximum engagement. The longitudinal component was considered to be dominant in impact force. Figure 5 shows body deformation of both vehicles after the crash. The magnitude of deformation was larger in the small

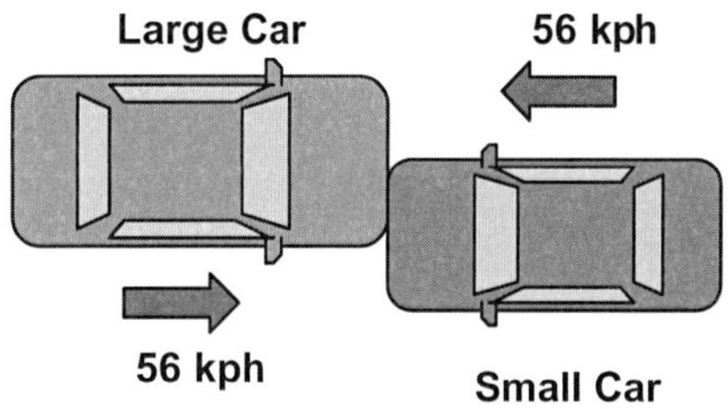

Figure 4.Car-to-Car Test Condition.

Small Car

LargeCar

Figure 5. Body Deformation of Two Cars After Crash.

vehicle extending to the right front-pillar. The lower part of the pillar was deformed possibly by a contact with the front-tire and the upper frame was bent upwards possibly induced by force through the hood-ledge. Local deformation around the right-front area of the floor suggested that the front-tire in the struck side moved rearwards and inwards then contacted the inner part of the body-side sill.

Impact force was estimated by the developed method for each vehicle. Figure 6 compares the time history curves of the calculated impact forces. Theoretically impact forces of two colliding vehicles should be identical with respect to time passage because their force interacts mutually. A good agreement was obtained between the two curves throughout the crash. Slight difference in single peak levels is possibly because of difference in vibration derived from eigenvalue of each vehicle. Validity of the method to estimate body deformation was examined by comparing measured and calculated velocity of two vehicles. By integrating body deceleration originated from the measured impact speed, velocity was calculated up to 150 ms for each vehicle. Figure 7 shows the velocity change in two vehicles. The calculated velocity was around 3.7 m/s at 150 ms while the final velocity was expected 3.3 m/s theoretically. Velocity change of the origin point was also calculated using the equation (4). It was found that the origin point moved at almost constant velocity throughout the crash. These results suggested that the developed method was valid to estimate body deformation of vehicles in a CTC crash. Another validation was made in terms of energy balance. Absorbed energy was calculated by integrating impact force with respect to body deformation while kinetic energy was simply obtained from velocity. As shown in Figure 8, total energy of the system was conserved throughout the crash. A small rise in the total energy around 50 ms possibly indicates slight overestimate of impact force or body deformation.

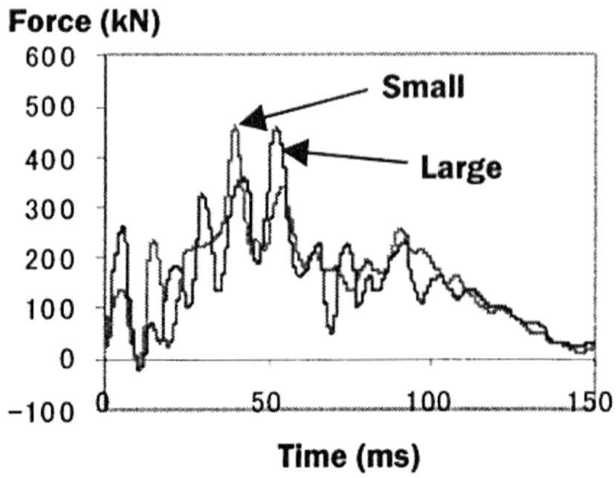

Figure 6. Estimated Crashing Forces.

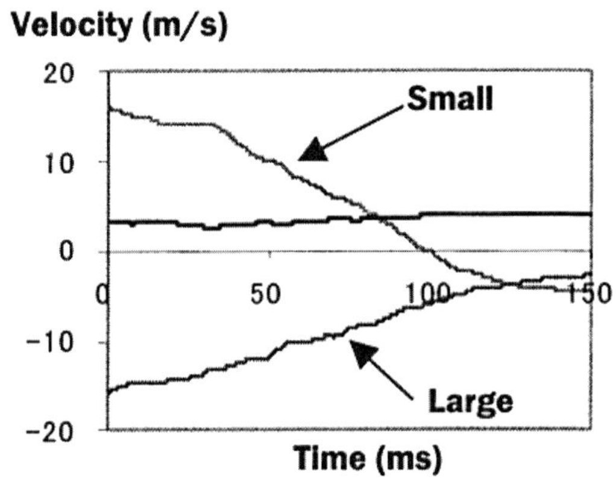

Figure 7. Velocity Change in Two Cars.

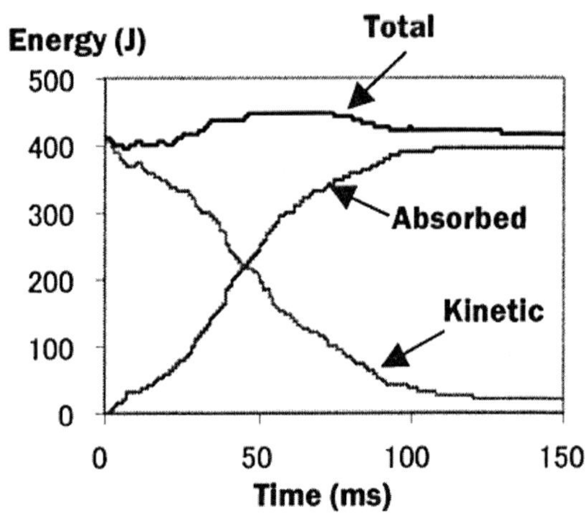

Figure 8. Energy Change in System.

MECHANISM OF CTC FRONTAL CRASH

A series of datasets was analyzed to understand the mechanism of CTC frontal crashes. Three passenger vehicles of different size were selected: Vehicle A is the smallest and lightest, Vehicle B is the middle, and Vehicle C is the largest and heaviest, as shown in Figure 9. CTC crash tests were conducted between Vehicle A and Vehicle C, Vehicle B and Vehicle C. The test conditions were basically the same with that explained before. The impact speed was 56 kph and the overlap ratio was 50% for the smaller vehicle in both cases. The same data acquisition system was used with that used in the previous tests. Impact force and body deformation were estimated using the developed method from measured deceleration. In addition, an offset deformable barrier test was conducted on each vehicle to examine a force profile in a single vehicle test. The impact speed was 80 kph for Vehicle A and B because overloading was expected on these vehicles when crashed against Vehicle C, while Vehicle C was tested at 64 kph. An EEVC deformable barrier was used in each test and its

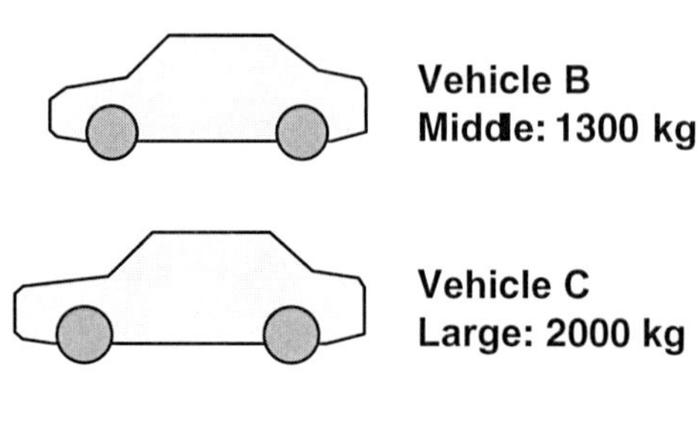

Figure 9. Tested Vehicles.

deformation was directly measured at the impact edge using a potentiometer. Body deformation was estimated by subtracting the barrier deformation from vehicle displacement. Figure 10 compares the impact force profiles between Vehicle A and Vehicle C. Estimated force-deformation curves of Vehicle A were plotted on the left side, while those of Vehicle C were drawn on the right side. On each side, impact force of the vehicle in a CTC crash test was compared with that in an ODB test. Figure 11 shows the impact force profiles of Vehicle B and Vehicle C, and the same comparison was made between a CTC crash test and an ODB test.

A possible mechanism of a CTC crash between two vehicles of different size can be explained by analyzing the force curves in Figure 11. The impact force increased as the body deformation grew in each vehicle. The force reached its maximum peak approximately at 700 mm of deformation on Vehicle B while it corresponded to around 600 mm on Vehicle C. It should be noted that maximum levels were almost the same between the two vehicles as they were in a relation of action and reaction. The gradient of force rising up to the maximum peak can be regarded as the stiffness of the vehicle body as the magnitude of deformation during the first half is primarily determined by this trend. The stiffness was found relatively steeper in Vehicle C. Beyond the maximum peak, the impact force still kept some level and generated further body deformation in Vehicle B, while the force gradually decreased in Vehicle C. The final magnitude of body deformation was relatively larger in Vehicle B as theoretically expected. Comparison with the force curves obtained from ODB tests clearly explains the features of force-deformation profiles in the CTC test. First, the stiffness in the CTC test is almost consistent with that in the ODB test for both vehicles. The maximum force peak in the CTC test is almost equal to that in the ODB test on Vehicle B, while it is lower than that in the single vehicle test on Vehicle C. The later force curves on Vehicle B are slightly different from each other but their average levels are almost the same. The later force curve on Vehicle C, on the other hand, suggests that the vehicle was basically unloaded after the peak though some local deformation might raise the final amount of deformation slightly. It is concluded that Vehicle C was overloaded in the CTC test while Vehicle C was loaded below its strength limit. The difference in stiffness and strength determined the energy sharing between the two vehicles.

The force-deformation curves in the CTC test between Vehicle A and Vehicle C, plotted in Figure 10, were also analyzed by comparing with those in the ODB tests. Although the same mechanism can explain some features, such as the maximum peak level and the magnitude of final deformation, significant difference was also found in the force curves for both vehicles. The maximum force peak on Vehicle A in the CTC test appeared later than that in the ODB test while the timing of the maximum peak on Vehicle C was relatively earlier

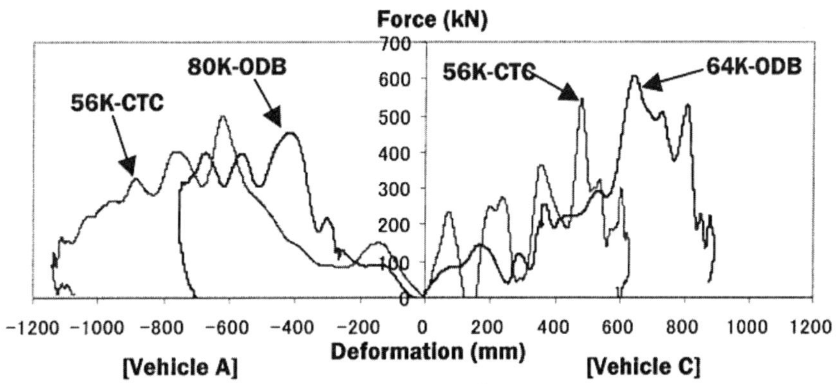

Figure 10. Force-Deformation Curves of Vehicle A and Vehicle C.

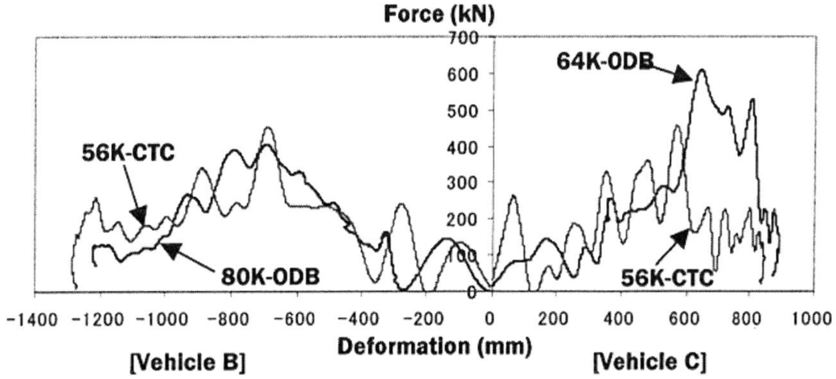

Figure 11. Force-Deformation Curves of Vehicle B and Vehicle C.

in the CTC test. The reason for the difference is possibly due to the interaction between the two vehicles. Since it is difficult to closely investigate the interaction in the crash test, especially during the first half of the crash phenomenon, a FEM study was conducted to simulate the CTC crash tests. An explicit finite element program 'PAM-CRASH' was used to perform the simulations. Figure 12 shows the front-end deformation of two vehicles when the impact force reached its maximum peak. Comparison from the bottom view well indicates the interaction among body structures and mechanical parts of two vehicles. In the CTC crash between Vehicle B and Vehicle C, as shown on the right side in the figure, the tire of one vehicle pushed against the other vehicle's engine block, as marked by (c) and (d). Such a loading condition is thought to be similar to that in an ODB test. When Vehicle C was impacted by Vehicle B, as shown on the left side in the figure, less interaction was observed among the mechanical parts. Since the tire of Vehicle A is relatively small, the engine block of Vehicle C was not loaded greatly, as marked by (a), as it was in the other case. The tire of Vehicle C was loaded by the engine block of Vehicle A but the timing was delayed compared with that in the other case because the engine block was relatively small and light, as marked by (b). The possible mechanism of interaction explained above was confirmed by comparing decelerations of mechanical parts. Figure 13 compares the decelerations of the engine block and the suspension frame, representing the tire, of Vehicle C between the CTC crashes with Vehicle A and Vehicle B. The maximum deceleration peak of the engine block was smaller when crashed with Vehicle A, and the maximum deceleration peak of the suspension frame was delayed when crashed with Vehicle A. Such a difference in interaction among mechanical parts resulted in the shift of the maximum force peaks as seen in Figure 10. Summarizing the analysis of the force-deformation curves, vehicle body stiffness and strength in a CTC test are primarily originated from the force profile of the subject vehicle in an ODB test, but are potentially affected by interaction among mechanical parts of two vehicles.

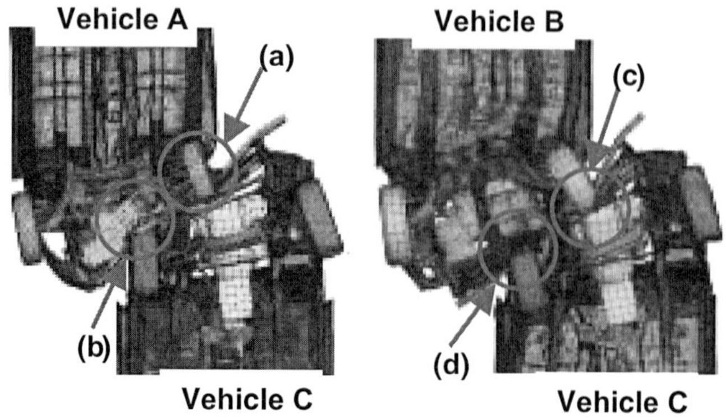

Figure 12. Front-End Interaction Between Tested Vehicles.

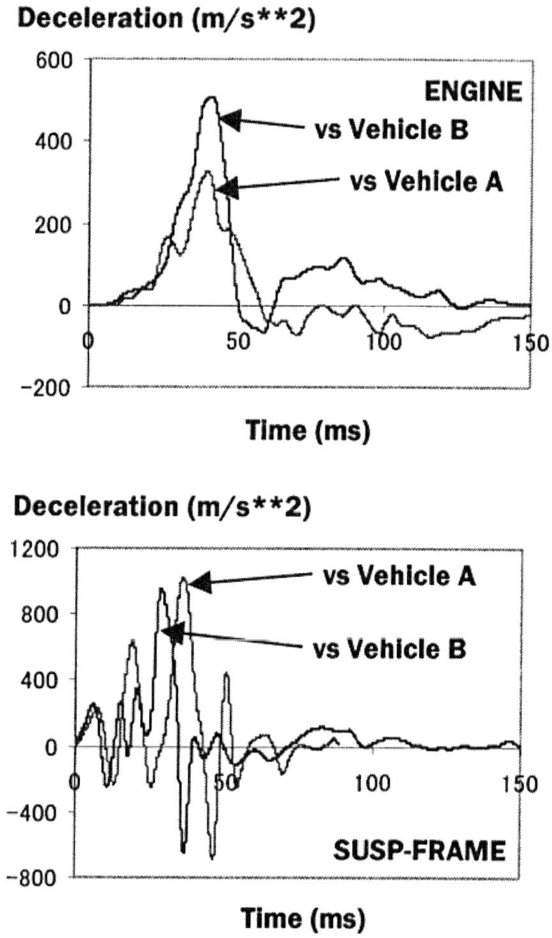

Figure 13. Deceleration of Mechanical Parts in Vehicle C.

DISCUSSION

In the past studies for compatibility, interaction problems have been pointed out from different aspects. Aggressivity is one of the issues in terms of partner protection. Large heavy vehicles are generally thought to be stronger than relatively small and light vehicles. Another feeling was that reinforced vehicles for self-protection would be potentially aggressive against older vehicles. The maximum force measured in a single vehicle test could be a possible indicator to assess aggressivity. Another issue is geometrical inconsistency especially in height among various vehicles. Typical results of inconsistency are known as over-riding and under-running. There is a proposal to control the height of the force center of vehicles within a specified area to improve consistency. The findings of this study suggest some other problems for compatibility assessment.

Basically every vehicle has a unique force-deformation profile, originating in its body structures and mechanical parts. Interaction with another vehicle, however, potentially changes the force profile in a CTC crash. Interaction among mechanical parts, which is dominant during the first half of a frontal crash, greatly affects the timing of the maximum force peak and possibly changes the stiffness of the vehicles. Energy sharing between two vehicles is primarily determined by the maximum force peak and its timing. Because of the difference in energy sharing, one of the vehicles has to absorb larger amount of energy generating body deformation. The force level following the maximum peak determines the final magnitude of body deformation. The trailing force level is also called 'end-of-crash force.' The interaction profile of a vehicle is, therefore, well described by the combination of the stiffness, the maximum force peak, and the trailing force level. The first problem for compatibility assessment is to obtain the interaction profile of the subject vehicle. Using the estimate method developed in this study, a force-deformation curve can be calculated even if load cells are not available. The obtained profile is symbolized as a polygon as shown in Figure 14. The stiffness, the maximum force peak, and the trailing force level can be easily evaluated in such a profile. The other problem is how to assess stability of the interaction profile for the vehicle. Desirably every vehicle should generate the same interaction profile regardless of opponent vehicles. A simple idea is to conduct CTC tests on the subject vehicle against other vehicles of different size but it is more reasonable to conduct a single vehicle test for simplicity and repeatability. Further studies are necessary to develop a method to assess stability of the interaction profile in a single vehicle test. In this study, possibilities of stability assessment using currently proposed barriers and test protocols are discussed based on the findings. Because the profile is greatly affected by interaction among mechanical parts, as explained before, force distribution during the first half of a crash could be helpful to assess stability. The area of a specified force level can be evaluated in addition to the maximum force level using a matrix of load cells. The number of load cells should be decided so that the force distribution shows enough resolution for assessment. It is assumed in the above method, however, that the front-end structures and mechanical parts have good shear connections among one another. In order to remove the assumption, shear deformation in lateral and vertical directions must be allowed for the subject vehicle. An offset crash test against a deformable barrier, like a current regulation test, is a good protocol for this purpose but a problem of bottoming out of the EEVC deformable barrier must be overcome. Increasing the thickness of the barrier may reduce the resolution in force distribution. By measuring deformation depth instead of force distribution on a progressively deformable barrier, such as the PDB, shear connections can be evaluated as well as aggressivity. Thus the indicator and the evaluation method must be carefully considered with respect to the barrier configuration for a better test protocol.

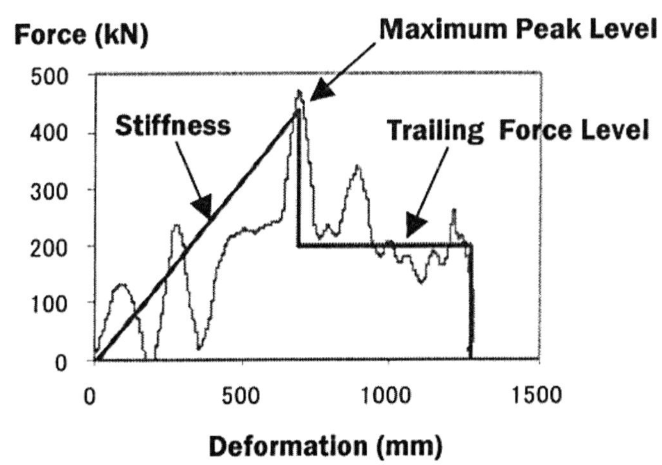

Figure 14. Symbolized Force-Deformation Profile.

CONCLUSION

1) A method has been developed to estimate impact force and body deformation from measured deceleration of two vehicles colliding into each other.

2) The validity of the developed method was confirmed by comparing the estimated force and the measured force, and by comparing the calculated velocity with the theoretical value for selected frontal crash tests.

3) A mechanism of CTC crashes was explained by analyzing the obtained force-deformation curves for two vehicles. Difference in stiffness and strength of the vehicle body well explained the energy sharing where a relatively small light car absorbs a larger portion of the crash energy generating body deformation.

4) The force-deformation profile of a vehicle can be characterized by the stiffness during the first half, the maximum force peak, and the trailing force level.

5) Since the force profile can change in CTC crashes due to interaction among mechanical parts, a vehicle can be both aggressive and vulnerable depending on the impact condition. It is desired to evaluate stability of the force profile in a single vehicle test.

6) The method described here can provide a force-deformation profile in CTC crashes, and should be considered as a future test protocol for vehicle compatibility assessment.

REFERENCES

1. Zobel, R., 'Demands for compatibility of passenger vehicles', Proc. 16th ESV Conference, Vol. 1, 98-S3-O-10, pp 729-741, 1998.
2. Appel, H., Deter. T., 'Crash compatibility of passenger cars achievable, but how?', Proc. of International Conference on Vehicle Safety 2000, held in London, England, 07/07-06/09/2000, pp 55-61, 2000.
3. Steyer, C., Dubos, A. 'Renault's proposal to improve compatibility', EEVC WG15 Document, Doc No. 31, 1997.
4. Edwards, M., Happian-Smith, J., Byard, H., Davies, H., Hobbs, C., 'Compatibility – the essential requirements for cars in frontal impact', Proc. of International Conference on Vehicle Safety 2000, held in London, England, 07/07-06/09/2000, pp 3-17, 2000.
5. Steyer, C., Delhommeau, C., Delannoy, P. 'Proposal to improve compatibility in head on collisions', Proc. 16th ESV Conference, Vol. 1, 98-S3-O-05, pp 682-692, 1998.
6. Delannoy, P., Diboine, A. 'Structural Front Unit Global Approach', Proc. 17th ESV Conference, Paper No. 199, 2001.

2001-06-0078

EFFECTS OF GEOMETRY AND STIFFNESS ON THE FRONTAL COMPATIBILITY OF UTILITY VEHICLES

Susan L. Meyerson
Joseph M. Nolan
Insurance Institute for Highway Safety
United States of America
Paper No. 91

ABSTRACT

The issue of vehicle incompatibility, especially between passenger cars and utility vehicles/pickup trucks, has received a lot of attention in recent years. Real-world crash data show that occupants of cars are much more likely to be injured in frontal crashes with utility vehicles and pickup trucks than with other passenger cars, even after controlling for vehicle mass. Factors in addition to mass that can influence compatibility are stiffness and geometry. In this paper, the effects of these factors on occupant injury measures and vehicle deformation patterns are examined. The Insurance Institute for Highway Safety conducted a series of car-to-utility-vehicle frontal offset tests with the Ford Taurus as a common collision partner. To vary stiffness, the Taurus collided with either a Mercedes ML320 or a relatively stiffer Isuzu Rodeo. To vary geometry, the test matrix included an ML320 at normal ride height, a Rodeo at normal ride height, an ML320 lowered 9 cm so the frame rails matched the ride height of the normal Rodeo, and a Rodeo raised 9 cm to match the ride height of the normal ML320. In each test, both vehicles were traveling at 48 km/h (30 mi/h). A Hybrid III 50th percentile male dummy was seated in the driver's seat of each vehicle. Vehicle deformation patterns, dummy injury measures, and vehicle accelerations were recorded and compared. Results indicated that, despite its lesser stiffness, the normally higher ride height of the ML320 produced a less 'compatible' crash than the Rodeo, whose front end is stiffer but normally lower. The benefits of lesser front-end stiffness were smaller and only apparent in comparisons of tests at a common ride height. These results indicate that there is little benefit of reduced utility vehicle stiffness without good geometrical alignment of vehicles' front structures.

INTRODUCTION

In the United States, the popularity of utility vehicles and pickup trucks has been growing over the years. From 1975 to 1997, registration of these vehicles has increased 35 percent (The Polk Company, 1997) and in 1998, they were 25 percent of all registered passenger vehicles. Coupled with their rising popularity, there has been growing concern with the aggressivity of these vehicles in crashes with other vehicles. Per registered vehicle, utility vehicles are four times as likely to kill occupants of the other vehicle in two-vehicle collisions as are cars (Insurance Institute for Highway Safety, 1998).

Much of this excess aggressiveness is due to increased mass—more than half of the utility vehicles and pickup trucks sold in 1999 (22 percent of total registered vehicles) weighed more than 1,815 kg (4,001 lb) while 87 percent of all cars sold in the same year (54 percent of total registered vehicles) weighed less than 1,590 kg (3,505 lb) (Lund et al., 2000). As shown in Figure 1, for cars, utility vehicles, and pickup trucks, the likelihood of deaths in other vehicles in two-vehicle collisions increases with their mass (Lund and Chapline, 1999). Thus, based solely on their greater average mass, utility vehicles would be expected to be more aggressive than cars in multiple-vehicle crashes.

Mass is clearly not the only factor. As figure 1 shows, at every weight category, utility vehicles and pickups account for more crash deaths in other vehicles than do cars (per registered utility vehicle, pickup, or car). Two other factors considered important in the incompatibility of vehicle-to-vehicle crashes are front-end stiffness and geometry. Some researchers believe stiffness to be the key factor (Klaner et al., 1998; Steyer et al, 1998). Conceptual models relating mass, stiffness, and front-end crush distance have been developed suggesting the need to soften the front ends of heavier vehicles and increase the stiffness of lighter ones (Zeidler et al., 1999; Zobel, 1999).

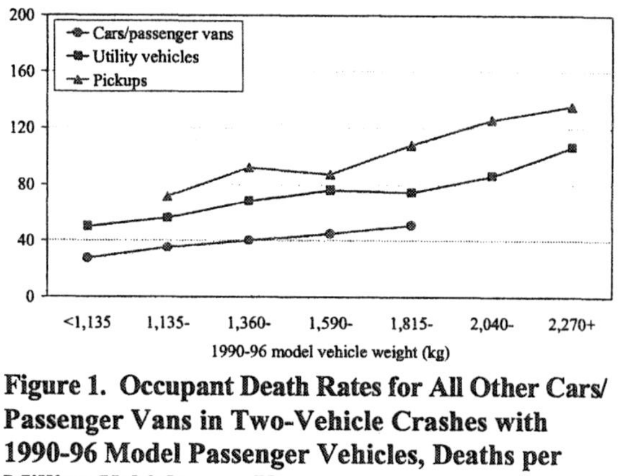

Figure 1. Occupant Death Rates for All Other Cars/ Passenger Vans in Two-Vehicle Crashes with 1990-96 Model Passenger Vehicles, Deaths per Million Vehicles per Year

Other researchers have argued that geometric effects may be even more important than stiffness (Faerber et al., 1998; Wykes et al., 1998). Research appears to support this position, at least with respect to side impacts. Early research at Transport Research Laboratory (Hobbs, 1989) and Renault (Provensal, 1981) suggested that the pattern of intrusion in side impacts was more important than the extent of intrusion in determining whether occupants would be injured. Front-end geometry, or the location of stiff structural elements, is more important in determining pattern of intrusion than is stiffness, which tends to affect the extent of intrusion.

This early research was confirmed in a series of side impact tests conducted by the Insurance Institute for Highway Safety (IIHS). In this series, a large American car, the Mercury Grand Marquis, was struck in the side by a Lincoln Town car and several Ford F150 4x2 or 4x4 pickups that varied in mass, stiffness, and ride height (geometry). The results confirmed that mass, stiffness, and geometry each affected overall physical damage to the struck vehicle, but the effect of vehicle stiffness on injury measures clearly depended on front-end geometry. As a result, the stiffest striking vehicle, the Ford F150 4x4, did not cause the highest risk of occupant injury. These results suggest that, while vehicle stiffness may be an important issue in crash incompatibility, front-end geometry can greatly affect the aggressiveness of increased stiffness in side impacts.

The results for side impacts may not be applicable to frontal impacts. In side impacts, vehicle front ends are inherently much stiffer than the side structures they strike. The relative subordination of stiffness effects in side impacts may reflect this fact. Therefore, another series of tests was conducted to assess the importance of front-end stiffness and geometry in frontal impacts. In this research, a passenger car is involved in head-on offset collisions with utility vehicles that vary in front-end stiffness and geometry (location of stiff structural elements).

TEST METHODOLOGY

A series of four tests was conducted to evaluate the effects of stiffness and geometry of a utility vehicle when crashed into a passenger car. In each test, the utility vehicle, either a 1995-96 Isuzu Rodeo or a 1998-99 Mercedes ML320, was crashed into a common partner, a 1996 Ford Taurus, with an offset equal to 50 percent of the Taurus front-end width. The masses of both utility vehicles were kept constant and equal to each other. The mass of the Taurus was also kept constant.

Front-End Stiffness

The ML320 was touted by the manufacturer as the "first SUV designed with 'crash compatibility' with smaller cars and other SUVs in mind" (Mercedes-Benz, 1998). Examination of load cell barrier data from the National Highway Traffic Safety Administration's New Car Assessment Program (NCAP), as well as testing conducted at Mercedes, indicated that the ML320 was indeed much less stiff than the Rodeo in the full-frontal rigid barrier crash test. Vehicle stiffness was assessed from force deflection curves derived from NCAP tests of these vehicles. In these flat, rigid barrier frontal crash tests, the barrier is fitted with load cells that measure the force exerted in 6 (in some cases 36) zones over time. Forces measured in the different zones were summed to obtain the total force at regular intervals during the crash. In addition, the vehicle's longitudinal acceleration measured at the B-pillars was integrated twice to estimate forward displacement of the vehicle's center of gravity at regular intervals after the crash began. Figure 2 gives a comparison of the force deflection curves for the ML320 (test conducted by Mercedes in 1995), the 1995 Isuzu Rodeo, and the 1996 Ford Taurus. It indicates that the Rodeo is considerably stiffer than the ML320, which in turn is stiffer than the Taurus. In 1998, Mercedes incorporated a structural change to strengthen the occupant compartment of the ML320. The ML320s that were tested by IIHS had these changes incorporated, while the vehicle tested in the NCAP-style rigid barrier test did not. To quantify the effects of the structural modifications, acceleration versus displacement plots from frontal offset deformable barrier tests, conducted by Mercedes, were compared (Figure 3). The figure shows that, due to structural changes, there was an increase in stiffness of the ML320 late in the crash event when the occupant compartment was beginning to be engaged. Because the ML320 occupant compartment was not as deformed in crashes with the Taurus as in the 64.4 km/h offset deformable barrier test, it was deemed appropriate to consider both the pre- and post-modification vehicles as having the same stiffness.

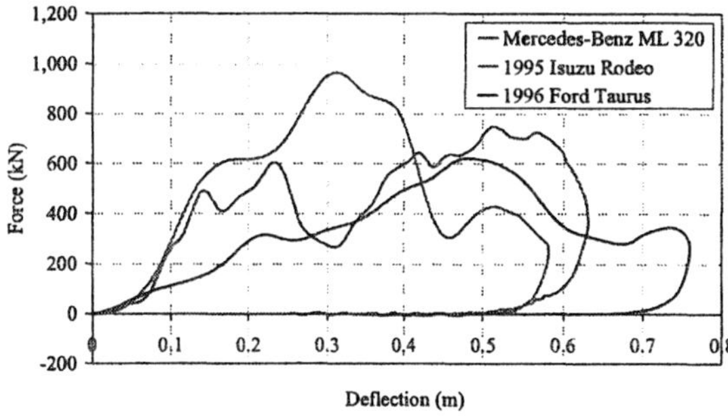

Figure 2. Force-Deflection Characteristics Frontal Compatibility Test Vehicles From NCAP Vehicle Acceleration and Load Cell Barrier Data

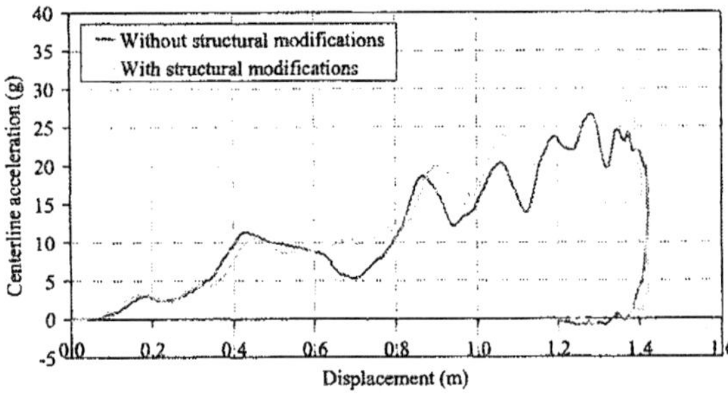

Figure 3. Comparison of Vehicle Acceleration of ML320 in Offset Deformable Barrier Tests Before and After Structural Modifications

Front-End Geometry

Although the side-view profiles of the two utility vehicles are very similar, the heights of the structural elements of the two vehicles are considerably different (Figure 4). The frame rails of the Rodeo are 9 cm lower than the frame rails of the ML320. The bumper element of the Rodeo is attached to the frame rails by weak brackets, thus suggesting that the primary load path to the frame rails is not through the bumper. Although the Rodeo appears more aggressive due to the location of this bumper element, its structural elements align more closely with those of the Ford Taurus. The horizontal centerline of the frame rails of the Rodeo is aligned with the lower edge of the Taurus front longitudinals, while the centerline of the ML320 frame rails is aligned with the upper edge of the Taurus front longitudinals. The first two crash tests were conducted with the utility vehicles in their normal configuration. To evaluate the effect of geometry, two additional tests were conducted; one with the ML320 lowered 9 cm to match the frame rail height of the Rodeo and one with the Rodeo raised 9 cm to match the height of the ML320. The test matrix is shown in Table 1.

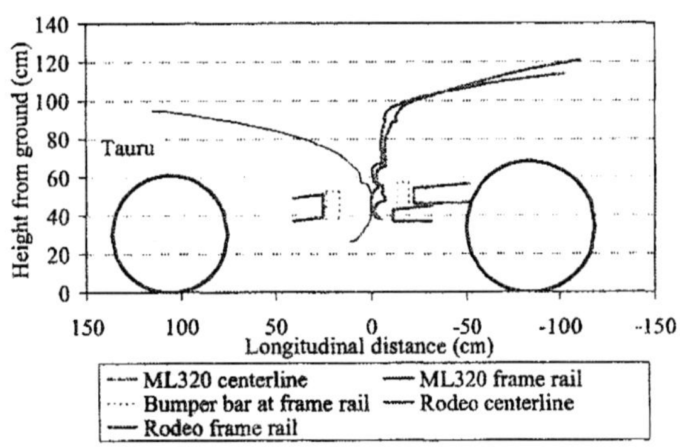

Figure 4. Comparison of Front Contours of the 1996 Ford Taurus to the 1999 Mercedes ML320 and 1995 Isuzu Rodeo

Table 1. Test Matrix

	Low Ride Height	High Ride Height
Lesser Stiffness	Lowered ML 320	Normal ML 320
Greater Stiffness	Normal Rodeo	Raised Rodeo

TEST PROCEDURES

The frontal offset tests were conducted with both vehicles moving at 48-49 km/h (30 mi/h), resulting in a 96 km/h closing speed. The frontal overlap was on the left side of the vehicles and was 49-50 percent for the Taurus, 55 percent for the Rodeo, and 50-51 percent for the ML320. The 50 percent offset for the Taurus was chosen to best replicate the loading condition seen in the 40 percent frontal offset deformable barrier test. Due to concerns about overwhelming the Taurus front structure, a 96 km/h closing speed was chosen for these tests, resulting in a slightly lower delta-V for the Taurus than seen in the 64 km/h offset deformable barrier test.

The mass of the Taurus was kept constant at 1660 kg (3660 lbs) and the masses of both utility vehicles were kept constant at 2180 kg (4806 lbs), which was the test weight of the first ML320 tested. The ML320 weighs approximately 164 kg more than its measured curb weight and the Rodeo weighs approximately 368 kg more than its measured curb weight. To achieve these weights, nonstructural components were removed from the rear of the vehicle or ballast was added in the floorpan of the passenger compartment.

To lower the ML320, the front torsion bars were let out all the way and the front shocks, which were already fully compressed, were unbolted at the lower end. In the rear, portions of the suspension springs were removed and the rear tires were replaced with smaller tires (from size 225/65 R16 to 225/555 R16). To raise the Rodeo, the torsion bars in the front of the vehicle were adjusted and the spring shackles in the rear were extended.

A Denton IDDAS was used for all dummy data acquisition. During the crash, all measurements were recorded at a sample rate of 10 kHz in accordance with SAE Recommended Practice J211/1 – Instrumentation for Impact Test – Part 1, Electronic Instrumentation MAR95. All filtering and subsequent calculations were executed using DSP Development Corporation's DADiSP Ver. 4.1 NI NK B07 (DSP Development Corporation, 1997). In addition to summary metrics for each of the recorded data channels, the vector resultant of the head acceleration, the head injury criterion (HIC), vector bending in legs, and tibia indices were calculated. All calculations comply with SAE Information Report J1727 – Injury Calculation Guidelines AUG96 (SAE, 1998).

RESULTS

Front-End Geometry

The height of the Rodeo frame rails is lower than that of most utility vehicles, including the ML320, and slightly lower than the frame rails of the Ford Taurus (Figure 5). The effect of this height difference is apparent upon examination of the structural crush of the Taurus after impact. In the test with the normal Rodeo, the Rodeo completely engages the frame rails of the Taurus, while in the test of the normal ML320, the front longitudinals of the Taurus are completely overridden. When the ML320 is lowered to the height of the normal Rodeo, the crash deformation of the Taurus is similar to that caused by the normal Rodeo. When the Rodeo is raised to the height of the normal ML320, the crash deformation to the Taurus is similar to that caused by the normal ML320 (Figures 6a-d and Figure 7).

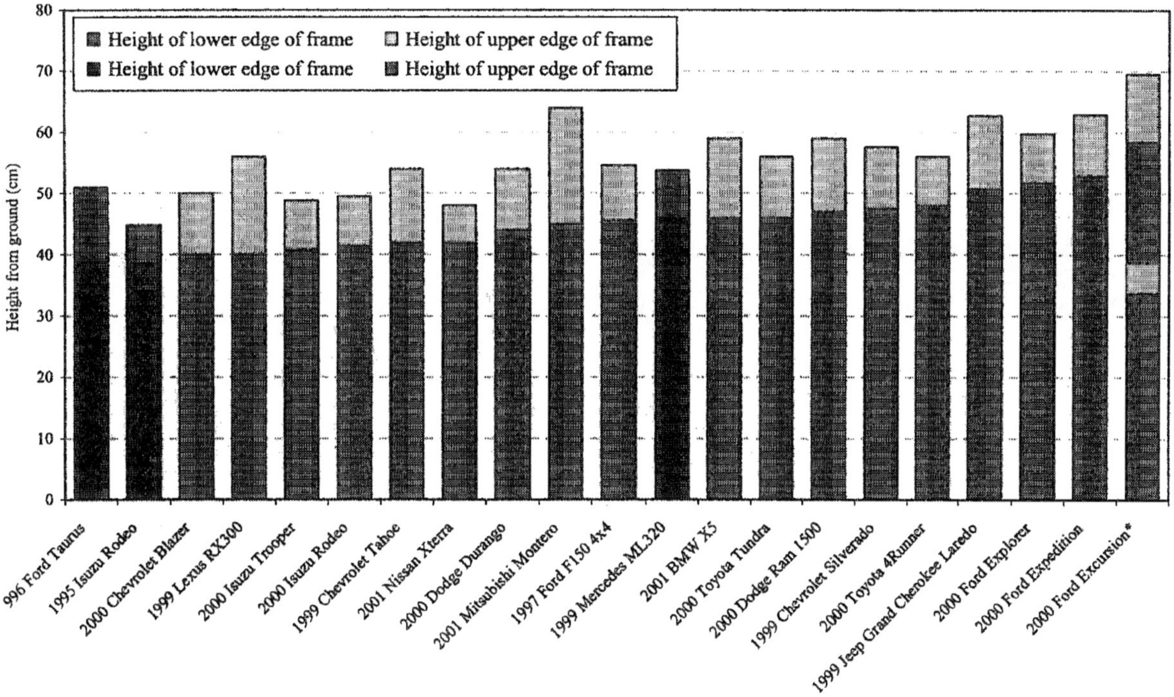

Figure 5. Comparison of Heights of SUV Frame Rails and Blocker Bars
*Lower orange section is BlockerBeam™, upper orange section is frame rail

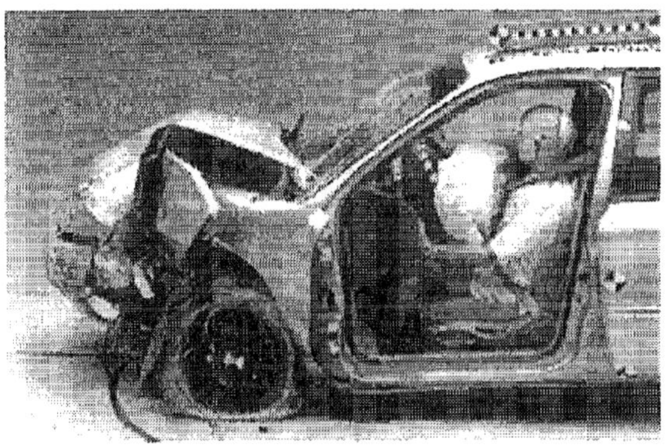

Figure 6a. Crush caused by Normal Rodeo

Figure 6b. Crush caused by Normal ML320

Figure 6c. Crush caused by Raised Rodeo

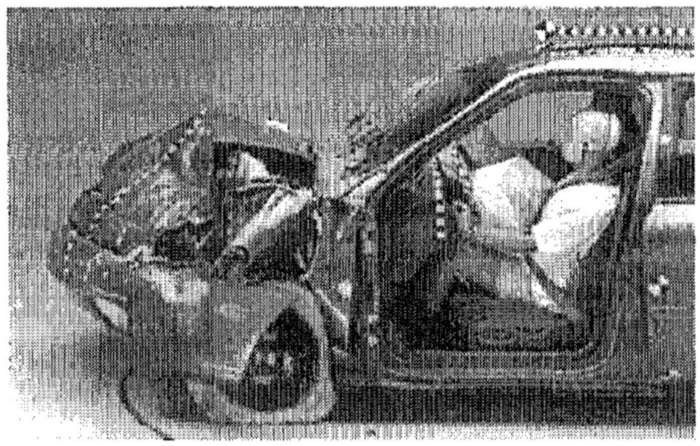

Figure 6d. Crush caused by Lowered ML320

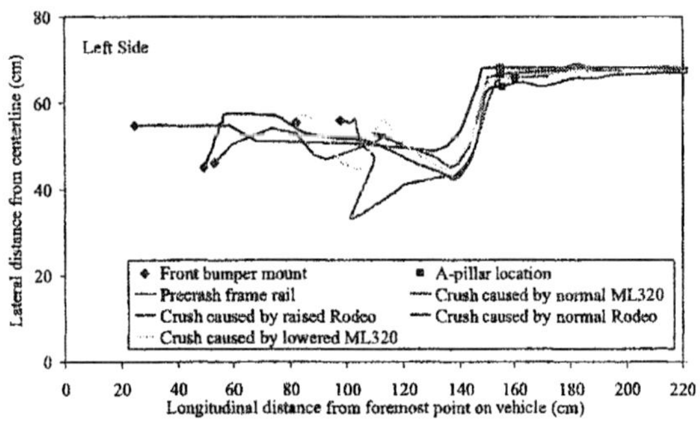

Figure 7. Underbody View of Taurus Left Frame Rail

The effects of the height differences can also be seen in the intrusion measures (Figure 8, Table 2). The normal Rodeo (low and stiff) causes the most intrusion in the Taurus footwell and some of the least intrusion at the instrument panel level (knee bolster, steering column, door aperture closure). When the Rodeo is raised and overrides the Taurus front longitudinal, the footwell intrusion decreases and there is an increase in intrusion at the instrument panel level. The same pattern exists, to a much lesser extent, for the ML320 crashes. The normal ML320 (high and soft) caused more intrusion at the instrument panel level than the lowered ML320.

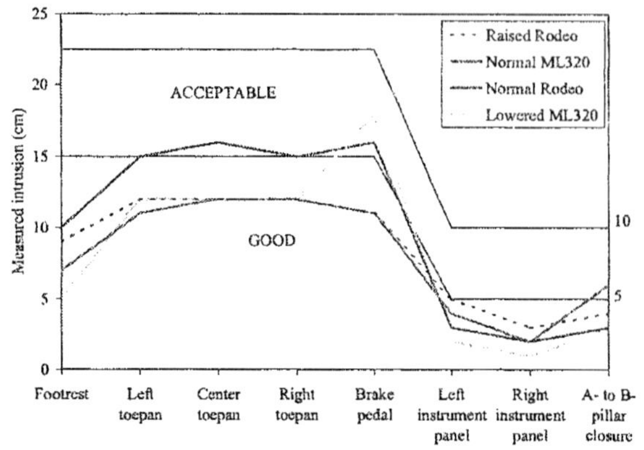

Figure 8. Intrusion Results - 1996 Ford Taurus

Table 2.
Upper Vehicle Intrusion in Ford Taurus

Vehicle Struck	Steering Column Rearward Movement (cm)	Steering Column Upward Movement (cm)	Door Aperture Closure (cm)
Normal Rodeo	2	8	3
Lowered ML320	0	5	3
Raised Rodeo	6	6	4
Normal ML320	3	6	6

Injury Values

Injury results for the Taurus are shown in Figures 9 and 10. Injury measures for the dummies in all vehicles tested were below injury assessment reference values. In normal conditions, the ML320 causes greater upper body (head, neck, chest, and femur) injury measures than the Rodeo. At the same ride height, the upper body injury values are similar for both vehicles, although the ML320 generally produced slightly greater injury measures than the Rodeo.

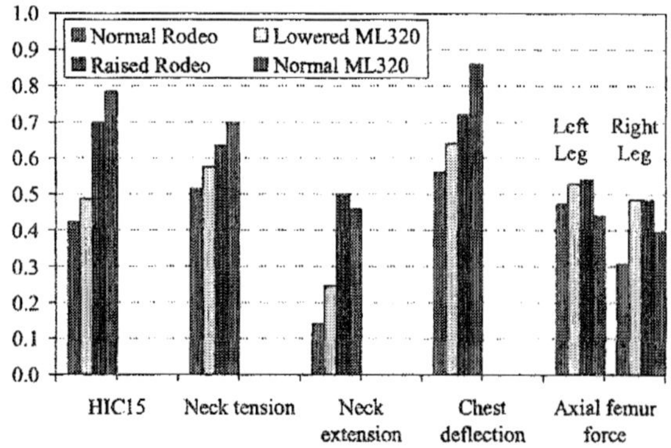

Figure 9. Frontal Compatibility Testing 1996 Ford Taurus - Upper Body Injury Measures

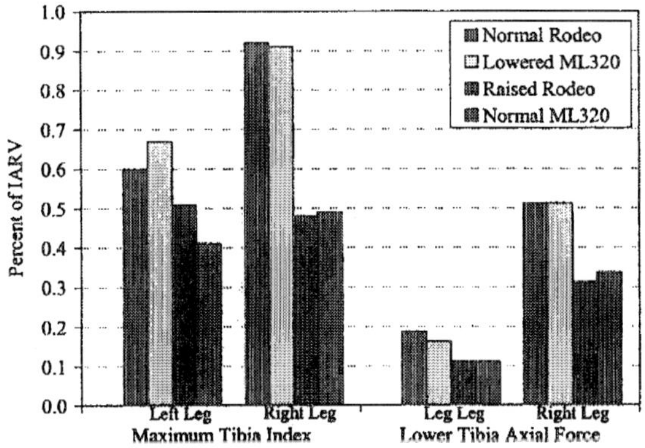

Figure 10. Frontal Compatibility Testing 1996 Ford Taurus - Lower Body Injury Measures

The results of the lower leg injuries in the Taurus indicate the opposite trend. The lower leg injury risk (maximum tibia index and tibia axial force) is greater in the low-ride height tests and is less in the high-ride height tests when the front longitudinal was overridden.

Front-End Stiffness

Initial assessments of the Rodeo and ML320 stiffnesses were confirmed by observations of bumper deformation in these tests. Figure 11 shows a top view of the bumper deformation of the utility vehicles. When compared at the same ride heights, the ML320 exhibited more crush than the Rodeo.

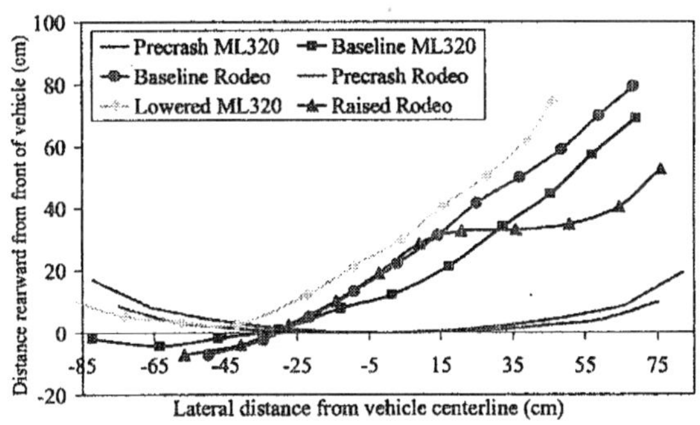

Figure 11. Bumper Crush Comparison of the 1999 ML320 and 1996 Isuzu Rodeo (Both Crashed into a 1996 Ford Taurus)

Results from the frame rail crush sustained by the Taurus (Figure 7) show similar results. The crush caused by the lowered ML320 is considerably less than the crush caused by the normal Rodeo. The crush caused by the vehicles at the raised ride heights cannot be compared, since the frame rails were overridden.

DISCUSSION

Results from this testing show that when the masses of vehicles are equal, frontal geometry, rather than stiffness, plays the greater role in vehicle compatibility. Although the ML320 is considerably less stiff than the Rodeo, this reduced the damage to the Ford Taurus only when the stiff structures of the cars and utility vehicles were geometrically matched. When the utility vehicles were in the higher configuration, the normal ML320 caused just as much upper vehicle intrusion to the Taurus as the stiffer Rodeo. More importantly, injury measures were clearly dependent on the ride height of the striking utility vehicle, while there was no consistent effect of vehicle stiffness on either upper or lower body injury risk.

Clearly the measured differences in Taurus intrusion patterns and injury measures are the result of the front longitudinal override that occurs with the high-ride height vehicles. The normal ML320 and raised Rodeo override the frame rail of the Taurus and therefore the crush is located higher on the Taurus than that caused by the lowered ML320 and normal Rodeo. The upper body injury measures, steering column intrusion, and door closure also were greater due to crush from these higher vehicles. Interestingly, the measures of upper body injury and door closure caused by the ML320 are slightly greater than those caused by the stiffer raised Rodeo.

In contrast, the normal Rodeo and lowered ML320 cause greater lower leg injury risk in the Taurus, because the frame rails of the utility vehicles interact with the frame rails of the Taurus and cause slightly greater intrusion in the toepan area. This result was also noted by Zobel.

At this crash energy level, the mismatch in frame rail and front longitudinal heights resulted only in slightly greater intrusion and injury risk. Despite good injury results for the Taurus in all tests, the exterior deformation suggests a problem. In the crashes where there was good interaction of frame rails, the Taurus had more structure available to absorb energy than in those crashes with poor frame rail interaction. This means the Taurus could absorb more energy in crashes with good frame rail interaction before catastrophic failure occurs.

CONCLUSION

When the ML320 debuted, Mercedes touted it as "the first SUV designed with 'crash compatibility' with smaller cars and other SUVs in mind," citing the height of the structure as an important factor. These test results confirm the importance of height, although Mercedes' marketing claim is placed in doubt. The height of the ML320 frame rails was found to be average compared with other utility vehicles. Since the differences in crash results due to geometry (frame rail/front longitudinal mismatch) are so dramatic, this would imply that most utility vehicles on the market would act similarly to the ML320 and completely override the Taurus under the same test conditions. Interestingly, when Isuzu changed the design of the new Rodeo in the 1998 model year, the frame rail height was raised several centimeters.

The results from side impact testing found that front-end geometry of the striking vehicle had a significant role in how the struck vehicle crushed and that the effects of stiffness were secondary to the geometry. These results, in conjunction with the offset frontal impact testing results discussed here, greatly strengthen the argument that the frontal geometry of vehicles needs to be equalized before any potential benefits from decreasing the stiffness of larger vehicles can be realized.

ACKNOWLEDGMENT

This work was supported by the Insurance Institute for Highway Safety.

REFERENCES

Faerber, E. et al. 1998. Improvement of crash compatibility between cars. *Proceedings of the 16th International Technical Conference on the Enhanced Safety of Vehicles*, 650-61. Washington, DC: National Highway Traffic Safety Administration.

Hobbs, C.A. 1989. The influence of car structures and padding on side impact injuries. *Proceedings of the 12th International Technical Conference on the Experimental Safety Vehicles*, 954-62. Washington, DC: National Highway Traffic Safety Administration.

Insurance Institute for Highway Safety. 1998. *Status Report Special Issue: Crash Compatibility*. 33(1). Arlington, VA. Available: http://www.highwaysafety.org/srpdfs/sr3301.pdf.

Klaner, W.; Felsch, B.; and van West, F. 1998. Evaluation of occupant protection and compatibility out of frontal crash tests against the deformable barrier. *Proceedings of the 16th International Technical Conference on the Enhanced Safety of Vehicles*, 693-702. Washington, DC: National Highway Traffic Safety Administration.

Lund, A.K. and Chapline, J.F. 1999. Potential strategies for improving crash compatibility in the U.S. vehicle fleet (SAE 1999-01-0066). Vehicle Aggressivity and Compatibility in Automotive Crashes (SP-1442), 33-44. Warrendale, PA: Society of Automotive Engineers

Lund, A.K.; O'Neill, B.; Nolan, J.M.; and Chapline, J.F. 2000. Crash compatibility issue in perspective (SAE 2000-01-1378). *Vehicle Aggressivity and Compatibility in Automotive Crashes 2000* (SP-1525). Warrendale, PA: Society of Automotive Engineers.

Mercedes-Benz of North America, Inc. 1998. 1999 Mercedes-Benz brochure. Montvale, NJ.

Provensal, J. and Stcherbatcheff, G. 1981. Identification of compatibility factors in side collisions. *Proceedings of the 8th International Technical Conference on the Experimental Safety Vehicles*, 587-99. Washington, DC: National Highway Traffic Safety Administration.

Steyer, C.; Delhommeau, M.; and Delannoy, P. 1998. Proposal to improve compatibility in head-on collisions. *Proceedings of the 16th International Technical Conference on the Enhanced Safety of Vehicles*, 682-92. Washington, DC: National Highway Traffic Safety Administration.

The Polk Company. 1997. National vehicle population profile 1997. Detroit, MI.

Wykes, N.; Edwards, M.; and Hobbs, A. 1998. Compatibility requirements for cars in frontal and side impacts. *Proceedings of the 16th International Technical Conference on the Enhanced Safety of Vehicles*, 667-81. Washington, DC: National Highway Traffic Safety Administration.

Zeidler, F.; Knöchelmann, F.; and Scheunert, D. 1999. Possibilities and limits in the design of compatible cars for real world accidents. SAE Technical Paper Series 1999-01-0068. Warrendale, PA: Society of Automotive Engineers.

Zobel, R. 1999. Accident analysis and measures to establish compatibility. SAE Technical Paper Series 1999-01-0065. Warrendale, PA: Society of Automotive Engineers.

2001-06-0109

FRONTAL OFFSET DEFORMABLE BARRIER CRASH TESTING AND ITS EFFECT ON VEHICLE STIFFNESS

Joseph M. Nolan
Adrian K. Lund
Insurance Institute for Highway Safety
United States of America
Paper No. 487

ABSTRACT

Since 1995, the Insurance Institute for Highway Safety (IIHS) has evaluated the crashworthiness of more than 120 new vehicle models in a 64 km/h (40 mi/h), 40 percent offset deformable barrier crash test. The offset test is especially demanding of the vehicle structure, requiring only 40 percent of the vehicle width to manage the crash energy. Many of the models originally tested have been redesigned and retested, with the majority producing better structural performance than their predecessors. Critics of such testing have suggested that these tests are forcing vehicle stiffness too high for compatibility with other vehicles and other crash modes. IIHS has studied the relationship between vehicle mass, stiffness, and front-end length to the structural rating in the offset test. IIHS then studied vehicle accelerations, deformation, and interior intrusion for eight pairs of vehicles whose structural performance changed in the offset test after redesign for evidence of front-end stiffness changes. The data indicate that there were no significant correlations between mass, front-end length, and stiffness to structural performance in the offset test. The data also indicate that for models that sustained catastrophic collapse of the vehicle structure in the offset test, an increase in overall stiffness was required for better structural performance. The majority of vehicles whose structural performance improved did so without significant alteration to the stiffness of the vehicle for the first half-meter of deformation. These vehicles have maintained essentially the same stiffness in the front crush zone but have rapidly increasing stiffness as the deformation approaches the occupant compartment.

INTRODUCTION

Frontal offset crash testing has gained acceptance worldwide as an assessment of the frontal crashworthiness of vehicles. In the offset test, 40 percent of the vehicle's front end strikes a deformable barrier mounted to a rigid barrier at 64 km/h (40 mi/h). The deformable barrier is intended to initiate deformation patterns seen in real car-to-car, 50 percent frontal offset impacts (Lund et al., 1995). Because the offset test forces a smaller portion of the vehicle's front end to manage crash energy, there is more localized deformation on the struck vehicle than seen in flat barrier tests such as the National Highway Traffic Safety Administration's New Car Assessment Program (NCAP).

To achieve good performance in the offset test it is necessary to minimize intrusion into the occupant compartment by absorbing the crash energy in the crush zone forward of the compartment. If a vehicle's occupant compartment is weaker than the front crush zone, the occupant compartment will be compromised before full use of the available energy-absorption capacity of the front crush zone has occurred. In this case, front-end stiffness needs to be decreased and occupant compartment stiffness increased. It is imperative to design vehicles with sufficient energy-absorption capacity in the crush zone and with an occupant compartment that can resist the forces of front-end crush without yielding. Zobel (1999) calls this strategy for occupant protection 'the bulkhead concept.'

Some are concerned that the bulkhead concept will drive vehicle front-end stiffness too high for compatibility with other vehicles (Zobel, 1999). This can be shown to be true when considering only simple models of vehicles in crashes. However, there are other possibilities to improve offset structural performance without increasing overall stiffness. In cases where occupant compartment failure occurs before full energy absorption of the front structure has been realized, a decrease in front-end stiffness is required. The front end also must be designed carefully to ensure that stiff front-end components such as engine cradles and longitudinal rails do not unnecessarily impinge on the occupant compartment before the crash energy has been absorbed.

Although not the focus of this paper, it should be noted that the effect of stiffness on injury causation in crashes remains unclear, particularly in side impact collisions (Nolan et al., 1999). The majority of research indicates that after controlling for mass, vehicle geometry is the dominant factor for compatibility in both frontal and side collisions (Meyerson and Nolan, 2001).

The Insurance Institute for Highway Safety (IIHS) has conducted more than 120 offset crashes of new vehicles available in the U.S. market since 1995, rating each vehicle's performance for consumer information. An instrumented 50th percentile male Hybrid III dummy is used to evaluate driver injury risk. Separate ratings are assigned to each vehicle in five rating categories: structural performance; head, neck, and chest injury risk; and occupant kinematics.

This paper includes two separate analyses. The first is a statistical analysis of the relationship between vehicle front-end stiffness, as calculated from full-frontal rigid barrier tests, to the structural performance of 80 vehicles in the offset test. Vehicle mass and front-end length also were included in the analysis to control for the separate effects of these physical variables.

The second analysis is an examination of the acceleration and displacement of the occupant compartment measured in the offset test for eight models that were retested for evidence of stiffness changes after a major redesign. The eight models had large changes in structural performance in the offset test after the redesign, with seven scoring better and one worse. These models provided an opportunity to study specific production vehicle changes and the influence of these changes on structural test performance and front-end stiffness.

METHOD

Each vehicle subjected to the IIHS frontal offset test is assigned a structural rating of good, acceptable, marginal, or poor based on its performance in the test (IIHS, 2000a,b). The structural rating is based primarily on the vehicle's ability to maintain its occupant compartment integrity during the crash. Vehicles with minimal intrusion into the occupant compartment are rated good, vehicles with excessive intrusion are rated poor, and the gradations between these extremes are rated acceptable or marginal. The structural rating also can be affected by structural performance not directly related to intrusion such as fuel leakage during the tests, door opening during the test, etc., but such downgrades are uncommon. Occupant compartment intrusion is quantified by measuring the precrash and postcrash coordinates of various parts of the vehicle interior. Intrusion is measured at the footrest, brake pedal, three locations on the toepan, and two locations on the instrument panel. The change in driver A-to-B-pillar distance also is measured, representing the door aperture closure as a result of the impact.

Statistical Analysis

The statistical analysis includes the 120 vehicles rated by IIHS. The data set includes 74 cars, 27 utility vehicles, 14 passenger vans, and 5 pickups. Recorded for each vehicle were its structural rating, test mass, and front-bumper-to-firewall distance (front-end length). Table 1 lists the vehicles included in the analysis by vehicle type. Stiffness values were calculated for 80 vehicles in the data set that were subjected to the NCAP test using the two methods described below.

Table 1
Summary of Test Vehicle Metrics

Model Year	Vehicle	Structural Rating	Mass (kg)	Wheelbase (cm)	Overall Length (cm)	Body Type	Front Bumper to Firewall (cm)
1995	Chevrolet Cavalier	Poor	1,362	264	458	Car	112
1995	Chevrolet Lumina	Good	1,645	273	510	Car	125
1995	Chrysler Cirrus LX	Poor	1,553	274	472	Car	110
1995	Ford Contour	Marginal	1,431	271	467	Car	105
1995	Ford Taurus	Good	1,565	269	488	Car	121
1995	Honda Accord	Acceptable	1,452	272	467	Car	120
1995	Mazda Millenia	Marginal	1,593	275	482	Car	111
1995	Mitsubishi Galant	Poor	1,459	263	475	Car	121
1995	Nissan Maxima	Acceptable	1,507	270	477	Car	118
1995	Saab 900S	Poor	1,489	260	464	Car	118
1995	Subaru Legacy	Acceptable	1,380	263	459	Car	114
1995	Toyota Camry	Acceptable	1,511	262	477	Car	123
1995	Volkswagen Passat	Marginal	1,557	262	461	Car	112
1995	Volvo 850	Acceptable	1,555	266	466	Car	117
1996	Ford Taurus	Good	1,651	276	502	Car	132
1996	Hyundai Sonata	Poor	1,485	270	470	Car	117
1996	Toyota Avalon	Marginal	1,584	272	483	Car	121
1997	BMW 540i	Good	1,876	283	478	Car	129
1997	Cadillac Seville	Poor	1,910	282	519	Car	138
1997	Dodge Neon	Marginal	1,308	264	436	Car	108
1997	Ford Escort	Acceptable	1,294	250	443	Car	108

Model Year	Vehicle	Structural Rating	Mass (kg)	Wheelbase (cm)	Overall Length (cm)	Body Type	Front Bumper to Firewall (cm)
1997	Honda Civic	Acceptable	1,244	262	445	Car	100
1997	Hyundai Elantra	Acceptable	1,355	255	442	Car	104
1997	Infiniti Q45	Acceptable	1,944	283	506	Car	134
1997	Infiniti Q45	Acceptable	1,944	283	506	Car	134
1997	Kia Sephia	Poor	1,316	250	436	Car	107
1997	Lexus LS 400	Good	1,886	285	500	Car	120
1997	Lincoln Continental	Acceptable	1,914	277	524	Car	137
1997	Mazda Protege	Acceptable	1,272	261	444	Car	104
1997	Mercedes E420	Good	1,802	283	481	Car	123
1997	Mitsubishi Mirage	Marginal	1,231	250	442	Car	109
1997	Pontiac Grand Prix	Acceptable	1,711	281	499	Car	133
1997	Saturn SL	Acceptable	1,240	260	449	Car	107
1997	Toyota Camry	Good	1,558	267	478	Car	117
1997	Volkswagen Jetta	Marginal	1,365	247	440	Car	106
1998	Honda Accord	Acceptable	1,526	272	480	Car	124
1998	Nissan Maxima	Acceptable	1,550	270	481	Car	121
1998	Nissan Sentra	Marginal	1,281	254	434	Car	103
1998	Toyota Avalon	Acceptable	1,680	272	488	Car	124
1998	Toyota Corolla	Acceptable	1,284	246	442	Car	109
1998	Volkswagen Beetle	Good	1,377	251	409	Car	105
1998	Volkswagen Passat	Good	1,576	271	468	Car	113
1999	Audi A6	Acceptable	1,822	276	488	Car	100
1999	Buick Park Avenue	Good	1,840	289	525	Car	122
1999	Cadillac Catera	Acceptable	1,842	273	493	Car	136
1999	Chevrolet Malibu	Acceptable	1,500	272	484	Car	126
1999	Chrysler LHS	Marginal	1,737	287	528	Car	123
1999	Daewoo Leganza	Poor	1,574	267	467	Car	119
1999	Hyundai Sonata	Marginal	1,544	270	471	Car	122
1999	Kia Sephia	Marginal	1,308	256	443	Car	108
1999	Lexus GS 400	Good	1,824	280	480	Car	122
1999	Mazda 626	Acceptable	1,416	267	474	Car	124
1999	Mazda Protege	Acceptable	1,302	261	442	Car	110
1999	Mitsubishi Galant	Acceptable	1,511	264	477	Car	122
1999	Pontiac Grand Am	Marginal	1,513	272	474	Car	128
1999	Saab 9-3	Acceptable	1,567	260	463	Car	125
1999	Saab 9-5	Acceptable	1,734	270	480	Car	117
1999	Volkswagen Jetta	Acceptable	1,455	251	438	Car	110
2000	BMW 328i	Good	1,620	272	447	Car	116
2000	Buick LeSabre	Good	1,728	285	508	Car	128
2000	Cadillac Seville	Good	1,916	285	510	Car	138
2000	Chevrolet Impala	Good	1,676	281	508	Car	134
2000	Dodge Intrepid	Acceptable	1,692	287	517	Car	124
2000	Dodge Neon	Marginal	1,319	267	443	Car	110
2000	Ford Taurus	Good	1,622	276	502	Car	127
2000	Lincoln LS	Good	1,837	291	492	Car	126
2000	Nissan Altima	Marginal	1,512	262	472	Car	116
2000	Nissan Maxima	Acceptable	1,624	275	484	Car	118
2000	Nissan Sentra	Acceptable	1,324	254	450	Car	108
2000	Saturn LS	Acceptable	1,565	270	484	Car	122
2000	Subaru Legacy	Good	1,600	265	468	Car	115

Model Year	Vehicle	Structural Rating	Mass (kg)	Wheelbase (cm)	Overall Length (cm)	Body Type	Front Bumper to Firewall (cm)
2000	Toyota Avalon	Good	1,677	272	488	Car	118
2000	Volvo S80	Good	1,715	279	482	Car	116
1996	Chevrolet Astro	Poor	2,131	282	482	Passenger van	74
1996	Dodge Grand Caravan	Acceptable	2,002	303	507	Passenger van	106
1996	Ford Aerostar	Poor	1,815	302	444	Passenger van	86
1996	Ford Windstar	Good	1,920	307	511	Passenger van	123
1996	Honda Odyssey	Poor	1,702	283	475	Passenger van	118
1996	Mazda MPV	Marginal	1,852	280	466	Passenger van	118
1996	Nissan Quest	Acceptable	1,862	285	482	Passenger van	119
1996	Toyota Previa	Marginal	1,874	287	475	Passenger van	89
1997	Pontiac Trans Sport	Poor	1,852	284	476	Passenger van	114
1998	Toyota Sienna	Good	1,928	290	492	Passenger van	120
1999	Ford Windstar	Acceptable	1,992	307	511	Passenger van	121
1999	Honda Odyssey	Acceptable	2,078	300	511	Passenger van	117
1999	Nissan Quest	Poor	1,936	285	495	Passenger van	120
2000	Mazda MPV	Acceptable	1,775	284	475	Passenger van	112
1998	Chevrolet S-10	Marginal	1,569	275	480	Pickup	120
1998	Dodge Dakota	Marginal	1,758	284	497	Pickup	119
1998	Ford Ranger	Marginal	1,584	283	476	Pickup	121
1998	Nissan Frontier	Acceptable	1,518	265	468	Pickup	111
1998	Toyota Tacoma	Acceptable	1,380	262	469	Pickup	109
1996	Chevrolet Blazer	Poor	2,013	272	461	Utility vehicle	124
1996	Ford Explorer	Acceptable	2,048	283	479	Utility vehicle	110
1996	Isuzu Rodeo	Poor	2,030	276	468	Utility vehicle	101
1996	Jeep Grand Cherokee	Acceptable	1,850	269	449	Utility vehicle	130
1996	Land Rover Discovery	Acceptable	2,134	254	452	Utility vehicle	127
1996	Mitsubishi Montero	Acceptable	2,162	272	470	Utility vehicle	119
1996	Toyota 4-Runnner	Acceptable	1,924	268	454	Utility vehicle	107
1997	Nissan Pathfinder	Poor	2,060	270	453	Utility vehicle	106
1998	Honda CRV	Acceptable	1,577	262	451	Utility vehicle	108
1998	Isuzu Amigo	Poor	1,758	246	427	Utility vehicle	108
1998	Jeep Cherokee	Marginal	1,701	258	425	Utility vehicle	106
1998	Jeep Wrangler	Acceptable	1,594	237	386	Utility vehicle	123
1998	Kia Sportage	Marginal	1,652	265	432	Utility vehicle	113
1998	Toyota RAV4	Acceptable	1,498	241	416	Utility vehicle	100
1999	Dodge Durango	Acceptable	2,312	294	491	Utility vehicle	108
1999	Jeep Grand Cherokee	Marginal	1,920	269	461	Utility vehicle	128
1999	Land Rover Discovery	Acceptable	2,314	254	470	Utility vehicle	126
1999	Lexus RX300	Acceptable	1,900	262	458	Utility vehicle	114
1999	Mercedes ML320	Good	2,125	282	459	Utility vehicle	122
1999	Mitsubishi Montero Sport	Marginal	1,990	272	453	Utility vehicle	100
1999	Subaru Forester	Acceptable	1,549	252	445	Utility vehicle	110
1999	Suzuki Grand Vitara	Acceptable	1,593	248	418	Utility vehicle	113
2000	Isuzu Rodeo	Good	1,946	270	451	Utility vehicle	107
2000	Isuzu Trooper	Poor	2,100	276	471	Utility vehicle	113
2000	Nissan Xterra	Good	1,998	265	452	Utility vehicle	108
2001	BMW X5	Good	2,168	282	467	Utility vehicle	126
2001	Mitsubishi Montero	Good	2,264	278	480	Utility vehicle	110

Stiffness Determination Method 1 The crush characteristics of vehicle front structures have typically been modeled as simple mass-spring systems based on the empirical data from full-frontal rigid barrier crash tests (Nash, 1987; Park, 1999; Prasad, 1990a,b). In these tests, the extent of frontal deformation is approximately a linear function of impact speed. This observed linear relationship suggests that vehicle front structure can be modeled as an energy-dissipating linear spring, as shown in Equation 1. Using this model, the initial kinetic energy of a vehicle of mass M moving at a velocity of V can be equated to the energy dissipated by a linear spring with a spring constant K and maximum crush x. This model allows for a simple computation of a front-end linear stiffness value based on data from crash tests.

$$\tfrac{1}{2}MV^2 = \tfrac{1}{2}Kx^2 \tag{1}$$

It should be noted that mass-spring models are not exact for vehicle front-ends in crashes. The model mass that is attached to the spring does not change during the impact, whereas in crash tests the moving mass changes as a function of frontal deformation. The observed linear relationship between impact speed and deformation in full-frontal crash tests is actually a combination of nonlinear stiffness and changing mass. The model approximates these two parameters with a single linear value. Despite their simplicity, mass-spring models continue to be the basis for estimating real-world crash severity (National Highway Traffic Safety Administration, 1981, 1999). These models, when appropriately applied, have been shown to produce reasonable estimates of crash severity for frontal offset deformable barrier crashes (O'Neill et al., 1996).

Values for stiffness were calculated using the NCAP published impact speed, test mass, and maximum dynamic displacement, as determined by double-integration of the rear-seat acceleration recorded in the NCAP test, and inputting the data into Equation 1.

Stiffness Determination Method 2 A second linear stiffness estimate was made for the same 80 vehicles subjected to the NCAP test. When Equation 1 is rewritten in its differential form, it is easily shown that the solution for x and $\dot{x}$ is sinusoidal (see Equation 2). This second method is slightly different from the first method in that it forces the actual velocity-time history to be sinusoidal to fit the model. Method 2 results in lower stiffness estimates than in method 1, because it does not account for the variable moving mass throughout the impact.

$$\tfrac{1}{2}M\dot{x}^2 = \tfrac{1}{2}Kx^2 \tag{2}$$

Solving Equation 2 for velocity yields the following general solution (Equation 3), where A is the initial speed of the vehicle and ω is the natural frequency of the mass spring system. Equation 4 is the natural frequency of the system, expressed in the form of the linear stiffness, K, and mass, M.

$$V = A\,Cos\,(\omega t + \beta) \tag{3}$$

$$\omega = \sqrt{K/M} \tag{4}$$

The second stiffness estimate was calculated by first determining the vehicle longitudinal velocity versus time for the 80 vehicles subjected to the NCAP test. Velocity was determined by integrating the data from longitudinal accelerometers located on the rear-seat crossmember. A quarter-cosine was fit to the data from the time of impact until the time when the vehicle velocity was zero (beginning of rebound). The actual test impact speed was fixed (A), and ω was varied to achieve the best fit using a least-square error criteria. The resulting values for ω were then used in Equation 4 to determine the curve-fit stiffness value. Table 2 lists the two calculated stiffness estimates for the 80 vehicles.

For the 120-vehicle data set (all vehicles), average mass and front-end length values were computed for each of the four structural rating categories (good, acceptable, marginal, poor), and a multiple regression was conducted to assess the combined effects of mass and front-end length.

A similar analysis was conducted using the 80-vehicle subset containing stiffness estimates. Average mass, front-end length, and stiffness values were calculated for each structural rating category, and a weighted regression analysis was conducted. A multivariate regression analysis also was conducted to isolate effects of stiffness from mass and front-end length.

To eliminate any possible problems with combining results from various vehicle types, the above analyses were repeated using cars only. There were 74 cars in the 120-vehicle data set and 47 cars in the 80-vehicle data set.

Table 2
Stiffness for 80 Vehicles Tested in NCAP

Model Year	Make	Model	Structural Rating	Stiffness Energy Method (kN/m)	Stiffness Curve Fit (kN/m)
1995	Chevrolet	Cavalier	Poor	706	658
1995	Chevrolet	Lumina	Good	536	513
1995	Chrysler	Cirrus	Poor	940	881
1995	Ford	Contour	Marginal	937	848
1995	Ford	Taurus	Good	793	707
1995	Honda	Accord	Acceptable	742	696
1995	Mazda	Millenia	Marginal	885	796
1995	Mitsubishi	Galant	Poor	651	619
1995	Nissan	Maxima	Acceptable	758	700
1995	Saab	900	Poor	668	630
1995	Subaru	Legacy	Acceptable	608	566
1995	Toyota	Camry	Acceptable	671	603
1995	Volkswagen	Passat	Marginal	884	813
1996	Chevrolet	Astro	Poor	1,108	1,108
1996	Chevrolet	Blazer	Poor	898	870
1996	Dodge	Grand Caravan	Acceptable	739	707
1996	Ford	Explorer	Acceptable	1,253	,107
1996	Ford	Taurus	Good	780	722
1996	Ford	Windstar	Good	777	689
1996	Honda	Odyssey	Poor	850	766
1996	Hyundai	Sonata	Poor	649	641
1996	Jeep	Grand Cherokee	Acceptable	1,050	989
1996	Land Rover	Discovery	Acceptable	1,015	1,005
1996	Mazda	MPV	Marginal	1,139	1,023
1996	Mitsubishi	Montero	Acceptable	1,715	1,394
1996	Nissan	Quest	Acceptable	976	932
1996	Toyota	4-Runner	Good	1,299	1,113
1996	Toyota	Avalon	Marginal	776	710
1996	Toyota	Previa	Marginal	1,385	1,299
1997	Dodge	Neon	Marginal	606	588
1997	Ford	Escort	Acceptable	688	667
1997	Honda	Civic	Acceptable	563	529
1997	Hyundai	Elantra	Acceptable	737	702
1997	Kia	Sephia	Poor	792	774
1997	Mazda	Protege	Acceptable	700	663
1997	Nissan	Pathfinder	Poor	1,124	979
1997	Pontiac	Grand Prix	Acceptable	645	602
1997	Pontiac	Trans Sport	Poor	784	687
1997	Saturn	SL	Acceptable	408	383
1997	Toyota	Camry	Good	810	713
1997	Volkswagen	Jetta	Marginal	852	809
1998	Chevrolet	S-10	Marginal	777	745
1998	Dodge	Dakota	Marginal	1,147	1,071
1998	Ford	Ranger	Marginal	1,011	806
1998	Honda	Accord	Acceptable	759	702
1998	Honda	CR-V	Acceptable	1,079	1,056
1998	Jeep	Cherokee	Marginal	1,125	1,014
1998	Jeep	Wrangler	Acceptable	861	667
1998	Nissan	Frontier	Acceptable	1,223	1,105

Model Year	Make	Model	Structural Rating	Stiffness Energy Method (kN/m)	Stiffness Curve Fit (kN/m)
1998	Nissan	Maxima	Acceptable	836	779
1998	Nissan	Sentra	Marginal	733	668
1998	Toyota	Avalon	Acceptable	928	851
1998	Toyota	Corolla	Acceptable	692	632
1998	Toyota	RAV4	Acceptable	1,125	1,049
1998	Toyota	Sienna	Good	940	862
1998	Toyota	Tacoma	Acceptable	1,336	1,112
1998	Volkswagen	New Beetle	Good	1,129	1,022
1998	Volkswagen	Passat	Good	929	820
1999	Chevrolet	Malibu	Acceptable	723	654
1999	Chrysler	LHS	Marginal	895	837
1999	Dodge	Durango	Acceptable	1,366	1,267
1999	Ford	Windstar	Acceptable	804	717
1999	Honda	Odyssey	Acceptable	833	762
1999	Jeep	Grand Cherokee	Marginal	1,394	1,308
1999	Mazda	626	Acceptable	754	711
1999	Mazda	Protege	Acceptable	782	748
1999	Mitsubishi	Galant	Acceptable	623	642
1999	Nissan	Quest	Poor	980	892
1999	Pontiac	Grand Am	Marginal	610	545
1999	Subaru	Forester	Acceptable	921	841
1999	Volkswagen	Jetta	Acceptable	1,062	942
2000	Buick	LeSabre	Good	784	710
2000	Chevrolet	Impala	Good	709	636
2000	Dodge	Neon	Marginal	760	717
2000	Ford	Taurus	Good	850	794
2000	Isuzu	Rodeo	Good	1,164	1,077
2000	Mazda	MPV	Acceptable	1,386	1,233
2000	Nissan	Altima	Marginal	654	620
2000	Nissan	Maxima	Acceptable	771	744
2000	Subaru	Legacy	Good	1,077	990

Vehicle Redesign Analysis

Included among the 120 vehicles are 16 models whose structural ratings changed in the offset test after the vehicle structure was redesigned. Of the 16 models retested, 14 of the redesigns scored better structural ratings than did their predecessors. Eight of the 16 redesigns had structural rating changes of two or more rating categories; seven were improvements, and one was a downgrade. These eight vehicles provided the opportunity to study specific examples of the stiffness changes required to obtain significantly better structural ratings in the offset test (or a downgrade in one case). The measured acceleration and calculated displacement in the offset test were compared for these eight models for evidence of stiffness changes. For this analysis, longitudinal vehicle acceleration was measured on the center tunnel in the rear seat, and displacement was calculated by double-integrating this acceleration. Unfortunately load cell barrier data was not available from the NCAP tests both before and after redesign for any of these models.

RESULTS

Statistical Analysis

There was no evident relationship between stiffness (using either method) and the offset structural performance when looking at all vehicles or cars alone (Figures 1 and 2). The effect of stiffness was estimated after controlling for mass and front-end length in multiple regressions, but still no significant correlation to test performance was found.

There also was no significant relationship between mass and test performance when looking at all vehicles or cars alone (Figure 3). However, there was a trend evident in the data toward longer front-end-length vehicles performing better in the offset test (Figure 4). This trend was most apparent (and statistically significant) when all vehicles were included in the analysis, but among cars alone there were some models with longer front ends that performed poorly.

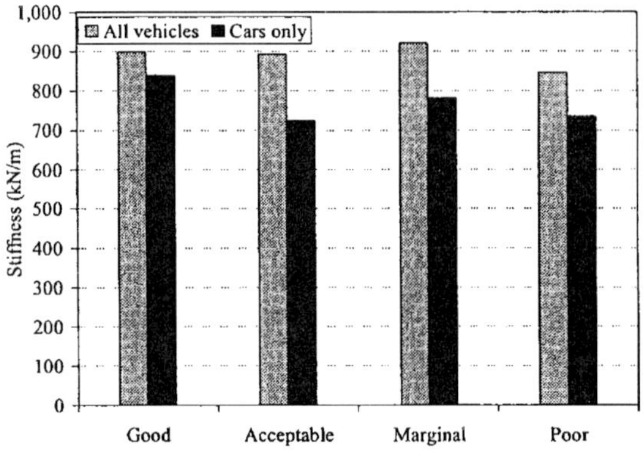

Figure 1. Average Stiffness (Method 1) by Rating Category

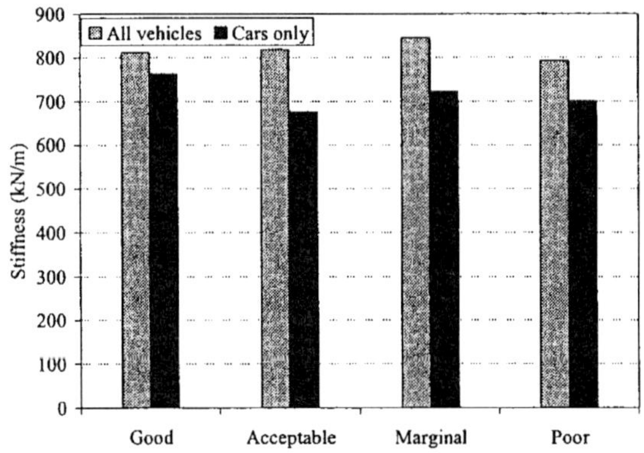

Figure 2. Average Stiffness (Method 2) by Rating Category

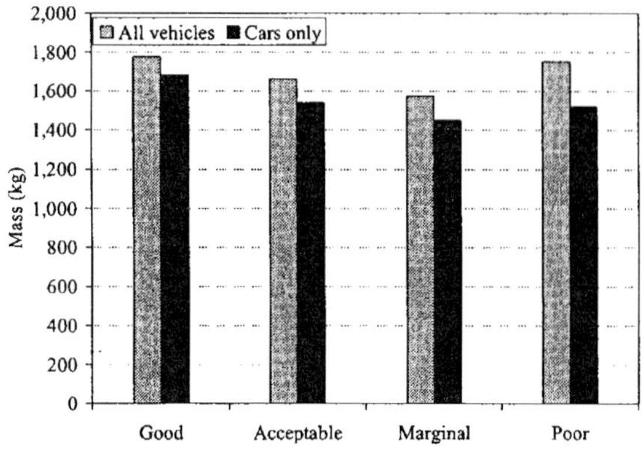

Figure 3. Average Mass by Rating Category

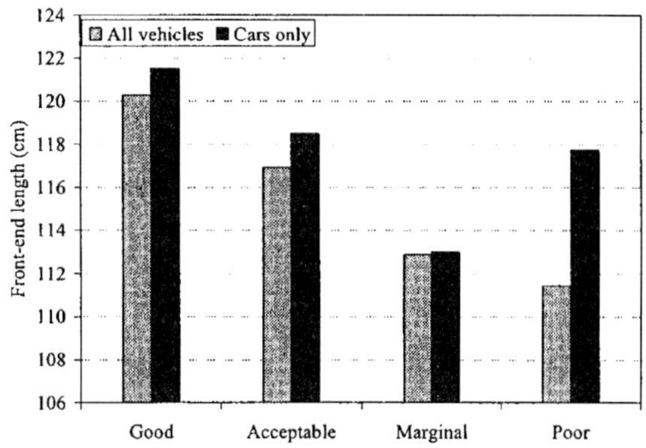

Figure 4. Average Front-End Length by Rating Category

Vehicle Redesign Analysis

Table 3 lists the 16 vehicle models that were tested before and after a major redesign. Listed for each vehicle is the maximum forward movement of the occupant compartment during the test, stiffness measures from the NCAP test, maximum footwell intrusion, and door aperture closure as a result of the crash. The eight models with significant structural rating changes after redesign are shaded. Figures 5-12 show acceleration versus displacement for the eight models that had large differences in structural ratings after redesign.

Table 3.
Vehicles Tested Before and After Redesign

Model Year	Make	Model	Structural Rating	Total Dynamic Crush (including Barrier) (cm)	Maximum Toepan Intrusion (cm)	Door Aperture Closure (cm)	Stiffness Method 1 (kN/m)	Stiffness Method 2 (kN/m)
1997	Cadillac	Seville	Poor	159	37	9	—	—
2000	Cadillac	Seville	Good	145	17	3	—	—
1996	Ford	Windstar	Good	186	18	9	777	689
1999	Ford	Windstar	Acceptable	155	22	3	804	717
1996	Honda	Odyssey	Poor	140	38	9	850	766
1999	Honda	Odyssey	Acceptable	147	22	0	833	762
1996	Hyundai	Sonata	Poor	140	31	17	649	641
1999	Hyundai	Sonata	Marginal	131	25	10	—	
1996	Isuzu	Rodeo	Poor	137	39	16	—	—
2000	Isuzu	Rodeo	Good	129	14	1	1,164	1,077
1996	Mazda	MPV	Marginal	154	35	11	1,139	1,023
2000	Mazda	MPV	Acceptable	129	29	7	1,386	1,233
1995	Mitsubishi	Galant	Poor	138	32	25	651	619
1999	Mitsubishi	Galant	Acceptable	136	25	5	623	642
1996	Nissan	Quest	Acceptable	139	32	11	976	932
1999	Nissan	Quest	Poor	150	40	19	980	892
1998	Nissan	Sentra	Marginal	130	28	11	733	668
2000	Nissan	Sentra	Acceptable	128	20	5	—	—
1995	Saab	900	Poor	140	34	21	668	630
1999	Saab	9-3	Acceptable	141	26	11	—	—
1995	Subaru	Legacy	Acceptable	137	30	8	608	566
2000	Subaru	Legacy	Good	128	18	1	1,077	990
1996	Toyota	Avalon	Marginal	125	26	9	776	710
1998	Toyota	Avalon	Acceptable	139	27	2	928	851
2000	Toyota	Avalon	Good	131	15	5	—	—
1995	Toyota	Camry	Acceptable	131	25	5	671	603
1997	Toyota	Camry	Good	131	12	2	810	713
1996	Toyota	Previa	Marginal	133	29	3	1,385	1,299
1998	Toyota	Sienna	Good	135	10	1	940	862
1997	Volkswagen	Jetta	Marginal	127	34	10	852	809
1999	Volkswagen	Jetta	Acceptable	127	25	3	1,062	942
1995	Volkswagen	Passat	Marginal	141	36	4	884	813
1998	Volkswagen	Passat	Good	147	16	7	929	820

— No measurement was recorded

The Cadillac Seville (Figure 5) went from a poor structural rating for the 1997 model to a good rating for the 2000 model. Both models have similar mass and available crush space, but the 1997 model experienced major collapse of the occupant compartment whereas the redesigned model did not. The accelerations recorded were similar until more than 100 cm of combined vehicle and barrier deformation had occurred (the deformable barrier is 54 cm deep in the offset test and is crushed completely in all tests). There was evidence of increasing acceleration in the 2000 model test as total deformation reached 110 cm. Maximum dynamic deformation was reduced by 14 cm, and intrusion into the toepan was reduced by 20 cm with the redesigned model.

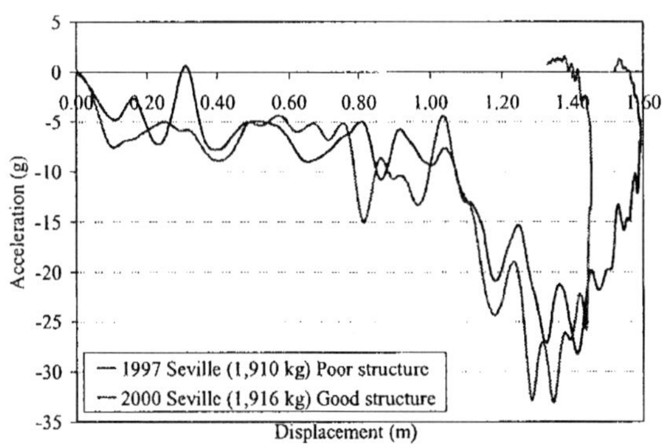

Figure 5. Acceleration vs. Displacement 1997 and 2000 Cadillac Seville

The Honda Odyssey (Figure 6) went from a poor structural rating for the 1996 model to an acceptable rating for the 1999 model. The 1999 model is significantly heavier than its predecessor (2,078 vs. 1,702 kg), but the front crush space is similar for both models. Despite allowing 7 cm more dynamic deformation, the redesigned model resulted in 16 cm less intrusion than its predecessor. Peak accelerations recorded were the same for both models but occurred much later in the redesigned model. Stiffness coefficients calculated from NCAP tests for the 1999 and 1996 models also showed the redesigned model to be less stiff.

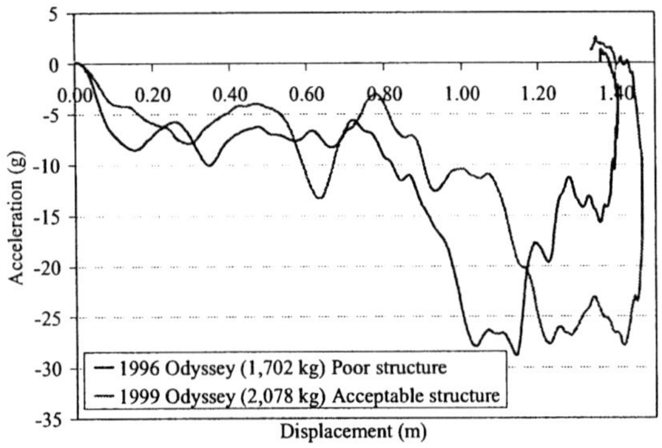

Figure 6. Acceleration vs. Displacement 1996 and 1999 Honda Odyssey

The Isuzu Rodeo (Figure 7) went from a poor structural rating for the 1996 model to a good rating for the 2000 model. The 2000 model has 6 cm additional front crush space and is slightly lighter. The accelerations recorded were similar until 75 cm of combined vehicle and barrier deformation had occurred. Peak accelerations were about the same, but the 2000 model peaked earlier than the 1996 model and had a second high peak at the end of the pulse. Maximum deformation was decreased by 8 cm, and intrusion into the toepan was reduced by 25 cm with the redesigned model. The reduction in intrusion was much greater than reduction in overall deformation.

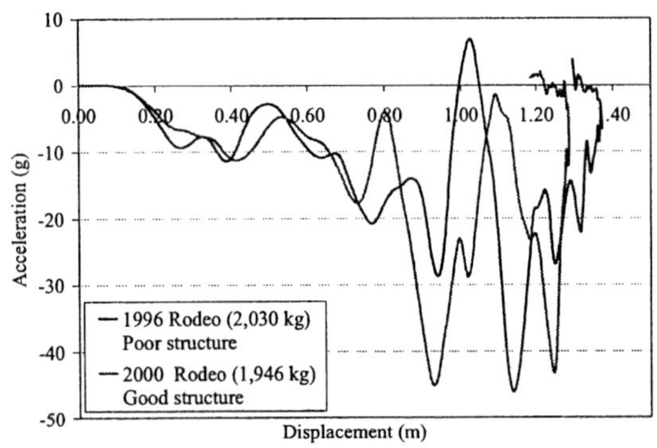

Figure 7. Acceleration vs. Displacement 1996 and 2000 Isuzu Rodeo

The Nissan Quest (Figure 8) went from an acceptable structural rating for the 1996 model to a poor rating for the 1999 model. This vehicle is the one example of a significant degradation in structural performance with a redesign. One major design difference between the two models is the addition of a left-side sliding door for the 1999 model. In the test of the redesigned model, the door structure buckled due to insufficient strength of the aperture. The stiffness calculated for both designs from the NCAP data were essentially the same. The accelerations recorded were similar until 70 cm of combined vehicle and barrier deformation had occurred. The 1996 model generally had higher accelerations than the 1999 model for the remainder of the crash. Maximum deformation was increased by 11 cm, and intrusion into the toepan was increased by 8 cm with the redesigned model.

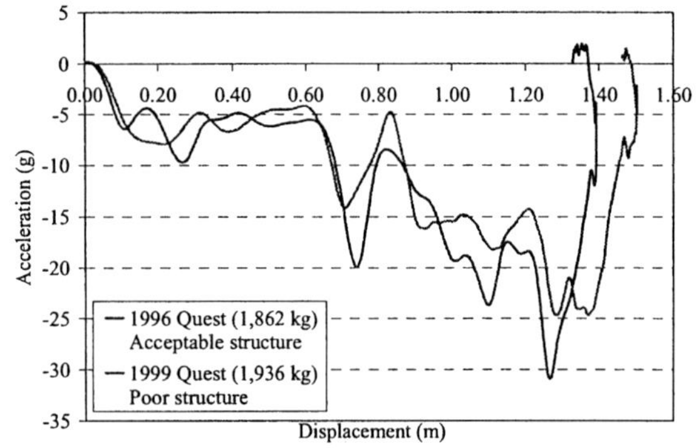

Figure 8. Acceleration vs. Displacement 1996 and 1999 Nissan Quest

The Saab 900 and Saab 9-3 (Figure 9) went from a poor structural rating for the 1995 900 model to an acceptable rating for the 1999 9-3 model. The accelerations recorded were similar until about 90 cm of combined vehicle and barrier deformation had occurred. Although both vehicles had the same maximum deformation, the 9-3 peak accelerations were higher and were sustained continuously as deformation neared the occupant compartment. The 900 acceleration fell off sharply between 120 and 130 cm of deformation, at which point collapse of the occupant compartment occurred. Maximum deformation was increased by 1 cm, and intrusion into the toepan was reduced by 8 cm with the redesigned model. The reduction in intrusion was much greater than reduction in overall deformation.

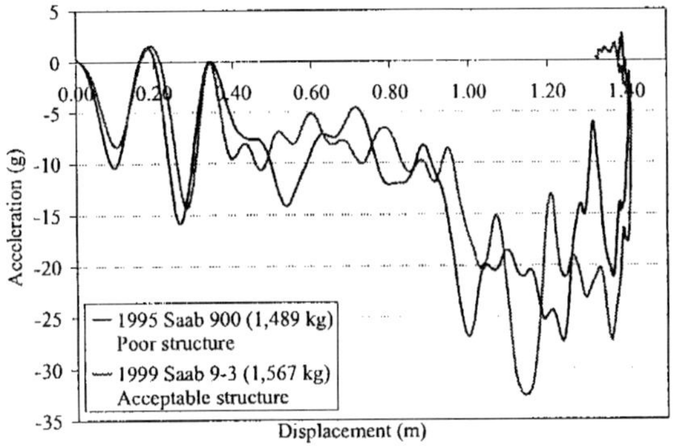

Figure 9. Acceleration vs. Displacement 1995 Saab 900 and 1999 Saab 9-3

The Toyota Avalon (Figure 10) went from a marginal structural rating for the 1996 model to an acceptable rating for the 1998 model and then to a good rating for the 2000 model. The 1996 model is 100 kg lighter than the 1998 and 2000 models, but they all have similar front crush space. Stiffness measures from the NCAP tests were available for the 1996 and 1998 models. These measures indicate the 1998 model is significantly stiffer, yet the data in Figure 10 would suggest the 1998 model is less stiff. The accelerations recorded were similar for all three models until 80 cm of combined vehicle and barrier deformation had occurred. Peak accelerations were about the same for all three models, with the best performer peaking first. The worst performer had the least mass and least total deformation. Maximum deformation was decreased by 8 cm, and intrusion into the toepan was reduced by 12 cm between the 1998 and 2000 models. The reduction in intrusion was somewhat greater than reduction in overall deformation.

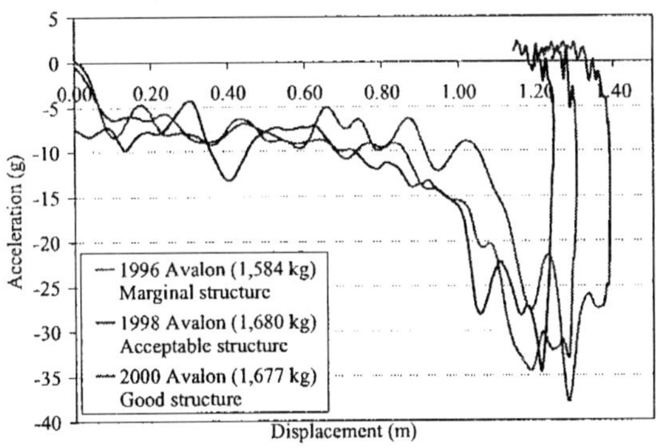

Figure 10. Acceleration vs. Displacement 1996, 1998, and 2000 Toyota Avalon

The Toyota Previa (Figure 11) went from a poor structural rating for the 1996 model to a good rating for the 1998 Sienna model. The Sienna is a completely new passenger van design that replaced the Previa, a snub-nosed mid-engine van. The Previa has only 89 cm of front crush space (one of the shortest, second only to the Ford Aerostar and Chevrolet Astro vans) compared with the Sienna's 120 cm. Despite similar dynamic deformation in the offset test, the Previa's occupant compartment failed catastrophically, whereas the Sienna had minimal occupant compartment intrusion. The stiffness calculated from the NCAP test indicated the Sienna is much less stiff than the Previa. The Sienna acceleration recorded was lower than that of the Previa until 120 cm of combined vehicle and barrier deformation had occurred. After this point, the Sienna acceleration increased rapidly to prevent intrusion into the occupant compartment. Maximum deformation was increased by 2 cm, and intrusion into the toepan was reduced by 19 cm with the redesigned model.

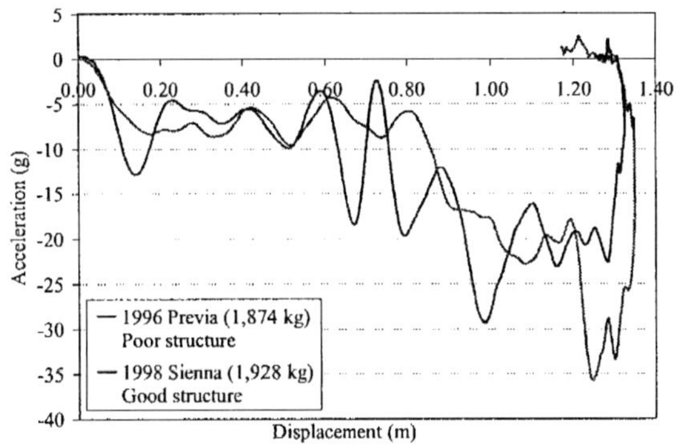

Figure 11. Acceleration vs. Displacement 1996 Toyota Previa and 1998 Toyota Sienna

The Volkswagen Passat (Figure 12) went from a poor structural rating for the 1995 model to an acceptable rating for the 1998 model. Both models have similar mass and available crush space. The NCAP test stiffness was slightly higher for the 1998 model using the first stiffness measurement method and slightly lower using the second method. The accelerations recorded generally were similar throughout the crash. Maximum deformation was increased by 6 cm, yet despite this, intrusion into the toepan was reduced by 20 cm with the redesigned model.

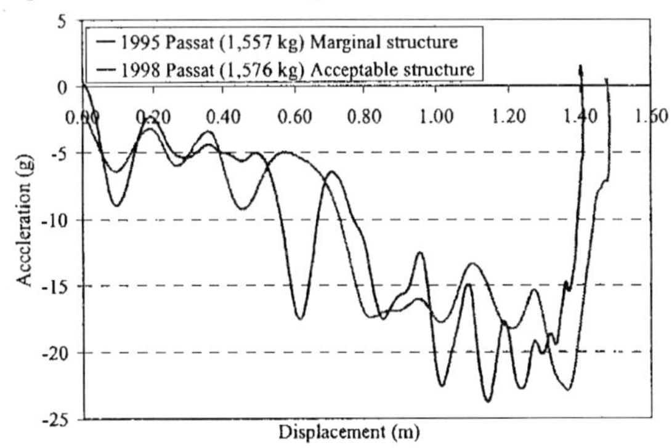

Figure 12. Acceleration vs. Displacement 1995 and 1998 Volkswagen Passat

DISCUSSION

Neither of the stiffness parameters calculated from NCAP test data correlated well with the IIHS structural evaluation in the offset test. No correlation was found even after controlling for mass, front-end length, and vehicle type. A general trend was observed between greater front-end length and better structural performance.

One possible reason for the lack of correlation is that the measures of stiffness are inappropriate. The extent of deformation that occurs in the NCAP test may be too low to adequately characterize what may occur in the offset test. In NCAP, the crash forces are distributed across the entire vehicle front end, and the occupant compartment is not challenged, resulting in little assessment of the occupant compartment strength relative to the crush zone strength. Neptune (1999) has compared the stiffness calculated from the NCAP test and the offset test and reported significant differences between the two for some vehicles.

Analysis of the eight models in this study indicates that those vehicles that exhibit signs of increased stiffness do so only late in the crash event, as deformation nears the occupant compartment (the 'bulkhead concept'). A good example of this is the Toyota Sienna shown in Figure 11; accelerations increase sharply after 120 cm of total deformation.

Seven out of eight vehicles tested before and after a redesign had improved structural performance after redesign. Four of these vehicles allowed more total occupant compartment movement during the crash than their predecessors while reducing intrusion dramatically. This suggests that these vehicles were designed to minimize occupant compartment intrusion, not by a simple increase in stiffness, but rather by careful placement of front structural elements and the matching of front-end stiffness to occupant compartment stiffness. The three other vehicles whose structural performance improved after a redesign allowed less total occupant compartment movement during the crash than their predecessors. These vehicles reduced occupant compartment intrusion by amounts greater than the overall reduction in occupant compartment movement, again suggesting careful front-end design is necessary for good test performance.

The Nissan Quest was the only vehicle to have significantly worse structural performance in the offset test after a redesign. The stiffness calculated from the NCAP test was essentially the same for both vehicles, and both have the same front-end length. The 1999 model had insufficient stiffness in the occupant compartment to sustain the front-end crash forces, resulting in excessive deformation in the occupant compartment without full use of the front energy-absorption capacity. Front-end stiffness must be well matched to occupant compartment stiffness, otherwise compartment failure will occur before complete energy absorption of the front structure is realized. This same phenomenon is apparent when comparing the IIHS tests of the 1995 Ford Taurus and 1995 Ford Contour. The Taurus front crush zone is used completely, and minimal intrusion into the occupant compartment occurred. In contrast, some of the energy-absorption capacity of the Contour remained unused, but the occupant compartment yielded, resulting in excessive intrusion into the occupant footwell by the engine cradle.

CONCLUSIONS

There was no significant correlation between stiffness (using either method) and the offset structural performance when looking at all vehicles or cars alone. Stiffness was analyzed after controlling for mass and front-end length in multiple regressions, but still no strong correlation to test performance was found.

Vehicles whose structural performance improved after a redesign do not show any evidence of increased stiffness for the first half-meter of vehicle deformation. All vehicles that improved after redesign had significant reductions in occupant compartment intrusion. These reductions typically were found to be much greater than reductions in overall movement of the occupant compartment during the crash. This suggests that careful design and placement of the front structural elements plus good matching of front-end stiffness to occupant compartment is required to improve offset test performance.

ACKNOWLEDGMENT

This work was supported by the Insurance Institute for Highway Safety.

REFERENCES

Insurance Institute for Highway Safety. 2000a. Crashworthiness evaluation offset barrier crash test protocol, ver. 8. Arlington, VA.

Insurance Institute for Highway Safety. 2000b. Crashworthiness evaluation guidelines for rating structural performance. Arlington, VA.

Lund, A.K.; Zuby, D.S.; and Nolan, J.M. 1995. Deformable barriers and occupant protection in offset frontal car-to-car crashes. SAE Technical Paper Series 950496. Warrendale, PA: Society of Automotive Engineers.

Meyerson, S.L. and Nolan, J.M. 2001. Effects of geometry and stiffness on the frontal compatibility of utility vehicles. *Proceedings of the 17th International Technical Conference on the Enhanced Safety of Vehicles* (CD ROM). Washington, DC: National Highway Traffic Safety Administration.

Nash, C. 1987. CRASH 3 – current status. SAE Technical Paper Series 870040. Warrendale, PA: Society of Automotive Engineers.

National Highway Traffic Safety Administration. 1981 (rev. 1982). *CRASH3 User's Guide and Technical Manual.* Washington, DC: U.S. Department of Transportation.

National Highway Traffic Safety Administration. 1999. *NASS/CDS Data Collection, Coding, and Editing Manual.* Washington, DC: U.S. Department of Transportation.

Neptune, J. 1999. A comparison of crush stiffness characteristics for partial-overlap and full-overlap frontal crash tests. SAE Technical Paper Series 1999-01-0105. Warrendale, PA: Society of Automotive Engineers.

Nolan, J.M.; Powell, M.R.; Preuss, C.A.; and Lund, A.K. 1999. Factors contributing to front-side compatibility: a comparison of crash test results (SAE 99SC02). *Proceedings of the 43rd Stapp Car Crash Conference* (P-350), 13-24. Warrendale, PA: Society of Automotive Engineers.

Nolan, J.M.; Preuss, C.A.; Jones, S.L.; and O'Neill, B. 1998. An update on the relationships between computed delta Vs and impact speeds for offset crash tests. *Proceedings of the 16th International Technical Conference on the Enhanced Safety of Vehicles,* 2:1234-41. Washington, DC: National Highway Traffic Safety Administration

O'Neill, B.; Preuss, C.A.; and Nolan, J.M. 1996. Relationships between computed delta V and impact speeds for offset crash tests. *Proceedings of the 15th International Technical Conference on the Enhanced Safety of Vehicles*, 1433-40. Washington, DC: National Highway Traffic Safety Administration.

Park, B., Hackney, J. et al. 1999. The New Car Assessment Program: Has it led to stiffer light trucks and vans over the years? SAE Technical Paper Series. 1999-01-0064. Warrendale, PA: Society of Automotive Engineers.

Prasad, A. 1990a. Energy dissipated in vehicle crush: a study using the repeated test technique. SAE Technical Paper Series 900412. Warrendale, PA: Society of Automotive Engineers.

Prasad, A. 1990b. CRASH3 damage algorithm reformulation for front and rear collisions. SAE Technical Paper Series 900098. Warrendale, PA: Society of Automotive Engineers.

Zobel, R. 1999. Accident analysis and measures to establish compatibility. SAE Technical Paper Series 1999-01-0065. Warrendale, PA: Society of Automotive Engineers.

Mass Effects

2003-06-0176

COMPATIBILITY FOR FRONTAL IMPACT COLLISIONS BETWEEN HEAVY AND LIGHT CARS

Shigeru Hirayama
Kazuhiro Obayashi
Tomosaburo Okabe
Nissan Motor Co., LTD.
Japan
Paper Number 454

ABSTRACT

Recently, frontal impact compatibility is discussed internationally and various procedures to assess compatibility and various measures to improve compatibility have been proposed. Considering the above, car to car tests between a heavy car and a light car were conducted to clarify the effect of homogenizing the front structure on compatibility. Then correlation between the results of the barrier impact tests proposed as the procedures to assess compatibility and the car to car test results and the requirements for the assessment procedure were discussed.

INTRODUCTION

The frontal impact performance is assessed by the impact tests of a car against a fixed barrier. These impact tests contributed to the reduction of the casualties in traffic accidents. Recently, frontal impact compatibility is recognized as an area where additional safety enhancements might be made. Through many studies in the world, it has become clear that the fundamental issues of frontal impact compatibility are (1)structural interaction, (2)frontal stiffness and (3)passenger compartment strength. The procedures to assess each issue have been proposed. Considering these things, car to car tests between a heavy car and a light car were conducted to clarify the effect of the front structure homogeneity on frontal impact compatibility. Then the barrier tests which are proposed as the test to assess frontal impact compatibility were conducted. From the comparison of the barrier test results and the car to car test results, the items which should be considered in case of deciding the assessment procedure of frontal impact compatibility were discussed.

CAR TO CAR TESTS

Structural interaction is one of the most important requirements for frontal impact compatibility. Therefore, a homogeneous front structure that can absorb the impact energy in a uniform manner in a wide variety of impact configurations is desired. It is assumed that not only the front structure homogeneity of heavy car but also that of light car is required to improve structural interaction in car to car tests. In order to confirm if this assumption is appropriate, car to car tests between a heavy car and a light car were conducted as shown in Table 1. The aim of mounting multiple load paths is to achieve the front structure homogeneity.

Table 1.

Car to car tests matrix

		Heavy car
		With multiple load paths
Light car	Without multiple load paths	X (Test No.:C2C01)
	With multiple load paths	X (Test No.:C2C02)

Test Conditions

Test details are shown in Table 2 and pre test position of both cars are shown in Figure 1 and 2.

Table 2.

Test details

		C2C01	C2C02
Test speed		56km/h	←
Overlap		50% of the smaller car's width	←
Mass	Heavy car	1539kg	←
	Light car	1125kg	←

Figure 1. Pre test side view.

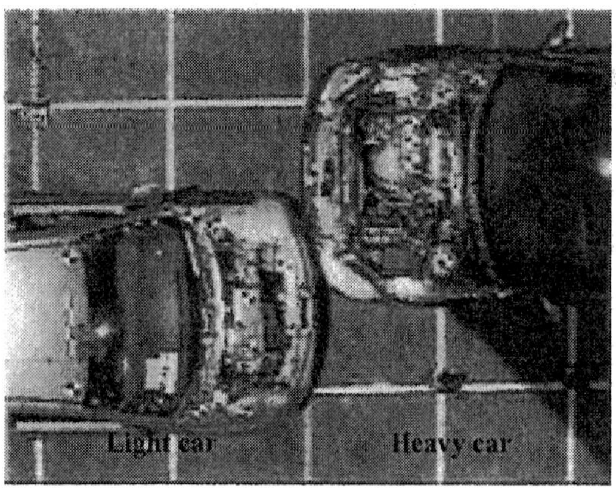

Figure 2. Pre test plan view.

Result of Car to Car Test between Heavy Car and Light Car without Multiple Load Paths

Figure 3 and 4 show the post test view of each car and Figure 5 shows the comparison of the firewall intrusion between two cars. The passenger compartment deformation of light car without multiple load paths is larger in comparison with that of heavy car. The result of this car to car test is not considered compatible.

Figure 3. Post test side view of light car in the C2C01 test.

Figure 4. Post test side view of heavy car in the C2C01 test.

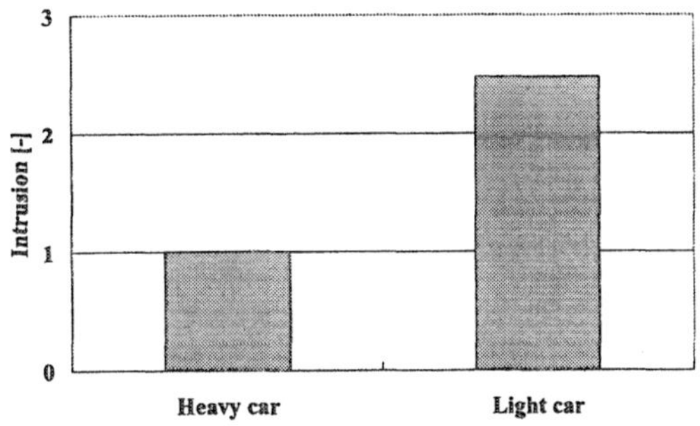

Figure 5. Comparison of firewall intrusion in the C2C01 test. (Each value is normalized by the intrusion of heavy car in the C2C01 test.)

Result of Car to Car Test between Heavy Car and Light Car with Multiple Load Paths

Figure 6 and 7 show the post test view of each car and Figure 8 shows the comparison of the firewall intrusion between two cars. The passenger compartment deformation of light car with multiple load paths in this test reduced significantly in comparison with that of light car without multiple load paths in the C2C01 test. On the other hand, the passenger compartment deformation of heavy car in this test slightly increased in comparison with that of heavy car in the C2C01 test. This car to car test is considered more compatible than the C2C01 test. Therefore, by comparing the results of these car to car tests, it indicates that the incorporation of multiple load paths in the light car improved frontal impact compatibility.

Figure 6. Post test side view of light car in the C2C02 test.

Figure 7. Post test side view of heavy car in the C2C02 test.

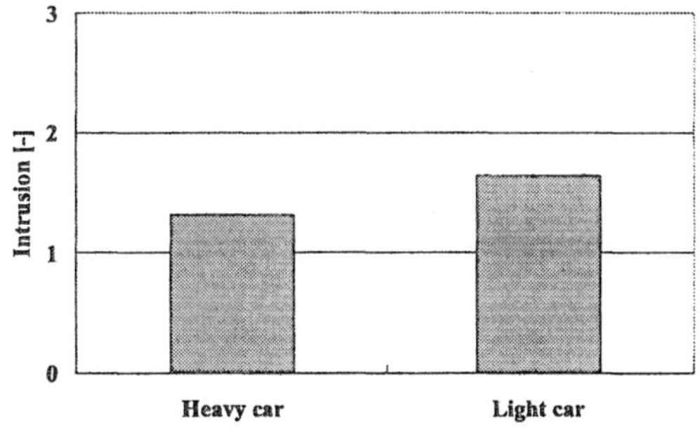

Figure 8. Comparison of firewall intrusion in the C2C02 test. (Each value is normalized by the

intrusion of heavy car in the C2C01 test.)

Comparison of Two Car to Car Tests

In frontal car to car impact between a heavy car and a light car the passenger compartment deformation of the light car is apt to be large in comparison with that of the heavy car because the passenger compartment strength of light car is weaker than that of heavy car, generally speaking. The C2C01 test illustrates this situation. However, in the C2C02 test the passenger compartment deformation of light car reduced significantly although the passenger compartment structure of light car was not varied. The reason is assumed as follows.

In the C2C01 test, where the light car did not have multiple load paths, the front structures of both cars did not interact with each other sufficiently. On the other hand, in the C2C02 test, both cars had multiple load paths, which resulted in the considerable improvement of compatibility performance.

It is often mentioned that the measures to achieve compatibility are (1)to homogenize the front structure of heavy car and (2)to increase the passenger compartment strength of light car. However, These test results show that it is effective for compatibility improvement to homogenize the front structure of light car as well as heavy car. In other words, it is not sufficient for achieving good compatibility that the front structure of only either one of two cars is homogeneous. These test results illustrate the importance that the front structures of both cars are homogeneous.

BARRIER TESTS

Various tests to assess the fundamental issues of compatibility are proposed as shown in Table 3 (refer to [1] and [2]). To verify the previous discussion on car to car tests results, the 80km/h ODB tests of the same cars as were used in car to car tests were conducted (see Table 4). In 80km/h ODB tests, passenger compartment strength is measured. And it is indicated as "End of crash force (refer to [3])".

Table 3.

Proposal of the tests to assess frontal impact compatibility

	Tests proposed by TRL	Tests proposed by Renault
Structural Interaction	56km/h Full width deformable barrier test	60km/h PDB test
Frontal stiffness	64km/h ODB test	
Passenger compartment strength	80km/h ODB test	60km/h ODB test

Table 4.

Barrier tests matrix

	Heavy car (With multiple load paths)	Light car (Without multiple load paths)	Light car (With multiple load paths)
80km/h ODB test	X	X	X

Results of 80km/h ODB Tests

Figure 9, 10 and 11 show the force vs. displacement curve of each car. In these figures, broken line shows "Mechanical force (refer to [1])" and chain line shows "Structural force (refer to [1])". Figure 12 shows the comparison of the "End of crash force" values. There is little difference in the "End of crash force" between the light car with multiple load paths and the light car without multiple load paths. On the other hand, there is a great difference in the "End of crash force" between heavy car and light cars.

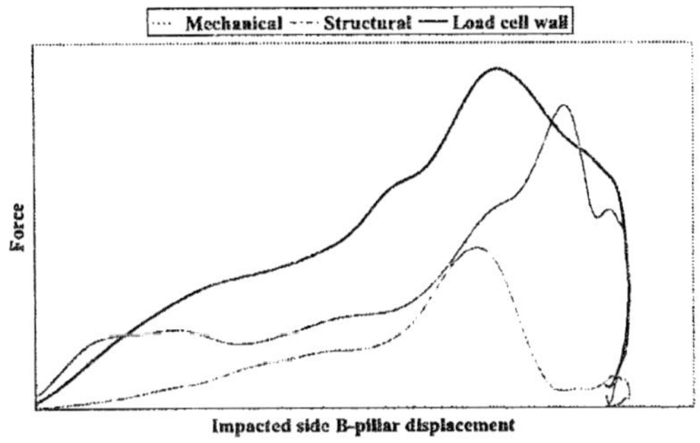

Figure 9. Force vs. displacement curve of heavy car in the 80km/h ODB test.

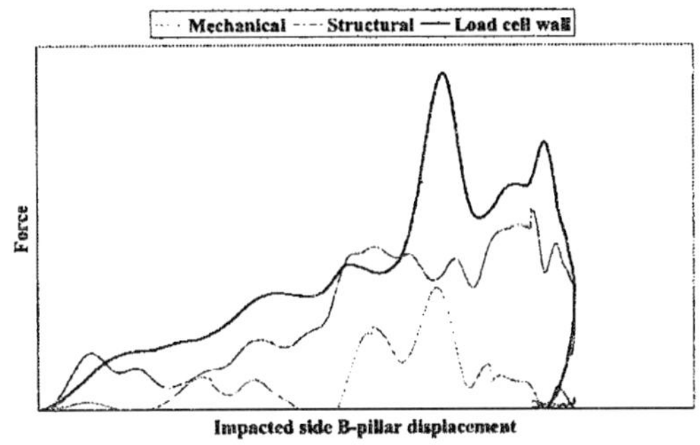

Figure 10. Force vs. displacement curve of light car without multiple load paths in the 80km/h ODB test.

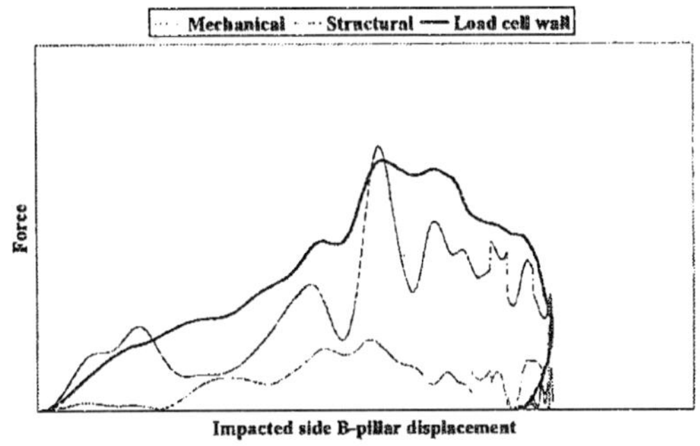

Figure 11. Force vs. displacement curve of light car with multiple load paths in the 80km/h ODB test.

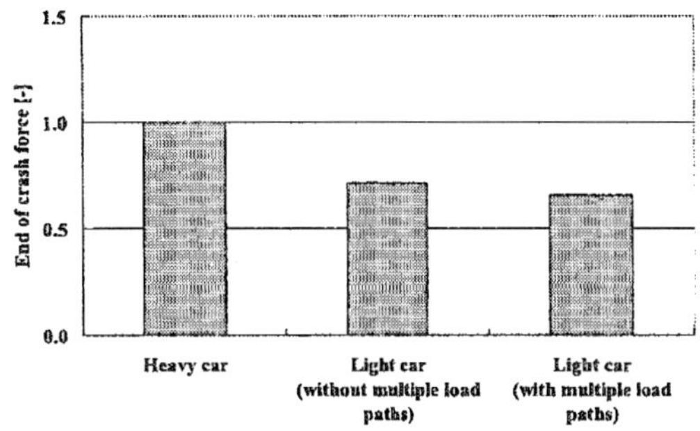

Figure 12. Comparison of "End of crash force". (Each value is normalized by the "End of crash force" of heavy car.)

Discussion

In case that there is a great difference in the "End of crash force" between heavy car and light car, consequently it is expected that the passenger compartment deformation of light car will be larger than that of heavy car when these cars collide each other (see Figure 13). However, the heavy car vs. light car with multiple load paths test resulted in compatible result although the heavy car vs. light car without multiple load paths test resulted in incompatible result.

The reason why the heavy car vs. light car with multiple load paths test resulted in compatible result is not that the passenger compartment strength increased, but that the energy absorbing capability of front structure was increased due to the improved structural interaction (see Figure 14).

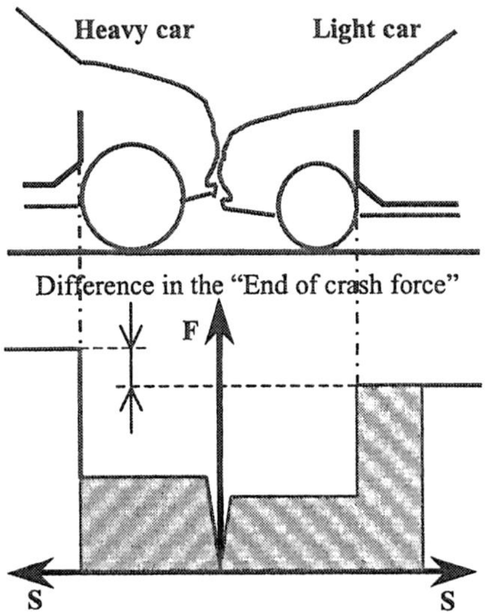

Figure 13. Force vs. deformation curve of each car (There is a great difference in the "End of crash force" between two cars.).

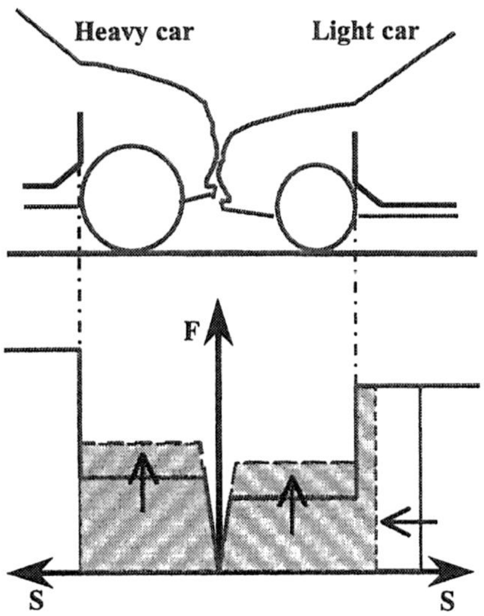

Figure 14. Force vs. deformation curve of each car (The energy absorbing capability is improved.).

CONCLUSIONS

Through the car to car test results and the barrier test results, the following have been shown. These findings should be considered in future development of test procedures for vehicle compatibility assessment.

(1) In order to achieve compatible result in a car to car impact between a heavy car and a light car, it is important to homogenize not only the front structure of the heavy car but also that of the light car.

(2) In case that compatibility of a light car is evaluated in the barrier tests, the front structure homogeneity as well as the passenger compartment strength should be analyzed.

REFERENCES

[1] Edwards, M., Hobbs, A., Davies, H. and Thompson, A. "Development of Test Procedures and Performance Criteria to Improve Compatibility in Car Frontal Collisions." Vehicle Safety 2002.

[2] Diboine, A. and Delannoy, P. "Improvements in car to car compatibility: Physics, design constraints and assessment test methodology and criteria." Vehicle Safety 2002.

[3] Steyer, C., Delhommeau, M. and Delannoy, P. "Proposal to Improve Compatibility in Head On Collisions." Sixteenth International Technical Conference on the Enhanced Safety of Vehicles.

986063

THE COMPATIBILITY OF MINI CARS IN TRAFFIC ACCIDENTS

Koji Mizuno
Janusz Kajzer
Nagoya University
Japan
Paper Number 98-S3-O-08

ABSTRACT

The compatibility problems of the mini car in car-to-car frontal collision and car-pedestrian accident are discussed using accident data and computer simulations.

In our analysis of the accident data in Japan, the number of fatalities was investigated using the vehicle masses and classes. It was found that the cars with identical mass are most compatible since the injuries per accident are minimal and injury risks to the driver in both cars are the same. The analysis of the car class indicated that the mini car and the sports utility vehicle are the most incompatible car types, with low and high aggressivity, respectively.

Our accident analysis in the present study shows that the safety of mini cars is the key point in achieving the compatibility in Japan. Computer simulations using MADYMO were carried out of crashes of mini car into a rigid wall and into a large car. It was found that either stiffening mini car with an optimized restraint system or modifying the large car with additional crush space can reduce the injury risk to the driver in the mini car.

The analysis of car-pedestrian accidents in Japan shows that the mini car has higher aggressivity in relation to the pedestrian than other bonnet-type cars. Computer simulation revealed that the head velocity of the pedestrian at impact is high since the pedestrian head contacts the body of the mini car in the early phase.

INTRODUCTION

In vehicle-to-vehicle collisions, the protection of all occupants in the subject and other vehicle should be considered. This compatibility problem was first discussed in connection with the Experimental Safety Vehicle (ESV) in 1970 and has not been solved yet. EC/EEVC has a leading responsibility for vehicle compatibility, which is one of the harmonized research activities of ESV [NHTSA 1996]. In the United States, the National Highway Traffic Safety Administration (NHTSA) has started a research program on this subject [Hollowell and Gabler 1996]. However, in Japan, there seems to have been little research on vehicle compatibility in the past decade. One reason is that it has been difficult to obtain accident data in Japan. Thus, the Institute for Traffic Accident Research and Data Analysis (ITARDA), established in 1992, provides some accident data.

Compatibility means that passenger vehicles of disparate size provide an equal level of occupant protection in car-to-car collisions [NHTSA 1996]. The field data show there are many vehicles which are incompatible with other vehicles. This incompatibility is induced by the difference in the mass, stiffness, geometry and structure of both vehicles [Buzeman 1997]. An incompatible vehicle induces high injury risks for the occupants in the other vehicle, which can be defined as aggressivity. The influences of mass, stiffness and geometry are combined and have aggressivity to other cars.

One purpose of the present study is to identify the compatibility problems of car-to-car and car-to-pedestrian collisions in Japan based on the accident data using vehicle masses and classes. The compatibility problem should be examined for each country because the traffic environment differs in each country in terms of vehicle size, population, velocity, travel distance and the road environment.

In car-to-car collisions, the injury risk to the occupants in a mini car is high due to high delta-V and large intrusion based on its small mass and size. To achieve the compatibility of this type of a car, a low mass vehicle (LMV) with a mass of 600-650 kg and length of 2.5-3.0 m was proposed [Waltz et al. 1991; Kaeser et al. 1992, 1995; Frei 1997]. The front structure of a LMV is designed to be stiff in order to reduce the intrusion into the passenger compartment. As the acceleration of a LMV becomes high due to the stiff structure, it needs a specially designed restraint system to ensure the occupant's safety. Optimum restraint system was analyzed using the crash victim simulation program MADYMO [Kaeser et al. 1995; Muser et al. 1996]. However, this analysis was conducted only for a crash with a deformable barrier. The analysis of the compatibility in a car-to-car collision is necessary to show the injury-reducing effect of high stiffness and the optimized restraint system of a LMV.

The safety of the occupant of a small car in a car-to-car frontal collision has been investigated in many studies by a mathematical model. The force–deformation characteristics of both cars were examined in order to decrease the acceleration and deformation of the small car using a simple spring-mass model [Ventre 1972, 1973; Appel et al. 1994; Tarrière 1994]. Finite element analyses of a car-to-car frontal collision were also conducted to describe the interaction of components of both cars [Maurer et al. 1996; Tarrière 1996]. These analyses of car-to-car collisions focused mainly on the car characteristics, and the model consists only of a vehicle without an occupant. Therefore, the influence of the car stiffness on the injury risk of the occupants by a combination of acceleration and intrusion in car-to-car frontal collision is not clear.

The compatibility of a mini car in a collision with a large car has to be achieved without increasing the injury risk to the driver of the mini car in a single-car crash. In the current study, the crash of a mini car into a rigid wall with full overlap, and the collision with a large car with a 50% overlap are simulated using MADYMO. The influences of front stiffness and the restraint systems of the mini car on the injury risk to the driver are studied to achieve the compatibility of the mini car in frontal collisions.

Since mini cars are mainly used in the city, car-pedestrian accidents are of great importance. The front geometry of the vehicle affects the pedestrian injury risk [Ishikawa et al. 1991]. Therefore the injury risk to the pedestrian struck by a mini car was examined by the analysis of the accident data and also computer simulation using MADYMO.

METHODOLOGY

Accident Analysis of Mini Car Crash

Distribution of Fatalities by Accident Type - The distribution of fatalities can be expressed for all types of accidents by fatalities related to the subject car [Appel 1996]. The fatalities can be distinguished as internal and external in relation to the car. Fatalities of the subject car in car-to-car collisions and in single car accidents are classified as internal fatalities, while those of other cars, motorcyclists, cyclists and pedestrians are classified as external fatalities. Fatalities in all types of accidents are estimated by the number of internal and external fatalities per registrations of the subject car.

Compatibility in Car-to-Car Frontal Collision - The goal of vehicle compatibility in car-to-car frontal collisions is to minimize the number of fatalities while the injury rates of the occupants in each car remain the same. Thus, in the current study to estimate compatibility in a car-to-car frontal collision, one method is employed to determine the total number of fatalities in both cars per accident when comparing the ratio of the fatalities occurring in each car.

To estimate the aggressivity in car-to-car frontal collisions, the following methods can be used:

1. (Number of fatalities in other cars) /(Number of fatalities in subject cars)
2. Percentage of fatalities in other cars
3. Number of fatalities in other cars per million subject car registrations

Methods 1 and 3 were suggested by Hollowell et al. [1996]. Using method 1, the aggressivity of a car without influence of human factors can be described. If the crash velocity of the subject car is high, the injury risk to the occupants in other cars as well as in the subject car is high. Thus, the influence of crash velocity on the aggressivity estimated by method 1 will be small. On the other hand, the aggressivity of the car including influence of crash velocity is estimated when the injury rate of the driver in the other cars is used in method 2. If the crash velocity of the subject car is high, the aggressivity obtained by method 2 will be higher because the number of fatalities in the other cars will increase. The aggressivity estimated by method 3 includes the effects of travel distances, vehicle velocities and accident rates, reflecting how they are used (Table 1.).

Table 1.
Effect of Crash Velocity and Accident Rate on Aggressivity Using Each Method

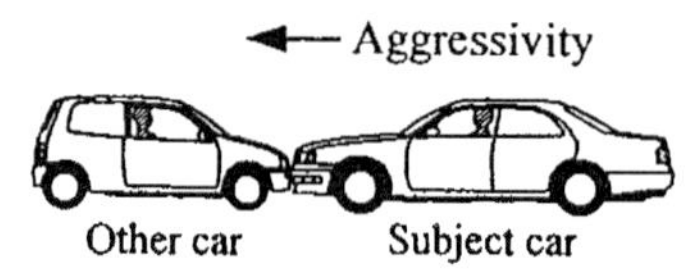

Method	Crash Velocity	Accident Rate
1	X	X
2	O	X
3	O	O

O = large effect X = small effect

The method of examining aggressivity depends on the problem under investigation. For example, car manufacturers can use method 1 to estimate aggressivity of cars because this method is related to the car itself. Method 2, which includes the velocity effect, is usable in studies dealing with driver behaviors. Method 3 expresses the aggressivity of each registered car, so it can be used when insurance problems are investigated. In the current study, only method 1 is used to discuss the aggressivity of the car itself.

The law enforcement accident data for the four years from 1992 to 1995 were used. The analysis is conducted only for cars with a model year of 1988 or later. In the current research, cars are categorized in eight classes: mini, sedan A, sedan B, sedan C, sports & specialty, wagon, 1BOX and Sports Utility Vehicle (SUV). Only injuries to drivers are examined to simplify the analysis.

Computer Simulation of Mini Car Crash

Car Model - The car model used in the computer simulations is based on a currently produced mini car. Figures 1 and 2 present a model of a car used to simulate crashes into a rigid wall and large car. The mass of the mini car is 700 kg, which is slightly larger than the average mini car (639 kg; Mizuno et al. 1997), and the mass of the large car is 1400 kg.

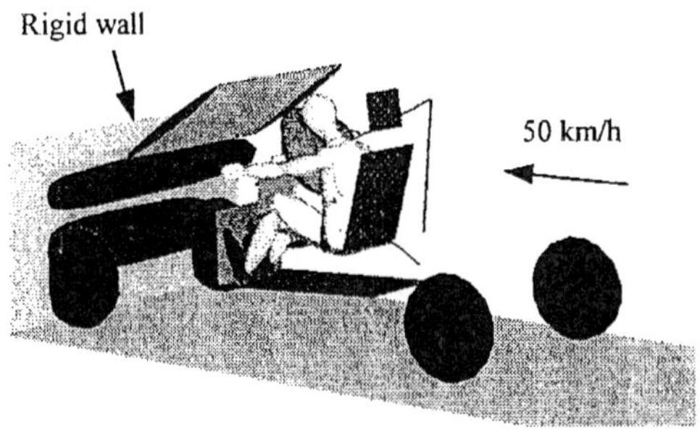

Figure 1. Simulation model of a mini car in a frontal crash into a rigid wall.

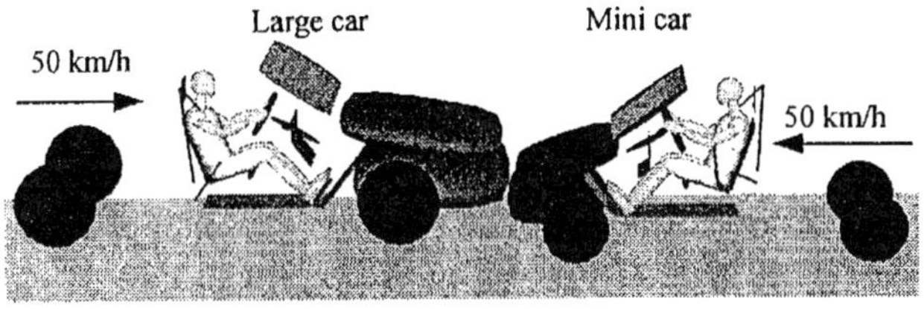

Figure 2. Simulation model for a frontal collision between a mini car and a large car with a 50% overlap for mini car.

Car front structures are represented by the ellipsoids. In the current model, the force–deformation characteristics of the mini and large cars are approximated by a straight line. The regression of the stiffness of the car (k) is expressed using car mass (m) as follows [Ishikawa 1990]:

$$k = 78m^{1/3} \text{ (kN/m)}. \qquad (1.)$$

Using Eq. (1), the stiffness of the current mini car (700 kg) is evaluated as 693 kN/m. The stiffness of the mini car is changed from 500 to 1000 kN/m in order to examine the effect of the stiffness in crashes into a rigid wall and into a large car. The stiffness of the large car (1400 kg) is calculated as 872 kN/m by Eq. (1), which is used in the simulation of a car-to-car collision.

In an offset crash, the damage profile of the car is classified into direct damage and induced damage. Due to the induced damage, the stiffness of the car in the offset crash increases by 30% per width of the car [Ishikawa 1995]. Thus, the front stiffness of the car in an offset collision is estimated as:

$$k_{offset} = 1.3k \times (overlap\ ratio) \qquad (2.)$$

where k_{offset} is the stiffness of the car in the offset crash and k is that in the full overlap crash. In car-to-car collisions, the forces acting on both cars have the same magnitude but a different direction. The deformation of each car is calculated using force–deformation characteristics according to this force.

According to Matsumoto et al. [1990], the intrusion of the firewall (x_{firewall}) can be approximated as:

$$x_{\text{firewall}} = 0.75(x - x_0) \qquad (3.)$$

where x is the deformation of the car and x_0 is the car deformation when the engine contacts the firewall. The deformation x_0 of the mini car is smaller than that of the large car due to its small size. In the current model, x_0 is estimated as 0.175 m for a mini car and 0.350 m for a large car. Based on the experimental results (Figure 3.), the longitudinal displacement of the steering column (x_{steering}) can be expressed by the intrusion of the firewall (x_{firewall}) as:

$$x_{\text{steering}} = 0.772(x_{\text{firewall}} - 0.0566). \qquad (4.)$$

In the model, the movements of the firewall and steering column are simulated as the displacement of translational joints based on Eqs (3) and (4). To express the intrusion of the firewall, the toe pan is designed to rotate first, and upon becoming perpendicular moves in the driver's direction.

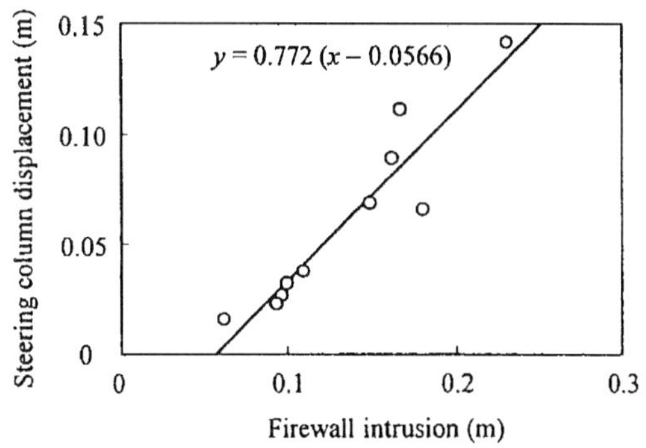

Figure 3. Relation between the firewall intrusion and longitudinal displacement of the steering column.

The HYBRID III database from MADYMO is used for the driver. The seat belt (10% webbing) and airbag (35l) is used for the basic restraint system for drivers in mini and large cars. This combination of restraint systems is commonly used in the current cars.

Mini Car Crashes into Rigid Wall and Large Car - The fatalities of the driver in single-car and head-on collisions account for a large portion of the driver fatalities in mini car accidents. Crashes into a rigid wall and car-to-car frontal offset collisions are representative of many cases of single-car and head-on collisions. In the present study, the simulations of the crash of the mini car into a rigid wall and the crash between mini and large cars with 50% overlap for the mini car were carried out.

According to the current regulation in Japan, the crash velocity of a mini car into a rigid wall is 40 km/h. In this simulation the crash velocity is put at 50 km/h, similar to the crash regulation of other types of cars. In the simulation of the car-to-car frontal collision, the crash velocity of each car is 50 km/h. The influence of the stiffness of the mini car on the injury risks to the driver in crashes into a rigid wall or a large car are examined. The injury criteria of the driver from the simulations are compared with threshold levels (HIC 1000, chest acceleration 60 g, chest deflection 0.075 m, femur force 10 kN).

Effect of Restraint Systems - The effects of restraint systems, including a seat belt force limiter, pretensioner (4 kN, 0.15 m), energy absorbing (EA) steering system (4 kN, 0.15 m), knee bolster and their combination, on the injury risk of the driver in a mini car are examined. The stiffness of the mini car is 1000 kN/m, which is larger than that of current mini cars to reduce the intrusion into the passenger compartment in a crash. The results are compared with those when a basic restraint system (airbag and seat belt) is used. The injury risks to the driver in the mini car in crashes into a rigid wall or a large car are studied when each restraint system or its combination is used with the basic restraint system.

Additional Crush Space of Large Car - When a large car has additional crush space designed for colliding with a mini car, the injury risk to the driver in the mini car may decrease. Tarrière et al. [1994] proposed a maximum force level 200 kN of a heavy car for compatibility with small car. Thus, in the present study, the additional crush length (c) of 0 to 0.4 m with a force level of 200 kN is simulated (Figure 4.) without changing the front length of the large car.

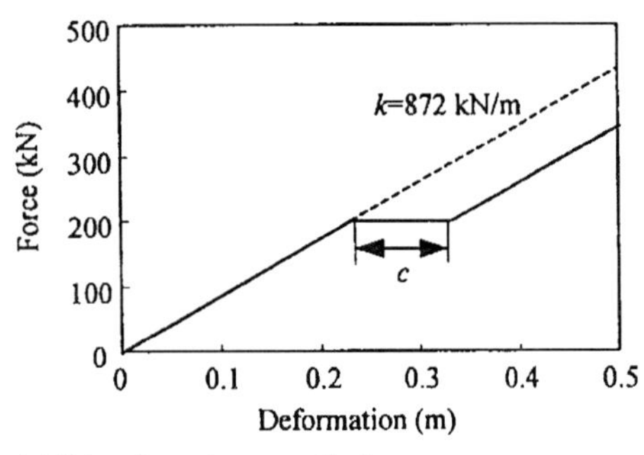

Figure 4. Additional crush space of a large car.

Car-Pedestrian Accidents

Accident Analysis - The compatibility in car-pedestrian accidents involves the mass, stiffness and geometry of the car. The car mass has little effect on a pedestrian injury because even the lightest car is much heavier than the pedestrian. The simulation demonstrated that the geometry of the car has a larger effect on the pedestrian injury than the stiffness [Ishikawa et al. 1991]. Thus, in order to clarify the influence of the geometry of the car, pedestrian accident data were examined in terms of car class since the cars have a similar geometry in the same car class.

Pedestrian accidents where the pedestrian was struck by the front of the vehicle were selected. To exclude the influence of impact speed of the vehicle, the accident data were used in which the velocity recognized to be dangerous was below 40 km/h. The velocity recognized to be dangerous is one of the items included among the accident data, which is defined as the velocity at the moment the driver perceives the danger of striking a pedestrian. It indicates the velocity before the driver brakes or steers to avoid the accident, and is compiled mainly from drivers' testimony. The distribution of the injuries according to the body region of the pedestrian were examined from accidents with fatal or severe injury to the pedestrian.

Simulation of Car-Pedestrian Accidents - Computer simulations using MADYMO were performed to examine the influence of vehicle geometry on the injury risk to the pedestrian (Figure 5.). Elderly people are frequently involved and injured in car-pedestrian accidents [Ishikawa et al. 1991]. Thus, the elder pedestrian model was made based on the average Japanese male aged 60 to 69, whose height and weight is 161.3 cm and 59 kg, respectively. The geometry, mass, moment of the inertia and center of the gravity of segments of this pedestrian model are generated by the GEBOD (Generator of Body Data). The joint characteristics and the stiffness of the ellipsoid of the pedestrian are based on the biomechanical data [Ishikawa et al. 1993; Yang 1998].

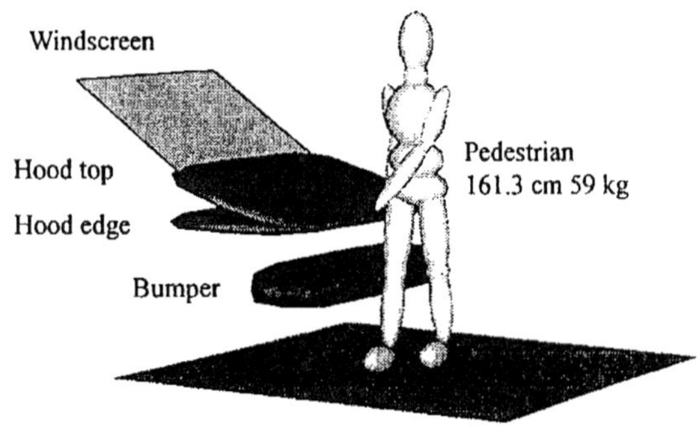

Figure 5. Pedestrian and vehicle model.

Three vehicle models representing mini car, sedan B and 1BOX were made to evaluate the injury risk to the pedestrian for each car. In examining the influences of vehicle shape on the pedestrian struck by a car, the model was designed so that the same part of the vehicle would have the same force-deformation characteristics in three vehicles. The crash velocity of the car is 40 km/h.

In order to estimate the injury risk to the pedestrian, the injury parameters [i.e., the HIC, chest, pelvis and femur accelerations (3 ms)] of the pedestrian in crash were evaluated for three car models. The results of the simulation were compared with those of statistical analysis of pedestrian accidents.

RESULTS

Accident Analysis of Mini Car Crash

Distribution of Fatalities by Accident Type - From accident data in Japan, the distribution of fatalities was calculated. This distribution was examined by the number of fatalities internal and external to the subject car in various types of accidents. Figure 6 shows the number of fatalities in relation to the subject car per million registrations.

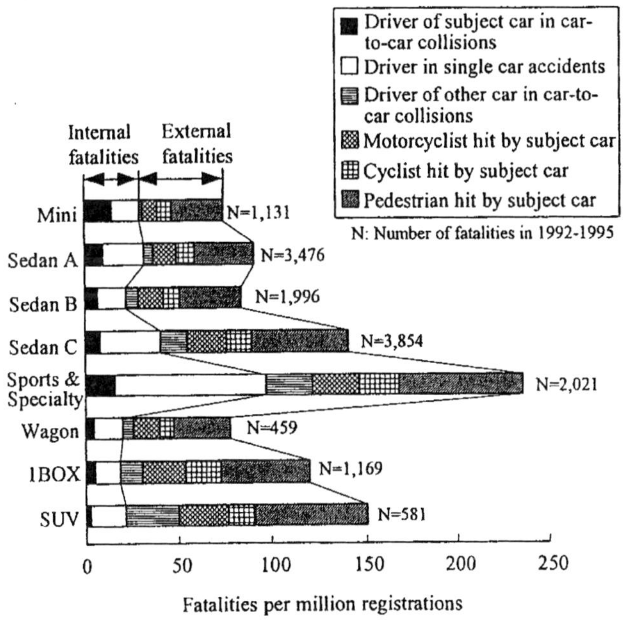

Figure 6. Internal and external fatalities of various subject cars for different types of accidents.

Sports & specialty, SUV, 1BOX and sedan C vehicles cause more external-type fatalities than any other type vehicles. SUV and sports & specialty car, in particular, cause the most fatalities in the other car in car-to-car collisions. Cyclists sustain more fatalities when struck by sports & specialty car and 1BOX vehicle, while more pedestrians die from accidents involving sports & specialty car and SUV.

From the analysis of distributions of fatalities, it is found that the total number of fatalities of mini car is lowest, so this car type could be considered as most compatible vehicle. But this conclusion cannot be drawn because the mini cars are used for short-distance travel at a relatively low velocity [ITARDA 1996a] and also the frequency of driver internal fatalities in car-to-car collision is high. It is also necessary in the analysis of the compatibility to exclude the influence of factors which are not related to the car itself, such as driver behavior, car velocity and accident rate.

The number of fatalities in single car accidents involving sports & specialty cars is especially large, reflecting their higher crash velocity and accident rate compared to any other car classes. As a result, the number of fatalities involving sports & specialty cars is large for all types of accidents.

Compatibility in Car-to-Car Frontal Collision - Car mass is one of the most significant factors affecting driver injury in car-to-car collisions. It is well known that the fatality rate of the driver decreases with car mass. Evans [1993] found that the ratio of the injury rate in a lighter car to that in a heavier one may be expressed by the power ratio of the car mass of the heavier car to that of the lighter car. In the present study, the individual injury rate is expressed by average car mass ratio.

According to Joksch [1993], the injury rate R (%) can be expressed by delta-V (Δv) as:

$$R = \left| \Delta v / \alpha \right|^k \tag{5.}$$

where α and k are parameters obtained by curve fitting. For many head-on collisions, delta-V is approximated for a central collision. Assuming the restitution coefficient is zero, the delta-V can be expressed using the average mass ratio as:

$$\Delta v_1 = \frac{m_2}{m_1 + m_2} v_c \tag{6.}$$

where Δv_1 is the delta-V of car 1, v_c is a closing speed, and m_1 and m_2 is the mass of car 1 and 2, respectively. Substituting Eq. (6) for (5), the injury rate of driver 1, R_1, is given by

$$R_1 = \left| \frac{m_2}{m_1 + m_2} \frac{v_c}{\alpha} \right|^k = a\left(\frac{m_2}{m_1 + m_2} \right)^k \tag{7.}$$

where $a = \left| v_c / \alpha \right|^k$. When Eq. (7) is applied to a real accident, the probability of serious and fatal injury to the driver of car 1 can be calculated as shown in Figure 7. The parameters k and a are calculated for seat belt wearing and injury severity as shown in Table 2. Based on this method [Mizuno et al. 1997], the parameter k is obtained as 2.64 for belted drivers sustaining fatal and serious injury. This value is almost the same as the 2.62 shown by Evans [1994]. However, he calculated the injury ratio of belted car drivers in heavier cars to those in the lighter cars, and considered all directions of impact.

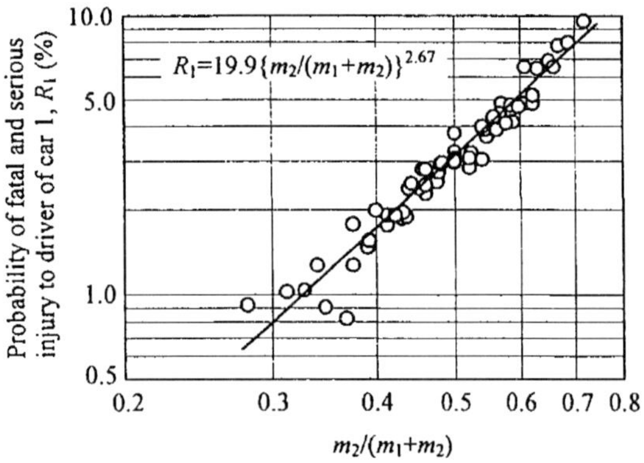

Figure 7. Average mass ratio and probability of driver injury in car 1 (Belted and unbelted driver).

Table 2.
Parameter *k* and *a*

Injury severity	Seat belt	k	a
Fatal and serious	Belted	2.64	13.9
	Unbelted	2.49	33.4
Fatal, serious, and minor	Belted	1.08	107.8
	Unbelted	0.955	123.5

The percentage of driver injuries in the subject car plus that in the other car corresponds to the driver fatalities per accident. From Eq. (7), we obtain

$$R_1 + R_2 = \frac{m_1^k + m_2^k}{(m_1 + m_2)^k} a . \qquad (8.)$$

R_1+R_2 has a minimum as $2^{1-k} a$ (k>1) when $m_1 = m_2$. Thus, cars with equal masses are most compatible in a collision since the injuries per accident are minimal and the injury rate is equal for both cars.

The percentage of driver fatalities in the subject and other car is shown in Figure 8. As the mass of the subject car increases, the fatality rate of the driver in the subject car decreases; on the other hand, that of the driver in the other car increases. The sum of the percentage of driver fatalities in the subject and the other car indicates the number of driver fatalities per accident where the subject cars are involved. When the car mass is 1150 kg, the number of fatalities per accident takes the minimum value, while the fatality rate of the subject car and that of the other car are almost the same. Then, the car with a mass of 1150 kg is considered most compatible in the current car population in Japan. The compatible car mass of 1150 kg is almost the same as the average mass of passenger cars in Japan, that is 1131 kg [Mizuno et al. 1997]. This is because there is a high possibility that the subject car with mass close to the average will crash into the other car with a small mass difference from the subject car.

When the mass of the subject car is in the range of 750 kg < m < 1350 kg, the number of fatalities per accident is small. However, when the subject car mass is less than 750 kg or greater than 1350 kg, the number of driver fatalities per accident increases. Thus, in order to decrease the total number of fatalities, it is necessary to design the lighter car so as to keep in mind the safety of the drivers in the subject cars, and to design the heavier car while keeping in mind the safety of the drivers in the other cars.

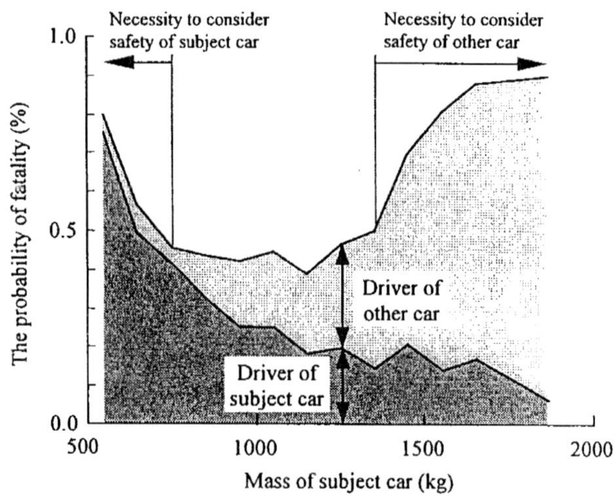

Figure 8. Car mass and the driver fatality of subject and other car in car-to-car frontal collisions.

Car classes have different mass, stiffness and geometry distributions. The effects of mass, stiffness and geometry are combined when the compatibility is analyzed by car class.

Figure 9 shows the number of driver fatalities in the subject and other car per thousand accidents. For SUV and 1BOX, the total number of driver fatalities is large and the proportion of the fatalities in other cars is large, so SUV and 1BOX can be considered incompatible cars. On the other hand, for mini cars, the number of fatalities in the subject car is the largest in all car classes. Therefore, mini cars cannot be said to be compatible in car-to-car frontal collisions. The wagon and sedan B are compatible cars in car-to-car frontal collision because the proportion of the number of fatalities in the subject to that in other cars is almost the same, and the total number of fatalities in the subject and other cars per accident is small. However, the number of registrations of the incompatible car types such as SUV and 1BOX is increasing, while that of sedan B, a compatible car type, is decreasing [Mizuno and Kajzer 1997].

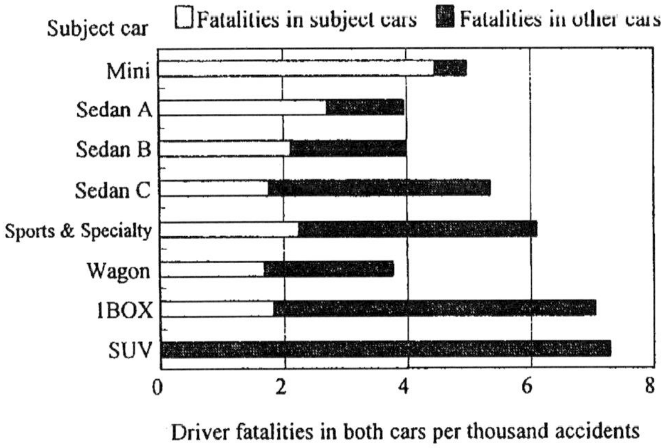

Figure 9. Car compatibility (N=531).

The aggressivity estimated by method 1 is shown in Figure 10. In this method, cars can be defined as aggressive when the aggressivity value is greater than one, because the number of fatalities in the other car is larger than in the subject car. Therefore based on Figure 10, the SUV, 1BOX, sedan C and sports & specialty can be described as aggressive. According to this analysis, the aggressivity ranking of the car itself is shown as:

Mini < Sedan A < Sedan B < Wagon
< Sports & Specialty < Sedan C < 1BOX < SUV.

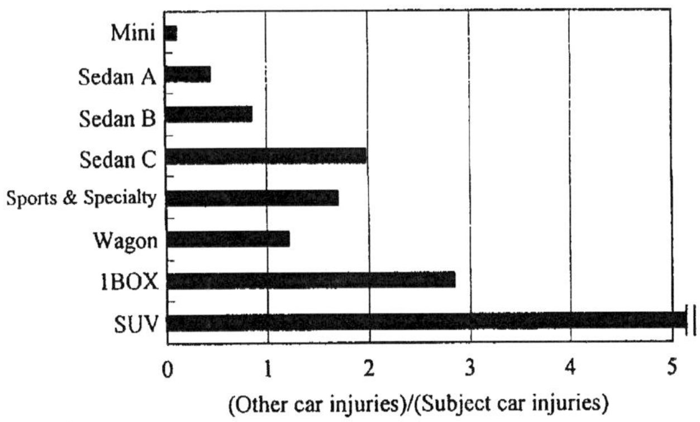

Figure 10. Car aggressivity calculated by method 1.

Computer Simulation of Mini Car Crash

Computer simulations of a crash of a mini car were performed in order to clarify the injury risk to the driver of mini car and to examine a compatible mini car.

Mini Car Crash into Rigid Wall - The crash of a mini car into a rigid wall at 50 km/h was analyzed in terms of different stiffness of the mini car (k). Figure 11 shows the variation of the acceleration, deformation of the car and the firewall intrusion with the stiffness of the mini car. The acceleration increases with the stiffness, while the deformation and the intrusion decrease. Thus when the stiffness increases, the driver is exposed to injury risk due to the high acceleration. On the other hand, when the stiffness decreases, the driver is exposed to injury risk due to the large intrusion.

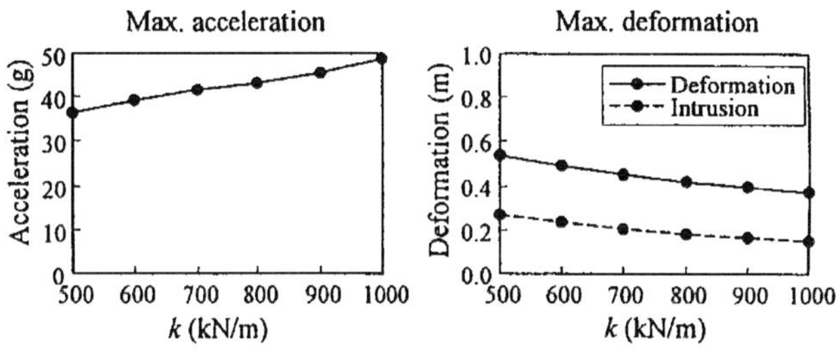

Figure 11. Maximum acceleration and deformation of mini car in crash into a rigid wall with varying stiffness of the mini car (k).

Figure 12 shows the driver behavior where the stiffness of the mini car (k) is 500 kN/m and 1000 kN/m, respectively. When the stiffness of the mini car is 500 kN/m, the intrusion is large but the acceleration of the car is small. As a result, the head and chest movement of the driver is less, but the foot rotation angle at 500 kN/m is greater than at 1000 kN/m.

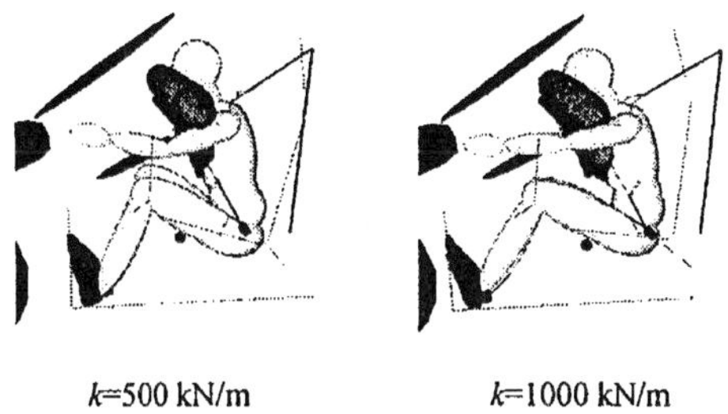

Figure 12. Kinematics of the driver in a mini car in a crash into a rigid wall (50 km/h).

Figure 13 shows the relation between the injury risk to the driver and the stiffness of the mini car (k). When k is lowest (500 kN/m), the HIC is 706, chest acceleration (3 ms) is 55.9 g and chest deflection is 0.042 m, all of which are less than the injury tolerance levels. The HIC and chest acceleration increase consistently with the stiffness of the mini car. On the other hand, the chest deflection, femur force, tibia axial force and moment do not change so much with the stiffness of the mini car, and its level is less than the injury threshold. These results suggest that in a mini car crash into a rigid wall, the risk of injury to the driver decreases when the front stiffness is low.

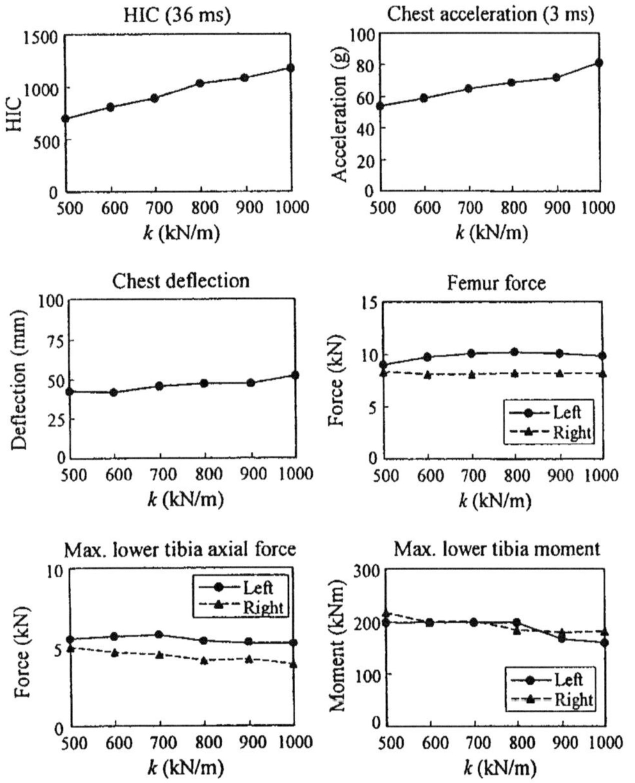

Figure 13. Driver injury risks in a crash into a rigid wall with varying stiffness of the mini car (*k*) (50 km/h).

The transition to serious injuries of the lower extremities (AIS 3 or more) appears when the intrusion exceeds 0.25 m [Morris et al. 1997]. As shown in Figure 11, the intrusion of the mini car firewall is less than 0.27 m in a crash into a rigid wall at 50 km/h. Therefore, in this type of crash configuration, the intrusion is a less important factor in determining the injury risk to the driver of a mini car, whereas the acceleration causes the majority of injuries.

<u>Mini Car Crash into Large Car</u> - Simulations of an offset frontal collision between mini and large cars were carried out. The overlap ratio of the mini car is 50% and that of the large car is 40%. Figure 14 shows the deformation of the car and intrusion of the firewall by the stiffness of the mini car (*k*). The acceleration level of the mini car in this type of crash is lower than in a crash into a rigid wall, whereas the car deformation and firewall intrusion of the mini car become large, especially when the stiffness of the mini car is small. Thus, in this type of collision, the effects of the acceleration and intrusion are combined, and the risk to the driver of the mini car becomes high.

As can be seen in Figure 15, when the mini car is less stiff (*k*=500 kN/m), the steering column, instrument panel and toe pan intrude and hit the chest, knee and foot of the driver, respectively.

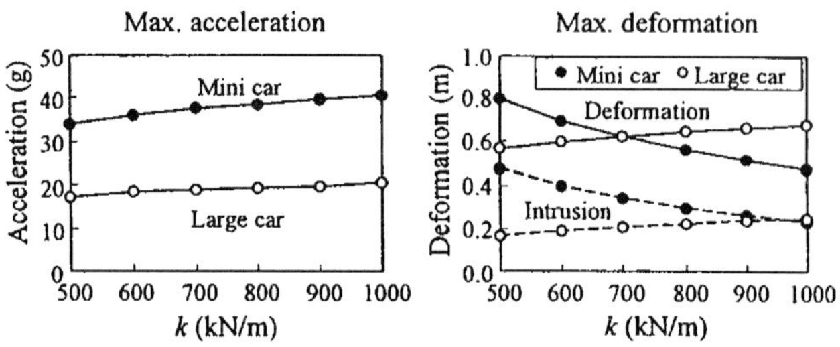

Figure 14. Maximum acceleration and deformation of mini and large cars in a car-to-car frontal collision with varying stiffness of the mini car (*k*).

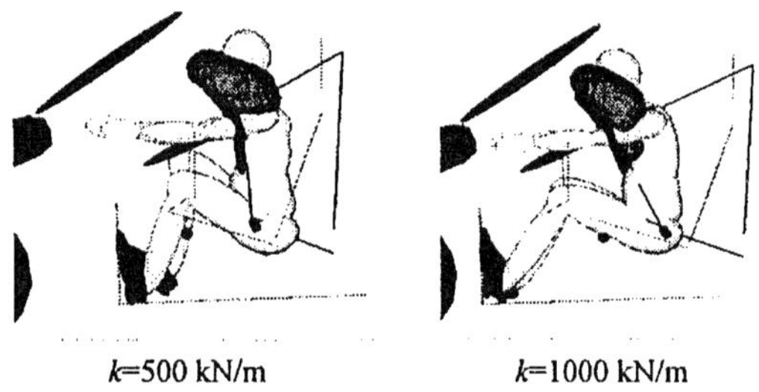

Figure 15. Kinematics of the driver in a mini car in a crash into a large car.

Figure 16 shows the variation of the injury risks to the drivers in mini and large cars with the stiffness of the mini car (*k*). When a comparison is made with Figure 13 and Figure 16, the HIC and chest acceleration of the driver in the mini car are lower in a crash with a large car than into a rigid wall, whereas the chest deflection, femur force, tibia force and tibia moment are higher. The chest deflection and tibia force are strongly affected by intrusion. Thus, in a crash of the mini car with a large car, the intrusion is an important factor in injuries. In addition, Figure 16 indicates that the injury risk of the driver in a mini car is higher than for the driver in a large car, irrespective of the stiffness of the mini car. This result corresponds to the findings from accident analysis that the injury risk to the driver in a mini car is high, while in a large car is low as shown in Figure 9.

When the stiffness of the mini car increases, there is a decrease in the risk to its driver as estimated on the basis of intrusion criteria such as chest deflection, maximum femur, tibia force and tibia moment. However, the HIC and chest acceleration of the driver in the mini car increase with the stiffness of the mini car because its acceleration becomes high.

As the stiffness of the mini car increases, the risk of injury to the driver in a large car tends to become greater. When the stiffness of the mini car is high, the chest acceleration, chest deflection, femur and tibia forces of the driver in the large car increase because both acceleration and intrusion of the large car become high. Nevertheless, the risk of injury to the driver of the large car is less than that of the driver of the mini car, and even less than the tolerance level of the relevant injury criteria.

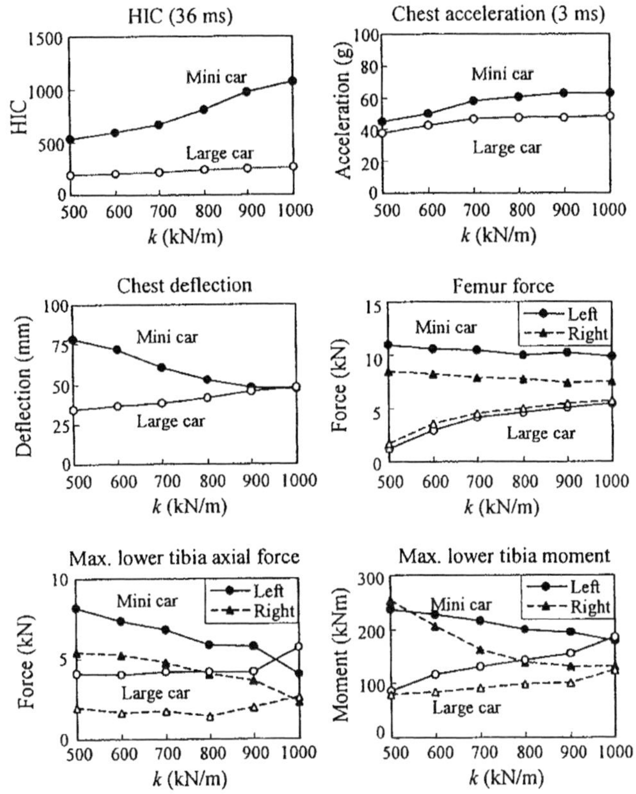

Figure 16. Driver injury risks in mini and large cars in a car-to-car frontal collision with varying stiffness of the mini car (k).

This analysis of collisions between mini and large cars demonstrates that the mini car should be stiff enough to prevent a large intrusion into the passenger compartment in a car-to-car frontal collision because greater intrusion means a higher risk of chest deflection and injury to the driver's lower extremities.

<u>**Effect of Restraint Systems**</u> - It is possible that the injury risk of the driver in mini car is reduced by the restraint system despite its high acceleration. In the current study, the stiffness of 1000 kN/m is applied for the mini car because the chest deflection, femur, tibia forces and tibia moment become low at this stiffness level as shown in Figure 16.

We evaluated the injury-reducing effect of the restraint systems (seat belt force limiter, energy absorbing steering system, knee bolster and their combination) on the driver of the mini car, in comparison with the results when an airbag and seat belt are simulated as a basic restraint system.

Figure 17 and Figure 18 show the effect of restraint systems on the injury criteria of the driver of a mini car in crashes into a rigid wall and a large car, respectively. The

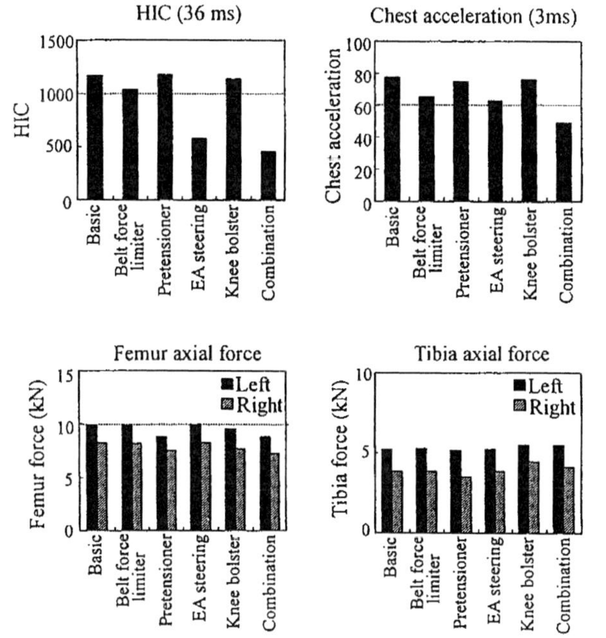

Figure 17. Effect of the restraint systems on the injury risks to the driver of the mini car in a crash into a rigid wall (k=1000 kN/m).

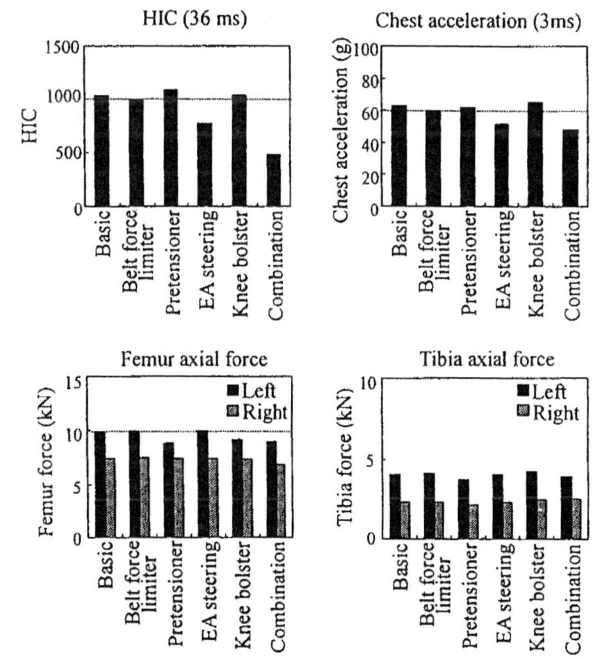

Figure 18. Effect of the restraint systems on the injury risks to the driver of the mini car in a collision with a large car (k=1000 kN/m).

injury-reducing effect of each restraint system for the driver of mini car differs between the two kinds of crashes. In the crash into the rigid wall, the seat belt force limiter effectively decreases chest acceleration and HIC by reduction of force transfer from the seat belt to the torso of the driver. Nevertheless, the force limiter has a small effect on the injury risks of the driver of the mini car in collision with a large car. In this type of crash, a large force is applied to the driver's chest by the steering wheel, not by the seat belt. Thus, the seat belt force limiter has a small effect on reduction of the chest acceleration in a collision with a large car.

The EA steering system is effective for both of the above-mentioned crashes. The movement of the steering column can decelerate the driver's head and chest by absorption of energy. The seat belt pretensioner reduces the femur force. The knee bolster also reduces the femur axial force, particularly in a crash with a large car. The restraint systems have little influence on the tibia force of the driver in a crash into a rigid wall or a collision with a large car. Thus, to reduce the tibia forces, the intrusion of the toe pan must be reduced.

When a mini car with high stiffness is equipped with restraint systems combining airbag, seat belt force limiter with pretensioner, EA steering system and knee bolster, the injury criteria levels for the driver are below the thresholds in either crashes into a rigid wall or a large car.

Additional Crush Space of Large Car - The additional crush space of the large car reduces the injury risk to the driver in the mini car due to reduction of acceleration and intrusion of the mini car in a collision. Figure 19 shows the chest acceleration, chest deflection, femur and tibia forces of drivers in the mini and large cars in terms of the length of additional crush space (*c*) of a large car (200 kN). The additional crush space reduces the chest acceleration and femur force of the driver in a mini car, when the stiffness of the mini car (*k*) is high. Particularly when *k* is small, the chest deflection and tibia force of the driver in the mini car decrease due to the small intrusion into the mini car, as the additional crush space of the large car increases. Therefore, the additional crush space of a large car has an injury-reducing effect on the driver of the mini car by reducing the acceleration and the intrusion of the mini car.

The chest acceleration of the driver in a large car decreases when the additional crush space of the large car is large due to the low acceleration of the large car. The chest deflection slightly increases by the additional crush space of the large car. The femur and tibia forces of the driver in the large car increase with the additional crash space of the large car, and have large values when the mini car is stiff. Thus, the analysis indicates that the additional crush space of the large car is effective in reducing injury risk to the driver of the mini car. However, when the mini car is stiff, the risk to the driver in the large car, especially for injuries to lower extremities, becomes high.

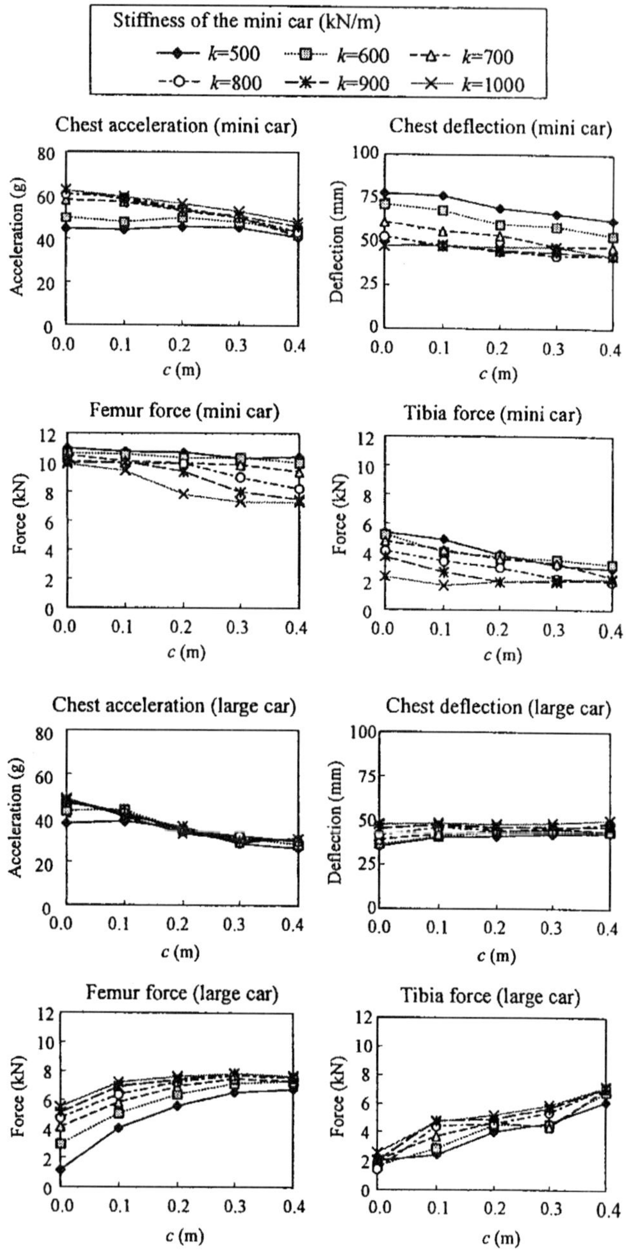

Figure 19. Driver injury risks in mini and large cars with the length of additional crush space (*c*) of the large car.

Car-Pedestrian Accidents

Accident Analysis - In car-pedestrian accidents, the vehicle geometry affects the injury risk to the pedestrian. Figure 20 illustrates the distribution of pedestrian injuries per thousand accidents by body region, injury severity and

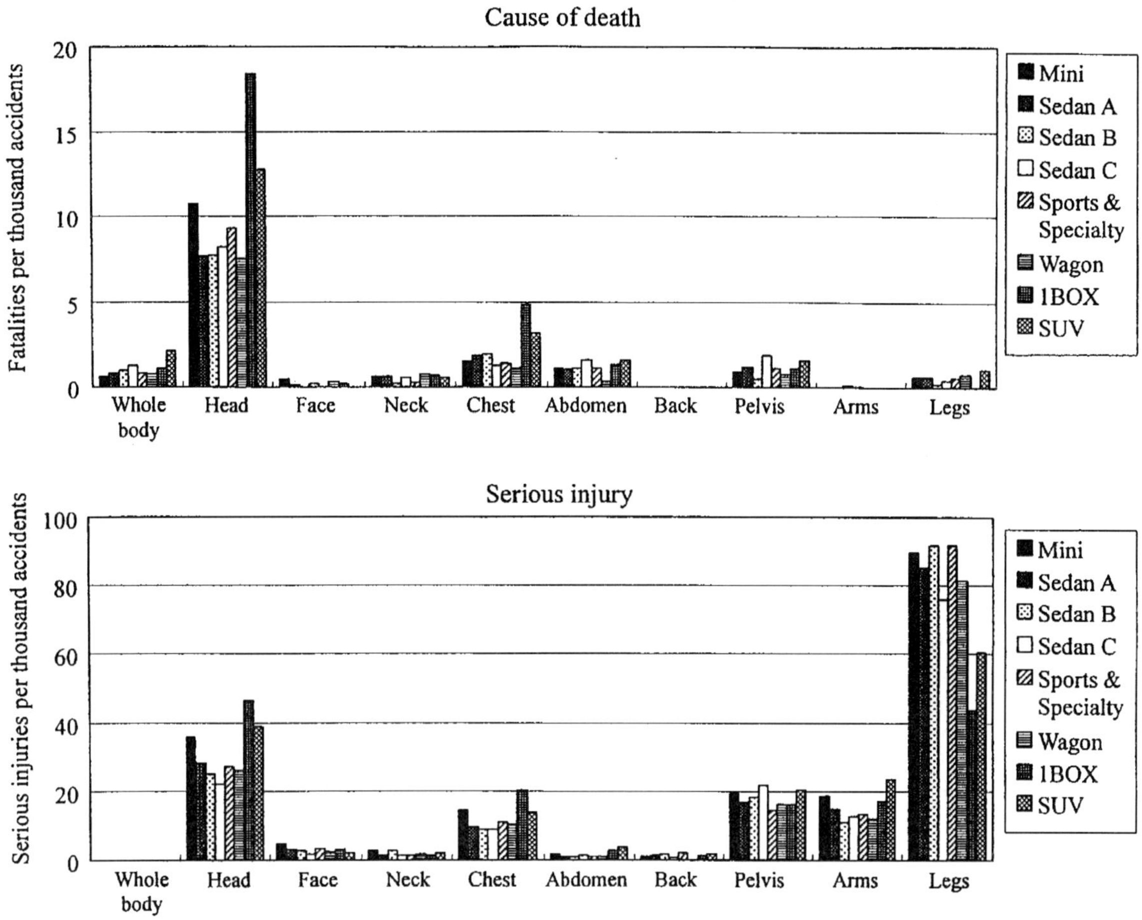

Figure 20. Distribution of pedestrian injuries per thousand accidents by body region, car shape and injury severity for the velocity recognized to be dangerous ≤ 40 km/h and pedestrian aged 13 or more (1994-1996).

car class. To exclude high-velocity collisions, especially between sports & specialty cars and a pedestrian, the velocity recognized to be dangerous was limited to 40 km/h or less.

The head is a dominant body region in fatalities for all car classes. The number of fatalities due to head and chest injuries is about two times larger for 1BOX than for bonnet-type cars (mini, sedan A, B, C, sports & specialty and wagon). In a serious injury, the number of head and chest injures which can be a cause of death is larger for 1BOX than for a bonnet-type car. Therefore, it is considered that the front shape of 1BOX is more aggressive for a pedestrian than that of a bonnet-type car. However, the shape of a bonnet-type car is aggressive in relation to pedestrian legs because the number of serious leg injuries is large in crashes with this type of car.

The risk of head injury to pedestrian when struck by a mini car is higher than for other bonnet-type cars. The SUV has a high aggressivity in relation to the head and chest of the pedestrian due to the height of the hood edge and bumper and the high stiffness of the vehicle body. As the body front shape affects the distribution of pedestrian injuries, the modification of the shape of the car front can be effective to increase the compatibility between car and pedestrian.

<u>Simulation of Car-Pedestrian Accidents</u> - Computer simulations were carried out for crashes of mini car, sedan B and 1BOX with a pedestrian. Figure 21 shows the pedestrian kinematics when the head of the pedestrian contacts the vehicle body. The kinematics differs when the pedestrian is struck by vehicles with different front shapes.

When a pedestrian is struck by a mini car or sedan B, the bumper hits the leg and the hood edge hits the thigh, then the upper torso of pedestrian rotates toward the hood of the car. The pedestrian's head contacts the windscreen when hit by a mini car, and the cowl area when hit by sedan B. With a 1BOX vehicle, the whole body of the pedestrian is struck by the vehicle front almost at the same time. The chest contacts the upper part of the front panel, the head contacts the lower part of the windscreen, and then the whole pedestrian body is projected ahead of the vehicle. The pedestrian kinematics when struck by a vehicle is influenced by the translational and rotational movement of the pedestrian. With a mini car and sedan B the rotational movement is dominant, while with a 1BOX the translational movement is dominant.

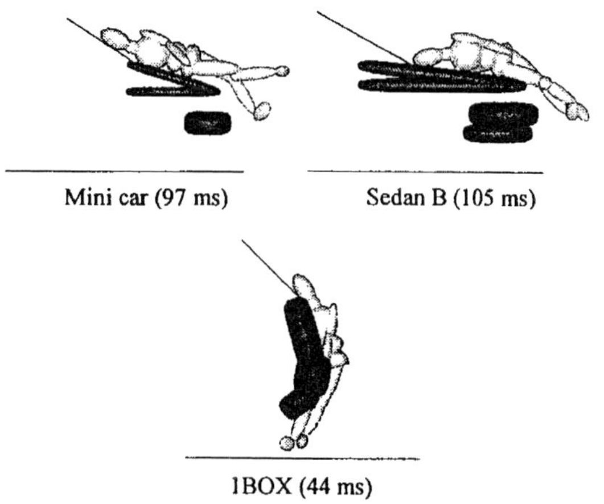

Figure 21. Pedestrian kinematics when the pedestrian head contacts the car body.

HIC is affected by the head resultant velocity at impact as well as the stiffness of the vehicle body area where the head makes contact. The time history of the head resultant velocity varies with vehicle shape. Figure 22 shows the head resultant velocity in relation to the respective car class. As shown in this figure, when the pedestrian is struck by a 1BOX, the head resultant velocity decreases consistently. When struck by the mini car and the sedan B, the head resultant velocity increases gradually due to the rotation of the upper body of the pedestrian, decreases gradually, and then drops abruptly after the pedestrian's head contacts the car body. For a mini car, the head contacts the windscreen in such an early phase that the head resultant velocity at impact is high.

The injury risk to the pedestrian body region differs with the shape of the car impacting it. The injury parameters of the pedestrian are HIC, chest, pelvis and thigh accelerations (3 ms) as shown in Figure 23. The HIC is the highest for sedan B due to the high resultant velocity at impact and the high stiffness of the cowl area where the pedestrian head makes contact. When the pedestrian is struck by a mini car, the HIC is lower than sedan B due to low stiffness of the windscreen. The chest and pelvis accelerations are highest when struck by 1BOX compared with other cars. The thigh acceleration is higher for sedan B and mini car than 1BOX. The distribution of injury risk to the chest and leg by car class agrees with those of real-world accidents (Figure 20.). However, in real-world accidents, the injury risk to the pedestrian head is highest when struck by a 1BOX, followed by a mini car and a sedan B. This differs from the simulation results.

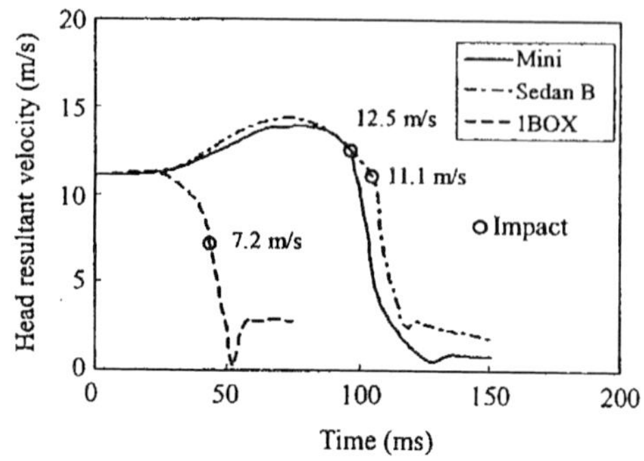

Figure 22. Head resultant velocity.

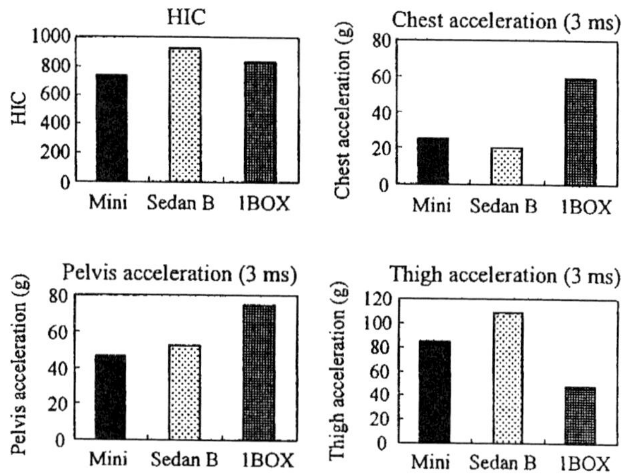

Figure 23. Injury risk to the pedestrian when struck by a car.

DISCUSSION

Mini Car Crash

The number of fatalities per registered mini car in real accidents is small, so the mini car may be considered compatible. However, in the method used in the present study, the injury risk to the driver in the mini car is underestimated since the accident rate and crash velocity of this type of car are low. When the injury risk to the driver is estimated by the probability of fatal injury, it is clear that the mini car is an incompatible car type because of the high fatality rate of the driver. Computer simulations of car-to-car frontal collisions also indicate that the injury risk to the driver in a mini car is far higher than in a large car.

This high injury risk to the driver of the mini car in a collision with a large car cannot be evaluated by the crash test into a rigid wall that is currently required by law. In a crash into a rigid wall, there is no influence of car mass on the injury risk to the driver and the influence of intrusion is small. However, in a collision with a large car, the driver of the mini car is at high risk of injury due to the high acceleration and large intrusion based on its small mass and size.

Two methods are considered to reduce the injury risk to the driver of a mini car. The first is to stiffen the mini car. Since the acceleration of this car tends to be high, optimized restraint systems combining airbag, seat belt force limiter, pretensioner, EA steering system and knee bolster is necessary. The stiff front structure and special restraint systems of the mini car can directly reduce the injury risk of the driver. In this method, no modifications of the large car are necessary to reduce the injury risk to the driver in the mini car. However, the aggressivity of a stiff mini car should be considered not only in car-to-car frontal collision but also in other types of collisions, such as side and rear-end collisions. In a low-velocity crash in which the airbag should not deploy, the risk of minor injury to the driver in the stiff mini car may increase due to high acceleration.

The second method is to provide a large car with additional crush space designed for a crash with a mini car. It is possible that by reducing the acceleration and the intrusion of the mini car, the injury risk to the driver in the mini car would be reduced. Thus, in both cases, whether the mini car is less stiff or stiff, this additional crush space in a large car is effective to reduce the injury risk to the driver of the mini car. On the contrary, the additional crush space of the large car causes intrusion into the large car, so the injury risk to the driver of the large car, particularly to the lower extremities, increases, when the mini car is stiff.

Car-Pedestrian Accidents

The accident data obtained in the present investigation demonstrate that the pedestrian has a high risk of head injury when struck by a 1BOX or a mini car. In the simulation, when struck by a 1BOX and a mini car, the HIC of the pedestrian is not so high because the head contacts the windscreen. The average elderly pedestrian aged 60-69 was used for the model. In the real mini car-pedestrian accidents, the head may contact the stiff part of the car such as windscreen frame and cowl area when the stature of the pedestrian is shorter than average or the velocity of the mini car is less than 40 km/h. As the width of the mini car is small, there is a high risk of the pedestrian striking A pillar of the car, or to be thrown away from the hood to the ground. When struck by a 1BOX vehicle, the pedestrian will often sustain a head injury because of being thrown to the ground following impact.

The simulation results show that the HIC of the pedestrian is small when his or her head contacts the windscreen. The mini car should be safety-engineered by designing the vehicle configuration so that should it strikes a pedestrian any contact of the head will be with the windscreen, no matter how large or small the pedestrian and for a wide range of impact velocity.

CONCLUSIONS

Mini car compatibility issues were discussed for car-to-car frontal collisions and car-pedestrian accidents using traffic accident data in Japan and computer simulations.

The results of the accident analysis for car-to-car frontal collisions are as follows.

1. A car with mass of 1150 kg is considered most compatible among the current car population in Japan. This compatible car mass coincides with the average mass of cars in Japan.
2. The SUV and mini car are the least compatible car types, with high and low aggressivity in relation to other cars, respectively.
3. Sedan B and wagon type cars are considered compatible. The proportion of the number of fatalities in the subject cars to that in other cars is almost the same, and the total fatalities in the subject and other cars are few.

Simulations of the safety of the driver in a mini car were performed using MADYMO for crashes into a rigid wall and into a large car. The following conclusions were obtained:

4. The crash test of a mini car into a rigid wall is insufficient to assure safety in a crash into a large car. This is because in a crash into a rigid wall the acceleration greatly influences the risk of injury to the driver of a mini car, whereas in a crash with a large car the effect of intrusion as well as acceleration is large.
5. The combination of the restraint systems in conjunction with high stiffness of the mini car provides good protection for the driver in either crashes into a rigid wall or a large car.
6. When a large car has additional crush space, it is effective for reduction of the injury risk to the driver in the mini car. However, if the mini car is stiff, the driver of the large car risks chest and leg injury due to the intrusion into the large car.

The accident analysis and simulation of pedestrian accidents were carried out. It is concluded that:

7. Incompatibility of the car geometry with a pedestrian has a large effect on the distribution of pedestrian injuries. Accident data show high injury risk to the pedestrian when struck by a mini car. These data could not be reproduced by computer simulation where a pedestrian is impacted only by the car without considering the head impact with the ground.

REFERENCES

Appel, H., Ni, J., and Deter, T., "Grenzen der Passive Sichrheit von leichten Fahrzeugen". VID Berichite NR. 1134. 1994. pp. 235-238.

Appel, H., "Compatibility Guideline for Passive Safety, for Active Safety and Even for the Total Transportation System". FISITA1996. 1996.

Buzeman D. G., "Car-to-Car and Single Car Crash Compatibility: Individual Effects of Mass, Structure, Stiffness and Geometry". Thesis for the Degree of Licentiate of Engineering, ISBN 91-7197-536-5, Chalmers University of Technology, 1997.

Evans, L., "Driver Fatalities Versus Car Mass Using a New Exposure Approach". Accident Analysis & Prevention. Volume 16. 1984. pp. 19-36.

Evans, L. and Frick, M. C., "Mass Ratio and Relative Driver Fatality Risk in Two-Vehicle Crashes". Accident Analysis & Prevention. Volume 25. 1993. pp. 213–224.

Evans, L., "Driver Injury and Fatality Risk in Two-Car Crashes Versus Mass Ratio Inferred Using Newtonian Mechanics". Accident Analysis & Prevention. Volume 26. No. 5. 1994, pp. 609–616.

Frei, P., Kaeser, R., Hafner, R., Schmid, M., Dragon, A., Wingeier, L., Muser, M. H., Niedere, P. F., Waltz, F. H., "Crashworthiness and Compatibility of Low Mass Vehicles in Collisions". SAE Paper 970122. 1997.

Hollowell, W. T., Gabler, H. C., "NHTSA's Vehicle Aggressivity and Compatibility Program". 15th ESV 1996.

IRF, "World Road Statistics". Geneva. International Road Federation. 1995.

Ishikawa, H., Yamazaki, K., Ono, K., Sasaki, A., "Current Situation of Pedestrian Accidents and Research into Pedestrian Protection in Japan". 13th ESV. 1991.

Ishikawa, H., "Kinematics of Automobile Collision — Analytical Formulas for One-Dimensional Impact". Journal of JARI. Volume 12. 1990. pp. 25-32 (in Japanese).

Ishikawa, H., Kajzer, J., Schroeder, G., "Computer Simulation of Impact Response of the Human Body in Car-Pedestrian Accidents". SAE Paper 933129, 1993.

Ishikawa, H., Digges K., Ennis, J., "Restitution Coefficients and Delta-V in Offset Frontal Collisions". Proceedings of JSAE Autumn Convention. 1995 (in Japanese).

ITARDA, "The Relation between Traffic Accidents, Driver and Vehicle", Research Report 1996a (in Japanese).

ITARDA "Report of Accident Investigation and Analysis in 1996". 1996b (in Japanese).

Joksh, H. C., "Velocity Change and Fatality Risk in a Crash – A Rule of Thumb". Accident Analysis & Prevention Volume 25. 1993. pp. 103–104.

Kaeser, R., Walz F. H., and Brunner, A., "Collision Safety of a Hard Shell Low Mass Vehicle". IRCOBI 1992.

Kaeser, R., Muser, M., Spiess, O., Frei, P., Guenat J., M., and Wingeier, L., "Passive Safety Potential of Low Mass Vehicles". IRCOBI 1995.

Matsumoto, H., Sakaida, M., Kurimoto, K., "A Parametric Evaluation of Vehicle Crash Performance". SAE Paper 900465. 1990.

Maurer, D., Thompson, G., M., Muller, R., Graffe, M., Weyersberg, A, "Safety Concepts for Very Small Vehicles, Example: Opel MAXX". 15th ESV. 1996.

Mizuno, K., Umeda, T., Yonezawa, H., "The Relationship Between Car Size and Occupant Injury in Traffic Accidents in Japan". SAE Paper 970123, 1997.

Mizuno, K., Kajzer, J., "The Compatibility Problems in Collisions Analyzed Using Accident Data". Proceedings of JSAE Spring Convention. 1997.

Morris, A., Thomas, P., Taylor A. M., Wallace, W. A., "Mechanisms of Fractures in Ankle and Hind-Foot Injuries to Front Seat Car Occupants". SAE Paper 973328. 1997.

NHTSA, "Report on International Harmonized Research Agenda", Proceedings of 15th ESV. 1996.

Sakurai, T. and Aoki, T., "On the Safety Body Structure of Finite Element Method Analysis". 12th ESV. 1989.

Tarrière, C., "Accident Research and Experimental Data Useful for an Understanding of the Influence of Car Structural Incompatibility on the Risk of Accident Data". 14th ESV. 1994.

Tarrière, "Compatibility Issues and Vulnerable Uses". Crashworthiness of Transportation Systems: Structural Impact and Occupant Protection. Kluwer Academic Publishers. 1997. pp. 139-170.

Ventre, P., "Homogeneous Safety Amid Heterogeneous Car Population". 3rd ESV. 1972.

Ventre, P., "Proposal for Test Evaluation of Compatibility Between Very Different Passenger Cars". 4th ESV. 1973.

Waltz, F., H., Kaeser R., and Nieder, P., "Occupant and Exterior Safety of Low Mass Cars (LMC)". IRCOBI 1991.

Yang, J., "Injury Biomechanics in Car-Pedestrian Collisions". Doctoral thesis. ISBN 91-7197-527-6. Chalmers University of Technology. Göteborg. 1997.

710861

Smaller Vehicle versus Larger Vehicle Collisions

Derwyn M. Severy, Harrison M. Brink, and David M. Blaisdell
Institute of Transportation and Traffic Engineering,
University of California, Los Angeles

Abstract

The research techniques of instrumented full-scale collision experiments were applied to evaluate relative crash performances of smaller passenger vehicles colliding with larger vehicles. The larger vehicle weighed from 1.5-4 times as much as the smaller vehicle. The structure-overriding tendencies of larger vehicles in a particular collision were found to greatly influence the severity of exposure to injury for occupants of the smaller vehicle; relative strength of structures was similarly important. The crash safety of a motorist is shown to depend more on the use of adequate restraining devices than on the smallness of his car. Mismatched sizes of vehicles were crashed head-on, as well as in rear-end and intersection-type exposures. Analytical relationships of post-impact displacements as well as transducer and photographic instrumentation data are presented.

Actual accident investigations were conducted which provided background preparation for this series of crash tests. Results of these investigations provide additional findings relative to the small versus larger vehicle collision injury problem.

IT'S DANGEROUS TO BE ALIVE, but the alternative makes this hazard an endurable burden for most of us. Similarly, in certain respects, it's more dangerous to drive a small car than a large car, but this hazard has alternatives that have caused more than half of the population seeking new cars to decide to endure this hazard. Briefly stated, the alternative of driving oversize passenger cars for "short trip" activities represents specific compromises which this majority population appear unwilling to accept. The initial price of larger cars is higher; their greater complexity necessitates more frequent and costlier maintenance, their larger size requires more fuel, often of higher premium gasoline. Even with the advent of regular and no-lead tolerant high cubic displacement engines, the increased rate of fuel consumption to accommodate antismog gadgetry has more than offset savings on purchases of the cheaper low-octane fuel. Larger cars are hastening the termination of our dwindling oil reserves, predicted by some as a reality around the turn of the century; larger cars contribute proportionately more to a predicted air pollution disaster, should corrective measures not be accomplished in time. Parking problems increase with the size of the car; additionally, the larger car's turning radius complicates its ease of handling, and its width reduces vehicle to vehicle clearance between lanes as well as the separation from opposing traffic. And then there is the youth image and similar intangible aspects that motivate people to purchase small cars.

For all the smaller car advantages, almost half of the United States population still select large cars; however, on the basis of related observations, the probability is that safety is not foremost among the factors governing a decision to buy big. Habit pattern still accounts for many such purchases, as does prestige, need to accommodate a larger family, or for carrying bulky equipment. Unquestionably, these factors compete to displace safety to a lower priority of consideration.

In this paper, the safety aspects of size will be shown to depend significantly on whether or not restraining devices are used by the small car motorist impacting a larger vehicle and that for some collision exposures, vehicle size has no bearing on injury exposure. Additionally, certain safety advantages of a small car are inherent in its design, such as maneuverability, increased roof crush resistance, absence of power-assist steering, and brake failures, all tending to make an alternative to the purchase of a large car an acceptable decision.

Background

The UCLA-ITTE Automotive Collision Research Project has conducted four series of head-on collision experiments, five series of read-end collision experiments, and other series involving passenger vehicle collisions corresponding to the more frequently occurring accidents. Generally, these experiments utilized medium-sized passenger vehicles owing to their preponderance during the 1950s and 1960s. The question of vulnerability of motorists to serious injuries because of vehicle size mismatch should not be directed exclusively to the subcompact car; the same potential for action occurs for a compact, medium, or large-size domestic vehicle when impacted by a correspondingly larger vehicle such as a truck. For this reason, the smaller vehicle for the two head-on collision experiments reported did not involve subcompacts but instead passenger vehicles of compact (Lark) and medium (Ford) sizes.

During a prior rear-end collision series involving 1967 Fords read-ended by 1967 Fords, data was assembled that was appropriate for comparative analysis with Experiment 107 reported in this paper on a 1969 VW (Bug) rear-ended by a 1967 Ford.

Other interesting relationships are presented based on previously reported findings by other researchers and by the authors as well as findings from investigations of actual accidents.

Facilities and Equipment

A fenced-in facility adjacent to the U.S. Naval Station, Long Beach, California has been used by UCLA-ITTE as their field station for the past 15 years. The facility includes an asphalt surface, mobile shops and buildings, as well as several towers to support photographic equipment for observation of collision experiments. Support facilities included four mobile vans consisting of electronics, photographic and machine shops, as well as a project control center.

An aluminum monorail guide track attached to the asphalt surface provided directional control for vehicles being powered to a crash (Fig. 1). The various techniques developed during prior years of study that were used for the experiments reported in this paper are described in prior publications cited in the bibliography.

Experimental Procedures

Variables Under Study - Prior UCLA studies involving intersection, rear-end, and head-on car-to-car collisions as well as mismatched weight collisions between passenger vehicles and pedestrains, motorcycles, trucks, and buses were conducted using a basic reference velocity of 30 mph. For this reason, the latest series of head-on collision experiments and the latest small/large car rear-end collision were likewise conducted at 30 mph. Alignment for these experiments was an inline impact with no vehicle offset. Location and size of occupants were selected to provide a match-up between vehicles and for comparison with findings from prior studies.

Where comparisons were made with prior experiments involving speeds other than 30 mph, adjustments were made on the basis of established trends so as to accurately reflect comparative data.

Instrumentation - Electronic transducers and high-speed motion picture cameras represent the principal systems used to obtain data. These systems along with related instrumentation are listed in Table 1. Other data collected include pre- and post-crash observations concerning the location of occupants and restraints, head supports, and instrumentation.

Photographic instrumentation ranged from high-G tolerant motion picture cameras to conventional motion picture cameras and from electronically actuated still cameras to high-speed motion picture cameras as depicted in Fig. 1.

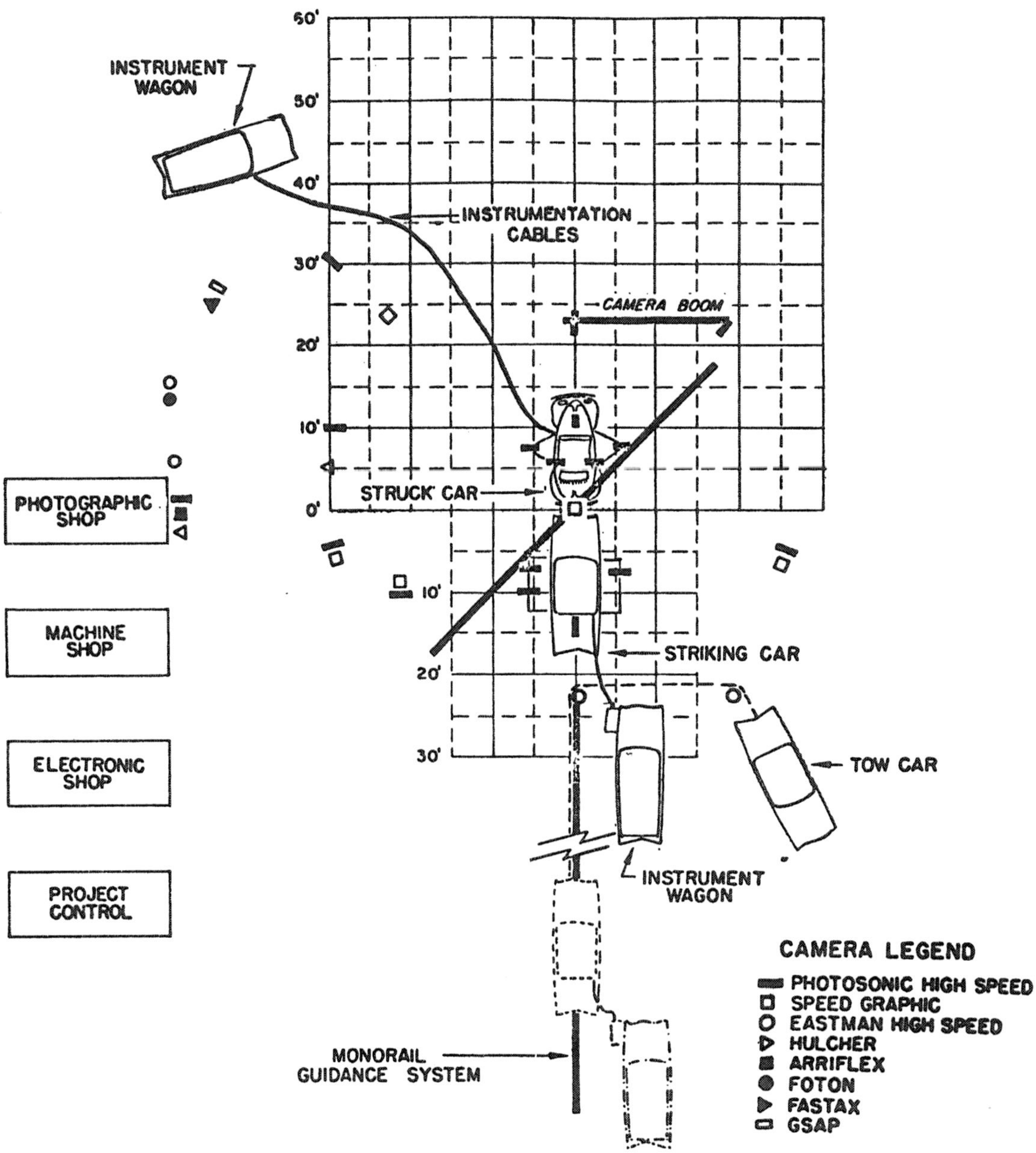

Fig. 1 - Vehicle control system, photographic and electronic facilities

Table 1 - Summary of Instrumentation

Device	To Provide	Location	Specifications
High-speed motion picture cameras	Vehicle collision dynamics and passenger kinematics	See impact center camera assignment, prior figure	2-Eastman, high speed; 2-Fastax WF-3; 16 mm, 600-1700 frames/s
Moderately high-speed cameras	Vehicle collision dynamics and passenger kinematics	See impact center and vehicle camera assignments prior figures	14-Photosonics, 1B, 1-GSAP MBH 200-26; 200 frames/s 16 mm
Standard motion picture cameras	General photographic coverage	Adjacent to impact center	1-K100, 16 mm; 1-Arriflex, 16 mm; 1-Bolex, 16 mm; 1-GSAP, 16 mm
Special motion picture cameras	Sequential, large format photographs of collision events	Strategic positions at impact center	2-Hulcher, Model 102, 20 frames/s 70 mm (camera rotated 90 deg to permit cine-reduction); 1-Bell & Howell Eyemo, 48 frames/s 35 mm; 1-Foton 5 frames/s 35 mm Bell & Howell
Still cameras	Precision-timed photographs	See impact center camera assignment, prior figure	5-Super Speed 4x5 in Graflex 1/1000 s
Calibrated references	Fixed ground reference points for micromotion analysis	Near the impact center	Reference markers at 5 ft intervals and vertical posts calibrated yellow and black at alternate 1 ft increments
Reference targets	Precision photographic references for micromotion analysis	At strategic positions on the vehicles and occupants	Diamond-shaped yellow and black targets
Electrical accelerometers	Acceleration measurements	Head, chest, and hips of occupants. Vehicle body	B & F-LF 50-50 and LF 100-75-120; Statham A 38a-60-350, A6-100-350, AJ43-100-350, A52-50-300, A5a-100-300, Ca-100-130; CEC 4-202-0001
Seat belt tensiometers	Measurement of belt loads	On belt webbing	Dual link and thread-through type, special UCLA construction

Device	To Provide	Location	Specifications
Recording oscillograph	Amplitude-time records of transducer signals	Carried by instruments recording vehicles for striking car and instrumentation trailer for struck car	2-18 Channel Consolidated Type 5-114 P2; 1-50 Channel CEC, Type 5-119, P3; 1-12 Channel CEC, Type 5-118
Electronic time delays	Precision timing for still photography	Electric pressure pads near impact center	100, 200, and 500 ms solid-state time delay units built by UCLA
Pulse generators	Timing for high- and moderate-speed cameras	Near cameras	Wollensak, 100 cps; 1000 cps and special 100 cps solid state devices
Photographic oscillographic synchronization units	Zero time (vehicle contact; flash bulb for film and pulse for oscillograph	Pressure switches between contacting surfaces of vehicles. Photocell and flash bulb on vehicle	SM flash bulb with photocell for oscillograph signal
Flash bulb timer	Timing at 0, 50, 100, 150, 200, 250, and 500 ms; flash bulb for film and pulse for oscillograph	Near impact center in view of all cameras	Solid-state timing device with SM flash bulbs and photocell for oscillograph signal
Auxiliary timer	Backup timing	Near impact center in view of all cameras	Rotating yellow-black drum; constant speed 1740 rpm motor
Fifth wheel	Accurate impact velocity	On tow car	Labeco, Model 5101, Weston Scale 0-60, 0-120 mph
Speed counter	Car speed data; time-displacement	On guide yoke	Induction pickup for oscillograph
Polar coordinate grid	Position data for vehicles from point of impact to positions of rest	On asphalt surface at impact center	Yellow traffic marker paint

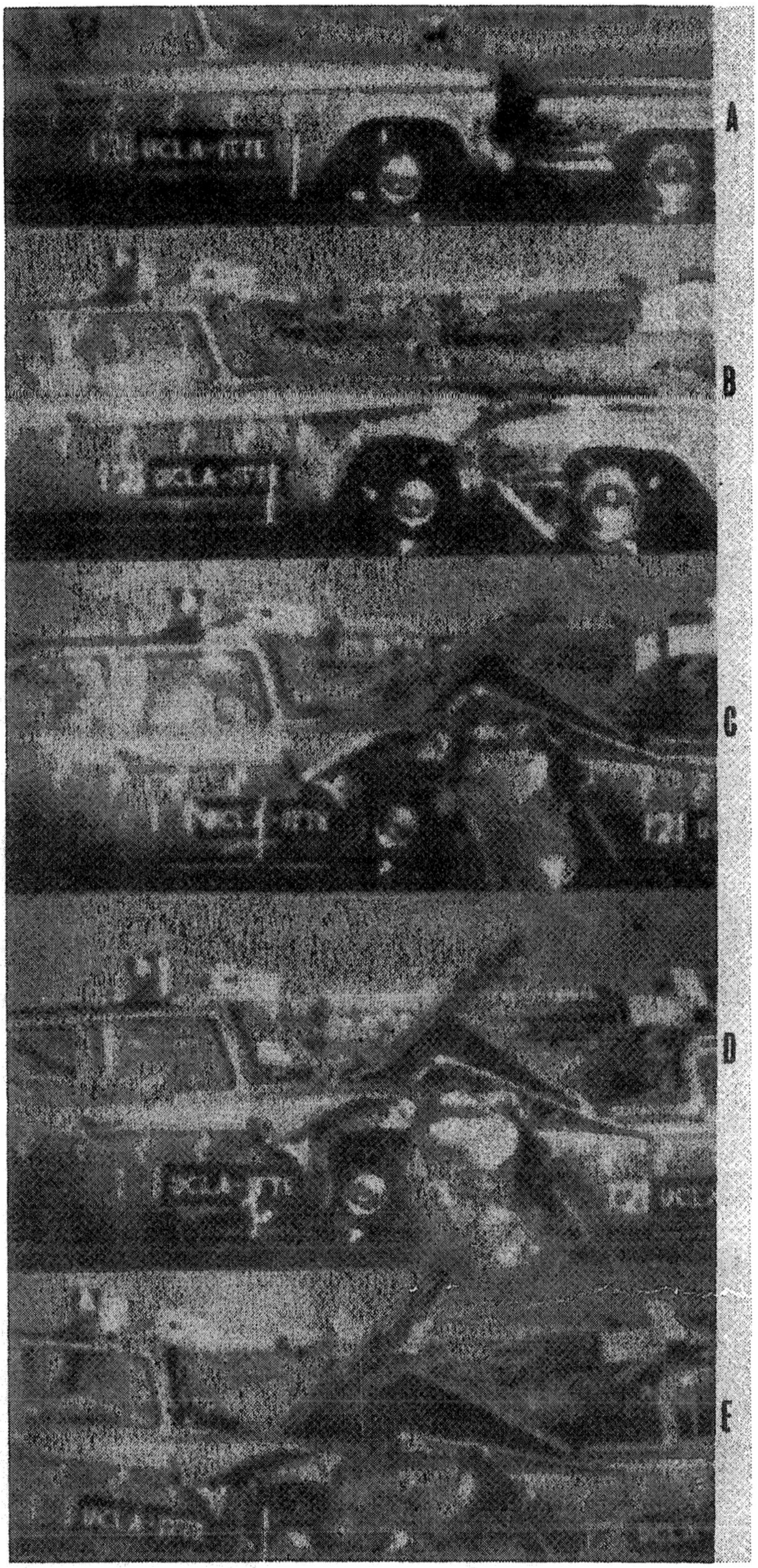

Fig. 2 - 30 mph head-on collision between Buick and Studebaker Lark; weight ratio 1.5:1 (Experiment 121)

Experiment 121 - Experiment 121 was an inline, head-on collision between a 1960 Studebaker Lark 2-door sedan weighing 2965 lb and a 1959 Buick 4-door sedan weighing 4880 lb with attached instrumentation; each vehicle was traveling 30 mph at impact. Gross weight of the Lark with four dummy occupants was 3795 lb, gross weight of the Buick was 5710 lb, for a weight ratio of 1.5:1.

The driver of each car was a 50th percentile adult male dummy and the right front seat passenger was a 95th percentile adult male dummy; each was restrained by a lap belt. They were seated in production bench seats with special anchorage provisions. The rear seat occupants were 95th percentile adult male dummies. The left rear seat occupants in each car had a three-point restraint; the right rear occupants were unrestrained except for the passive restraint of the anchored seatback ahead of them.

Collision Dynamics - The Studebaker impacted the front of the Buick with the bumpers squarely matched at the instant of initial contact (Fig. 2A). Touch point of the Buick was 12 in onto the Studebaker's side of the zero reference line. After contact, the Studebaker crushed forward 20 in before coming to a momentary stop (Fig. 2C). Following maximum collapse, the Lark rear-end commenced to elevate (Fig. 2E), and as rebound occurred, the rear end reached a maximum elevation of 1-1/2 ft, accompanied by the rear tires leaving the pavement (Fig. 3). After maximum compression of the two vehicles, total rearward displacement of the Studebaker was 52 in ; thereafter, owing to suspension elastic rebound, the Studebaker shifted forward again approximately 4-1/2 in to its final position of rest. The Buick, after impact, crushed forward 2.5 ft and then appeared to come to a momentary stop, before continuing forward an additional 0.6 ft. As the Buick front end collapsed, the passenger compartment continued forward a distance of approximately 22 in before the front seat passengers showed evidence of commencing any significant forward movement relative to

Fig. 3 - Rear tires of Lark leave pavement as rebound occurs (Experiment 121)

Fig. 4 - Buick passenger compartment continues forward 22 in before occupants begin to shift forward (Experiment 121)

the vehicle interior (Fig. 4). The heavier Buick follow through, after bumper contact, to its final position of rest was only 3.1 ft (Fig. 5).

Driver - Lark - Following onset of impact, the driver shot forward with his hinged seatback following him; he impacted the hub of the steering column with his upper chest, after his mid-chest had deflected the lower rim of the steering wheel forward. This action caused the upper portion of the steering wheel rim to strike the driver's mouth. The two-spoke steering wheel rim was fractured free from the hub as a result of driver chest impact (Fig. 6). His chest then rotated counterclockwise (looking down on the driver) as he was thrown forward past the hub, striking the instrument panel with his head. Holes were chopped in the clothing and chest of the driver as a result of his severe impact with the steering hub. The peak chest resultant deceleration was 71 G, occurring at 85 ms after initial contact. Forward excursion of his torso amounted to 10 in before his knees intruded the instrument panel sufficiently to prevent further stretching of his seat belt; thereafter, he rebounded to the center of the front seat and slumped to the right against the front seat passenger.

Driver - Buick - During the collision, the steering column shaft extended 3-1/2 in further within the passenger cabin as a result of front-end crush; the 50th percentile driver, wearing a lap belt, was struck in the chest by the steering wheel hub after the lower rim had been deflected by contact with his lower chest; this action projected the upper rim against his face (Fig. 7). As the Buick displacement reached 2-1/2 ft following contact, the lower rim of the steering wheel contacted the driver's sternum (mid-chest). His chest peak resultant deceleration was 49 G at 95 ms and his longitudinal head deceleration was 31 G at 125 ms. Following impact with the steering wheel assembly, he rebounded back into the seat, assuming a slumped seated posture.

Right Front Passenger - Lark - The front seat passenger was thrown forward, jackknifing about his lap belt, bringing his head into forceable contact with the

Fig. 5 - Final positions of rest (Experiment 121)

windshield and nonpadded instrument panel; he sustained a head impact of 82 G at 110 ms. This impact only slightly distorted the gradual contour of the sheet metal panel. His peak chest impact of 68 G at 45 ms was a short duration pulse apparently the result of the right rear passenger's knee-impact through the restrained seatback. At the time the front seat passenger's head impacted the windshield and instrument panel, the Studebaker had rebounded only 3 in from the most forward position it obtained. Thereafter, he rebounded rearward, violently whiplashing over the low seatback and striking the chest of the right rear passenger with his head. Secondary rebound from the seatback was not significant and he remained in a seated, semireclined posture.

Right Front Passenger - Buick - This 95th percentile dummy continued forward at impact, maintaining an erect seated posture as he took up the slack in his lap belt. Forward excursion of his torso was 10 in when his knees impacted the instrument panel sufficiently to prevent further stretching of his belt. The lap belt peak load was 1875 lb at 90 ms, and his chest sustained a 73 G (resultant) at 135 ms after initial vehicle contact. Thereafter he rebounded to a seated position.

Right Rear Passenger - Lark - The right rear seat passenger continued forward

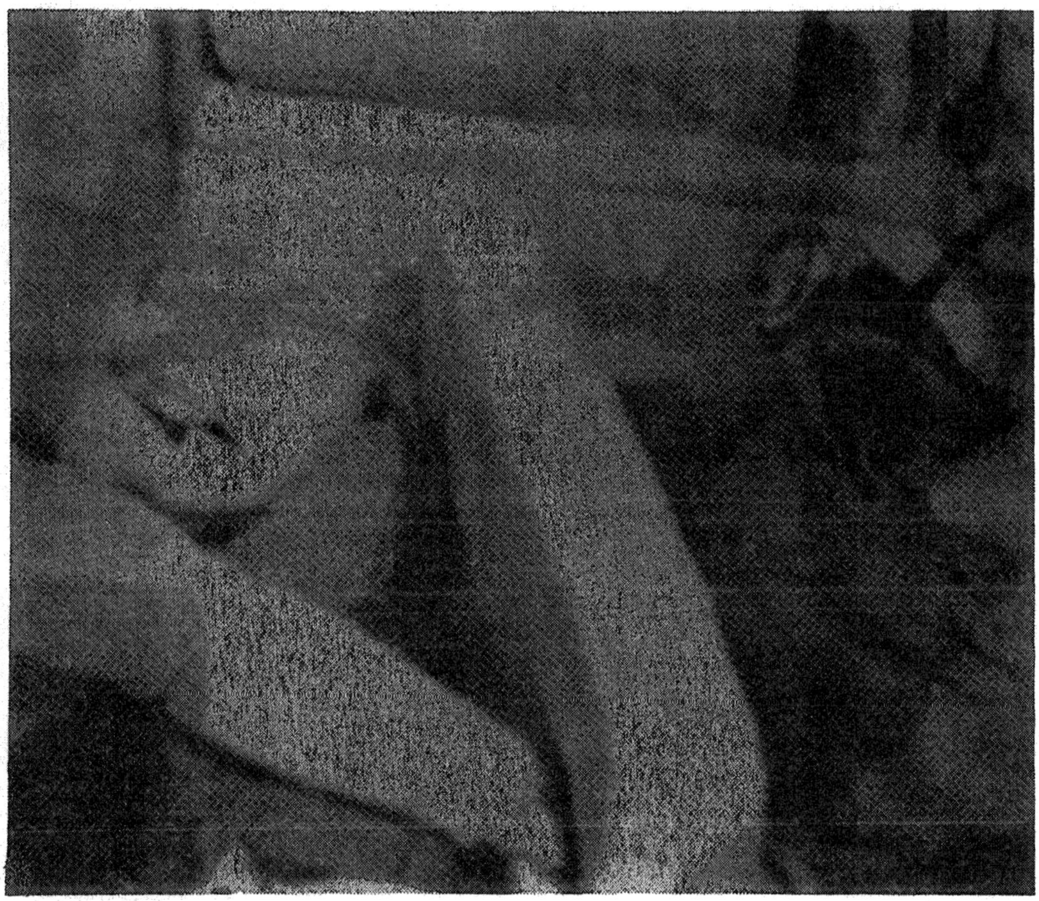

Fig. 6 - Lark steering wheel rim is fractured free of hub during driver impact (Experiment 121)

in his seat during initial vehicle deceleration, maintaining a rather normal posture, as his knees plunged into the backrest of the restrained right front seat, bowing the seatback substantially forward. His chest then struck the top edge of the restrained seatback, bowing it forward additionally; this specially anchored backrest provided a rather effective passive restraint by distributing forces across his entire torso (Fig. 8). A resultant chest deceleration of 63 G was sustained at 115 ms.

The right rear passenger's head pivoted over the low seatback and slammed into the front cushion of the backrest, sustaining a head impact of 178 G at 145 ms. This whip-like action demonstrated the desirability of having a high-back seat so as to provide a protective decelerative structure for the head instead of the observed injury-producing action as the head and neck flexed violently over the top of the low backrest. Moreover, an adequately padded front-seat integral head support provides the primary head protection for front seat occupants during a rear-end collision. The right rear passenger pitched forward, relative to the interior of the passenger compartment, 3-1/2 ft; at this point of maximum pocketing, the seatback had been distorted forward to approximately the wind-

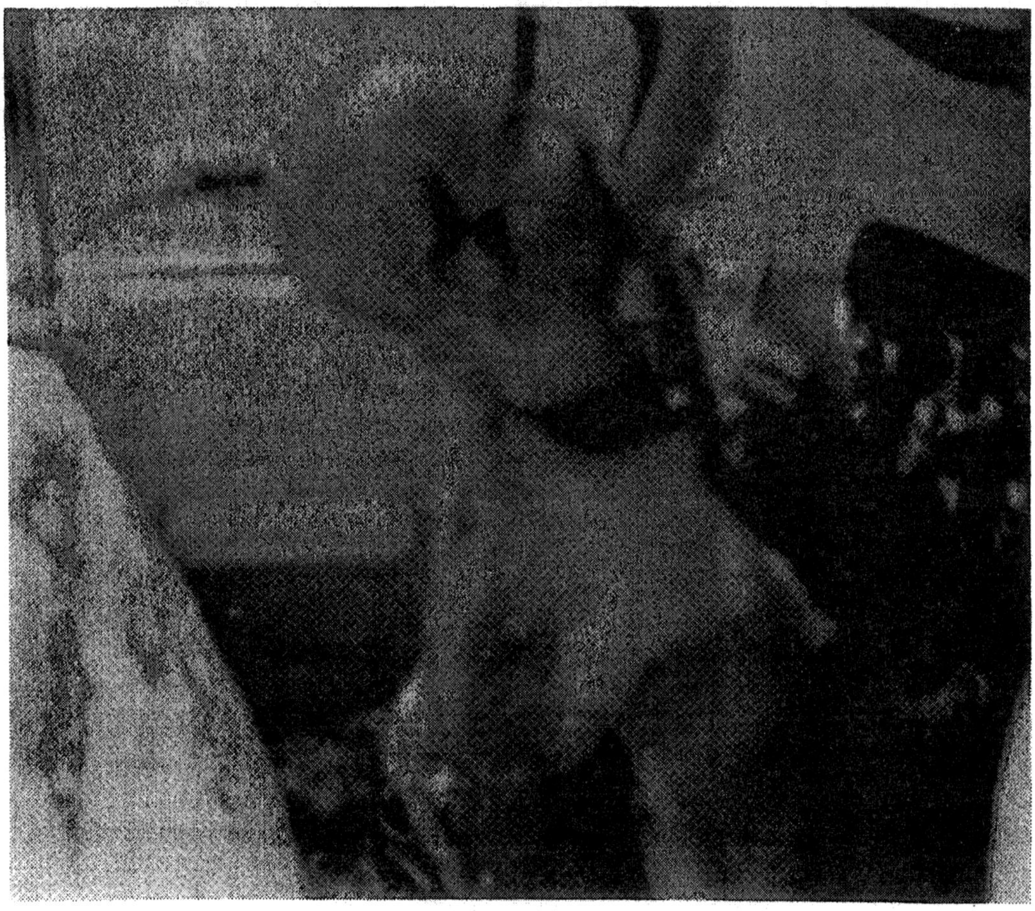

Fig. 7 - Lap-belted Buick driver slams against steering wheel during this 30 mph collision (Experiment 121)

shield "A"-post. Thereafter, he rebounded to his original seat space and attained a semiprone posture at rest over his dislodged seat cushion.

Right Rear Passenger - Buick - The 21 in front seat backrest was reinforced against forward movement and provided effective restraint for this unrestrained passenger. His peak resultant chest deceleration was 25 G at 135 ms. Thereafter, his head flexed over the seatback and pocketed into the top of the backrest, sustaining a peak resultant deceleration of 56 G at 170 ms; better protection would be provided by a high backrest, as previously indicated. Elastic rebound from the front seatback was minor and he remained slumped over the seatback at his position of rest.

Left Rear Passenger - Lark - This passenger's three-point restraint was somewhat loose at the hips, accounting for his lower torso submarining sufficiently to allow his knees to follow the driver's seatback as it pitched forward. The upper torso cross-chest restraint provided the principal restraining action for this left rear seat passenger; the shoulder strap was provided with a friction device to attenuate peak forces to a relatively constant value between 65-130 ms. For this exposure the restraint force reached a peak of 1000 lb at 70 ms and sustained no value below 740 lb throughout this 65 ms time period. The peak resultant chest deceleration was 27 G at 120 ms. The peak resultant head deceleration of 82 G at 130 ms was primarily in the vertical axis (76 G vertical, 30 G longitudinal). Additionally, the forward movement of the driver's hinged seat-

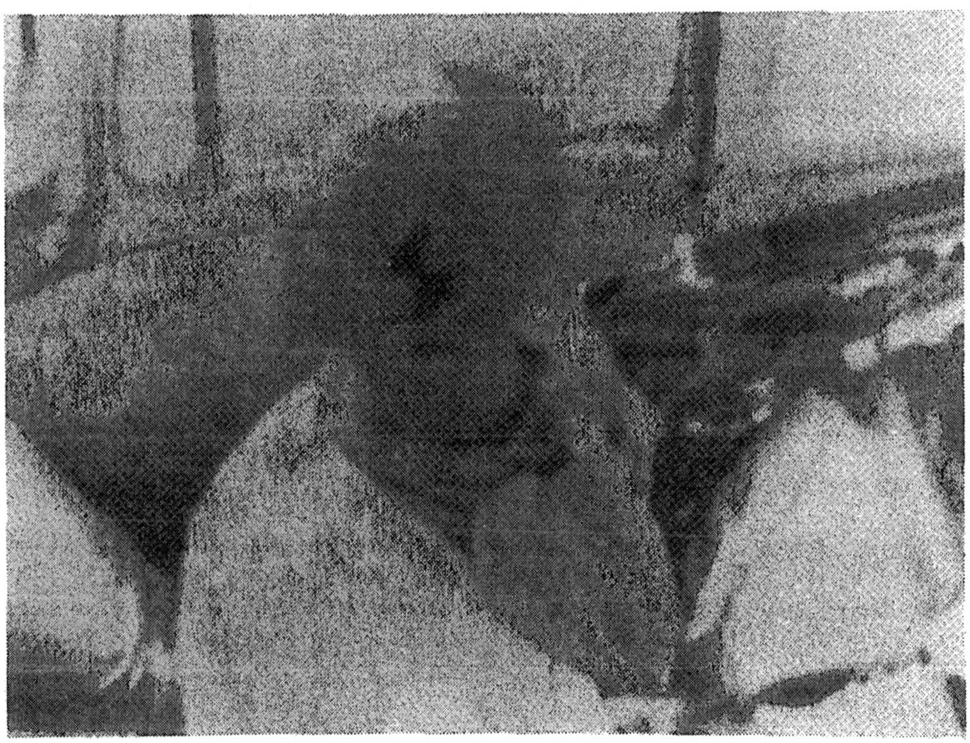

Fig. 8 - Right rear passenger in Lark impacts restrained seatback with his chest; higher padded back-rest could also provide passive restraint for head (Experiment 121)

back provided increased space for more effective use of this occupant's restraint. After fully loading his three-point belt, the left rear passenger was pulled rearward by elastic rebound to a normal seated posture and then slumped to his right.

Left Rear Passenger - Buick - Following contact, the left rear passenger pitched forward, maintaining the same natural posture until his knees struck the restrained backrest of the front bench seat; his chest peak deceleration was 34 G at 110 ms. As his knees struck the seatback, his head flexed downward against the top of the seatback, sustaining a peak deceleration of 134 G at 145 ms. The 34 in of forward movement by the left rear passenger brought his head into position to impact the top of the front seat backrest even though he was restrained by lap and cross-chest belts. His shoulder strap registered a peak of 765 lb at 120 ms. The Buick displacement reached 3 ft beyond initial contact before the left rear passenger commenced to rebound from the elastic action of his restraint system to a normal seated posture.

Post-Collision Observations - Lark - The 1960 Studebaker 2-door sedan had an overall length of 14 ft 7 in. It was pushed rearward by the heavier Buick 30 in from its original contact position to its position of rest. The damaged front ends of the Buick and Studebaker at rest were approximately 37 in apart. The deformation to the front of the Studebaker was about 22 in ; it had a uniform flat-like crush, similar to a barrier collision (Fig. 9). Although the left front fender

Fig. 9 - Front end of Studebaker Lark sustained permanent deformation of 22 in (Experiment 121)

was pushed rearward and to the outside of the door, it could still be opened; the right fender, however, had been pushed rearward sufficiently to jam the door. The left front tire was deflated by collapse damage and both front wheels were pushed rearward to the inner fender pan at the firewall. The engine deformed the firewall into the passenger compartment about 6 in. Both frame members were buckled downward to the pavement at a point behind the front wheels. When the driveshaft failed at the front end, it bent on contact with the pavement, causing the rear wheels to lock up.

The gasoline tank was torn loose from its single-bolt attachment on the right side, but remained attached on the left side where it was fastened by two bolts.

The Lark passenger compartment did not receive major collision intrusion; however, the lap-belted driver impacted the crash-extended steering column so severely that the steering wheel was separated from the spokes; the spokes were bent downward and forward around the steering column; the steering column shaft was extended approximately 2 in relative to its housing.

The right front seat was restrained at the top of the backrest and at the track adjuster by tensiometer links. Although the seatback failed at the left hinge point, the top reinforcing attachments held and continued to provide the intended passive restraint for the passenger seated behind it.

Post-Collision Observations - Buick - The 1959 Buick 4-door sedan had an overall length of 18 ft, 1 in. After contact this vehicle's intact passenger compartment crushed forward 3 ft, 1 in to its position of rest. The front-end collapse was approximately 22 in in the central portion and 20 in at the headlights (Fig. 10). Although the front wheels were pushed rearward to the floor pan at the firewall, the tires remained inflated and the front bumper was wrapped about them. When the gasoline tank attachment straps failed, the tank became jammed between the differential housing and the pavement. Additionally, the enclosed drive shaft buckled and locked up the rear wheels. Notwithstanding the 1.5 heavier weight advantage of this vehicle, its front end moved forward only 15 in beyond the initial contact point.

The Buick passenger compartment received moderate damage. The steering wheel and shaft were pushed into the compartment about 1.5 in.; the lap-belted driver deformed the steering wheel and column to an almost vertical position. The horn button rim was fractured and became a sharp, knifelike object. The lap-belted right front passenger deformed the glove compartment door 0.5 in with his knees and his head impacted the windshield 5 in below the header. The unrestrained right rear occupant deformed the reinforced seatback 12 in with his knees and the restrained left rear occupant did not cause any significant damage to the seatback. The right doors were jammed and inoperable; the left front door could be opened, but not relatched; and the left rear door worked in a normal fashion.

Experiment 122 - was an inline head-on collision between a 1967 Ford Custom 4-door sedan and a 1962 Ford F350 pickup truck; each vehicle was traveling 30 mph at impact (Fig. 11). The truck weight with instrumentation and simulated load was 10,050 lb; the pickup bed was loaded in a manner to prevent load shift during impact. The truck with two dummies grossed 10,430 lb; the Ford sedan with instrumentation weighed 4065 lb and grossed 4895 lb with four dummies. The ratio of gross weights of these two vehicles was 2.1 to 1.

The driver of each vehicle was a 50th percentile adult male dummy and the passengers were 95th percentile adult male dummies. Both occupants in the truck had lap-belt restraints; the sedan driver and right front occupant had lap belts, the right rear occupant was unrestrained except for the passive restraint afforded by the tethered front seatback. The left rear occupant had a three-point restraint with inertial reel.

Collision Dynamics - After initial bumper-to-bumper contact, the sedan crushed forward 1.5 ft before the front seat occupants commenced significant forward movement, relative to the vehicle interior. The sedan's forward displacement was 24 in in the first 60 ms and it reached a 28 in maximum just before 100 ms. The sedan thereafter reversed its direction and the heavier truck pushed rearward about 2 in at 100 ms (Fig. 12). After 150 ms the sedan rearward shift had reached 15 in as the truck started to override the sedan. Thereafter, the sedan was pushed rearward 13.2 ft to its position of rest and the truck moved forward 18 ft to its position of rest (Fig. 13).

Driver - Ford Sedan - The driver lagged behind the right front passenger by 4 in as he traveled forward after initial contact to strike the front of the passenger compartment; the steering column elevated out of his path before he

Fig. 10 - Post-collision Buick front-end collapse is more pronounced in center owing to impact by the narrower Lark front end (Experiment 121)

Fig. 11 - During this 30 mph head-on collision, the smaller vehicle partially underrides the truck front end structure (Experiment 121)

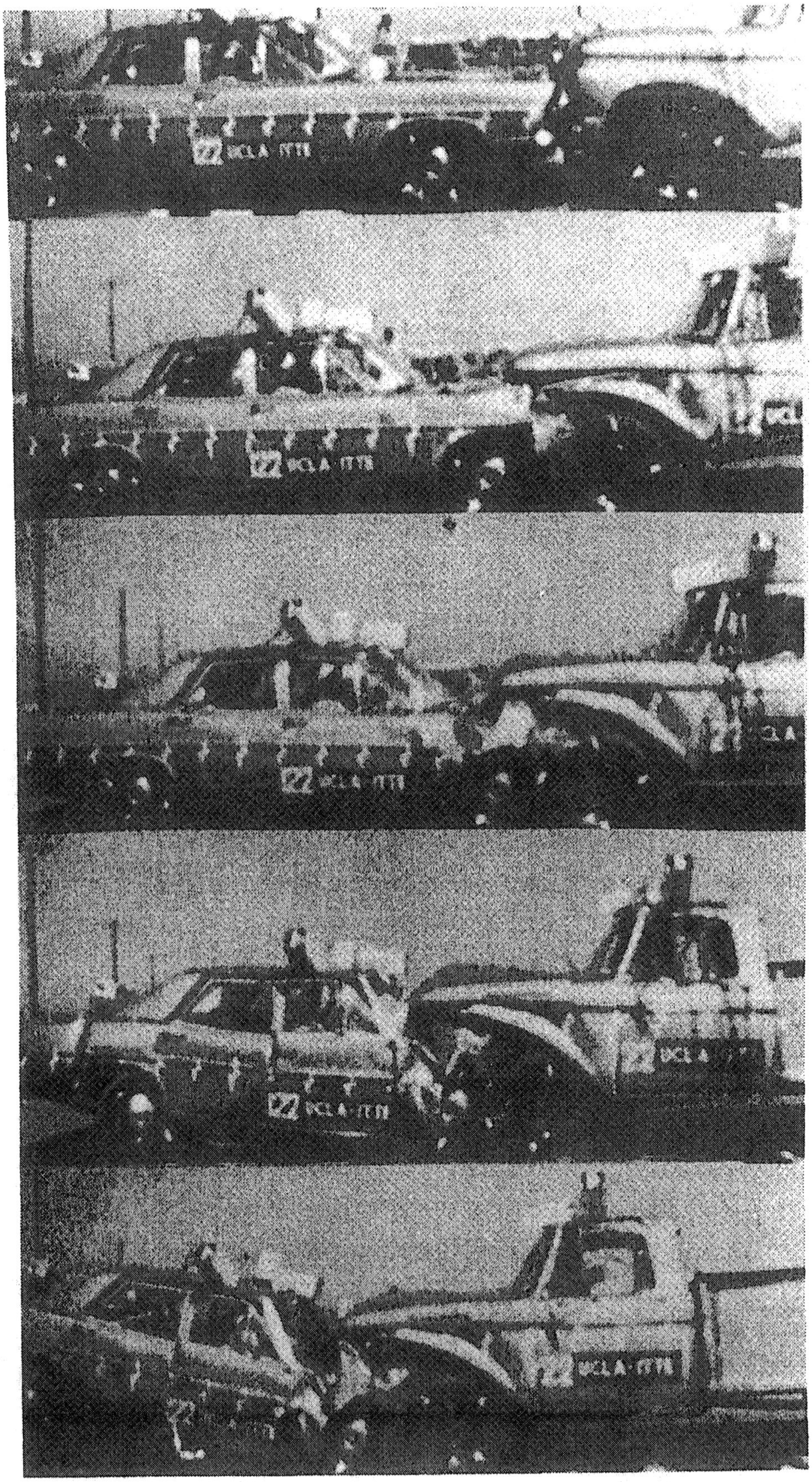

Fig. 12 - 30 mph head-on collision between Ford truck and sedan; weight ratio 2.1:1 (Experiment 122)

reached it, so his impact with the steering column was not directly to the hub. This unusual steering column response occurred because of the extreme front-end collapse to the left front of the passenger vehicle as a result of truck encroachment. The elevation of the steering column and steering wheel continued until it struck the upper left side of the windshield, fracturing it with a typical impact pattern. The driver's head sustained a peak deceleration (longitudinal) of 102 G at 105 ms. After impacting the steering wheel and column, the driver was struck by the unrestrained right rear passenger who rebounded from the instrument panel. The driver was forced to the left side of his seat and subsequently slumped to the floor with his upper torso on his seat cushion.

Driver - Truck - After initial contact, the truck proceeded forward about 3 ft before the lap-belted driver showed any response and thereafter he abruptly shifted forward about 3 ft relative to his vehicle's interior.

The driver had assumed an erect seated posture during his travel from his seat toward the steering column; his seatback tore loose and followed him as he shifted forward. When the pickup reached 4 ft of forward movement following contact, the driver had reached his furthest forward position in the driver's compartment; he jackknifed about his lap belt and crushed against the steering wheel. His peak head deceleration in the longitudinal axis was 48 G, occurring at 120 ms. As the truck reached 5 ft of forward movement following contact, the driver started to rebound relative to the interior of the compartment, ending up in a seated posture leaning against his door.

Right Front Passenger - Ford Sedan - The right front seat passenger was lap belted and as the sedan decelerated, he was pitched forward maintaining a normal seated posture; thereafter, his right knee crushed against the right side of the instrument panel. As he continued forward, still maintaining an erect posture, his knees tended to deflect upward on the instrument panel as they crushed

Fig. 13 - Final positions of rest (Experiment 122)

further into the panel. A peak lap belt load of 4250 lb was sustained at 80 ms; thereafter, his chest sustained a peak resultant deceleration of 59 G at 90 ms. The right front passenger's head flailed forward and downward as he rotated about his lap belt and slammed against the top of the instrument panel. His head impacted the windshield at 100 ms as the vehicle reversed its direction and he sustained a peak head deceleration of 100 G at 105 ms. The right front passenger impacted the instrument panel and was momentarily held against it by his seatback as the right rear passenger crushed forward. He rebounded rearward with his seatback to a normal seated posture. He did not whiplash over his low seatback since the car was being pushed rearward almost as fast as his rebound velocity.

Right Front Passenger - Truck - The right front seat passenger in the Ford truck was lap belted and as he started to shift forward during collision, the seatback followed him forward, having torn free from its mount, exposing the fuel tank positioned behind the front seat (Fig. 14). His knees penetrated the lower section of the instrument panel and his head impacted the top of the instrument panel; then his shoulders contacted the instrument panel, but not forceably. His peak longitudinal chest deceleration was 15 G at 125 ms. The lap belt prevented this right front seat passenger's head from striking the windshield and prevented his torso from contacting the instrument panel forceably; the lap belt tensiometer registered a peak force of 2200 lb at 100 ms. He then

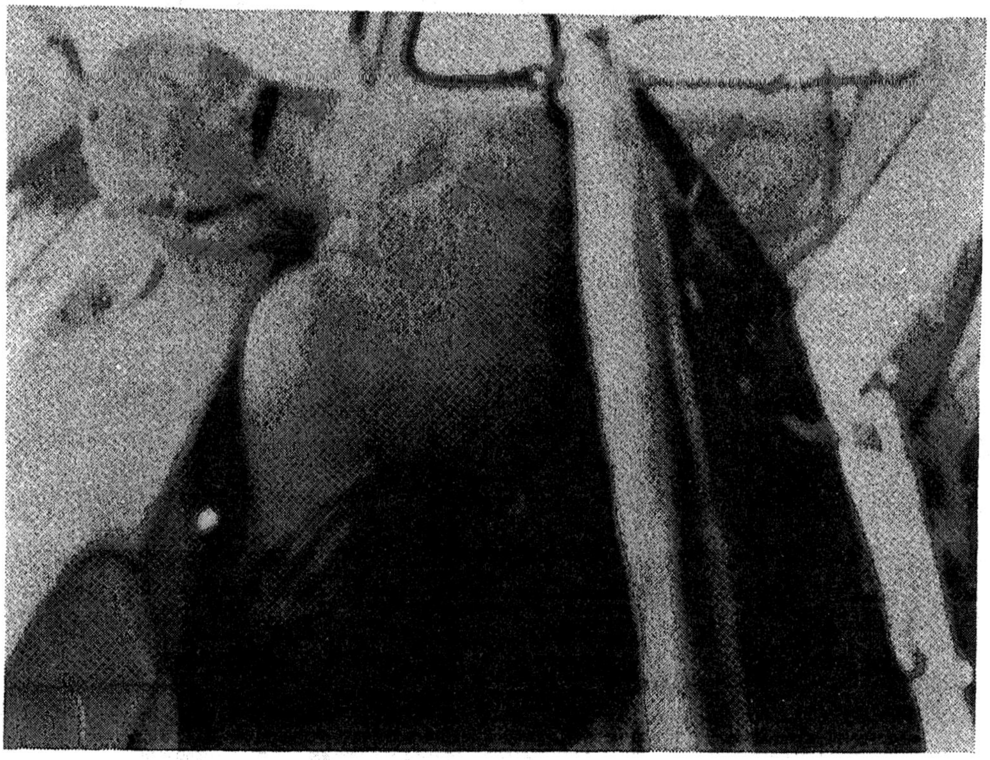

Fig. 14 - Lap-belted truck passenger slams head into instrument panel as seatback flails forward (Experiment 122)

rebounded rearward from the instrument panel and arched over the inclined plane formed by his torn-loose seatback cushion. He remained restrained by his lap belt as he returned to a seated posture, slumping slightly to his left at rest.

Right Rear Passenger - Ford Sedan - The 21 in bench-type front seat was reinforced to resist forward movement of the backrest. The unrestrained right rear seat occupant was thrown against it, pocketing his knees deeply. This restrained front seat backrest partially retained his torso in a normal seated posture as he approached his maximum forward displacement; however, the right-side seat restraint failed at 1860 lb, 115 ms after initial bumper contact. Thereafter, the seatback was forced against the back of the right front seat passenger; the right rear seat passenger's head remained erect until the two were close to touching one another. Then the right rear seat passenger's head flailed forward and downward over the left shoulder of the right front seat passenger and struck the instrument panel.

After striking the instrument panel, the unrestrained right rear passenger's torso was momentarily horizontal; as he rebounded rearward, he struck the driver and subsequently struck the restrained left rear passenger with his buttocks. He then fell forward to the floor space behind the driver's seat. The seatback, as a result, was extremely collapsed and distorted following this impact.

Left Rear Passenger - Ford Sedan - The 95th percentile dummy in this location was wearing a three-point restraint with an inertia reel on the cross-chest strap. An apparent delay in the reel's snubbing action allowed his torso to shift forward initially at the same rate as the unrestrained right rear seat companion. He reached a position about half the distance to the front seatback still in a normal seated posture before the lap belt caused his upper torso to commence rotating about the hips. The shoulder belt partially restrained him as he sustained a peak load of 1260 lb at 115 ms; the peak chest deceleration occurred slightly earlier, 67 G at 95 ms. This excessive torso movement allowed his head to strike the restrained front seatback, 146 G at 120 ms. At this time, his cross-chest strap bottomed out and violently rebounded him rearward to his original seat position. The force of rebound was sufficient to cause his head to whiplash over the top of the rear seatback, followed by a head impact with the header over the rear window. Thereafter he slumped to the right on the rear seat cushion.

Post-Collision Observations - Ford Sedan - The 1967 Ford 4-door sedan was pushed rearward by the heavier truck 13 ft, 3 in from the impact line (Fig. 15). The permanent deformation was about 46 in to the front end sheet metal in addition to major structural damage. The transmission and drive shaft were forced to the pavement as well as the left rear rocker panel and frame member; both front tires were deflated by collapsed structures and were pushed rearward against the firewall (Fig. 16). All four doors were pushed rearward approximately 1/2 in and were jammed closed.

Fig. 15 - Position of rest of Ford sedan showing compressive buckling to left rear door—all doors were jammed shut after impact (Experiment 122)

Fig. 16 - Sedan has extreme front-end collapse caused by override (Experiment 122)

The passenger compartment was not directly crushed by the truck, but it did sustain major damage from occupant impacts. The steering column was pushed rearward and upward to within 30 deg of vertical; the steering wheel rim impacted the windshield 2 in below the header (Fig. 17). The front seat occupants, although lap-belted, deformed the instrument panel with their knees. The front seatback and seat adjustment track were reinforced by tensiometer links and provided some passive restraint action for the unrestrained right rear passenger even though the right reinforcement failed during collision. The upper seatback horizontal frame member received a permanent bow of 12 in at its center, from the right rear occupant impact.

Post-Collision Observations - Ford Truck - The 1962, F350 Ford truck had an overall length of 18 ft, 2 in. with a 9 ft pickup bed. The truck crushed onward 18 ft after initial contact with the sedan before reaching its position of rest. The truck and the sedan remained interlocked after the collision and the truck received about 12 in of deformation to the bumper and lower front end sheet metal sections (Fig. 18). The truck frame had minor frame buckling ahead of the front wheels. The hood showed slight damage as it did not directly contact the opposing vehicle. The front suspension leaf springs and shackles remained connected; however, the main spring leaves bent ahead of the rear anchor point to the frame. Both doors remained functional as fender collapse did not progress significantly rearward.

Fig. 17 - Lap-belted front seat occupants in sedan did not impact windshield—star fracture pattern, upper left corner, caused by steering wheel impact as it bent upward (Experiment 122)

The passenger compartment received minor damage; direct collision contact forced the steering shaft and wheel only 0.5 in into the compartment. The steering wheel rim was bent forward 2 in at the top of the four-spoke wheel, as the driver flexed over his lap belt. The passenger brushed the windshield with his head but did not fracture the glass or significantly damage the top of the unpadded instrument panel. The seatback came loose and revealed the presence of the gasoline tank located behind it. For this exposure, the tank did not rupture or spill fuel; however, as shown by prior research (1)*, fuel system components should not be placed within the passenger compartment to expose occupants to incineration in the event that fuel is spilled during collision or subsequent vehicle upset.

Rear-End Collision

Experiment 107 - Experiment 107 was an inline collision involving a 1969 Volkswagen 2-door sedan, weighing 1700 lb, rear-ended by a 1967 (full-size) Ford 4-door sedan weighing 3570 lb; gross weight of the VW was 2440 lb with three adult dummies and a 13-year-old; gross weight of the Ford with four adult dummies was 4400 lb for a weight ratio of 1.8 to 1. The VW was stationary when rear-ended at 30 mph by the Ford.

---

*Numbers in parentheses designate References at end of paper.

Fig. 18 - Truck front-end collapse was approximately 12 in at bumper elevation (Experiment 122)

The driver and right front seat passenger of the rear-ended VW were unrestrained, 95th percentile, adult male dummies; the left rear seat 13-year-old male dummy was unbelted as was the 50th percentile adult dummy in the right rear seat.

Collision Dynamics - After the initial bumper-to-bumper contact, the Ford crushed forward, overriding the VW bumper until its hood contacted the VW below the rear window (Fig. 19). This hood contact occurred at 20 ms after the bumpers touched and without any appreciable VW forward displacement, even though the Ford had encroached into the rear of the VW. The VW did not move forward appreciably until 50 ms after contact; during this period, energy was elastically stored in the collapsing VW rear end and in the deforming front end of the Ford. Then at 50 ms, the VW commenced to spring forward in response to the rear-end collision forces. The maximum mutual penetration of 30 in was reached at 95 ms and at this instant the momentary common velocity for the two vehicles was 19 mph (28 fps) (Fig. 20). At 95 ms, the VW had been displaced one foot forward and the Ford had advanced 3-1/2 ft beyond the location of initial bumper contact. Thereafter, vehicle to vehicle partial rebound took place as the vehicles, still mechanically interlocked, continued their roll-out for a total of 156 ft before reaching their positions of rest. Except for the overriding action of the bumpers, significant force resistive structure of the VW

Fig. 19 - Ford hood encroaches almost to VW rear window before significant forward movement of VW occurs (Experiment 107)

was not encountered until the VW engine-transmission was struck, which resulted in a 3/4 in distortion of the engine-transmission mount. Consequently, the nominally nonyielding engine structure of the VW forced the Ford front end to yield beyond the amount it would ordinarily sustain from encountering the more common, readily crushable rear end. However, the combination of the higher stiffness associated with the larger (Ford) vehicle and the higher stiffness identified with the VW and other rear-engine vehicles, in addition to the lighter weight of the VW, accounted for the abruptness of the VW acceleration, which prevented the striking car from encroaching into the passenger compartment of the smaller vehicle during this relatively severe, 30 mph rear-end collision.

The collision accomodation process of the (subcompact) Volkswagen rear end is comparable to the resistance to passenger compartment intrusion by the rear end of a 1967 Ford if each of these vehicles was to be exposed to a rear-end collision by a 1967 Ford. The more crushable rear end of the 1967 Ford is likewise collapsed for this 30 mph impact nearly to the base of the rear window, for UCLA Experiments 93, 95, and 96 (2). Because the weight ratio is 1:1 for the Ford-Ford rear-end collision and is 1:1.8 for the VW- Ford rear-end collision, the lighter VW is accelerated more abruptly than the Ford where both of the vehicles were rear-ended by the same type Ford (Fig. 21). Inasmuch as the slopes of the steepest parts of the V-T curves of Fig. 21 provide the peak acceleration attained during these rear-end collisions, the steeper VW slope (25 G) accounted for almost three times the peak acceleration sustained by the Ford rear-ended also by a Ford.

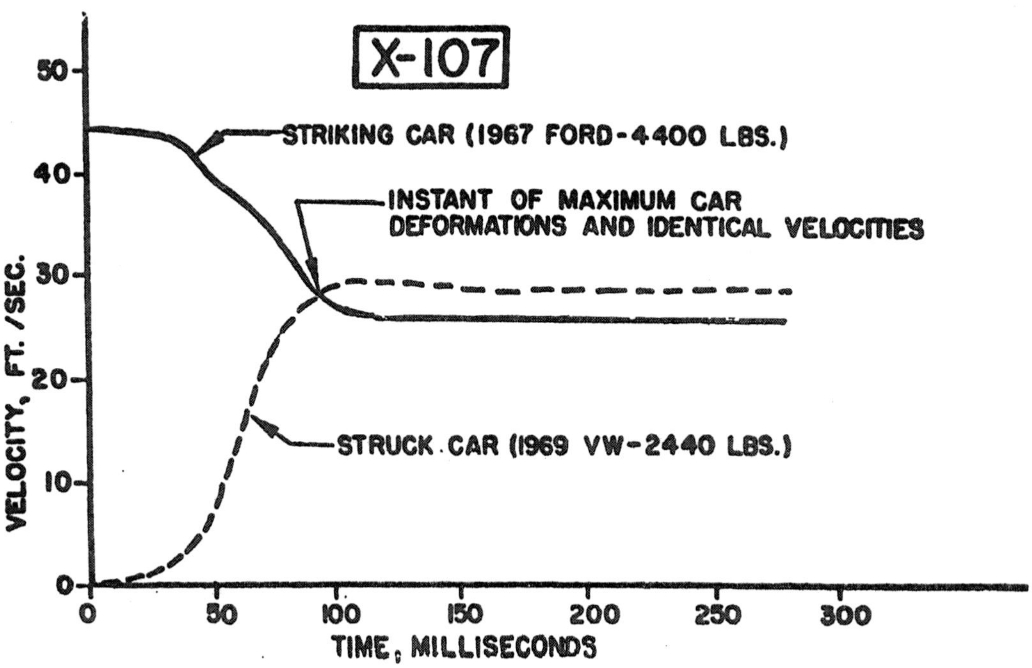

Fig. 20 - Velocity changes during a 30 mph rear-end collision; cars remain mechanically interlocked for their 156 ft of roll-out.

Volkswagen Driver - The occupant seat assignment, the forced-posture and peak acceleration of time of collision, the type of seat and the instrumentation for the rear-ended Volkswagen are all shown in Fig. 22. The illustration is sufficiently detailed that only one example, the driver, will be described. Before collision, the driver was sitting with a natural posture against his seatback which had a normal 20 deg slope (relative to vertical). During the abrupt acceleration, the inertial loading of the driver's torso against the backrest caused it to fail; the torso followed the seatback rearward as it slid 8-10 in up the plane of the backrest and the head of the dummy continued rearward in a normal posture relative to the driver's shoulder-neck alignment until the backrest struck the knees of the rear seat passenger. The driver's head received a severe impact of 120 G at 155 ms as it struck the rear passenger's chest; he thereafter rebounded to a rest position partially laying on his permanently deformed seatback.

Ford Driver - As was provided for the description of the VW occupants, the forced body postures for each of the seated locations along with the type dummy, his seat assignment, his peak accelerations, the type of seat and instrumentation are detailed for the striking Ford in Experiment 107 (Fig. 23). The 95th percentile driver was restrained by a lap belt to his bucket seat. The seat had been strengthened to provide support needed and to withstand the unrestrained rear seat passenger's impact against it.

Following impact, the driver pivoted about his lap belt and struck his abdomen against the steering wheel's lower rim at 50 ms. As he crushed forward, his chest struck the center pad of the steering column resulting in a 10 G blow at 101 ms. Next his head struck the upper portion of the steering wheel rim sus-

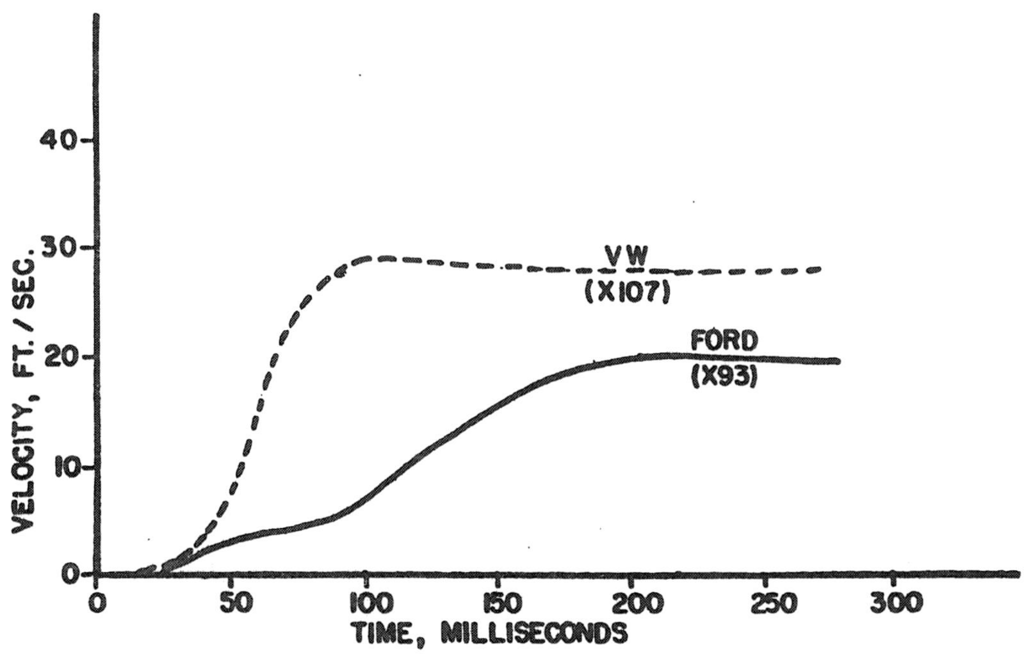

Fig. 21 - Velocity changes of a 1969 VW and a 1967 Ford, each rear-ended by a 30 mph Ford

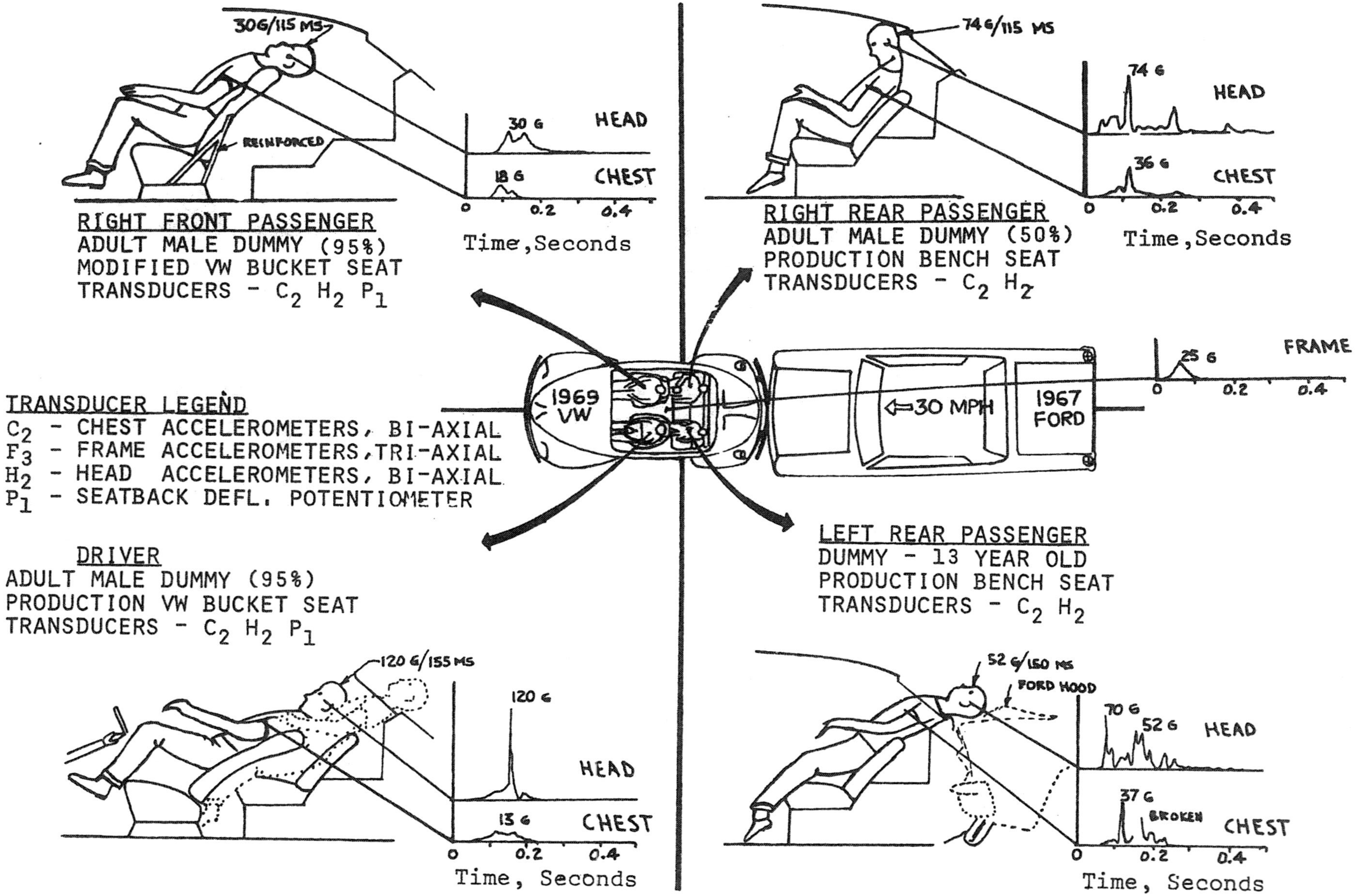

Fig. 22 - Transducer patterns shown with occupant posture and restraint condition during impact for each seated position, Volkswagen (Experiment 107)

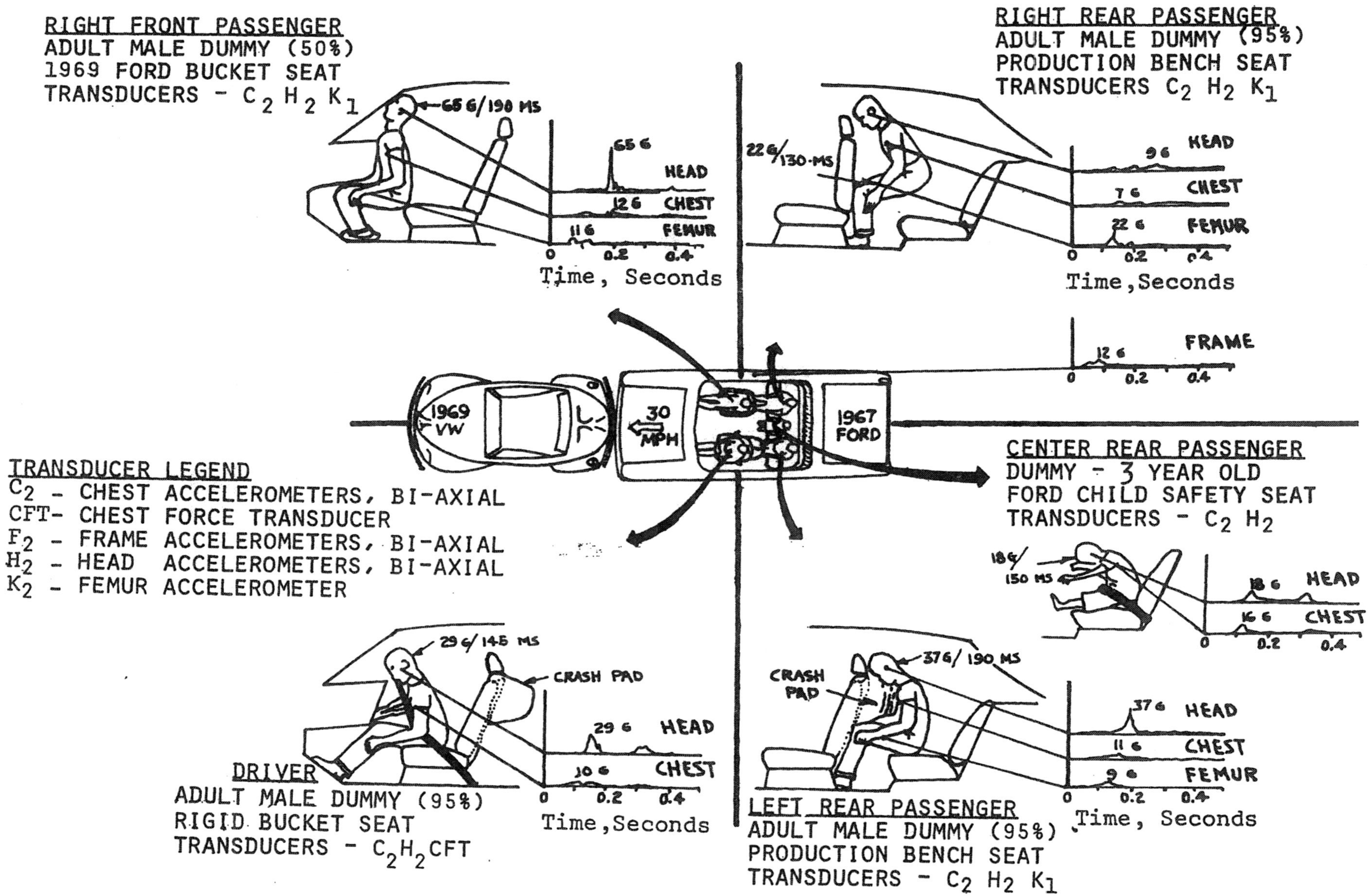

Fig. 23 - Transducer patterns shown with occupant posture and restraint condition during impact for each seated position, striking Ford (Experiment 107)

taining 29 G at 147 ms. Although initially snug, the dynamic factors contributing to slack in the belt during collision allowed the knees of this dummy to be displaced forward against the instrument panel. The findings for the other occupants of this Ford can be derived by reviewing Fig. 23.

Analysis and Findings

Vehicle Displacement Rates During Collision - Experiments 121 and 122 - The speed of a vehicle, measured in feet per second, represents its displacement with respect to time, referenced to some fixed object. A graph of this displacement-time relationship is a straight line with a fixed slope, ds/dt, when the vehicle is traveling at a constant speed.

The vehicles in Experiments 121 and 122 prior to collision were traveling 30 mph or a constant 44 fps, as shown by the straight line on Fig. 24. Any slowing on the part of the approaching vehicle, except for "collision-slowing" would not be evident for the limited period presented by Fig. 24. Even if the vehicle locked up all four wheels and could derive full decelerative action from such skidding independent of the collision, the departure from the precrash displacement line would not be appreciable during the time period shown. For this reason, the somewhat pronounced arching of the displacement curve away from the straight line, even for the heaviest vehicle (the truck), attests to its rapid ve-

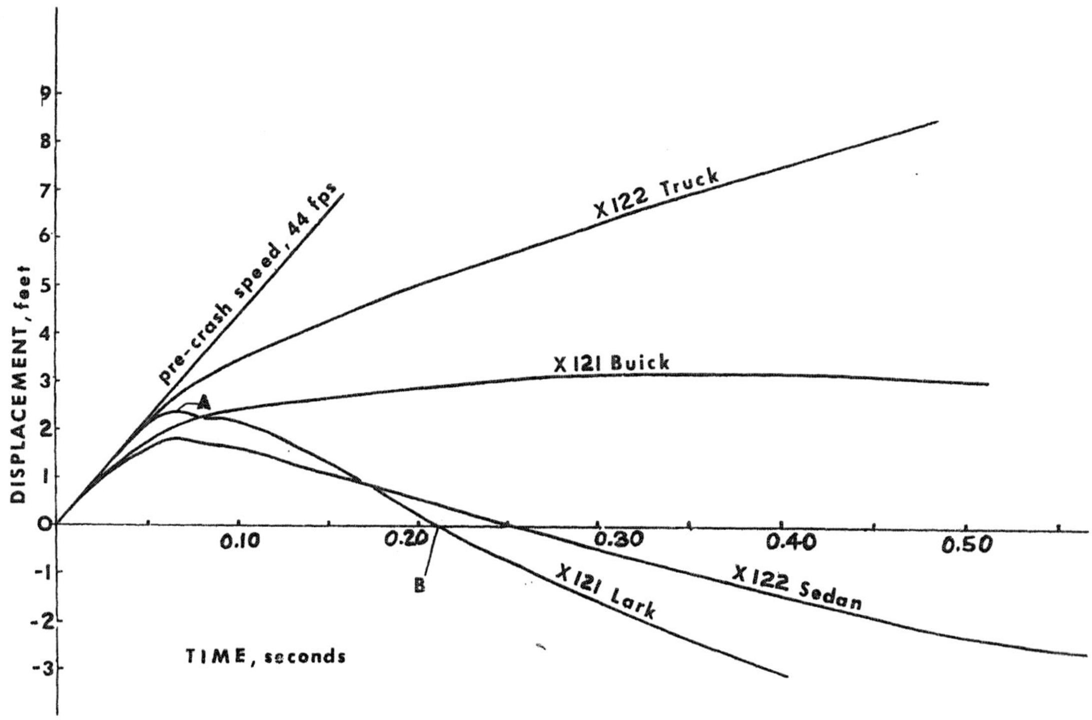

Fig. 24 - Passenger vehicle displacement time patterns during head-on collisions (Experiments 121 and 122)

locity change during collision. Far more sudden, however, was the change in displacement per unit time by the Ford sedan which was impacted head-on by the truck. The speed of the Ford sedan rapidly diminished as the heavier truck crashed head-on into it until no further forward movement of its intact passenger compartment and rear structure occurred (Point A of Fig. 24). Thereafter, the Ford sedan commenced to be shoved in the opposite direction until its intact structure reached the location it had passed as its front bumper initially struck the opposing truck (Point B). The greater kinetic energy of the heavier truck continued to shove the Ford sedan rearward until it came to rest, 13.2 ft rearward of its initial contact location.

The weight ratio of the Buick-Lark head-on (1.5:1) was lower than that of the truck-Ford sedan ratio (2.1:1) which accounts for the fact that the Buick's intact structure was nearly stopped after one-tenth of a second. The more abrupt initial reversal of the Lark displacement was attributed to a shorter, more compact hood and lack override. The pronounced reverse direction displacement of the Lark notwithstanding the observation that the Buick displacement was small after 1/10 s is attributed to structural restitution or elastic rebound.

Although passenger compartment forward displacement during the critical 1/10 s after initial contact was markedly less for the lighter vehicle in each head-on collision, the forward displacements were somewhat larger than those previously measured by Fredricks (3) for barrier collisions involving three sizes of vehicles (Fig. 25).

Velocity Patterns During Head-on Collision - The extent of exposure to injury from rapid changes in velocity depends upon the degree of motorist restraint and the rate of vehicle deceleration. In most head-on collisions, the vehicle has arrested its forward movement before unrestrained occupants impact their vehicle interior. Any helpful crash moderation that may have occurred during the crushing stop of the vehicle is lost to the unattached motorist who continues to move forward as his car stops; his impact with the car interior can be as severe as though he had been catapulted from the car against a rather rigid structure. "Ride-down" refers to the extent to which motorist deceleration conforms to vehicle deceleration. Motorist ride-down can never be precisely synchronized with vehicle deceleration but this is a worthy goal, one nearly obtainable. A uniform deceleration pattern provides a better opportunity for motorist ride-down than one representative of delayed onset followed by an abrupt deceleration.

Passenger vehicle velocity patterns for four head-on collision experiments are shown by Fig. 26. Two collisions (Experiments 121, 122) concern smaller/larger car impacts and two collisions (Experiments 38, 40) (see Ref. 4) concern identical vehicles crashing head-on (Fig. 27). Only one composite vehicle curve each is shown for Experiments 38 and 40 because these velocity-time patterns were taken from almost identical individual vehicle deceleration curves by integration.

The more uniform and nearly horizontal the velocity pattern, the lower the

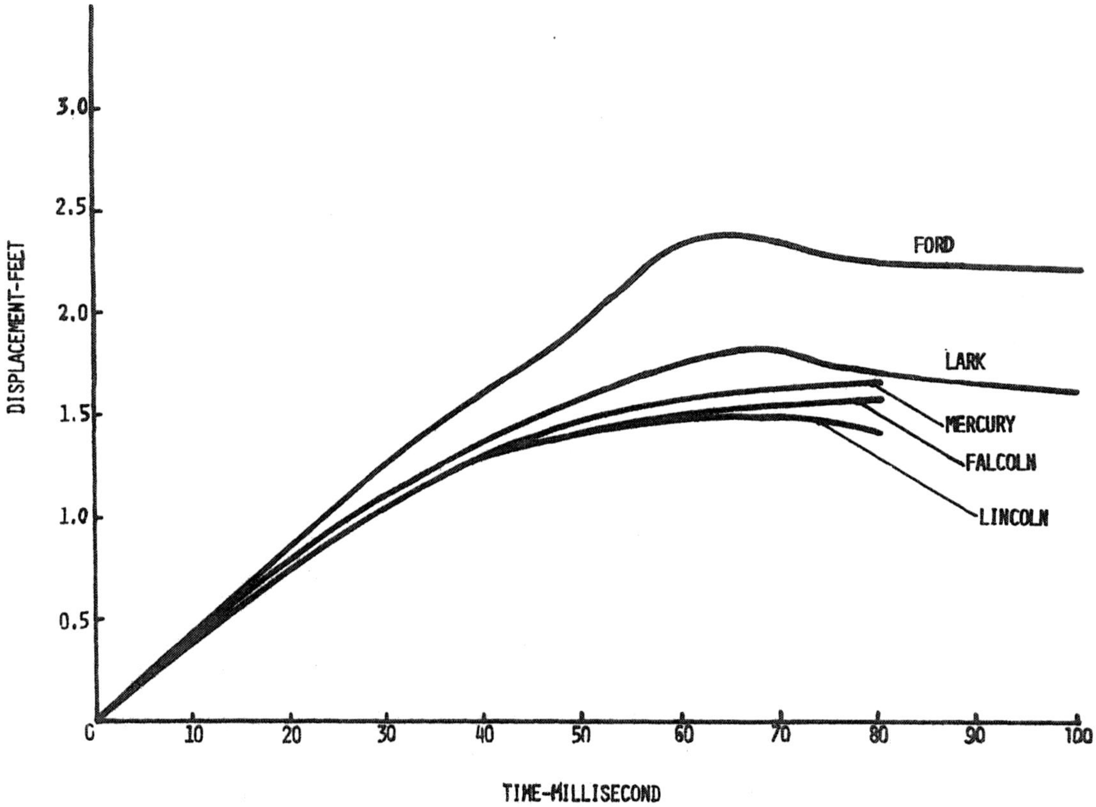

Fig. 25 - Displacement-time curves of the lighter vehicles in head-on collisions, Lark Experiment 121, Ford sedan Experiment 122, are compared to three barrier crashes of different weight vehicles, Fredricks (3)

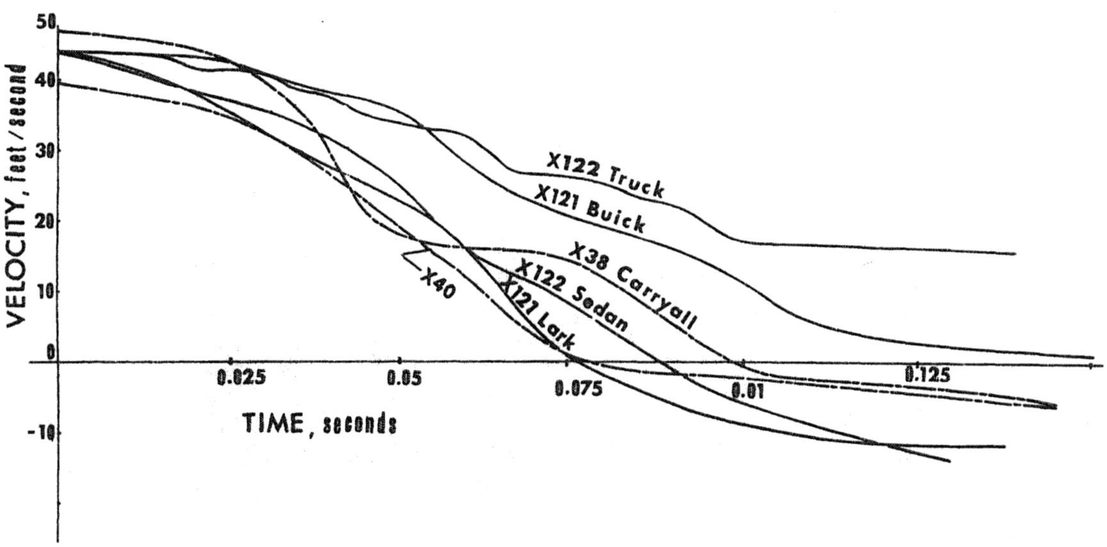

Fig. 26 - Passenger vehicle velocity patterns during head-on collision (Experiments 38, 40, 121, and 122)

crash deceleration. The Ford truck, Experiment 122, was 2.1 times the weight of its opposing Ford sedan and provided the mildest crash velocity pattern, with the Buick, Experiment 121, (1.5 : 1 weight ratio) closely following the truck's velocity pattern through the critical 100 ms after contact. Their respective opposing vehicles, the Ford sedan and the Lark, slowed at a faster rate but this rate is as favorable as for the equally matched Experiments 38 and 40 vehicles, and is slightly less severe than for barrier collisions at similar impact speeds. The velocity time patterns for the previously mentioned barrier collisions involving three vehicle sizes are compared to values obtained from Experiments 121 and 122 (Fig. 28).

Of the four steeper sloping curves of Fig. 26, the Experiment 122 Ford sedan, which was head-on impacted by the truck, is representative of the general performance of this group. Referring to Fig. 29, this same Ford sedan velocity time pattern has been repeated to provide a reference basis for discussing the "ride-down" performance of the driver and left rear passenger. The driver, restrained by a lap belt, continued forward in a normal seated posture and as the Ford crushed to a stop, he took up the slack of his lap belt and pivoted his head and chest forward until he struck the steering column assembly and instrument panel at 55 ms. At this instant, car velocity had diminished to half of the driver's velocity, indicating a relatively poor "ride-down." This injury-producing situation can be minimized, not by relying solely upon the current energy absorbing steering columns but by use of the diagonal chest strap. The authors' observations of the relatively ineffective action of many of the energy absorbing steering columns was described in another paper (5).

Characteristically, the rear seat occupants are more protected during frontal collisions; however, the left rear seat passenger of the Ford sedan represented in Fig. 29 appears to have sustained an exceptionally high rate of change of veloc-

Fig. 27 - Head-on collision between identical Chevrolet carryalls, each traveling 32 mph at impact (Experiment 38)

ity, considering the presence of a lap and cross-chest belt; it was found that the inertia reel for the cross-chest strap apparently did not function correctly, allowing the left-rear passenger to flail forward, pivoting at his hips, until he slammed against the front seat backrest. Then the late-acting shoulder strap snapped him to a stop and caused him to be rebounded violently into his seat.

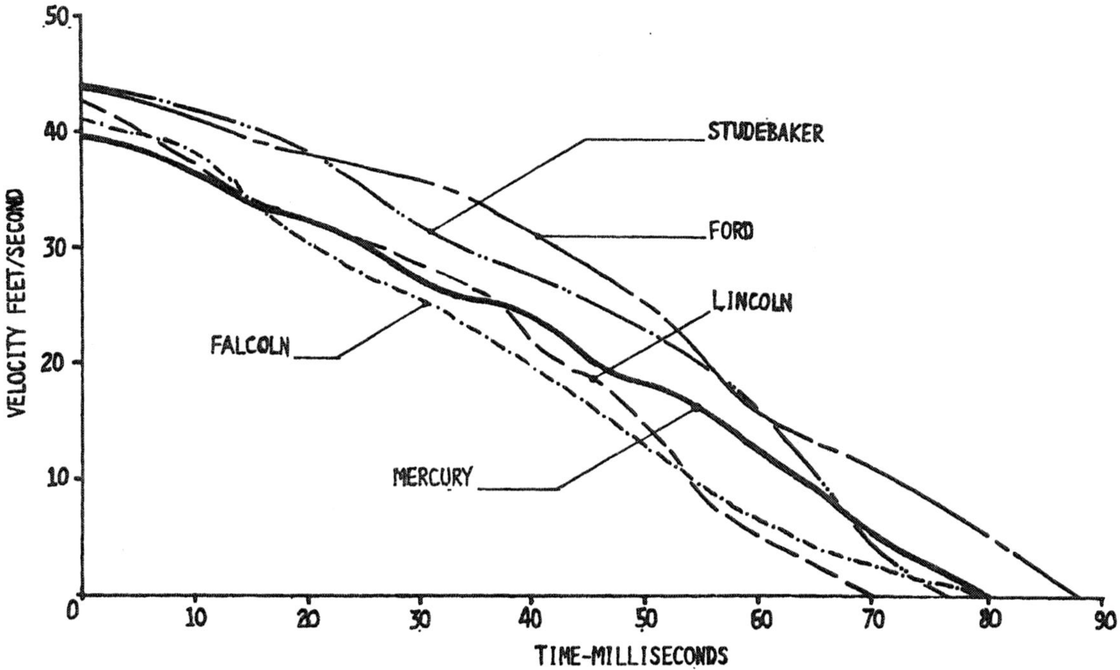

Fig. 28 - Velocity-time patterns for lighter vehicles in head-on collisions, Lark Experiment 121 and Ford sedan Experiment 122; note similarity to three barrier crashes of different weight vehicles, Fredricks (3)

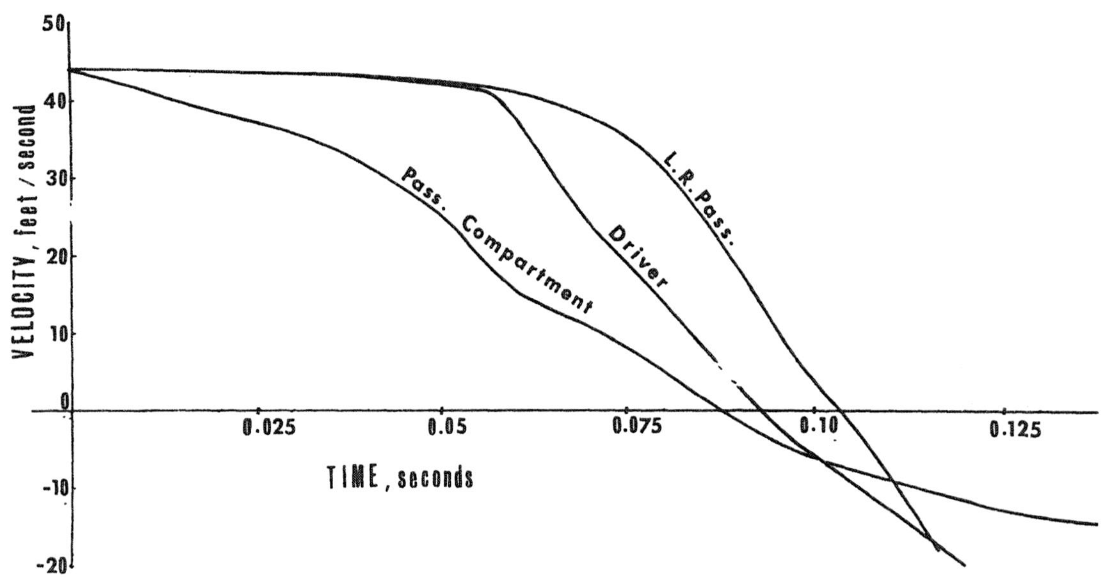

Fig. 29 - "Ride-down" performance for the restrained driver and L. R. passenger of the Ford sedan (Experiment 122)

Peak Deceleration versus Impact Velocity–Experimental and Theoretical - The deceleration patterns registered by instrumentation attached to intact sections of crashing vehicles appear as an erratic curve primarily owing to the nonuniform crush response and intermeshing action of the opposing structures. With appropriate instrumentation, the peak acceleration attained for each collision exposure has been found to indicate, to a reasonable degree of accuracy, their relative crash intensities. In Fig. 30, the peak accelerations labeled "experimental" for the UCLA Head-on Collision Series II (4) tend to increase proportionately with the speed of impact*. Each deformation value corresponds to a car's maximum (centerline) deformation. Maximum deformation is the total deformation occurring during impact and therefore is elastic plus permanent deformation. For the maximum deformation and impact speed for each ex-

---

*Impact velocity represents the speed of each car; both cars had the same speed at impact making their closing velocity double that shown on the graph.

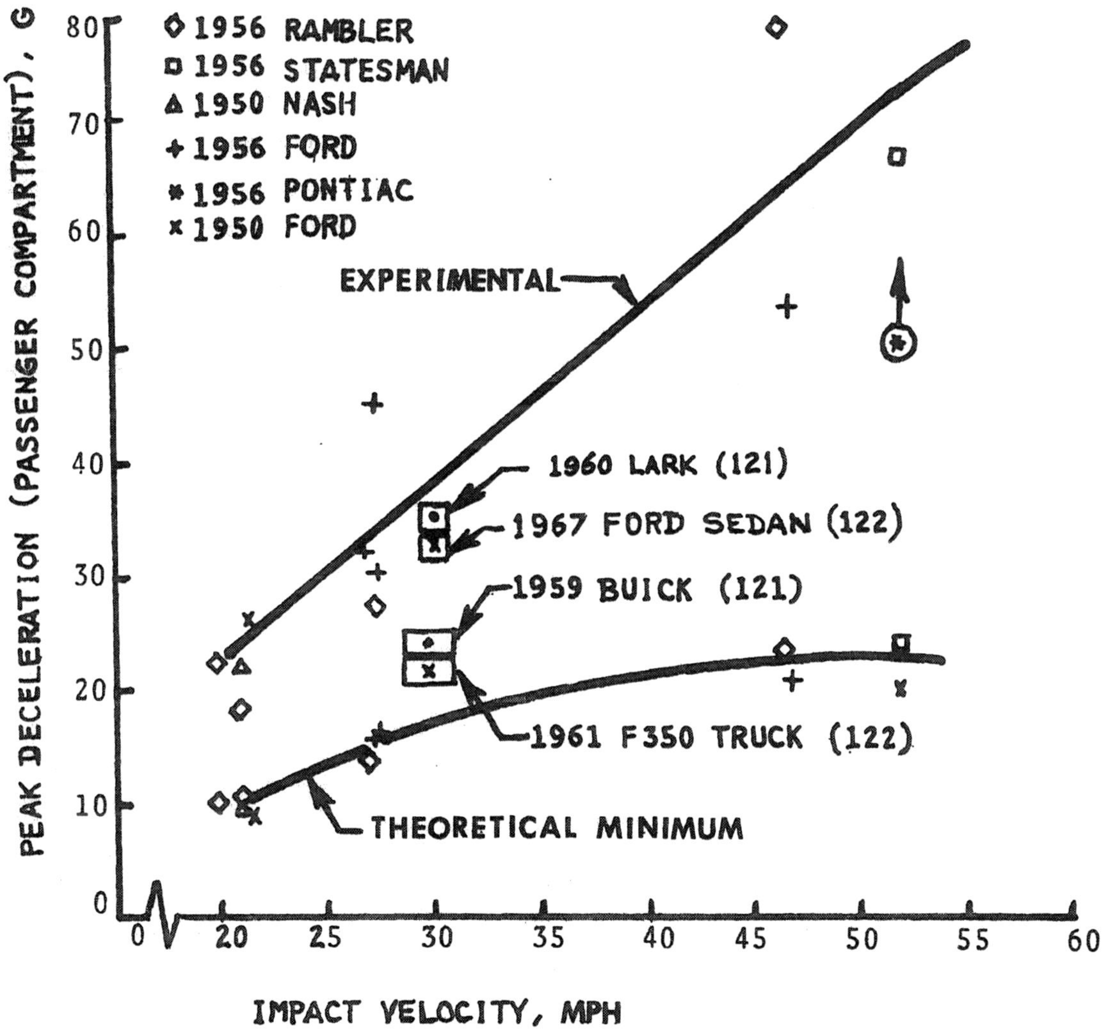

Fig. 30 - Peak declerations, passenger compartment, compared with theoretical minimums

periment, there may be calculated a theoretical minimum peak deceleration value which would result if the car decelerated uniformly throughout the entire collapse distance up to its maximum deformation. These "theoretical minimum" values of peak deceleration are represented by the lower curve of Fig. 30. Comparison of these two curves provides the following information:

1. At 21 mph, the experimentally determined peak decelerations of the automobile are more than double their theoretical minimum value and at 52 mph triple their theoretical minimum. Collision performance decreases with increases in impact speed from 20-50 mph.
2. The experimental peak decelerations appear to increase uniformly with increases in impact velocity, whereas the theoretical minimum decelerations appear to reach a plateau at about 47 mph, because of the proportionately larger maximum deformations occurring for impact speeds of 47 and 52 mph.

The data for Fig. 30 concerns the peak accelerations obtained for vehicles crashing head-on; the opposing vehicles are of equal weight except for the two 30 mph collisions which are further identified by the rectangular enclosures of the data points. These vehicles were mismatched by 1 : 1.5 for the Lark-Buick and 1 : 2.1 for the Ford sedan-truck collision.

Further evidence of the insensitivity of small cars to sustaining intensified head-on collision decelerations over those that would be obtained if the small car were as large as the opposing vehicle is readily apparent in Fig. 30. The Lark and Ford sedan were substantially smaller than their opposing Buick and Ford truck, yet sustained peak decelerations less than the average for previously conducted head-on collision experiments involving equal-weight opposing vehicles. This seeming incongruity in collision performances is attributed to the self-compensating factor that, as the differences in weights for opposing vehicles increase, so does the mismatch of their principal force-resistive structures, necessitating greater mutual penetration before a maximum encroachment is reached. Although mismatched head-on collisions at speeds higher than 30 mph were not conducted, they would be expected to show a trend toward greater passenger compartment encroachment and collapse for the small car head-on impacted by a substantially larger car than was observed for equal weight cars crashed at speeds through 52 mph, as shown in Fig. 30. Adequately restrained small car operators impacting large passenger vehicles appear to have no serious impact deceleration handicap for head-on collisions through 30 mph and probably approaching 40 mph. In the range of 40 mph and above, the collapse distance ahead of the front seat location is sufficiently short that direct encroachment with the passenger compartment would be expected to occur to a greater extent for the small car struck by a larger car than for equally weighted opposing vehicles.

The relative influence of mismatch in car size can be illustrated further by considering a more extreme weight ratio, that of a 17,500 lb, 60 passenger school bus and a 4400 lb passenger vehicle, each stationary when rear-ended by a 60 mph, 4400 lb passenger vehicle. The bus reached a peak acceleration of 10 G; the peak acceleration for the passenger car weighing one quarter the bus weight was 20 G (Fig. 31). The peak acceleration obtained for rear-ended vehicles struck by equal weight vehicles is shown by the dotted trend line of Fig. 31. The actual peak acceleration curve is shown as a solid line. On the basis of the trend line and considering the rather orderly collision performances observed for other series of experiments involving a range of impact speeds, the trend line for the 17,500 lb bus is shown for a rear-end collision by a 4400 lb passenger vehicle. Although the bus weighed four times as much as the passenger vehicle, which was subjected to a similar rear-end collision exposure, the passenger car peak acceleration was only twice as high as the bus acceleration.

The lack of effective structural opposition to underride intrusion by the passenger car into the rear of the bus substantially extended the deceleration distance with attending metal shearing replacing a portion of the more force-resistive front-end compression action. As with the previously discussed head-on collision performance for mismatched weights of vehicles, this offsetting feature of mismatch of heights and widths compensates very effectively for the abusive aspects of a smaller car impacting a larger vehicle, except where the speed is suf-

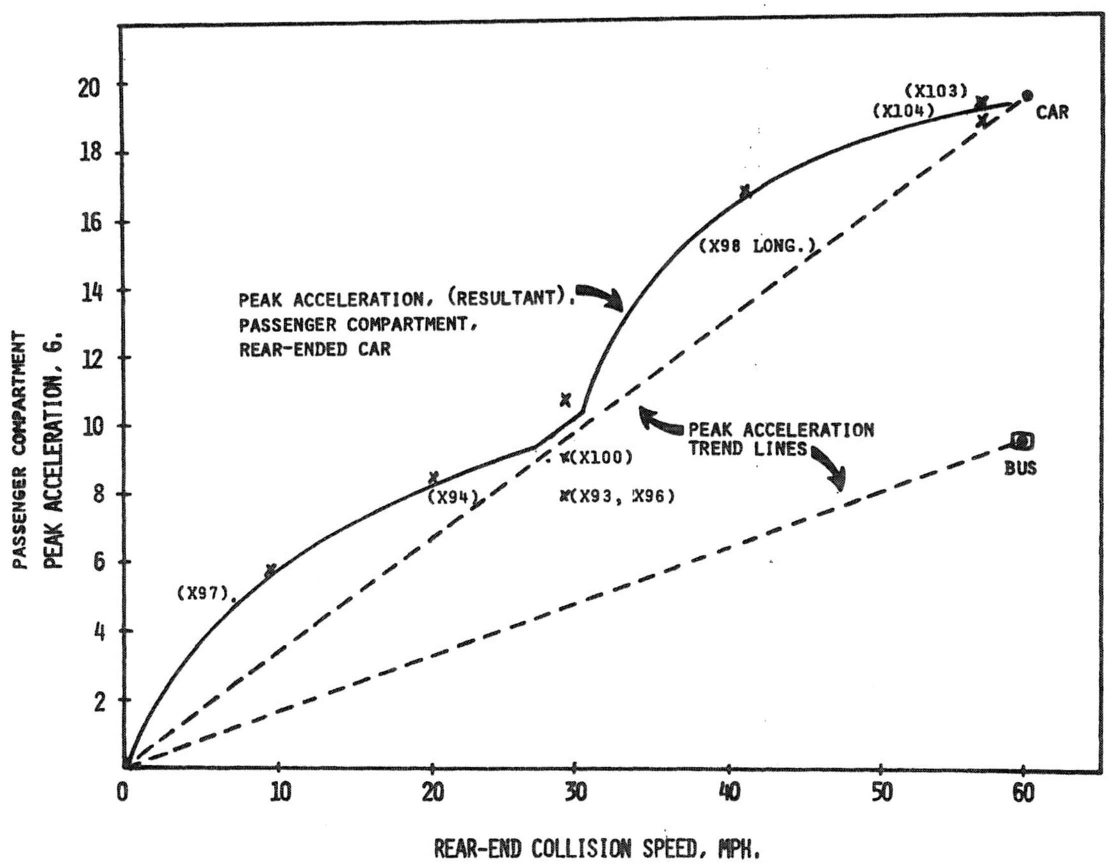

Fig. 31 - Peak acceleration values for rear-ended vehicles struck by a standard size passenger car

ficiently high to develop in the small car, direct passenger compartment encroachment as occurred for the smaller vehicle of Fig. 32. Passenger compartment deceleration was only 18 G for this 60 mph impact into the rear of a school bus, but the severe encroachment into the front passenger compartment represented a greater exposure to serious injury than if the opposing vehicle structure had matched sufficiently to eliminate passenger compartment encroachment, even though deceleration of the striking car passenger compartment would be substantially increased.

Impact Response Characteristics as a Function of Size - The four principal factors that account for the severity of a direct impact sustained by a vehicle are its size and weight, the relative closure speed of impact, the relative collision performances of the vehicles, and the size and weight of the opposing vehicle. This "bag" of variables provides an unlimited number of combinations, but such combinations operate within rather specific constraints.

For purposes of discussing impact response characteristics of small cars relative to larger vehicles, speed of impact will be held to a constant 30 mph. Further, in order to avoid the confusion of considering a double variable, the size of the striking vehicle will be held constant, and will involve a medium-size domestic passenger car grossing out at about 4000 lb.

Fig. 32 - Direct passenger compartment intrusion occurs to the smaller vehicle during this 60 mph rear-end collision (Experiment 87)

The simplest, most practical, and universal vehicle man regularly uses is his shoes; regrettably about one out of every 25 new cars manufactured will, before being retired to a salvage facility, strike a pedestrian, sometimes so forceably as to eject him from those shoes. Pedestrian impacts are generally devastating to the pedestrian but nearly always cause no physical injury to the motorist. The striking vehicle weighs, on the average, 25 times as much as the pedestrian, as contrasted wtih an average weight difference of 2 for a small car colliding with a large passenger vehicle. Pedestrians, for example, the three-year olds impacted by a 30 mph medium-size passenger vehicle were subjected, on an average, to a peak acceleration of 100 G (Fig. 33). The striking car velocity is not noticeably affected by a pedestrian impact. A motorcyclist impacted by this 30 mph reference vehicle has an exposure potentially as serious as that of the pedestrian since the motorcycle-car weight ratio is approximately 10 and this does not represent a significant improvement over the 25 value for pedestrians. The high incidence of death or grossly crippled, paralyzed, and disfigured motorcycle accident victims attest to their extreme vulnerability during collisions with larger vehicles.

The small car has sufficient weight that our medium-size reference car is no more than twice its weight. The small car acceleration is 25 G, one-quarter that of the pedestrian (Fig. 33); this acceleration level can be sustained without injury for certain collision exposures providing restraining devices are used. The referenced striking car is decelerated at nearly half the small car acceleration, or 12 G.

By doubling the size of the vehicle from small size to equivalent that of our reference vehicle, the impact acceleration is more than halved (25 G to 10 G). This difference would not be quite as much if the small car had its engine in front instead of in the rear, since by necessity rear-engine vehicles have structure at their rear that is less collapsible than the rear-ends of the front-engine vehicles*. As the weight disparity widens further, as for example, to 4:1, the (heavier) struck vehicle impacted by our reference vehicle has become a special purpose passenger vehicle of large proportions. In collision experiments with school buses, it was found that the bus, weighing four times as much as the reference passenger vehicle, was substantially taller and wider, accounting for collision performance being modified as the striking car underrode the bus and bored into the bus with a relatively small impact frontal area. Consequently, much of the front end of the striking reference passenger vehicle was not directly opposed by bus structure leaving adjacent structure to sustain combinations of compression and shear stresses; these operate at a reduced force level accounting for the nominal adjusted peak of 5 G**.

---

*The effect of rigid rear structures on peak accelerations is demonstrated by the 17 G peak sustained by the rear-ended 1960 Chevrolet in the 30 mph, Exp. 88.

**Adjustment to 30 mph based on previously established trends (2,4).

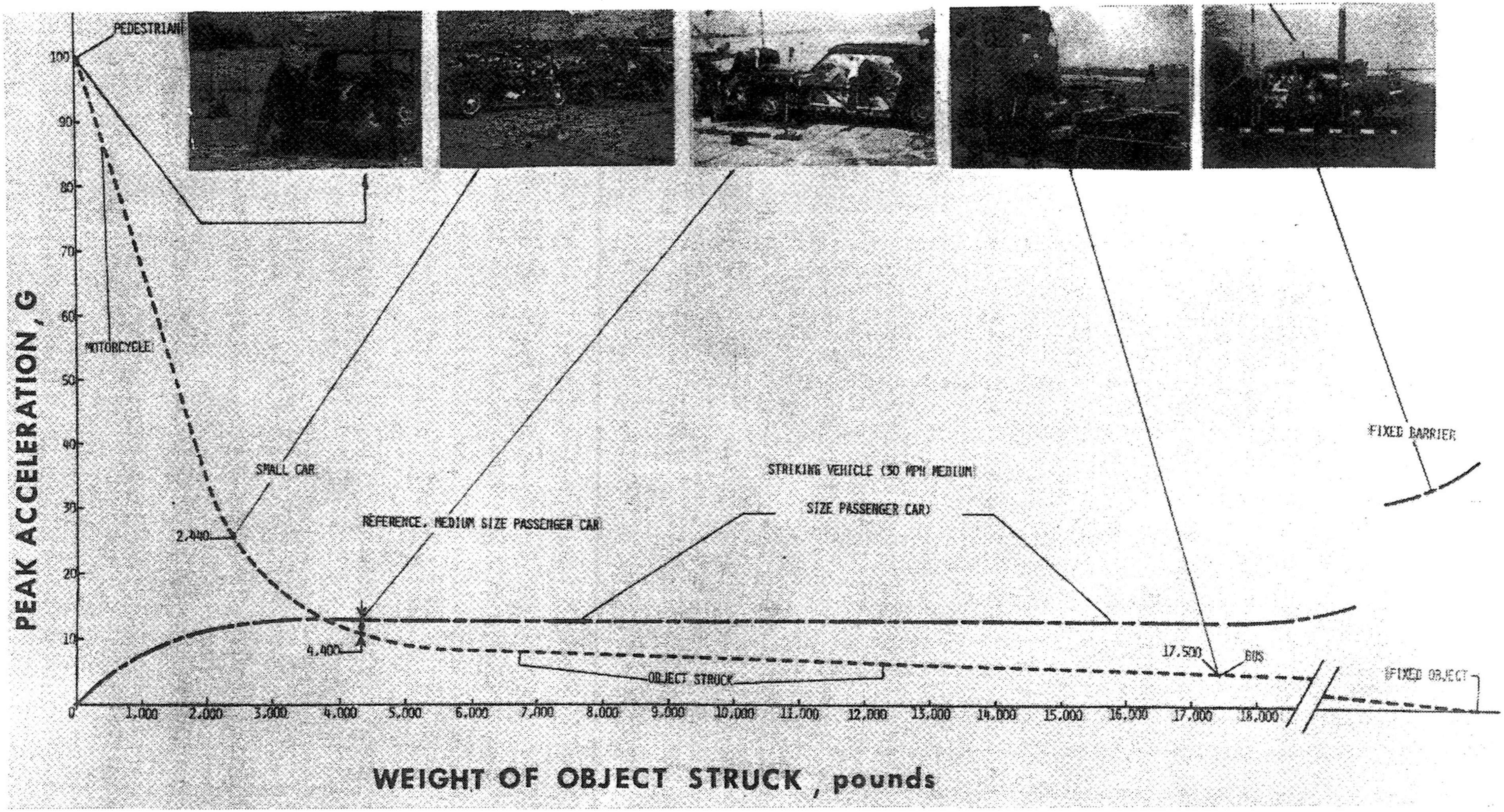

Fig. 33 - Accelerations sustained for increasing weights of struck object. Where required, data values were adjusted to a 30 mph impact by a medium size passenger vehicle

As the struck vehicle or object increases in weight, and as this object (more effectively) directly opposes the impacting reference car, the deceleration of the reference car becomes more dependent on its own collapse action to moderate its collision deceleration. The extreme or limit of resistance is represented by the fixed barrier. The nonyielding barrier by definition sustains no acceleration but our reference passenger vehicle sustains its maximum peak deceleration of 38 G.

Based on the previous discussion of the representations of Fig. 33, these conclusions appear appropriate:

1. Relative to the average vehicle weight likely to be encountered during normal usage, a small car sustains higher collision accelerations than a large car, impacted under otherwise similar conditions; the magnitude of such differences, although a function of their weights, does not represent a serious increase in exposure when compared with an alternative mode of travel such as the motorcycle.
2. Collision deceleration for a small car and for a larger car acting as the striking vehicle from studies reported in this paper and by others (3) is not usually greatly affected by the size of the opposing vehicle, as illustrated by the "Striking Vehicle" curve of Fig. 33. This is attributed to the built-in crushability of all vehicles which functions to an increasing degree as the resistance to displacement is increased for the opposing structure.
3. The curves shown by Fig. 33 are representative trendlines based on the specific data indicated and while useful for estimating responses of vehicles and conditions intermediate or adjacent to the data points, the curves should not be treated as absolute without recognition of the differences in data characteristics between these values and any comparison situation under evaluation.

When a small car strikes a larger vehicle headon at the same speed as the opposing larger vehicle, the heavier vehicle's excessive kinetic energy brings the lighter vehicle to a momentary halt substantially before the larger vehicle's forward movement is arrested. However, the time required to stop the smaller vehicle correlates with previously determined time required for two medium-size passenger vehicles, to wit, 75-90 ms. For the two head-on collision experiments reported in this paper, the smaller vehicles were stopped by their larger opposing vehicles at 75 ms (Lark) and at 88 ms (Ford sedan) (Fig. 34). However, because of the unspent momentum of the heavier vehicle, the lighter vehicle reversed direction and was pushed rearward by the still slowing heavier vehicle. Most of this rearward displacement occurred after the dummy occupants had already sustained their peak decelerative forces and this displacement, therefore, did not seriously increase their exposure to injury. Occupant rebound within the passenger compartment was, for some exposures, reasonably well synchronized with vehicle rebound so that the overall effect of this vehicle direction reversal was not significant. In order to minimize occupant injury for this exposure, restraint systems should be designed to afford a "ride-down" efficiency that

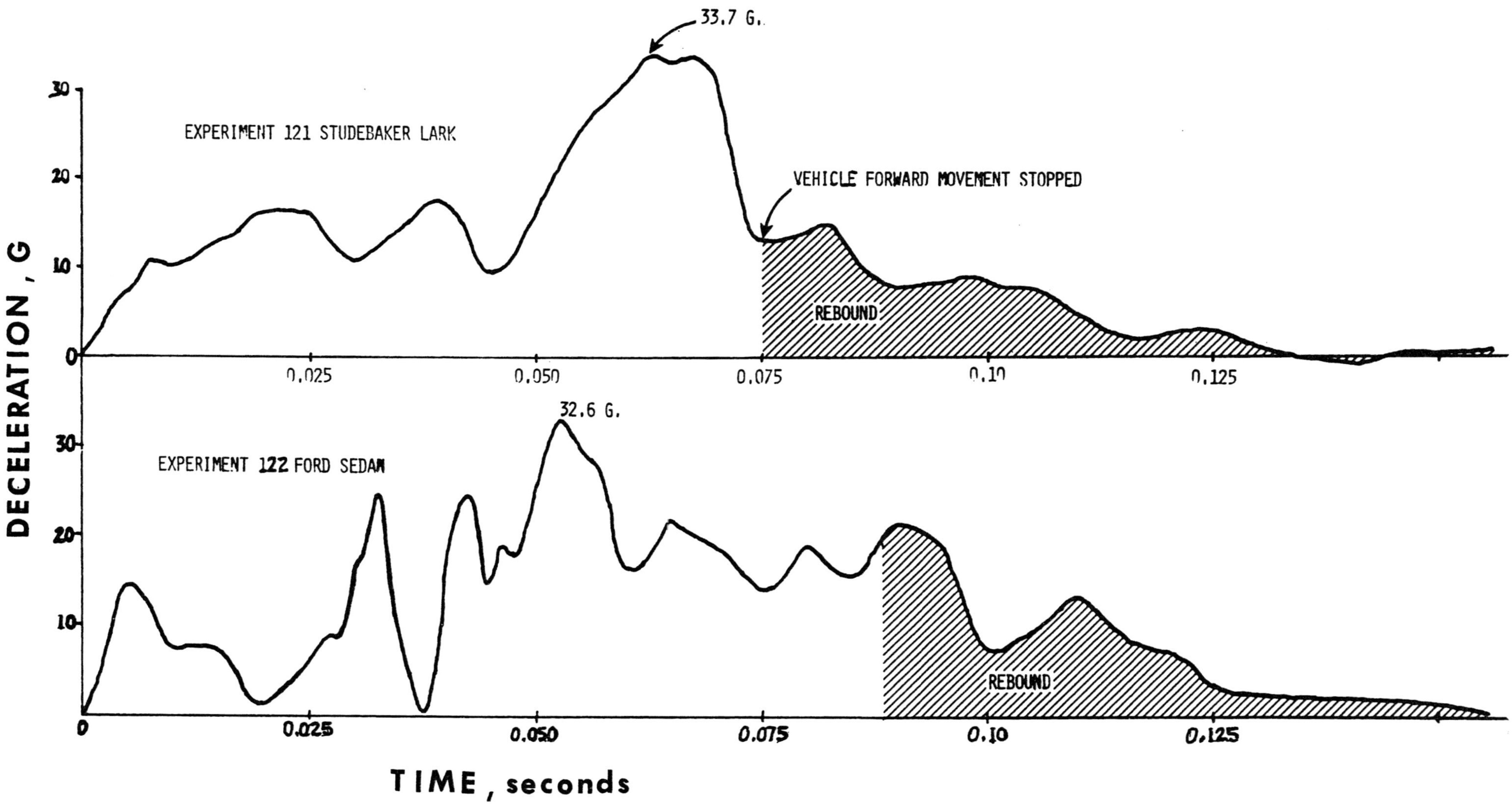

Fig. 34 - Deceleration curves for passenger compartment of lighter vehicles (Experiments 121, 122); shaded area denotes extended deceleration from forced-rearward movement

brings the motorist to zero ground velocity within the critical 70-100 ms time period where the vehicle has reached its maximum forward movement.

For the reasons outlined above, the exposure of the motorist was not appreciably greater for having been in the smaller vehicle than it would have been if the motorist were in a vehicle equally matched in size. The prolongation of the crash deceleration of approximately 60 ms caused by vehicle rebound was of a level that could easily be accommodated by motorists in the smaller car, assuming they were properly restrained. There is a somewhat self-compensating protective feature for motorists in small cars striking larger vehicles head-on which concerns the similarity between the two vehicles in collapse and passenger compartment forward displacement during the first 1/10 s before commencement of the follow-through movements of the heavier vehicle. This self-compensating feature becomes reduced in effectiveness as the relative speed of collision increases to the point where significant intrusion to the smaller car motorist compartment takes place.

It is essential for the small car motorist to be wearing restraining devices if they are going to achieve the "ride-down" similarity during the initial phase of the collision. Also, because small cars have relatively compact interiors, these vehicles provide the greatest opportunity for incorporation of passive restraints, conforming with the passenger safety concept of motorist minimum space (5).

Small-car and large-vehicle crash safety designs share the common relationship of providing a three-stage protective barrier for isolating the passenger from direct confrontation with the excessive impact forces of the hostile external environment, either from an opposing vehicle or fixed object. The first stage relates to the crushable structure ahead, behind, to the sides, and below the motorist which serves to deform and resist passenger compartment intrusion. Where forces and the penetrating qualities of opposing structures are excessive, encroachment into the passenger compartment will occur. The padded surfaces, energy absorbing controls, and interior trim of the passenger compartment represent the second stage protective barrier serving to reduce the intensity and abusiveness of crash forces encountered by the vehicle exterior. A padded steering wheel spoke as contrasted with a dagger-like fractured steering wheel spoke, a padded and gently contoured instrument panel in place of the knife-like projections of earlier instrument panel designs, recessed or deformable window cranks and door handles with padded arm rests in place of rigid side panel structures all exemplify the second stage protective barrier of the modern passenger car. Although significant progress has been made in the two stages of barrier protection for the motorist, considerable additional improvement is readily obtainable with the current level of understanding of collision phenomenon.

The third stage protective barrier, although of higher priority on the list of need for further development than the other two protective barriers, has received relatively little attention. The third protective barrier is provided by inertial restraints positioned about the motorist for the purpose of directing and attenuating the forces of the second collision, the impact exposure of the motorist with his vehicle interior. The integrated seat (6) provides the most effective third stage protective barrier presently evaluated.

Considering the importance for occupants of small cars to commence ride-down in time to take maximum advantage of the 70-100 ms vehicle forward displacement period during severe frontal collisions, it would appear desirable that installation and refinement of active restraint systems be continued for the use of knowledgeable motorists who desire maximum protection for themselves and their family. Such systems should continue to be available until at least air-bags or other activated-passive restraint systems are developed that can be fully deployed by the time cabin deceleration becomes significant; onset of cabin deceleration occurs at approximately 10 ms and is significant by 20 ms.

For motorists driving smaller cars who do not wish to take advantage of the added protection of active restraints, the incorporation of "structured" passive restraints in the form of effective head and back supports, steering wheel crush-pads, extended (and retractable) instrument and subinstrument panel padded crash guards, seat dividers, and similar improvements can provide passive restraint and compartmentization of a more practical nature than the concept of crash-inflatable air bags. Both the structured interior passive restraint concept and the air-bag activated-passive restraint concept require an integrated seat for effective motorist protection; the added refinements of passenger compartment design necessary to complete a passive restraint system represent a nominal investment compared with inflatable restraint systems.

The cost difference will be even more remarkable when the manufacturer liability aspect is understood. For example, sleepy drivers who run off the road and into an embankment, a tree, or bridge abutment will have a new excuse for loss of control; sorting out the actual "air-bag" malfunctions from the driver malfunctions could of itself become a monumental task, unless an indication is built into the system establishing whether or not external forces triggered the unit.

Accidental discharge, crash-malfunction, passenger mispositioning during actuation, motorist tampering during vehicle modifications, and salvage hazards all add to the problems inherent with an explosive device. A review of the accidental firings, malfunctions, and related injuries by the highly refined military aircraft explosive ejection seat suggests the type of problems normal to this system's approach.

Undoubtedly there are many collision configurations for which refined inflatable restraint systems can provide more effective protection than that afforded by a nonactivated passive restraint system. However, until crash-actuated systems can provide protection from impact at any angle, for multiple collision impacts and from collisions followed by upset, active and structured passive restraint systems will be required if maximum passenger protection is to be achieved.

The Intersection Collision - During side-impact collisions directly involving the passenger compartment, motorists on the impacted side are subjected to the most severe injury exposures to be found for any collision at corresponding impact speeds, including the head-on collision (4,8). Even for large passenger cars, inadequate energy absorbing structure exists between the near-side struck car occupant and the front end of the striking car. The concept of adding steel corrugated beams to the door panel represents an approach for limiting intrusion of the striking car by distributing the forces over a larger area of the vehicle; however, severe accelerative forces can still be applied to the occupants. The principal injury-producing factors for the side-impact collision exposure are passenger compartment intrusion and struck-vehicle lateral acceleration. Intrusion develops localized impact acceleration (and shear-strains) applied to the occupant that can greatly exceed the acceleration of the intact vehicle structure. Intrusion may carry an added penalty for those individuals sitting nearest the impacting vehicle who are following the proper procedure of wearing their restraining devices, because such devices tend to limit the range in which their bodies can be forceably displaced laterally by the intruding structure; this observation primarily relates to restraints that attach at the floor rather than directly to the seat. Floor-attached safety belt systems may provide severe constraint opposing the sideways displacement of a motorist rather than allowing his body to be displaced out ahead of the intruding striking vehicle. Earlier UCLA studies (7) confirmed the need for integrated seats with sufficient side structure to minimize the effect of side-impact intrusion.

The acceleration of the struck vehicle by the striking vehicle at a given speed and position of impact is dependent upon the relative masses of the vehicles and, to a lesser extent, upon the particular crush properties of those specific vehicles. For a specific speed of impact by the striking car, the effect on the struck vehicle according to whether it is a small (light weight) or large (heavy weight) vehicle is illustrated by considering the lateral acceleration of 20 G sustained by a bus weighing 17,500 lb struck at 60 mph by a 4400 lb vehicle as contrasted with the extrapolated value of 35. From this comparison, it can be seen that although the weight of the larger bus is four times the weight of the passenger vehicle, their lateral accelerations varied by a factor of two or less. The lateral displacement and partial roll of the bus during impact somewhat offset the effectiveness of its fourfold weight advantage.

Further tests are needed to provide data on a small car struck by a small car at the same speed as a small car struck by a larger car. Such data will probably show no greater differences in the extent of intrusion owing to the following offsetting factors:

1. The larger striking car has a larger frontal area and therefore distributes its forces over more of the small car structure, thereby developing less intrusion.
2. For the same speed of impact, the inherent resistive inertial forces of the impacted small car are relatively smaller than those sustained by a large car, allowing it to be pushed aside more easily and tending to minimize intrusion.

These relationships tend to remain valid until the size difference has become so great that the smaller car becomes overridden rather than pushed away.

As was found for the rear-end collision, acceleration of the smaller vehicle will be higher when struck by a large vehicle than when struck by a smaller vehicle, where speed and other factors remain constant. This higher acceleration presents an increased exposure to injury, a danger not easily overcome by improved restraint systems. When conditions of grossly mismatched opposing structure occur that allow the larger vehicle to crush over and into the small car's passenger compartment, occupant survival is unlikely, and unless provisions can be made to prevent this too commonly observed condition of excessive override with corresponding intrusion, it will matter little to occupants of the smaller vehicle involved in such a collision whether or not they are adequately restrained.

Conclusions—Head-On Collisions

1. For head-on collisions up to the speed where cabin intrusion occurs, the similarity of collapse and forward displacement of unequal weight vehicles during during the initial 1/10 s of collision provides the opportunity for considerable improvement in smaller car safety if adequate "ride-down" restraint systems are provided.
2. Restraint systems should be designed to provide a sufficiently effective "ride-down" enabling the motorist to stop close to the critical 70-100 ms instant following contact corresponding to the time of maximum forward movement of the vehicle.
3. A uniform deceleration pattern provides a better opportunity for motorist ride-down than one representative of delayed onset, followed by abrupt deceleration. The greater the differences in weights for vehicles colliding head-on, the greater the mismatch of their principal force-resistive structures and the lower their peak deceleration; this mismatch produces greater mutual penetration before a maximum encroachment is reached.
4. The noncollapsed portions of the passenger compartment of the smaller vehicle in the head-on collisions reported by this paper continued their forward displacement after initial bumper contact to a maximum forward position comparable to the forward displacements attained by equal weight vehicles in prior experiments.
5. Peak deceleration rates for smaller vehicles opposing larger passenger vehicles in head-on collisions are slightly less severe than values these smaller cars would sustain in barrier collisions at similar impact speeds.
6. During head-on collisions between unequal weight vehicles, occupants of the heavier vehicle have a greater deceleration distance available tending to reduce their collision injury exposure; restrained smaller car occupants, how-

ever, are not necessarily subjected to a significantly more severe collision than they would sustain for an impact with a vehicle of similar weight.

7. In most head-on collisions, the vehicle has arrested its forward movement before unrestrained occupants impact their vehicle interior. Therefore, the vehicle front-end, force-moderating properties provide no advantage for the unrestrained motorist as he continues forward after his car has stopped.

8. Because small cars have relatively compact interiors, these vehicles provide a greater opportunity for incorporation of passive restraints, consistent with the concept of motorist compartment minimum space.

9. During this series of head-on collisions head whip-like action of rear seat occupants over the front seatback reemphasizes the desirability of having an energy absorbing high-back seat in addition to the strengthened backrest so as to provide a protective decelerative structure for rear seat occupants.

10. Proper design of restraints and of vehicle collision structural performance can provide significant passenger protection for small car head-on collisions, comparable with larger vehicles, unless the collision is so violent that the passenger compartment integrity is violated.

11. Properly restrained small car operators appear to have no serious impact deceleration handicap for head-on collisions with larger passenger vehicles through 30 mph and probably approaching 40 mph. In the range of 40 mph and above, the collapse distance ahead of the front seat location is not sufficient to prevent direct encroachment of the passenger compartment by the larger car; for equally matched vehicles this condition did not occur until speeds of 50 mph each was reached.

Conclusions—Rear-End Collisions

1. A rear-engine vehicle presents a stiffer or less crushable structure to a rear-ending vehicle than would be encountered for a front-engine vehicle. The reduced intrusion of rear-ended vehicles is especially desirable for smaller cars because higher speed rear-end collisions can be accommodated without danger of passenger compartment encroachment. The price paid for this protection occurs in the form of higher acceleration forces to the rear-ended occupants but this presents a problem only if the vehicle is not equipped with an adequately structured full head and back support system.

2. The combination of the relatively higher stiffness common to the front-end of a larger striking vehicle, the higher rear-end stiffness identified with rear-engine vehicles as well as the lighter weight of a smaller car accounted for the abruptness of the small car (VW) crash acceleration.

3. For most rear-end collisions where compartment integrity is maintained, forces imparted to the smaller struck vehicle are within the range of occupant survivability without permanent injury provided seatback yield is not excessive.

4. For the rear-ended small car, the motorist injury exposure is especially dependent on the adequacy of the seatback restraint. Smaller cars must have

reinforced seatbacks of sufficient height to provide effective support for the head, neck, and torso as the vehicle sustains excessively abrupt accelerations.

5. As was also identified for head-on collisions, rear-end collisions involving vehicles of substantial weight differences may sustain accelerations comparable with accelerations sustained for rear-end collisions with matched vehicles. Mismatching of opposing structures provides less effective crush resistance, thereby allowing a greater proportion of collision damage to involve relatively superficial shearing of structure.

Conclusions—Side-Impact Collisions

1. In side-impact collisions the acceleration of the struck vehicle for a given speed and location of impact by the striking vehicle is dependent upon the relative masses of the vehicles, and, to a lesser extent, upon the particular crush properties of those specific vehicles (Fig. 35).

2. Acceleration of the smaller struck vehicle during a side-impact collision will be higher when struck by a large vehicle than when struck by a smaller vehicle, where speed and other factors remain constant; however, unlike the rear-end collision exposure, this higher acceleration is a danger not easily overcome by improved restraint systems.

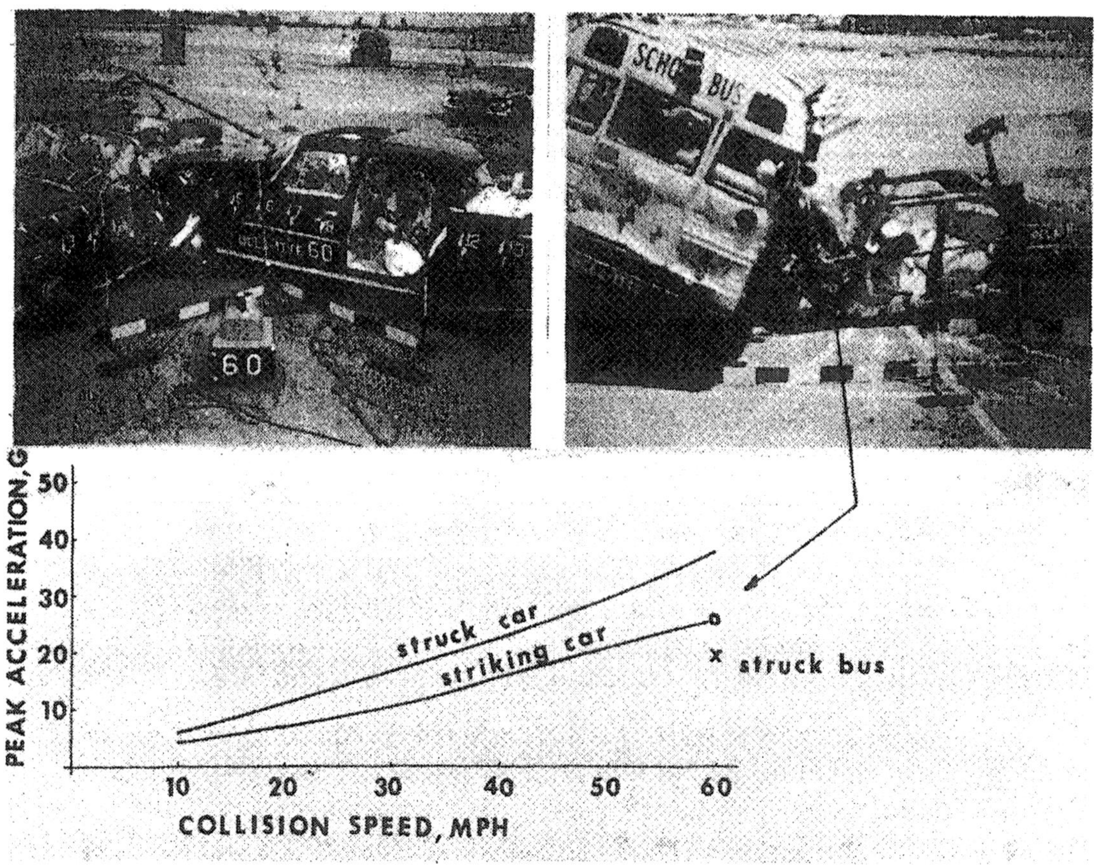

Fig. 35 - Peak acceleration values as a function of speed; side impact collisions

3. During side-impact collisions where the motorist's compartment is impacted directly rather than at locations forward or rearward of the compartment area, motorists on the impacted side are subjected to the most severe injury exposures to be found for any collision at corresponding impact speeds, including the head-on collision. Even for large passenger cars, inadequate energy absorbing structure exists between the side of the near occupant of the struck vehicle and the front end of the striking car.

4. For side-impact collisions involving a grossly mismatched striking vehicle, the larger vehicle can crush over and into the small car's passenger compartment; until provisions can be made to prevent this condition of excessive override with its corresponding intrusion, it will matter little to occupants of the smaller vehicle involved in this type collision whether or not they are adequately restrained.

General Conclusions

1. Considering the average striking vehicle weight likely to be encountered during normal usage, a small car when struck generally sustains higher collision accelerations than a larger car impacted under otherwise similar conditions; however, this difference does not necessarily represent a serious increase in exposure if adequate restraint systems are employed.

2. Crush resistant qualities of the passenger compartment can be improved by such procedure as standardization of bumper heights and by providing vertically oriented frame crash guards at front and rear ends (within the body sheet metal) and by providing intrusion resisting structures and designs to the passenger compartment doors and side panels.

3. Findings from these experiments show that in the majority of those collision exposures where the integrity of the passenger compartment is maintained, smaller vehicle occupants who are adequately restrained are protected from injury nearly as effectively as motorists in larger passenger cars undergoing comparable collision exposures.

Acknowledgments

Acknowledgment is made for Jack Baird's engineering liaison, which greatly facilitated these experiments; the excellent engineering and design assistance rendered by Hans Jakob, staff engineer; the dedicated assistance of Bernard McGuire for photography; Jean Belsha for data reduction; technicians Virgil Sibell, Bernard Tai, and Edmund Cababa for their assistance in preparing for and conducting the experiments; Glen Severy for engineering services during preparation of this paper; Lillian Nelson and Naomi Raffel for clerical support. Acknowledgment is made for assistance by the Institute of Transportation and Traffic Engineering, and the School of Engineering, UCLA, for student assistance and test site facilities.

Experiment 107 was conducted as part of the UCLA 104CD Senior Engineering Student Course under the instruction of Derwyn Severy, Harrison Brink, and Jack Baird. This student project was facilitated by the staff and equipment of Automotive Collision Research, UCLA; students participating in this study were under the student leadership of John Kerkhoff.

References

1. Derwyn M. Severy, Harrison M. Brink, and Jack D. Baird, "Vehicle Design for Passenger Protection from High-Speed Rear-End Collisions." Proceedings of Twelfth Stapp Car Crash Conference, P-26, paper 680774. New York: Society of Automotive Engineers, Inc., 1968.
2. Derwyn M. Severy, Harrison M. Brink, and Jack D. Baird, "Backrest and Head Restraint Design for Rear-End Collision Protection." Paper 680079 presented at SAE Automotive Engineering Congress, Detroit, January 1968.
3. R. H. Fredricks, "Progress in Safety Vehicle Design." Proceedings of Fifth Stapp Car Crash Conference. Minneapolis: University of Minnesota, 1961.
4. Derwyn M. Severy, J. H. Mathewson, and A. W. Siegel, "Automobile Head-on Collisions - Series II." SAE Transactions, Vol. 67 (1959).
5. Derwyn M. Severy, Harrison M. Brink, Jack D. Baird, and David M. Blaisdell, "Active Versus Passive Motorist Restraints." 1970 International Automobile Safety Conference Compendium, P-30, paper 700424. New York: Society of Automotive Engineers, Inc.
6. Derwyn M. Severy, Harrison M. Brink, Jack D. Baird, and David M. Blaisdell, "Safer Seat Designs." Proceedings of Thirteenth Stapp Car Crash Conference, P-28, paper 690812. New York: Society of Automotive Engineers, Inc., 1969.
7. Derwyn M. Severy, Harrison M. Brink, and Jack D. Baird, "Collision Performance, LM Safety Car." SAE Transactions, Vol. 76 (1967), paper 670458.
8. Derwyn M. Severy, John H. Mathewson, and Arnie W. Siegel, SP 232, "Automobile Side-Impact Collisions, Series II." Society of Automotive Engineers, Inc., March 1962.

Appendix

Small Car, Large Car Accident Case Study: 1964 Ford–1964 Volvo–Head-On Collision–Vehicle Weight Ratio: 1:1.5 - A head-on collision occurred between a large domestic passenger car and a foreign sports car, both traveling at highway speed on a 24-ft wide rural highway. The Ford was eastbound and the Volvo was westbound; pavement gouge marks indicated that the Volvo crossed the centerline stiking the Ford head-on with approximately one-half car off-set, biased to the left front and sides of both vehicles.

After collision, the Volvo was rotated 180 deg counterclockwise and was near the point of impact in the westbound lane. The Ford had rotated about 150 deg counterclockwise and ended up off the road, 12 ft south of the highway and approximately 54 ft past the point of impact. No braking skid marks were found at the scene.

The Ford driver (unbelted) was killed but the cross-chest lap-belted driver of the Volvo survived. The Volvo driver had heavy bruising from the diagonal strap, a fractured clavicle, and a broken left foot. Both he and his doctor attributed survival to his three-point restraint system. A surviving passenger of the Ford (seated on the nonimpact side) said that the Volvo came across the dividing line and that the Ford driver could not take effective evasive action.

The larger vehicle, with its greater crush distance and greater resistance to crush, nevertheless carried the only fatality. The "second collision" by the Ford driver was not controlled by use of restraining devices. This accident serves as an example of observations reported in this paper that although the size of the vehicle is an important contributing factor to the survivability or nonsurvivability of a specific collision, the use, or failure to use, restraining devices can be a more decisive factor than the size of the vehicle. There is little or no advantage to having a large crush distance if, as the vehicle collapses to a stop, the unrestrained motorist within is continuing at the same precrash speed toward his own (second) collision with his vehicle interior after it has completely stopped.

Geometry Effects

2003-06-0206

VEHICLE FRONT STRUCTURE IN CONSIDERATION OF COMPATIBILITY

Shinji Fujii, Masanobu Fukushima, Akiko Abe, Shigeru Ogawa, Hideharu Fujita, Takayuki Sunakawa, and Yukiko Tanaka
Mazda Motor Corporation
Japan

Paper No.518

ABSTRACT

A structure which effectively improves compatibility in a vehicle-to-vehicle frontal impact has been considered focusing on sub-frame structure that disperses applied force with multiple load paths. Evolved sub-frame structure has been studied by CAE with RADIOSS to search the possibility to reduce aggressivity and to improve self-protection at the same time.

Vehicle models used for this compatibility study were a large saloon car with sub-frame and a small family car without sub-frame. The large saloon car had three different front structures: original, forward-extended sub-frame, and original with 25%-stiffness reduced structures. The types of collision contained four different crash modes in a combination of lateral overlap rate difference and side member height difference. With these three different structures in four different crash modes, crash simulations were conducted to evaluate aggressivity and self-protection based on front structure and compartment deformations, energy absorption amount, and Average Height of Force (AHOF).

As a result, it was found that the front structure with forward-extended sub-frame improved both aggressivity and self-protection by preventing override effect through structural interaction enhancement.

INTRODUCTION

In the United States, with the growing popularity of light trucks and vans (LTVs), the aggressivity of LTVs as an issue of concern is growing. According to field data analysis by NHTSA based on Fatality Analysis Reporting System (FARS) data, driver fatality ratio in frontal-frontal LTV-to-passenger car crashes ranges from 1:2.6 to 1:6.2 [1]. The aggressivity of LTVs is obvious and undeniable. Highly possible factor of aggressivity is geometric difference, in particular, height differences of structural stiff parts like side members.

Recent studies on crash compatibility between vehicles have shown that the factors influencing crash compatibility performance are vehicle mass, stiffness, and geometry. The majority of the studies has concluded that geometry is the most dominant factor [2, 3, 4, 5, 6, 7, 8]. And of the geometric incompatibilities, height difference of stiff structural parts is a major concern. Height difference of structural parts leads to override and/or underrun effects, where energy absorption efficiency of both vehicles is impaired, generating additional compartment intrusion. When a vehicle is overridden, the crash energy is absorbed only by the upper body, generating a significant upper body intrusion in cowl and instrument panel areas of the overridden car compartment [4, 6, 8], compounding injury and fatality risks to the occupants.

For compatibility improvement, structural interaction to minimize override potential and effect, therefore, is most important. Thus, this study attempts to reduce aggressivity and to enhance self-protection by controlling override effect through structural interaction enhancement.

VEHICLE-TO-VEHICLE FINITE ELEMENT SIMULATIONS

Crash Models

Vehicle-to-vehicle frontal crash simulations were conducted by FEM to isolate and identify the vehicle structures that contribute to improvement of compatibility performance. Vehicle models used for the study were a large saloon car with sub-frame shown in Figure 1 and a small family car without sub-frame in Figure 2. These models were highly correlated with rigid barrier frontal impact. The large saloon car had three different types of frontal geometry:

- Original structure, a base structure (hereinafter described as "original type")
- Forward-extended sub-frame structure for override prevention shown in Figure 3 ("extension type")
- Original structure with 25%-stiffness reduced by side member and sub-frame material thickness reduction ("low stiffness type")

The weight of large saloon car, 1,326kg, was equalized to that of small family car so that weight factor in this study can be eliminated. This is the average weight of passenger cars in Japan: 1,150kg plus two Hybrid III 50th percentile male dummies. Initial crash velocity of both vehicles is also the same, 56km/h.

Table 1 summarizes vehicle model characteristics of large saloon and small family cars. RADIOSS software was used to make calculations during the period of first 150ms after the collision.

Figure 1. FEM model of large saloon car with sub-frame.

Figure 2. FEM model of small family car without sub-frame.

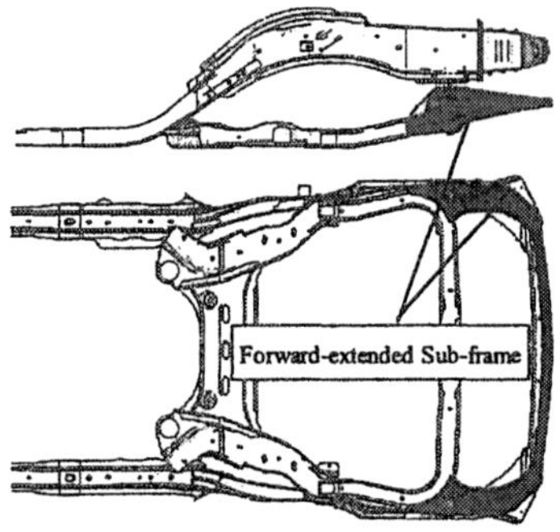

Figure 3. Forward-extended sub-frame structure of large saloon car.

Table 1.
FEM model characteristics

	Large saloon car			Small family car
	Original type	Extension type	Low stiffness type	
Nodes	265,000	267,000	265,000	231,000
Elements	270,000	273,000	270,000	261,000
Length	4,670 mm			3,925 mm
Width	1,780 mm			1,680 mm
Mass	1,326 kg			1,326 kg

Simulation Matrix

There were four vehicle-to-vehicle frontal crash modes. Figures 4 and 5 depict two different lateral lap amounts, 50% and 100%, and two different height differences, 0mm and 100mm, respectively. In the case of 100mm-height difference, the side member of the large saloon car is higher than that of the small family car. Combinations of these overlap amounts and height differences were adopted as crash modes. The simulation test matrix contains 12 cases shown in Table 2, combining these four test modes and three different vehicle front geometries of the large saloon car.

Under the combination, toeboard and A-pillar intrusions on driver side on both models were measured to evaluate compatibility performance.

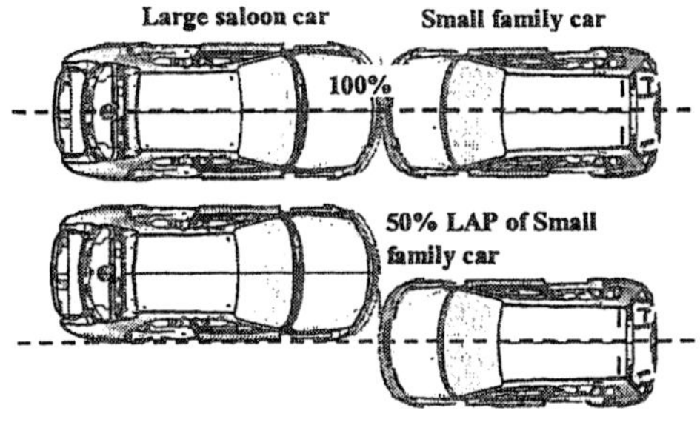

Figure 4. Lateral overlap.

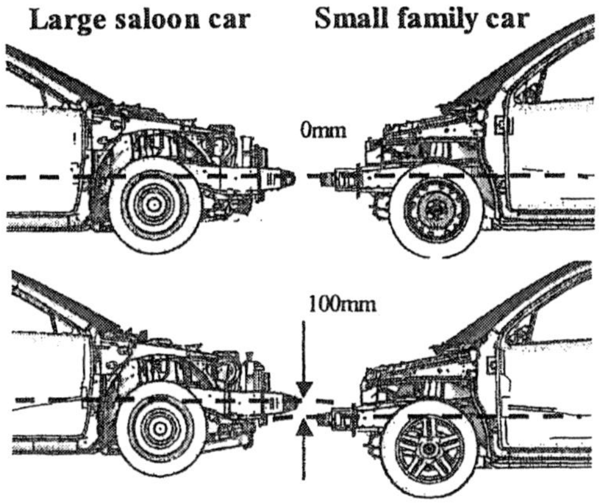

Figure 5. Side member relative height difference.

Table 2.
Simulation matrix

Crash mode		Front end structure	
Lateral lap	Side member height of large saloon car in comparison with small family car	Large saloon car	Small family car
100%	0mm	Original Extension Low stiffness	Original
100%	100mm	Original Extension Low stiffness	Original
50%	0mm	Original Extension Low stiffness	Original
50%	100mm	Original Extension Low stiffness	Original

RESULTS AND DISCUSSION

Front Structure Deformation

This section describes vehicle behavior varied by difference in vehicle structure and crash modes. Figures 6 through 9 illustrate deformation modes of original and extension types of large saloon cars and small family cars at 40ms with 100mm-height difference. The greatest difference in deformation mode appeared at 40ms and no outstanding mode change was seen after 40ms. Therefore, deformations at 40ms were evaluated. Figure 6 shows original type with 100% lateral lap, and Figure 7, extension type. Figure 8 exhibits original type with 50% lateral lap, and Figure 9, extension type. Figures 8 and 9 depict right side members. The results show a conspicuous structural interaction difference between original and extension types. In the case of original type in Figure 6, the side member of the large saloon car overrode that of the small family car. This left some portion of the small family car's side member undeformed. Meanwhile, in the case of extension type in Figure 7, the side member of the small family car was sandwiched between the extended sub-frame and side member of the large saloon, preventing the small family car from underrunning the large saloon. Further, the frontal area of the small family side member also deformed, proving that there had been good interaction. Similar results were seen in the 50% lateral lap as shown in Figures 8 and 9. However, no difference was seen in the vehicle movement between low stiffness and original types. Thus, it follows that the extension type can prevent the opponent vehicle from underrunning by improving interaction even when the side member height is different.

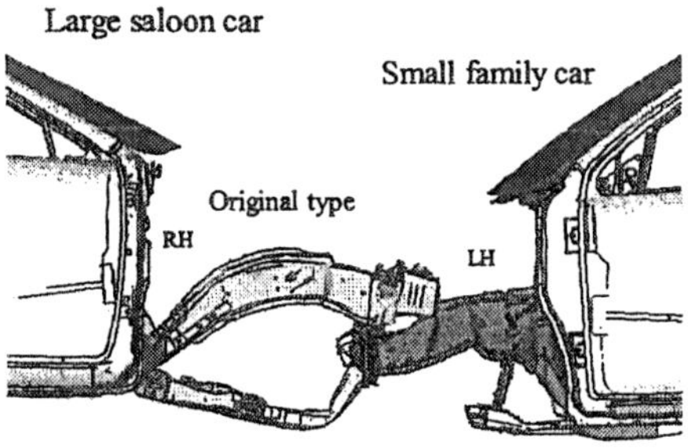

Figure 6. Deformation mode in 100%-lap - original type.

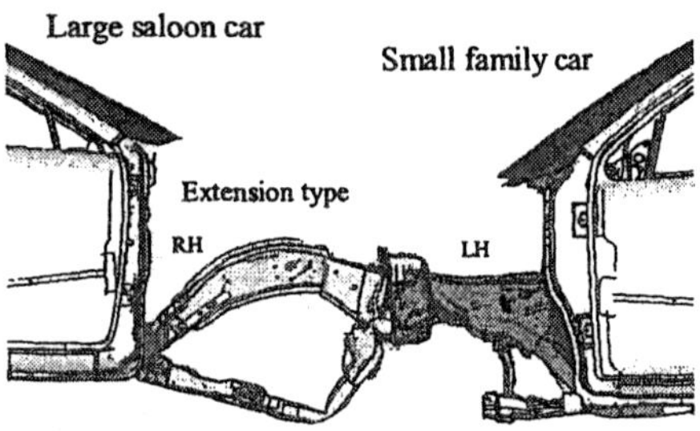

Figure 7. Deformation mode in 100%-lap - extension type.

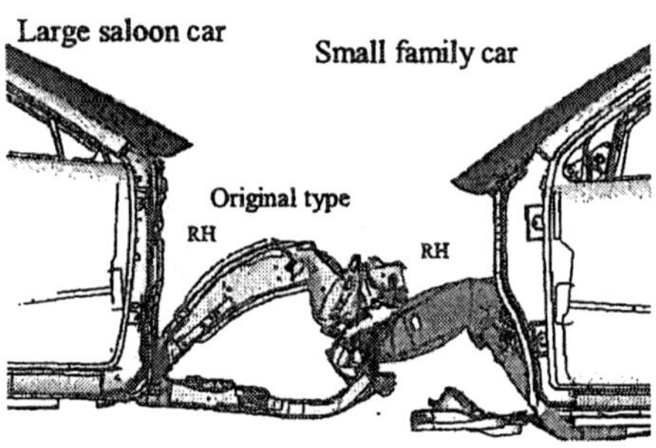

Figure 8. Deformation mode in 50%-lap - original type.

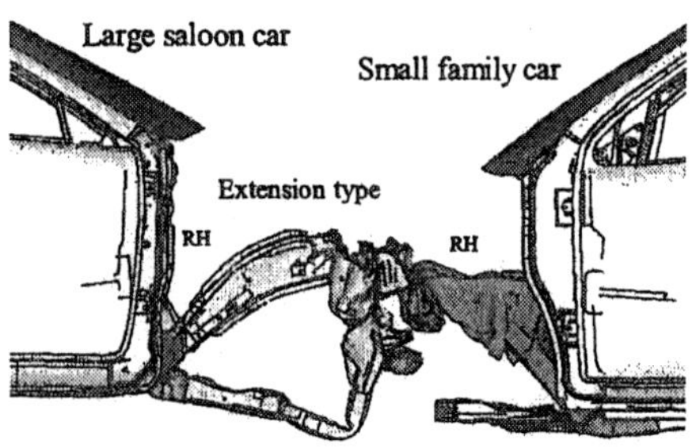

Figure 9. Deformation mode in 50%-lap - extension type.

A comparison of side member and sub-frame deformation was made quantitatively, and the characteristics of extension type were investigated. Figure 10 indicates measuring points on the front areas of the large saloon and the small family cars. Note that the front end of the extended sub-frame was not measured and therefore not illustrated for the extension type. Connecting lines are drawn between these points, and from a side view of the lines were made the comparisons.

Figures 11 through 14 show deformation modes of original and extension types of large saloon cars and small family cars at 40ms. These figures reveal two notable deformation features when the extension type of large saloon car is collided with the small family car. One is that the sub-frame of the extension type bends downward at the center more stably compared with the original type. The bending points are marked with "A" in Figures 10 through 14. The other feature is that the extended sub-frame holds and deforms the side member of the small family car, preventing the small family car from underrunning as marked with circles in Figures 12 and 14. The deformation of the side member of the small family car contributes to the smaller deformation of its compartment.

As far as low stiffness type is concerned, it did not show any improvement in interaction as illustrated in Figure 15.

The extension type of large saloon car has the following features:

- It stabilizes deformation mode of its own side member and side frame.
- It prevents the small family car from underrunning.
- It deforms the side member of small family car effectively.

The extended sub-frame has the potential to reduce aggressivity by improving structural interaction, especially when there is geometric incompatibility between two vehicles.

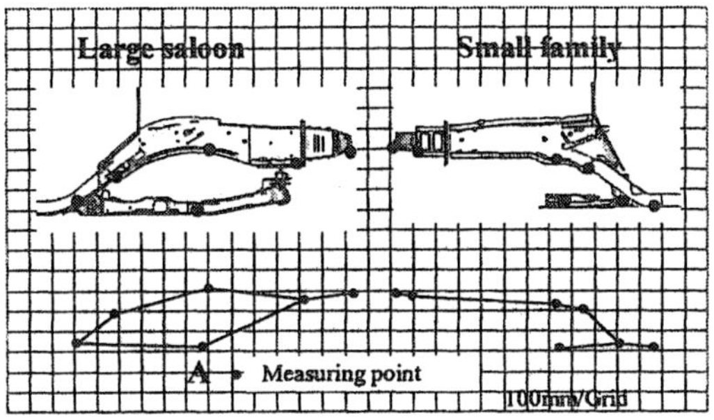

Figure 10. Measuring points for deformation mode comparison.

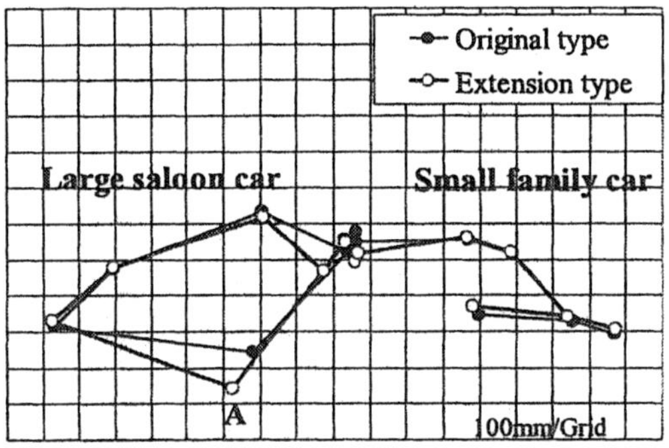

Figure 11. Deformation comparison between original and extension types in 100%-lap and 0mm-height difference.

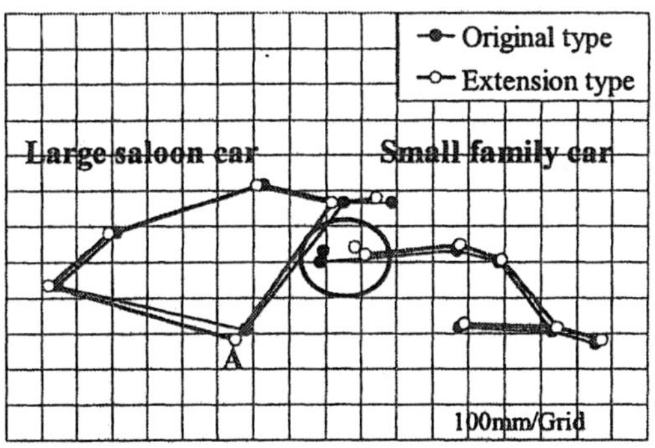

Figure 12. Deformation comparison between original and extension types in 100%-lap and 100mm-height difference.

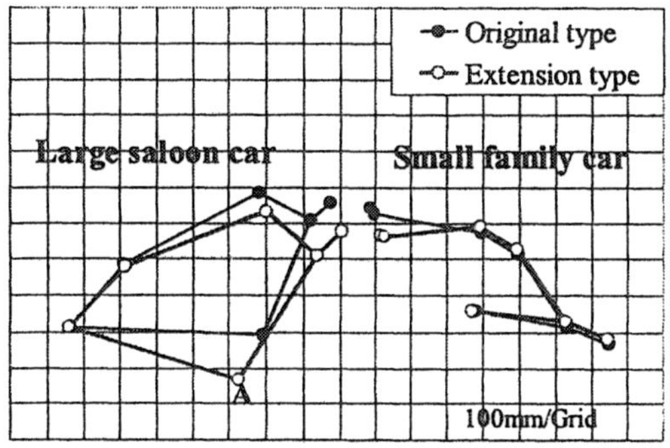

Figure 13. Deformation comparison between original and extension types in 50%-lap and 0mm-height difference.

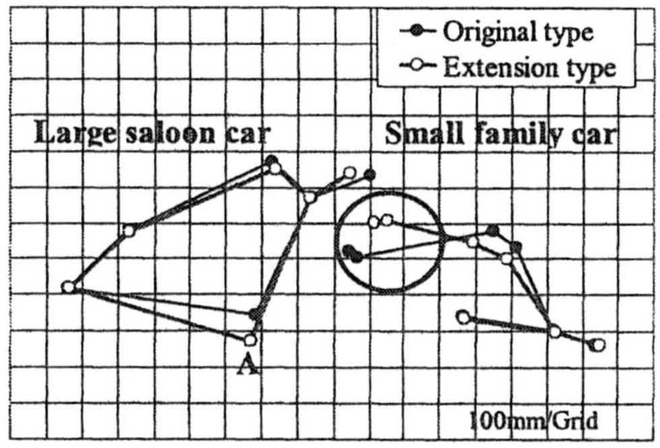

Figure 14. Deformation comparison between original and extension types in 50%-lap and 100mm-height difference.

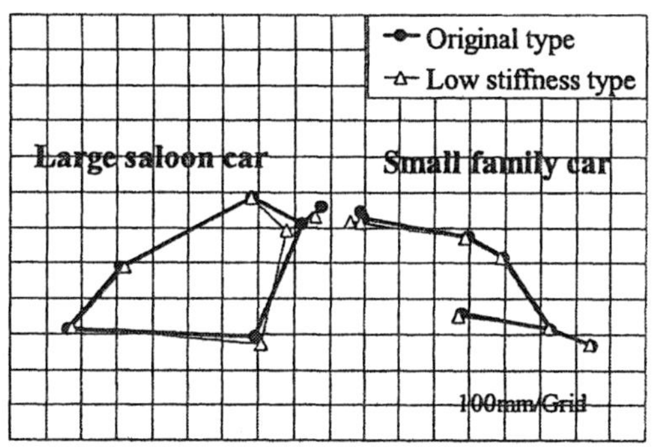

Figure 15. Deformation comparison between original and low stiffness types in 50%-lap and 0mm-height difference.

Compartment Deformation

Vehicle deformation amount was compared to evaluate compatibility performance. For the compatibility evaluation, toeboard and A-pillar intrusions were chosen to measure the damage on compartment. The large saloon car was defined as subjective vehicle and the small family car as opponent vehicle. When the intrusion amount of the large saloon car was smaller than that of original type of the large saloon car, it was judged as effective for self-protection. And if the intrusion amount of the small family car was smaller than that of the small family car collided with original type, it was judged as effective for aggressivity reduction.
Figure 16 shows the toeboard intrusion on driver side. With regard to the extension type, toeboard intrusions of both the large saloon and small family cars are reduced compared to the original type in every crash mode. Figure 17 shows A-pillar intrusion on the driver side. Since the A-pillar intrusion of the large saloon was too small, we only evaluated intrusion of the small family car. In the case of extension type, A-pillar intrusion of the small family car was reduced compared to the original type in every crash mode. Therefore, it follows that the extension type can reduce toe-board and A-pillar intrusions for both vehicles.
This means that the extension type improved self-protection and reduced aggressivity at the same time. As mentioned earlier, this is because the extension type has the possibility of improving interaction and it deforms the front area of both vehicles efficiently.
For the low stiffness type, intrusion amounts of toeboard and A-pillar are not necessarily smaller than that of the original type. It is theoretically considered that reduced stiffness should increase the deformation amount of its own vehicle. However, because the geometry influence is so dominant in a vehicle-to-vehicle crash, stiffness reduction effect was not observed as expected.

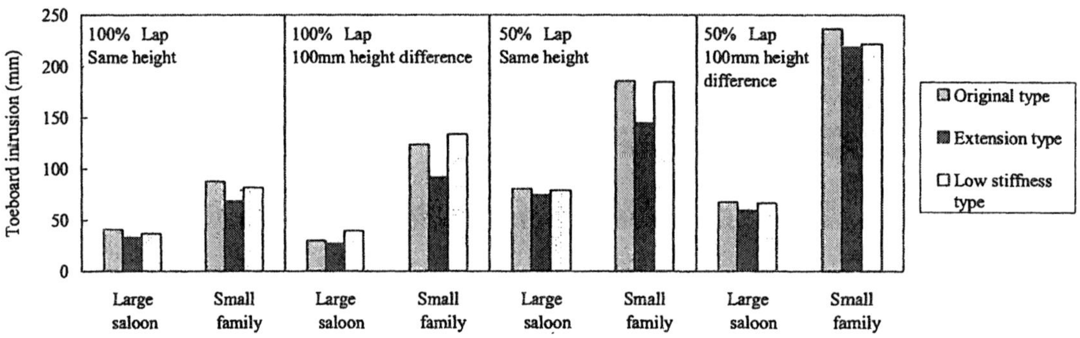

Figure 16. Comparison of toeboard intrusion.

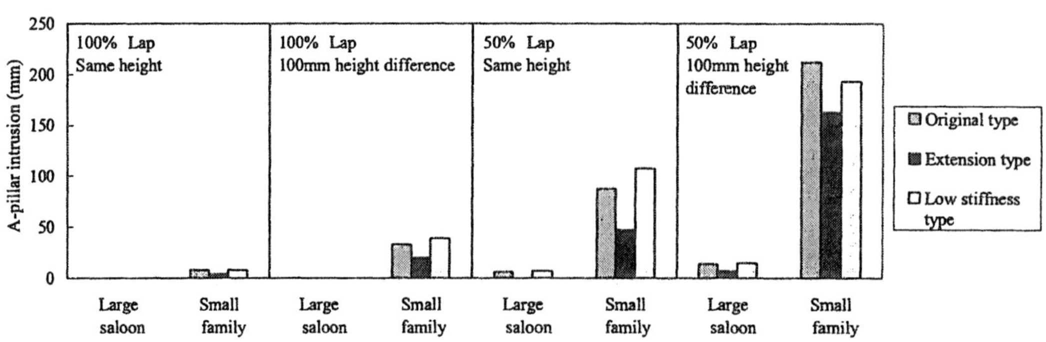

Figure 17. Comparison of A-pillar intrusion.

Energy Absorption

From standpoints of deformation on front structure and compartment, compatibility was so far investigated. The results show that the extension type has the possibility of improving aggressivity and self-protection. In this section, compatibility is discussed in terms of energy absorption. The energy absorption amount means a total value of energy absorbed by every component of a vehicle model calculated by computer simulation. Figure 18 shows energy absorption share of large saloon car. To calculate the energy share, the energies absorbed by the large saloon car and by the small family car were added. The total energy absorbed by both vehicles was always reserved as 100%.
Compared to the original type, the extension type of large saloon car increases its own energy absorption share and decreases the energy absorption share of the opponent small family car, which is good for aggressivity reduction. This is one of the reasons why the vehicle deformation amount of the small family car was decreased (See Figures 16 and 17). Further, despite increased energy absorption share of the large saloon, toeboard and A-pillar intrusion amounts did not increase. We consider that the sub-frame and side member of the extension type of the large saloon car absorbed energy effectively without passing the energy on to its compartment, which enhances self-protection.

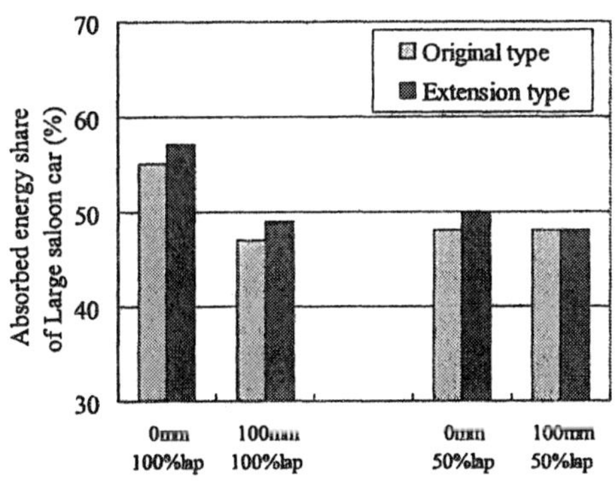

Figure 18. Energy absorption share of large saloon car.

Evaluation of Average Height of Force (AHOF)

This section explains the relation between AHOF and compatibility improvement effect displayed by the extension type in a vehicle-to-vehicle collision.
Vehicle-to-rigid barrier full-lap frontal impact test procedure to calculate AHOF is now under discussion by IHRA. In the present study, original type and extension type of large saloon car were impacted against a rigid barrier with load cell wall, as illustrated in Figure 19, under the following test specification by FEM:

- Test speed: 56km/h
- Aluminum honeycomb: without
- Load cell size: 125 x 125mm
- Load cell wall matrix: 16 x 10
- Load cell wall location: 125mm above the ground line

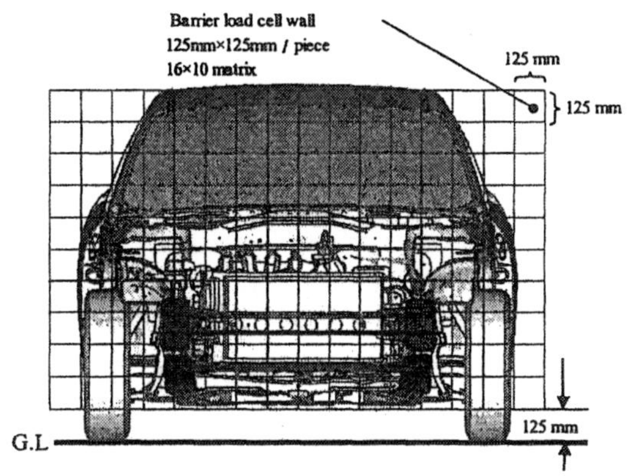

Figure 19. Relation of load cell wall and vehicle.

Figure 20 illustrates AHOF of original type and extension type. AHOF of the extension type within the first 25ms is lower than that of the original type. Specifically, at the initial stage of crash, AHOF of the extension type is lower approximately by 70mm compared to the original type.
As for the total load applied to the barrier load cell in Figure 21, total load of the extension type is higher than that of the original type until 15ms. Load distribution at 10ms is shown in Figure 22 for the original type and in Figure 23 for the extension type. Regarding load distribution of the extension type, load was applied in a lower and wider range of load cell walls compared to the original type. This is because extended sub-frame collided against the bottom row of load cell wall,

which made AHOF of the extension type lower than that of the original type.

The extension type improves vehicle front structural interaction and controls override/underrun effects even when there is height incompatibility. As for the relation between AHOF and improved interaction, it showed close relation during the initial crash, namely, within the first 25ms. Accordingly, in order to evaluate structural interaction in a vehicle-to-vehicle crash by means of a vehicle-to-rigid barrier test, the use of AHOF during the initial stage of the collision is practical. It is important to reduce the AHOF difference during the first stage of crash for compatibility improvement.

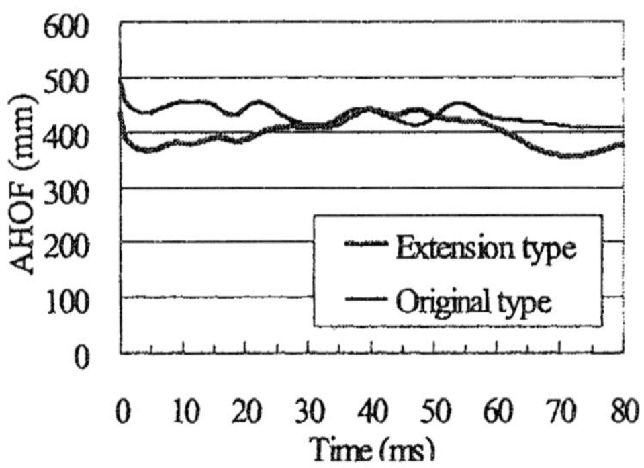

Figure 20. AHOF comparison.

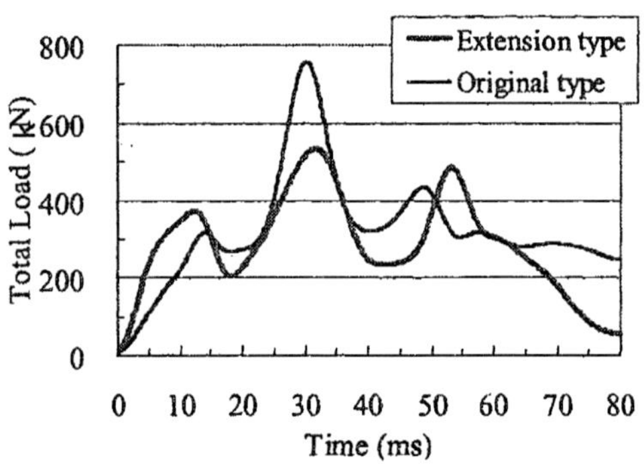

Figure 21. Load cell wall total load.

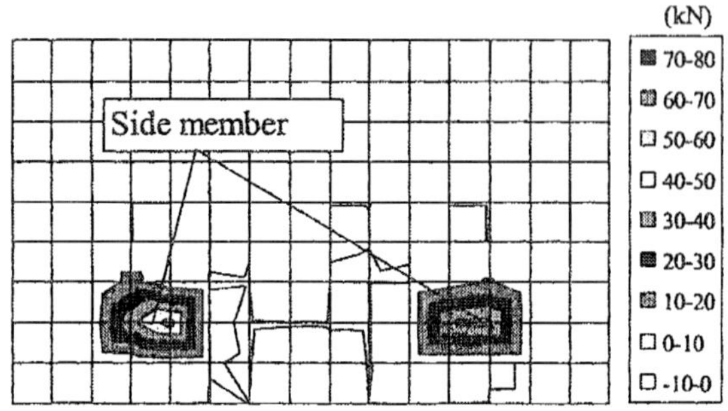

Figure 22. Load distribution of original type at 10ms.

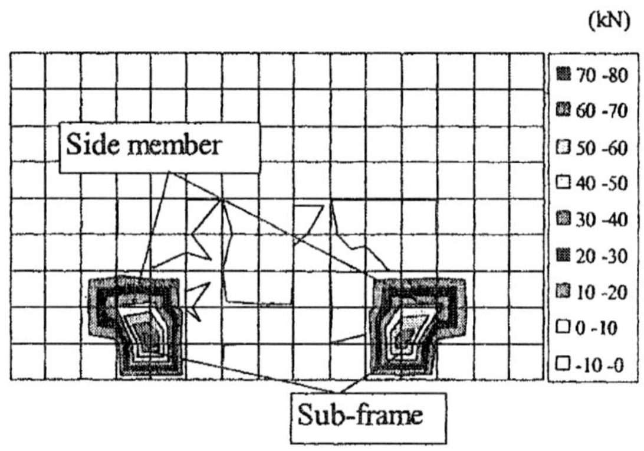

Figure 23. Load distribution of extension type at 10ms.

CONCLUSIONS

1. Forward-extended sub-frame structure has a possibility to improve compatibility for aggressivity and self-protection because of structural interaction enhancement and prevention of override.
2. Forward-extended sub-frame lowers the AHOF in a vehicle-to-rigid barrier full-frontal impact.

This study was conducted only on limited vehicles with and without sub-frame by CAE. The results are not applied to all cases and to every vehicle. Further study needs to be done on various vehicles because structure tends to depend on vehicle packaging that varies by vehicle.

REFERENCES

[1] SUMMERS, S. M., PRASAD, A., HOLLOWELL, W. T., National Highway Traffic Safety Administration, "NHTSA's Research Program for Vehicle Aggressivity and Fleet Compatibility", Paper Number 249, ESV2001.

[2] FAERBER, E., EEVC WG15, "EEVC Research in the Field of Improvement of Crash Compatibility Between Passenger Cars", Paper Number 444, ESV2001.

[3] O'REILLY, P., IHRA, "Status Report of IHRA Vehicle Compatibility Working Group", Paper Number 337, ESV2001.

[4] EDWARDS, M., HAPPIAN-SMITH, J., DAVIES H., BYARD, N., AND HOBBS, A., TRL, "The Essential Requirements for Compatible Cars in Frontal Collisions", Paper Number 158, ESV 2001.

[5] MONNET, S., PSA Peugeot Citroën, "PSA's Views on Compatibility: A Potential Two-step Approach to Improve Compatibility among the Vehicle Fleet", Paper Number 343, ESV 2001.

[6] MAYERSON, S. L., Nolan, J. M., IIHS, "Effects of Geometry and Stiffness on the Frontal Compatibility of Utility Vehicles", Paper Number 91, ESV 2001.

[7] BARBAD, S., Li, X., PRADAD, P., Ford Motor Company, "Evaluation of Vehicle Compatibility in Various Frontal Impact Configurations", Paper Number 97, ESV 2001.

[8] BARBAD, S., Li, X., PRADAD, P., Ford Motor Company, "A Comparative Analysis of Vehicle-to-Vehicle and Vehicle-to-Rigid Fixed Barrier Frontal Impacts", Paper Number 31, ESV 2001.

[9] DIGGES, K., EIGEN, A., George Washington University, "Examination of Car to Light Truck Compatibility in Frontal Crashes", Paper Number 2001-01-1165, SAE 2001.

2001-06-0146

INFLUENCE OF BODY INTRUSION AND DECELERATION ON OCCUPANT INJURIES IN FRONTAL COLLISIONS BETWEEN PASSENGER CARS

Tomiji Sugimoto
Hideki Suzuki
Honda R&D Co. Ltd. Japan
Yoshiji Kadotani
American Honda Motor Co., Inc. U.S.A.
Paper Number 433

ABSTRACT

In this paper, the results from a matrix of tests performed to evaluate the response of the occupant injuries in collisions between passenger cars at 55 km/h are reviewed. Various crash tests were conducted with different vehicle weights and stiffness to investigate the effect of body intrusion and vehicle deceleration on occupant injuries.

Some cases showed high intrusion which resulted in high occupant injury. Conversely some cases exhibited severe body deceleration which resulted in high occupant injury. In the cases with severe body deceleration, the body intrusion was almost the same as a 64 km/h ODB test but the injuries occur because the body deceleration is much greater in the car-to-car collision than the ODB test.

To assess vehicle compatibility, an MDB test method is proposed which is one of the representative test methods of real world car-to-car accidents.

BACKGROUND

Research is continuing in countries around the world to improve vehicle compatibility, but the current situation is that concrete analysis and evaluation methods have not yet been clarified.

Research on compatibility begins with understanding the problems of compatibility, which is done through analysis of accident surveys. Of course, traffic environments differ according to the country, due to differences in the mass ratio and model configuration between the crash vehicles, and also the traffic environment and driver characteristics. These differences in environment and other factors are thought to result in different crash directions and speeds. At the same time, the progress of existing safety standards concerned with crash safety also differs between countries. This difference in safety performance between environments is one reason for the increasing complexity of this problem. That is to say, analysis of vehicle compatibility also faces the issues concerned with the establishment of a global standard.

INTRODUCTION

The authors are also involved in this research, and reported the results of comparative analysis of offset crash test methods at the previous '98 ESV. The offset deformable barrier (ODB) test method approximates only car-to-car crashes between vehicles of the same limited weight. However, the honeycomb bottoming out phenomenon, which does not occur in actual car-to-car crashes, sometimes occurred when the weight or stiffness of the test vehicle exceeded certain values.

The moving deformable barrier (MDB) test method which is being studied by the National Highway Traffic Safety Administration (NHTSA) has the problems of MDB over-ride onto the test vehicle, and also the honeycomb bottoming out phenomenon such as in ODB tests.

This paper added the perspective of occupant injury which was not touched upon in the previous comparative analysis of offset crash test methods. It also investigated the relationship between body deformation, deceleration and occupant injury severity. Methods for evaluating and testing compatibility are also discussed.

In particular, car-to-car tests crash were conducted with vehicles of different weights, and these results were used to verify changes in occupant injury severity and investigate characteristic factors which may be able to be used for compatibility evaluations.

In addition, it was found that the newly proposed evaluation method may be able to reproduce the body deformation and deceleration observed in actual car-to-car crashes by adjusting the honeycomb characteristics to prevent the above-mentioned MDB over-ride phenomenon.

Table 1 lists the features of current safety standards from the perspective of compatibility.

Table 1.
Features of Compatibility Evaluation Methods

	Test Procedure	Car to Car	MDB	Full Lap	ODB
	Crash Pulse	*stiff*	*stiff*	*stiff*	*soft*
	Intrusion	*high*	*high*	*low*	*high*
Compatibility Factors	Mass	*base*	O	×	×
	Stiffness	*base*	△	△	△
	Geometry	*base*	△	△	△

O can evaluate
× can not evaluate
△ difficult to evaluate

Load cell barrier is required to evaluate stiffness and geometry.

Frontal full lap crashes generate strong body deceleration, and these tests are considered suitable for improving the performance of restraint devices such as seatbelts and airbags. On the other hand, the ODB crash tests that are currently widely used could be said to focus on the evaluation of cabin integrity with respect to uneven input to the body. Also, it should be noted that the crash conditions are between vehicles of the same weight and model.

These crash tests are improving vehicle safety performance for the crash formats concerned. In other words, it can be said that the self-protection performance of vehicles is increasing. However, there is room for doubt as to whether improving the self-protection performance of each vehicle will clearly contribute to improving compatibility.

The usual method for improving compatibility is to improve factors that are thought to cause incompatibility. In other words, steps are taken to optimize each of the three factors which are generally related to incompatibility, namely: (1) Mass, (2) Stiffness and (3) Geometry. However, there is currently a lack of methods and criteria for evaluating these factors. Methods for simultaneously optimizing multiple factors have not been established.

This research performed a number of tests using vehicles of varying mass ratios in order to:

1) Clarify the position of current evaluation criteria.
2) Discover characteristic relationships between body deformation, deceleration and occupant injury severity.
3) Investigate the possibility of resolving issues concerned with improvement technologies.
4) Propose new evaluation methods.

TEST RESULTS

The test vehicles used were commercially available 1996 to 2000 year model passenger cars. The test conditions were a frontal offset of 50% with reference to the width of the lighter vehicle, and the two crash speeds of 50 km/h and 55 km/h for both vehicles. In addition, the tests were conducted with a colinear crash angle so that the center axes of the vehicles were parallel. This test method is the same as that described in the authors' previous paper. The dummies used were HY-III dummies, and the test vehicles were equipped with airbags.

Fig. 1 shows the test matrix for the vehicles tested.

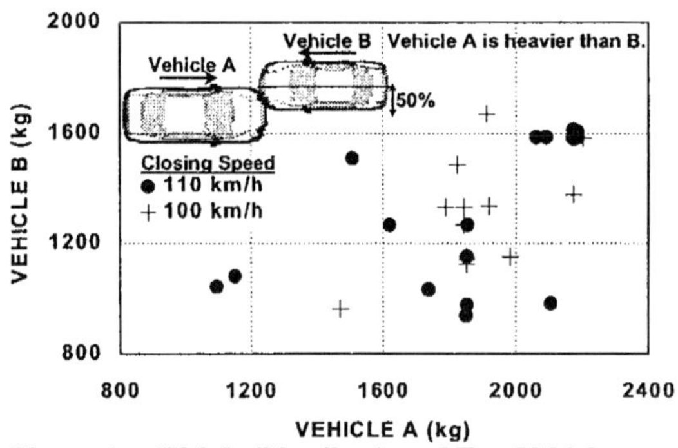

Figure 1. Weight Distribution of Test Vehicles.

Figs. 2 to 7 show the dummy injury severity measurement values for the driver. Reference value is a ratio of one to average injury number.

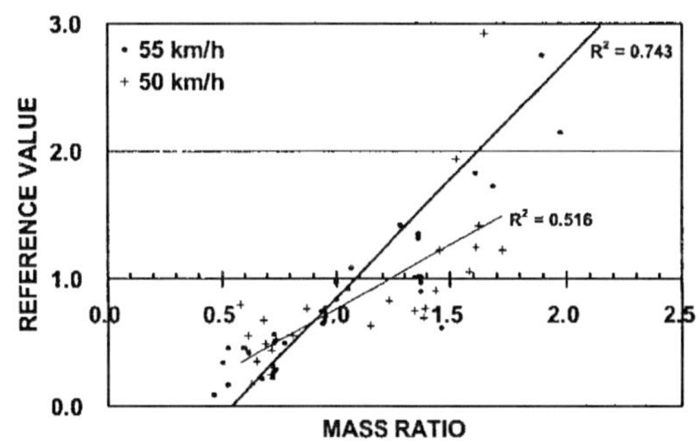

Figure 2. Relationship between Driver HIC and Mass Ratio.

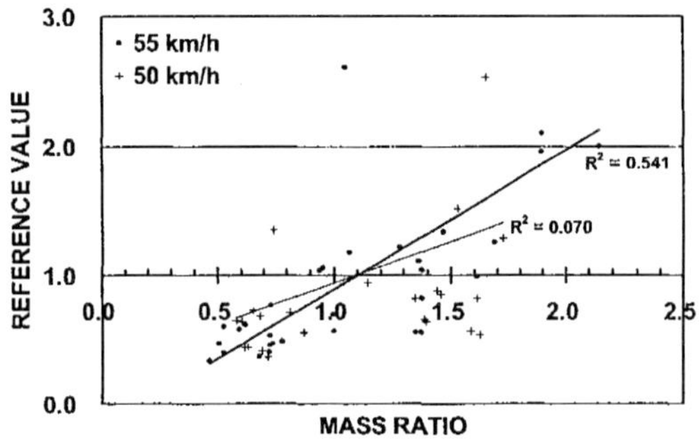

Figure 3. Relationship between Driver Nij and Mass Ratio.

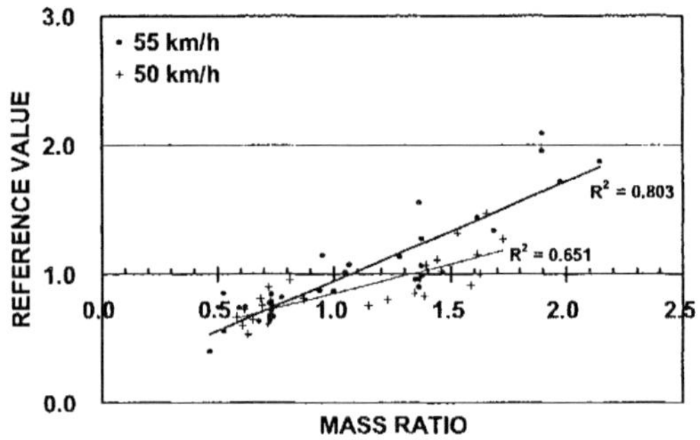

Figure 4. Relationship between Driver Chest G and Mass Ratio.

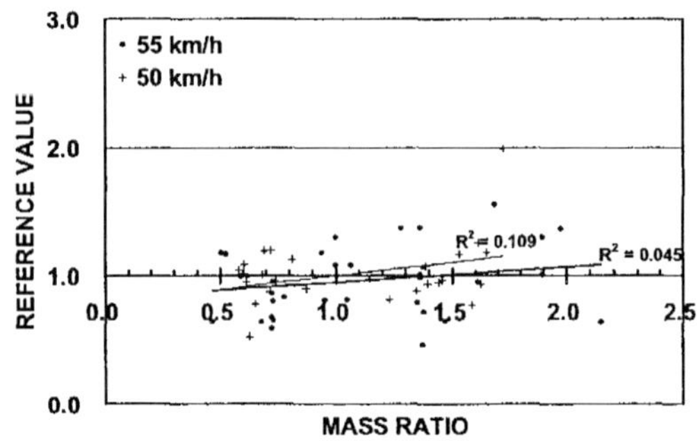

Figure 5. Relationship between Driver Chest Disp. and Mass Ratio.

The head injury (HIC), neck injury (Nij), Chest (G, displacement), femur load and tibia index (T.I.) characteristics from the perspective of mass ratio are as follows.

• HIC exhibited a large slope and also a high correlation relative to mass ratio. When the mass ratio exceeded 1:1.1, it exceeded reference value in some cases.

• Nij showed similar characteristics to HIC.

• Although Chest G showed strong correlation, it exceeded reference value even at mass ratios near 1:1 in some cases.

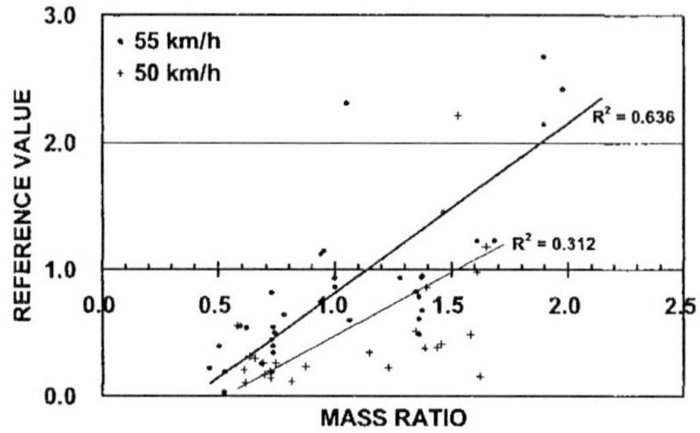

Figure 6. Relationship between Driver Femur Load and Mass Ratio.

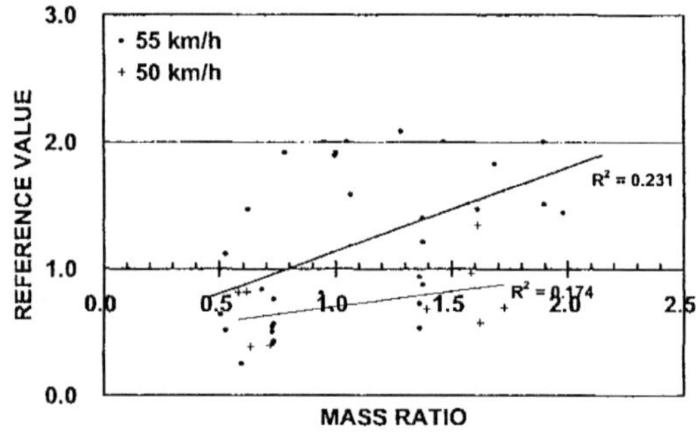

Figure 7. Relationship between Driver Tibia Index and Mass Ratio.

• Chest displacement did not show good correlation relative to mass ratio.

• Femur load exhibited correlation relative to mass ratio, and showed high value only when the vehicle experienced significant deformation. On the other hand, tibia index (T.I.) exceeded reference value in many cases for all mass ratios.

Figs. 8 to 11 show the average body deceleration, the maximum speed difference (delta V) and dummy injury severity.

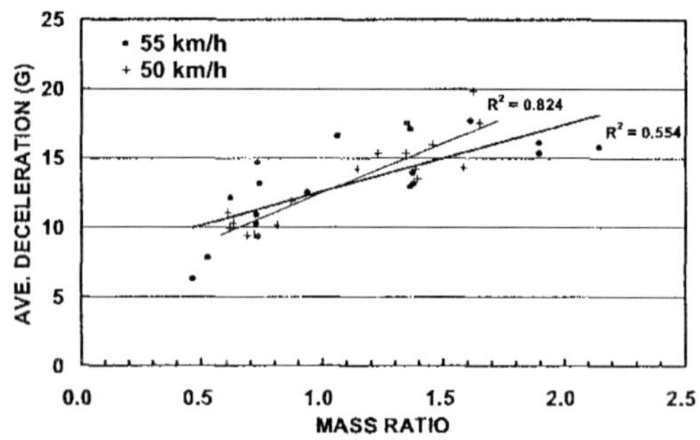

Figure 8. Relationship between Body Deceleration and Mass Ratio.

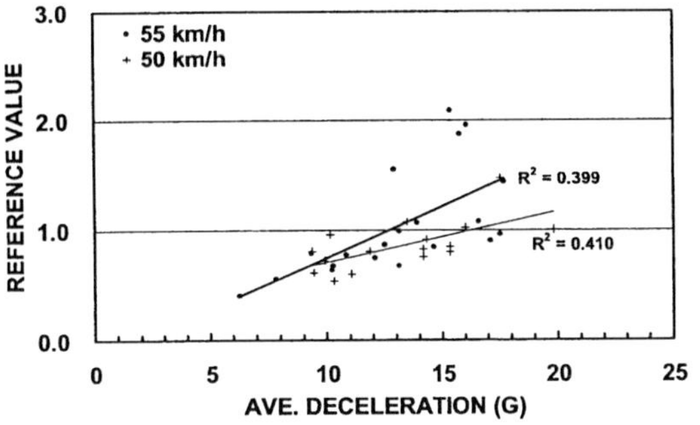

Figure 9. Relationship between Body Deceleration and Occupant Injury Severity (Driver Chest G).

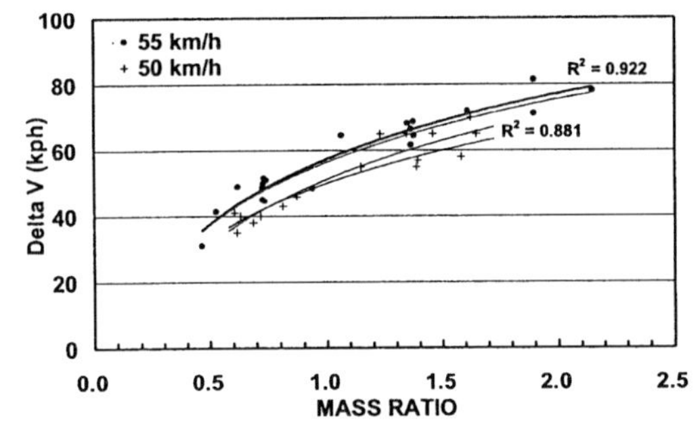

Figure 10. Relationship between Delta V and Mass Ratio.

Dotted lines are theoretical delta V curves of 50 and 55 km/h.

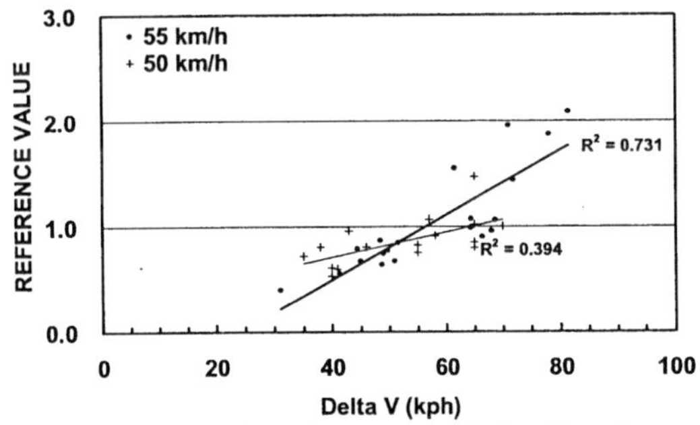

Figure 11. Relationship between Delta V and Occupant Injury Severity (Driver Chest G).

The average body deceleration was observed to increase as the mass ratio became larger, and Chest G was found to worsen as the body deceleration increased. The delta V also showed same phenomenon.

Figs. 12 and 13 show the amount of toe board, steering shaft intrusion and dummy injury severity.

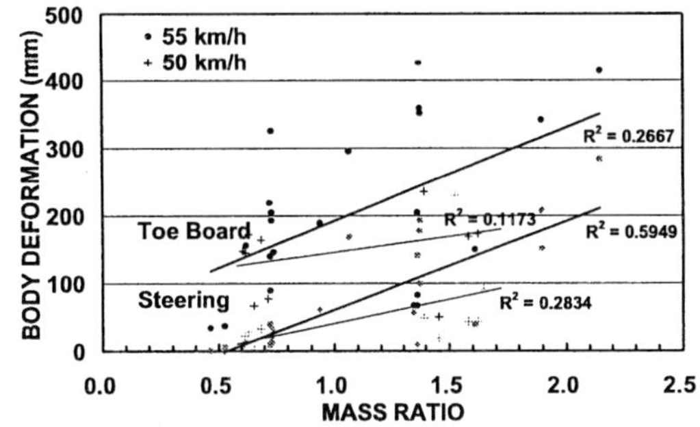

Figure 12. Relationship between Body Deformation and Mass Ratio.

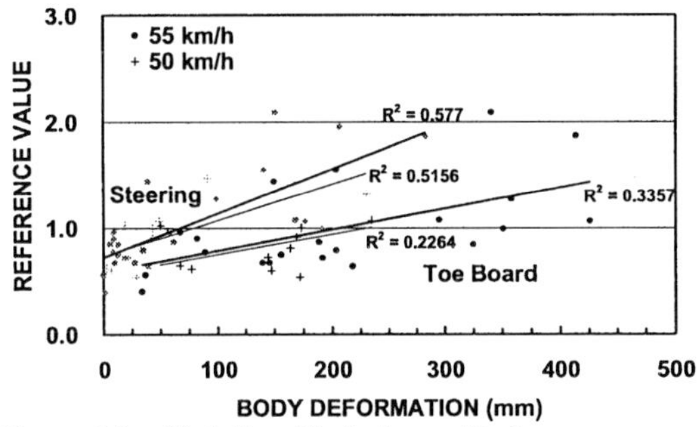

Figure 13. Relationship between Body Deformation and Occupant Injury Severity (Driver Chest G).

The amount of toe board intrusion tended to increase overall as the mass ratio increased.

Likewise, the amount of steering shaft intrusion also tended to increase in accordance with an increase of mass ratio.

It was observed that when the amount of toe board and steering intrusion increase, Chest G which is a major factor in injury severity tended to worsen.

ANALYSIS

The test results clearly show that at the current performance level, the HIC and chest G values exceed the reference value when the crash mass ratio is 1:1.1 or more. (for a crash speed of 55 km/h)

Table 2 shows the Chest G-Displacement characteristics. The factors influencing G-Displacement characteristics are body deceleration, cabin intrusion, and the performance of the occupant restraint system.

The body deceleration characteristics and occupant restraint characteristics in car-to-car crashes were compared for fixed barrier crash test. This method is effective for improving self-protection performance and ODB crash test performance and is suitable for thoroughly evaluating body and restraint system performance in conditions where the loading input to the body is uneven.

Table 2.
Relationship between Body Deceleration and Deformation and Occupant Injury Severity by Crash Format

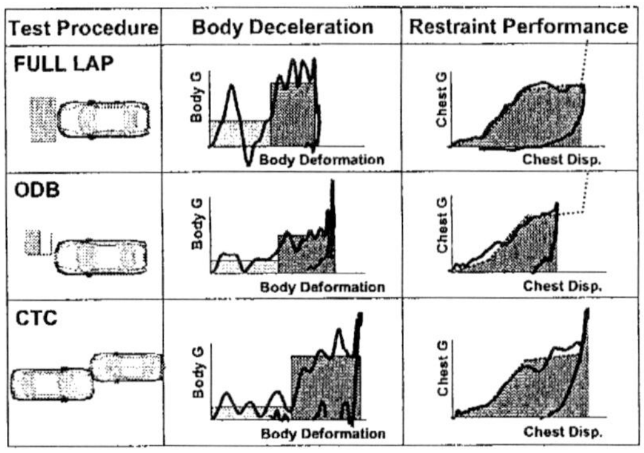

As previously mentioned, full lap barrier crashes generate the highest average body deceleration, and for this reason, also the largest relative energy absorption by the occupant. On the other hand, ODB crashes exhibit an extremely low deceleration due to the effects of the offset ratio and energy absorption by the honeycomb. As a result, the crash energy translated the occupant was found to be less than that for a full lap crash.

The outcome of a car-to-car crash depends on the characteristics of the two opposing vehicles. Although car-to-car crashes have a low initial deceleration similar to ODB crashes, they also produce extremely high input loads into the cabin and high G levels in the second half of the crash. As a result, the crash energy translated to the occupant by car-to-car offset crash is actually equal to or greater than that for a full lap barrier crash. However, the cabin survival space is significantly reduced compared to the other two tests, and there are many cases where this causes secondary collisions with interior parts which worsen the occupant injury severity.

Thus, avoiding secondary collisions with the vehicle interior through the use of appropriate restraining devices while not increasing body deceleration and maintaining occupant survival space becomes an issue. It is extremely important to achieve a balance between the crash vehicles so that the body deceleration of small lightweight vehicles is not raised to levels higher than is necessary to ensure occupant survival space.

An examination of evaluation test methods that combine these characteristics shows the above-mentioned moving deformable barrier (MDB) crash test to be ideal.

The MDB test method is currently one of the test method which models car-to-car crashes from the dual perspective of body deceleration characteristics, which control occupant injury severity, and occupant survival space. The proposed car to MDB test method was presented in a previous paper, which presents specifications for preventing MDB over-ride in order to further improve the characteristics of deformable barriers in Figure 14.

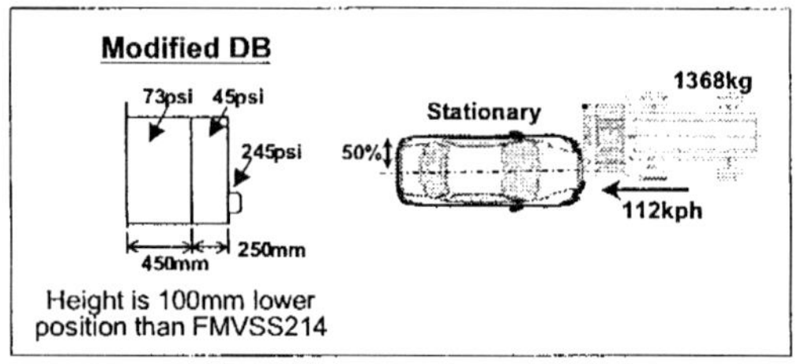

Figure 14. Modified Deformable Barrier and Test configuration.

*See the 16th ESV Paper, Technical Session 1, No. 98-O-08

MDB over-ride is thought to be caused by the interactive forces occurring between the front of the vehicle and the deformable barrier and the inertia characteristics of the vehicle and MDB.

In particular, the over-ride phenomenon is related to the vertical interaction forces between the front of the vehicle and the MDB. A crash between an MDB and car produces less interaction than that which occurs in an actual car-to-car crash. This reduced interaction, which is a result of the inertial characteristics of the deformable barrier, allows greater relative motion between the MDB and car. The deformable barrier and bumper heights were adjusted to avoid this phenomenon, and it was confirmed that over-ride was reduced. Test results show that the body deformation and deceleration of the MDB and the test vehicle closely approximated those in an actual car-to-car crash. The test results also show that although there was significant body deformation to the top of the cabin when MDB over-ride occurred, this did not produce high body decelerations.

Figs. 14 to 16 show body deceleration, deformation and occupant injuries compare with over-ride MDB test result.

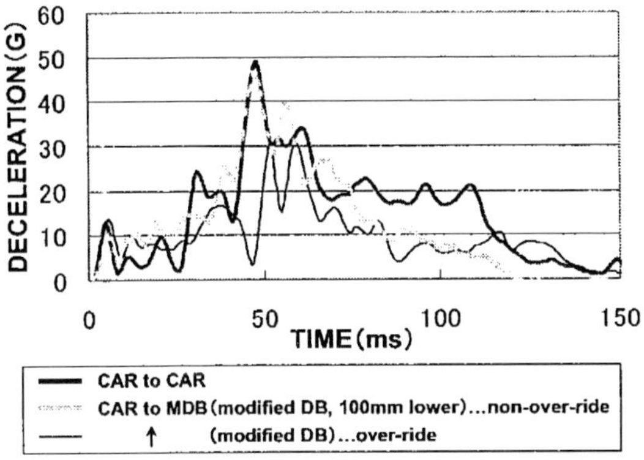

Figure 15. Body Deceleration of Modified MDB and Car-to-Car Test.

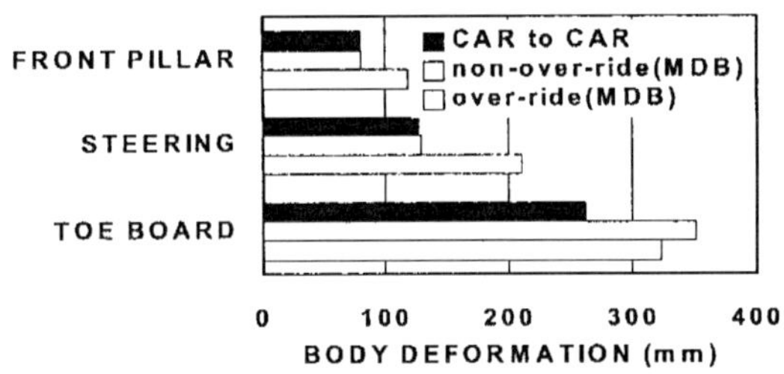

Figure 16. Body Deformation of Modified MDB and Car-to-Car Test.

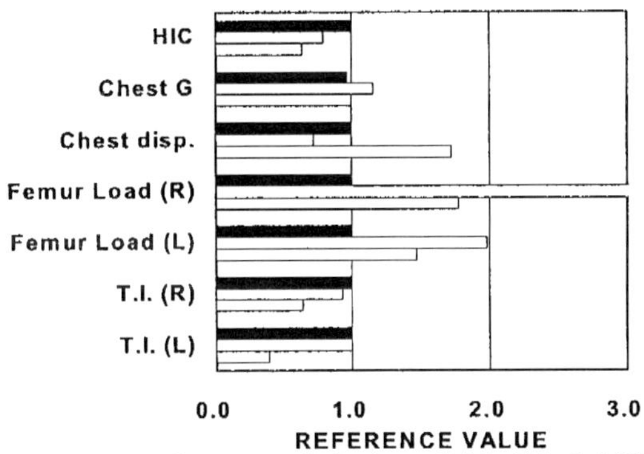

Figure 17. Occupant Injuries of Modified MDB and Car-to-Car Test.

These tests were conducted as research to investigate whether the MDB to car test can be substituted for actual car-to-car crashes between mid-size passenger cars. As a result, it was found that car-to-car crashes may be better reproduced using an MDB to car test as opposed to a conventional fixed barrier test.

It is thought that these MDB characteristics will allow realistic evaluation of compatibility by the MDB test method by using the cumulative average weight of marketed vehicles as the weight and by having the crash angle, that is to say the crash format, conform to market accident formats, etc.

DISCUSSION

When discussing occupant injuries, the concept of body deceleration is thought to be extremely important. In addition, of the three compatibility factors, it is impossible to accurately reflect mass and stiffness to the test method as long as fixed barriers are used.

In order to improve compatibility, it is necessary to establish test methods that can evaluate the various criteria without becoming enmeshed in existing test methods. From the viewpoint of being able to more realistically evaluate the impact input to occupants, the MDB test method is thought to be a suitable method. The proposed MDB tests should be repeated while varying the crash angle, speed, honeycomb characteristics and other factors in consideration of the regional characteristics of each country. Then, these results should be used to investigate and determine vehicle specifications that can improve occupant injury severity and indices for the future evaluation of compatibility.

Finally, the clearest target is to tentatively set characteristic targets for compatible vehicles. Furthermore, it is thought that crash tests with mass ratios that most closely reproduce actual car-to-car crashes are necessary in order to improve incompatibility as quickly as possible.

CONCLUSION

It was confirmed that the test vehicles (up to 2000 year models) used in this research do not achieve compatibility from the viewpoint of occupant injury severity for certain combinations of mass, stiffness and geometry. In particular the influence of steering intrusion and rise caused HIC and Chest G to significantly exceed the reference values, indicating that these are extremely important factors compared to other injury criteria. On the other hand, the tibia index exhibited high values overall, indicating that future analysis such as comparison with ODB tests is necessary to determine whether this is rooted in the body deformation characteristics of car-to-car crashes. Occupant injury is a compound phenomenon caused by body deformation, deceleration and secondary collisions, so the simultaneous improvement of both body countermeasures and restraining devices is thought to be the key to improving compatibility in the future.

The bottoming out and over-ride problems of the MDB test method were improved in this specific test. The resulting body deformation, body deceleration and occupant injury severity of the test vehicles matched closely with actual car-to-car tests. However, as mentioned above this moving deformable barrier (MDB) test method reproduced only crashes between mid-size passenger cars, and there are plans to also investigate test methods that can evaluate compatibility in terms of aggressivity and self-protection performance while varying the specifications to match real world accidents.

REFERENCES

Hampton C. Gabler, William T. Hollowell "NHTSA's Vehicle Aggresivity and Compatibility Research Program" Enhanced Safety Vehicle Paper No. 98-S3-O-01, June 1998, Winsor, Ontario CANADA

Christian Steyer, Marc Delhommeau, Pascal Delannoy "Proposal to Improve Compatibility in Head on Collisions " Enhanced Safety Vehicle Paper No. 98-S3-O-05, June 1998, Winsor, Ontario CANADA

Stephen Summers, Aloke Prasad, Willam T. Hollowell "NHTSA's Vehicle Compatibility Research Program", Society of Automotive Engineers Paper No. 1999-01-0071. March 1999

M J Edwards, J Happian- Smith, N Byard, H C Davies, and C A Hobbs "Compatibility-the essential requirements for cars in frontal impact", Vehicle Safety 2000 Paper No. C567/011/2000 June 2000 London, UK

K H Digges and A M Eigan "Load-cell barrier measurements of geometric compatibility", Vehicle Safety 2000 Paper No. C567/060/2000 June 2000 London, UK

James Holgate "Frontal Offset Test Conditions", SAE Frontal Offset Crash Testing and Lower Extremity Injuries TOPTECs January 2000.

Koji Mizuno "Current Situation and Problem of Vehicle Compatibility" Japan Society of Automotive Engineers Paper No. 9942628

Tomiji Sugimoto, Yoshiji Kadotani, Shigeru Ohmura "The Offset Crash Test-A Comparative Analysis of Test Methods" Enhanced Safety Vehicle Paper No. 98-S1-O-08, June 1998, Winsor, Ontario CANADA

Test Procedures and Test Tools

2004-01-1171

New Barrier Test and Assessment Protocol to Control Compatibility

Pascal Delannoy
Teuchos- Snecma Group

Jacques Faure
Renault S.A.

Donat Coulombier and Richard Zeitouni
PSA Peugeot-Citroën

Tiphaine Martin
UTAC Laboratory

ABSTRACT

After a lot of researches in the field of compatibility through testing and accident analysis, the current knowledge allows us to propose a new test protocol able to control the two issues of compatibility: partner and self protection.
In other words, the procedure would be able to assess the most relevant parameters: structural interaction (geometry / stiffness) and compartment strength (stiffness). The most effective way to meet this target is to control both of them at the same time, thus avoiding many test regulations.

The new final protocol will be a mix of two very well known test configurations:

- current frontal regulation in Europe (ECE 94) to assess compartment strength and structural integrity through vehicle and dummy criteria
- Progressive Deformable Barrier test (PDB) to assess and control force distribution through barrier deformation.

The aim of this paper is to clearly describe the spirit of the regulation approach, to present the test procedure and to propose assessment rules.

INTRODUCTION

The past few years have shown that compatibility cannot be treated separately without taking into account the self protection of other road users or without detriment to vehicle occupants. Non-aggression towards other vehicles extends to the side collision configuration. Such constraints taken separately could seriously compromise passive safety for some configurations and/or road users.
The experience of new requirements over the past few years, associated with a real awareness for safety, is suggesting approaching front-unit design from a global point of view in terms of partner and self protection.
Whenever it is possible, in order to ensure a maximum of passive safety when renewing vehicle populations, front-unit design must as much as possible take into account previous generation vehicles.
Accident research conducted over the last few years clearly shows the gains of such improvements.
This paper is based on internal research studies into passive safety, conducted since 1995, as well as papers presented all around the world, international working groups and current and future regulations.
The aim of this document is to propose a better assessment procedure in a regulation approach for improving compatibility to answer real world accident observation.

The goal of the test procedure is:

- to improve partner protection without reducing self protection
- to assess partner protection
- to assess self protection
- to decrease the compartment strength difference between heavy and light cars

- to optimize front end unit in terms of pressure (pushing surface and force distribution)
- to harmonize test procedures.

Furthermore, this document is a guide for performing partner and self protection test at the same time. This test is based on current ECE R94 frontal impact; three parameters are changed in the test configuration:

- velocity (56 to 60 kph),
- overlap (40% to 50%)
- obstacle (frontal current EEVC barrier to PDB – Progressive Deformable Barrier).

Some aspects regarding assessments and test data processing analyses are not yet fully defined. The procedure is opened to additional improvements specially regarding partner assessment formula.
However, no change regarding the basic principle (deformable elements that avoids bottoming out, overlap test and assessment are based on barrier deformation) is considered.

REAL WORLD ACCIDENT TREND

Concerning fatalities in vehicles involve in car to car, two main reasons can be put forward. The first is associated with a relatively low level of self-protection. The second results from a high level of aggressiveness of the recent models. There is here one aspect regarding self-protection (against other cars and fixed obstacles), and another aspect regarding protection of the others.
European data shows that in most cases, occupants suffer serious or fatal injuries due to passenger-compartment intrusion.

CURRENT FRONT STRUCTURE DESIGN

Self protection

One priority in vehicle design was to protect the vehicle occupants against fixed obstacle collisions. This is today well applied and well known, and this is what assesses the ECE R94 regulation in Europe and ADR 73 in Australia. The test is carried out at 56 kph against a deformable element (EEVC barrier) and 40% overlap and 0° impact angle.

Since intrusion was identified as the first cause of fatal injuries (far before injuries due to deceleration), all car manufacturers have made significant improvements in car body design to decrease it.

Due to the specific nature of self-protection tests (more stringent for large vehicles due to their own mass), these improvements have driven manufacturers to increase the stiffness not only of their small vehicles, but of larger ones also. Large vehicles which due to their design are stiffer already ... In effect, the quest for similar intrusion performance, whether for a small or a large vehicle, leads naturally to increase front unit and passenger compartment stiffness. Note that large vehicles - even if they are for the most part longer - require higher level deformation loads.

This higher stiffness has been already proven, by French studies, to be dangerous for older-generation vehicles, and as well as for car-to-car frontal and lateral compatibility of the same generation vehicles.
Reduce the intrusion level provokes increasing front-end and passenger compartment stiffness.

Possible contradiction with partner protection issue

As we have seen, it would be rather easy to design a car offering excellent protection against rigid and plane obstacles. But this could be to the detriment of other vehicles. Real life safety in that case would have nothing to gain.

To address protection against fixed obstacles demands greater front unit stiffness to limit intrusion. Without considering partner protection, the deformation forces can only be limited by the restraint system and may be proven aggressive for other vehicles due to its geometry, layout and stiffness. This is why it is today essential to include the notion of partner protection in modern vehicle development.

To ensure a correct compatibility level, the two vehicles should present a homogeneous –stiffness and geometry- contact surface (for frontal and lateral impacts), as well as high compartment strength.

Lateral compatibility

European studies undertaken on this issue show the necessity for homogeneous load distribution on the impacted vehicle. Unlike frontal impacts, overlapping of structures may load the occupants directly and reach dramatic proportions. Fortunately, front end requirements to meet lateral compatibility are very closed to frontal one's.

Geometry - architecture

As seen before, whenever a rigid obstacle is concerned, front structures will be in all cases well deformed. The need to address protection against rigid and fixed obstacles involves the sought of a greater stiffness in order to absorb maximum energy for the lowest deformation. But, it is certainly not recommended for partner protection. However, high front unit force is not incompatible with partner protection, it depends on the force distribution over front area.
That is why frontal and lateral compatibility requires a homogeneous pushing surface directly link to the geometry. The ideal pushing surface leads to a multiplication of the load paths, with a high degree of interconnectivity. This allows to limit the degree of overlapping in frontal impacts, and the lower structures of the impacted vehicle allow very early loading in lateral impacts.

COMPATIBLE FRONT UNIT DESIGN

Physical approach

The most important formula to know is:

Pressure = Force / Surface

Designing a vehicle around acceleration and force requirements is not sufficient any more. Today, contact pressure notion must be taken into account to develop vehicle front end.
The pressure is directly linked to the force and the contact surface.
That's why compatibility evaluation procedures must be able to detect without any doubt the pushing surface created by the geometry (horizontal and vertical links among load paths).

Ideal front unit design to meet partner protection

As it was explained in previous paper (ESV 2001), multiple front end load transfer paths with strong connections and moderate localised forces is the right answer. Furthermore, a progressive longitudinal increase in stiffness from the front to the compartment is better.

NEW ASSESSMENT PROTOCOLE

Aims of the test procedure

Only a regulation that must be applied by all manufacturers can answer the problem.
The final target is to control and assess partner and self protection with only one test at the same time.
If regulation ECE R94 in Europe already exists for self-protection, it cannot be used for partner one due to the barrier used.
It is therefore up to the vehicle to offer a sufficient pushing surface for deformation. This phenomenon corresponds precisely to what is seen in a vehicle / vehicle collisions.
The self protection assessment is based on the same criteria using dummy criteria and car deformation (future EEVC WG 16 proposal).
The partner protection will be based on barrier deformation which is supposed to represent the better possible, actual deformation of a car.

The procedure must be in line with:
- Current regulation (no rupture in the fleet)
- Light and heavy cars
- Real life accidents

The procedure must be able to check:

- Load path position and stiffness
- Structural and rigid link between them

The procedure must be able to generate:

- Local and global longitudinal / horizontal shears
- Large deformation depth / light car overloading

Current self protection test

This test is certainly the best known. Today, it uses an offset deformable barrier in Europe but also in others countries.
The low stiffness deformable barrier makes it possible for all vehicles to use the wall behind for structural collapsing. It is thus closed to a rigid wall test – figure 2.
The energy absorbed by the barrier is practically the same for all vehicles, which makes it more severe for the heaviest (that have higher EES –Equivalent Energy Speed- is increased) – figure 1.

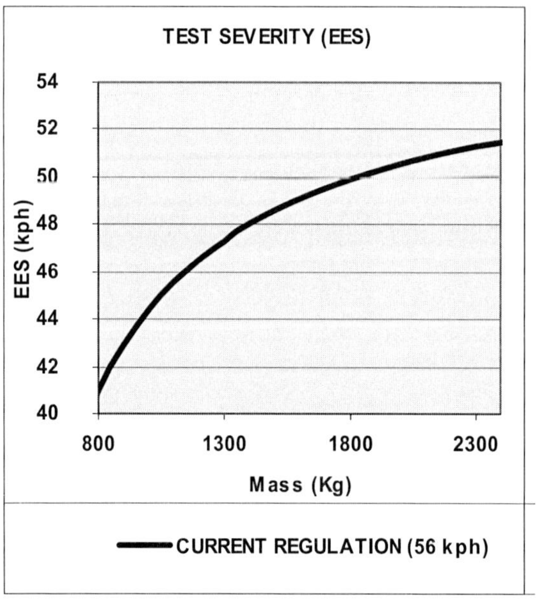

Figure 1: Severity for a vehicle vs its mass

This test is not the right one for measuring the level of aggressiveness of the front unit. Only an obstacle which behaves like a car could answer this question.
We may consider that the current self-protection test serves as the basis for assessment of self protection
But if this configuration allows for a good assessment of the load onset on the passenger-compartment, it does not provide enough indication about its stability – capacity to sustain force- under high intrusion (a mechanism frequently observed in real world accidents).

Test at higher violence levels (60 kph), could provide information on the stability of the passenger compartment, especially for the light cars.

For this reason, we propose filling this gap with a new test configuration.

Figure 2: Deformation of current barrier

TEST PROTOCOL PRESENTATION

The proposal is intended to replace the soft current EEVC barrier from current R94 regulation by the PDB one. This proposal is one of the most effective ways to assess partner protection without deeply changing anything. It could be very quickly applied because this test configuration is very well known all around the world.

Furthermore, this proposal could generate higher compartment deceleration pulse and could be able to combine both acceleration and intrusion. Another advantage of the test proposed is to be able to combine rigid wall full overlap test (high deceleration) and EEVC barrier test (high level of intrusion). That's why it could be helpful to have a test able to control both acceleration and intrusion on the same time because it is what we observe on the road. However, further researches are needed to validate the approach.

Changes compared with current R94 regulation

This test protocol is close to ECE R94 except speed, overlap and a new barrier.

The three main changes:

-1-
Obstacle :
EEVC barrier => PDB barrier

-2-
Test speed
56 kph => 60 kph

-3-
Overlap
40% => 50 %

Obstacle to replace the current one

The deformable element proposed and used is the Progressive Deformable Barrier (PDB) developed in association with AFL Honeycomb Company. Barrier is built with two layers of aluminium honeycombs (figure 3).
This barrier offers a progressive increase in stiffness in depth, and two height independant stiffnesses (figure 4)

The barrier allows for checking the front pushing surface of the vehicle, and links among transfer load paths (cross member, pendant etc…). Its dimensions and stiffness make the bottoming-out very unlikely. In real life car to car collision there is no bottoming out, thus the barrier has been designed in such a way as to prevent this phenomenon.

Furthermore vertical car force distribution is also represented; the lower front load path is usually stronger than the upper one.
The PDB concept close to another vehicle is presented as a tool to assess partner protection.

The PDB approach and validation have been fully explained in previous papers (ESV2001, SAE 2003…).

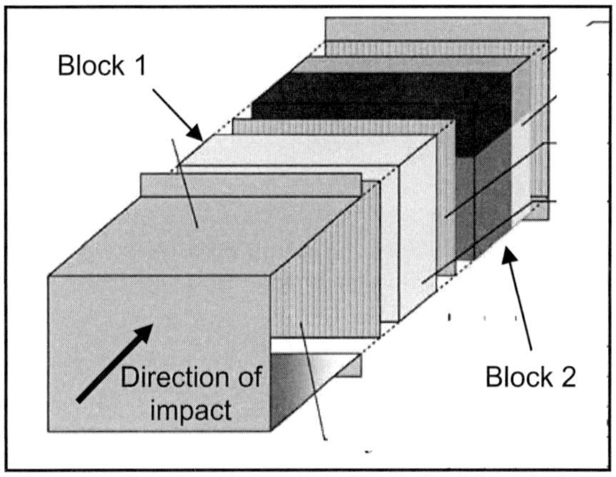

Figure 3: Barrier details: 4 Aluminium sheets and 2 deformable elements

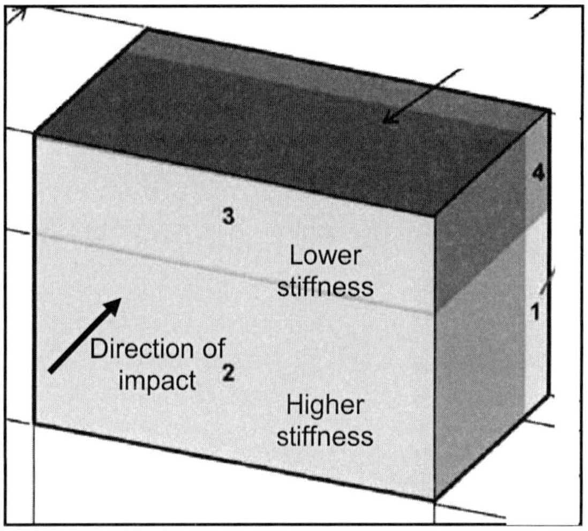

Figure 4: Progressive Deformable Element Bloc (Block 2)

Barrier Positioning and mounting

The lower edge of the barrier is at 150 mm from the ground and aligned on the load cells.
The barrier is mounted on a high resolution load cell wall for research purposes. Further researches are needed to validate and control the accuracy of a force measurement behind a deformable element.

Figure 5: Side view of the barrier positioned on the Load Cell Wall

So, barrier deformation, force distribution and global force are measured. Only deformation pattern is used for the criteria calculation.

Test mass

The mass of the vehicle as it is usual in regulation approach is the worst case. Two front instrumented dummies are used.

Overlap and angle

There was a need to increase the current overlap to 50%. On one hand it represents the average overlap in real world accident and, on the other hand it is commonly accepted and used by scientific community for car to car test. Furthermore, the increase in overlap could generate higher deceleration and contribute to the severity increase for passenger restraints.

Test speed

The impact test speed is 60 kph +/- 1 kph and fixed for all vehicle mass range. The draft frontal European regulation are 60 kph (proposed by WG16) and our proposal is linked to the fact that the improvement of the compatibility cannot be done to the detriment of self protection (specially true for heavy vehicles).
In order to deal with light cars, it is not suited to set higher test speed. It would conduct to increase the compartment strength difference between heavy and light vehicles, which is not in line with compatibility self protection requirement.

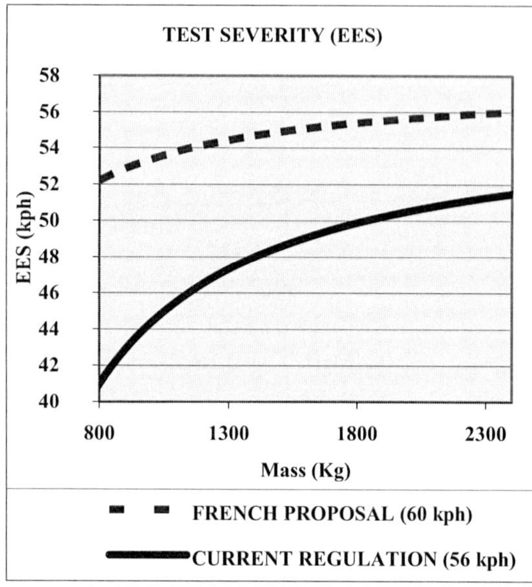

Figure 6: Test severity compared with the current ECE R94 regulation

Self protection

The self protection assessment is based on current ECE 94 regulation specifications. Passenger compartment and dummies criteria give information about compartment strength and stability -figure 7-. Specific aspects of the assessment may be found in the regulation 94.

Figure 7: Self protection level is assess from vehicle intrusion and dummies criteria

Partner protection

Partner protection assessment is less well known, however, validation of the PDB approach has been fully explained in papers SAE 2002 and ESV 2003. Even if it has been presented in previous publications, the process is going to be explained in more details.

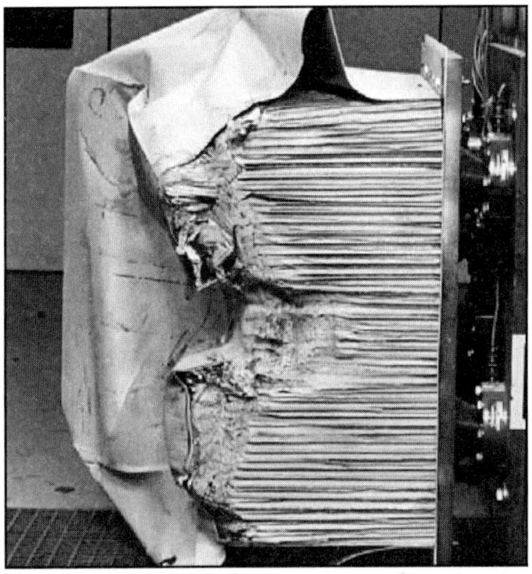

Figure 8: PDB deformation after test

After a test, if the barrier present deep and inhomogeneous deformation the conclusion can be withdrawn is that it was provoked by a very aggressive vehicle.

Assessment summary

Test configuration presented in this paper deals with the low aggressiveness requirements, and takes into account the introduction of new generation of vehicles.
All criteria and investigations are based on the deformation pattern of the barrier; the force distribution measurement is recorded only for research. It is similar car to car accident or crash test analysis, except that in this case, the barrier deformation pattern is investigated instead of the car. An aggressive vehicle would be identified by deep deformation of the barrier and / or substantial local penetration and / or a high overall load and / or a centre of forces application that is too high.

The PDB barrier is able to detect local stiffness but also transversal and horizontal links among load paths. The test is able to detect the positive effects of an optimized vehicle structure design as: the type of front cross member, lower cradle subframe and pendants linking that improve vehicles compatibility.
The final barrier pattern gives information about front unit homogeneity as a combination of the force distribution and the pushing surface.

After having digitizing the barrier shape, an assessment is defined. Based on the deformation, three main parameters are identified as very important:

- deformation in depth –x- representativity of the stiffness
- height of the deformation – z- representative of the geometry
- surface area

Two limits are considered in the assessment: geometric in height and stiffness in depth. The assessment formula describes each colour surface weighted by its height (Z position) and its deformation (X position).

For each surface (Si) :

Average height: Zi
Average deformation: Xi

$$\sum_{i=1}^{13}\left(\frac{Zi}{Z\lim}\right)^4 * \left(\frac{\mathrm{Xi}}{\mathrm{Xlim}}\right)^2 * \mathrm{Si}$$

$$\textbf{Compatibility mark} = \frac{0{,}52}{10}\mathbf{R}^{0{,}55}$$

Aggressiveness scale

1	2	…	6	7	…	11	12

NON AGGRESSIVE → AGGRESSIVE

The distance Zi represents the addition of the measurement in the barrier referential and the ground clearance (150 mm).

Z lim = 420 mm corresponding to average longitudinal height in Europe.
Xlim = 300 mm corresponding to a certain level of force (300 kN) that represents an average front unit force -50% overlap-.
Both limits can be adapted with the considered fleet (vehicles stiffness, size and geometry of the car population in-use in a continent) and what kind of cars that must be protected.

The geometric parameter (Zi / Zlim) is higher weighted than the stiffness parameters (Xi / Xlim), identified as more important.
An aggressive car will be identified by a large mark (above 10). Up to know, the mark of fifty tested vehicle are included between 2 and 11.

This assessment is more complicate than the Average Height Of Force assessment proposal.
However, this approach is testing and trying with barrier deformation pattern as
- Average Depth Of Deformation for stiffness parameters
- Average Height Of Deformation for geometric parameters.

This proposal could be a first step before using the corridor approach.

Barrier digitization

Only the front surface is digitized. Several methods can be used. The reference point (origin) is located on the opposite side of the deformation: on the left for a left hand drive, on the right for a right hand drive.
Several type of file can be generated. The surface is modelled as a CAD surface. The characteristics of the file usually used are:

- Type : IGES or VDA
- number of nodes : 40000
- number of elements: 80000
- mean distance: 12 mm.

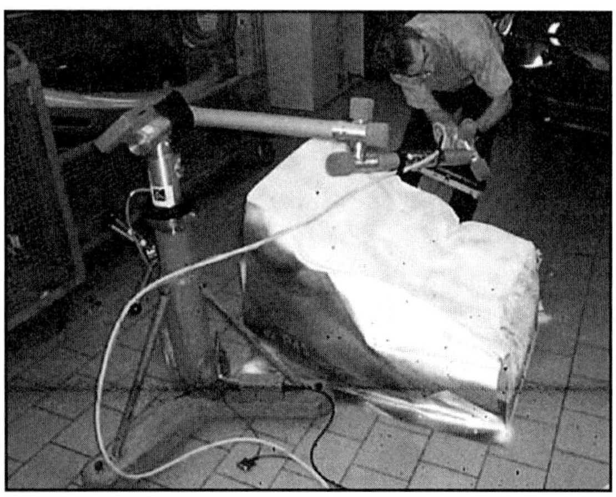

Figure 9: Barrier digitization in progress

Figure10: Referential used for digitization

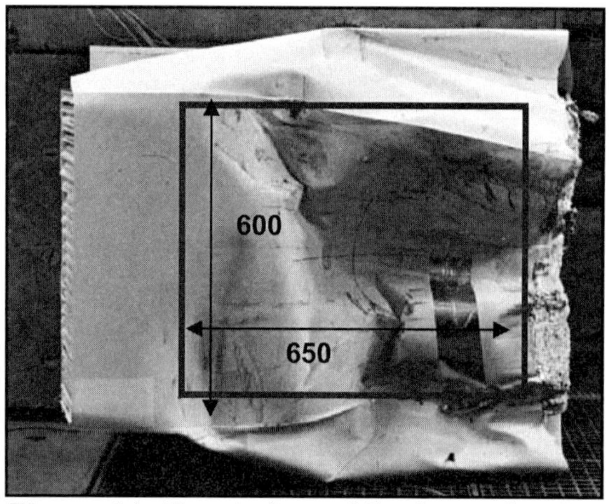

Figure 11: Front view of the investigation area

Investigation area and graphic representation

Several colour corridors have been chosen to represent barrier local deformation. The barrier is cut in 13 corridors. Each colour represents a 50 mm corridor (figure 12). To avoid boundary effect the deformation map is resized to keep a map of 600*650mm that create the investigation one (figure 11).
For each colour corridors, the surface (Si) is calculated with its average height (Zi) and average deformation (Xi).

An automatic assessment is developing to generate the aggressiveness mark. This software will be available in june 2004.

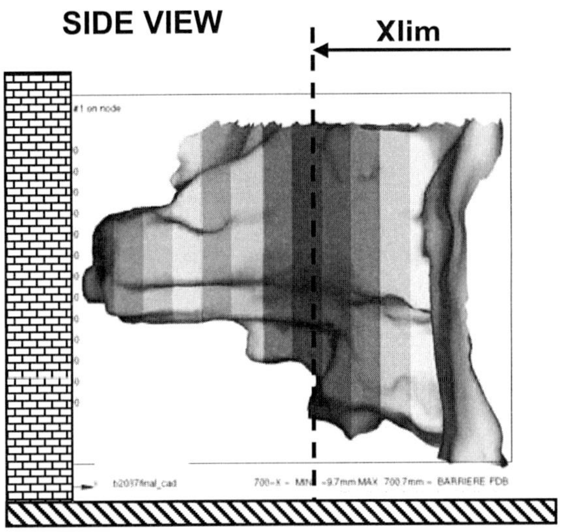

Figure 12: Side view of the barrier deformation: 50 mm colour corridor and X lim representation

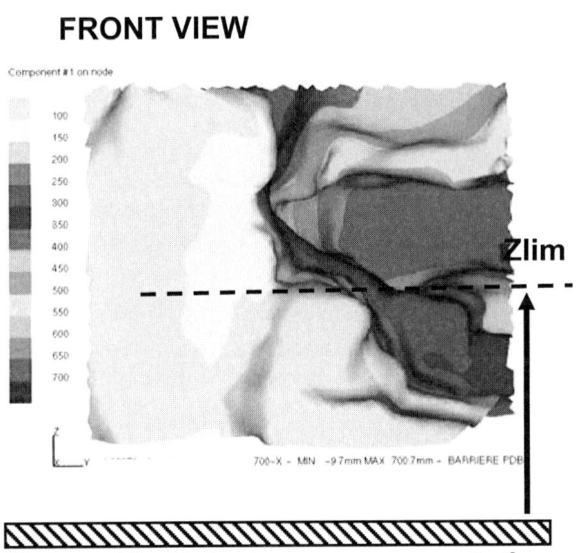

Figure 13: Colour map representation and Zlim representation

We will keep this kind of representation as a basis.

Area example

2: 100mm ➔150mm ☞ GREY

...

6: 300mm ➔350mm ☞ BLUE

...

13: 650mm ➔700mm ☞ RED

CONCLUSION

The aim of this paper was essentially to clearly explain our global approach in terms of self and partner protection,

Self-protection driven vehicle structure design could quickly become potentially aggressive for other road users.

A regulation already exists for self-protection and is very well known. A partner protection measurement method is today possible with a slight change: replace the barrier. The PDB concept representatives another vehicle has been chosen. This change allows for measuring the front aggressiveness.

One test is able to do two assessments: the first to measure the level of self-protection by vehicle analysis, and the second the level of aggressiveness by barrier analysis.

Furthermore, the increase in speed allows for a good compromise: overloading light cars without becoming too much severe for heavy cars, due to PDB energy absorption capacity.

References

1- P.Delannoy, J.Faure "Compatibility Assessment Proposal Close To real Life Accident", ESV 2003

2- NHTSA Research Program, "Initiative to Address Vehicle Compatibility", 2003

3- EEVC WG 15 and IHRA "Vehicle Compatibility - Working Group Activities", Nagoya 2003

4- P.Delannoy, A.Diboine "Compatibility: Causes, Constraints, Improvements and Evaluation Proposal" SAE World congress 2002

5- EUCAR project: "Development of criteria and standards for vehicle compatibility", 2000

1999-01-0720

Application of Load Cell Barrier Data to Assess Vehicle Crash Performance and Compatibility

K. Digges, A. Eigen and J. Harrison
The National Crash Analysis Center
The George Washington University

ABSTRACT

The National Highway Traffic Safety Administration (NHTSA) routinely measures the force exerted on the barrier in crash tests. Thirty-six load cells on the face of the rigid barrier measure the force. This study examines the load cell barrier data collected during recent years of NCAP testing to determine how it can be used to assess vehicle structural crash characteristics and vehicle compatibility in car to car crashes. To illustrate the value of the data, the load cell measurements for an SUV are compared with a small car.

Several aggressiveness metrics are proposed for different crash modes. The proposed metrics for frontal crash modes are the force at 250 mm of crush, the linear stiffness at various levels of crush, and the height of the center of force at 250 mm of crush.

For front-to-side vehicle crashes, some additional metrics are proposed. The force distribution when the loading is sufficient to cause intrusion of the side door is proposed as the basis for a metric. A high percentage of force on the lowest rows is indicative of sill loading, which should be favorable. A high percentage of force on the highest rows of load cells is indicative of intrusion in the region of the occupant's thorax, which should be unfavorable. The presence of loading in the upper row of load cells at any time during the crash is indicative of a high hood, which could be the source of head impacts.

VEHICLE AGGRESSIVENESS RESEARCH BY NHTSA

Gabler summarized the results of NHTSA's vehicle compatibility and aggressiveness program [Gabler, 1998]. The reported research used data from the GES and the FARS national crash data systems for the years 1991-94 to assess the relative aggressiveness of vehicles. Table 1 shows the rate of driver fatalities in the struck vehicle based upon the striking vehicle. The rate expressed in Table 1 is deaths per 1000 police reported crashes. All crash directions are included.

Gabler also reported analysis of FARS 1992-1996, which examined the fatality risk to drivers of passenger cars by crash direction. The results are shown in Table 2. The reported results are the number of fatalities in passenger cars for each fatality in the striking vehicle. In Table 2, the passenger car is always the struck vehicle and the listed vehicle is the striking vehicle. All occupant restraint conditions are included in the data. Driver ages are restricted to the 26-55 range.

A major objective of NHTSA's research program on vehicle aggressivity and compatibility is to understand the relationships among the vehicle factors which contribute to the large difference in fatality risks shown in Tables 1 and 2.

Table 1. Driver Fatalities in Struck Vehicle per 1000 Police Reported Crashes (FARS/GES 1991-94)

Striking Vehicle	Fatality Rate in Struck Vehicle
Full Size Van	2.47
Full Size Pickup	2.31
SUV	1.91
Small Pickup	1.53
Minivan	1.46
Large Car	1.15
Midsize Car	0.70
Compact Car	0.58
Subcompact Car	0.45

Table 2. Fatally Injured Drivers in Passenger Cars per Fatally Injured Driver in Striking Vehicle

Class of Striking Vehicle	Frontal Crash Ratio	Side Crash Ratio
Car	1.0	6
Full Size Van	6.0	23
Full Size Pickup	5.3	17
Utility Vehicle	4.1	20
Minivan	3.3	16
Small Pickup	1.6	11

VEHICLE CRASH INCOMPATIBILITY METRICS

The crash incompatibility between vehicles has been attributed to three vehicle factors: (1) mass incompatibility, (2) stiffness incompatibility, and (3) geometric incompatibility [Gabler, 1998]. The measurement of vehicle mass is relatively straightforward. However, the measurement of stiffness and geometric compatibility needs further definition.

Gabler suggested the height of the forward-most load-bearing member of the vehicle structure as a metric for geometric incompatibility. Since this element has no precise definition, Gabler used the rocker panel height as a geometric metric.

For a stiffness metric, Gabler used the linear stiffness based on the vehicle crush at the maximum barrier force in a 35-mph crash into a rigid barrier.

NHTSA's crash test program produces additional measurements, which can contribute to assessing stiffness and geometric characteristics of vehicle frontal structures. For most of the 35-mph crash tests conducted under the NCAP program, the time history of the distribution of force applied by the vehicle to the barrier was measured. These measurements indicate the geometric location of "hard spots" and the amount of force the vehicle imparts to a rigid barrier. This data permits the calculation of local stiffness and of load paths at various heights.

Different aggressiveness metrics may be applicable to different crash modes. The efficacy of any proposed metric would need to be verified using on-the-road crash and injury data. However, a number of metrics can be proposed and developed from the available NCAP test data.

For a front to side impact, the front of the striking vehicle may crush only 5 to 10 inches. The force developed in this intermediate crush range and the height of the force may be the critical parameters.

For a frontal-offset crash, the force and geometry of only the left or right half of the car may be applicable.

The use of barrier force data permits a finer discrimination of vehicle stiffness and geometry that can be further investigated as appropriate aggressiveness metrics. The purpose of this paper is to propose metrics from barrier test data that may be used to assess vehicle aggressiveness in frontal and side crashes.

BARRIER INFORMATION

The barrier used in the New Car Assessment Program (NCAP) is a rigid, fixed barrier with 36 force measuring load cells on its surface. The load cells array consists of 4 rows of 9 cells, as shown in Figure 1. The rows are designated by letters A through D, with A at the bottom. The columns are numbered 1 through 9, starting at the left, facing the barrier. The array is subdivided in 6 groupings, 1 through 6, numbered left to right, and beginning with lower left grouping (see Figure 1).

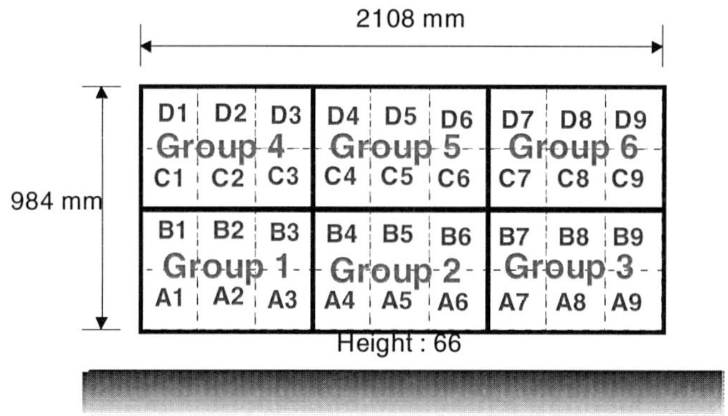

Figure 1. Configuration of Load Cells on Barrier

The array of load cells provides the opportunity to assess the distribution of forces that the vehicle imposes on the barrier during the crash. In this study, the relationship between barrier forces, and their geometric location are of particular interest. In offset crashes, the left or right side of the structure principally deforms and absorbs energy. In centerline impacts with narrow objects, the center response is primary. In car to car crashes the vertical location of the high stiffness regions is important.

To address these various requirements, the barrier measurements have been used to graphically present the local stiffness as measured by all 36-load cells. The linear stiffnesses are calculated at three levels of vehicle deformations – 125 mm, 250 mm, and 375 mm. The stiffness is calculated by dividing the force measured by the load cell at a particular time by the calculated vehicle crush at that time. The vehicle crush is determined by double integration of the longitudinal acceleration measured on a structural member close to the vehicle's center of gravity.

To quantify the height of the structural loading, a center of impact force was calculated for each column of cells. The left column contained the 1 and 4 groupings, the center column the 2 and 5 groupings, and the right the 3 and 6 groupings. In addition, the height of the center of force for the total loading was calculated. The center of force was calculated for three values of vehicle crush – 125 mm, 250 mm, and 375 mm. For each crush value, the force on each of the cells was assumed to be uniformly distributed. The height of the center of the force was calculated by applying static equilibrium relationships similar to the calculation for a center of pressure.

Typical results of the barrier loading are shown in Figures 2 through 7. The figures compare the loading by the Jeep Grand Cherokee and that of the Dodge Neon. These two vehicles were selected to illustrate the range of footprint geometry and force which exists in new on-the-road vehicles. Figures 2 and 4 show the loading footprints at 125 mm of crush. Figures 3 and 5 show the footprints at 250 mm of crush. These plots show the lines of equal loading as registered by each load cell. The total load on all cells is 100%, and the iso-lines are for equal percentages of the total barrier force. The total barrier force is indicated in the figure title. Figures 2 though 5 graphically illustrate the geometric differences between two vehicles with quite different geometric force patterns.

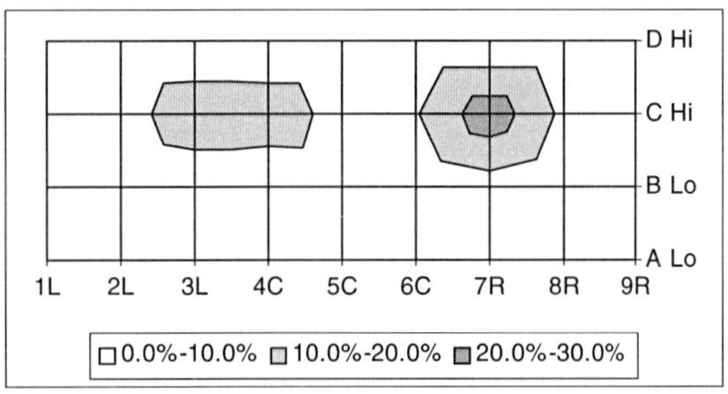

Figure 2. Distribution of Barrier Force, Jeep Grand Cherokee – 125 mm of Crush; Total Force = 172 kN.

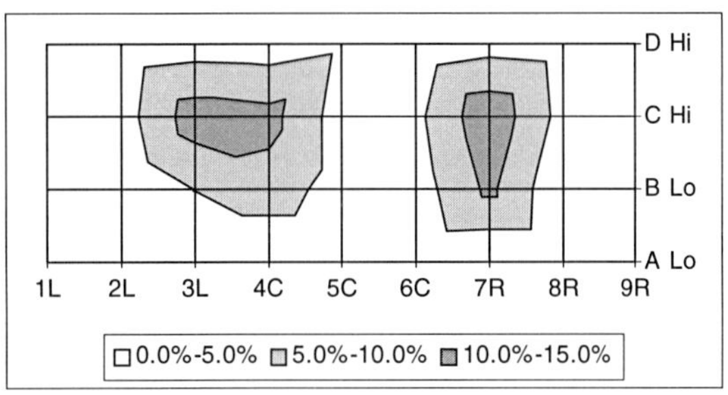

Figure 3. Distribution of Barrier Force, Jeep Grand Cherokee – 250 mm of Crush; Total Force = 438 kN

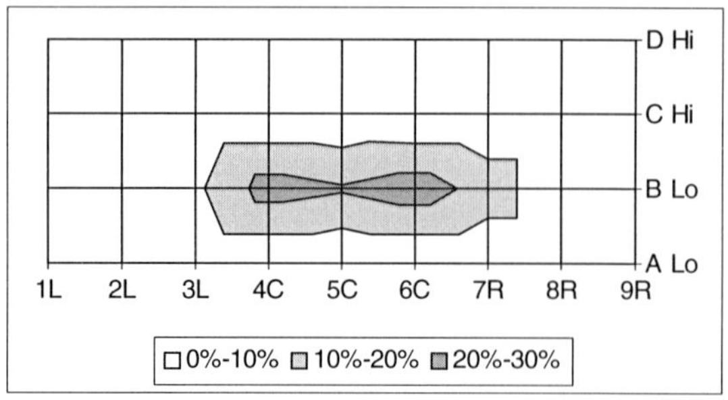

Figure 4. Distribution of Barrier Force, Dodge Neon 125 mm of Crush; Total Force = 108 kN.

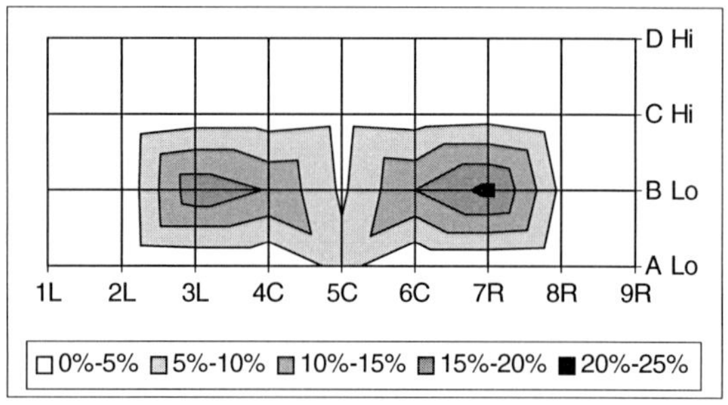

Figure 5. Distribution of Barrier Force, Dodge Neon 250 mm of Crush; Total Force = 244 kN.

The forces registered by each load cell in the array are plotted in Figures 6 and 7. The height of the array is plotted on the x-axis, and the width on the y-axis. The figures are viewed from the top left corner of the barrier. The nomenclature for the cells corresponds to that shown in Figure 1. The extent of crush for these figures is 250 mm.

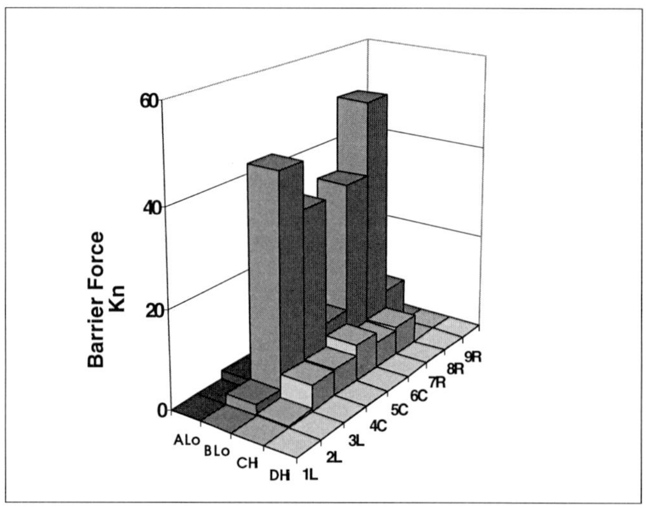

Figure 6. Distribution of Barrier Force, Dodge Neon – 250 mm of Crush; Viewed from Top Left

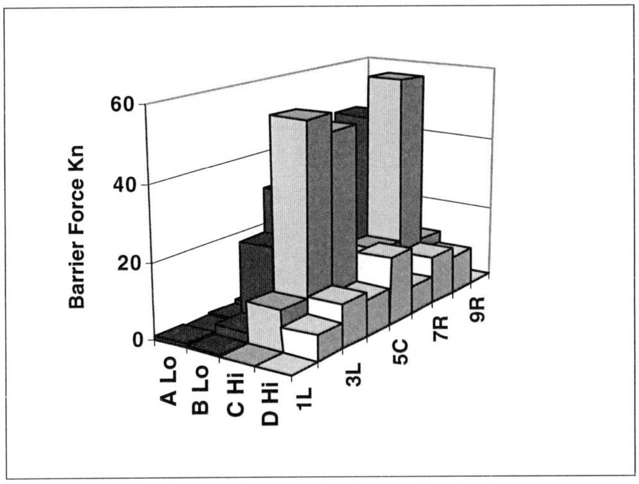

Figure 7. Distribution of Barrier Force, Jeep Grand Cherokee - 250 mm of Crush: Viewed from Top Left

A quantification of vehicle geometric and stiffness data is shown in Tables 3 and 4. The load cell force, and the height of the center of force are shown in the tables. The nine columns load cells are divided into three groups, and separate metrics are calculated for the left, right, and center columns. In addition, the total value for all load cells is shown.

Table 3. Force and Height of Force Center for Jeep Grand Cherokee – 250 mm of Crush

	Left	Center	Right	Total
Total Force, kN	114	174	150	438
Force Ctr, in.	25.5	24.9	24.7	25.0
Force Ctr, mm.	649	633	626	635

Table 4. Force and Height of Force Center for, Dodge Neon – 250 mm of Crush

	Left	Center	Right	Total
Total Force kN	55	120	69	244
Force Ctr, in.	17.7	16.9	17.6	17.3
Force Ctr, mm.	448	430	446	439

The way in which the height of the barrier forces change with crush distance is shown in Figures 8 and 9. In these Figures, the total force on each row of cells is plotted for three levels of crush.

It is evident that the Grand Cherokee maintains the higher load path than the Dodge Neon. This higher load path gives rise to geometric incompatibility in frontal vehicle-to-vehicle crashes.

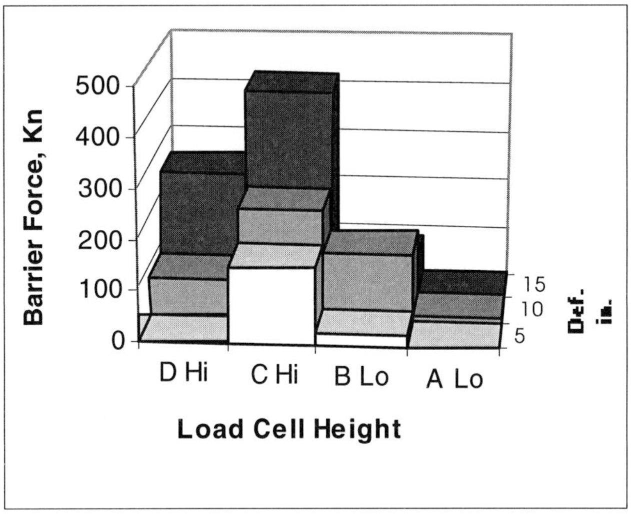

Figure 8. Barrier Force Distribution vs. Crush Jeep Grand Cherokee

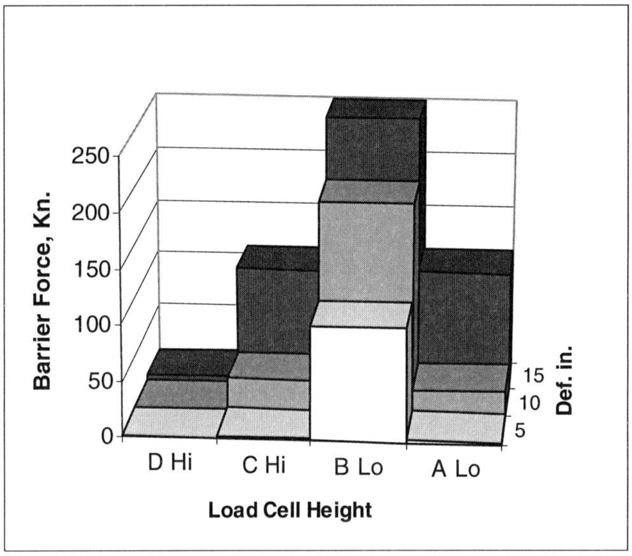

Figure 9. Barrier Force Distribution vs. Crush Dodge Neon

A comparison of the force levels in Figures 8 and 9 shows that even if the geometry of the force were aligned, there remains a large difference in force developed at equal levels of crush. Figure 10 compares the linear stiffness of the two vehicles, based on three levels of crush. At a crush of 250 mm, the Jeep Grand Cherokee exerts almost twice as much force as the Dodge Neon. This difference in stiffness will result in a higher extent of crush for the Dodge Neon in a frontal crash involving the vehicles. This difference illustrates the stiffness incompatibility between the two vehicles.

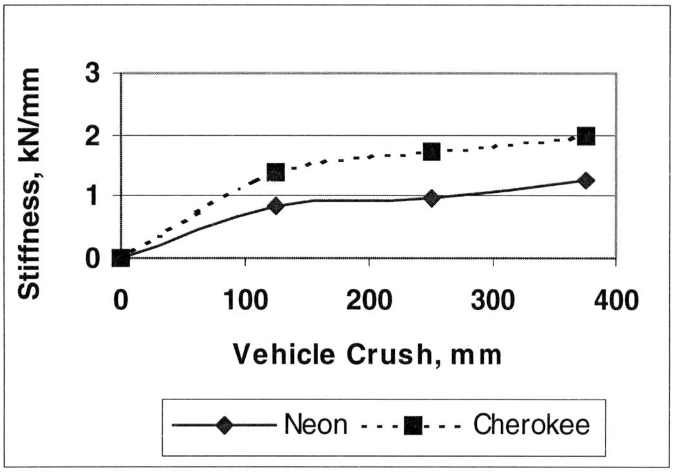

Figure 10. Total Frontal Stiffness vs. Crush

An expected consequence of the observed barrier measurements would be structural override/underride in frontal collisions. In view of the predominant force from the two longitudinal structural members, the override might be expected in crashes that engage primarily the left or right frontal structure, such as offset and angular offset collisions.

SIDE IMPACT CONSIDERATIONS

In vehicle-to-vehicle front-to-side impacts, the occupants of the vehicle struck in the side are at a disadvantage, as shown in Table 2. The disadvantage increases substantially when the striking vehicle is a light truck, van or SUV. The application of barrier force data may provide some insight into the cause of the difference.

In examining front-to-side compatibility, additional metrics may be required, compared with front-to-front compatibility. Leading researchers have reported that side impact protection is reduced when the intrusion at the chest level, precedes the intrusion at the pelvic level [Hobbs 1989, Dalmotas 1991]. These observations suggest the desirability of a low load path from the striking vehicle.

The height of the striking vehicle's hood is another significant parameter in side impact injury causation. Observations of side impact crashes at the William Lehman Injury Research Center, indicate that head impacts with the front or hood of the striking vehicle are common sources of severe injury to occupants of vehicles impacted in the side by larger vehicles. These types of head strikes do not occur in the FMVSS 214 test because the top barrier face is lower than the top of the hoods of many light trucks, vans and SUV's.

The stiffness of the side of a passenger vehicle is much less than its frontal stiffness. Further, the vehicle sill contributes to the side stiffness. If the striking vehicle does not engage the side sill, the load must be transferred primarily by the door, seat, and upper pillars. A low load path is desirable to engage the sill and to control the door intrusion profile so that the occupant's chest is not impacted earlier and more severely than the pelvic.

In a past research program, NHTSA measured the side door stiffness of seven small cars from the early 1980's time period [Willke, 1982]. The measurements were made by statically loading the door with a device that simulated the front of a stiff vehicle. The test device did not engage the sill. The range of results are summarized in Figure 11. The static load vs. deformation for all seven vehicles fell between the upper and lower curves in Figure 11.

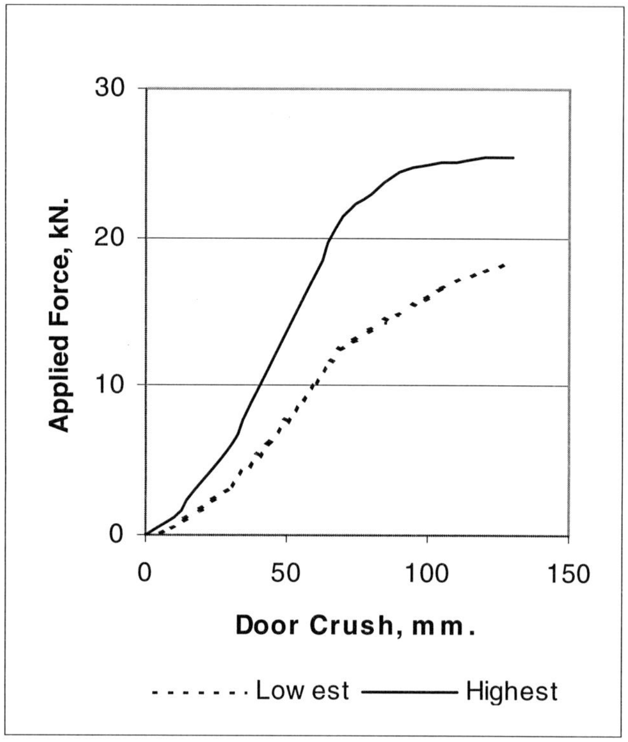

Figure 11. Load vs. Door Crush from Static Tests of 7 MY 1980 Vehicles

In many real world crashes that engage primarily the door, the striking vehicle impacts the struck vehicle at an angle other that 90 degrees. Consequently, only the left or right frontal structure may engage the side of the struck vehicle. Figure 12 shows the forces measured by the load cells on the left side of the barrier for the Grand Cherokee. The longitudinal frame rails primarily generate this force. A comparison of the force magnitudes from Figure 11 and 12 indicates that engagement of only a portion of a high striking vehicle can lead to large door intrusions. The large difference in the stiffness of front and side components of the vehicle structure suggests that the frontal characteristics at relatively low values of frontal crush govern the interaction in side impacts.

One proposed metric for assessing side impact compatibility is the height of the barrier force when the frontal force level is sufficient to cause severe door intrusion. A representative value for this force needs to be examined further, based on analysis of available test data. For the Grand Cherokee, the force generated by each frame rail reaches about 38 kN at 125 mm of deformation. This level appears to be in the range of force that can cause door intrusion in a side impact.

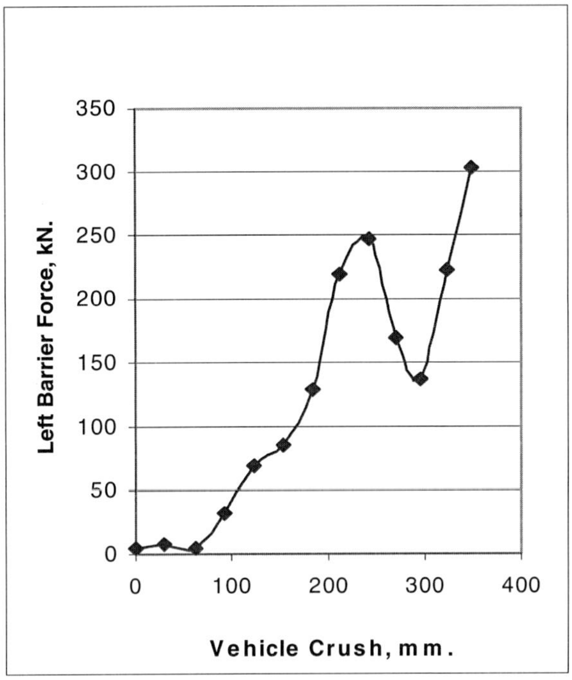

Figure 12. Barrier Force vs. Crush for Load Cells on Left Side of Barrier - Jeep Cherokee

For the side impact compatibility, the distribution of force at a crush of 125 mm is a candidate metric. Table 5 shows this distribution for the Grand Cherokee. Similar data is shown for the Dodge Neon in Table 6. The presence of a load path low enough to engage the sill of the struck vehicle may reduce the door intrusion at the level of the occupant's chest. Consequently, a metric to represent the extent of sill loading could be based on the load percentages in the lowest rows of load cells. Conversely, a high percentage of loading in the higher rows of load cells could induce unfavorable intrusion, as discussed earlier.

The risk of head impact with the striking vehicle may be related to its hood height. This metric can be measured directly on the vehicle. The barrier force data provides information on whether or not the striking vehicle loaded the top row of cells at any time during the crash. In the case of the Grand Cherokee, 33% of the barrier force was registered by the top row of cells, late in the crash. This force distribution is shown in Figure 8.

Table 5. Distribution of Total Load Cell Force by Row at 125 mm of Crush– Grand Cherokee

Load Cell Row	% of Load
D-Hi	0%
C-Hi	88%
B-Lo	12%
A-Lo	0%

Table 6. Distribution of Total Load Cell Force by Row at 125 mm of Crush– Dodge Neon

Load Cell Row	% of Load
D-Hi	0%
C-Hi	10%
B-Lo	80%
A-Lo	10%

RESULTS

The results of the barrier data provide useful insights into the geometry and height of the stiffest portions of the vehicle structure in a barrier crash. By developing metrics for these properties it may be possible to more precisely quantify vehicle compatibility. The performance of vehicles in frontal offset collisions and front to side collisions may be better understood using metrics of the type developed in this report.

The proposed metrics need to be further evaluated. The evaluation should include the assessment of a large number of vehicles and an assignment of proposed compatibility metrics based on barrier crash test data and physical measurements. The resulting metrics should be evaluated by determining the extent to which they explain the aggressiveness characteristics observed in the on-the-road crash data.

Several enhancements to the barrier array would assist in more accurately measuring the force distribution. One enhancement would be additional rows of smaller load cells to provide better discrimination of the footprint. The desirability of additional discrimination is illustrated by the force distribution in Figure 8. As the crush increases from 10 to 15 inches, the Grand Cherokee exhibits a large increase in the force on the C row, but a decrease in the force on the B row. It is probable that the maximum load at 10 inches of crush was close to the border of the B and C rows. At 15 inches, the maximum load moves to the C row. The high magnitude of the force on the top of the B row at 10 inches may indicate that the load path is lower than it actually is. Additional rows of cells on the barrier face would provide more a precise location of the center of force.

A second barrier enhancement would be a higher row of load cells to record the force from very high vehicles, particularly late in the crush. As reported earlier, 33% of the force from the Grand Cherokee is on the top row of cells at 15 inches of crush. It is probable that some present vehicles will exhibit loading above the D row.

CONCLUSIONS

On-the-road crash data indicates large differences in the aggressiveness of vehicles. An analysis of the distribution of loads that vehicles impart to the barrier in crash tests provides valuable data to assess the stiffness and geometric aggressiveness characteristics of vehicles.

Different metrics are proposed for different crash modes. For frontal-offset and frontal oblique, the force distribution from only the left portion or the right portion of the load cells may be needed. The proposed metrics for frontal crash modes are: (1) the barrier force at 250 mm of crush, (2) the linear stiffness at various amounts of crush, and (3) the center of force at 250 mm of crush.

For front-to-side vehicle crashes, some additional metrics are proposed. The force distribution when the loading is sufficient to cause intrusion of the side door is proposed as the basis for a metric. A high percentage of force on the lowest rows is indicative of sill loading, which should be favorable. A high percentage of force on the highest rows of load cells is indicative of intrusion in the region of the thorax, which should be unfavorable. The presence of loading in the upper row of load cells at any time during the crash is indicative of a high hood, which could be the source of head impacts.

ACKNOWLEDGMENTS

The authors would like to express their appreciation to NHTSA for sponsoring this research.

REFERENCES

1. Dalmotas, D., et.al., “Prospects for Improving Side Impact Protection Based on Canadian Field Accident Data and Crash Testing”, SAE 910321, 1991
2. Gabler, H., and Hollowell, W., “The Aggressivity of Light Trucks and Vans in Traffic Crashes”, SAE 980908, 1998.
3. Hobbs, A., “The Influence of Car Structure and Padding on Side Impact Injuries”, SAE 896115, 1989.
4. Willke, D., and Monk, “Side Interior Stiffness Measurements”, DOT HS 806 708, September 1984.

2003-06-0013

DEVELOPMENT OF TEST PROCEDURES AND PERFORMANCE CRITERIA TO IMPROVE COMPATIBILITY IN CAR FRONTAL COLLISIONS

Mervyn Edwards
Huw Davies
Adrian Hobbs
TRL Limited (Transport Research Laboratory)
United Kingdom
Paper Number 86

ABSTRACT

Compatibility is now recognised as the next major step forward for improved car occupant secondary safety. The work reported here forms part of a research project that was undertaken to understand the current compatibility problems in car to car collisions and develop crash evaluation procedures that are suitable for consumer and legislative testing. The research performed to date has focused on the structural performance of vehicles in order to provide a safe environment for the protection of the occupants. This should also provide sufficient space to allow intelligent restraint systems of the future to operate effectively. This paper outlines the present understanding of compatibility for frontal impact collisions and reports the current state of development of three possible test procedures to address the fundamental issues, namely: structural interaction, frontal stiffness matching and passenger compartment strength. The development of a new deformable barrier face and revised performance criteria for the full width test to assess structural interaction are described. Analysis of the load cell wall data collected in EuroNCAP tests, to address the frontal stiffness problem, is reported. Performance criteria are suggested and future work necessary to help set performance limits outlined. Initial work to investigate the repeatability of the passenger compartment strength test and possible performance criteria are described. This research is being carried out in co-operation with the European Enhanced Vehicle-safety Committee (EEVC) and the International Harmonisation of Research Activities (IHRA) Working Groups, and is funded by the UK Department for Transport (DfT).

INTRODUCTION

Compatibility is now recognised as the next major step forward for improving car occupant safety and reducing road casualties. Since 1995, research carried out by TRL on behalf of the Department for Transport (DfT) has changed focus from frontal impact to compatibility. This ongoing work is being used to support the European Enhanced Vehicle-safety Committee (EEVC) and the International Harmonisation of Research Activities (IHRA) compatibility working groups.

Initially this research was aimed at gaining an understanding of compatibility and the factors that affect it. Having achieved this, more recent research has focussed on developing test procedures able to measure the most important characteristics that influence compatibility. Prior to this research, the conventional wisdom was that compatibility problems were limited to crashes between cars of different masses, where mass ratio had the dominant influence. Now it is clear that it is the effect that mass has on frontal stiffness that is primarily responsible for this effect. Furthermore, the importance of good structural interaction between impacting cars has been highlighted (1). This aspect of compatibility plays a part in virtually every road crash. Without good structural interaction, the energy absorbing capability of the frontal structure is compromised, leading to compartment intrusion in severe accidents. Once good structural interaction has been achieved, frontal stiffness matching between vehicles, combined with strong passenger compartments, should ensure that the impact energy is absorbed with minimal passenger compartment intrusion. Beyond this, there is scope for better optimisation of the car's deceleration pulse to minimise restraint induced deceleration injuries. With good compatibility, cars should perform in a more predictable manner over a range of impact configurations, enabling the meaningful development of advanced restraint systems.

For the year 2000, Great Britain's national accident statistics show that two-thirds of the road accident casualties were in cars or light goods vehicles (2). Occupants of these vehicles accounted for about half of the fatalities and the seriously injured, which is typical for recent years. Using values calculated by the DfT for preventing a road accident casualty, the cost to society of these casualties was about £6.3 billion in 2000. Although the improved structural interaction aspects of compatibility are relevant for virtually all car frontal impacts, the main benefits from stiffness matching are expected in car impacts with another vehicle. The GB national data show that over sixty percent of fatal and seriously injured car occupant casualties in 2000 were in frontal impacts. Approximately two-thirds of these occurred in an impact with at least one other vehicle, the type of accident in which improved compatibility should offer most benefit (Table 1).

Table 1.
Distribution of car occupant casualties by first point of impact for Great Britain in the year 2000

First point of impact	Fatalities		Seriously injured	
	No	(%)	No	(%)
Front	1002	(60)	11931	(66)
Side	560	(34)	4123	(23)
Rear	65	(4)	1527	(8)
Other	38	(2)	473	(3)
Total	**1665**	**(100)**	**18054**	**(100)**

It is expected that improved vehicle compatibility will result in far better occupant compartment integrity in frontal impact accidents but precise assumptions about how a compatible car will perform are very difficult to make. A benefit analysis has been performed for GB, which assumed that improved vehicle compatibility would either, pessimistically, eliminate injuries related to contact with intruded parts of the vehicle interior, or optimistically, eliminate injuries related to contact with the vehicle interior whether it had intruded or not. It was then assumed that removal of these injuries from the existing accident data would quantify the benefits for the applicable occupant population. With the additional assumption that compartment integrity would be maintained for all impact severities, it was predicted that improved compatibility could save between 37 and 55 percent of car occupant fatalities in frontal impacts. This prediction can be regarded as an upper estimate as it unlikely that compartment integrity could be maintained for high speed impacts. Repeating the analysis using the assumption that compartment integrity would be maintained for impact severities up to 56 km/h Equivalent Test Speed, predicted that between 14 and 26 percent of fatalities would be saved. It is expected that the actual benefit lies somewhere between these predictions.

This paper summarises the current frontal impact compatibility problems and describes the current development status of three test procedures to address the fundamental issues, namely, structural interaction, frontal stiffness matching and compartment strength.

CURRENT COMPATIBILITY PROBLEMS AND PROPOSED TEST PROCEDURES

Structural Interaction

In rigid wall crash tests, the wall controls the way the impact deformation is distributed across the car's front, irrespective of the car's stiffness distribution, so ensuring good structural interaction. Cars designed with limited numbers of frontal load paths that have few interconnections can obtain good results in these tests. Unfortunately, when such cars impact each other the chances of their stiff structures interacting is very limited. This reduces the chance that the impact energy can be efficiently absorbed in the frontal structure in the designed manner. In severe accidents this leads to excessive compartment intrusion and subsequent occupant injury. This poor interaction can manifest itself as the lateral fork effect, where the stiff members of one vehicle penetrate the soft areas of the other vehicle, due to lateral misalignment, or the over-riding of one car's structure by that of the other. With no control over the height of car structures, geometrical mismatches can give rise to over-riding from static misalignment. Even when structures are aligned statically, dynamic over-riding may occur.

An example of the static misalignment problem was seen in a collision between a BMW 3 series and a VW Sharan. The higher stiff structure of the Sharan overrode the BMW lower rails leaving them virtually undeformed. It directly loaded the less stiff area of the BMW and subsequently pushed the engine onto the firewall (Figure 1). This prevented the BMW absorbing its share of the impact energy efficiently and probably resulted in an undesirable 'back loaded' compartment deceleration pulse shape (3), i.e. a low deceleration at the start of the impact leading to a high deceleration at the end of the impact, as the engine loaded the firewall.

The sensitivity of structural interaction with current cars has been demonstrated previously (3). A 100 mm variation in ride height, in an impact between two identical cars, resulted in significant over-riding of the raised car over the lowered one. The energy absorption capability of both cars was compromised, as the structures were not loaded in the intended manner. This resulted in greater intrusion for the lowered car at facia level and in the raised car at footwell level. Subsequent EUCAR simulation modelling indicated that over-riding can occur with a height difference of only 25 mm, with identical cars (4). Even where structures are aligned vertically, dynamic pitch or bending of the structure during the impact can lead to misalignment.

Figure 1. Comparison of damage to BMW (top) to Sharan (bottom) showing that Sharan has overridden BMW.

Detailed accident case studies performed using the UK CCIS database found that structural interaction is a major problem, with less than 2 percent of the 162 car to car frontal impact cases examined showing good interaction.

In order to achieve good interaction, it is important that the structures of each car meet something substantial on the other car to react against. Current views are that this is best achieved by utilising multiple load paths, with good links between them. These links may take the form of frontal inter-connections or of shear connections set back from the front. Such structures should provide a more homogeneous front against which the other car's structure can react. In addition to the provision of a homogeneous front, it is important that there is adequate vertical alignment to ensure, for example, that a low sports car could interact with the front of a high off-road vehicle.

The Offset Deformable Frontal Impact test was intended to encourage manufacturers to increase the number of load paths and links between them. Unfortunately, so far, few manufacturers have taken advantage of the weight saving opportunities of this approach. Most have simply increased the stiffness of the car's main rails. However, for load spreading, all cars now have substantial crossbeams between the main rails but few other frontal connections have been improved. No cars currently have effective lateral connections, at the bonnet latch platform level, and few have any significant vertical connections between the lower load path and any upper load path.

In summary, these aspects of compatibility are general to all frontal impacts, not just car to car impacts. Achieving good interaction will allow a car to perform more predictably in accidents, in terms of energy absorption and compartment deceleration. Apart from the resulting reduction in intrusion this would help advanced restraint systems to perform correctly and predictably.

A Full Width Test for Structural Interaction has been proposed that uses a load cell wall (LCW) to assess and control the car's frontal stiffness distribution. The logic behind this is that cars with more homogeneous fronts should offer greater potential for good structural interaction. However, there are some issues that generate problems when a rigid faced load cell wall is used:

- The parts of the car that first impact the wall are decelerated instantaneously giving rise to large inertial forces. Such forces are not present in impacts with deforming structures, such as other cars.
- Similarly, when the engine impacts the wall, it is brought to rest very rapidly generating high inertial forces. In a car to car impact, the engine can rotate or move slightly out of the way of the other car's engine, so reducing its deceleration.
- Localised stiff structures can form preferential load paths to the wall and reduce the loading from adjacent structures which are slightly set back. This does not occur in impacts with other cars.
- No relative shear is generated in the front structure to exercise any shear connections between load paths.

In order to overcome these problems, a deformable barrier face is fitted to the front of the load cell wall. As the test is also intended to function as a high deceleration test for frontal impact, care has been taken to ensure that the overall car deceleration has not been significantly affected by the addition of the deformable face.

Frontal Stiffness

All current frontal impact crash tests place direct or indirect controls on energy absorption and deceleration of the car. If there is inadequate energy absorption in the frontal structure intrusion occurs which, at some level, will be detected by the instrumented dummies. Similarly, the dummies are sensitive to the car's deceleration, which is detected through such things as chest loading from the seat belt. However, there are currently no requirements controlling the frontal stiffness of the car. Indeed, the tests encourage heavier cars to be stiff, in comparison with lighter cars. As all the tests place a limit on the car's deceleration, through control of dummy loading, all cars tend to have similar stopping distances in the tests. The dummy's experience of deceleration is totally independent of the mass of the car it is travelling in. Data from EuroNCAP tests show that most cars, irrespective of size, have an overall ride-down distance of 1200 (+/- 200) mm (Figure 2). This includes the depth of the deformable barrier face of 540 mm. As most manufacturers aim to limit the length of the front structure, for a variety of reasons, crush depths tend to be kept to the minimum.

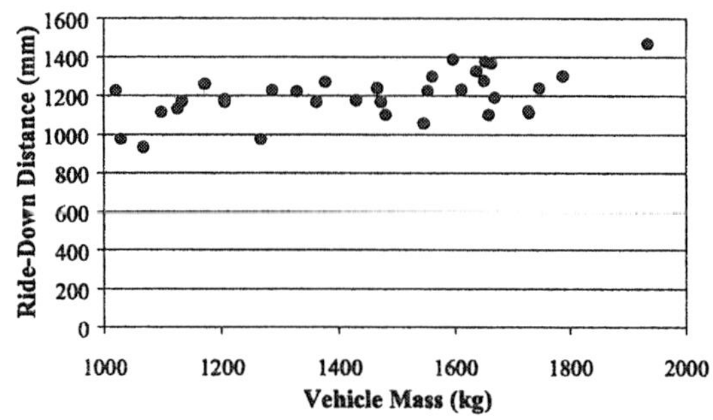

Figure 2. Ride-down distances recorded from EuroNCAP tests showing little variation with increasing mass. Note: Ride-down includes barrier depth of 540 mm.

With the energy absorbed being the integral of force against distance, the only way to maintain the same crush depth, whilst at the same time absorbing the car's kinetic energy, is for the frontal stiffness to increase with vehicle mass. This means that, even without other influences, current frontal crash tests lead to a stiffness incompatibility between cars of different mass.

Detailed accident case studies were performed using the UK CCIS database to quantify the size of frontal stiffness / compartment strength incompatibility problem. This was achieved by identifying cases where there was a significant intrusion difference between the colliding cars from the 162 car to car frontal impact cases examined. There were 78 cases where at least one of the vehicles had intruded and therefore it was possible to identify an intrusion difference. A significant intrusion difference was identified in 68 percent of these cases indicating that stiffness / compartment strength incompatibility problem is large. However, it should be noted that the extent to which poor structural interaction contributed to this problem is unknown.

In order to overcome this aspect of compatibility, it is necessary to control frontal stiffness by limiting the force imposed by a car on its opponent, in an impact. This may be less of a problem than it might at first appear. Data from EuroNCAP tests is indicating that the stiffness of some small cars has increased and becoming more in line with that of larger cars.

In setting a force limit requirement for cars, there are a number of factors to be considered:

- Whatever the force level is set to be, it will be necessary for the passenger compartments of all cars to be strong enough to resist this force without suffering significant intrusion.
- If the force level is set to be low, heavy cars will have to increase their available crush depth and may require longer front structures.
- If the force level is set to be high, light cars will have to become stiffer and the requirements for passenger compartment strength will also be high.
- A limit on how high the force level can safely be set will come from the potentially increased risk of deceleration induced injuries from the restraint system. A worst case situation would be where a low mass car had a full width frontal impact with a high mass car and the occupants were frail or elderly. For these occupants, the velocity change and deceleration of their cars will be high and there will be a limit to the ability of even advanced restraint systems to provide an adequate ride down.
- With a high force limit, the need to understand the influence of deceleration pulse shape, in combination with advanced restraints, will become more urgent.

A 64 km/h Offset Deformable Barrier (ODB) test for frontal stiffness has been proposed that uses a LCW to assess and control the force generated by the car. This requirement could simply be added to the current EuroNCAP type test. As previously reported (5), the load measured is a combination of the force coming from the deceleration of the passenger compartment (structural component) and the force coming from the deceleration of the mainly rigid masses ahead of the firewall (mechanical component), a large proportion of which is due to the engine and gearbox. In setting a limit for this force, it is necessary to consider the extent to which the engine force needs to be taken into account. In a car to car impact some of the engine load directly acts on the engine of the other car and has little effect on the structure. The remaining load does act on the structure, either directly or indirectly. The deformable face can attenuate the force to decelerate the engine and this may allow the maximum total force measured by the load cell wall to be used.

There may also be a need to set a minimum force level for the car front, so producing a range for the acceptable forces. This would prevent the design of small cars with excessively soft fronts, where the deceleration pulse might have to increase rapidly, when the front structure bottoms out on the strong passenger compartment. Such deceleration pulses are known to be injurious. It is unlikely that a minimum force requirement would come into play for larger cars, as there is no indication that any manufacturer has an interest in producing a long soft fronted car.

Compartment Strength

The frontal stiffness test effectively sets a limit for the force that one car can impose on its opponent. However, it gives no guarantee that the passenger compartment can sustain the load imposed by another car. For example, where a car, which generated a force well below the limit in the frontal stiffness test, impacted one which generated a force near to the limit, there could be no confidence that its passenger compartment would survive. Furthermore, any slight variation in the impact configuration might affect the force levels. For these reasons, it will be necessary to have a requirement for the strength of the passenger compartment, ensuring that it can resist forces greater than those used to control frontal stiffness.

It is clear that the strength of the passenger compartment is dependent upon the load paths used to transmit forces to it. In a frontal impact the most important load paths are the main rails, the upper rails, the engine subframe, via the road wheel to the sill and via the engine to the firewall. The upper rails and / or engine subframe may or may not be present. The way the loads are distributed between these load paths is dependent upon the car design, the impact configuration and the characteristics of the object hit. As the distribution of loads between the load paths varies, so the effective strength of the passenger compartment also varies. In order to ensure survival of the passenger compartment, cars should be designed to be tolerant of the distribution of the impact load. In principle this could be achieved by having a passenger compartment which is strong enough, irrespective of some variation in load path use, or by having a frontal structure that controls the way loads are distributed to the various load paths. The indications are that good structural interconnections control adjacent load paths to deform together and help to achieve this.

An 80 km/h ODB test to measure the compartment strength has been proposed that uses a LCW to assess and control the force generated by the car. It should be pointed out that there is no intention to require that cars provide a survivable performance for the occupants, at this severity. The test is simply designed to measure the strength of the passenger compartment.

If the passenger compartment becomes unstable in the impact, it will be necessary to ensure that the strength measured is prior to any major intrusion occurring. Once the passenger compartment becomes unstable, the measured load can be expected to reduce but it might again increase if subsequent structural blocking occurs. However, with conventional car designs this is unlikely.

CURRENT DEVELOPMENT STATUS OF TEST PROCEDURES AND ASSOCIATED PERFORMANCE CRITERIA

Full Width Deformable Barrier Test

This section describes the development of a new deformable barrier face, the modification of a current car to improve its structural interaction compatibility and the revised performance criteria for this test.

<u>Development of a new deformable barrier face.</u> Initial development work for this test was performed with a deformable face that consisted of 150 mm deep aluminium honeycomb element with a longitudinal crush strength of 0.34 MPa. A series of full width tests using this face were performed with current cars varying in size from small family to an off-road vehicle, using an impact velocity of 56 km/h. High resolution load cell wall measurements were recorded using a wall, which consisted of 128 load cells of size 125 mm by 125 mm arranged in a 16 by 8 matrix. These results have been reported previously (6).

Unfortunately, the results of some of these tests have shown that localised stiff structures on the car can form preferential load paths, which dramatically reduced loading from adjacent structures leading to an incorrect assessment of the stiffness homogeneity of the car. An example of this effect was seen with a family sized car, which has several such structures, namely a towing eye and radiator mount brackets located on the engine subframe (Figure 3).

Figure 3. View of family sized car structure showing towing eye and radiator mount bracket protruding structures.

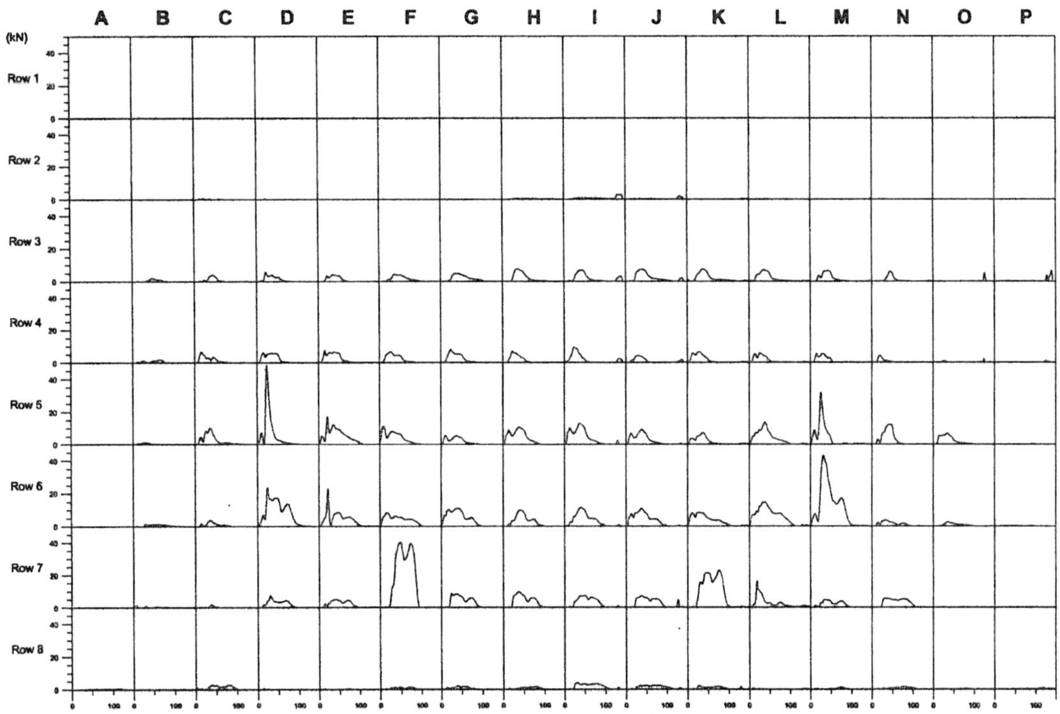

Figure 4. Load (scale 0 –50kN) / time (scale 0-150ms) curves for complete load cell wall for 0.34 MPa barrier face. Row 7 shows loading from engine subframe crossbeam with substantially higher loads recorded on cells in columns F and K caused by preferential loading of these cells by radiator mount brackets.

Examination of the deformed car and barrier face showed that the front crossbeam of the engine subframe applied load to row 7 of the load cell wall (LCW) with over 50 percent of this load being applied to the two cells in columns F and K (Figure 4). This was caused by the radiator mount brackets penetrating the deformable barrier face to make direct contact with the LCW to form preferential load paths. These unloaded the adjacent crossbeam structure. Unfortunately, this load distribution was not representative of the stiffness homogeneity of the crossbeam structure.

In order to resolve this problem the barrier face was redesigned. Following several iterations, the new barrier face design has two layers, each 150 mm deep. The front layer consists of a 0.34 MPa crush strength honeycomb element, the same as the original face, and the rear layer consists of a 1.71 MPa honeycomb element. The rear layer is segmented into individual blocks the same size as the load cells so that each block can be aligned with a load cell behind the barrier face. The reason for segmenting this layer was to reduce its shear strength to prevent the bridging of recessed load cells. Bridging can cause a redistribution of the load measured on the LCW leading to an incorrect representation of the stiffness homogeneity of the car being assessed.

It was proposed to repeat the test with the family sized car that showed the preferential load path problem using this new barrier face. However, due to technical difficulties in the manufacture of this face, which were later overcome, a face with a rear layer depth of 85 mm instead of 150 mm was used to repeat the test. The LCW results from this test are shown in Figure 5. Comparison of the LCW results from this test with those from the original single layer face (Figure 4) show that the revised face gives a much more even force distribution along the row which was loaded by the engine subframe crossbeam (row 7). This load distribution is a much better representation of the stiffness homogeneity of the crossbeam structure. Examination of the barrier face following the test showed that the radiator mount brackets had penetrated the stiffer rear layer of the face but had not made direct contact with the LCW, which allowed the rest of the crossbeam to load the wall.

One of the requirements of the barrier face is that, compared to a rigid wall test, it should have a

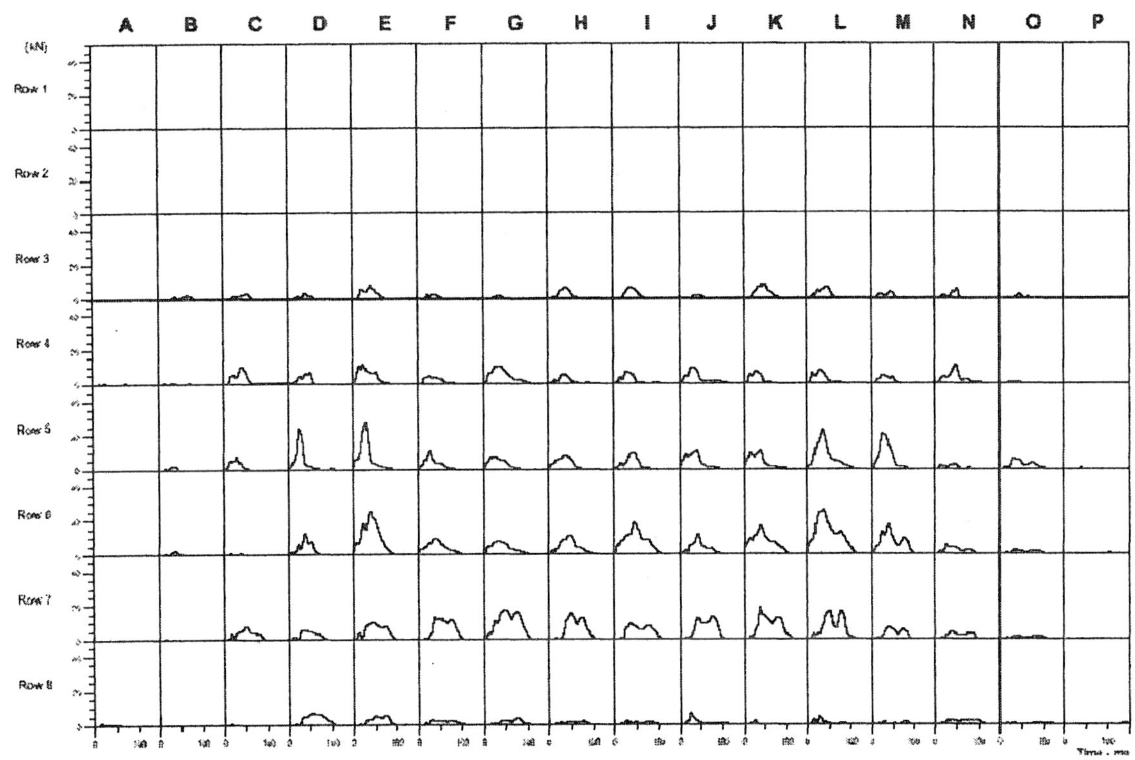

Figure 5. Load (scale 0–50kN) / time (scale 0-150ms) curves for complete load cell wall with double layer barrier face with 85 mm deep segmented rear layer showing even distribution of load on row 7 from engine subframe crossbeam.

minimal effect on the occupant compartment deceleration pulse so that the test could be used as a high deceleration frontal impact test similar to US FMVSS 208. A comparison of the compartment deceleration pulses from equivalent rigid wall and deformable barrier tests shows that the new face meets this requirement (Figure 6).

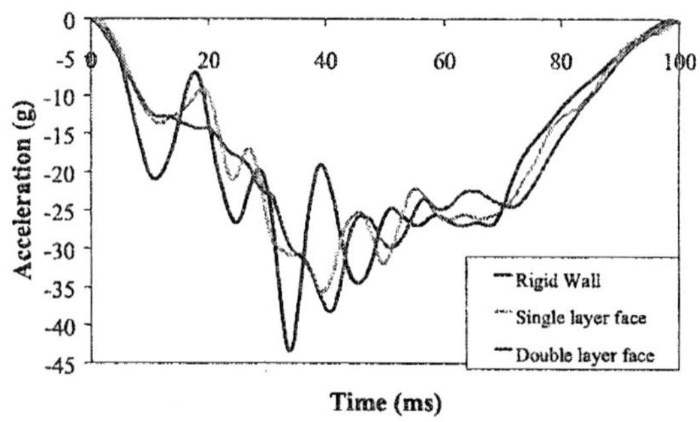

Figure 6. Comparison of compartment decelerations from rigid wall and deformable barrier tests showing minimal effect of deformable face. Note: single layer face (150mm, 0.34 MPa), and double layer face (1st layer 150mm 0.34 MPa, 2nd layer 85 mm 1.71 MPa, segmented).

In summary, a new barrier face was developed which overcame the preferential load path problem while still meeting the requirements of the original face. These were, compared to a rigid wall test, a) to attenuate the initial high decelerations at the front of the car and reduce the magnitude of the engine loading to make the test more representative of a vehicle to vehicle impact and b) to minimise the effect of the face on the compartment deceleration pulse for the reason mentioned above.

Modification of a car to improve its compatibility

A current European mid-sized car was modified to improve its compatibility and hence demonstrate the principles of how a car's compatibility performance, in particular its structural interaction potential, can be improved. The approach taken was to identify a suitable car to be modified, that not only exhibited poor structural interaction in a car to car crash test, but also would allow for additional load paths and vertical and lateral connections to be added to its base structure to improve its compatibility. A repeat car to car crash test with modified cars would then be performed and compared to the test with unmodified cars to demonstrate the improved crash performance. The car chosen for modification had lower rails which extended significantly further forward than the other load paths (Figure 7). This type of single level load path design is much more susceptible to under / override problems than a multi level one because of its smaller frontal interaction area.

Figure 7. Side, front and bottom view of mid sized car structure showing forward extent of lower rails compared to other load paths.

The modifications made were the addition of a lower load path and additional frontal vertical and horizontal connections to provide a structure with a more homogeneous stiffness distribution to improve structural interaction. The additional lower load path was made by attaching an extension to the engine subframe. The extension consisted of two arms with a deep front crossbeam horizontal connection. It had a large thin walled box section design to maximise the proportion of energy absorbed in axial crush compared to bending (Figure 8).

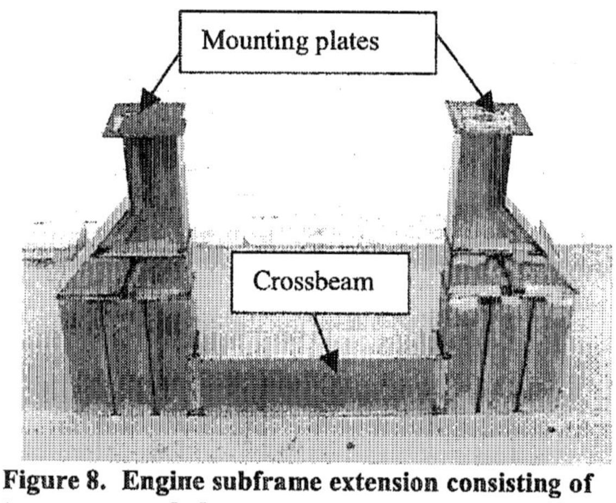

Figure 8. Engine subframe extension consisting of two arms and front crossbeam. Extension was attached to engine subframe via mounting plates.

The subframe extension was attached to the lower rails by vertical shear connections, which were formed from metal plate with additional ribs to increase their shear capacity (Figure 9). The modifications added a total mass of 16 kg to the car. The mounting plates accounted for 6 kg of this.

Initially, it was also proposed to include additional shear connections between the lower and upper rails. However, this was not done because supporting finite element analysis work showed that this would most likely increase the load through the upper rails, which could not be supported by the door aperture.

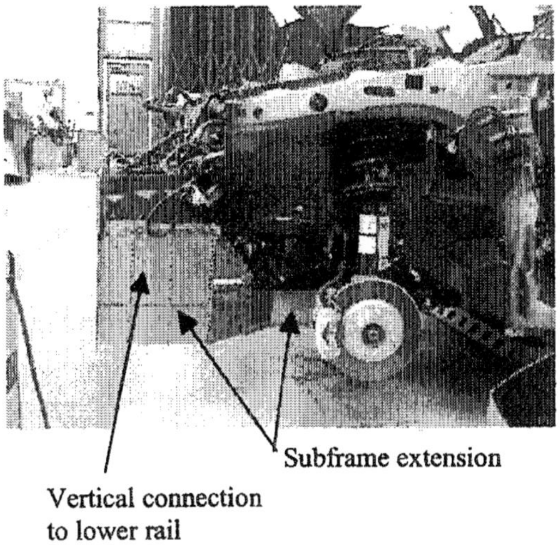

Figure 9. Modified car showing subframe extension and vertical shear connection to lower rail.

The results from a car to car impact test between unmodified cars were compared with those from an identical test with modified cars. The test configuration was 50 percent overlap with a 112 km/h closing speed. Also one car was raised to give a 60 mm ride height difference between the cars to promote any under / override that was likely to occur. The deformation of the lower car from each test shows that the occupant compartment deformation was substantially less for the modified car especially above waist rail level, indicating that less under / override has occurred with the modified cars (Figure 10). For the raised cars the modified car had less compartment deformation than the unmodified one but the difference was not as great as for the lower cars (Figure 11). Furthermore, the modified car had had less floorpan and sill deformation even though it had an additional load path at this level.

Unfortunately, the structural interaction performance for the modified cars was not improved as much as expected because the vertical shear connections between the engine subframe extension and the lower rails failed on the impacted side of the car for both the lower and raised cars. This was caused by spot weld failure.

In summary the crash performance of the modified cars was better than the unmodified ones because of the improved structural interaction and increased energy absorption of the frontal structure given by the additional load path and frontal connections.

Figure 10. Comparison of lower cars from car to car impact tests showing reduced occupant deformation of modified car.

Figure 11. Comparison of raised cars from car to car impact tests showing the reduced sill deformation between the B and C pillars for the modified car.

Development of revised performance criteria. Full width deformable barrier tests using the new face were conducted with an unmodified and modified car to check that the test and assessment technique could correctly distinguish the improved stiffness homogeneity of the modified car given by its additional load path. Unfortunately, it was found that although a subjective assessment of the load cell wall results correctly ranked the modified car as having a better homogeneity than the unmodified car the Coefficient of Variance assessment technique reported previously (6) incorrectly ranked the cars. Because of this failure a new technique to objectively assess the force homogeneity was developed.

The revised homogeneity criterion is based on the difference between peak cell loads and an ideal or target load level over a specified assessment area or footprint. The calculation of the homogeneity consists of the following steps, which are described, in more detail below:

- LCW data smoothing.
- Determination of peak cell loads.
- Definition of assessment area.
- Calculation of target load level.
- Calculation of homogeneity criterion.

A small variation in the test configuration, such as the alignment of the car with the barrier, may change the number of load cells that a main structure, such as a lower rail, loads. This can alter the homogeneity assessment causing problems with the repeatability and reproducibility of the test. To overcome this problem the entire LCW data set is smoothed. The smoothing process involves averaging the output of four adjacent load cells and allocating this value to the centre point of these cells in a stepwise manner for each of the load cells at each time step. Following this, the maximum load recorded on each cell of the smoothed data set is determined.

The methodology to determine the assessment area is still under development. To ensure good structural interaction, this area should be defined to include the main structure of the vehicle and to ensure that there is adequate geometrical overlap, both vertically and horizontally, for all vehicles. Currently, to define the area vertically, a methodology based on defining a minimum vertical extent and controlling the average height of force (AHOF) is being investigated.

A simple way to determine the target load level to achieve homogeneity within the assessment area would be to sum the peak cell forces within the area and divide by the number of cells in the area. However, to encourage all loads applied to the wall to be applied within the assessment area the target load was defined as the sum of all the smoothed data set peak cell forces (whether inside or outside the area) divided by the number of load cells in the assessment area.

The homogeneity criterion is based on the square of the difference between the peak smoothed data set cell load levels and the target load level. The reason for this was to additionally penalise loads that varied significantly from the target load level. The homogeneity is calculated for cells, rows and columns of the smoothed data set as defined in the equations below. The row and column measures are included to assess the force homogeneity between rows (vertically) and columns (horizontally), respectively. The reason for this was that although the cell assessment provides an indication of the global cell force homogeneity, it did not take into account the relative position of the cells. The overall homogeneity criterion is calculated by summing the individual cell, row and column assessments. However, if, in the future, it is determined that one assessment measure was more important than the others were, then weighting could be applied to bias the homogeneity criterion appropriately.

Cell homogeneity

$$V_c = \frac{\sum_{i=1}^{n_C} (L - f_i)^2}{n_c}$$

Row homogeneity

$$V_r = \frac{\sum_{j=1}^{n_r} \left[\frac{L.n_{cl} - \sum_{i=1}^{n_{cl}} f_{ij}}{n_{cl}} \right]^2}{n_r}$$

Column homogeneity

$$V_{cl} = \frac{\sum_{j=1}^{n_{cl}} \left[\frac{L.n_r - \sum_{i=1}^{n_r} f_{ij}}{n_r} \right]^2}{n_{cl}}$$

where:
L = Target load level.
f = Peak cell force.
n_c = Number of cells in the assessment area.
n_{cl} = number of columns in the assessment area.
n_r = number of rows in the assessment area.

This revised homogeneity assessment technique has been shown to correctly rank the unmodified and modified cars described previously (Figure 12). The assessment area was determined from a subjective examination of the LCW data. It is interesting to note that the modified car's additional load path results in a much improved row (vertical) homogeneity assessment.

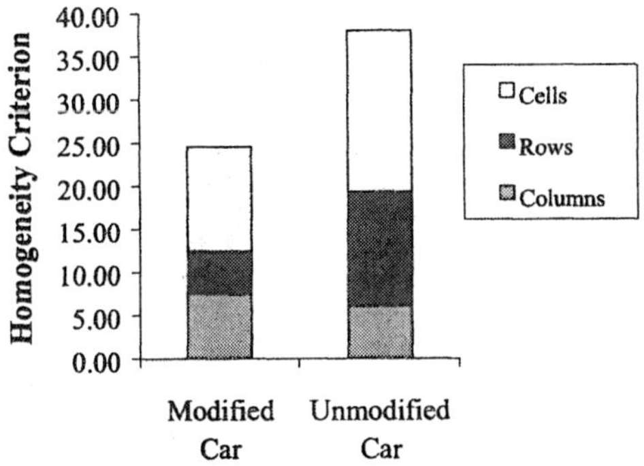

Figure 12. Homogeneity Criterion for unmodified and modified European mid sized car broken down into cell, row and column assessment.

Frontal Stiffness Test

The potential of controlling a car's stiffness by using the peak LCW force measured in a 64 km/h ODB test has been demonstrated and reported previously (6). A 50 percent overlap car to car test, with a closing speed of 112 km/h, was conducted between two small cars with a mass ratio of 1.01. Intrusion measurements showed that the car, which had recorded a lower peak load cell wall measurement (240 kN c.f. 310 kN) in the 64 km/h ODB test, suffered relatively more intrusion in the car to car test than in the ODB test.

As mentioned previously, the LCW force is a combination of the force coming from the deceleration of passenger compartment (compartment component) and the force coming from the deceleration of the mainly rigid masses ahead of the firewall (mechanical component), a large proportion of which is due to the engine and gearbox. For a typical car, the 'mechanical' component is relatively constant, as the engine and gearbox decelerate gradually, as the car deforms the barrier (Figure 13).

However, in a small number of cases, the magnitude of the mechanical force component increases significantly towards the end of the impact, which increases the peak LCW force recorded (Figure 14). This is caused by the engine 'bottoming out' the deformable barrier face and directly loading the wall. It would be more difficult for cars that exhibit this behaviour to comply with a LCW force maximum limit. However, it is not believed that this would be detrimental as, generally, these cars have little structure ahead of the engine, making them more aggressive. On the other hand, small cars could use this approach to help comply with a minimum force requirement. Further work is necessary to determine if this could be a significant problem.

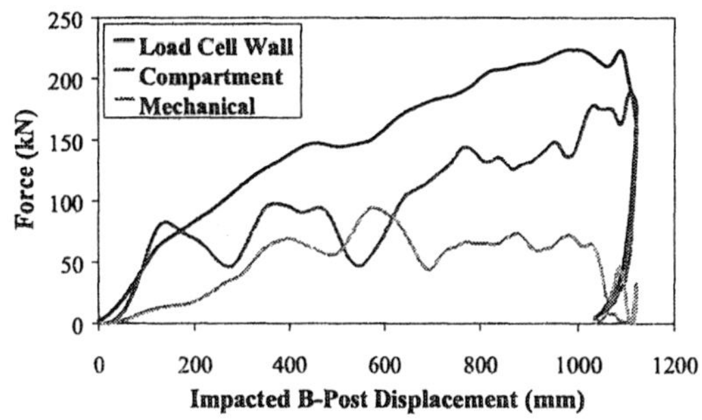

Figure 13. Load Cell Wall force showing passenger compartment and mechanical components for a typical car in a 64 km/h ODB impact.

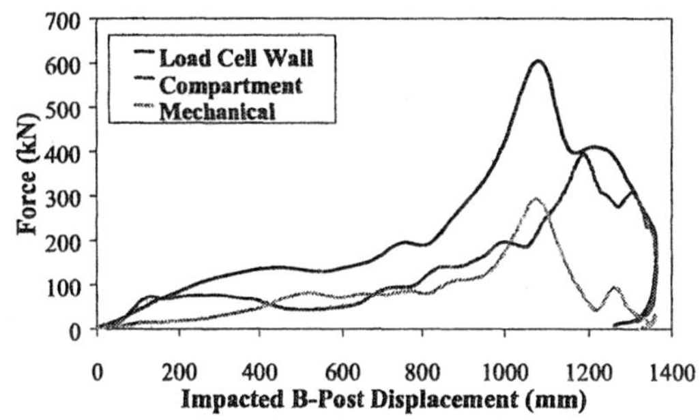

Figure 14. LCW force measured for a 64 km/h ODB test of large family car showing large mechanical force component towards end of crash.

As part of the continuing development of this test procedure, LCW peak force measurements have been taken for many recent EuroNCAP tests (Figure 15). Examination of the data shows that the peak forces lie in the range from 200 to 500 kN. From this information a first estimate for a maximum force limit could be 400 kN and for a minimum 300 kN. To determine whether these suggested values are appropriate and practicable much further work is necessary to address the issues mentioned previously. These are passenger compartment strength, deceleration pulse and the need to increase the crush depth in heavier cars.

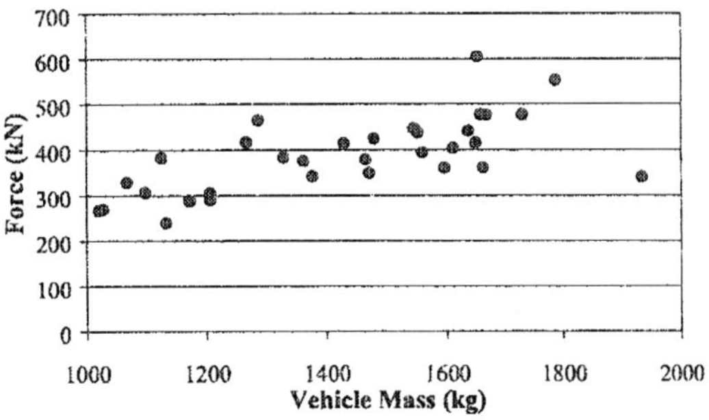

Figure 15. Peak Load Cell Wall measurement for EuroNCAP tests.

Compartment Strength Test

Typically in an 80 km/h ODB test, towards the end of the impact the engine has 'bottomed out' the deformable barrier face and stopped decelerating so that the LCW force consists mainly of the passenger compartment force component (Figure 16). The LCW force at this point is termed the 'end of crash force,' a phrase first used by Renault (7). This force represents the load imposed on the compartment and hence can be used as an indication of a minimum load that the compartment can withstand for this loading configuration. From the limited number of tests performed, it appears that the time at which the difference of the compartment deceleration and average engine deceleration records a maximum can be used to determine the time at which the 'end of crash force' should be measured. It is possible that the end of crash force requirement may be achieved with just one load path, for example via the road wheel to the sill. To ensure this does not occur, intrusion limits may also be necessary, in particular at waist rail level. Further work is necessary to address this issue.

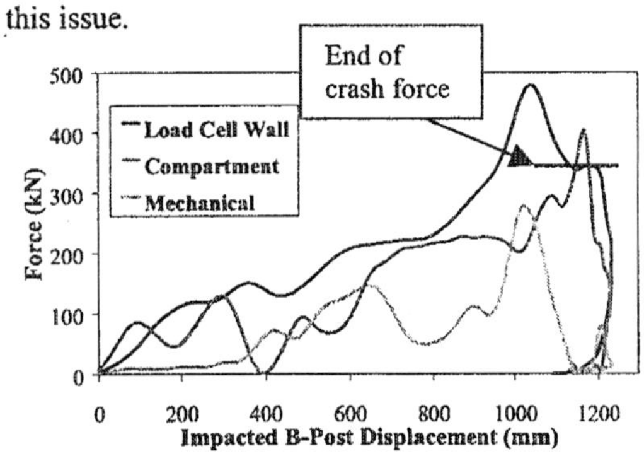

Figure 16. LCW force showing compartment and mechanical components for 80 km/h ODB test.

Some concern has been expressed about the possible repeatability of this test especially if the passenger compartment becomes unstable (4). Two similar tests have been performed for a super mini size car with an impact speed of 80 km/h. These show good repeatability for the LCW force (Figure 17) and the car's deformation (Figure 18) even though the load path through the door beam has become unstable.

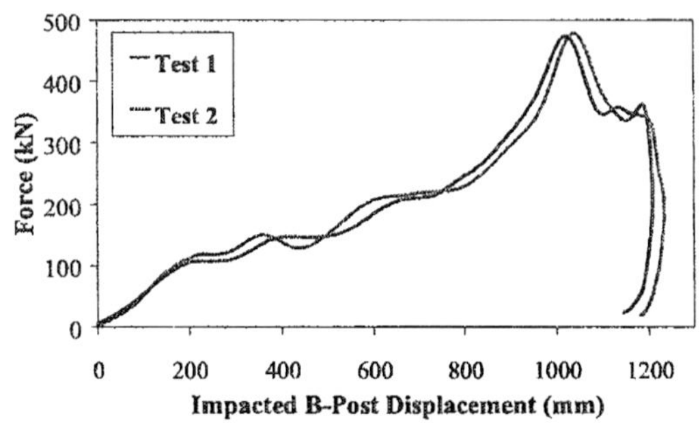

Figure 17. Load Cell Wall force for two 80 km/h ODB tests with a super mini size car showing good repeatability.

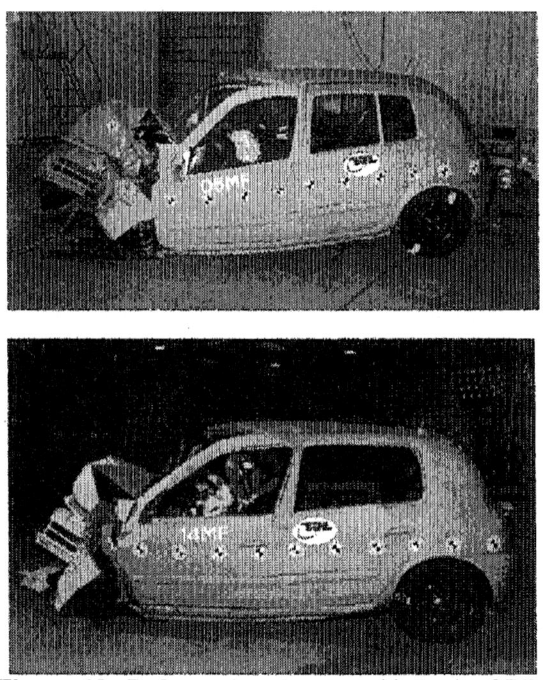

Figure 18. Deformation for two 80 km/h ODB tests with a super mini size car showing good repeatability.

The compartment strength measured in this test will be dependent on the load paths used. One possible concern, which requires further investigation, is that the wheel to sill load path is used more in this test configuration than it would be in accidents.

If a maximum force level of 400 kN is set for the frontal stiffness test, a suggested minimum limit for the end of crash force to control the compartment strength may need to be somewhat greater, say 450 kN. However, it may be possible to set the limit lower and still allow a sufficient safety margin. In a car to car impact some of the engine load acts directly on the engine of the other car and does not act on the passenger compartment. It is possible that not all of the engine component of the LCW force measured in the frontal stiffness test acts on the passenger compartment in a car to car collision, so this may allow a sufficient safety margin to set the limit lower. One advantage of a lower limit would be to minimise the risk of car designs where the deceleration pulse might have to increase rapidly when the front structure bottomed out on the strong passenger compartment. Further work is required to check that the suggested value of 450 kN is appropriate and practical.

CONCLUSIONS

The first requirement for frontal impact compatibility is to ensure good structural interaction. It helps to address problems seen in all frontal impacts and without it any control of stiffness would have limited effect. With good structural interaction, it will then be possible to control frontal stiffness and passenger compartment strength. An inevitable consequence of these actions to reduce passenger compartment intrusion is that car deceleration will increase along with associated injuries, unless they are mitigated by improved restraint systems. Although any increase in injuries from deceleration is likely to be small compared with the decrease due to improved passenger compartment survival, there is going to be a growing need to understand the importance of and potentially control the shape of the deceleration pulse.

To address these issues and improve compatibility three test procedures are under development. It should be noted that two of these are modifications of current tests. The tests are:

- A full width test at 56 km/h (the current US NCAP test) with a deformable barrier face and high resolution load cell wall (LCW) to assess and control structural interaction. This will be achieved by controlling the force distribution measured on the LCW, to encourage the development of structures that behave in a more homogeneous manner.
- A 64 km/h Offset Deformable Barrier (ODB) test (the current EuroNCAP frontal test) with a high resolution LCW. From the load cells, the car's frontal stiffness could be controlled by specifying that the peak force recorded should lie within a specified range. In the future, some control of the pulse shape could be used to manage the passenger compartment deceleration and restraint loading.
- An 80 km/h ODB test with a LCW to assess the strength of the passenger compartment. This test would not require instrumented dummies.

One advantage of this set of tests for frontal impact is that the full width test would generate a 'hard' deceleration pulse on the vehicle and restraint system, whereas the 64 km/h ODB test would generate a 'soft' pulse. This would ensure that restraint systems are better able to deal with both types of pulse. Another advantage is that, assuming the full width and 64 km/h ODB tests are specified for frontal impact world wide, only one additional test would be required for compatibility.

The current state of development of these test procedures has been reported, covering issues such as the development of a new deformable barrier face and a revised homogeneity assessment technique for the full width test. A car has been modified to demonstrate the principles of how a car's compatibility performance, in particular its structural interaction potential, can be improved. It has been shown that the full width test procedure with the new barrier face and revised assessment technique correctly ranks the better structural interaction of the modified car above that of the unmodified car. Some performance limits for the frontal stiffness and compartment strength tests have been tentatively suggested. However, further work is required to ensure that these suggestions are appropriate and practicable.

In addition, the benefit of improving compatibility for the UK has been estimated and shown to be substantial.

This is an evolving area and much further work is required to complete the development of these procedures to a level suitable for consumer and / or legislative use.

ACKNOWLEDGEMENTS

The work described in this paper forms part of a Department for Transport (DfT), Vehicle Standards and Engineering Division, funded research project, being carried out by the TRL Limited. The authors gratefully acknowledge the contribution of the Nissan Motor Company Ltd and UTAC.

REFERENCES

1. Edwards M, Happian-Smith J, Byard N, Davies H, and Hobbs A, (2001). The Essential Requirements for Compatible Cars in Frontal Impacts, Seventeenth International Technical Conference on the Enhanced Safety of Vehicles, Amsterdam 2001.

2. Road Accidents Great Britain: 2000, The Casualty Report, DTLR National Statistics, available from 'The Stationery Office (TSO)', see http://www.TSO-online.co.uk

3. Wykes N, Edwards M AND Hobbs C, (1998). Compatibility Requirements for Cars in Frontal and Side Impact, Sixteenth International Technical Conference on the Enhanced Safety of Vehicles, Windsor 1998.

4. EUCAR (2001). Development of Criteria and Standards for Vehicle Compatibility, Vehicle Compatibility Workshop, Wolfsburg, Germany, Feb 13th – 14th 2001.

5. Edwards M, Happian-Smith J, Byard N, Davies H, and Hobbs A, (2000). Compatibility – The Essential Requirements for Cars in Frontal Impact, Vehicle Safety 2000 conference, London, June 2000.

6. Edwards M, Happian-Smith J, Byard N, Davies H, and Hobbs A, (2001). The Essential Requirements for Compatible Cars in Frontal Impacts, Seventeenth International Technical Conference on the Enhanced Safety of Vehicles, Amsterdam 2001.

7. Steyer C, Delhommeau M, and Delannoy P, (1998). Proposal to Improve Compatibility in Head On Collisions, Sixteenth International Technical Conference on the Enhanced Safety of Vehicles, Windsor 1998.

2001-06-0240

TEST PROCEDURES TO EVALUATE VEHICLE COMPATIBILITY

Koji Mizuno
Traffic Safety and Nuisance Research Institute
Kazumasa Tateishi
Japan Automobile Research Institute
Yukihiro Ezaka
Ministry of Land, Infrastructure and Transport
Japan
Paper Number 127

ABSTRACT

Test procedures for evaluating vehicle compatibility were investigated based on accident analysis and crash tests. This paper summarizes the research reported by Japan to the IHRA Compatibility Working Group.

Passenger cars account for the largest share of injuries in head-on collisions in Japan and were identified as the first target for tackling vehicle compatibility in Japan.

To ascertain situations in collisions between vehicles of different sizes, we conducted crash tests between minicars and large cars, and between small cars and large cars. The deformation and acceleration of the minicar and small car is greater than that of large car.

ODB, Overload and MDB tests were performed as procedures for evaluating vehicle compatibility. In overload tests, methods to evaluate the strength of the passenger compartment were examined, and it is found that this test procedure is suitable for evaluating the strength of passenger compartments. The MDB test is a procedure taking into account the effects of vehicle mass in evaluating vehicle compatibility. The MDB tests with three different barrier faces indicated problems with MDB override and bottoming-out, which have a great effect on the behavior of the test vehicles.

INTRODUCTION

Compatibility is defined as the ability to protect not only the occupants, but also other road users as well. Analyses of global accident data of car-to-car collisions from various countries have indicated that there are vehicles with low compatibility, such as cars with poor self-protection and cars with high aggressivity with respect to other cars. The aggressivity of sport utility vehicles (SUV) has become an issue in the United States [1] and Australia, as has the self-protection of small cars in Europe. In Japan as well, vehicle sizes vary widely, and compatibility is considered an important problem. It is therefore necessary both to ascertain the current state of compatibility issues in Japan and to evaluate and improve compatibility performance in crash tests.

Test procedures for evaluating and improving the compatibility of passenger cars are currently under discussion in the International Harmonized Research Activities (IHRA) Compatibility Working Group. Japan considers the activities of the IHRA to be significant activities that will inform future legislation and regulation, and has conducted research with the aim of making an active contribution to these activities. This report summarizes the results of fleet studies, structural surveys, accident analysis and crash tests that Japan has conducted and reported to the IHRA Compatibility Working Group.

METHODOLOGY

Fleet Studies

Vehicle compatibility is strongly related to fleet composition, and fleet studies can identify which vehicle types, vehicle mass and other such properties of the vehicles in a country's fleet should be subject to compatibility.

The four-wheel vehicles used in Japan were classified into different types and the proportion of each among registered vehicles was surveyed. To examine further passenger cars with the greatest number of registrations, passenger cars were divided into classes and the number of vehicles in each class ascertained. From this analysis, we identified those vehicle types with high exposure.

Vehicle mass are factors with great effects on compatibility. The distributions of vehicle curb mass in Japan were examined on the basis of the 113 passenger car models newly registered in 1998. Such data were provided by the Japanese IHRA Side Impact Working Group. Total registrations of these models accounted for 70% of the new four-wheel vehicle registrations in 1998.

Structural Surveys

The dimensions and locations of the members of a vehicle have a great effect on the structural interactions between them in a crash. To survey the dimensions of members of current cars, we used the data on the 113 passenger car models (i.e., the same data as used in vehicle mass distribution analysis). The dimensions used were the top and bottom edge heights of the front end of the longitudinal members, cross member bottom-edge height and shock absorber height. Since interactions with the longitudinal member in a crash are especially significant, the distributions of the top- and bottom-edge heights of longitudinal members were determined on the basis of the data.

Accident Analysis

To address the issue of vehicle compatibility in Japan, it is necessary to identify priority target vehicles and their numbers. We therefore used global accident data (1995) to find the proportion of head-on collisions in all accidents and further identify the vehicle type with the largest proportion of such collisions.

Passenger cars were grouped into classes, and the ratios of fatal and serious injuries of the drivers in subject and other cars were examined. Using this method, the effect of velocity is not a major factor, and the compatibility of the car itself can be evaluated. This analysis reveals which classes of passenger car present significant vehicle compatibility issues in Japan.

Crash Tests

To examine the car deformation and the injury risk to drivers in crashes, offset frontal car-to-car crash tests were performed between cars of different sizes. Moving deformable barrier (MDB) tests and overload tests were also conducted to examine tests as part of our attempt to find fitting methods to evaluate compatibility. Table 1 shows the test matrix.

Table 1.
Test matrix

Test	Car model	elocity (km/h)	Mass (kg)	Overlap ratio	Remarks
Car-to-Car	Car A (small car)	55.9	1180	50%	with driver dummy
	Car C (large car)	55.9	1595		with driver dummy
Car-to-Car	Car B (minicar)	55.9	929	50%	with driver dummy
	Car C (large car)	55.9	1595		with driver dummy
MDB test	Car B (minicar)	0	1263	40%	with driver and passenger dummy
	MDB (FMVSS 214)	111.6	1367		FMVSS 214 barrier face
MDB test	Car B (minicar)	0	1258	40%	with driver and passenger dummy
	MDB (FMVSS 214)	112.4	1369		ECE R94 barrier face
MDB test	Car B (minicar)	0	1260	40%	with driver and passenger dummy
	MDB (FMVSS 214)	112.3	1368		2-stage barrier face
Overload	Car A (small car)	80.0	1095	40%	without dummy ECE R94 barrier face
Overload	Car B (minicar)	80.0	845	40%	without dummy ECE R94 barrier face

<u>**Car-to-car crash tests**</u> - To ascertain the present condition and problems in car-to-car crashes, we conducted frontal crash tests between small car (car A) and large car (car C) and between minicar (car B) and large car (car C). Overlaps were 50% for smaller cars, and the impact velocity of each car was 56 km/h. We compared the deformation, acceleration and dummy injury parameters of the two cars. The results were then compared with those for the overload tests and MDB tests and examined the validity of the tests and related problems.

<u>**Overload tests**</u> - Obviously, the structural integrity of the passenger compartment secures survival space in collision, is important for the effective operation of restraint devices and is fundamental to self-protection[2]. To prevent the collapse of the passenger compartment in a car-to-car collision, the strength of the passenger compartment (*Fmax*) should be greater than the final strength (*Fend*) of the front structures of the other vehicle [3].

Overload tests were conducted for a small Japanese car, and for a minicar to evaluate the strength of the passenger compartment. The tests were performed at 80 km/h and with 40% overlap on the ECE R94 barrier to completely collapse the passenger compartment. For measuring passenger compartment strength, the barrier force and the acceleration of the bottom of the passenger-side B pillar were used. The barrier forces observed in the overload tests (*Fmax*) were compared with those in the offset deformable barrier (ODB) tests (*Fend*) at 64 km/h to determine the possible relationship between *Fmax* and *Fend*.

MDB tests - MDB tests are one way of evaluating vehicle compatibility while accounting for the effects of vehicle mass. Since the in-depth accident data showed that the proportion of the collision with impact angle zero and 30-40% overlap is the largest, we performed collinear MDB tests at 40% overlap. To reproduce car-to-car collisions at a velocity of 56 km/h, we crashed the MDB (FMVSS trolley) into stationary small cars (car A) at 112 km/h.

To survey the effects of the barrier, we performed experiments using ECE R94 and FMVSS 214 barrier faces. And to prevent MDB override and bottoming-out, an MDB test was conducted with a two-stage barrier face[4] 700 mm depth mounted 180 mm above the ground (Figure 1). The results were compared with those from car-to-car crash tests.

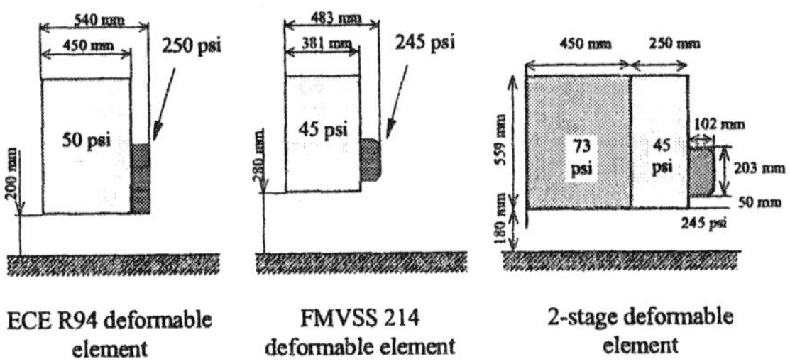

Figure 1. Deformable elements of MDB.

RESULTS

Fleet Studies

The total number of four wheel-vehicle registrations in Japan was 69,528,110 at the end of August 1999 (Figure 2). Among them, passenger cars accounted for 50,601,149 (72.8%) and trucks for 18,690,426 (26.9%). The number of passenger cars increases year by year, while the number of trucks shows no significant change. Therefore, given the number of registrations, the passenger car is the most important to improve compatibility in vehicle fleets in Japan.

Passenger cars were grouped into classes, and the number of registrations in each class is shown in Figure 3. The proportions of small and medium sedans tend to fall from year to year, and those of minicars, station wagons and minivans to rise. In 1998 SUVs were 5.0% of total registrations, which form a lower proportion in Japan than in the United States and Australia.

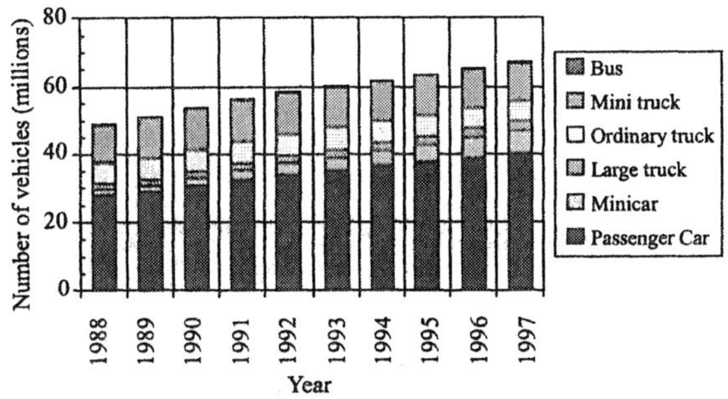

Figure 2. Vehicle registrations in Japan (1988-1997).

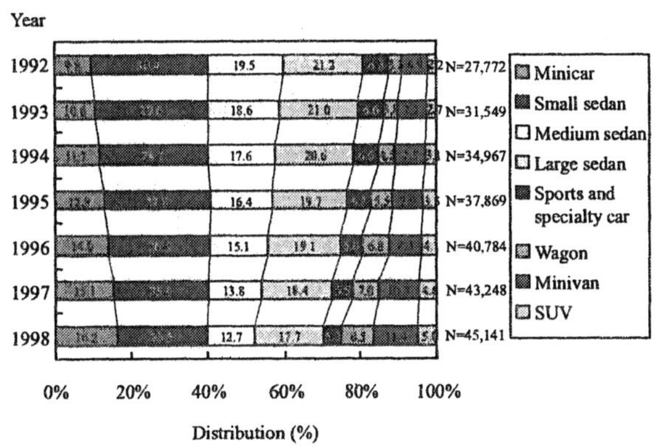

Figure 3. Registration of passenger cars by car class.

From the data on the 113 models of passenger car newly registered in 1998, the distributions of curb mass of cars in Japan were obtained (Figure 4). Curb mass is distributed from 700 kg to 1,800 kg (95% interval), and the average mass is approximately 1,150 kg (1998). This represents an increase of approximately 70 kg from the 1991 average of 1,080 kg. Reasons for this increase in mass appear to be the larger size of vehicles used in Japan such as minivans and SUVs as well as full frontal impact test that became mandatory in 1994, and side impact test mandatory as of 1998.

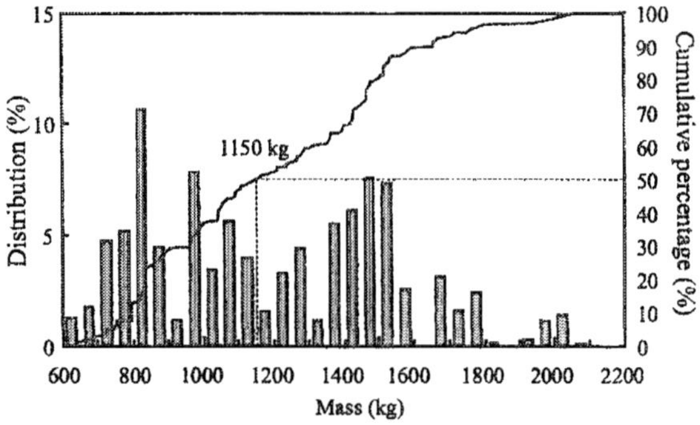

Figure 4. Vehicle mass distribution (1998).

Structural Surveys

The front-end dimensions of the 113 passenger car models newly registered in 1998 were investigated, and the weighted averages for heights of members were obtained from the number of vehicles sold. The average dimensions of the principal members are as follows:

Longitudinal member top-edge height	504 mm
Longitudinal member bottom-edge height	376 mm
Cross member bottom-edge height	256 mm
Front shock absorber top-edge height	788 mm

These figures are similar to those given in European vehicle survey data. A 1991 survey found cross member bottom-edge height to be 242 mm, so the ground clearance of vehicles has increased somewhat.

Figure 5 shows the distributions of the top-edge and bottom-edge heights of the front end of longitudinal members. The average of the top-edge height is 507 mm and the average bottom-edge height 381 mm. Although the longitudinal member heights of many cars are nearly average, many SUV models have higher longitudinal members than the average, and exhibit great variation.

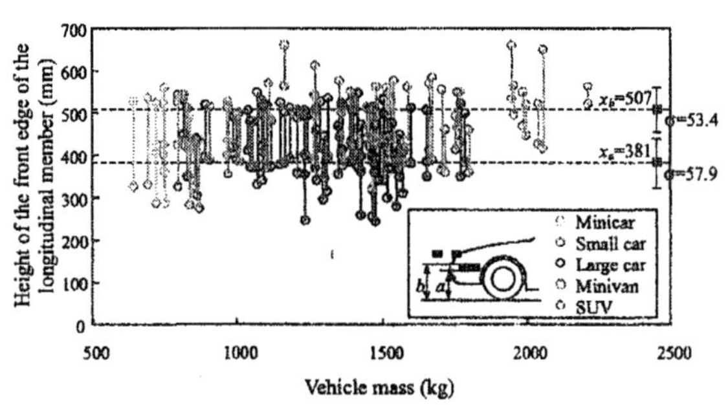

Note: The averages weighted by vehicle registrations for top and bottom heights of longitudinal member edges are 504 and 376 mm

Figure 5. Distributions in height of front edges of longitudinal members.

Accident Analysis

In the present study, accidents in Japan were analyzed on the basis of global data in 1995. Figure 6 shows the number of fatal or serious injuries of drivers in car-to-car collisions by type of accident. Of 13,157 fatal or serious injuries of drivers, the largest number of 4,372 (33%) was in head-on collisions. Figure 7 classifies these drivers in head-on collisions broken down by type of vehicle. The largest number of 680 was in collisions between passenger cars, and 390 were in collisions between passenger cars and minicars. The total of these two groups is 1,075 (44%). Therefore, from the results of accident studies there is a plurality of fatalities and injuries in car-to-car collisions in Japan in head-on collisions between passenger cars.

Global data (1993-96) were employed to divide passenger cars into classes, and the ratio and the numbers of fatal and serious injuries to the drivers in the subject cars in comparison to the other cars were obtained (Table 2). While minicars exhibit a high fatality and a serious injury rate and a large number of accidents, SUVs exhibit high aggressivity but relatively few accidents. Therefore, the self-protection of minicars is another important issue in accidents in Japan.

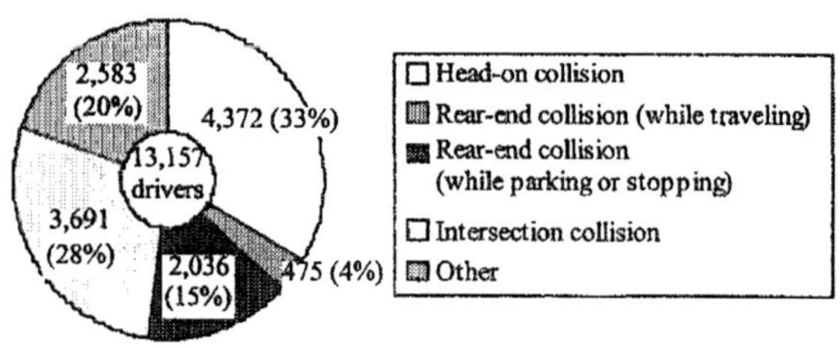

Figure 6. Drivers sustaining fatal or serious injuries by type of vehicle-to-vehicle collisions (1995).

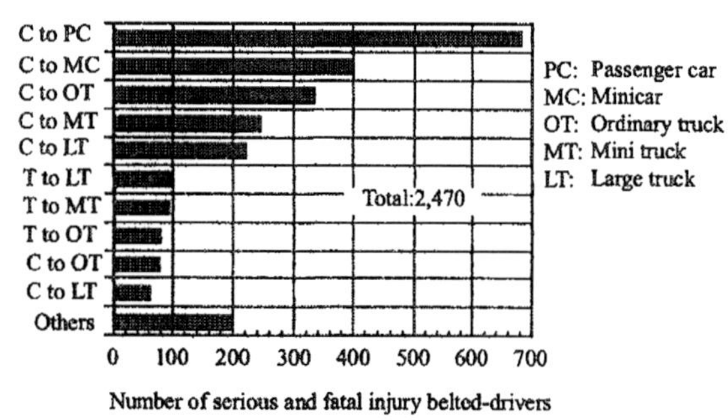

Figure 7. Number of fatal and serious injuries to belted drivers by type of head-on collision (1995).

Table 2.

The ratios of serious or fatal injuries in subject cars to other cars in head-on collisions (1993-1996)

Cars in collision	All drivers	Belted drivers
Minicar vs. Sedan	5.3:1.0 (553:104)	5.4:1.0 (328:61)
Sedan vs. Sedan	1.0:1.0 (1938:1938)	1.0:1.0 (1084:1084)
Mini van vs. Sedan	0.62:1.0 (117:189)	0.77:1.0 (73:95)
SUV vs. Sedan	0.25:1.0 (35:142)	0.25:1.0 (20:81)

( ): Number of injuries

Crash Test

<u>Car-to-car crash test</u> - Offset frontal car-to-car crash tests were performed with an overlap of 50% and an impact speed of 56 km/h between a small car (car A) and a large car (car C), and between a minicar (car B) and a large car (car C).

In crash between small and large cars, the longitudinal member of each car impacted the other's bumper beam; there was no underriding, and interaction was good (Figure 8). There was extensive deformation to the passenger compartment of car A, but hardly any to that of car C. The acceleration-time histories of both cars are shown in Figure 9. The level of the acceleration in car A is slightly higher than that in car C. However, the deformation of car A is larger than car B as shown in Figure 10. The steering column of car A was moved over 100 mm, and toe board intrusion was over 250 mm.

Figure 11 shows injury parameters for the drivers. Injury parameters related to intrusion such as femur force and tibia index grew with the extent of passenger compartment intrusion in car A; the tibia index for car A exceeds the reference value (1.3).

Car A Car C

Figure 8. Car A and car C after impact.

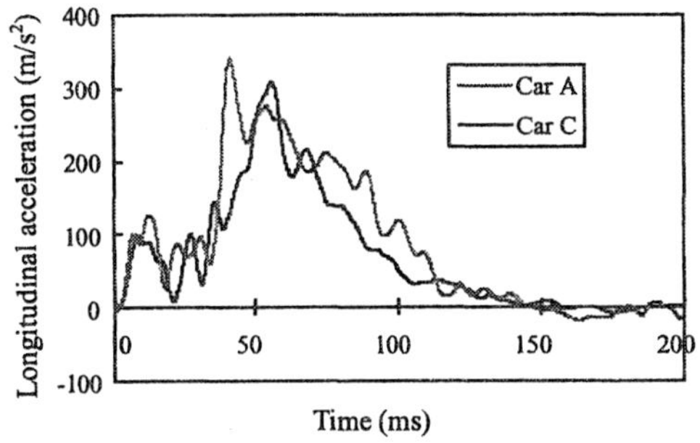

Figure 9. Acceleration-time histories for car A and car C.

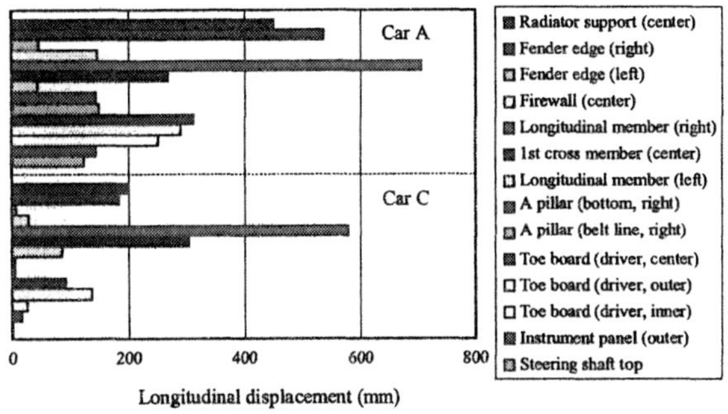

Figure 10. Deformation of car A and car C

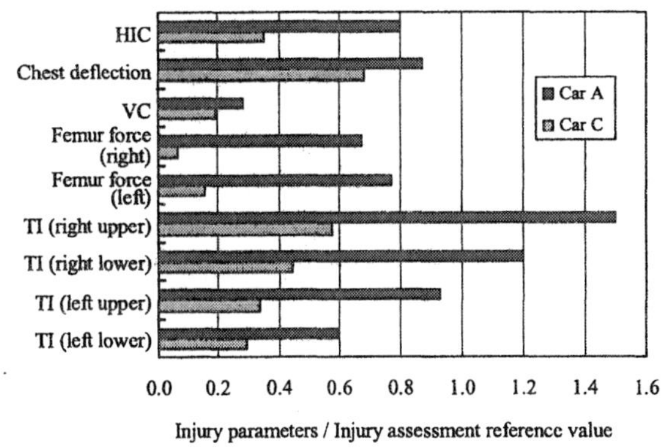

Figure 11. Driver injury parameters in crash between car A and car C.

In crash between minicar and large car the deformations were not uniform due to poor interaction (Figures 12 and 13). The minicar exhibited underriding of the large car. The longitudinal member of each car crashed into the space between the other's longitudinal member and tire, and longitudinal members showed no extensive deformation. However, the tire of car B crashed into the bumper beam of car C, and the fork effect of the longitudinal member of car C was diminished. This lateral mismatch resulted in local deformation of the wheel house and extensive deformation of the (right-side) toe board of car C.

Figure 14 shows the acceleration-time histories. The acceleration of car B is high, whereas that of car C is low. The deformation of the car is shown in Figure 15. Due to high strength of the passenger compartment, the intrusion of car B is small. Since car B had high acceleration and intrusion into the passenger compartment, the injury parameters related to both acceleration and intrusion are high; HIC is 962, and the tibia index exceeds the reference value (see Figure 16). However, the tibia index of the driver in car B is not necessarily higher than that in car A because the intrusion of the car B is smaller than car A. Though the injury parameters for car C are low, the tibia index exceeds the reference value (1.3) due to local deformation of the wheel house.

Figure 12. Car B and car C after impact.

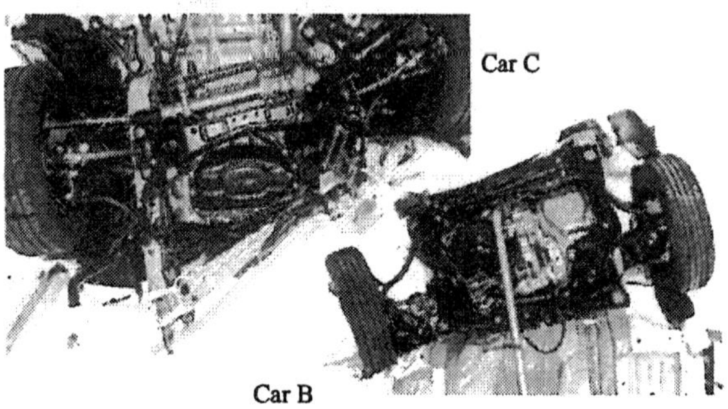

Figure 13. Lateral mismatch of car B and car C.

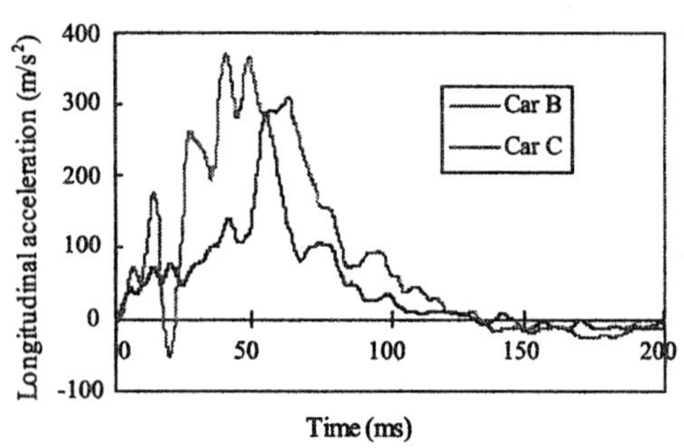

Figure 14. Acceleration-time histories for car B and car C.

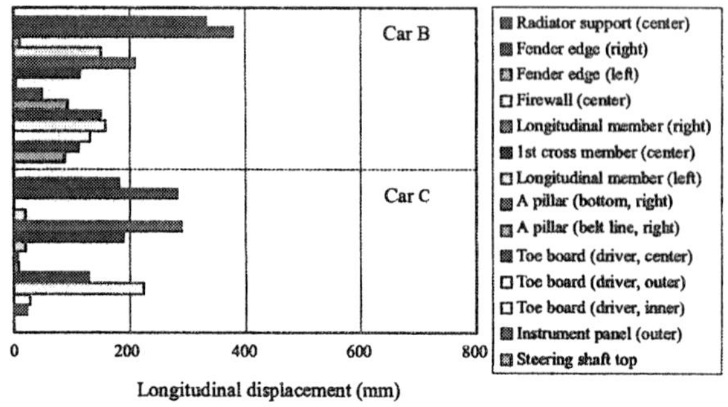

Figure 15. Deformation of car B and car C.

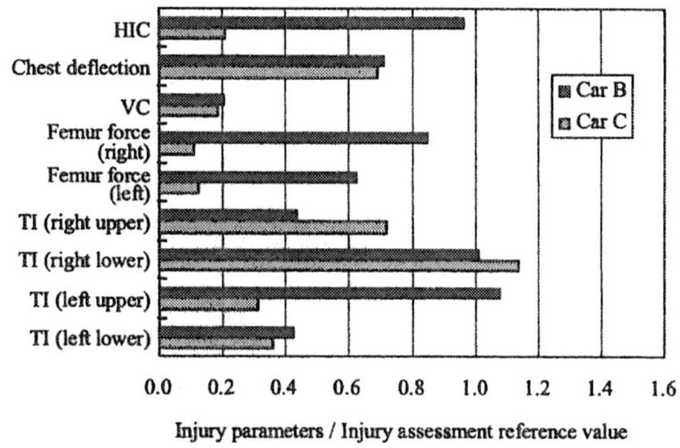

Figure 16. Driver injury parameters in crash between car B and car C.

When comparing the crash of car A and car B into car C, the injury risk is not only related with car mass. The high strength of the passenger compartment of car B is effective even with the poor interaction.

Overload Test - Overload test of the small car (car A), and that of minicar (car B) were conducted. Both car A and car B showed side sill bending, large intrusion of the instrument panel and large displacement of the steering column upwards and to the rear. The test showed no more survival space in the passenger compartment due to the crash.

Figure 17 shows the force-time histories derived from the barrier load cells and B pillar acceleration on the passenger side that suffered little deformation. To derive force from acceleration, the product of longitudinal acceleration and car mass was used. The curve indicates that barrier load was greatest at 66 ms and load from the B pillar greatest at 74 ms. Examination of vehicle deformation at each of these times (Figure 18) revealed that the instrument panel intrusion into the passenger compartment began after 66 ms. After 74 ms the side sill buckled and the force derived from B pillar acceleration on the passenger side decreased.

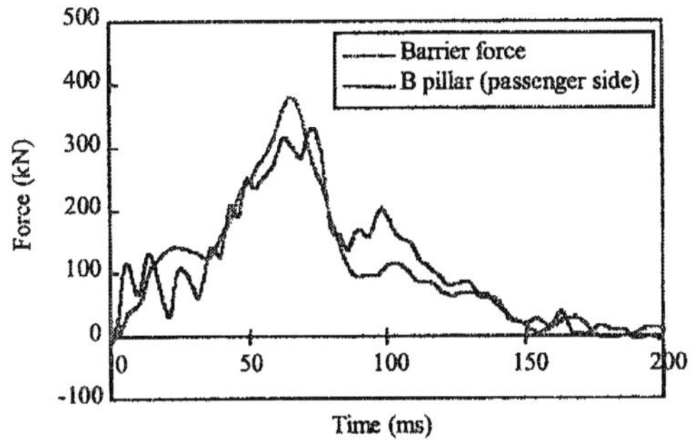

Figure 17. Force-time histories of car A in overload test.

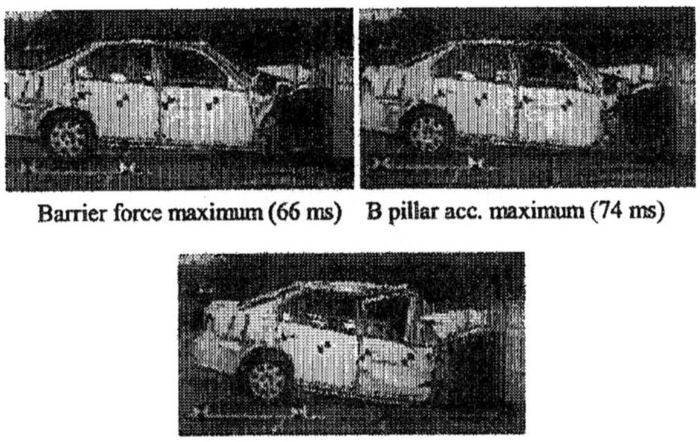

Barrier force maximum (66 ms) B pillar acc. maximum (74 ms)

Deformation almost maximum (100 ms)

Figure 18. Deformation of car A in overload test.

Figure 19 shows the force-time histories for car B. Barrier load was greatest at 50 ms, and B pillar load greatest at 68 ms. Like car A, the instrument panel intrusion into the passenger compartment began at the time of maximum barrier load, and the bending from the center of the side sill began after the force obtained from the passenger-side B pillar reached the maximum (Figure 20).

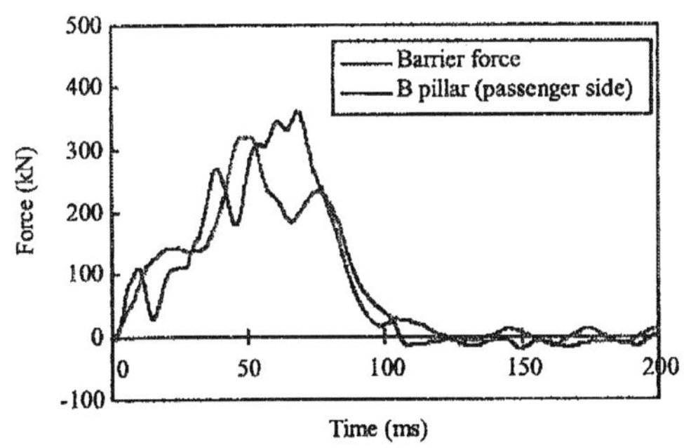

Figure 19. Force-time histories of car A in overload test.

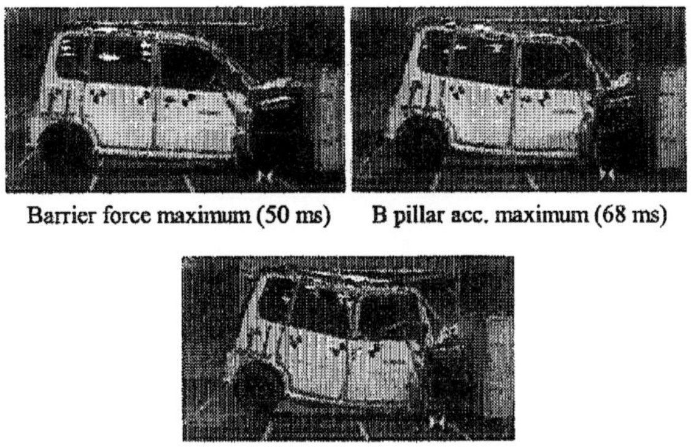

Barrier force maximum (50 ms) B pillar acc. maximum (68 ms)

Deformation almost maximum (80 ms)

Figure 20. Deformation of car B in overload test.

Evaluation of the strength of the passenger compartment varies with load measurement location and procedures. Table 3 gives the characteristics of passenger compartment strength derived from barrier load cells and passenger-side B pillar acceleration, and the car A and car B passenger compartment strength derived from these. In car-to-car collisions the force of impact is transmitted through the impact plane. Therefore, passenger compartment strength measured by barrier force in overload tests may be directly used to predict the passenger compartment collapse in car-to-car collisions.

Passenger-side B pillar acceleration reaches maximum immediately prior to the buckling of the side sill, after which time force is inadequately transmitted through these members. Therefore, a measurement of passenger compartment strength derived from B pillar acceleration may be considered to represent the strength of these members and the survival space in the final stage.

Table 3.
Evaluation of compartment strength

Measurement locations	Features	Compartment strength	
		Car A	Car B
Barrier force	• Evaluation of crash force on the interface • Maximum when instrument panel intrudes • Effect of inertial forces of engine is large	381 kN	321 kN
B pillar (acceleration)	• Evaluation of collapse force level at side sill • Depending on calculated mass • Great influence of fluctuation of the acceleration pulse	331 kN	360 kN

Figures 21 and 22 show the force-displacement characteristics of car A and car B, broken down by impact velocity. Force-displacement characteristics describe the similar curve, regardless of velocity, and the deformation mode at 80 km/h may be considered to match those at lower velocities. The maximum force of car A in the overload test and in the 64 km/h ODB test are near matches of around 330 kN. However, the energy absorption in this force level can be obtained from the overload test. For car B, on the other hand, the maximum force is 360 kN in the overload test, which is higher than that in the ODB test (329 kN). Therefore, whereas car B did not reach its maximum passenger compartment strength at 64 km/h, the overload test makes it possible to determine the actual passenger compartment strength.

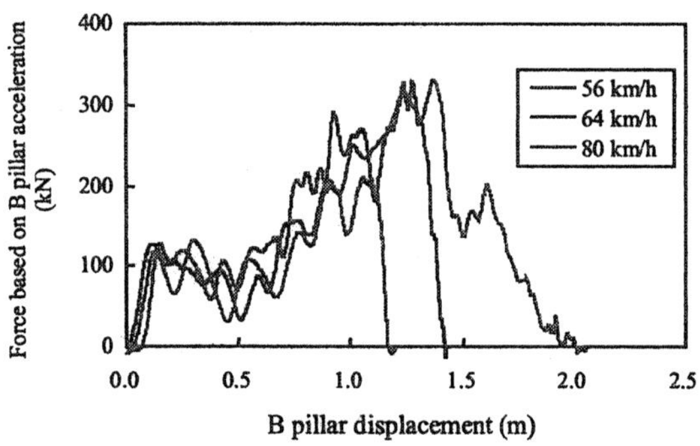

Figure 21. Force-displacement characteristics for various impact velocities (car A).

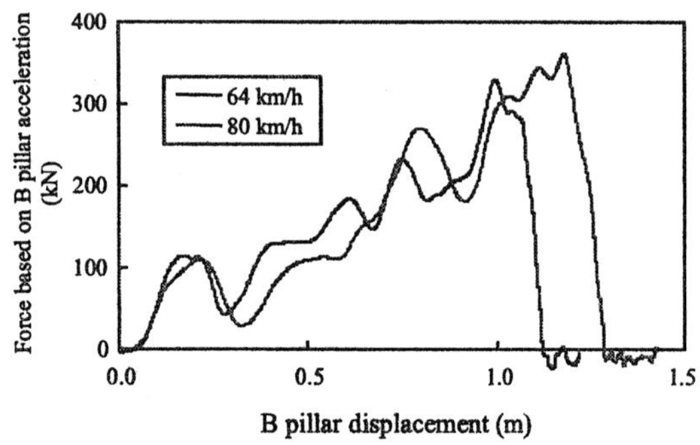

Figure 22. Force-displacement characteristics for various impact velocities (car B).

The maximum barrier forces in overload test and 64 km/h ODB test were compared in Figure 23. In ODB tests, maximum barrier force increases with vehicle mass. Comparing these values with the maximum force in overload tests of car A and car B, we find that they correspond to the maximum loads in 64 km/h ODB test of a 1,620 kg car A and a 1,920 kg car B. Therefore, we may infer that the passenger compartments of car A and car B will not collapse before vehicle deformation reaches the extent observed in 64 km/h ODB test. The figure also indicates that car A and car B passenger compartments would not collapse in collisions with car C, and in actual car-to-car crash test, the passenger compartments of car A and car B were preserved.

Deformation is less uniform in actual car-to-car collisions than in an overload test. Since various deformation modes occur in car-to-car collisions, it is necessary to evaluate passenger compartment strength in overload tests on the basis of various parameters such as instrument panel intrusion and side sill strength. The differences in vehicle deformation observed in the overload test and car-to-car crash test also require further study.

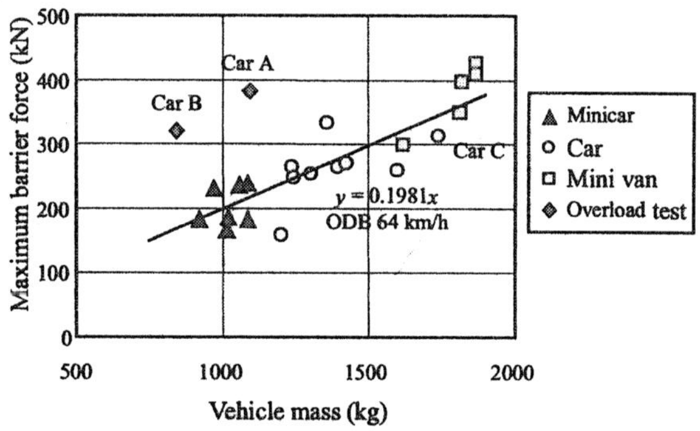

Figure 23. Barrier forces in overload test and 64 km/h ODB test (JNCAP).

<u>**MDB test**</u> - In theory, the MDB test is one way of reproducing car-to-car collisions in terms of acceleration and deformation. We conducted an MDB test with three barrier faces: the ECE R94 barrier face, the FMVSS 214 barrier face and a two-stage barrier. We compared the results with those from car-to-car crash tests.

Figure 24 shows acceleration-time histories. Since overlap was 50% in the car-to-car collisions and 40% in the MDB test, acceleration in this test was lower than in the car-to-car collisions. Override of the MDBs occurred for all barriers. In the MDB test with ECE barrier face, the acceleration of the car was high due to bottoming-out of the barrier face. Since the FMVSS barrier ran fully on top of the test vehicles, acceleration fell from 50 ms. Acceleration with the two-stage barrier was lower than in the car-to-car collisions, but its crash pulse was closest to the car-to-car collisions.

Figure 25 compares deformation in the MDB tests. Tests with the two-stage barrier face exhibited A pillar deformation, steering column movement and toe board intrusion most resembling those in the car-to-car collisions. In tests with the ECE and FMVSS barriers, MDB override resulted in greater deformation of the fender edge than of the longitudinal member.

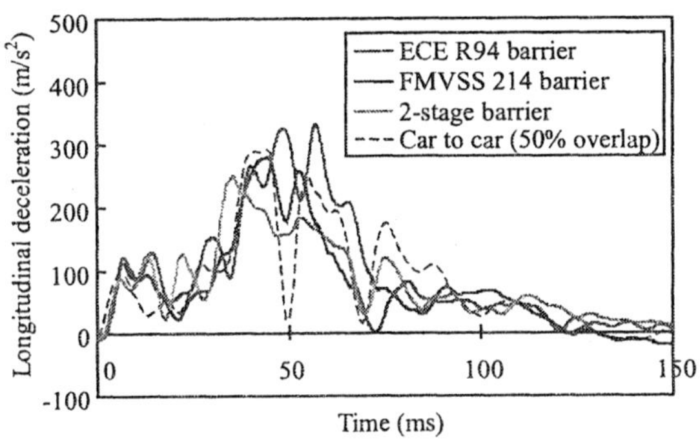

Figure 24. Car acceleration in MDB test with various barrier faces.

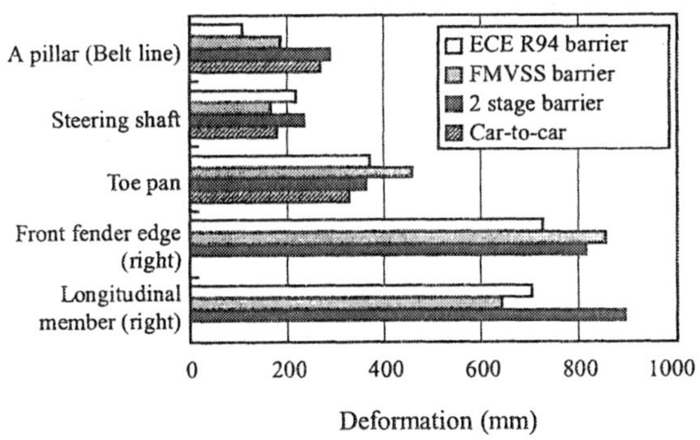

Figure 25. Car deformation in the MDB test with various barrier faces.

Vehicle behavior in an MDB test using a two-stage barrier is presented in Figure 26. With this barrier MDB override occurs past 60 ms, but acceleration already starts falling by this time (Figure 24) and did not appear to have a great effect on impact phenomena. With the ECE R94 and FMVSS 214 barriers, bottoming-out and override was starting at 40 ms, greatly affecting acceleration and deformation.

Considering acceleration, deformation and MDB overriding, crash tests with the two-stage barrier (180 mm ground clearance) reproduce car-to-car collisions more closely than with the other barrier face and ground clearance examined in the tests.

Figure 26. Vehicle behavior in MDB test (2-stage barrier).

DISUSSION

On the basis of fleet studies, accident studies and crash tests, we examined test procedures for the evaluation of compatibility. Many studies reported in the IHRA Compatibility Working Group confirm that good interaction is essential in preventing override and vertical mismatch of longitudinal members to ensure energy absorption for both cars [(5)]. Thus, in order to improve compatibility, it is necessary to control the absorption of energy by the vehicle front structure and maintain the passenger compartment integrity in good structural interaction. Test procedures must be selected with these considerations in mind. Further, test procedures must improve vehicle compatibility without lowering levels of self-protection. Therefore, we consider it important to use the following combination of three test procedures:

(1) Full frontal impact test against rigid walls with high resolution load cell measurement
(2) ODB test
(3) Overload test

Remarks and some possible criteria for these tests are as follows:

(1) We did not examine the load distributions in full frontal tests in our research, but it will be effective for good interaction to evaluate the homogeneity from the force distribution of barrier load cells. To allow homogeneous evaluation, a center of force and coefficient variation can be among the criteria.

(2) ODB test is capable of evaluation of energy absorption in the front of vehicles, acceleration pulse and injury risk to occupants. To evaluate the self-protection and aggressivity of the car, the maximum force level in this test can be also one of the criteria.

(3) For correct measurement of passenger compartment strength, both acceleration of the passenger compartment and barrier force may be necessary. Further research is required on impact velocity, methods of measuring passenger compartment strength, thresholds and differences in deformation modes in car-to-car collisions.

This combination of tests allows control of vehicle interaction, the extent of acceleration and intrusion, and is highly likely to lead to greatly improved vehicle compatibility. However, the crash conditions and criteria that should be used in the individual tests require more detailed study.

It is possible that MDB test may reproduce the acceleration and vehicle deformation observed in car-to-car collisions. However, MDB override and barrier bottoming-out occurred in the three tests we performed, greatly affecting the behavior and deformation mode of the test vehicles. More detailed research is required to resolve these problems, and at this time we are considering tests with fixed deformable barriers (FDB) based on three tests described above to be more capable of evaluating vehicle compatibility.

CONCLUSIONS

This paper summarizes the results of fleet studies, structural surveys and crash tests reported from Japan to the IHRA Compatibility Working Group. Our conclusions are as follows:

1. From the number of vehicles registered as well as accidents in Japan, we find that passenger cars are the priority target for vehicle compatibility in the Japanese transportation environment.
2. We surveyed front-end vehicle dimensions of car models and identified the average dimensions of such members as the longitudinal member and cross member.
3. Vehicle compatibility requires the evaluation of passenger compartment strength. Overload tests were performed using a minicar and a small car. This kind of high-velocity test can be one of the methods to evaluate the passenger compartment strength.
4. MDB tests were performed with three different barrier faces. In spite of the benefit of the MDB test to reproduce acceleration and deformation, there are problems with MDB barrier override and bottoming-out, and these proved to greatly affect test vehicle behavior and deformation.
5. Vehicle compatibility requires control of energy absorption by the vehicle front structures and intrusion into the passenger compartments in good structural interaction between the cars. In this light, we consider the combination of full frontal impact tests against rigid walls to measure load distribution, ODB test and overload test to be appropriate for the evaluation of vehicle compatibility.

REFERENCES

1. Gabler, C. H., Hollowell, T. W., "NHTSA'S Vehicle Aggressivity and Compatibility Research Program", 16th ESV, Paper No. 98-S3-O-01, 1998.
2. Steyer, C., Diaboine, A., Delhommeau, M., Delannoy, D., "Taking Compatibility Aspects into Account When Designing Vehicles", VDA Technical Congress, Frankfurt, 1999.
3. Zobel, R., 'Accident Analysis and Measures to Establish Compatibility', SAE Paper 1999-01-0065, SAE Congress, 1999.
4. Sugimoto, T., Kadotani, Y., Ohmura, S., "The Offset Crash Test–A Comparative Analysis of Test Methods", 16th ESV, Paper No. 98-S1-O08, 1998.
5. Edwards, J., Happian-Smith, J., Byard, N., Davies, C., Hobbs, A., "Compatibility–the Essential Requirements for Cars in Frontal Impact", IMechE, Conference Transactions, Vehicle Safety 2000, pp.3-17, 2000.

986008

THE OFFSET CRASH TEST -A COMPARATIVE ANALYSIS OF TEST METHODS

Tomiji Sugimoto
Yoshiji Kadotani
Shigeru Ohmura
Honda R&D Co., Ltd.
Japan
Paper Number 98-S1-O-08

ABSTRACT

This research will discuss the issue of how the currently used frontal crash tests correlate to actual accidents. The following data will be presented in relation to this:
1. Results of offset crash tests now being conducted, and results of vehicle-to-vehicle crash tests, especially results of crash tests in which the vehicles have different weights.
2. Why do such differences occur?
3. Differences between the results of tests with moving deformable barriers (MDB) which are being studied by the National Highway Traffic Safety Administration (NHTSA) and results of vehicle-to-vehicle crash tests.
4. Results of modifications to test methods
The following aspects of the above mentioned issues will be discussed:
1. Important items and information to be considered in studying crash test methods to be used in the future.
2. Information which needs to be taken into consideration in developing cars in the future.

INTRODUCTION

In response to the need to improve crashworthiness, various countries have proposed and implemented a variety of test methods in order to provide regulations and safety information. Recently, offset crash tests have come into widespread use in addition to full frontal crash tests or oblique impact tests. In actual accidents, chassis deformation and intrusion into the cabin has been observed in many cases. In addition, passenger deaths have been reported in conjunction with chassis and cabin deformation. Therefore, with the primary objective of securing cabin space and thereby reducing passenger deaths, a great deal of research has been conducted on offset crash tests, as well as on the body frame structure in order to improve passenger survivability. Full frontal crashes are considered useful for evaluating the performance of safety devices which restrain passengers during a crash. Offset crashes are considered appropriate for evaluating cabin deformation caused by the impact loads on the vehicle during a crash. As has already been described in a wide range of literature on the subject, in a certain sense, these two test methods involve evaluating mutually contradictory phenomena. This is an extremely serious and difficult problem for automobile development engineers who are attempting to improve crashworthiness. Issues which will be critical in discussions of vehicle crashworthiness in the future are:
(1) Does each of these evaluation techniques provide methods and criteria which are suitable for increasing vehicle crashworthiness?
(2) Which of these test methods is useful in developing and evaluating a vehicle?

A variety of configurations and conditions have been proposed, especially for offset crashes, so further research and discussion are needed.

An area which is currently a main focus of concern is the types of considerations that are needed for vehicle designs which will provide compatible crashworthiness for both small cars and large cars. This issue is especially important for vehicles which are evaluated with these methods.

This research seeks to verify how crash test methods, either full frontal or offset frontal crashes, are associated with actual accidents. This research also discusses what needs to be done in the future.

BACKGROUND

Among actual accidents, deaths of passengers riding in vehicles may be classified as shown in Figure 1 for Japan and the U.S.

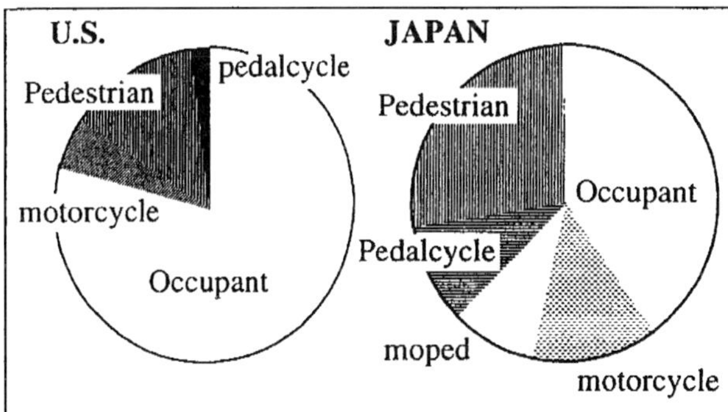

Sources: US; FARS 1995
JAPAN; The Japanese police traffic accident statistics since 1995

Figure 1. Fatalities in traffic accidents

Fatalities of passengers riding in vehicles may be further categorized by the type of accident. There are two general classifications: single-vehicle accidents and vehicle-to-vehicle accidents. The breakdowns for these classifications are shown in Figure 2. As shown in the figures, about half of the accidents are single-vehicle accidents and the other half are vehicle-to-vehicle accidents.

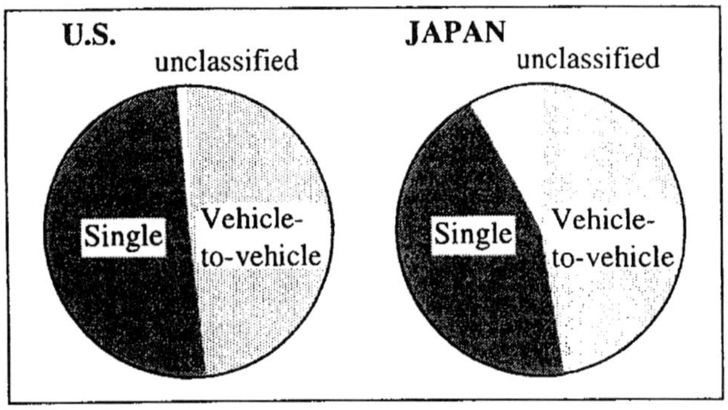

Figure 2. Classification of fatal collisions

Figure 3 presents the numbers of cumulative fatalities and the corresponding barrier equivalent speeds.

Approximately 90% of the cumulative fatalities occur at speeds of 50-55km/h or less.

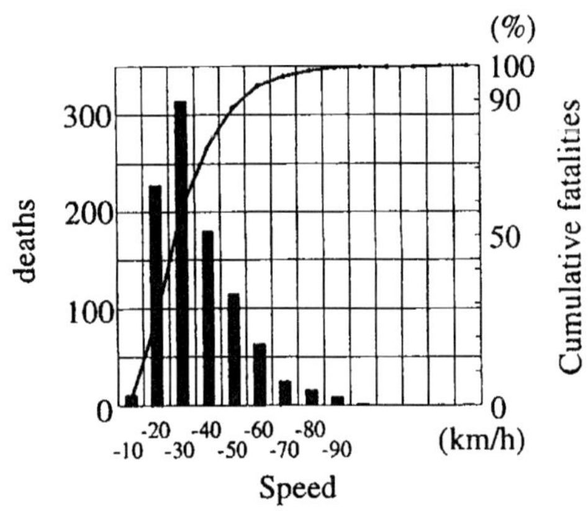

Source: NASS CDS 1991-1995
head on collisions with belted

Figure 3. Barrier equivalent speed

The conditions for the tests currently being conducted were established based on such information.

We will now consider which types of actual accidents each of the test methods is applicable to. The discussion will be simplified in order to maintain a comprehensive focus on current problems and future trends. For further information on the detailed verifications, the reader is referred to the results of research conducted by various researchers in the course of establishing each of the crash test methods. The frontal crash test methods which are currently used in Japan, the U.S., Canada, Europe, and Australia are listed in Table 1.

Table 1. Test configurations

	JAPAN	U.S.	EUROPE	AUSTRALIA
Regulation	Full flat 50km/h	Full flat 48km/h R/L 0°~30° 48km/h	ODB 56km/h	Full flat 48km/h
Consumer safety information test	Full flat 55km/h	Full flat 56km/h ODB 64km/h	ORB 55km/h ODB 64km/h	Full flat 56km/h ODB 64km/h

The common types of full frontal crash tests into a flat, rigid barrier, are the regulation tests used by the NHTSA in the U.S., Transport Canada in Canada, the Federal Office of Road Safety (FORS) in Australia, and the Ministry of Transport in Japan. This same type of test is also used in the New Car Assessment Program (NCAP), which serves to provide consumer safety information and incorporates some changes (e.g., a higher crash speed). These test methods will now be considered in relation to actual accidents. In vehicle-to-vehicle accidents, vehicles of the same weight may collide head on with almost no offset. In single-vehicle accidents, the vehicle may collide head-on into an object such as a structure. In actual accidents where the vehicle collides into a structure, vehicles may collide into trees, utility poles, or experience under-ride impact into trucks in addition to colliding into flat objects. At the present time it is very difficult to narrow down correlation with macro data. It is difficult to postulate the exact extent to which this test method covers actual accidents. However, it is possible to infer from the statistics on cumulative fatalities that there are cases in which passengers are subjected to rather strong impacts during collisions.

In light of such considerations, the full frontal rigid barrier crash test methods seem extremely useful for evaluating life saving capabilities which would reduce passenger injuries during extremely strong actual impacts. This test method is advantageous in that it allows evaluations under conditions in which driver and passenger impact severities are nearly identical. The offset crash tests conducted in the past few years may be broadly divided between offset rigid barrier (ORB) crash tests and offset deformable barrier (ODB) crash tests.

In the case of offset rigid crashes, offset crashes between vehicles of the same weight during vehicle-to-vehicle accidents and offset crashes into structures during single-vehicle accidents are covered by this method and are considered applicable. However, as in the case of full frontal crashes, there is not a clear association between offset rigid crashes and collisions into trees or utility poles, or under-ride impact into trucks.

In the case of offset deformable crashes, the results of experiments replicating vehicle-to-vehicle accidents have been used to establish test conditions, such as collision speed and the specifications of the honeycomb (a deformable device), as has been referred to in European Experimental Vehicle Committee (EEVC) and Insurance Institute of Highway Safety (IIHS) research reports. However, there has been little in the way of verification under conditions in which the vehicles involved have different weights. Therefore, in this study we would like to compare the results of such offset deformable crashes with the results of vehicle-to-vehicle tests based on vehicles with different weights. The need for verification using offset crash tests is to determine how well passenger space in the vehicle cabin is protected. This test serves to evaluate cabin deformation, and resistance to intrusion as a result of the collision. Thus, this method can be used to verify how well the cabin and frame in the engine room compartment are able to absorb the impact energy from the collision and distribute the impact forces. As reported in the for EEVC and IIHS research reports, the specifications of the offset deformable barrier (honeycomb) which is used with this test simulates the stiffness of the structure at the front of a vehicle of nearly average weight (normally called a mid-size vehicle). In terms of actual vehicle-to-vehicle accidents, this test seems to simulate vehicle-to-vehicle collisions involving vehicles of average weight or less.

TEST RESULTS

Actual vehicle crash tests were conducted under these offset conditions. Figure 4 illustrates the vehicle deformation results of offset rigid collisions. The offset rigid crash test method (ORB) was used by Auto Motor Sport, a German magazine. Figure 5 illustrates the vehicle deformation results of offset deformable collisions.

Figure 6 compares the deformation results of vehicle-to-vehicle offset crash tests in which both vehicles weighed approximately 1500kg. The vehicle-to-vehicle crash test conditions were a speed of 56km/h for both vehicles and an offset of 50%

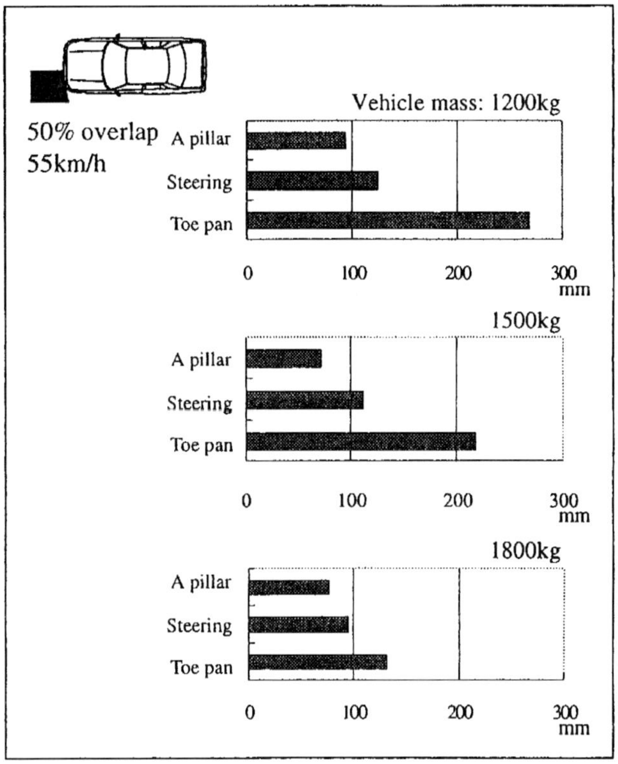

Figure 4. Vehicle deformation (ORB)

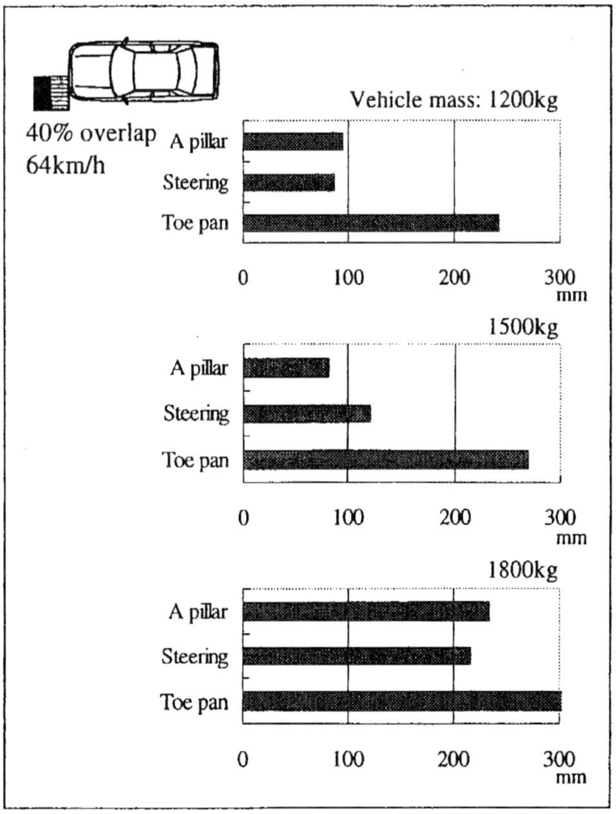

Figure 5. Vehicle deformation (ODB)

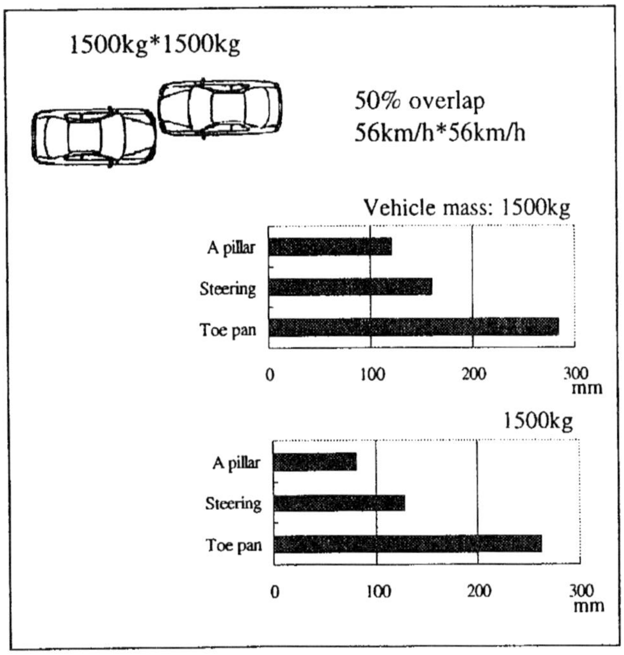

Figure 6. Vehicle deformation (vehicle-to-vehicle)

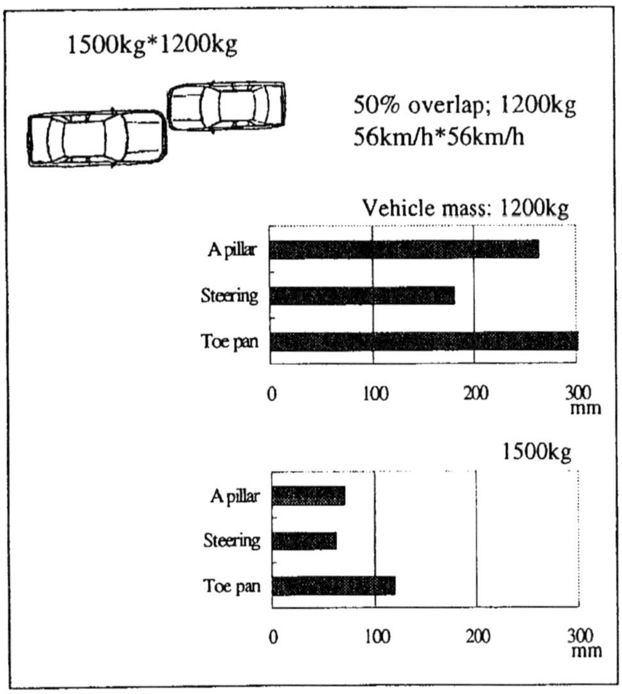

Figure 7. Vehicle deformation (vehicle-to-vehicle)

Figure 7 compares the deformation results of the same type of vehicle-to-vehicle offset crash tests in which one of the vehicles weighed approximately 1200kg, and the other approximately 1500kg.

Figure 8 compares the deformation results of the same type of vehicle-to-vehicle offset crash tests in which one of the vehicles weighed approximately 1200kg, and the other approximately 1800kg.

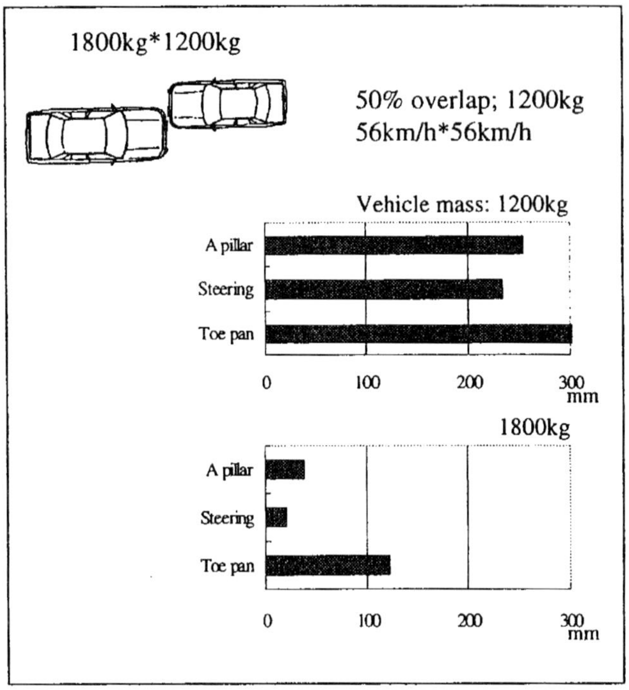

Figure 8. Vehicle deformation (vehicle-to-vehicle)

ANALYSIS

As illustrated above, the results for vehicle-to-vehicle offset crash tests in which both vehicles weighed approximately 1500kg were consistent with the offset deformable crash test results. When the vehicles had different weights, there is a significant difference between the vehicle-to-vehicle crash test results and the barrier crash test results.

As mentioned above, offset rigid barrier collisions simulate collisions between vehicles of the same weight, or collisions into structures. In contrast, offset deformable barrier collisions are essentially offset collisions between vehicles of average weight. However, the results of the offset deformable barrier crash tests indicate that if the colliding vehicle weighs more than average, (e.g.,1800kg) a bottoming out phenomenon will occur due to the characteristics of the deformable barrier (i.e., the honeycomb). As a result it would seem that an actual vehicle-to-vehicle crash is not simulated in such cases. Similar problems have already been pointed out among researchers; this will remain a topic for future study.

Nonetheless, this cannot be set aside as a simple "issue". In other words, vehicles which are developed in order to obtain good evaluation results using such test methods may create a number of problems under actual road conditions.

One such problem is an increase in vehicle weight. It is inevitable that weights will increase as a result of improvements in crashworthiness. Unfortunately, excessive increases in vehicle weight remain a significant problem. Specifically, vehicles whose structures are designed based on test conditions and evaluation criteria which are significantly different from actual accident conditions will not contribute appropriately to efforts to improve crashworthiness under actual road conditions. Also this is a problem of compatibility in vehicle-to-vehicle crashes. Along with the need to protect the vehicle of a person driving in mixed traffic, it is also necessary to protect the other vehicle in an accident. This capability may be an important issue in the future. Among actual accidents, total fatalities are divided approximately evenly between single-vehicle accidents and vehicle-to-vehicle accidents. It is necessary to protect passengers in both of these types of accidents. In particular, during vehicle-to-vehicle collisions, it is necessary to consider the safety of the other driver -- not just the driver in the car which is being designed. Results from the vehicle-to-vehicle crashes of different weights specifically show this problem. Figure 9 illustrates this phenomenon graphically. As vehicle weight increases, the stiffness of the vehicle front increases.

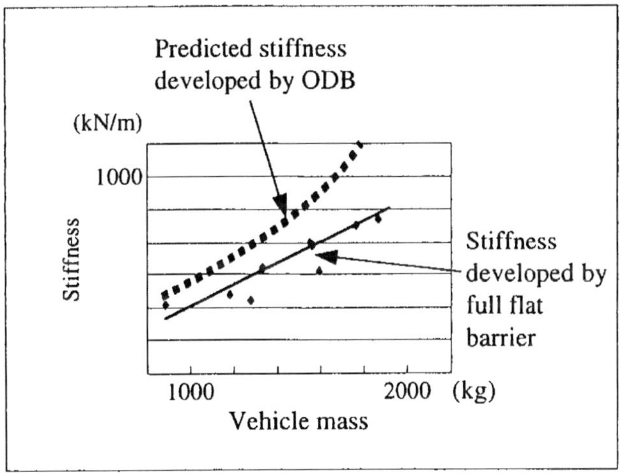

Figure 9. Prediction of stiffness

The test results done on each vehicle in the U.S. with an NCAP full frontal barrier show a strong correlation between vehicle weight and stiffness. In other words, an increase in vehicle weight can be inferred to lead to an increase in aggressiveness toward the other vehicle. As used here, the term 'vehicle stiffness' is defined as the slope of the load on the chassis as derived from an accelerometer attached to the cabin floor on the chassis. Figure 10.

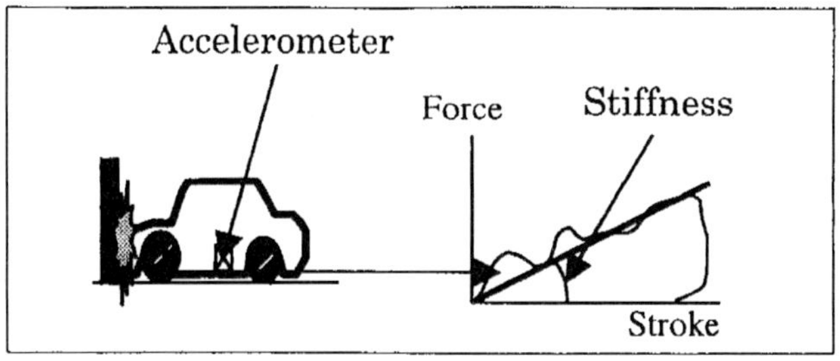

Figure 10. Definition of vehicle stiffness

Next we used same-weight vehicles as described above to verify the recent test method of the U.S. NHTSA, which is being researched based on vehicle-to-vehicle crashes. The 56km/h vehicle-to-vehicle crash results are shown in Figure 11.

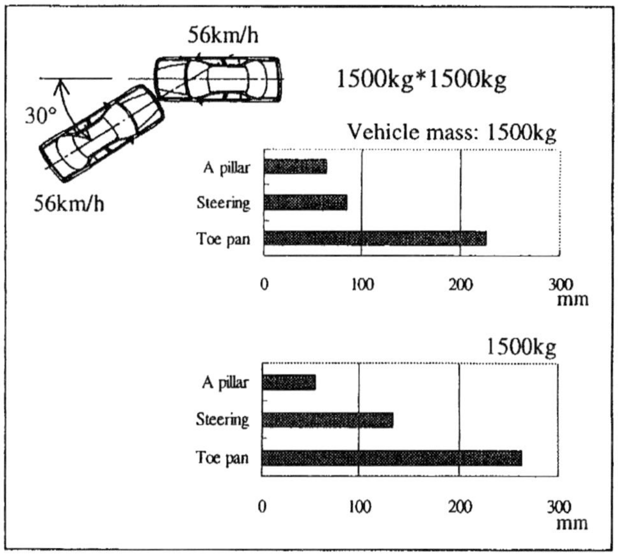

Figure 11. Vehicle deformation (vehicle-to-vehicle)

Figure 12 illustrates the results of a stationary vehicle crash test using a 112km/h moving deformable barrier (MDB).

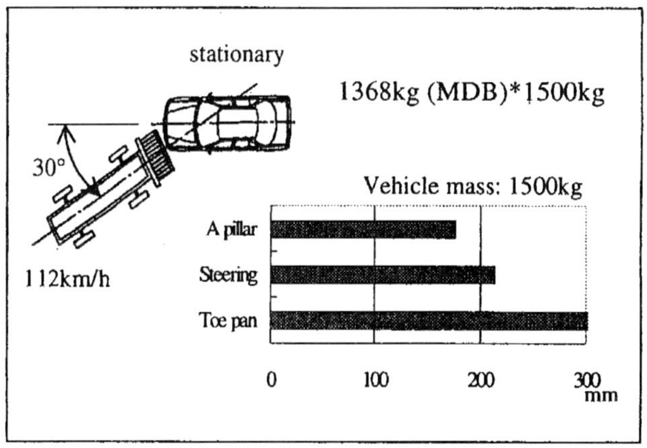

Figure 12. Vehicle deformation (vehicle-to-vehicle)

As indicated by the diagram, this test method clearly involves a vehicle-to-vehicle type of accident. One way it is different from the frontal offset crashes discussed thus far is that the offset is oblique. The second is to use an MDB. The MDB weighs 1368kg, which is the average vehicle weight in the U.S. This method appears to be based on the type of accident which is likely to occur frequently under actual road conditions. Note that the NHTSA research results should be checked for details regarding what types of actual accident situations are covered. As illustrated in Figures 11 and 12, a comparison of deformation amounts in the vehicle which is collided into shows that deformation for MDB and a vehicle is much greater than deformation between one vehicle and another.

One reason for this can be clarified by comparing the amount of deformation in the deformed area on the colliding vehicle. This comparison shows that there are problems in the characteristics of the barrier, i.e., the honeycomb, similar to the results for the offset deformable crash tests. As in the EEVC and IIHS tests, this problem seems to be due to honeycomb bottoming out, i.e., the stroke is significantly different than that of actual vehicles. Figure 13 illustrates the force (deceleration) vs. displacement characteristic in an actual vehicle compared to the results obtained in a test using a honeycomb.

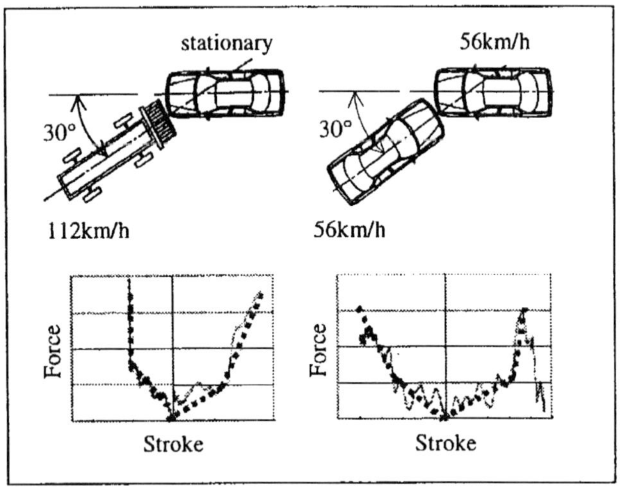

Figure 13. Comparison of Force-Stroke characteristics

Figure 13 also indicates a clear difference between the results. Assuming the collision speed simulation parameter is physically and theoretically correct, the honeycomb characteristics are a definite problem with this test method. This test method has other problems as well: reproducibility and practicality. Since this test method involves an oblique crash test, there is inconsistency in the amount of offset. And it is almost impossible to conduct the high MDB test speed in an ordinary indoor laboratory, so it is not well suited to third-party evaluation tests, including compliance. Then a test method which would theoretically solve the problems discussed above was devised. This test method, illustrated in Figure 14, was developed with consideration for reasonableness, faithfulness, reproducibility, practicality, and aggressiveness evaluations.

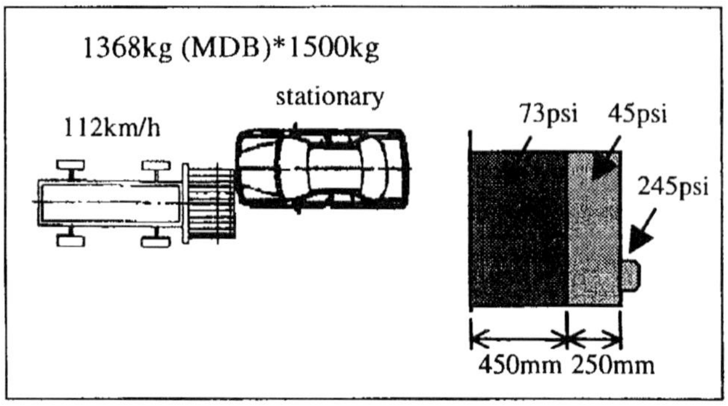

Figure 14. Test method of MDB-to-vehicle

Like NHTSA, for the MDB we selected the average weight which was most likely to be encountered under actual road conditions. We tried using a compound honeycomb consisting of a honeycomb which is average or has a hardness that is nearly the same as the stiffness of the engine rooms of vehicles which are commonly sold in the U.S., plus a honeycomb with stiffness characteristics similar to cabin stiffness. A relative MDB speed between 100km/h and 120km/h would simulate vehicle-to-vehicle collision speed of approximately 56km/h. In this test we used a speed of 112km/h. In order to minimize inconsistency in the data caused by the test method, we decided to make the collided vehicle stationary in a frontal offset collision. Some evaluations may consider an oblique collision to have a better correlation to actual road conditions, but oblique collisions were not used in this test. The test results are shown in Figure 15.

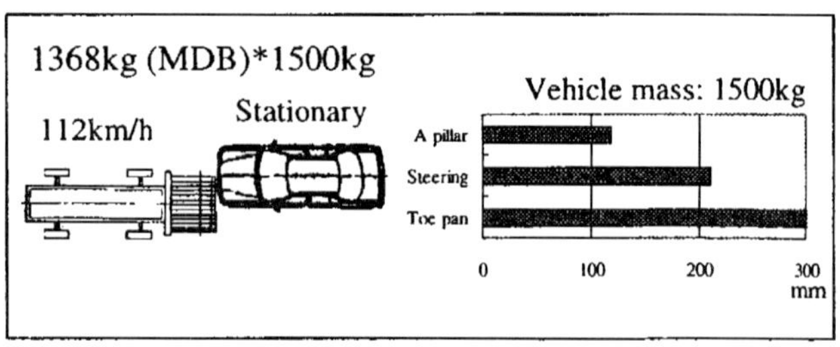

Figure 15. Vehicle deformation (MDB-to-vehicle)

The vehicle deformation approaches the test results in Figure 15, but the amount of deformation for the steering wheel is still larger. This seems to be due to the fact that the MDB rose onto the collided vehicle. There are two further problems with this test method. One is that it does not solve the difficulty of conducting the test in an ordinary indoor laboratory. The second problem is that the use of a compound honeycomb comprising two different honeycomb types makes it necessary to verify whether the method is acceptable in terms of production technology (including reproducibility), and whether the desired characteristics can be obtained.

CONCLUSION

The offset deformable barrier tests currently conducted using honeycombs are suitable for evaluating vehicle safety in vehicle-to-vehicle accidents involving vehicles which weigh approximately 1500kg or less. However, the results are not necessarily consistent with actual accidents in cases where the vehicles weigh more than approximately 1500kg. This is due to a problem with the specifications of the honeycomb, which is the deformable device. Specifically, the force-stroke characteristic of currently used honeycombs is not suitable for vehicle-to-vehicle crash tests with vehicle weights of 1500kg or greater. Our results reconfirm the recognition and observation by others that this is a bottoming out problem. In addition, it was learned that vehicle-to-vehicle offset collisions involving an MDB are not necessarily consistent with actual accidents in terms of what actually happens (e.g., the MDB rises onto the test car). Therefore further research is needed.

DISCUSSION

As described above, test methods involving deformable barriers have been proposed and are used to simulate vehicle to vehicle accidents. However, based on these tests results, the barrier characteristics do not always seem to replicate actual accidents. In cases where there is a difference in weight between vehicles, as is commonly found in vehicle-to-vehicle accidents, the heavier vehicle will suffer less deformation than the lighter vehicle. This has been confirmed experimentally, so test methods which provide different results are clearly problematic in a number of respects. Specifically, there is the problem of the collision speed, which is not related to the vehicle weights, and the related honeycomb characteristics. In the future we believe it will be necessary to establish appropriate test methods based on further research.

Another problem which may arise is that vehicles which are developed using such problematic test methods may not be suitable in terms of compatibility in vehicle to vehicle collisions -- an issue which is expected to be important in the future. In particular, heavier than average vehicles which are sold in each market have the potential to increase aggressiveness toward small and lighter-weight vehicles.

This paper presents research on vehicle to vehicle tests involving an MDB, and compares these tests to ODB crash tests which are currently used. Further research will be needed in the future on criteria for evaluating vehicle aggressiveness.

Collision accidents are extremely complex. For this reason, it is necessary to have a number of methods for evaluating crashworthiness -- not just one method. In particular, it is impossible to use a single test method to evaluate mutually contradictory phenomena (i.e., single vehicle crash protection evaluations and securing cabin space in vehicle-to-vehicle crashes). Therefore, care must be taken in publishing test results supposedly serving as safety information.

REFERENCES

1. EEVC Working Group 11 Report on the Development of a Front Impact Test Procedure. R W Lowne on behalf of EEVC Working Group 11 (TRL) 1994 ESV 94-S8-05
2. Offset Frontal Impacts - A Comparison of Real-World Crashes with Laboratory Tests. Brian O'Neill, Adrian K. Lund, David S. Zuby, Charles A. Preuss(IIHS) 1994 ESV 94S-S4-O-19
3. An Examination of Different Test Procedures for Frontal Offset Crashes. Sheldon L. Stucki, William T. Hollowell (NHTSA)

2003-06-0168

A STUDY OF COMPATIBILITY TEST PROCEDURE IN FRONTAL IMPACT

Satoshi Takizawa
Tomiji Sugimoto
Hideki Suzuki
Honda R&D Co., Ltd. Tochigi R&D Center
Japan
Paper Number 437

ABSTRACT

The purpose of this study is to examine compatibility test procedures proposed in the IHRA Vehicle Compatibility Working Group. Various crash tests were conducted with different vehicle weights and stiffness in our previous study, and each of the compatibility problems, namely mass; stiffness and geometric incompatibility were identified in these tests. In order to improve the compatibility, it is necessary to evaluate and control relevant vehicle characteristics of compatibility in test procedures.

According to the IHRA study, relevant aspects for compatibility in frontal impact are

- Good structural interaction
- Frontal stiffness matching
- Maintaining passenger compartment integrity
- Control the deceleration time histories of impacting cars

Possible candidate test procedures to evaluate four items given above are as follows:

1) A full width Load Cell Barrier test in which the load cell data was analyzed to evaluate structural interaction of vehicles, and some geometric indices such as Average Height Of Force, interaction area, and stiffness indices were measured. It was found from the analysis that several candidate metrics could be identified given their high correlation with laboratory vehicle-to-vehicle crash tests.

2) An MDB-to-vehicle test, which allows the mass ratio to be taken directly into account. Potentially it can generate a realistic delta V and vehicle acceleration pulse. It has been recognized that the MDB could be used as a representative of an actual vehicle, and it provides more flexibility in compatibility test procedures. MDB-to-vehicle tests were conducted to confirm the reproducibility of vehicle-to-vehicle tests; the test results and analysis are reported in this paper.

INTRODUCTION

Crash compatibility can be defined as the ability of a vehicle to help protect not only its own occupants, but other users as well. Compatibility in vehicle-to-vehicle crashes has become an issue in countries with various vehicle compositions; thus compatibility is strongly affected by fleet composition. The National Highway Traffic Safety Administration (NHTSA) reported the aggressivity issue of sport utility vehicles (SUV) and light trucks and vans (LTV) in the U.S. fleet [1]. In Japan, the sales volume of small and medium sedans has been decreasing over the past decades, and the sales of mini-cars, station wagons and minivans have been increasing. As the range of the vehicle composition has been spread widely in Japan, the concern with crash compatibility has been growing for the last several years.

After the International Harmonized Research Activity (IHRA) was set up, a number of studies have been made on compatibility between vehicles in the IHRA Vehicle Compatibility Working Group [2]. Referring to the IHRA work, various crash tests were performed with different vehicle weights and stiffness in our study, and each of the compatibility problems, namely mass; stiffness and geometric incompatibility were identified in these tests.

A major focus of compatibility research is a development of a laboratory test procedure to evaluate compatibility. Within the IHRA Compatibility WG, a number of possible candidate test procedures have been proposed for a frontal impact test. We have studied each of the proposed test procedures, and the present development activity is being focused on both a full width barrier test with load cells and an MDB test. In order to improve compatibility, it is necessary to evaluate and control relevant vehicle characteristics of compatibility in test procedures. According to the IHRA study, relevant aspects for compatibility in frontal impact are

- Good structural interaction
- Frontal stiffness matching
- Maintaining passenger compartment integrity
- Control the deceleration time histories of impacting cars

The work described in this paper provides a comparative analysis of the vehicle-to-vehicle test and vehicle-to-fixed/moving barrier test. The full width barrier test with load cells is mainly used to evaluate the aggressivity and the MDB test is intended to evaluate the self-protection. The barrier load cell data obtained from full width rigid barrier is analyzed to evaluate the aggressivity of a SUV/LTV, and the MDB-to-Car impact was compared to the SUV-to-Car impact.

SUV/LTV-TO-CAR CRASH TESTS

Test configuration

Figure 1a,b and Table 1 show the selected vehicles and their associated weight used in SUV/LTV-to-Car impacts. The target vehicle was a compact four-door sedan representing a small passenger car in the US fleet. SUV and LTV with different mass and geometry were selected for striking vehicles in SUV/LTV-to-Car impacts. SUV/LTV-to-Car crash tests have been performed with these cars impacted at 50 km/h each. The offset ratio for this test was 50% width of the compact sedan. Although a Hybrid III 50th percentile male dummy for the driver and a Hybrid III 5th percentile female dummy for the passenger was used in these SUV/LTV-to-Car impacts, however we limit the discussion to the structural behavior of these vehicles.

Figure 1a. SUV-to-Car impact.

Figure 1b. LTV-to-Car impact.

Table 1
Selected vehicles and test weight

Vehicle type	Vehicle test weight	Mass ratio
Compact Sedan	1310 kg	1.0
SUV	2205 kg	1.68
LTV	2680 kg	2.04

Figure 2a,b and Table 2 show the geometrical alignment of the two bullet vehicles compared to the target vehicle. Regarding the overlap of their front longitudinal member, there is no clear difference between the SUV-to-Car and LTV-to-Car, however the SUV has an engine sub-frame as an extra load path of the impact. In terms of structural interaction, the sub-frame could reduce the structural mismatch between the SUV and the compact sedan.

Figure 2a. SUV-to-Car.

Figure 2b. LTV-to-Car.

Table 2
Dimension of vehicle front-end structures

	Car	SUV	LTV
Longitudinal Top Height (mm)	510	550	590
Longitudinal Bottom Height (mm)	412	460	479
Sub-Frame Height (mm)	181	312	
Longitudinal Overlap (mm)		50	31

Crash test results

The deformation mode of the target vehicle is shown in figure 3a,b In the SUV-to-Car impact, the main element in the front structure of both vehicles absorbed kinetic energy and the passenger compartment integrity of the compact sedan was maintained. No over-riding phenomenon was seen in the SUV-to-Car impact. There was good structural interaction between the front longitudinal member of the passenger car and the sub-frame of the SUV.

However, in the LTV-to-Car impact the lack of the sub-frame of the LTV caused the structural mismatch between the two vehicles, thus it was a disadvantage for the vehicle aggressivity. Overriding phenomenon was seen in the LTV-to-Car impact. This overriding was caused by structural mismatch between the two vehicles. The LTV caused significantly more intrusion in the passenger compartment of the target vehicle compared to those caused by the SUV-to-Car impact. It was clear in the LTV-to-Car impact that there was little structural interaction due to geometrical difference that caused the LTV to override the target vehicle.

Figure 3a. Deformation of the target vehicle in the SUV-to-Car impact.

Figure 3b. Deformation of the target vehicle in the LTV-to-Car impact.

FULL WIDTH BARRIER LOAD CELL DATA ANALYSIS

Geometrical Compatibility

The LTV-to-car test result revealed that a poor structural interaction between two vehicles led to overriding. The difference in height between the vehicle front-end structures of the passenger car and the LTV is a good example to illustrate the geometric incompatibility. The geometric incompatibility must be examined in more detail.

Basically, the crash incompatibility is characterized by three vehicle factors:
(1) Mass Incompatibility
(2) Stiffness Incompatibility
(3) Geometric Incompatibility
Several studies have been made on a measurement of stiffness and geometric compatibility [(3), (4)]. NHTSA used a full width barrier test data with 36 load cells to evaluate stiffness and geometric compatibility [(5), (6)]. The full width rigid barrier test has been conducted in New Car Assessment Program (NCAP) in the US.

In this study, the load cell data of the NCAP test (36 Load Cells Barrier) and that of our in-house test (180 Load Cells Barrier) was used to assess the geometric and stiffness compatibility. The barrier load cell wall configurations are displayed in Figure 4a,b.

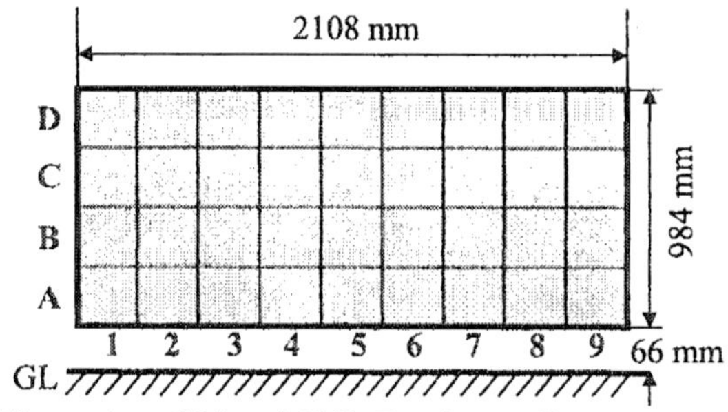

Figure 4a. 36 Load Cells Barrier configuration.

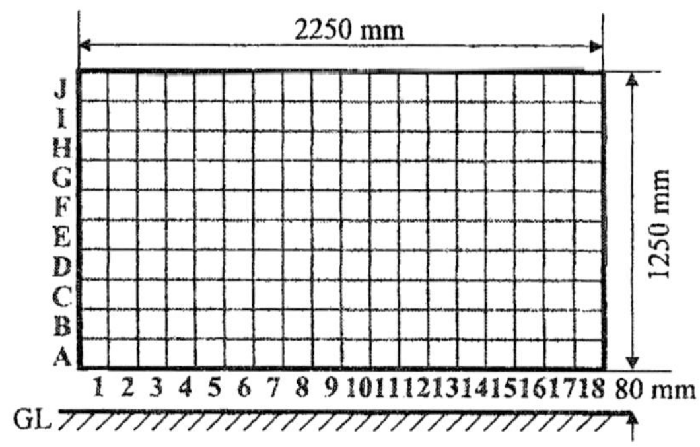

Figure 4b. 180 Load Cells Barrier configuration.

In order to assess the structural interaction, Average Height Of Force (AHOF) derived from the barrier load cell data was analyzed. The vehicle deflection was obtained by double integration of the acceleration derived from the accelerometer placed on the center of the vehicle. The barrier load cell data and the acceleration data were filtered with CFC 60 according to SAE J211 standard.

The Height Of Force (HOF) and the Average Height Of Force (AHOF) used in this study are computed as follows.

$$HOF(d) = \frac{\sum_{1}^{cell} Fi \times Hi}{\sum_{1}^{cell} Fi}$$

$$AHOF = \frac{\sum_{0}^{d} HOF(d) \times F(d)}{\sum_{0}^{d} F(d)} \quad \text{(d)=deflection}$$

The HOF at each deflection is computed from balancing the moments acting on each load cell with the total moment acting on the barrier. The AHOF is then averaged using the F (d) data as a weighing function. The reason why F (d) data was used as the weighing function in this study is that the height of force could be strongly affected by the Force-Deflection characteristics in the full width barrier test. Especially, the contact force of engine with load cell wall could act the important role for the AHOF analysis. The crush based AHOF is helpful to realize the vehicle structure and the location of the engine.

The AHOF for the LTV was compared with that for SUV as a predictable parameter of the overriding phenomenon. Figure 5 shows a plot of vehicle weight versus AHOF. The AHOF for the SUV and the LTV is generally higher than for the passenger cars. However, the AHOF for the LTV indicated similar value to that for the SUV. Thus the AHOF could not discriminate the overriding phenomenon of the LTV.

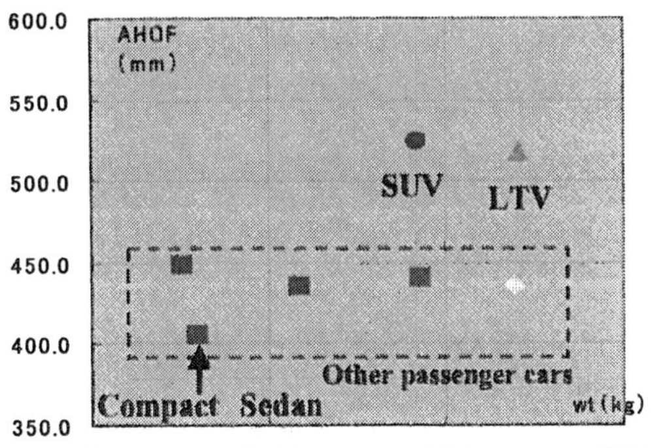

Figure 5. Vehicle weight vs AHOF

The variation on the HOF-Deflection curve was analyzed in detail. The HOF for the passenger car, SUV and LTV in the full width rigid barrier test are shown in Figure 6 as a function of deflection.

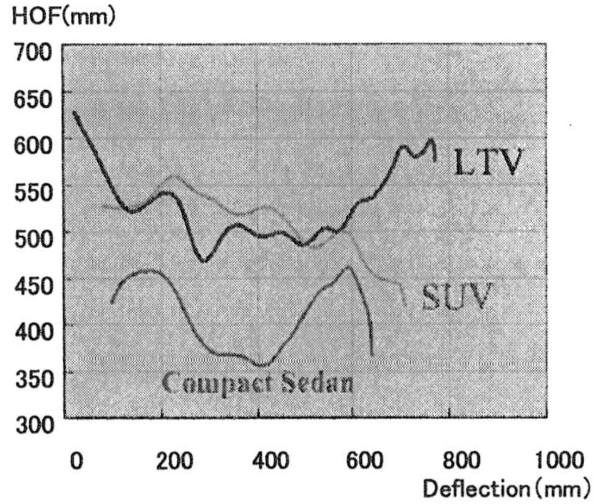

Figure 6. HOF-Deflection curve

Higher HOF values are indicated for the SUV within 200-500 mm at the HOF-Deflection curve. In order to investigate the relationship between the overriding and the HOF, each load cell output at 200-500 mm was investigated. The force distribution derived from the barrier load cells at 300 mm and 400 mm on HOF-D curve is shown in Figure 7. The test data for the LTV was from 36 load cells, and that for the SUV was from 180 load cells in this study. An important point to emphasize is that the 36-barrier load cell is about twice as large as 180-barrier load cell. If it sees roughly, similar force distribution was seen in the load cell output for both vehicles. The force distribution for the LTV at 300 mm demonstrates two of the main load paths clearly; these are the two longitudinal members.

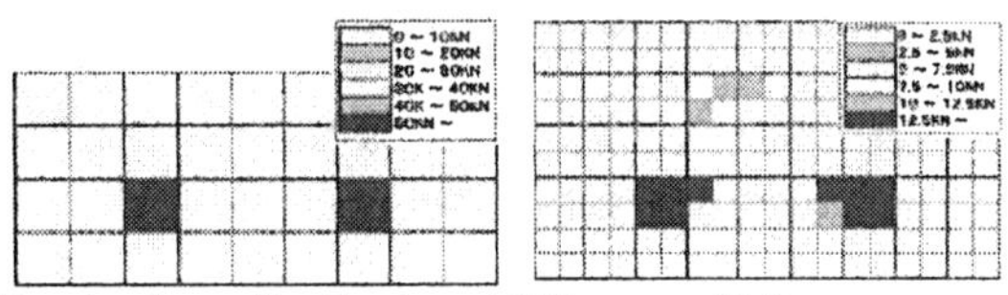

Barrier force distribution at 300 mm on F-D curve

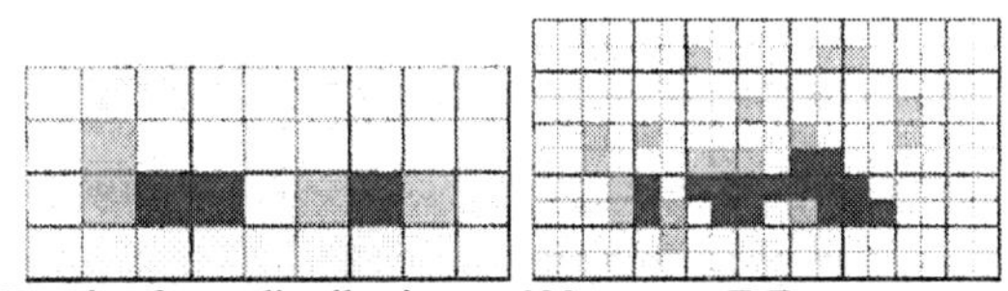

Barrier force distribution at 400 mm on F-D curve

Figure 7. Barrier force distribution for LTV and SUV.

Figure 8 shows the vehicle front structure of the LTV. The vertical distance from the bottom of the longitudinal member to the ground was 479 mm. The crush force of the longitudinal member of the LTV is measured by the second row from the bottom of 36 barrier load cells, whereas it is measured by fourth row of 180 barrier load cells. In that case, the HOF computed by the 36 load cells is different from that of the 180 load cells in that the moment arms for HOF is derived from the center height of the load cell.

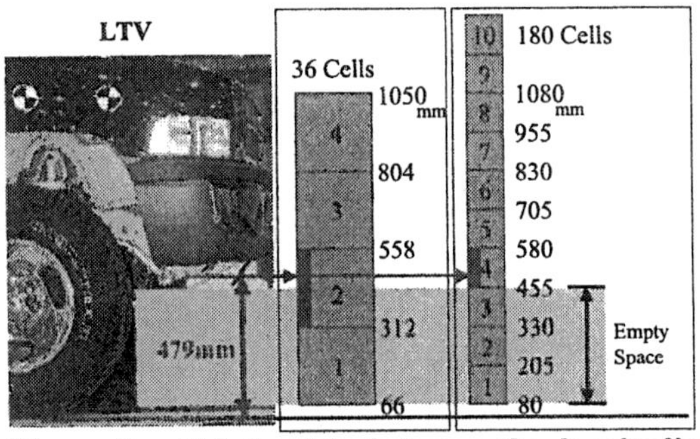

Figure 8. Relationship between the longitudinal member of the LTV and barrier load cell height.

In order to examine the AHOF, we have to clarify the influence of the load cell size first. By getting adjacent 4 load cells together in the 180 barrier load cells, imaginary 45 barrier load cells were created. (Figure 9) The total barrier load of the imaginary 45 load cells is exactly same as that of the 180 load cells. The HOF of the SUV computed by 45 load cells was compared with original HOF. It was found from the result that there was a considerable variation of the HOF that comes from the different load cell size. The result shows that the load cell size affects greatly the HOF because of the difference in the center height of the load cells. (Figure 10)

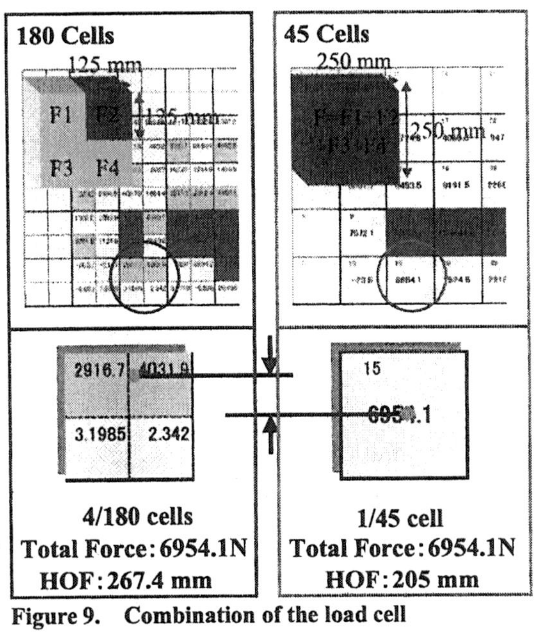

Figure 9. Combination of the load cell

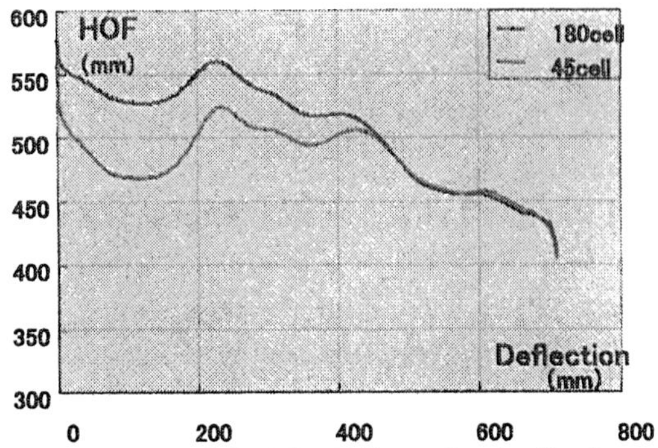

Figure 10. HOF of the different load cells size.

Let us now return to the discussion of the overriding phenomenon. It is important to note that the difference of load cell size affects the AHOF analysis, as stated above. Since 36 load cells are about twice as large as 180 load cells, 36 load cells were divided into imaginary 144 load cells tentatively. The measured barrier forces for the LTV from the 36 load cells were assigned to the imaginary 144 load cells. Assuming that the main reason for the overriding is caused by the structural mismatch between the main load paths for energy absorption, barrier load analysis was focused on the longitudinal member of the LTV. The bottom of the longitudinal member of the LTV was 479 mm in height; therefor it is reasonable that the 3rd row from the ground in the imaginary 144 load cells does not contact with the longitudinal member. The barrier force distribution was reflected according to the above reason, and the result is shown in Figure 11.

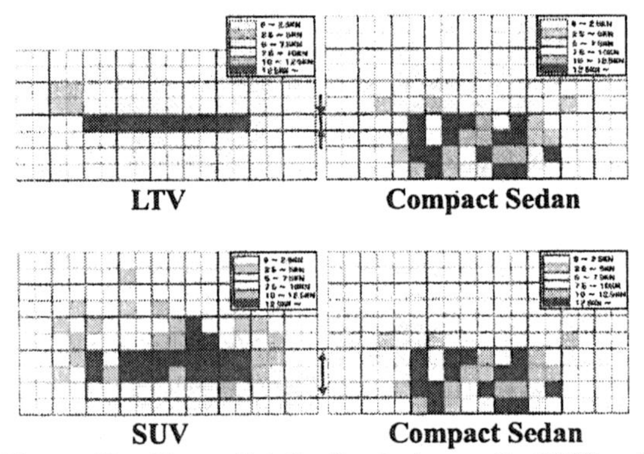

Figure 11. Force distribution between the LTV and the passenger car at 400 mm crush.

Compared with the SUV, only the two longitudinal members of the LTV place any significant force on to the load cell wall, and this was very localized. This was far from ideal and indicates a potentially aggressive structure. The SUV had a sub-frame as an extra load path, which provided a much more homogeneous force distribution. The force distribution of the LTV in height indicates that the amount of overlap for the structural interaction between the LTV and passenger car was only approximately 1 load cell row. It is quite likely that this poor interaction area for the interface force leads to the overriding of the LTV. Therefore, the vertical width of interaction area should be considered to be a geometric compatibility index.

Generally speaking, as the AHOF of the LTV becomes high, the structural interaction would decrease between LTV and passenger cars in frontal impact. From this point of view, one may say that the AHOF could be a geometric index for compatibility. However no overriding phenomenon was seen in the SUV-to-Car impact with reasonable interaction area between two vehicles, thus the interaction area could also be one of the important factors to prevent overriding. The SUV/LTV-to-vehicle impacts in this study revealed that the reasonable extent of the interaction area should be taken as a geometric compatibility index. (Figure12a,b)

Figure 12a. AHOF and Interaction Area.

Figure 12b. AHOF and Interaction Area.

Ideally, the AHOF and the interaction area should be matched completely to absorb energy, however it is necessary to study the reasonable extent of the interaction height and area. The geometrical alignments of the bumpers and the longitudinal members of passenger cars in the US fleet are derived from the US bumper standard. Additionally, a lot of passenger cars in other countries are having the similar longitudinal members in height, because they have a US model. This concentration of the longitudinal member in height provides an opportunity to encourage the resolution of the interaction area. SUV/LTV should respond to support this bumper height range that prevent partner cars from being overridden.

Stiffness Compatibility

The force concentration causes a local penetration into partner vehicle, which is far from an ideal and indicates a potentially aggressive structure. Vehicle front-end structures must be homogeneous to avoid a "fork effect", and must absorb a sufficient energy to reduce a passenger compartment intrusion. In order to assess the stiffness compatibility, full with barrier load cell data was analyzed by using the coefficient of variance (CV) [7].

$$CV = \text{Standard Deviation} / \text{mean}$$

The localized barrier forces were seen in the LTV test, however the number of 36 load cells used in the NCAP test is not acceptable for the CV analysis, therefor an example of another case was selected to assess stiffness compatibility. The lateral structural mismatch between vehicles can be often seen in the vehicle-to-vehicle impact that has a single load path for energy absorption. Here is another example of single load path car shown in Figure 13a.

Figure 13a. Single load path Car A after vehicle-to-vehicle impact test.

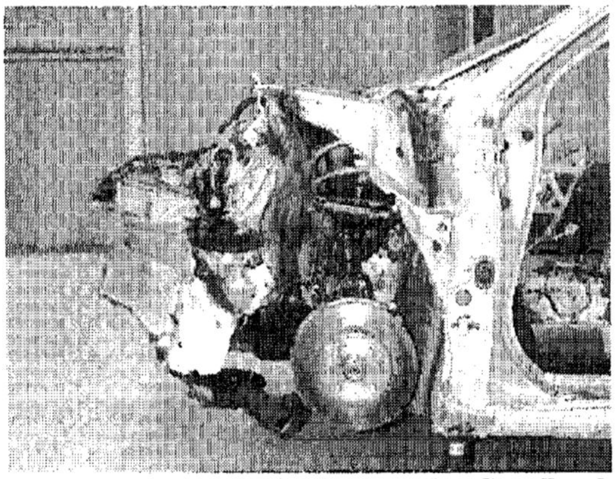

Figure 13b. Multiple load paths Car B after vehicle-to-vehicle impact test.

Figure 13b shows a Car B that has a multiple load path for energy absorption and more homogeneous structure in vehicle front. A subjective visual inspection of force distribution on each load cell shows a force concentration of the higher load for the car A caused by the single load path. The CV was computed for the Car A and the Car B and shows substantially higher values for the Car A.

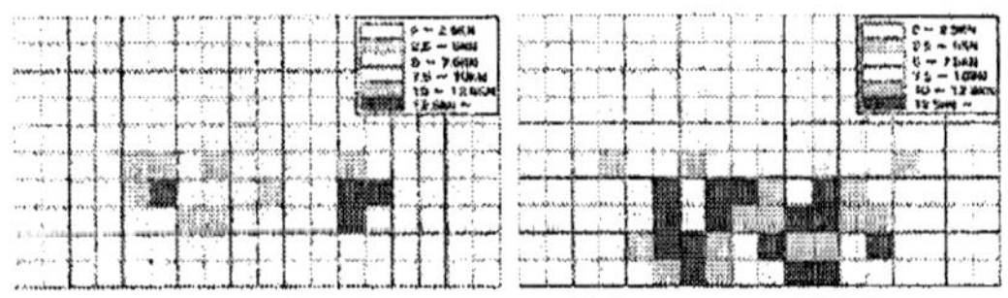

Barrier force distribution at 400 mm on CV-D curve

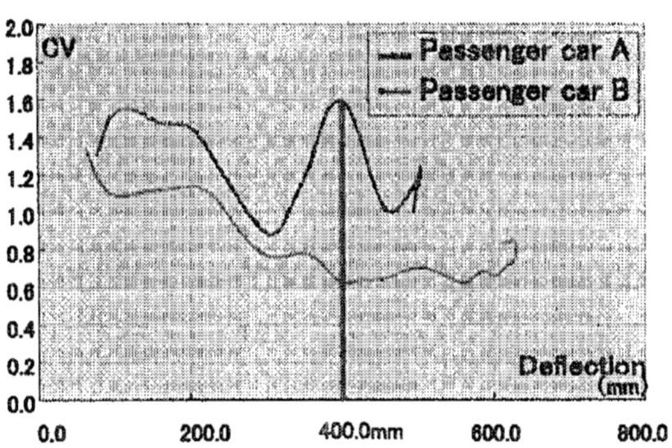

Figure 14. The CV for the Car A and the Car B

To demonstrate the potential of this analysis, the load cell data for more compatible family cars have been compared to those for less compatible vehicle. To make direct comparison of the results of the CV possible, overall average was computed using the deflection data as a weighing function.

As shown in Figure 15, Passenger cars and Minivans were found to be the similar homogeneous vehicle category with a CV from CV = 0.75 to CV = 1.2. The CV of SUV was obviously higher and ranged from CV = 1.25 to CV = 1.3. The difference in these values demonstrates the potential of this parameter as an index to evaluate the stiffness compatibility of vehicle front structure.

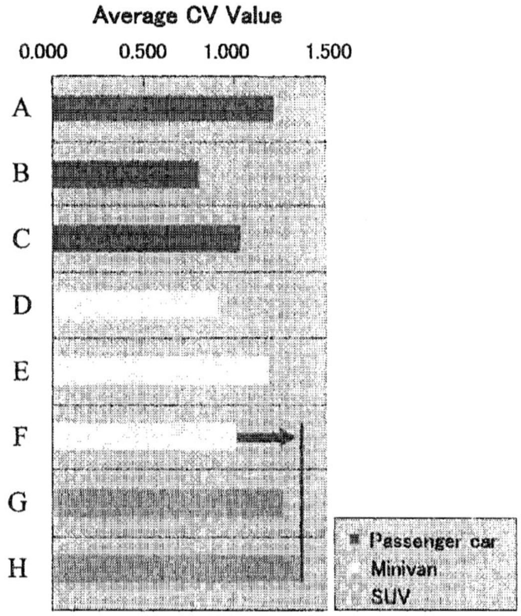

Figure 15. Average CV for various vehicles.

Here the question now arises: the SUV-H and the Minivan-F have almost same framework in the vehicle front structure (Figure 16), however SUV-H with CV = 1.3 was much higher than that of Minivan-F with CV = 1.0. Assuming that the similar framework provides the similar CV value, the question is why those two CV values were very different. Since the major difference between the two vehicles is a ride height, the influence of the ride height was investigated in the first place.

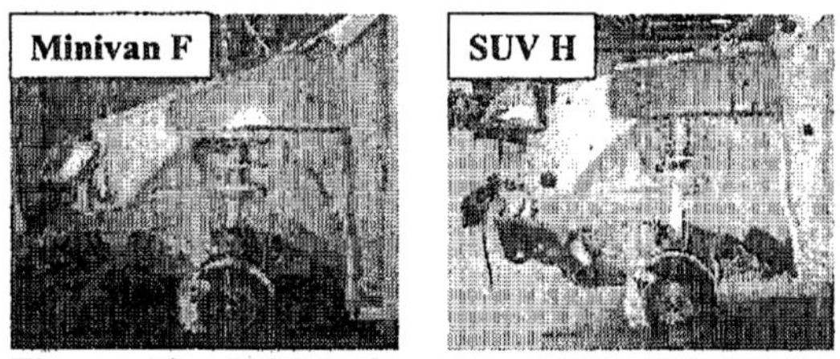

Figure 16. Vehicle front structure of Minivan F and SUV H.

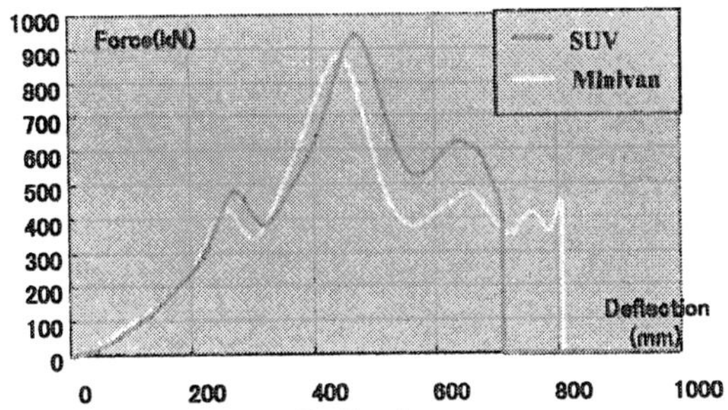

Figure 17a. Force-Deflection curve.

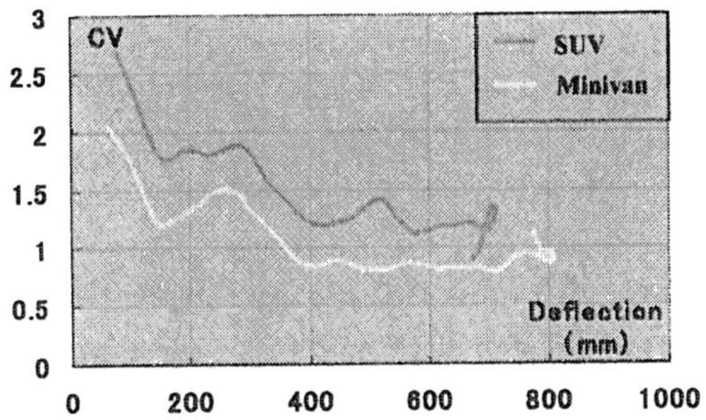

Figure 17b. CV-Deflection curve.

Figure 17a,b shows the F-D and CV-D curve for the SUV-H and the Minivan-F. The F-D curves for both vehicles showed similar trend in early stage of the impact on the ground that they have the similar framework, whereas the CV for both vehicles showed considerable difference. The difference of the CV value has resulted from the assessment area that was used for computing the CV. The assessment area is identified with the area that at least one load cell in the outside row or column was indicated a peak force greater than 5 kN in the impact. The difference of the assessment area was derived from the different ride height; namely the higher SUV has the larger assessment area. The mean value of the barrier force of the SUV-H was lower than that of the Minivan-F because of its larger assessment area. The low mean value of the barrier force causes the high CV value. That is the reason why the CV for the SUV-H has become higher than that for the Minivan-F as a consequence.

The CV is too sensitive when the mean value of the force is low in early stage of impact, because the gap between the mean force and higher force causes large standard deviation. The CV for the SUV-H and the Minivan-F were computed again with the assessment area that was defined at each deflection, then the results appear in Figure 18. By using this method, the CV value of the SUV-H became similar to that of the Minivan-F. In other words the homogeneity of the vehicle front structure was assessed more precisely.

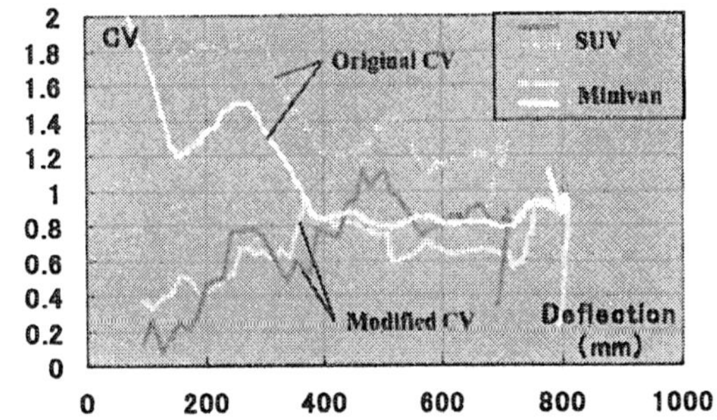

Figure 18. CV computation for each deflection.

A local penetration issue would be controlled by homogeneity assessment. However, in addition to homogeneity assessment, there is one other thing that is important for stiffness compatibility. There is no doubt about the fact that heavier vehicles have a higher global stiffness than lighter vehicles, because heavier vehicles must absorb their kinetic energy in approximately same crash distance in a fixed barrier test. The stiffness of the vehicle is somewhat related to its mass. In order to assess the stiffness matching between vehicles, some candidate test procedures have been proposed. A further research direction of this study will be to examine those test procedures to clarify reasonable global stiffness. As discussed in geometrical compatibility, it is important for stiffness compatibility that the stiffness within the interaction area would be investigated carefully with regard to frontal impact.

MDB-TO-CAR TEST

Various test procedures are being used to evaluate the self-protection of occupants provided by production vehicles in frontal crashes. The offset deformable barrier (ODB) test is designed primarily to reproduce the deformation patterns seen in real world crashes, accordingly addressing intrusion induced injuries. In contrast, the full width barrier test is designed to provide information relating to deceleration pulse induced injuries.

The NHTSA has been developing an MDB test to evaluate both self-protection and the compatibility of the vehicle [8]. A fixed mass MDB test allows the mass ratio effect to be taken into account, and it can generate a realistic delta V and vehicle deceleration pulse. One of the goals to use the MDB is to reproduce a vehicle-to-vehicle crash response, deformation of the vehicle and occupant kinematics seen in the real world type crashes.

One of the concerns of the MDB testing is its proneness to overriding issue, and IHRA WG report says that the overriding was greater when a vehicle was stationary [(2), (9)]. Offset frontal vehicle-to-vehicle test with same two passenger cars was carried out to confirm this issue according to the test configuration shown in Figure 19. In this test, the initial velocity of striking vehicle was chosen at 112 km/h and the struck vehicle was stationary.

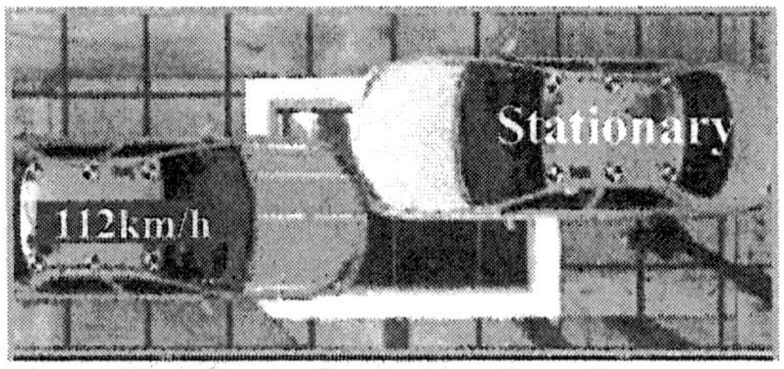

Figure 19. Car-to-Car test configuration.

In this test, injury measures of the dummies in the struck vehicle was the same level as those in the striking vehicle, therefore the injury measures did not indicate either vehicle being overridden. The result of the intrusion indicated that the deformation of the struck vehicle was slightly greater than that of the striking vehicle.

Figure 20a. Deformation mode of the striking car (front).

Figure 20b. Deformation mode of the struck car (front).

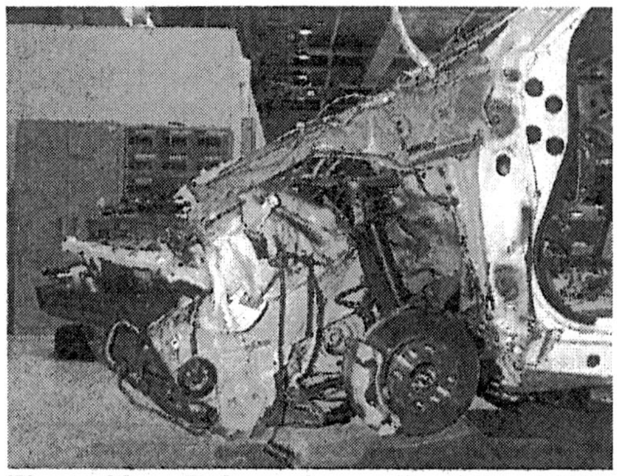

Figure 20c. Deformation mode of the striking car (side).

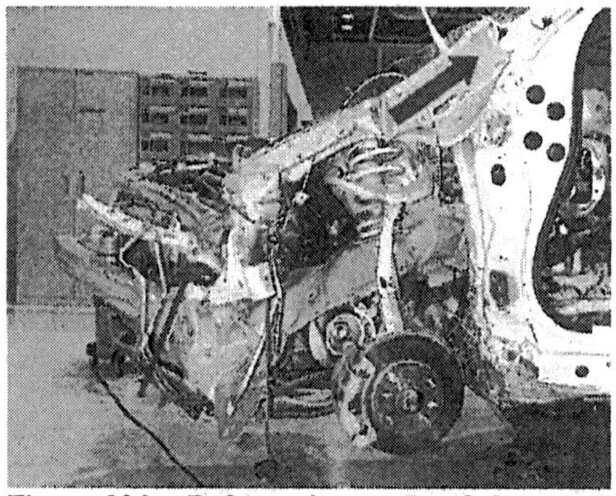

Figure 20d. Deformation mode of the struck car (side).

The different bending mode of the bumper beam was observed in both vehicles; to put it briefly, the bumper beam of the striking vehicle bent downward, and that of the struck vehicle bent upward to the contrary. (Figure 20a,b) The upper rail of the struck vehicle was more deformed than that of the striking vehicle and the lower rail of both vehicles demonstrated a different failure mode. (Figure 20c,d) There was no obvious over-riding phenomenon during the test, however the deformation profile of the struck vehicle showed the signs of being over-ridden. It was found from the result that the proneness of overriding to the stationary vehicle holds true, consequently the MDB-to-Car test has been conducted with both vehicles moving to prevent overriding.

Another issue for the MDB test is the bottoming out of the defomable face. The Federal Motor Vehicle Safety Standard (FMVSS) 214 MDB was studied in our previous study and the honeycomb bottoming out phenomenon was observed. There is no bottoming out in vehicle-to-vehicle impact; therefore the barrier must be design to avoid this phenomenon.

A Progressive Deformable Barrier (PDB) has a

depth of 700 mm and was developed to avoid bottoming out [10]. Judging from the above, the PDB was selected as one of the candidates of deeper honeycomb, and MDB-to-Car test was carried out with the deformable face of PDB. Figure 21 shows the test MDB-to-car test configuration. The 1362 kg MDB was used to strike the compact sedan in 40% offset collinear with both vehicles traveling at a speed of 56 km/h. In the MDB-to-Car test, Hybrid III 50th percentile male dummies were used to evaluate the injury levels for the driver and passenger.

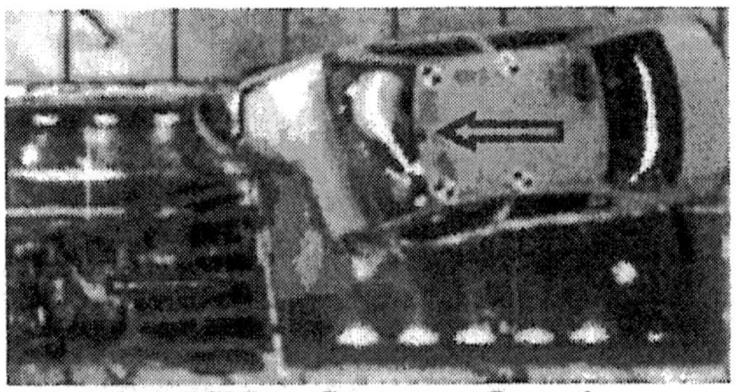

Figure 21. MDB-to-Car test configuration.

This preliminary MDB test was conducted to compare the vehicle deformation and injury levels in an equivalent vehicle-to-vehicle test. Target compact sedan was subjected to three energy equivalent crash tests shown in Table 3, and results were compared.

Table 3. Test matrix

Test	Test speed	Total Kinetic Energy	Offset Ratio
Car-to-Car	50 km/h	4451 kN·m	50%
MDB-to-Car	56 km/h	4312 kN·m	40%
ODB-to-Car	64 km/h	4262 kN·m	40%

The deceleration pulses of the target vehicle in the vehicle-to-vehicle collinear test, in the MDB test, and in the ODB test are shown in Figure 22.

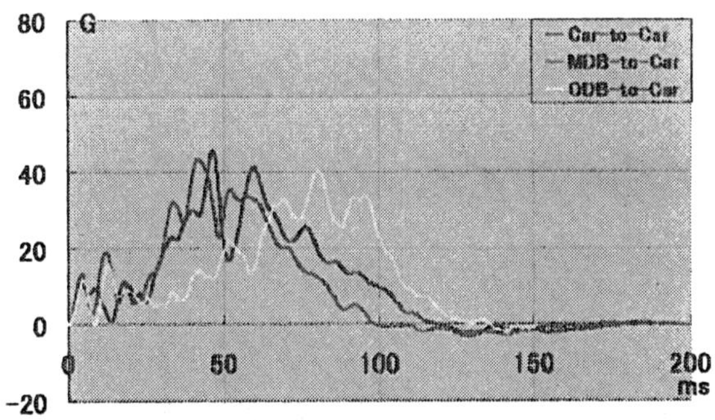

Figure 22. Comparison of Deceleration Pulses.

The overall deceleration pulse shape and timing in the MDB 40% test was generally similar to that in the vehicle-to-vehicle 50% test, in contrast the peak deceleration occurred approximately 40 millisecond later in the ODB 40% test. This figure shows that the MDB test matched the vehicle-to-vehicle response very well, whereas the crash pulse resulting from the ODB test does not appear similar to the vehicle-to-vehicle crash pulse.

Figure 23 and 24a,b show the comparison of dummy responses for the three crash test conditions for the compact sedan. Injury Assessment Reference Values (IARV) were used to normalize the injury measures. These reference values are defined in FMVSS 208 except for the tibia index that uses a reference value of 1.3, as defined by the EU96/79 standard.

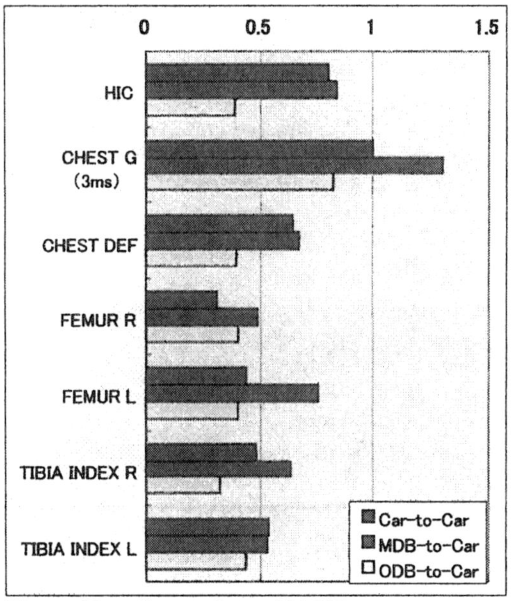

Figure 23. Comparison of IARV

A very good match was observed between the Car-to-Car dummy responses and the MDB test. This MDB test is slightly more severe than corresponding Car-to-Car test, resulting in higher vehicle response and higher dummy head, chest and femur response. The ODB test responses was compared to the Car-to-Car response and MDB-to-Car test. Comparing dummy responses from Figure 23 shows much lower dummy responses for all major injury measures, even though the ODB is at similar impact energy.

This holds especially true for the head and chest response comparisons. (Figure 24a,b.) This match is expected since the MDB test was designed to reproduce the vehicle-to-vehicle test mode.

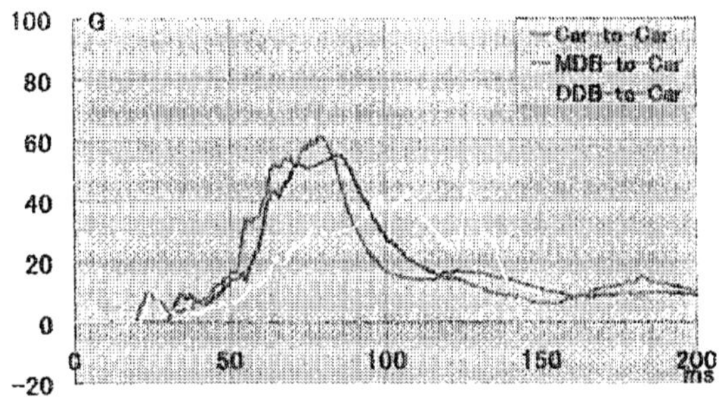

Figure 24a. Comparison of Driver Head Deceleration Pulses.

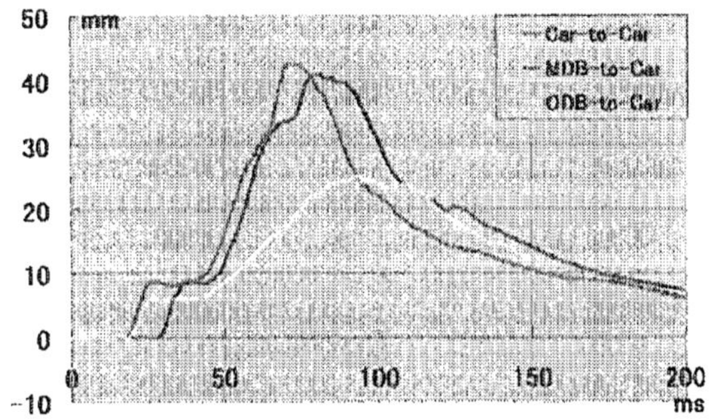

Figure 24b. Comparison of Driver Chest Deflection Pulses.

Figure 25 and 26 show the overall vehicle deformation. A very good reproducibility was observed with regard to the vehicle deformation as well as dummy responses. These results lead us to the conclusion that Car-to-Car crashes can be better reproduced using an MDB-to-Car test. It is thought that this MDB test will allow realistic evaluation of compatibility.

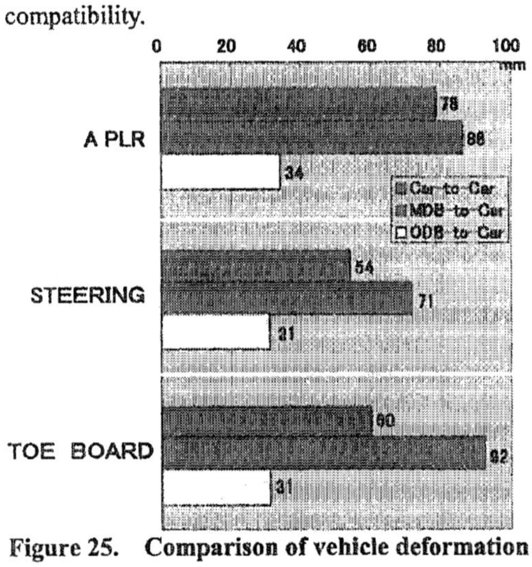

Figure 25. Comparison of vehicle deformation

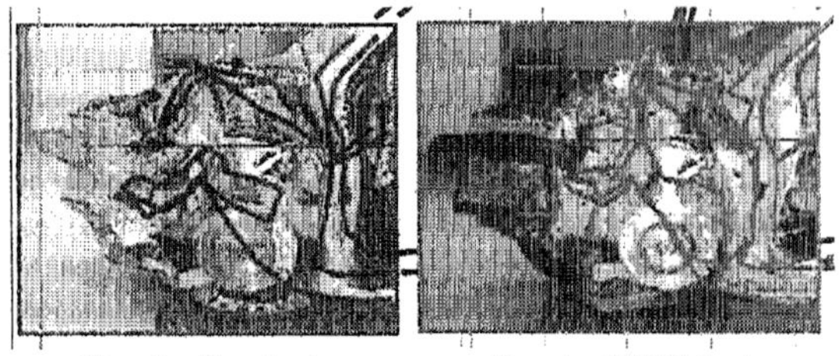

Car-to-Car test **Car-to-MDB test**

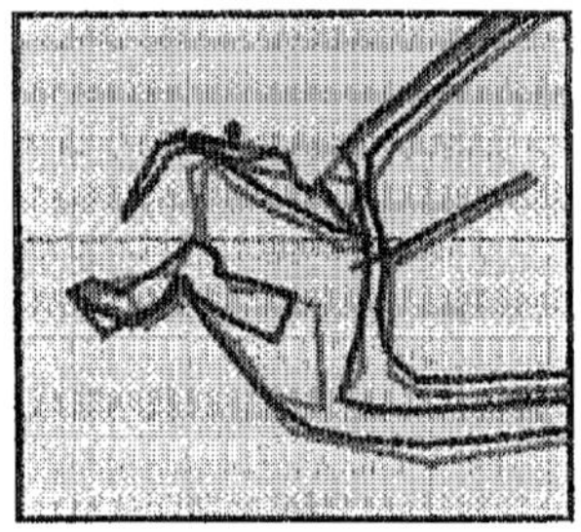

—Car-to-Car
—Car-to-MDB

Figure 26. Comparison of 3-D measurement data of the Compact Sedan.

The result of Car-to-Car crash depends on the characteristics of the two opposing vehicles. Although Car-to-Car crashes have a low initial deceleration similar to ODB crashes, they also produce extremely high input loads into the cabin and high deceleration levels in the second half of the crash. As a result, the crash energy translated to the occupant by Car-to-Car offset crash is actually equal to or greater than that for a full width barrier test. However, the cabin survival space is significantly reduced compared to the other two tests, and there are many cases where this causes secondary collisions with interior parts which worsen the occupant injury severity [11]. (Table 4)

Table 4.

Test Procedure	Body Deceleration	Restraint Performance
FULL LAP	Body G / Body Deformation	Chest G / Chest Disp.
ODB	Body G / Body Deformation	Chest G / Chest Disp.
CTC	Body G / Body Deformation	Chest G / Chest Disp.

Relationship between Body Deceleration and Deformation and Occupant Injury Severity by Crash test

An examination of evaluation test procedure that combine these characteristics shows the moving deformable barrier (MDB) crash test to be ideal. The MDB test procedure is currently one test method which models Car-to-Car crashes from the dual perspective of body deceleration characteristics, which control occupant injury severity, and occupant survival space. Test results show that the body deformation and deceleration of the test vehicle closely approximated those in an actual Car-to-Car crash. As a result, it was found that vehicle-to-vehicle crashes might be better reproduced using an MDB-to-Car test as opposed to a conventional fixed barrier test.

DISCUSSION

The barrier load data analysis calls for further investigation. The full width rigid barrier test is too sensitive to the inertia force when hard structures hit with the rigid wall. The load cell data in the rigid barrier test should be compared with that in the full width deformable barrier test proposed by the Transport Research Laboratory (TRL) with respect to the AHOF and the CV [12].

In the case of SUV/LTV-to-Car offset crashes, the target vehicle experienced a lower deceleration level in the early stage of the impact (up to 30 ms) and a higher deceleration in the later stage of the impact. The lower deceleration pulse can be reproduced by the ODB test and the higher deceleration pulse can be reproduced by the full width barrier test. However, the combination of initial low deceleration and the following higher deceleration, which causes severer injury levels, will not be reproduced by those test procedures. Nothing will reproduce the realistic deceleration pulse in the vehicle-to-vehicle impact better than that in the MDB test. A further research direction of this study will be examining if the MDB characteristics will allow realistic evaluation of compatibility in various countries. MDB test procedure should be considered varying the crash angle, speed, offset ratio and other factors of the vehicle fleet in each country. Then, these results should be used to determine vehicle specifications that can improve occupant injury severity and metrics for the future evaluation of compatibility.

CONCLUSION

This report has examined the compatibility between SUV/LTV and passenger cars in vehicle-to-vehicle crash tests.

1. The LTV overrode the passenger car during the LTV-to-Car frontal offset crash. Significant deformation and dash intrusion were observed in the target vehicle in the LTV-to-Car impact compared to that in the SUV-to-Car impact. This was caused by vertical structural mismatch between the two vehicles. In the studied frontal SUV/LTV-to-Car impacts, it was found from the result that good structural interaction was a fundamental requirement to crash compatibility.
2. In order to improve compatibility, it is necessary to develop test procedures that can evaluate the various criteria for both aggressivity and self-protection. Our development activity is being focused on two test procedures, namely one is a full width barrier test with load cells and the other is an MDB test.
3. The load cell data in the full width barrier at 56 km/h was analyzed to assess the structural interaction. The AHOF could become a geometric compatibility index and interaction area should also be taken as a compatibility index at the same time. As one of the indices for stiffness compatibility, the CV could evaluate homogeneity of the vehicle front structure. However, further study for the calculation method of the CV is needed to assess the homogeneity more precisely.
4. It was clear that the barrier load cell size affects seriously the result of the AHOF and the CV. Therefore, barrier load cells should be smaller and identical size among various laboratories in order to assess these compatibility indices correctly.
5. The issues of bottoming out the honeycomb and overriding of the MDB were improved in this study. It was found that the MDB test could reproduce the body deformation and deceleration observed in actual vehicle-to-vehicle impact by using a PDB as a deformable face. The resulting vehicle deformation, vehicle deceleration and occupant injury severity matched closely with actual vehicle-to-vehicle tests. The MDB would address deceleration-induced injuries in addition to contact-induced injuries.
6. Combining the full width barrier test and the MDB test has a possibility to provide an improved compatibility. This combination of the test procedures can evaluate the vehicle characteristics from vehicle front-end homogeneity to passenger compartment strength. The MDB provide more flexibility to reproduce vehicle-to-vehicle crash, hence the MDB test would offer the best overall coverage of real world accidents.

REFERENCES

1. Stephen M. Summers, Aloke Prasad, William T. Hollowell "NHTSA's Research Program for Vehicle Aggresivity and Fleet Compatibility" 17th International Conference on the Enhanced Safety of Vehicles Paper No. 249, Amsterdam, the Nether lands, June 2001

2. Peter O'Reilly "Status Report of IHRA Vehicle Compatibility Working Group" 17th International Conference on the Enhanced Safety of Vehicles Paper No. 337, Amsterdam, the Nether lands, June 2001

3. K H Digges and A M Eigan "Analysis of Load Cell Barrier Data to Assess Vehicle Compatibility", Society of Automotive Engineers Paper No. 2000-01-0051. March 2000

4. K H Digges and A M Eigan "Measurement of Stiffness and Geometric Compatibility in Front-to-Side Crashes", 17th International Conference on the Enhanced Safety of Vehicles Paper No. 349, Amsterdam, the Nether lands, June 2001

5. Stephen Summers, Aloke Prasad, Willam T. Hollowell"NHTSA's Compatibility Research Program Update", Society of Automotive Engineers Paper No. 2001-01-1167. Detroit, March 2001

6. Stephen Summers, Willam T. Hollowell, Aloke Prasad"Design Considerations for a compatibility Test Procedure", Society of Automotive Engineers Paper No. 2002-01-1022, March 2002

7. Mervyn Edwards, Julian Happian- Smith, Huw Davies, Nigel Byard and Adrian Hobbs "The Essential Requirements for Compatible cars in Frontal Collisions", 17th International Conference on the Enhanced Safety of Vehicles Paper No. 158, Amsterdam, the Nether lands, June 2001

8. Carl L. Ragland "Offest Test Procedure Development and Comparison" Enhanced Safety Vehicle Paper No. 98-S1-O-03, Windsor, Ontario CANADA, June 2001

9. Koji Mizuno, Kazumasa Tateishi, Yukihiro Ezaka "Test Procedure to Evaluate Vehicle Compatibility" 17th International Conference on the Enhanced Safety of Vehicles Paper No. 127, Amsterdam, the Nether lands, June 2001

10. Alain Diboine, Pascal Delannoy "Improvements in car to car compatibility: Physics, design constraints and assessment test methodology and criteria " Vehicle Safety 2002 , London, UK, May 2002

11. Tomiji Sugimoto, Hideki Suzuki, Yoshiji Kadotani "Infuluence of Body Intrusion and Deceleration on Occupant Injuries in Frontal Collisions between Passenger Cars" 17th International Conference on the Enhanced Safety of Vehicles Paper No. 433, Amsterdam, the Nether lands, June 2001

12. Mervyn Edwards, Adrian Hobbs, Huw Davies, and Alex Thompson "Development of Test Procedures and performance Criteria to Improve Compatibility in car in Frontal Collisions" Vehicle Safety 2002 , London, UK, May 2002

2004-01-1165

Significant Factors in Height of Force Measurements for Vehicle Collision Compatibility

M. K. Verma, R. Nagappala, Y. J. Tung, M. R. Zimmerman, M. Murugan and T. A. Bernstein
General Motors Corporation

ABSTRACT

The concept of height of force has been suggested by some researchers as one possible parameter defining the structural interaction probability between vehicles of different sizes. This proposed parameter was defined as the vertical centroid of forces exerted on a flat barrier surface when a vehicle crashes into the barrier. It is therefore measured as a function of elapsed time since crash.

In this paper, the height of force is obtained from theoretical calculations and also measured in crash tests at 56 km/h against barriers instrumented with an array of load cells.

It is observed that the measured values of height of force
have significant errors which are dependent on factors other than the crash conditions and the properties of the vehicle's structure and geometry. These factors need to be taken into account in future discussions of using the height of force or the average height of force as an indicator of vehicle compatibility.

INTRODUCTION

The issues related to collision compatibility between vehicles of different sizes have been discussed by several authors in the past (1-4). Most of the societal harm associated with collision compatibility can be generally divided (2) into the broad categories of (a) front to front impact between two vehicles, and (b) front (of a larger vehicle) to side (of a smaller vehicle) impacts. The potential countermeasures for improving front-to-front compatibility between two vehicles have generally been discussed as countermeasures for the larger vehicle.

It is in this context that the 'height of force' (HOF) and the 'average height of force' (AHOF) have been proposed (5,6) as potentially representing the location where a vehicle effectively applies forces during a collision with another vehicle. The definitions of these parameters are in the above-mentioned references and, though this concept is similar to that of center of gravity, there are major differences. For example, whereas the gravitational forces can be considered constant and invariant with time for ground-based objects, the forces applied in a crash vary greatly both as a function of time as well as of spatial co-ordinates. In addition, these forces are dependent on many other factors, e.g. the speed and the direction of impact, the mass of the vehicles, the relative dynamic properties of the impacting vehicle structures, etc. Therefore, these variables (height of force and average height of force) are not 'independent variables' in the mathematical sense but rather, they must be thought of as statistical abstractions. The usefulness of these abstract variables therefore needs to be examined on the following issues:

- Are there strong correlations between these statistical variables and safety of vehicle occupants in crashes so that performance measures based on these variables are likely to improve overall safety?

- Are these variables uniquely representative of the vehicle's dynamic properties in a given crash so that performance measures based on these variables will definitively modify vehicle's properties in the desired direction?

The issue of correlation requires examination of the relationship of these variables to vehicle safety (both in self-protection and in partner-protection) and is not included in the present study.

This paper examines the issue of these variables being representative of a vehicle's dynamic properties in a crash from theoretical calculations, finite element simulations and test results on vehicles.

MEASUREMENT OF HEIGHT OF FORCE

Errors are present whenever a continuous variable is measured using a discrete measurement device. Such errors are sometimes referred to as 'discretization errors' if they are dependent only on the resolution of the measurement device. However, there can be other sources of such errors inherent to the definition of the variable or of the measurement method. This 'inherent error' in height of force at any instant can be calculated by simple equations for simple shapes of the impacting structures or by techniques such as finite element methods for more complex shapes (8).

The force applied on a rigid barrier by a vehicle varies as a function of time. For a frontal impact, the initial point of contact is usually the front face of the bumper. For high-speed ($\geq$30 mph) impacts, the bumper is fully crushed within the first few (< 5) milliseconds at usually much smaller loads than those applied later in the event .The subsequent application of force is mainly by the primary structure (which is usually referred to as 'frame rails' or 'motor compartment rails') of the vehicle. In all currently known automotive designs, this primary structure dissipates most of the kinetic energy of the crash by means of structural deformation. Thus, in the initial phases of the crash sequence (after the bumper crush), the location of this primary structure is the location of application of crash forces on the rigid barrier. The error in measuring the height of force at any instant of time can then be calculated as follows.

Consider the case where the rigid barrier is instrumented with load cells of size d x d and that the primary structure of the impacting vehicle is a rectangular section of depth r as shown below in figure 1. If the lower edge of this section is at height x above the ground at any instant, then the actual height of force is at the vertical

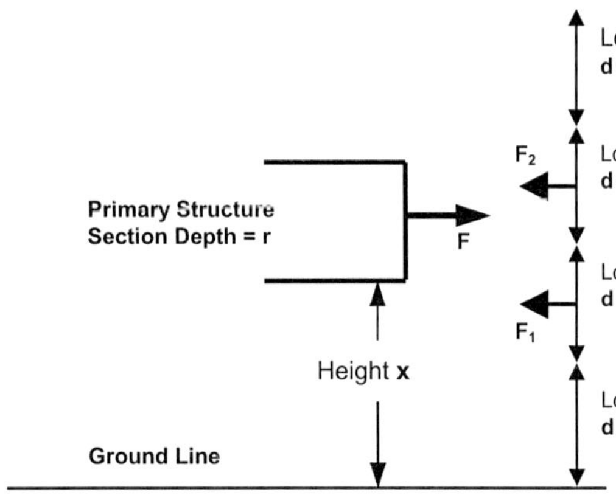

Figure 1: Height of Force Calculation

center of this section, i.e.

actual height of force = x + ½ r

The total force F applied by the vehicle's primary structure is distributed across the cross section and, in the case shown above, is resisted by forces F_1 and F_2 in the two load cells if the primary structure spans two load cells vertically.

F_1= F.(2d-x) / r; F_2 = F.(r-(2d-x)) / r; and

Calculated height of force = $({}^{3}/_{2}\,d.\,F_1 + {}^{5}/_{2}\,d.\,F_2) / F$

Thus,
Measurement error = Calculated HOF - Actual HOF

By substituting above values for F_1 & F_2,
Measurement error = ${}^{5}/_{2}\,d - 2\,d^2/r + d.x/r - x - \tfrac{1}{2}r$

Numerical estimates are presented below for several cases. Two values of primary structure depth are used below - 80mm and 160 mm, respectively. These depths are representative of vehicle structures on the road today. Error estimates are generated for two load cell sizes of 125 mm and 50 mm respectively. These load cells are single axis cells so that a load applied anywhere on the surface of the cell is measured as being at the center of the cell.

<u>Case 1:</u> Load cell height is the same as LTV structure depth; d = r. In this case, it can be shown that

Error in measuring HOF = 0

<u>Case 2:</u> Load cell height (125 mm) is more than the depth of the vehicle structure (80mm) and the bottom edge of the structure is aligned with bottom edge of load cell. In this case, the vehicle structure section spans only one load cell in depth, i.e. d = 125 mm, r =80 mm. Then,

Actual HOF = 540 mm,
Measured HOF = 562.5 mm
Error in measuring HOF = **+**22.5 mm

<u>Case 3:</u> Same as above except the vehicle structure section is raised by 4.5 cm (the top edge of the structure is aligned with top of load cell array); i.e. d=125 mm, r=80mm. Then

Actual HOF =585 mm,
Measured HOF = 562.5 mm.
Error in measuring HOF = **-**22.5 mm

Thus, it is possible to change the height of the vehicle structure by 45 mm and not detect any change in the measured height of force using 125 mm load cells.

The measurement error can be similarly calculated for other sizes of load cells. For example, when using load cell size of 50x50 mm, the error in measuring the height of force of a primary structure with a depth of 80 mm is found to be 3.75 mm when bottom of the rail is aligned with the bottom of a load cell array.

It is therefore evident that errors are inherent in measuring the height of force with an array of discrete single-axis load cells whenever the impacting section depth is not the same as the load cell height. This error in measurement of the height of force is then a function of :
i) size of the load cell,
(ii) size of the structure impacting the load cell, and
(iii) the relative position of the impacting structure with respect to the load cell geometry.

Thus, this error is inherent to the definition and measurement of the height of force. In the case when the impacting sections are more complex or when many parts of the vehicle are impacting the barrier, the calculation of error will become more complex but will lead to the same conclusions regarding the measurement error and the significant variables affecting the height of force calculation.

FINITE ELEMENT MODEL FOR HOF

A finite element model was created for calculating the height of force when the vehicle structure is more complex than what can be evaluated by the simplified technique in the previous section. For this purpose, the DYNA-3D program from Livermore Software Technology Corporation was used to simulate the crash of the vehicles into the barrier with instrumented load cells.

Shown below are results from this simulation of the frontal impact (fully frontal impacts, no offset) of the vehicle at speed of 56 km/h into the barrier surface at the TRL facility with load cells of approximately 123x123 mm size, with approximately 2 mm gap between individual cells on each edge. The bottom edge of the lowest row of load cells is located approximately 175 mm from the ground. Each load cell is covered by an individual layer of thin, deformable material which can be replaced prior to a test if necessary. The load cells are arranged in a 16x8 matrix (horizontal x vertical) covering the entire impact area for these test vehicles.

SIMULATION RESULTS WITH 125mm LOAD CELLS

There are significant difficulties in reliably calculating the height of force in the first few milliseconds after the impact. This is due to several factors such as the large values of acceleration at the moment of impact, the noise and the cross-talk in the measured data and the distortions caused by any band-pass filter that is traditionally used in processing crash data. It was therefore necessary to make several assumptions in order to get the desired data, e.g. the calculations were not begun until the load on the barrier exceeded a certain value, thereby discarding the data for a few milliseconds after t=0. Both the simulation results and the test data were subjected to the same set of processing constraints, such as using SAE180 filter.

As observed in figure 2, the finite element simulation of the height of force as a function of time (measured from the beginning of the crash event) shows generally good correlation with the test data for the time period of approximately 5 to 30 msec. This is usually the time period during which most of the primary vehicle structure interaction with the barrier occurs. The correlation of the simulation results to the test data beyond this time period is not as good. In order to improve this correlation, it may be necessary to enhance the representation of both the barrier as well as of other parts of the vehicle.

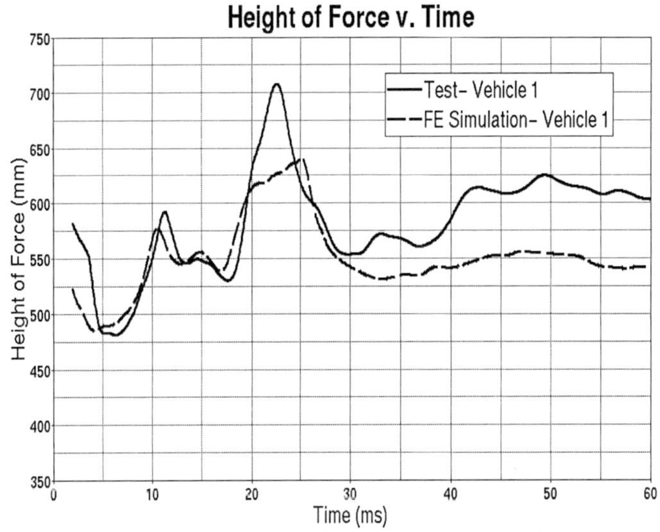

Figure 2- Comparison of FE and test results, HOF versus Time, 125 mm Load cells

This is also seen from figure 3 where the height of force is shown as a function of the crush of the front end of the vehicle. The results from the finite element simulation as well as those from the crash test are compared in this figure and a good agreement is seen between these two results in the range of vehicle crush of 300 mm.

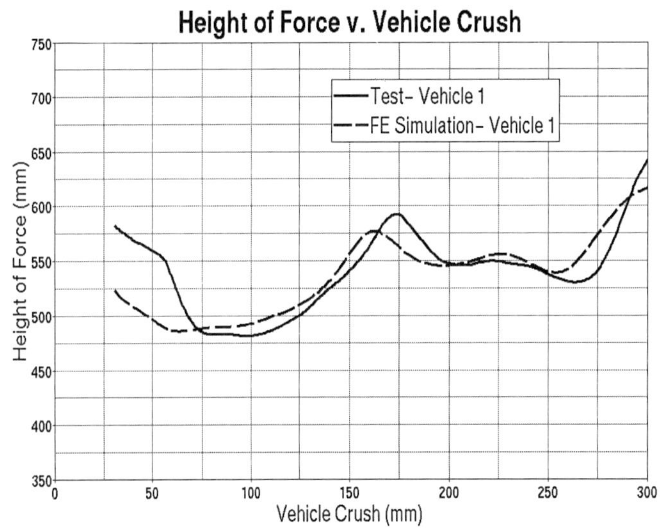

Figure 3- Comparison of FE and test results, HOF versus Crush; 125 mm Load cells

The average height of force (AHOF), which is the time-averaged value of the height of force at any instant (6,7) was also calculated for these vehicles. The comparison of the AHOF calculation from the finite element simulation and from the tests is shown in figure 4 as a function of time and in figure 5 as a function of vehicle crush. As expected from the HOF results, the simulation results show generally good agreement with measurements in the same range of time and vehicle crush as mentioned above.

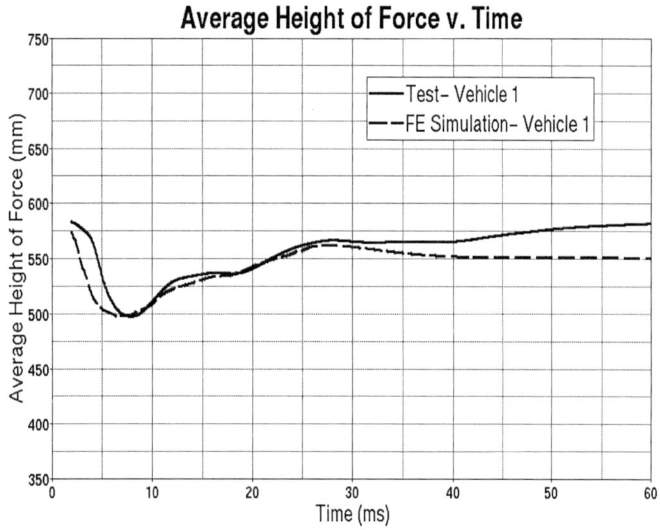

Figure 4- Comparison of FE & Test Results, AHOF versus Time, 125mm Load cells

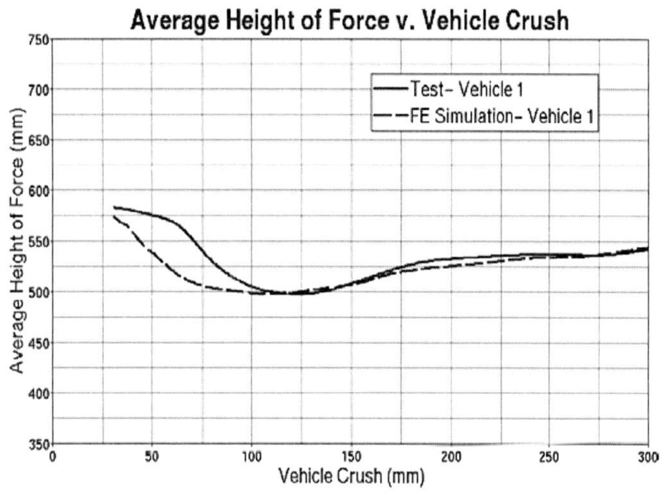

Figure 5- Comparison of FE and Test Results, AHOF versus Crush, 125mm Load cells

It is therefore reasonable to expect that numerical simulations such as the finite element examples shown above can yield indications for the height of force and the average height of force in most of the range where the crush of the vehicle is due to the initial deformation of the vehicle's primary structure. As the previous figures have shown, this FE simulation can be used to estimate the Height of Force variation for time duration of approximately 30 msec and vehicle crush of approximately 300 mm in this case. It may be possible to extend the range of validity of the finite element simulation with further refinement of the FE model and with better representation of other parts of the vehicle that impact the barrier.

TEST RESULTS WITH 125 mm LOAD CELLS

The first set of two crash tests were conducted at the barrier test facility of TRL (UK) with the load cells arranged as described in the preceding section. Two vehicles referred to as VEHICLE1 ("baseline") and VEHICLE2 ("lowered vehicle") were used in these tests. These vehicles were identical within manufacturing variations except that VEHICLE2 was lowered by 38 mm from the configuration of VEHICLE1.

VEHICLE1 (Baseline Vehicle):
Mass = 2372 kg , Test Speed = 55.9 km/h

VEHICLE2 (Lowered by 38 mm from VEHICLE1):
Mass = 2372 kg, Test Speed = 55.7 km/h

DISCUSSION OF TEST RESULTS WITH 125MM LOAD CELLS

The test results for the height of force and the average height of force are shown in figures 6-9. Also shown in these figures are lines indicating the upper and lower surfaces of the forward-most point of the bumper and of the primary structure (frame) of VEHICLE1 prior to the impact. The vehicle position with respect to load cells on the barrier face was analyzed and it was observed that with the configurations of the vehicles and the test barrier as defined, the primary energy absorbing structure (frame rail) of VEHICLE1 lines up with the third row of load cells of the barrier face and the frame rail of VEHICLE2 overlaps the top portion of the load cells on the second row of the barrier face.

Figure 6 shows the height of force vs. time plot for both

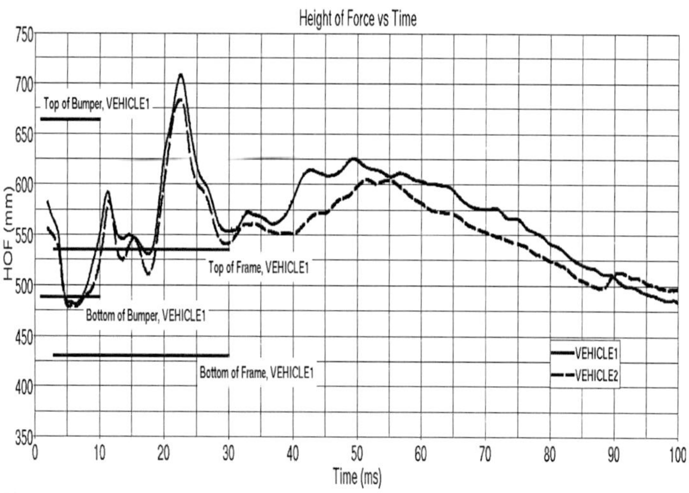

Figure 6 – Height of force vs. Time with 125mm load cells

the VEHICLE1 and the VEHICLE2 for the first 100 msec. The initial point for each vehicle represents the point of contact of its bumper with the load cells. The measured height of force at the first point in the plot was 581 mm for VEHICLE1 and 556 mm for VEHICLE2, showing a difference in height of 25 mm versus the actual change of 38mm. Thus, the error due to the load cell size is approximately 13 mm at this point. Since the VEHICLE2 had been lowered by 38 mm as compared to VEHICLE1, the measured results in the ideal case would have been expected to show this amount of difference throughout the test. However, this is not the case here and even after ignoring the results beyond 80 msec (since the vehicles may be in rebound), the difference between the two vehicles is found to vary as a function of time from the initial value of 25 mm to a value of 0 at 5 sec and other values at varying time steps.

Figure 7 shows the height of force vs. vehicle crush, for VEHICLE1 and VEHICLE2 and leads to the same conclusions as before. The plots for the height of force remain within the upper and lower surfaces of the primary structure for only the first 150 mm or so and the correlation to the centerline of the vehicle frame (its primary structure) is inconclusive.

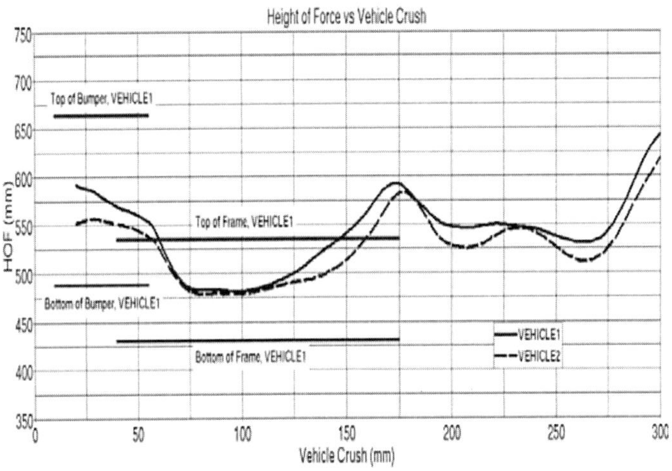

Figure 7 – HOF vs. Crush, 125mm Load cells

Shown in Figure 8 are the average height of force plots calculated from the test data for VEHICLE1 and VEHICLE2. The average height of force of VEHICLE2 is seen to be lower than that of VEHICLE1 throughout the plotted range but not by the correct amount. The average height of force is seen to be always in error in this case, by as much as 35 mm.

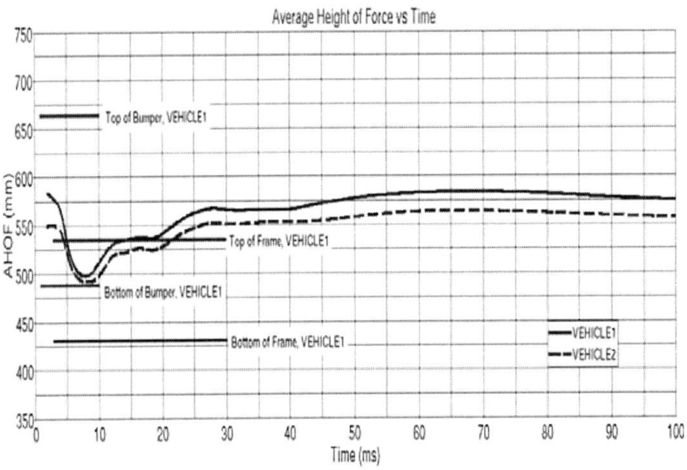

Figure 8 – AHOF versus Time with 125mm load cells

Figure 9 shows the average height of force plotted as a function of the crush of the front end of the vehicles, for both VEHICLE1 and VEHICLE2 and leads to the same conclusions as before.

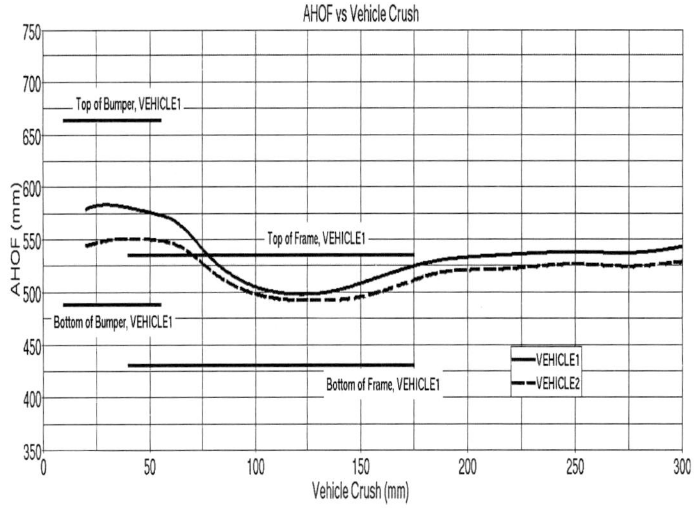

Figure 9 – AHOF versus Vehicle Crush

It is also concluded that average the height of force is not correlated to the geometry of the primary crash structure of the vehicles for most of the time duration shown here.

TEST RESULTS WITH 50mm LOAD CELLS

The second set of two tests was conducted at the test facility of MIRA (UK) which has a flat rigid barrier instrumented with 50x50 mm load cells over most of the barrier, with larger load cells at the periphery (beyond the impact surface of the vehicle). The gap between individual load cells is very small (~1 mm). The load cells are covered first by a thin aluminum sheet and then by a thicker plywood sheet. These sheets are continuous surfaces, covering all of the 50 mm load cells.

Two vehicles (substantially different structurally from the those used for tests with 125 mm load cells) were used for this study:

VEHICLE3 (Baseline vehicle)
Mass = 2524 kg, Test Speed = 56.7 km/h

VEHICLE4 (Lowered from VEHICLE3 by 43 mm)
Mass = 2521 kg. Test Speed = 56.9 km/h

Again, these two vehicles were identical to each other within production variations except for the lowered height of VEHICLE4. The results from these tests are shown in figure 10 (height of force as function of time) and figure 11 (average height of force versus time). Also shown on these plots are lines indicating the top and the bottom surfaces of the bumper and the top and the bottom surfaces of the primary structure (frame) for the VEHICLE3 at the moment of impact. The respective lines for VEHICLE4 (not shown in these plots) will be 43 mm lower than that for VEHICLE3.

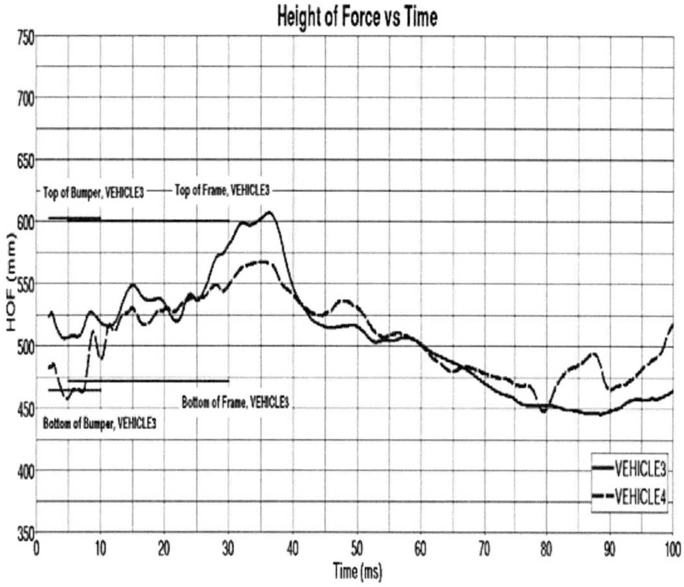

FIGURE 10- Height of Force vs. Time with 50mm Load cells

As observed from the above figures, the initial values of the height of force of the two vehicles represent the midpoint of their bumper profiles (the first point of contact with the load cells). Numerically, the height of force of VEHICLE4 (482mm approximately) is lower by 42 mm from that of VEHICLE3 (524 mm approximately) at this instant. Thus, the measurement error introduced by 50 mm load cells is much smaller than those introduced by the larger 125mm load cells. Similar conclusions apply to the initial value of the AHOF results for these two vehicles.

However, the conclusions to be derived from the height of force results are again time-dependent - the change ranging from a maximum 50.5 mm lowering at 7.4 msec

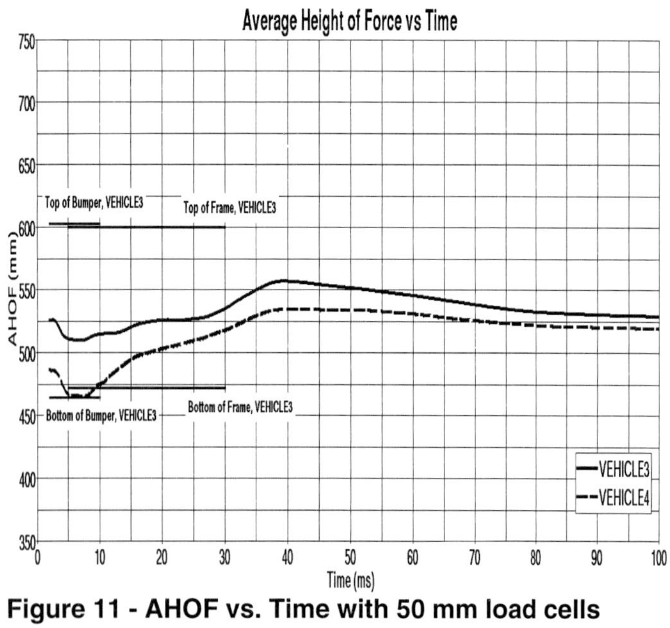

Figure 11 - AHOF vs. Time with 50 mm load cells

to 8mm raising at 22 msec and 0 at several other time steps. The measured height of force for these vehicles shows a much closer relationship to the location of the vehicles' primary structure for the first 25 msec. The data beyond 80 msec should not be considered since the vehicle may be in the rebound phase at that time.

The average height of force results for these two tests are more reliable in showing that one vehicle has been lowered. However, the amount of lowering inferred from these tests again varies with time – approximately 42 mm at initial time, a maximum of 45mm at 8 msec to a minimum of 18 mm at 25 msec to a minimum of 12mm at 80 msec (later data being ignored). The location of the average height of force is also close to the centerline of the vehicles' primary structure for the first 30 msec and remains within the upper and lower surfaces of the primary structure for the entire period shown here.

CONCLUSIONS

The measured height of force is an implicit function of several variables;

Measured HOF = ***f*** (**vehicle structure geometry; load cell size; location of structure relative to load cell boundary; time*)***

Thus, the height of force (and the average height of force) is not a property of the vehicle uniquely but is a time-dependent function of interaction among several variables.

As seen from the data presented above, the height of force and the average height of force results have significant errors in indicating the change in the actual height of a vehicle and of its structure. The error introduced by the load cell size is dependent on the dimension of the load cell used for measurement. It is observed that the 50mm load cells provide better resolution in the measured results. However, there are other errors also which appear inherent to the definition of the variables and the measurement procedure. In all cases presented above, it is only in the first few (<5) milliseconds after the initial impact that the height of force reliably determines whether the vehicle height has changed and the amount of such change. Similar conclusions apply to the average height of force. Overall, in the cases studied here, the load cell measurements have less fidelity in measuring the change in height of a vehicle than a simple measurement of a well-defined reference point on the vehicle structure.

The relationship of the measured height of force (and the average height of force) to the location of the vehicle's primary structure was not found to be unique and different conclusions would be reached in the two series of tests mentioned above. Since automobiles are designed to dissipate most of the energy of frontal crash by deformation of their primary structure, it may be expected that the location of this primary structure is a better indicator of the likely location of crash load on another vehicle. More studies are needed to determine the possibility of constraining the time period of measurement such that the height of force or the average height of force are reliable indicators of the location of the primary structure.

It is important that additional investigations be conducted to establish any relationships that may exist between these parameters and actual safety in vehicle-to-vehicle collisions. But, in order to examine such correlations, it is necessary to use the correct height of force (or average height of force) and not values with large inherent errors.

ACKNOWLEDGEMENTS

We wish to express our thanks to David van Egeren for writing the software for processing the results reported in this paper. Thanks are also due to the members of the TRL (UK) team for helping to set up and conduct the tests as well as to the members of the MIRA (UK) team who helped conduct the tests and process the data for the 50mm load cell tests.

REFERENCES

1. Summers S., Hollowell, W.T., Prasad, A., *"NHTSA's Research Program for Vehicle Compatibility"*, 18th International Conference on the Enhanced Safety of Vehicles, Paper 307, May 2003.
2. Verma, M.K., Lavelle J., Lange R.C., *"Perspectives on Vehicle Crash Compatibility and Relationship to Other Safety Criteria"*, 18th International Conference on the Enhanced Safety of Vehicles, Paper 412, May 2003.
3. Reilly, P.O., *"Status Report of IHRA Compatibility and Frontal Impact Working Group"*, 18th International Conference on the Enhanced Safety of Vehicles, Paper 402, May 2003.
4. Gabler, H.C., *"The Evolution of Side Crash Compatibility Between Cars, Light Trucks and Vans"*, Society of Automotive Engineers, Paper 2003-01-0899, March 2003.
5. Summers S., Hollowell, W.T., Prasad A., *"Design Considerations for a Compatibility Test Procedure"*, Society of Automotive Engineers, Paper 2002-01-1022, March 2002.
6. Jerinsky, M., Hollowell, W.T., *"NHTSA's Review of High Resolution Load Cell Walls' Role in Designing for Compatibility"*, 18th International Conference on the Enhanced Safety of Vehicles, Paper 393, May 2003.
7. Edwards, M., Davies, H., Hobbes, A., *"Development of Test procedures and Performance Criteria to Improve Compatibility in Car Frontal Collisions"*, 18th International Conference on the Enhanced Safety of Vehicles, Paper 86, May 2003.
8. Turley, J., Sheldon, A., *"Vehicle Compatibility – Influence of Load Cell Size on Average Height of Force & CV"*, SAE Government Industry Meeting, May 2003.

Design Strategies

946089

Car size and safety:
a review focused on identifying causative factors

Leonard Evans

General Motors NAO R&D Center
Automotive Safety and Health Research Department
Warren MI 48090-9055
USA

Paper No. 94 S4 W 28

ABSTRACT

In the last few years a number of additions to the technical literature on relationships between car size or mass and occupant risk of fatality or injury have appeared. This new information is reviewed, synthesized and used as the basis for additional calculations aimed at better identifying causal factors. It is concluded that if a car crashes head-on into a 12 000 kg truck, the car driver is 36% more likely to be killed in a 900 kg car than in an 1800 kg car solely as a result of differing Newtonian kinematics. Five studies from two countries consistently support that when cars of similar mass crash head-on into each other, driver risk is inversely related to the common car mass. Size is the dominant causative factor in this relationship, and in the higher rollover risk in lighter cars. Mass and size are causal factors in single-car nonrollover crashes. Mass exercises a dominant causal effect on car driver risk in crashes between vehicles whose masses differ by more than about 10%. All cars becoming lighter/smaller would have only a small effect (most likely an increase) on net driver risk in two-car crashes, such crashes accounting for 22% of car-occupant fatalities. As 70 % of car-occupant deaths occur in crashes involving only one car, and lighter/smaller cars increase driver risk in all of these, a smaller/lighter fleet leads to increased casualties regardless of any interactive effects in the two-car crash component. Because mass is a dominant causal factor in crashes that account for over 50% of car-occupant fatalities, mass reductions (even if size remained unchanged) would lead to casualty increases. Any measure that reduces the mass of cars, even if car size remains unchanged, will increase car-occupant fatalities.

INTRODUCTION

One of the most firmly established effects in traffic safety is that when crashes occur, occupants of smaller lighter vehicles are at greater injury risk than are occupants of larger heavier vehicles. The early studies of Campbell and Reinfurt [1973], Negri and Riley [1974], Joksch [1976], and Grime and Hutchinson [1979] established that when cars of unequal weight crash into each other, fatality and injury risk is greater in the lighter car. Higher risks in lighter, smaller cars are found in data and investigations from many countries: : Germany [Kramer, Scholpp and Otte 1993;

Ernst et al. 1993]; France [Thomas et al. 1990]; the United Kingdom [Jones 1976; Grime and Hutchinson 1979]; Sweden [Koch, Kullgren and Tingvall 1993; Folksam 1990]; Finland [Roine and Kulmala 1994]; and the United States [Evans 1986; Partyka 1988; Partyka and Boehly 1989; Klein, Herts and Borener 1993; Kahane 1993; Insurance Institute for Highway Safety 1987].

Despite an extensive literature, many important aspects of the relationship between occupant risk and vehicle size remain uncertain, particularly identification of causative factors. The size and mass (or weight) of vehicles are closely related, so that any observed relationship between occupant risk and vehicle size does not establish vehicle size as a causative factor; a similarly systematic relationship with mass would have emerged if mass had been chosen as the independent variable. This paper aims at further clarifying such questions by synthesizing results from studies published in the last few years, and by performing some new analyses.

For most purposes mass is the most convenient independent variable because it is, conceptually at least, a fairly unambiguous physical characteristic of a vehicle. It additionally tends to be known, and is often part of the legal description of the vehicle for taxation purposes. In contrast, size is a more nebulous characteristic not quantified by one measurement. Many measures of vehicle size have been used, including overall length, wheelbase, trackwidth (the distance between left and right wheels) and interior volume. In this paper, we examine driver risks versus vehicle mass, but later discuss the findings in terms of relationships between vehicle size and mass, and probable causative factors.

Much of the material to be presented is for fatalities, because reliable information for this highest severity outcome is available from the Fatal Accident Reporting System [National Highway Traffic Safety Administration 1991]. Figure 1 shows that 45 percent of fatally injured car occupants are killed in single-car crashes, and an additional 25 percent are killed in crashes with vehicles other than cars (trucks, vans, etc.). Thus for 70 percent of the car-occupants killed, the only car characteristics related to their deaths are the characteristics of the car in which they were traveling; the mass of the other cars on the road is irrelevant. However, in two-car crashes the mass of the other involved car is important. Much of this paper is devoted to two-car crashes even though they account for far fewer deaths than single-car crashes. This is done because various methods that are not applicable to other types of crashes allow us to focus on detailed aspects of two-car crashes. For single-car crashes, traffic crash data provide information on the striking car, but nothing on the struck object (most commonly a tree). However, in a two-car crash, the struck object is another car, so that important characteristics of it are available in the data set. In many cases precise relationships are obtained which offer insight into the effect of mass in other types of crashes. Physical mechanisms identified in two-car crashes offer possible explanations of mechanisms in other types of crashes.

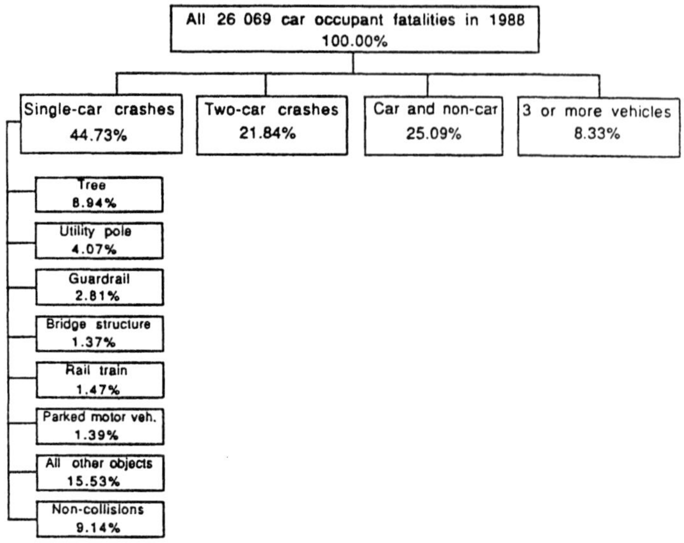

Fig. 1. Distribution of car occupants killed according to the number of vehicles involved. The values in all cells indicate the percents of the total 26 069 car occupant fatalities coded in the 1988 FARS file. From Evans [1991, p 47].

TWO-CAR CRASHES

Crashes between cars of unequal mass

For every crash between two cars of known mass, we can define a mass ratio, μ, as

$$\mu = \frac{\text{Mass of the heavier car } (m_2)}{\text{Mass of the lighter car } (m_1)} \tag{1}$$

and a driver fatality risk ratio, R, as

$$R = \frac{\text{Probability of a driver fatality in lighter car}}{\text{Probability of a driver fatality in heavier car}} . \tag{2}$$

<u>**Empirical findings**</u> - Using FARS data, Evans and Frick [1993a] examined R versus μ. Figure 2 shows results for head-on crashes (regardless of driver belt use, car model year, driver age, etc.) The relationship fitted to the empirical data is

$$R = \mu^u . \tag{3}$$

Values of the parameter u, and its associated standard error, are given in Evans and Frick [1993a] for 30 cases (head-on compared to all impact directions, belted compared to unbelted drivers, urban compared to rural, night compared to day, drivers of same sex and similar age, etc.). All the results are inferred from FARS data in a largely assumption-free calculation. The effect of mass ratio is large. For example, Fig. 2 indicates that if one car crashes head-on into another 20 percent heavier, then the driver in the lighter car is twice as likely to be killed as is the driver in the heavier car. If one of the cars is twice as heavy as the other, the driver in the lighter car is 13 times as likely to be killed as is the driver in the heavier car. The finding of so large and robust an empirical relationship invites a search for understanding at a more fundamental and theoretical level, especially with a view to identifying underlying causative mechanisms.

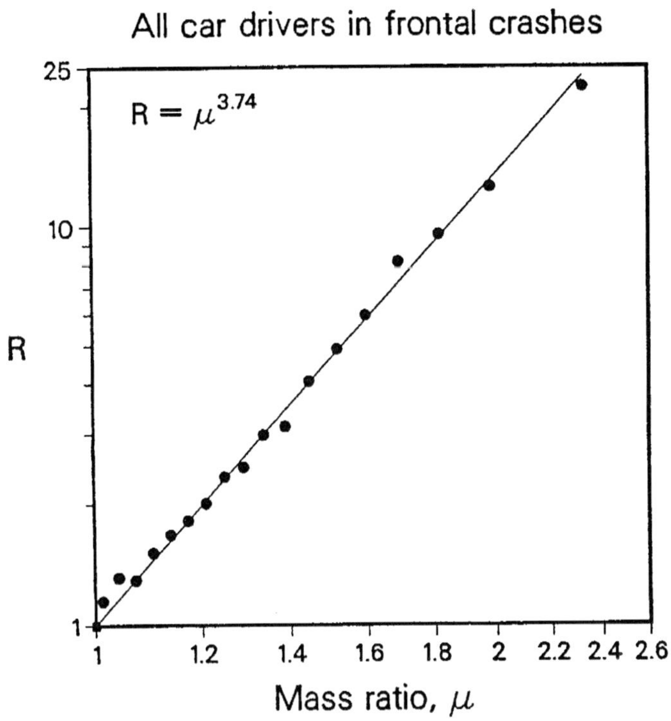

Fig. 2. The ratio, R, of driver fatalities in the lighter compared to in the heavier car versus the ratio, μ, of the mass of the heavier to the mass of the lighter car for frontal crashes (both cars with principal impact point at 11, 12 or 1 o'clock). 1975-1989 FARS data. From Evans and Frick 1993a.

<u>An explanation based on Newtonian Mechanics</u> - Evans [1994] considers two cars traveling in opposite directions crashing head-on into each other, and assumes that the crash is inelastic -- that is, after the crash the vehicles remain in contact as one combined object. Because of the assumed symmetry about the common axis, the combined mass will also travel along this axis. The assumption of an inelastic crash is a reasonable approximation because in most cases crashed vehicles remain relatively close to each other after the crash. By applying the law of conservation of linear momentum, the change in speed, Δv, that each car undergoes as a result of the crash can be computed in terms of the initial speeds and masses of the two cars. The ratio of the changes in speed depends only on the mass ratio, and is given by

$$\Delta v_1/\Delta v_2 = m_2/m_1 = \mu, \tag{4}$$

where Δv_1, Δv_2, are the speed changes of the lighter and heavier cars, respectively.

The speed change a vehicle undergoes in a crash is the best available measure of crash severity for vehicles that have not been specially instrumented for crash testing. It is a measure that has been widely used and studied [Ricci 1980; Roberts and Compton 1993; Joksch 1993]. Figure 3 shows the probability of a driver fatality in a frontal impact crash versus Δv obtained from National Accident Sampling System (NASS) data for 1982-1991 [National Highway Traffic Safety Administration 1988]; corresponding figures for belted and unbelted drivers, but for all crashes regardless of impact direction, are given in Evans [1994]. The curve fitted is the *rule of thumb* suggested by Joksch [1993] with the general form

$$P = (\Delta v/\alpha)^k, \tag{5}$$

where P is the probability of fatality and α and k are parameters determined by a least squares linear regression fit to the log-transformed data. The relationship applies provided $\Delta v \leq \alpha$; for $\Delta v > \alpha$ we assume P = 1. If P_1 and P_2 are the fatality risks to two drivers involved in the same crash, then the ratio of their fatality risks, R, is given by

$$R = P_1/P_2 = P(\Delta v_1) \div P(\Delta v_2) = (\Delta v_1/\alpha)^k \div (\Delta v_2/\alpha)^k = (\Delta v_1/\Delta v_2)^k; \tag{6}$$

Substituting eqn 4 gives the result that

$$R = \mu^k. \tag{7}$$

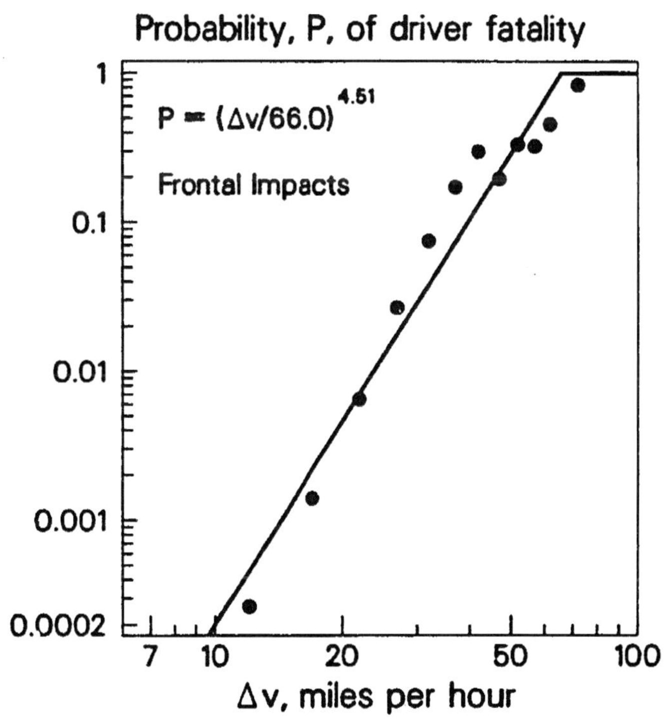

Fig. 3. Probability of a driver fatality in a frontal impact nonrollover crash versus the speed change, Δv. The points plotted are computed from 1982 - 1991 NASS tabulations provided by Susan Partyka.

Thus, the eqn 5 relationship between fatality risk and Δv, combined with simple Newtonian mechanics, generates the empirically derived relationship given by eqn 3 and shown in Fig. 2. The parameter k = 4.51 derived from the NASS data is, given the approximate and general nature of the calculation, in reasonably good agreement with the value u = 3.74 from the head-on two-car crash data in Fig. 2. (For all crashes, regardless of direction of impact, k = 3.80 compared to u = 3.53.)

The comparison presented here is for all driver fatalities, regardless of belt use. The corresponding analyses are given in Evans [1994] for severe injuries (AIS ≥ 3) as well as fatalities, and in each case for unbelted and belted drivers. Separating by belt use so reduced sample sizes that all crashes, regardless of impact direction, were included. The effect of mass ratio for injuries, although still considerable, was substantially less than for fatalities (for unbelted drivers, k = 2.22 for injuries compared to k =3.54 for fatalities; the corresponding values for belted drivers were 4.57 and 2.62). The lower values for injuries finds support in Wood [1993a], who finds an exponent of 2.5 for injuries, based on a calculation using equations relating dynamic and static crush characteristics of cars to injury risk. If one car crashes into another twice its mass, k = 3.54 for fatalities implies that the driver in the lighter car is 12 times as likely to be killed as the driver in the heavier car; k = 2.2 for injuries implies that the driver in the lighter car is 5 times as likely to be injured as is the driver in the heavier car.

Thus simple Newtonian mechanics, together with a relationship between fatality risk and Δv, implies that the risk in the lighter compared to in the heavier car increases as a power function of the mass ratio, as observed directly from data. This agreement definitively establishes that mass causes a large difference in risk, and identifies mass as the main causative factor in the relationship in Fig. 2. Such identification does not exclude the possibility that other factors, such as size, may also be important.

Crashes between cars of the same wheelbase but different mass

Another approach to isolating causative factors in the relationship between relative fatality risk and mass ratio is to examine what happens when cars with identical wheelbases, but different masses, crash into each other. Results of such an analysis [Evans and Frick 1992] are shown in Figure 4. Each point represents crashes between cars coded in the FARS data as having identical wheelbase but different mass. Such a stringent selection criterion leads to few data and large standard errors.

Fitting eqn 3 to the data gives u = 3.8, a value not materially different from u = 3.5 for all crashes regardless of wheelbase. Thus, when cars of the same size, as indicated by wheelbase, crash into each other, the effect of mass is not distinguishably different from the case when no restrictions are placed on size. Size, as indicated by wheelbase, cannot contribute to the effect, so that mass is unambiguously identified as the causative factor. This result fits the explanation above in terms of Newtonian mechanics -- two cars of identical wheelbase crashing into each other will have Δv ratios in inverse proportion to their masses. This and the previous result suggests that increasing the mass of a vehicle reduces the risk its occupants face in a crash, regardless of the source of the additional mass (provided it is not free to move inside the vehicle). Thus adding cargo or restrained passengers reduces your risk when you drive, but increases the risk of those in vehicles into which you crash.

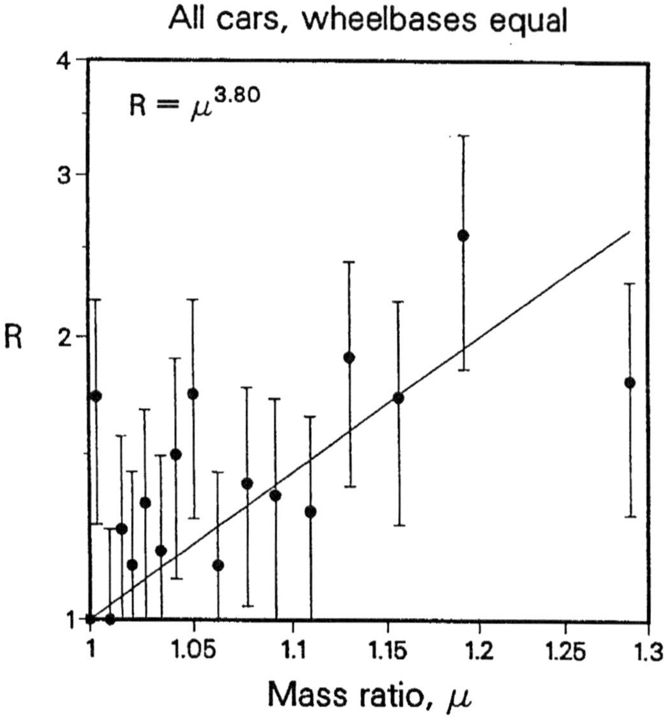

Fig. 4. The ratio, R, of driver fatalities in the lighter compared to in the heavier car versus the ratio, μ, of the mass of the heavier to the mass of the lighter car for crashes between cars with <u>identical wheelbase</u>, based on cars of all model years in 1975-1989 FARS data. From Evans and Frick [1992].

Crashes between cars of the same (or similar) mass

The relationship between driver risk and car mass when cars of the same (or similar) mass crash head-on into each other has been examined using five independent data sets (Fig. 5). The FARS results [Evans and Wasielewski 1987] are for driver fatalities, with exposure being estimated using the *pedestrian exposure approach* described later in the discussion of single-car crashes. The NC and NY results [Evans and Wasielewski 1987] are driver injuries per registered car from data for North Carolina [Campbell and Reinfurt 1973] and New York [Negri and Riley 1974]. The other two sets of results are from Ernst et al. [1993] showing driver serious injuries per driver involved in any type of crash in the German State of North-Rhine/Westphalia, supplemented by Federal data. The Ernst et al. [1993] data are given for rural crashes, and for crashes in built-up areas, giving the two relationships in Fig. 5. All five data sets are scaled using the procedure in Evans and Wasielewski [1987] in which two 1800 kg cars crashing head-on into each other defines a risk of 1.0.

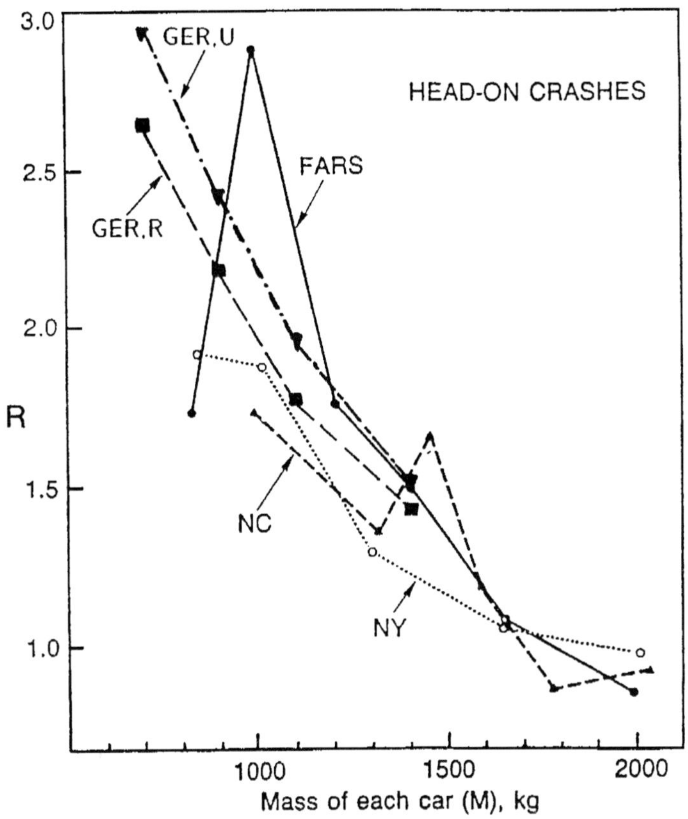

FARS	Fatalities [Evans and Wasielewski 1987]
NC	Injuries in North Carolina [Evans and Wasielewski 1987]
NY	Injuries in New York State [Evans and Wasielewski 1987]
GER.R	Injuries on rural roads in Germany [Ernst et al. 1993]
GER.U	Injuries on roads in built-up areas in Germany [Ernst et al. 1993]

Fig. 5. Relative risk, R, of driver injury or fatality when cars of similar mass crash head-on into each other versus M, the common mass of each car. In all cases the data are scaled so that a regression fit passes through R = 1.0 at M = 1800 kg. The data of Ernst et al. [1993] added to the figure on p 70 of Evans [1991].

All five data sets consistently support that when two 900 kg cars crash head-on into each other, the risk of driver death or injury is about twice as great as when two 1800 kg cars crash head-on into each other. Although Fig. 5 plots the relationships as a function of mass, mass cannot be the causative factor. In terms of the Newtonian mechanics of the crashes, mass is irrelevant when two cars of the same mass crash into each other.

Wood [1993a] investigated injury risk when identical cars crashed head-on using equations relating injury risk to the dynamic and static crush characteristics of cars. He finds that when cars of the same mass crash head-on into each other, driver injury risk increases in inverse proportion to mass; a relationship he notes fits well the same five sets of empirical data displayed in Fig. 5.

Another part of the Evans and Frick [1992] study was an examination of crashes between cars of similar mass, but different wheelbase. While no effect above the noise in the data was observed, the results were compatible with the interpretation that risk increased with decreasing wheelbase in accord with Fig. 5 when the relationship between mass and wheelbase discussed later is taken into account. The similarity between the relationships for injuries and fatalities is supported by the data of Thomas et al. [1990], even though these data, because of exposure limitations, do not support more specific conclusions [Evans 1990].

Two-car crash system-wide effects

Because a driver's risk in a two-car crash depends not only on the mass of the car in which that driver is traveling, but also on the mass of the other involved car, many questions have arisen regarding the overall effect on safety of a general change in the size of cars. Any such discussions should keep firmly in mind the information in Fig. 1 which shows that such considerations are irrelevant for crashes that kill 70 percent of car occupants.

When a driver changes from a heavier to a lighter car:

1) That driver's risk increases

2) The risk that driver's lighter car poses to all other drivers declines.

If the magnitude of the first of these is larger than the second, then the traffic system will be less safe, and vice versa. In order to answer the question whether system safety increases or decreases, we need to know not only the magnitudes of items 1 and 2, but we need to know them with higher precision than the expected small difference between them.

Figure 5 shows that replacing a fleet of identical mass cars with another fleet in which every car was 10 percent lighter would increase casualties by 10 percent. Although this result is simple and relatively reliable, it cannot be generalized in any simple way to actual fleets that contain a wide mix of car sizes.

Klein, Hertz and Borener [1993] applied logit analysis to two-car crashes in Texas and Maryland to estimate the change in casualties that resulted from the reduction in the average mass of a car from 3700 pounds in 1970 to 2700 pounds in 1982 (1678 kg to 1224 kg, a reduction of 454 kg). In their calculations they reduced the masses of all cars by the same proportion to achieve this 1000 pound reduction, so that the same relative mix of cars by mass was maintained. They find that such a change in the average mass of cars in the fleet would increase casualties by 14.3 percent (based on the Texas data) and by 4.3 percent (based on the Maryland data).

Using the *pedestrian exposure approach* described later, Evans and Frick [1993b] estimated driver fatality risk as a function of the mass of each of the two cars involved. They found that replacing any car by a lighter one leads to an increase in net risk for nearly all two-car crashes. The effect was in the opposite direction for some large values of mass ratio, an effect that now appears related to the smaller cars in their sample preferentially benefiting from crashworthiness improvements [Evans and Frick 1994]. However, averaged over the mix of cars into which any particular car might crash, Evans and Frick [1993b] concluded that replacing any car in the fleet by a lighter one always led to a net increase in fatality risk.

While the relationship between risk and Δv in Fig. 3 was useful for investigating large mass-ratio effects corresponding to driver risk ratios that vary by over an order of magnitude, it is unsuitable for examining the small effects discussed here. The Fig. 3 relationship estimates that when cars of identical mass crash into each other, the outcome is mass-independent, in clear contradiction of Fig. 5. If we proceed by making the incorrect assumption that the probability of injury versus Δv is independent of car mass (equivalent to ignoring size effects), then we can calculate the effect of substituting one car with another of different mass. If a crash involving two equal mass cars is replaced by a crash in which one of the cars is 10 percent lighter than the other, the mass ratio change is computed to increase the fatality risk to the driver of now lighter car by 26 percent, but decrease the risk to the driver in the now heavier car by 22 percent. The calculated net effect is that the calculated sum of the risks for the drivers combined increases by 2 percent. If the original mass ratio differs from one, then, in general, decreasing the mass of the heavier car reduces net risk, while reducing the mass of the lighter car increases net risk. As mass ratio becomes larger, the estimated change in combined driver risk increases, so that at some point it will exceed the bias. If vehicles are grossly dissimilar in mass, decreasing the mass of the lighter one generally increases that driver's risk more than it decreases the risk to the driver in the much heavier vehicle. An exception occurs if the crash severity is so great that the driver of the lighter vehicle is killed ($\Delta v > \alpha$); making that vehicle still lighter cannot increase this risk, but will reduce the risk to the driver in the heavier vehicle, for a net decrease in risk.

SINGLE-CAR CRASHES

While single car-crashes account for 45 percent of car-driver deaths, it is not possible to perform the diverse collection of analyses that have been applied to two-car crashes. A central difficulty is that of exposure, made all the more difficult by additional biases that apply specifically to single-car crashes. The probability that a collision with any object immobilizes a car decreases as the mass of the car increases. If the car driver is the only one to observe the crash, and the car is driveable, the crash may never be reported. Accordingly, such measures as fatalities per police-reported crash systematically underestimate how much more likely a driver of a small car is to be killed compared to a driver of a large car in single car crashes.

The *pedestrian exposure approach* [Evans 1991, p 72; Evans 1984a] circumvents these problems. When a car crashes into a tree (the most commonly struck object -- Fig. 1), this crash will generally not be recorded in FARS. However, if a car crashes into a pedestrian, this crash will be in FARS if the pedestrian is killed. If we assume that cars crash into pedestrians at the same rate that they crash into objects such as trees, then the number of pedestrian fatalities will estimate exposure provided that the probability that a pedestrian is killed when struck is independent of car mass. This is likely to be fairly correct, because the lightest car is so much heavier than the heaviest pedestrian.

Figure 6 shows the number of drivers killed in cars of a given mass divided by the number of pedestrians killed in crashes involving cars of the same mass. This is interpreted to indicate the relative probability that a driver is killed in a crash, other factors being equal. Supporting evidence that what is shown is a physical effect related to the car is that the phenomenon is essentially independent of driver age, a factor that has a large influence on crash rates. The effect of car mass in Fig. 6 is substantial -- the driver in a 900 kg car is 2.4 times as likely to be killed as is a driver in an 1800 kg car. Such a difference is much larger than any likely uncertainties in the assumptions on which the exposure approach rests, although these assumptions likely overestimate the effect. In terms of the earlier

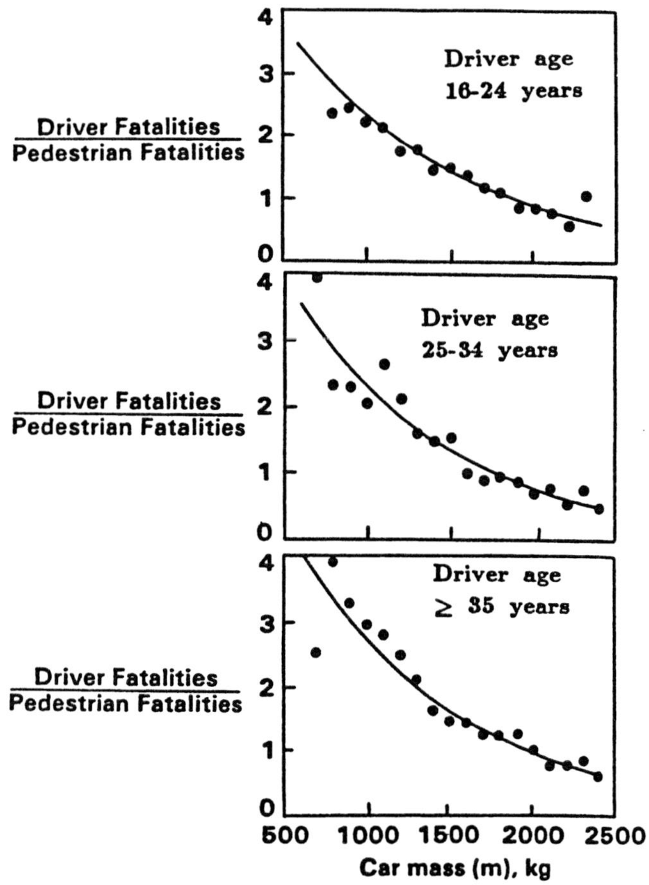

Fig. 6. The ratio of drivers killed in single-car crashes to the number of crashes in which pedestrians were killed versus the mass of the involved cars (FARS 1975-1980). This ratio is approximately proportional to the probability that a driver is killed given that a single-car crash occurs, so the graphs are interpreted to reflect driver fatality risk in single-car crashes versus car mass. From Evans [1984a; 1991, p 73].

discussion of the Newtonian mechanics of crashes, the Δv sustained by a 80 kg pedestrian in an inelastic (not too plausible assumption) collision with an 1800 kg car is 4 percent greater than if struck by a 900 kg car. Applying eqn 6 as a very rough approximation associates a 20 percent greater pedestrian fatality risk with this Δv, compared to an assumption of no effect.

Rollover crashes

We consider single-car crashes as either rollover (account for 18 percent of car occupant fatalities) or nonrollover (27 percent of car occupant fatalities), because different physical mechanisms apply to each. Fig. 7 shows the number of occupants killed in single-car rollover crashes per registered vehicle [Partyka and Boehly 1989]. Although the large effect (an occupant fatality is 1.8 times as likely in a 900 kg car as in a 1800 kg car) is plotted versus car mass, it is not mass, but other vehicle properties correlated with mass, that generate the physical mechanisms. Kahane [1993] writes "Short light cars usually have less directional stability than long, heavy ones. Narrow cars have less rollover stability than wide cars. Since 'small' cars are shorter, lighter and narrower than full-sized cars, they tend to have lower directional and rollover stability." Other factors being equal, the lower the center of gravity the greater is the resistance to rollover. The center of gravity is determined by vehicle function. While geometrical properties of a vehicle are related to resistance to rollover, all other factors being equal, the influence they exercise is small compared to the much larger influences of driver behavior and use patterns. British double-decker busses have far lower rollover rates than sports cars. (My searches failed to uncover a single case of a double-decker bus rolling over -- if any reader knows of such a case, I'd be grateful if they could provide information).

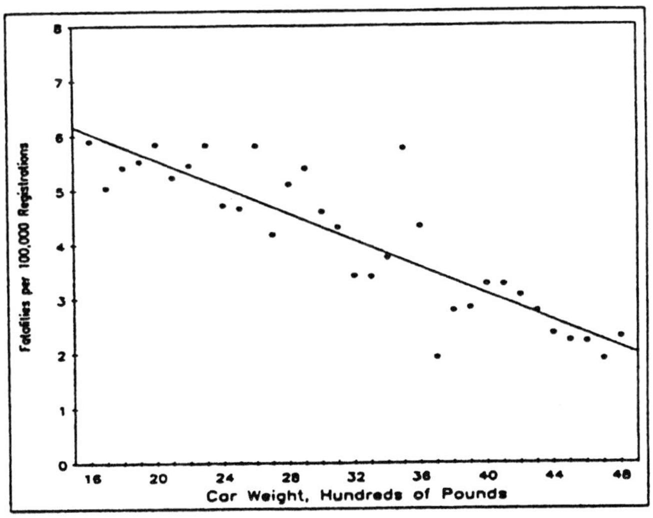

Fig. 7. Occupant fatalities per 100 000 registered cars in single-car rollover crashes. From Partyka and Boehly [1989].

This paper addresses how vehicle size and mass affect risk of death and injury, given that a car crash occurs. The rate *driver deaths per registered car* reflects many properties of the driver as well as of the vehicle. Smaller vehicles are more likely to be driven by younger drivers [Evans 1985a]. Younger drivers have higher crash rates, but are less likely to be killed in a crash of given severity [Evans 1988; 1991], so that these factors will exercise influences in opposite directions. There are indications [Evans 1984b] that smaller cars (corrected for driver age) are associated with lower crash involvement rates, though such an effect is not apparent in insurance claims [O'Neill 1985]. Rothengatter [1988] suggests that larger cars may be associated with higher travel speeds. Even small increases in average speed would have large effects on fatality rates, because fatality rate increases as the fourth power of speed [Evans 1991, p 154]. I consider that the net effect is that *driver deaths per registered car* somewhat underestimates the effect of car mass, given that a crash occurs, especially in view of the results the same measure gives for nonrollover crashes.

Nonrollover crashes

Partyka [1989a] examined driver deaths per registered car versus car mass and found a statistically significant increasing risk with decreasing mass (equivalent to a driver in a 900 kg car being 15 percent more likely to be killed than a driver in a 1800 kg car). The corresponding difference for occupants rather than drivers was not statistically different from zero [Partyka and Boehly 1989]. Concentrating on drivers provides results that are easier to interpret because larger cars have higher occupancy rates, so examining occupant fatality rates adds a non-vehicular confounding factor.

Partyka and Boehly [1989] also applied multivariate methods to the NASS data, and obtained a relationship between car mass and moderate driver injury rate per towaway crash indicating that the risk in 900 kg cars is 30 percent higher than in 1800 kg cars. Quite similar results were obtained in the same study using police-reported data from Michigan, Maryland and North Carolina. The study by Klein, Hertz and Borener [1993] also examined single-car nonrollover crashes, and concluded that a reduction of average car mass from 3700 pounds to 2700 pounds would increase driver fatalities by 9.8 percent, with 95 percent confidence limits from 0.5 percent to 19.9 percent.

CRASHES BETWEEN CARS AND VEHICLES OTHER THAN CARS

The relative risks when cars crash into other vehicles, obtained in a largely assumption-free way from FARS data [Evans and Frick 1993a], show remarkably consistency. For every row the values decrease in every step from left to right, implying a systematic decrease in the relative risk to the car driver as the car's mass increases from light, to medium, to heavy, regardless of the other vehicle involved. The data in Table 1, being for relative risk, do not estimate quantitatively differences in risk faced by drivers of small and large cars (or the differences in risk they impose on others). However, they do lead to the clear conclusion that in crashes with a wide spectrum of other vehicles, drivers of heavier cars are at lower risk than drivers of lighter cars. Thus lowering the average masses of cars increases the net risk to car drivers in crashes with other vehicles. As the great majority of car-driver deaths in crashes with other vehicles that are not cars involve crashes with trucks, we examine such crashes in more detail.

Table 1: Relative driver risks when cars crash into other vehicles. For example, when light cars and motorcycles crash into each other, the motorcycle driver is 42 times as likely to be killed as is the car driver; when light cars and heavy trucks crash into each other, the car driver is 44 times as likely to be killed as is the truck driver. Extracted from the table in Evans and Frick [1993a], which gives all combinations of the listed nine vehicles.

Other vehicle ↓	Mass class* of car		
	Light	Medium	Heavy
Moped	1/139	1/202	1/205
Motorcycle	1/42	1/85	1/153
Light car	①	1/3.1	1/7.7
Medium car	3.1	①	1/2.4
Heavy car	7.7	2.4	①
Pick-up	7.1	2.7	1.3
Van	9.3	3.3	1.7
Medium truck	34	18	12
Heavy truck	44	29	22

**Definitions*

Light car: Mass from 655 kg to 1227 kg, mean 1014 kg
Medium car: Mass from 1227 kg to 1599 kg, mean 1428 kg
Heavy car: Mass from 1600 kg to 2606 kg, mean 1833 kg
Medium truck: Gross Vehicle Weight (GVW) > 10 000 pounds (4536 kg), but not a heavy truck
Heavy truck: GVW > 26 000 pounds (11 794 kg), or tractor-trailer combination, or with cargo trailer(s), or truck tractor pulling no trailer.

Car-Truck Crashes

Klein, Hertz and Borener [1993] examined cars crashing into trucks of gross vehicle weight rating greater than 10 000 pounds (4536 kg). They concluded that a reduction of average car mass from 3700 pounds to 2700 pounds would increase net driver fatalities by 11.0 percent, with 95 percent confidence limits from 0.5 percent to 22.7 percent. Applying the *pedestrian exposure approach* to car-truck crashes showed that car-driver fatality risk was related to car mass not too differently from the Fig. 6 relationships for single-car crashes [Evans 1984a].

The analysis using Newtonian mechanics applied to two-car crashes [Evans 1994] can be extended as follows. Consider a head-on crash between a car of mass m_1 and a truck of mass m_T. The speed change for the car, Δv_1, is given by

$$\Delta v_1 = (v_1 + v_T)\, m_T/(m_1 + m_T)\,, \tag{8}$$

where v_1 and v_T are the pre-crash speeds of the car and truck. Let us imagine this same crash repeated with all factors the same, except that the mass of the car is increased to m_2. The resulting smaller speed change, Δv_2, is given by substituting $m_1 = m_2$ in eqn 8. The ratio of the speed changes for the two cars is given by

$$\Delta v_1/\Delta v_2 = (m_2/m_T + 1)/(m_1/m_T + 1)\,. \tag{9}$$

Comparing 900 kg and 1800 kg cars crashing into a heavy truck of mass 12 000 kg gives a speed change ratio for the cars of 1.070, which, when substituted into eqn 6 indicates that the driver in the lighter car is 36 percent more likely to be killed than is the driver in the heavier car. This effect is due exclusively to the mass-induced difference in the speed change given by eqn 9. In analytical relations giving driver fatality risk as a function of the mass of each of the cars involved, the risk of death in one of the cars still increased with decreasing car mass even when the mass of the other car approached infinity [Evans and Frick 1993b], a case that is conceptually somewhat equivalent to a car crashing into a very large truck. As the mass of the truck becomes infinite, the mass of the car becomes irrelevant in eqn 9. However, the heavier car has safety advantages not reflected in eqn 9. The heavier car, if larger, will provide greater crush space. Because the truck is not a perfectly rigid body, the heavier car will penetrate further into it, thereby providing a longer stopping distance.

Car mass also can affect car-driver risk in sever crashes with much lighter vehicles. If a 200 kg motorcycle and a 900 kg car each traveling at 60 mph crash head-on into each other, the probability that the car driver is killed (from eqn 9 and Fig. 3) is about 1 percent, compared to a fifteen times smaller risk for a driver of an 1800 kg car.

THE RELATIONSHIP BETWEEN SIZE AND MASS

Consider a hypothetical construct (as in Evans and Frick [1992]), in which all cars are made from material of the same density. If cars were of identical shape, differing only by a scale factor, then mass would be proportional to any linear dimension to the power three. However, regardless of their size, cars must be of sufficient height to accommodate seated humans. If all cars had the same height, so that only length and breadth varied, then mass would be proportional to length (or breadth) to the power two. Real cars are likely to be intermediate between these two hypothetical cases, suggesting a relationship between mass and wheelbase of the form

$$M = \lambda W^{\beta}\,, \tag{10}$$

where M is the mass of the car, W is a linear dimension (other than height), and λ and β are constants, with β expected to be between 2 and 3. Figure 8 shows mass versus wheelbase for 3824 unique mass/wheelbase combinations available from the FARS data for cars of model year 1980 or later [Evans and Frick 1992]. The fit to eqn 10 gives

$$M = (6.47\ W)^{2.51}, \tag{11}$$

thus validating our intuitive understanding that β should be between 2 and 3.

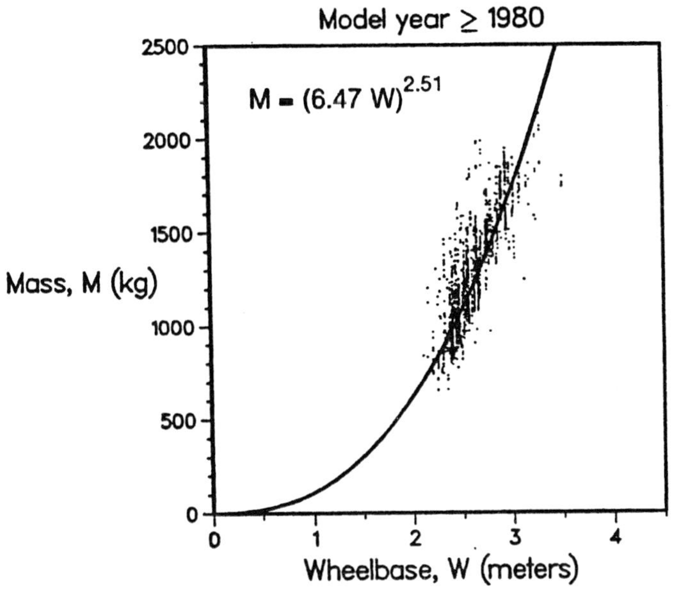

Fig. 8. Mass versus wheelbase based on for MY > 1980 cars coded in 1975-1989 FARS data. From Evans and Frick [1992].

Wood [1993a] fitted data for overall length, L, versus mass for European cars, and obtained an equation that can be expressed as

$$M = (3.89\ L)^{2.48}, \tag{12}$$

where, consistently with eqn 11, L is in meters and M is in kg. The β coefficients in eqns 11 and 12 agree closely. The constant terms agree provided the wheelbase is equal to 3.89/6.47 times the overall length; that is, the wheelbase is 60.0 percent of the overall length. Wood [1993b] computed the ratio of the wheelbase to the overall length for his data and obtained 59.5 percent.

Additional data on the close relationship between measures of size and weight are provided by Kahane [1993] who finds correlation coefficients between curb weight and wheelbase, curb weight and trackwidth, and wheelbase and trackwidth of 0.93, 0.92 and 0.91 respectively. Any observed empirical relationship with any one of these quantities is going to provide a similarly good relationship with any of the others. In multivariate analyses, simply as a result of eqn 11, parameters associated with length will be about 2.5 times as large as those associated with mass. It seems possible that inappropriately attaching more meaning than this to larger parameters in regressions on size compared to mass could have contributed to suggestions [O'Neill, Joksch and Haddon 1974] that size was a more important causative factor than mass.

SUMMARY AND DISCUSSION OF RESULTS

Table 2 summarizes the results found, in all cases derived for car drivers. Although occupants have different risks depending on seating location [Evans and Frick 1988; 1989], passenger risk is sufficiently closely proportional to driver risk that the results may be assumed to apply, with little additional error, to all occupants in the car. Apart from the 8 percent of car occupants killed in crashes involving three or more vehicles, Table 2 includes all car occupants killed in crashes.

Causative factors

For rollover crashes and crashes between cars of similar mass, size is the dominant causative factor. For all other crash types, mass causally affects outcome, and is the dominant causal factor in crashes between two vehicles whose masses differ by more than about 10 percent. This is to be expected because of Newtonian mechanics. Indeed, the results presented here show that even in crashes between cars and large trucks, kinematic effects due to the mass of the car have a larger influence on fatality risk than was generally assumed previously. For example, if a car crashes into a heavy truck of mass 12 000 kg, the car driver is about 36 percent more likely to be killed in a 900 kg car than in an 1800 kg car solely as a result of the slightly larger speed change the lighter car undergoes.

Mass plays a role in the dynamics of crashes beyond that captured in the speed change calculations. Essentially all objects into which cars crash may break, bend, dent, distort or tear. Each of these effects reduces the forces on occupants. The heavier the striking vehicle, the more likely is a struck object (such as a tree -- the most common fixed object in fatal crashes) to break (assuming car force deflection characteristics are independent of car mass). If the vehicle uproots the tree or breaks it, rather than bouncing back from it, risk to occupants is dramatically reduced. Even the proverbial brick wall from which a light car will bounce back may still be penetrated by a heavier vehicle, with beneficial effects to its occupants. Many of the objects struck have some dynamical properties in common with some members of the wide spectrum of vehicles for which we have definitive information (from mopeds through large trucks). All the mechanisms by which the striking car may move or damage the struck object reduces the speed change the car undergoes, and thereby lower driver fatality and injury risk; small changes in speed change, or the time over which the speed change takes place, translate into large changes in occupant risk.

In nearly all crashes, increased mass, as such, reduces occupant risk. In barrier crash tests, mass is rendered irrelevant. The design goals for the barrier are that it be equivalent to an object of infinite hardness and infinite

Table 2: Summary of how the mass of the car in which a driver is traveling affects that driver's risk in a crash.

Crash Type	Main Causative Factor	Quantification / Comment
Two-car crashes		**22% of car-occupant fatalities**
Cars of dissimilar mass	Mass	$R = \mu^{3.7}$: If one car is twice as heavy as other, driver in lighter car about13 times as likely to be killed as driver in heavier
Cars of same wheelbase	Mass	$R \approx \mu^{3.8}$: If masses differ by 10%, driver in lighter car about 40% more likely to be killed than driver in heavier car
Cars of same (or similar) mass	Size	$R \propto 1/M$: If common mass, M, is halved, risk doubles
System-wide effects	Mass and size	Likely small increase in system-wide two-car crash risk if cars become lighter/smaller; difficult to quantify
Single-car crashes		**45% of car-occupant fatalities**
Overall	Size and mass	Halving car mass multiplies driver fatality risk by about 2.4; exposure method likely overestimates effect
Rollover	Size (length and track width; height of center of gravity)	Halving car mass multiplies driver fatality risk by about 1.8; exposure method likely underestimates effect
Nonrollover	Mass and size	Halving car mass increases driver injury risk by about 30%.
Crashes between cars and non-car vehicles		**25% of car-occupant fatalities**
Overall	Mass and size	Drivers of lighter cars are at greater risk in crashes with motorcycles, vans, trucks, etc.
Car-truck crashes	Mass and size	Halving car mass multiplies driver fatality risk by about 2.4; exposure method likely overestimates effect

Definitions:

R = risk to driver in lighter car divided by risk to driver in heavier car

μ = mass of heavier car divided by mass of lighter car

mass, properties possessed by very few objects into which vehicles actually crash.

The focus of this paper has been the relationship between risk, given that a crash occurs. We have not addressed any changes in driver behavior or use patterns that additionally may be associated with different size cars [Evans 1991; 1985b]. While drivers who chose higher horsepower cars have higher crash rates [Robertson 1993; Evans 1993], horsepower cannot affect injury risk given that a crash occurs (ignoring that, other factors being equal, a heavier engine will in fact reduce risk in a crash). The relationships are for vehicles on the roads, so they may not apply to vehicles using different design concepts. For example, Niederer et al. [1993] describes a low mass vehicle designed with a rigid body belt aimed at preventing vehicle crush, in contrast to the usual design goal of using vehicle crush to absorb crash energy in order to provide the occupants with as much stopping distance as possible.

Total system-wide effects

The least certain relations relate to how reducing the mass of all cars affects the average risk from two-car crashes. If all cars are the same mass, then the lighter this mass, the greater the risk. If all cars become lighter by some percentage, then average risk increases. However, the uncertainty is too great to rule out the possibility that some distributions by mass may lead to lower risk than other distributions with the same average mass.

Of all car occupants killed, 70 percent are killed in crashes involving only one car. For all of these crashes, occupants in lighter, smaller cars are at greater risk than occupants in larger heavier cars. For more than 50 percent of these occupants, mass is an important casual contributor to outcome. These effects are so large and clear that no conceivable effect in the opposite direction in the two-car crash case could come even close to canceling them. The conclusion is unmistakable -- if all other factors are held constant, a lighter car fleet leads to more car-occupant deaths and injuries.

Effect of occupant protection devices

The relationships in this paper were derived mainly for unbelted drivers of cars without airbags. A number of studies allow us to address how different these relationships would be if all occupants were belted and/or had airbags. The computed relative effect of mass on injuries and on fatalities was greater for belted than for unbelted drivers [Evans 1994]. Empirical fatality data suggest (weakly) an effect in the same direction [Evans and Frick 1993a]. No systematic car-mass effects dependent on belt use were found in earlier fatality studies using methods unlikely to uncover small differences. Evans and Frick [1986] and Evans [1985c] found no systematic relationship between belt effectiveness and car mass, whereas Partyka [1989b] found higher effectiveness for very light cars. The more recent finding [Evans 1994] of a larger mass effect for belted drivers implies that if all drivers were belted, the relative disadvantage to drivers of lighter cars would become greater. The finding of Zador and Ciccone [1993] (see also Insurance Institute for Highway Safety [1993]) of a higher effectiveness of airbags for larger cars likewise implies that if all cars had airbags, the relative disadvantage in smaller cars would increase. Although occupant protection devices appear to increase the relative disadvantage in small cars, they of course reduce the risks to drivers in cars of any size. The interpretation is that they reduce the risk in the heavy car by a somewhat larger proportion than in the light car.

IMPLICATIONS

While it has been well recognized that policies which lead to lighter cars increase casualties [Crandall and Graham 1989; Graham 1992], the results presented here demonstrate more definitively, and in finer detail, than hitherto that reducing the mass of cars increases the number of car occupants killed and injured. Increased use of occupant protection devices does not diminish the strength of the relationships between car mass and relative risk. Such devices reduce total casualties, in the same sense that other measures such as reduced drunk driving or converting rural two-lane roads to freeways reduce total casualties, but occupant protection devices do not preferentially reduce the risks in small cars -- if anything, the reverse. By analogy, improvements in safety design or the development of new materials, if equally applied to cars of all sizes, cannot be expected to alter the relative safety disadvantages in the smaller car.

Although there is considerable scientific interest in the question of causative factors, such questions seem less central to policy decisions. For any set of manufacturing materials, a larger car will always weigh more than a smaller car if both cars are similarly designed. If new lower density materials became available, size and mass would still be intrinsically related, but by a different quantitative relationship than applies presently. There is no reason to expect the relative driver risk relationships reported here would be particularly different for cars made from lower density material. If all cars remained the same size, but with reduced mass, rollover risk and risks when two cars made from the lower density material crashed into each other would remain similar. However, car occupant risk in nonrollover single-car crashes, and crashes with other vehicles, which together account for more than half of car occupant deaths, would increase. Most car occupant deaths resulting from crashes with vehicles other than cars are crashes with trucks. As cargo capacity is a crucial determinant of total truck weight, it is not possible to substantially reduce the weight of laden trucks without reducing the flow of goods (or increasing the number of trucks). Thus, reducing the masses of cars leads to an increase in the number of car occupants killed in crashes with trucks even if the size of cars remained unchanged.

Additional mass, as such, increases protection in any crash in which the struck object can move, bend, break, distort, or tear. Increased mass of the striking car increases damage to the struck object, thereby increasing the distance over which the car comes to rest, with consequent reduction in risk to the car's occupants. Reducing the mass of the car fleet by even small amounts substantially increases total traffic fatalities and injuries.

REFERENCES

Campbell, B.J.; Reinfurt, D.W. The relationship between driver crash injury and passenger car weight. Chapel Hill, NC: Highway Safety Research Center, University of North Carolina; 1973.

Crandall, R.W.; Graham, J.D. The effect of fuel economy standards on automobile safety. Journal of Law and Economics 32:97-118; 1989.

Ernst, E.; Bruhning; E; Glaeser, K. P; Schmidt, M. Compatibility problems of small and large passenger cars in head on collisions. Proceedings of the 13th International Technical Conference on Experimental Safety Vehicles, Paris, France, 4-7 November 1991, Document DOT HS 807 990, Washington DC, Vol. 1, p 103-110, July 1993.

Evans, L. Driver fatalities versus car mass using a new exposure approach. Accident Analysis and Prevention 16:19-36; 1984a.

Evans, L. Accident involvement rate and car size. Accident Analysis and Prevention 16:387-405; 1984b.

Evans, L. Driver age, car mass and accident exposure -- a synthesis of available data. Accident Analysis and Prevention 17:439-448; 1985a.

Evans, L. Driver behavior revealed in relations involving car mass. In: Evans, L.; Schwing, R.C., editors. *Human behavior and traffic safety.* New York, NY: Plenum Press, p. 337-352; 1985b.

Evans, L. Fatality risk for belted drivers versus car mass. Accident Analysis and Prevention 17:251-271; 1985c.

Evans, L. Car size and safety: results from analyzing US. accident data. Proceedings of the Tenth International Technical Conference on Experimental Safety Vehicles, Oxford, UK; 1-4 July 1985. National Highway Traffic Safety Administration, report DOT HS 806 916. Washington, DC, p. 548-556; February 1986.

Evans, L. Risk of fatality from physical trauma versus sex and age. Journal of Trauma 28:368-378; 1988.

Evans, L. Discussion of "The problem of compatibility in car-to-car collisions" by Thomas et al. Proceedings of the 34th Annual Meeting of the Association for the Advancement of Automotive Medicine, p 269-273, Scottsdale, AZ, 1-3, October 1990.

Evans, L. *Traffic safety and the driver.* New York, NY: Van Nostrand Reinhold; 1991.

Evans, L. Evans responds (Letter to the editor). American Journal of Public Health 83: 769-770; 1993.

Evans, L. Driver injury and fatality risk in two-car crashes versus mass ratio inferred using Newtonian Mechanics. Accident Analysis and Prevention, in press, 1994.

Evans, L.; Frick, M.C. Safety belt effectiveness in preventing driver fatalities versus a number of vehicular, accident, roadway and environmental factors. Journal of Safety Research 17:143-154; 1986.

Evans, L.; Frick, M.C. Seating position in cars and fatality risk. American Journal of Public Health 78: 1456-1458;1988.

Evans, L.; Frick, M.C. Relative fatality risk in different seating positions versus car model year. Accident Analysis and Prevention 21:581-587; 1989.

Evans, L.; Frick, M.C. Car size or car mass -- which has greater influence on fatality risk? American Journal of Public Health 82:1105-1112; 1992.

Evans, L.; Frick, M.C. Mass ratio and relative driver fatality risk in two-vehicle crashes. Accident Analysis and Prevention 25:213-224; 1993a.

Evans, L.; Frick, M.C. Driver fatality risk in two-car crashes: Dependence on masses of driven and striking car. Proceedings of the 13th International Technical Conference on Experimental Safety Vehicles, Paris, France, 4-7 November 1991, Document DOT HS 807 990, Washington DC, Vol. 1, p 83-93, July 1993b

Evans, L.; Frick, M.C. Car mass and fatality risk -- has the relationship changed? American Journal of Public Health 84:33-36; 1994.

Evans, L.; Wasielewski, P.F. Serious or fatal driver injury rate versus car mass in head-on crashes between cars of similar mass. Accident Analysis and Prevention 19:119-131; 1987.

Folksam. Folksam car model safety rating 1989-1990, Folksam Research and Development, 1990

Graham, J.D. The safety risks of proposed fuel economy legislation. Risk: Issues in health and safety 3:95-126; 1992.

Grime, G.; Hutchinson, T.P. Vehicle mass and driver injury. Ergonomics 22:93-104; 1979.

Insurance Institute for Highway Safety. Small passenger vehicles a problem. Status Report 22(2), Feb. 28 1987.

Insurance Institute for Highway Safety. Air bags in perspective. Status Report 28(11), October 9 1993.

Joksch, H.C. Analysis of the future effects of the fuel shortage and increased small car usage upon traffic deaths and injuries. Report DOT-TSC-OST-75-21; January 1976.

2003-01-0902

Changes in Vehicle Designs from Frontal Offset and Side Impact Crash Testing

Adrian K. Lund and Joseph M. Nolan
Insurance Institute for Highway Safety

ABSTRACT

The Insurance Institute for Highway Safety (IIHS) has been conducting frontal offset crash tests of new passenger vehicles and providing comparative crashworthiness information to the public since 1995. This program has resulted in large improvements in frontal crashworthiness largely because vehicle structures have been redesigned to prevent significant collapse of the occupant compartment. In late 2002, IIHS began a side impact crash test program in which the side-impacting barrier has been designed to simulate the geometry of the front ends of SUVs and pickups, which pose a much larger threat in side impacts than the lower front ends of cars. It is anticipated that this program, too, will result in changes in vehicle structure, in this case the structure of the vehicle side pillars and door hardware. Good performance in the side impact test also is likely to require installation of side airbags (or comparable system) to protect the head and/or chest.

These improvements in vehicle crashworthiness for both frontal and side impacts will reduce incompatibility in motor vehicle crashes by improving self protection without adversely affecting the consequences for occupants of "partner" vehicles in two-vehicle crashes. Evidence for this is that the structural changes seen so far in response to the frontal offset test do not appear to be increasing vehicle aggressivity. Similarly, the vehicle changes expected in response to the side impact test procedure seem unlikely to have any effect on aggressivity in two-vehicle crashes.

INTRODUCTION

The Insurance Institute for Highway Safety (IIHS) has conducted crash tests of new passenger vehicles to provide the public with comparative crashworthiness information since 1995. IIHS's frontal crash test program consists of 64 km/h (40 mi/h) offset tests into a deformable barrier. In late 2002 testing was expanded to add side impacts, consisting of 50 km/h (31 mi/h) perpendicular impacts with a moving deformable barrier designed to simulate the front end of a sport utility vehicle (SUV) or pickup. IIHS's frontal test program already has prompted significant design changes, especially to the front ends and occupant compartment structures of many recent model passenger vehicles. We expect the side impact program also to produce changes to side restraint systems and structures.

There is no question that these design changes will substantially improve the protection vehicles provide to their occupants — sometimes referred to as "self" protection — in serious front and side crashes. However, it has been suggested that if these improvements result from vehicles becoming stiffer, the improved occupant protection may come with some negative consequences for occupants of other (or "partner") passenger vehicles involved in crashes with vehicles that have the improved self protection.

This paper reviews in some detail vehicle design changes that have resulted from IIHS's frontal crash test program and projects structural changes that could result from the newer side impact program. Such changes are considered in the context of self and partner protection.

VEHICLE CHANGES WITH OFFSET CRASH TESTING

IIHS's frontal offset crash test, in which 40 percent of a vehicle's width overlaps the barrier, simulates a real-world crash in which two similar vehicles meet head on at high speed (closing speed of about 120 km/h) but with only partial overlap of their front ends. For vehicles with poor structural designs, such crashes often cause large intrusion of the occupant compartment. This loss of occupant compartment space means there is less opportunity for restraint systems to protect occupants from serious injury. Indeed, research in the late 1980s and early 1990s had shown that most serious injuries to restrained occupants occurred in the presence of massive occupant compartment intrusion [1] and that these collisions often were offset frontals [2-3]. The frontal offset crash test was developed to assess front-end and occupant compartment structural designs.

The offset test highlights an aspect of occupant protection that differs from the U.S. New Car Assessment Program (USNCAP), which employs a full-width, frontal, flat-barrier test that causes less vehicle intrusion for similar impact speeds. In USNCAP tests, crush depth also is less, and occupant decelerations are correspondingly higher. Figure 1 shows the improvements in USNCAP performance that occurred between 1979 and the mid 1990s, which primarily reflect improvements in restraint systems to manage occupant energy with these high vehicle accelerations [4]. Many vehicles that were good USNCAP performers in 1995 still had relatively poor structural designs, as reflected in IIHS's first offset frontal crash tests of midsize car models in 1995 (Figure 2).

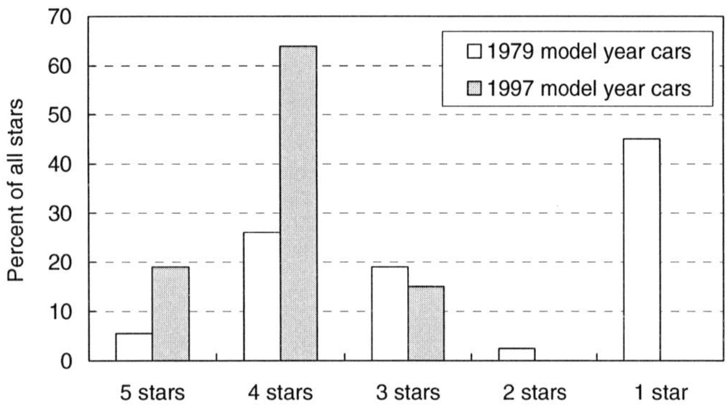

Figure 1. NCAP star ratings over time

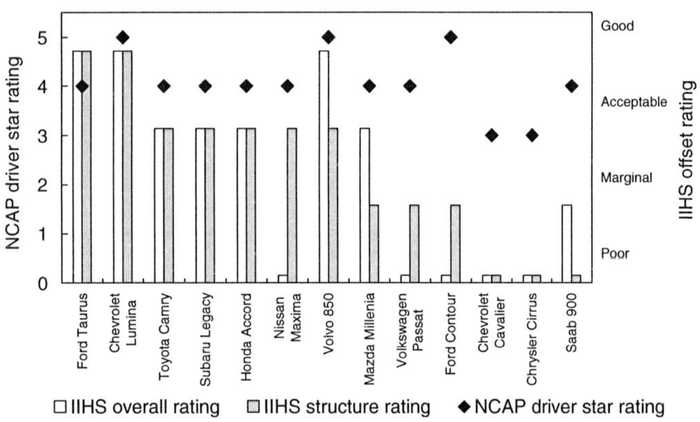

Figure 2. NCAP scores and offset frontal ratings, 1995 model cars

As a result of offset crash testing by IIHS (and other consumer test organizations around the world), this situation has changed dramatically. Just as happened earlier with the USNCAP program, there have been major improvements in offset crash test performance. Among the 1995-98 model year passenger cars and vans tested, 22 percent achieved good overall crashworthiness ratings on IIHS's four-point rating scale (good, acceptable, marginal, or poor). Among 1999-2001 models, 52 percent achieved good overall ratings [5]. More recently, the results have been even more impressive: Of 18 tests of 2002-03 model year cars and minivans, 16 vehicles have been rated good overall and two acceptable. None has received a marginal or poor overall rating. For pickups and SUVs, the results also are encouraging: Among 11 vehicles, 6 have been rated good, 3 acceptable, and only 2 marginal. Figure 3 shows the trend in ratings for all vehicles by year.

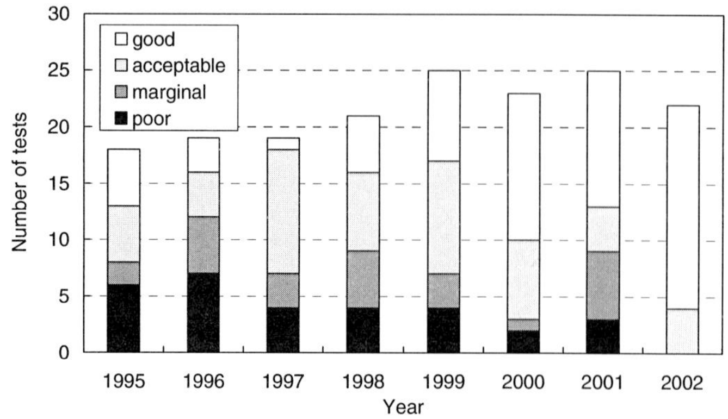

Figure 3. Crashworthiness evaluation ratings by calendar year of test

The improvements in offset crash test performance have been achieved in nearly all cases through improved designs of front vehicle structures and occupant compartments to reduce intrusion into the compartments. When testing began, there were many vehicles with front structural elements that did not bend or crush in the offset test but, instead, pushed back into the occupant compartment, which was not strong enough to resist this intrusion. Design changes to reduce the intrusion have entailed both strengthening of the occupant compartment where structural elements make contact and strategic *weakening* of some structural elements. Other design changes have focused on making sure stiff structures that move rearward make contact with parts of the safety cage that are capable of withstanding the forces those stiff structures bring. For example, we rarely see front wheels that turn inboard to apply force on the wheel well in later model vehicles; instead the wheel movement tends to be controlled during front-end crush so that the wheel applies force to the stronger rocker panel. Other changes have involved careful consideration of the location of stiff structures in the engine compartment and making sure these stiff structures do not stack up and, thereby, reduce front crush space. In simple terms, front ends now are more effective crush zones, and occupant compartments are stronger safety cages.

Some have argued that the structural improvements required to achieve good offset crash performance can increase incompatibility in crashes involving multiple vehicles. Specifically, there is concern that good offset crash performance requires a stiffer vehicle, which crushes less in collisions with other vehicles [6-7]. Assuming that overall vehicle length does not change, then, in theory, this contention can be true — if a vehicle's occupant compartment does not crush, it is stiffer at that point than a vehicle whose occupant compartment does crush. At the same time, a more effective crush zone may be less stiff than one that does not effectively absorb crash energy. In any case, self protection requires that the safety cage be very stiff. Good occupant protection requires that a vehicle's safety cage be capable of resisting crash forces.

But is it true that this increased occupant compartment stiffness, necessary for self protection, is offset by an increase in aggressivity against other vehicles? IIHS has examined the accelerations recorded for some of the vehicles that have shown dramatic improvements in offset crash test performance [8]. The acceleration versus deflection curves do not support the contention that vehicles are becoming significantly more aggressive in their interactions with other vehicles. For example, through a series of design changes, the Cadillac Seville went from a poor structural evaluation in 1997 to a good evaluation in 2000. Figure 4 shows the acceleration versus deflection curves for these two vehicles in the offset test.

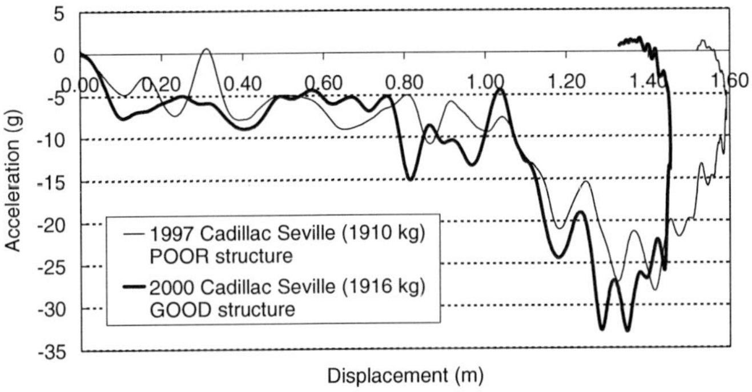

Figure 4. Longitudinal occupant compartment acceleration vs. displacement in 64 km/h offset tests, 1997 and 2000 Cadillac Seville

It is apparent that during the early part of the crash tests after the deformable barrier has been crushed (0.54 m), the force exerted by the two vehicles is quite similar; it is when the crush begins to approach the occupant compartment that the redesigned vehicle appears to be stiffer. The point of this comparison is that during the early stages of a crash involving two vehicles, the old and new Seville designs will exert similar forces on their crash partners. This is because the redesign involved not only strengthening of the occupant compartment but also ensuring that front structural elements bend and crush early in the crash.

Comparisons of the acceleration versus deflection of other vehicles that have shown striking improvements in structural performance in offset crash tests reveal similar changes; the increased stiffness occurs as crush reaches the occupant compartment (Figures 5-8). In the case of the Honda Odyssey, the acceleration versus deflection curve suggests that the improved vehicle structure may be even less aggressive in interactions with other vehicles. The redesigned Odyssey achieved a good structural rating in 1999. In comparison with the 1996 version of the Odyssey, which had a poor structural performance, we see not only that the accelerations are low during the early part of the redesigned vehicle's crash test but also that the crush length actually has been extended to manage the energy of the heavier, redesigned vehicle (Figure 8).

These analyses are limited to accelerations recorded in the 64 km/h offset crash test. Nolan et al. [9] also found no clear relationship between stiffness as measured from loadcell barrier data in USNCAP tests and offset test ratings (Figure 9).

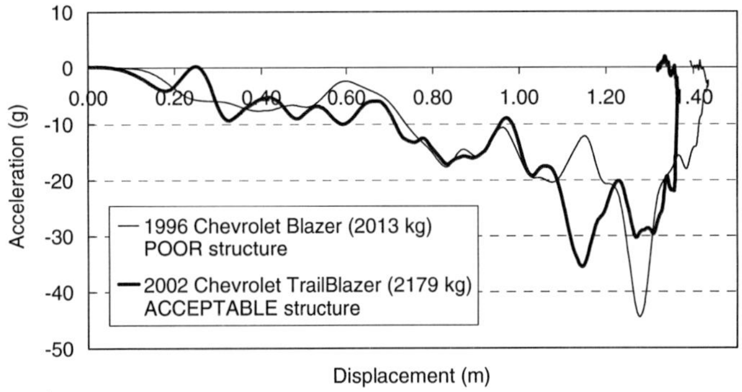

Figure 5. Longitudinal occupant compartment acceleration vs. displacement in 64 km/h offset tests, 1996 Chevrolet Blazer and 2002 Chevrolet TrailBlazer

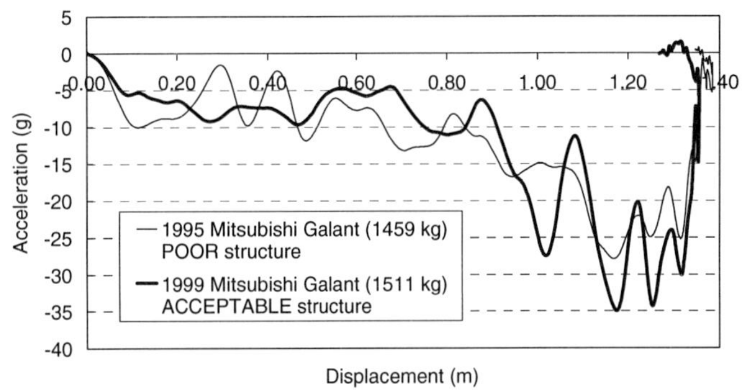

Figure 6. Longitudinal occupant compartment acceleration vs. displacement in 64 km/h offset tests, 1995 and 1999 Mitsubishi Galant

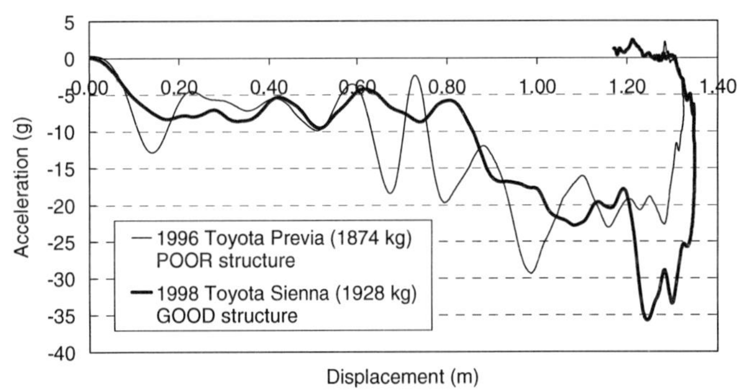

Figure 7. Longitudinal occupant compartment acceleration vs. displacement in 64 km/h offset tests, 1996 Toyota Previa and 1998 Toyota Sienna

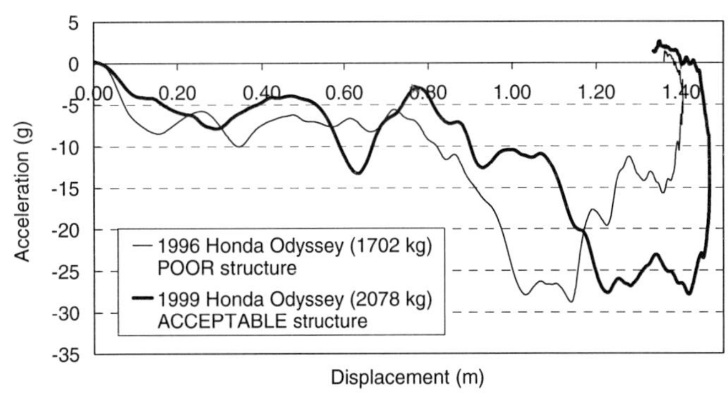

Figure 8. Longitudinal occupant compartment acceleration vs. displacement in 64 km/h offset tests, 1996 and 1999 Honda Odyssey

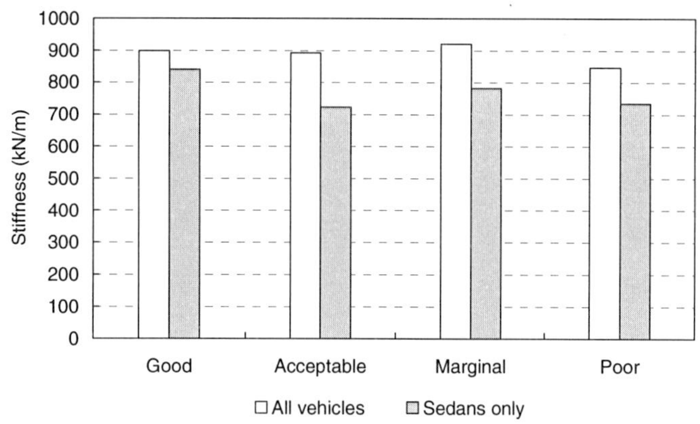

Figure 9. Average stiffness by rating category, NCAP loadcell barrier analysis

Thus, nothing about offset crash tests requires a vehicle manufacturer to significantly increase a vehicle's aggressivity in crashes with other vehicles. In fact, there is no convincing evidence that there have been increases in stiffness that would greatly increase aggressivity. This conclusion is supported by the results of IIHS crash tests, comparing the injury risk to occupants of a Ford Taurus with an Isuzu Rodeo or a Mercedes M-class [10]. The Rodeo and M-class differ in stiffness, as measured by force deflection curves from USNCAP tests. They also differ in the heights of their structural elements, with the Rodeo being somewhat lower and matching up with the front structure of the Taurus. The results indicated only marginal evidence that the softer front structure of the M-class was beneficial, and then only when the M-class had been lowered so the height of its structure matched that of the Rodeo. The primary observation of these tests was that Taurus occupants fared much worse in collisions with either the M-class or Rodeo when those vehicles were at a higher ride height. This study demonstrated that even in frontal collisions, a major factor in the incompatibility of collisions between SUVs and cars is the mismatch in positions of their structural elements.

ANTICIPATED CHANGES IN VEHICLES FROM IIHS SIDE IMPACT TESTS

In November 2002, IIHS began a new side impact crashworthiness evaluation program, intended to assess how well vehicles protect near-side occupants when struck by pickups and SUVs [11]. Figure 10 shows the side view of the IIHS side impact barrier and a Ford F-150 aligned in test position against the side of a Hyundai Elantra. Figure 11 shows an overhead view of the IIHS side impact test condition, including barrier mass, and impact conditions.

This test is intended to help address a key aspect of vehicle crash incompatibility — the increased risk of fatality for side-struck occupants when the striking vehicle has the higher front profile of an SUV or pickup [12]. The presumption is that this increased incompatibility reflects not only the aggressivity of SUVs and pickups but also the vulnerability of occupants in vehicles with side structures and restraint systems that have not been designed

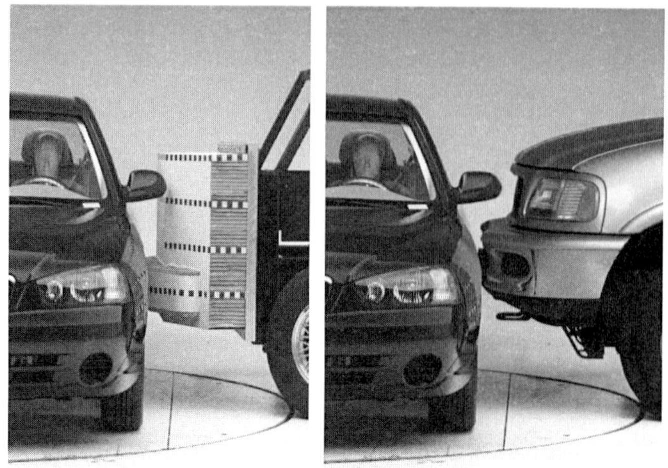

Figure 10. IIHS barrier and Ford F-150 4x4 with Hyundai Elantra

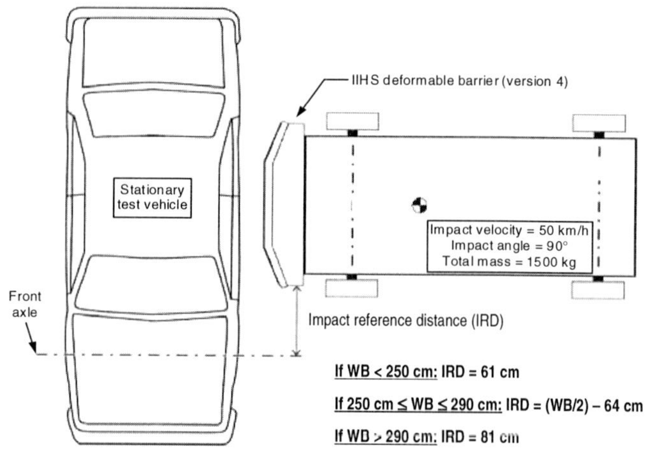

Figure 11. IIHS side impact test configuration

for side collisions from taller vehicles. The purpose of the new side impact test is to reduce incompatibility in such crashes by focusing attention on vehicle designs that can reduce the vulnerability. Incompatibility in side impacts always has been influenced by the fact that vehicle front ends are substantially stiffer than side structures but, with more and more SUVs and pickups in the vehicle mix, geometric mismatch has become even more important. It seems probable that improvements in side impact protection will make greater contributions to reducing side impact crash incompatibility than changes to the front ends of striking vehicles.

Because the side impact test program is just beginning, it is too early to know precisely how automakers will redesign side structures and restraint systems to improve protection. However, one aspect of side impacts from taller vehicles, compared with typical passenger cars, is an increased risk of head contact, and hence serious head injury, with the intruding structure of the striking vehicle front. Figures 12-13 show the increased risk of head contact from striking vehicles with higher ride heights in side impact crashes [9]. It is anticipated that the increased risk of head injury will be addressed through airbags that deploy between the heads of occupants and the striking vehicles. Preliminary test data suggest this countermeasure will be effective; in every IIHS test conducted with side airbag head protection so far (a total of 9 vehicles), forces on the head have been low.

Figure 12. Lincoln Town Car striking the side of a Mercury Grand Marquis

Figure 13. Ford F-150 striking the side of a Mercury Grand Marquis

The IIHS side impact test also increases the measured forces on occupants' torsos and pelvic regions. Because the barrier is higher off the ground than the standard FMVSS 214 barrier, the B-pillar and doorsill of the struck vehicle have less strength to resist the intrusion of the barrier. To reduce this intrusion, or at least limit the velocity at which it occurs, it is expected that side structures will be strengthened. This increased strength is likely to require reinforcement of not only the sill and B-pillar but also the doors themselves in order to prevent the deformable barrier face from intruding. One manufacturer already has shared data with IIHS demonstrating that, without sufficient strengthening of doors, the barrier will deform around a very stiff pillar, punching in the softer doors to make forceful contact with the occupants.

What are the implications of these vehicle changes for the issue of crash incompatibility? Clearly, the anticipated countermeasures are aimed at reducing the vulnerability of side struck occupants; in this way the changes, if they are effective, will reduce the incompatibility of side impacts between SUVs and pickups, on the one hand, and passenger cars, on the other. Of course, simultaneously the potential increase in side structure stiffness means that crashes will become more severe for the occupants of striking vehicles in side impacts. This is an *increase* in incompatibility from their point of view, though certainly the increase is minor compared with the potential reduction in the vulnerability of the side-struck occupants.

DISCUSSION

There is no question that IIHS's frontal crash test program has produced major changes in vehicle designs, some of which have resulted in increased stiffness. The side impact program also may increase side stiffness and may add weight. But as this paper argues, it is not at all clear that these stiffness changes, which predominantly affect compartment stiffness, will result in any significant adverse consequences to occupants of partner vehicles. In fact, improved crush zone designs could improve not only self protection but also protection in partner vehicles. It also should be noted that improved self protection does not necessarily require increases in vehicle stiffness and consequent potential adverse effects in partner vehicles. Instead front ends could be increased in length so there would be more crush space to more effectively manage crash energy. Manufacturers may claim that this design alternative is not possible, due to other design considerations. But what this really means is that occupant crash protection, both self and partner protection, is something that is considered after styling and other vehicle design constraints are imposed.

So far, this discussion has considered the effects of crash testing on incompatibility in two-vehicle crashes. Any full discussion of possible trade-offs between improvements to self protection in serious frontal and side crashes and consequent deterioration of protection in partner vehicles involved in crashes with the improved vehicles should recognize that the benefits of self protection accrue in all kinds of crashes — single-vehicle, two passenger vehicles, more than two vehicles, crashes with large trucks, etc. In contrast, possible adverse consequences for occupants in partner vehicles occur principally in two-vehicle crashes. In the United States about 20 percent of car occupant deaths occur in crashes with other cars, and another 15 percent occur in crashes with SUVs or pickups [13]. Thus, there is no partner passenger vehicle in substantially more than half of the car crashes with occupant deaths. For car occupants with nonfatal (MAIS 2+) injuries, the proportion involving crashes with other passenger vehicles is somewhat higher (38 percent other cars and 13 percent SUVs or pickups), but still many crashes producing occupant injuries do not involve partner passenger vehicles [13].

Because of the terms "self" and "partner" protection that have been adopted in this debate, it is easy to think of self protection as a selfish concept and partner protection as an ethical necessity. This viewpoint would have some limited merit if the advocates of improved self protection were urging that it be achieved solely by building much bigger and heavier passenger vehicles. But this is not the case. Crash test programs such as those conducted by IIHS are aimed at producing self-protection improvements within a vehicle size and weight class. In this latter case any adverse consequences resulting from vehicle stiffness increases and/or slight increases in vehicle weight could, at worst, have only slight negative consequences for partner protection (most of whose deaths and serious injuries occur in other than two-vehicle crashes).

In summary, increasing self protection within a vehicle size and weight class can involve trade-offs in partner protection. However, it does not have to, and there is little evidence that it has in frontal crashes. Most important, because improvements in self protection will work in a wide range of crashes, while partner protection is relevant only in crashes involving two passenger vehicles, the balance always should favor improving self protection.

ACKNOWLEDGMENTS

This work was supported by the Insurance Institute for Highway Safety.

REFERENCES

1. Green, R.N.; German, A.; Nowak, E.S.; Dalmotas, D.; and Stewart, D.E. 1994. Fatal injuries to restrained passenger car occupants in Canada: crash modes and kinematics of injury. *Accident Analysis and Prevention* 26:207-14.

2. Lowne, R.W. 1994. EEVC Working Group II report on the development of a front impact test procedure. *Proceedings of the 14th International Technical Conference on Enhanced Safety of Vehicles*, 1334-51. Washington, DC: National Highway Traffic Safety Administration.

3. O'Neill, B.; Lund, A.K.; Zuby, D.S.; and Preuss, C.A. 1994. Offset frontal impacts: a comparison of real-world crashes with laboratory tests. *Proceedings of the 14th International Technical Conference on the Enhanced Safety of Vehicles*, 649-70. Washington, DC: National Highway Traffic Safety Administration.

4. Park, B.T.; Hackney, J.R.; Morgan, R.M.; Chan, H.; Lowrie, J.C.; and Devlin, H.E. 1999. The New Car Assessment Program: Has it led to stiffer light trucks and vans over the years? (SAE 1999-01-0064). *Vehicle Aggressivity and Compatibility in Automotive Crashes* (SP-1442), 7-20. Warrendale, PA: Society of Automotive Engineers.

5. Insurance Institute for Highway Safety. 2001. Crashworthiness improvements. *Status Report* 36(3). Arlington, VA.

6. Zeidler, F. and Knöchelmann, F. 1997. The influence of frontal crash test speeds on the compatibility of passenger cars in real-world accidents. Report No. 1354. Verlag des Verein Deutscher Ingenieure GmbH, Dusseldorf, Germany.

7. Verma, M.K.; Lange, R.C.; and General Motors Corporation. A manufacturer's view of vehicle compatibility in crashes. Presented at the SAE Government Meeting, May 2002, NHTSA Docket Submission 3332.

8. Nolan, J.M. and Lund, A.K. 2001. Frontal offset deformable barrier crash testing and its effect on vehicle stiffness. *Proceedings of the 17th International Technical Conference on the Enhanced Safety of Vehicles* (CD ROM). Washington, DC: National Highway Traffic Safety Administration.

9. Nolan, J.M.; Powell, M.R.; Preuss, C.A.; and Lund, A.K. 1999. Factors contributing to front-side compatibility: a comparison of crash test results (SAE 99SC02). *Proceedings of the 43rd Stapp Car Crash Conference* (P-350), 13-24. Warrendale, PA: Society of Automotive Engineers.

10. Meyerson, S.L. and Nolan, J.M. 2001. Effects of geometry and stiffness on the frontal compatibility of utility vehicles. *Proceedings of the 17th International Technical Conference on the Enhanced Safety of Vehicles* (CD ROM). Washington, DC: National Highway Traffic Safety Administration.

11. Insurance Institute for Highway Safety. 2001. Head protection in side impacts. *Status Report* 36(1). Arlington, VA.

12. Arbelaez, R.A.; Dakin, G.J.; Nolan, J.M.; Dalmotas, D.J.; and Tylko, S.. 2002. IIHS side impact barrier: development and crash test experience. *IMechE Conference Transactions: International Conference on Vehicle Safety 2002*, 73-88. London, United Kingdom: Professional Engineering Publishing Ltd.

13. Lund, A.K.; O'Neill, B.; Nolan, J.M.; and Chapline, J.F. 2000. Crash compatibility issue in perspective. SAE Technical Paper Series 2000-01-1378. Warrendale, PA: Society of Automotive Engineers.

2003-06-0074

INNOVATIVE BODY STRUCTURE FOR THE SELF-PROTECTION OF A SMALL CAR IN A FRONTAL VEHICLE-TO-VEHICLE CRASH

Masuhiro Saito
Tetsuya Gomi
Yoshinori Taguchi
Takeshi Yoshimoto
Tomiji Sugimoto
Honda R&D Co., Ltd. Tochigi R&D Center
Japan
Paper Number 239

ABSTRACT

Preservation of passenger compartment space during a frontal vehicle-to-vehicle collision is extremely significant for the self-protection of small cars.

It is well known that crash speed, mass, stiffness and geometric interaction all have an influence on the intrusion of the passenger compartment in a frontal impact between vehicles. This paper reports on a new enhanced body structure to reduce passenger compartment intrusion in a crash between large and small cars. The test discussed in this report set the crash speed of both cars at 50kph, the mass of the large car at almost twice that of the small car, and the small car over lap at 50%. The proposed innovative body structure for the front end of small cars achieved a previously unavailable level of efficiency of energy absorption and was able to maintain cabin integrity.

INTRODUCTION

In recent years the use of stationary barrier crash tests as a method of evaluation of crash safety performance has increased internationally. This has been very effective in improving vehicle crash safety performance and reducing the number of casualties in traffic accidents.

However, in the case of frontal collisions between small cars and large cars in the real world accident, it is said that the risk of injury to the small car's occupants is higher than that to occupants of the large car. This is caused by incompatibility between 'mass' 'stiffness' and 'geometry' in vehicle-to-vehicle collisions. A collision in which the mass and stiffness ratios of the vehicles are large is equivalent to an extremely high speed stationary barrier crash for a small car.

Small cars which receive good evaluations in full lap and offset frontal crash barrier tests are therefore not always sufficiently safe in a small car to large car collision in the real world accident. And it begins to be pointed out the necessity to have an another manner to evaluate it in the collisions with relatively different sized vehicles.

In particular in the case of narrow offset collisions in which overlap distance is relatively small in the direction of the vehicle' s width and collisions with differing bumper height, it is very difficult for conventional body structures to maintain crash safety performance. It is therefore necessary to propose innovative body structures based on new design concepts. Of course it goes without saying that it is extremely significant for compatibility not only to consider the progress of self-protection but also partner-protection (for opposite vehicles). This research reports on the possibility to improve the self and partner protection which are discussed in the society by modifying the body structure such as one of the unique technique.

THE SAFETY IMPROVEMENT FOR SMALL CARS

In frontal vehicle-to-vehicle collisions it has been discussed that mainly 'mass' 'stiffness' and 'geometry' are the factors for incompatibility. In this part we report on some subjects and solutions in the view of the self-protection for small cars.

The portion of the velocity change before and after collisions is influenced by mass ratio as follows.

$$Vr = 1/Mr \qquad (1.)$$

$Vr = \Delta V_2/\Delta V_1$ ΔV_i: velocity change of Vehicle i $Mr = M_2/M_1$ M_i: mass of Vehicle i

The lighter vehicles are forced to a higher deceleration level than the heavier vehicles. As a result the risk of injuries in small cars is higher than in large cars. We will be able to solve such mass incompatibility developing superior restraint systems for small cars and to reduce mass in large cars.

The portion of energy absorption in vehicle deformation depends on stiffness ratio of two vehicles as follows.

$$Er = 1/Kr \qquad (2.)$$

$Er = E_2/E_1$ E_i: energy absorption of Vehicle i

$Kr = K_2/K_1$ K_i: stiffness of vehicle i

The problem is how to balance several vehicles' stiffness. To be realistic we need to improve the stiffness for small cars.

Certainly as stated above the difference between mass and stiffness is a problem for the compatibility, however we should first improve geometrical compatibility.

In recently years the level of self-protection for small cars has advanced greatly because of the adoption of offset deformable barrier testing. However the deformable barrier is very uniform in stiffness. But in reality, the front of vehicle in vehicle-to-vehicle collisions lack uniform stiffness, which causes severe damage to small cars. After this we focus on the improvement of interaction in frontal structures.

NEW DESIGN CONCEPT

To improve the offset and full lap frontal crash performance, conventional body structures are generally designed with two main frames located on each side of the engine compartment to absorb vehicle energy and to control vehicle deceleration.

However, in the case of vehicle-to-vehicle collisions such as narrow offset collisions in which there is less overlap distance in the direction of vehicle width and collisions between a passenger cars and a Sports Utility Vehicle (SUV) in which the bumper beam heights of the vehicles differ (as shown in Figure 1), this body type allows structural penetration into the engine compartment.

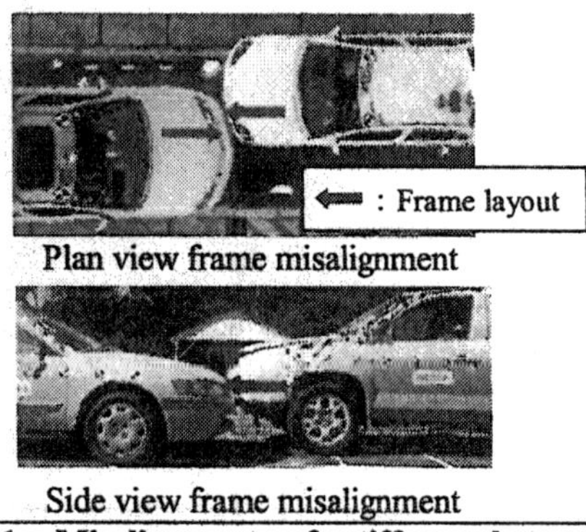

Figure 1. Misalignment of stiffness between vehicles in vehicle-to-vehicle collision.

In particular when there is a significant difference in vehicle weight, the bumper beam and main frame of the large car passes into the frame of the small car without sufficient energy absorption and deceleration, penetrating the weaker part of the small car. As a result, deformation extends to the small car's passenger compartment, increasing injuries to occupants from secondary collisions in the passenger compartment, as shown in Figure 2.

It is important for the improvement of the level of protection offered by small cars to prevent structural penetration of major frontal components, increase the homogeneity of strength distribution and improve energy absorption in the engine compartment in the event of vehicle-to-vehicle collisions between vehicles with misalignment of stiffness. The design concept will be expected not only to achieve progress for self-protection but also the effect of partner-protection for compatibility.

Figure 2. Large deformation of the passenger compartment after narrow offset vehicle-to-vehicle collision.

STRUCTURAL OUTLINE AND CRASH PERFORMANCE

The proposed structure consists of three components, as shown in Figure 3.

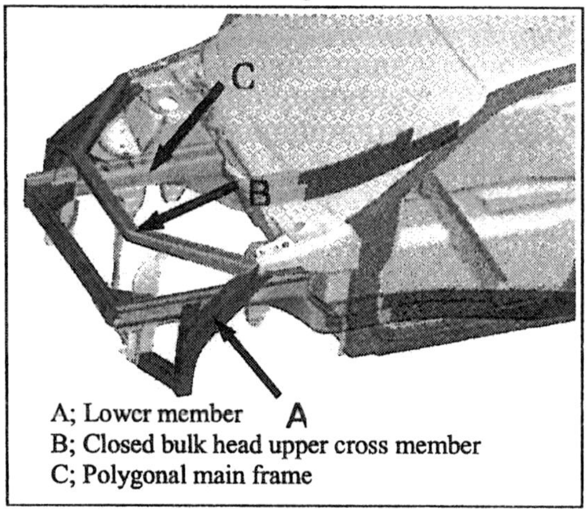

Figure 3. Structural outline.

These components are A; A lower member to prevent penetration. B; A closed bulk head upper cross member to assist energy absorption in the upper part of the engine compartment. and C; A polygonal main frame enabling high efficiency energy absorption. This paper will offer a structural outline of the lower member system and discuss the predicted effectiveness of this system as determined by computer simulations.

The new 'lower member' was positioned in front of the tires extending from the wheel house upper member, and was connected to the main frame and bulk head cross member. This prevents the penetration of the frames of the respective vehicles in narrow offset collisions and collisions between vehicles with differing bumper heights. (See Figure 4). On impact, the lower member makes contact with the front structure of the other vehicle and deforms, thus achieving a high level of energy absorption (as shown in Figure 5).

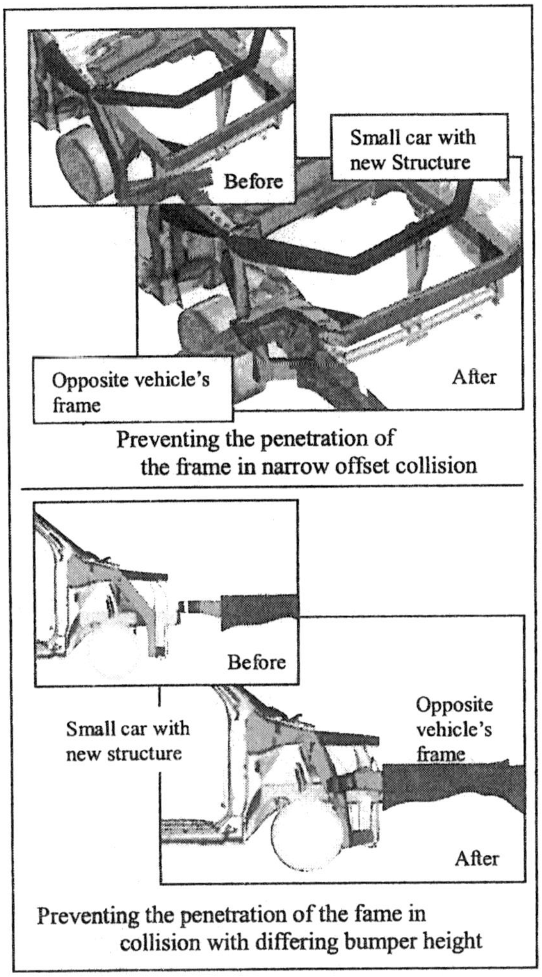

Figure 4. Preventing the penetration of the Frame.

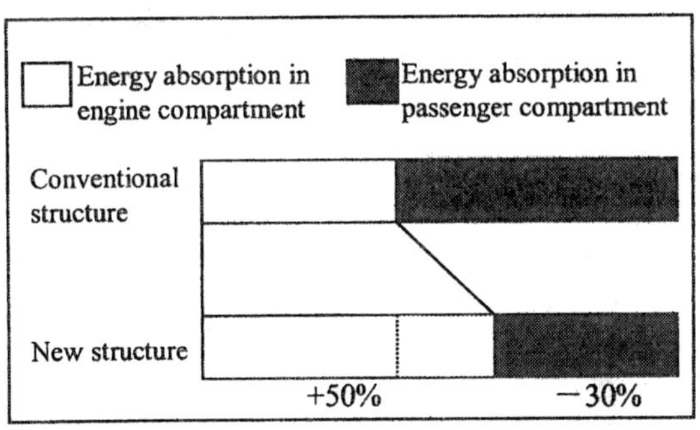

Figure 5. Increasing energy absorption in the engine compartment from vehicle-to-vehicle collision.

This leads to a significant reduction in the proportion of energy absorption in the cabin and of the degree of passenger compartment intrusion. The effectiveness of the new structure in reducing passenger compartment intrusion is shown in Figure 6.

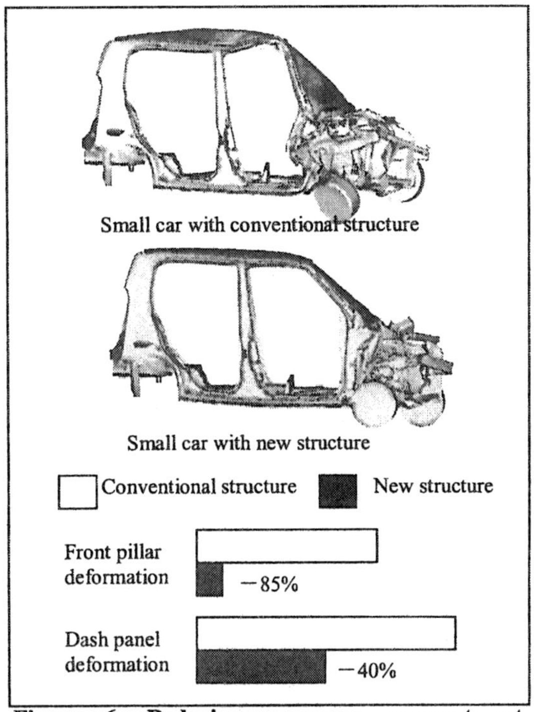

Figure 6. Reducing passenger compartment intrusion in vehicle-to-vehicle collision.

The new structure also enables reduction of the strength and weight of the main frame, allowing a more homogeneous distribution of strength in the front end structure.

It is predicted that this structure will be effective in reducing passenger compartment intrusion for various overlap distances in the direction of vehicle width (as shown in Figure 7), for the difference of bumper height (as shown in Figure 8) and for angle of approach (as shown in Figure 9).

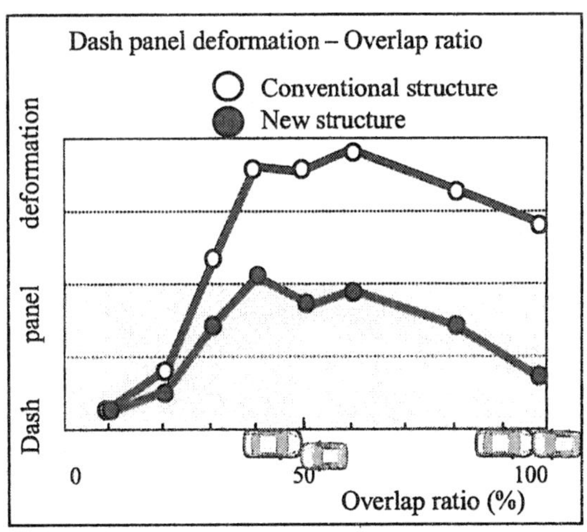

Figure 7. Reducing passenger compartment intrusion in various offset vehicle-to-vehicle collisions.

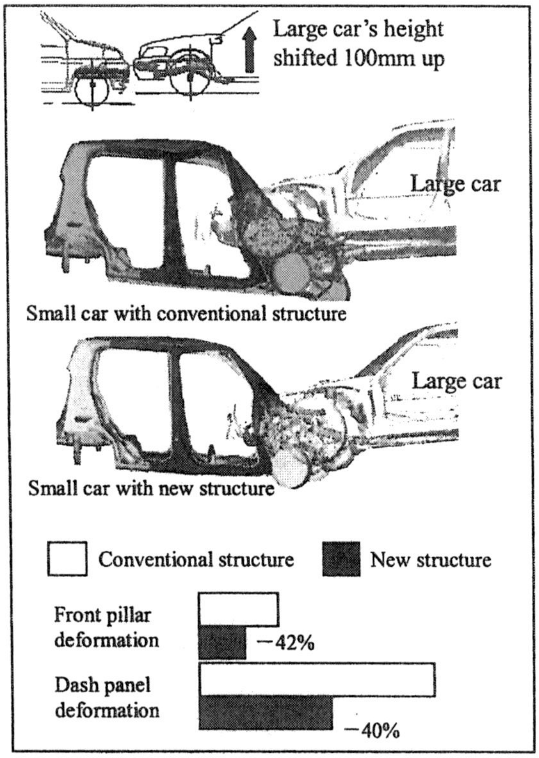

Figure 8. Reducing passenger compartment intrusion for vehicle-to-vehicle collisions with differing bumper height.

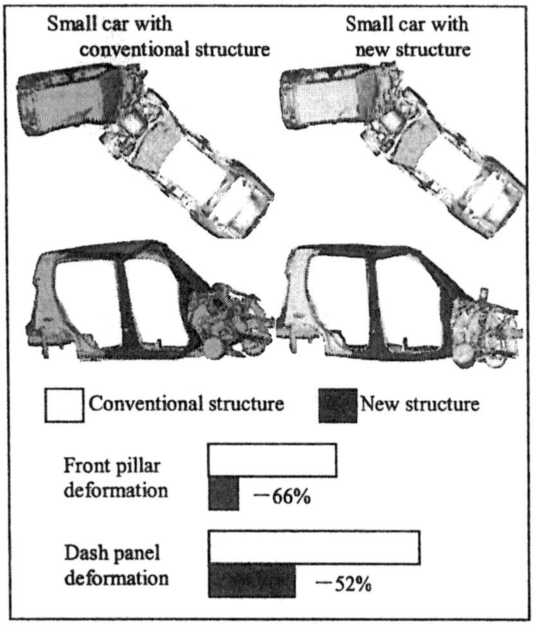

Figure 9. Reducing passenger compartment intrusion for angle of approach in vehicle-to-vehicle collision.

AGGRESSIVENESS

Secondly we give consideration to aggressiveness. As a result of above-mentioned vehicle-to-vehicle simulations, energy absorption in the engine compartment has increased slightly and the energy absorption in the cabin is decreased in large cars with conventional body structures by means of the effect from the new structure. It can therefore be predicted that the new body structure will not increase aggressiveness towards the partner vehicle in a collision (as shown in Figure 10).

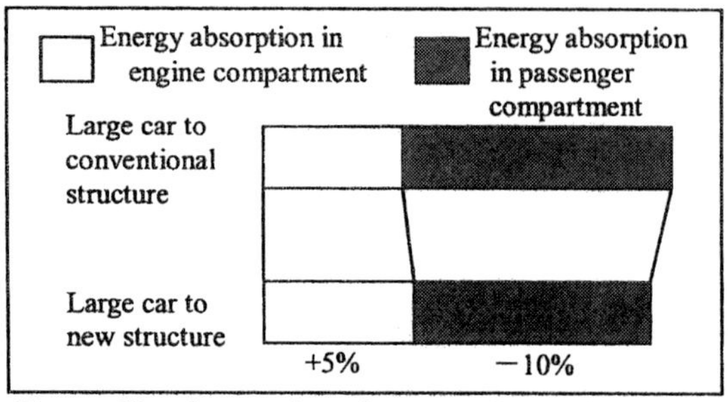

Figure 10. Improving of energy absorption.

Further we analyzed aggressiveness towards small cars as follows. We describe the simulation results in a frontal collision between similar small cars A and B with conventional structure (case 1), and between a small car A with conventional structure and a small car C with the new proposed structure (case 2)in Figure 11. Opposite car's intrusion in the passenger compartment was approximately similar in both cases. We could assess that our proposed new structure greatly improved self-protection, and doesn't increase the aggressiveness towards the small car, and we could find the possibility for compatibility.

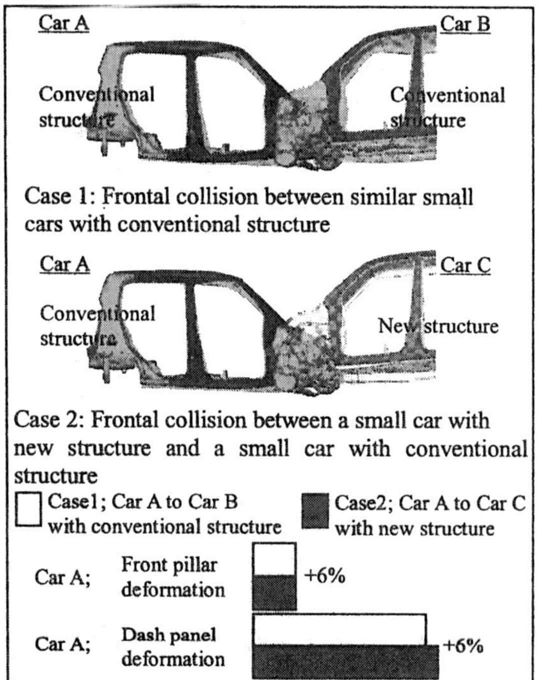

Figure 11. Not increasing aggressiveness to small cars.

ACCIDENT ANALYSIS

This time the condition in a vehicle-to-vehicle crash test was based on the accident analysis in Japan. Some accident data were provided from the Institute for Traffic Accident Research and Data Analysis (ITARDA) established in 1992.

Almost half of the number of occupant deaths is in the case of frontal collisions as shown in Figure 12[1]. About 90% of the frontal collision deaths were at speeds lower than 50kph as shown in Figure 13 [1]. And when the opposite vehicles are heavier than the subject vehicles, the driver deaths in frontal collisions are about 75% as shown in Figure 14 [1]. With relation to overlap in frontal collisions, 30% and 50% overlap cases are the primary overlap conditions for offset collisions (as shown in Figure 15) [1].

For reduction casualties, we would get effective test results from test condition based on accidents analysis.

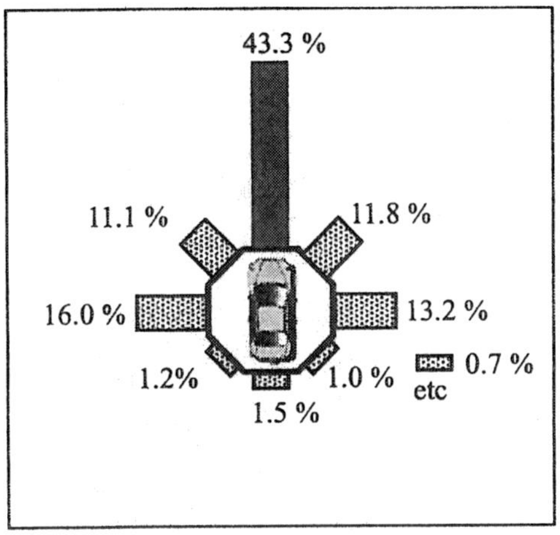

Figure 12. The collision direction in the fatal accidents for vehicle-to-vehicle collisions.

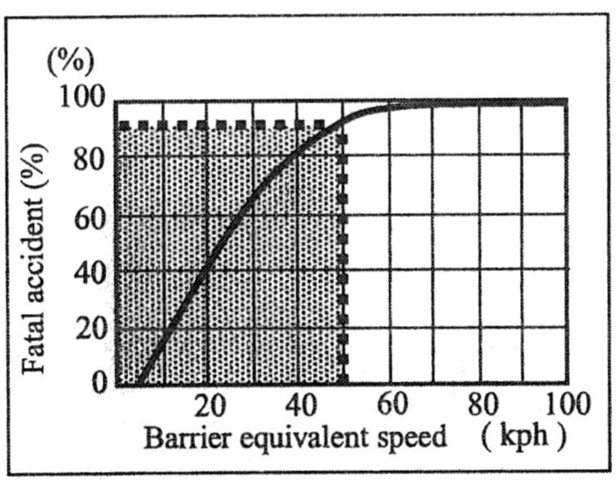

Figure 13. Fatal accident speed and percentage in frontal collisions.

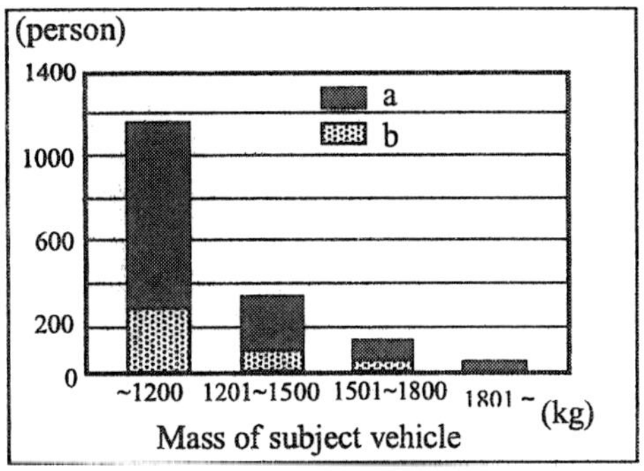

Figure 14. Vehicle mass and the driver fatalities of subject vehicle in frontal collisions.
; a) The opposite vehicle is heavier than the subject vehicle.
; b) The opposite vehicle is lighter than the subject vehicle.

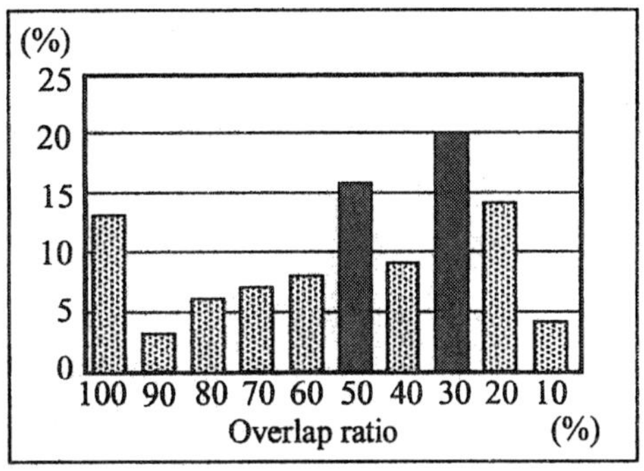

Figure 15. Overlap ratio in frontal collisions.

TEST CONDITIONS

The target in this test is the verification for a small car' s self-protection in the conditions based on real world accidents, especially the improvement of geometrical interaction in the frontal structure.

Small car: Prototype model
(This time only the 'Lower member system' was added on.)
Large car: Conventional model
Speed: 50kph per car
Mass : Small car 985kg Large car 1855kg
Mass ratio: 1.9 (Large car／Small car)
Overlap ratio: 50% of small car

The 50% overlap case is the reason that the risk of injury is higher than 30%. (as shown in figure 7)
Impact angle is 0°to the car's longitudinal axis.
Figure 16 shows the two cars before the crash test.
The main structural layouts of the cars are shown in Figures 17.

Figure 16. Two cars before the crash test.

Bumper heights are approximately the same.

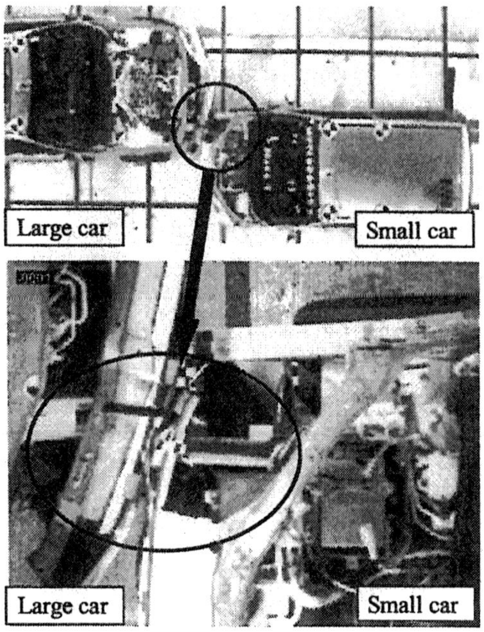

The main frame of the vehicles did not overlap in the direction of width.

Figure 17. Main structural layout of two cars.

In this test the main frames of the cars did not overlap in the direction of the cars' width, and the bumper heights were approximately the same.

Dummy: HybridIII

Restraint system: airbag and seat belt pretensioner with load limiter.

The restraint specification is similar to a small car without suitable modification for the change of the vehicle's deceleration characteristic.

TEST RESULTS

Speed: Small car 50.0kph
Large car 49.9kph
Overlap ratio: 48% of Small car

Figure 18 shows two cars after the crash test.

The structural deformation of the small car and the large car is shown in Figures 19 and 20 respectively.

The mode of structural deformation in the engine compartment of the small car during the crash is shown in Figure 21. Figure 22 shows deformation in the engine compartment of the small car after the crash.

Each part of deformation in both cars is listed in Table 1. Figure 23 shows the velocity change of both cars.

Figure 18. Two cars after the crash.

Figure 19. Deformation of small car with the new structure after the crash.

Figure 20. Deformation of large car after the crash.

0msec 10msec 38msec

Figure 21. Deformation mode of small car's new structure during the crash.

Before the crash After the crash

Figure 22. Deformation of small car in the engine compartment after the crash.

Table 1. Deformation

		Small car with new structure	Large car
Deformation	Front pillar(mm)	13	9
	Dash panel(mm)	114	162

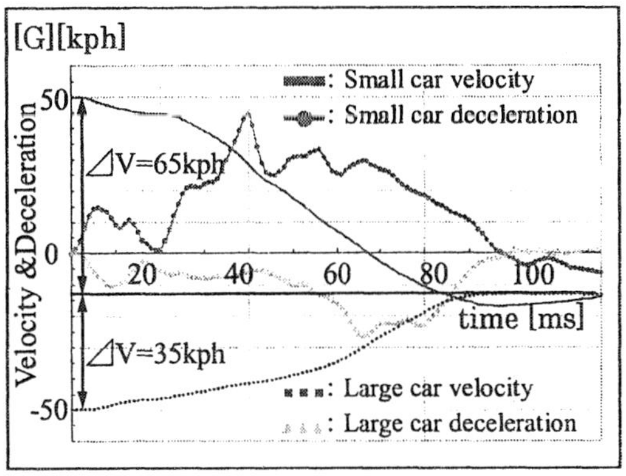

Figure 23. Velocity & Deceleration of both cars.

The vehicle-to-vehicle crash test confirmed the effectiveness of the new design in increasing the level of self-protection of small cars. The mode of deformation on impact confirmed that the bumper beam and main frame of the large car collided with the new lower member fitted in the small car.

The lower member restrained intrusion into the small car by making contact with the tire and wheel. Intrusion of the passenger compartment was therefore significantly reduced and the integrity of the cabin was maintained for occupants. The large car deformations in each part were approximately similar with the small car deformations.

The risk of injury to the small car's occupants were generally low by the prevention of secondary collisions in the passenger compartment.

The test results show the small car driver's injury risk is higher than the large car driver's injury risk. The reason is that the small car's Δ V[65kph], (namely the velocity change of before and after impact) is higher than the large car's Δ V[35kph], due to the influence by mass ratio.

CONCLUSION

A small car to large car crash test confirmed that the new body structure is one of the ways in increasing the level of self-protection of small cars.

It is very difficult for small vehicles with conventional body designs to maintain cabin integrity in narrow offset collisions and collisions with SUV because of the intrusion of the frame of the other vehicle

The innovative body structure proposed in this research reduces passenger compartment intrusion and occupant injury by restraining frame intrusion and enabling a high level of energy absorption in the engine compartment.

The structure improves the level of self-protection in small cars, it is also expected to improve the level of partner-protection offered by large car. As a further step we are going to research the aggressiveness for large cars based on the proposed new structure in this report.

Therefore, new design concept in making vehicles isn't an individualistic one and doesn't aim for only superior self-protection. Rather, the concept is harmony with the society of automobiles.

Finally it is hoped that the proposal of this new structure will trigger further research on body structures enabling reduction of traffic accident casualties in the future.

REFERENCES

[1] The Annual Reports of Traffic Accident Research and Data Analysis.1996.ITARDA.

2002-01-1022

Design Considerations for a Compatibility Test Procedure

Stephen Summers, William T. Hollowell and Aloke Prasad
National Highway Traffic Safety Administration

ABSTRACT

A major focus of the National Highway Traffic Safety Administration's (NHTSA) vehicle compatibility and aggressivity research program is the development of a laboratory test procedure to evaluate compatibility. This paper is written to explain the associated goals, issues, and design considerations and to review the preliminary results from this ongoing research program. One of NHTSA's activities supporting the development of a test procedure involves investigating the use of an mobile deformable barrier (MDB) into vehicle test to evaluate both the self-protection (crashworthiness) and the partner-protection (compatibility) of the subject vehicle. For this development, the MDB is intended to represent the median or expected crash partner. This representiveness includes such vehicle characteristics as weight, size, and frontal stiffness. This paper presents distributions of vehicle measurements based on 1996 fleet registration data. While there is still considerable work to be done to develop meaningful aggressivity metrics that relate to real-world crash performance, this paper summarizes NHTSA's work to date in this area.

INTRODUCTION

For several years, NHTSA has been conducting a vehicle compatibility research program to study the problems, trends, and evolving issues in vehicle-to-vehicle crashes. Much of this research has been statistical in nature, trying to define the nature and extent of the compatibility problem within the US automotive fleet. This research, conducted by NHTSA and many others, has led to a growing interest in the development of test procedures to evaluate the compatibility of individual vehicles. In fact, the major aim of the International Harmonised Research Activity (IHRA) on compatibility has been to develop internationally agreed test procedures designed to improve compatibility of vehicle structures in front-to-front and front–to-side impacts[1].

Within the IHRA compatibility working group, a number of possible candidate test procedures have been identified for a frontal test. These include a full frontal barrier test with load cells, with or without a thin (150 mm) deformable element; an offset deformable barrier (ODB) test with load cells; and an overload test using the ODB to evaluate the passenger compartment integrity; a test using barrier elements to explore shear (e.g., the progressive deformable barrier test being developed by the French[2]); and a moving deformable barrier (MDB) test with load cells behind the deformable element.

While NHTSA has been studying each of the identified test procedures, the focus of the developmental activity has been on the full frontal barrier test with load cells and on the MDB test. This paper updates NHTSA's previously reported efforts[3] evaluating the use of a load cell MDB for use in a potential compatibility test. The use of a load cell MDB allows a test procedure to account for the mass of the vehicle, which will affect the change in velocity for the struck vehicle and therefore the severity of the collision. It is expected that a compatibility test will provide a self protection challenge for smaller vehicles, while for larger vehicles, the challenge will be not to "overload" the measured forces on the MDB face. It has not been shown that the forces measured on an MDB or rigid barrier face have any correlation to vehicle compatibility; however, this is an area of active research which seems to have some potential[4]. While this work is continuing, there are still the many questions regarding how a MDB should be designed for use in a potential compatibility test. Any compatibility test is intended to evaluate a vehicles' performance in vehicle-to-vehicle crashes. Federal motor vehicle safety Standard (FMVSS) 214 testing evaluates a vehicles' vulnerability to side crashes. Here the MDB is designed to be the aggressor. In contrast, an aggressivity test would require a more vulnerable MDB to evaluate the force and acceleration environment in a frontal-frontal crash. Both aggressivity and vulnerability could be simultaneously evaluated by developing an MDB to represent an average vehicle or the most likely collision partner. Here the vehicle under test must protect its occupants while providing an acceptable crash environment for the MDB. Regardless of the test objectives, proper MDB design requires some understanding of vehicle characteristics within the U.S. fleet. This paper will analyze available data on the US fleet and develop distributions of vehicle characteristics that are relevant to the design of an MDB.

DATA SOURCES

This analysis utilizes the R.L Polk Company's National Vehicle Population Profile (NVPP) to determine the composition of the US fleet based on annual registrations. The registration data are analyzed by grouping corporate twins and siblings into car groups that span vehicle design years[5]. These car groups were developed using changes in the vehicle wheelbase to identify which model years can be grouped together. This technique is generally, but not completely consistent with the significant vehicle design changes reported in NHTSA's New Car Assessment Program (NCAP) and compliance programs. The registration data extracted from Polk NVPP are used to weight and distribute vehicle characteristics obtained from NHTSA's test programs and other data sources. The test data is primarily from the frontal NCAP program, but where available static vehicle measurements from side NCAP, FMVSS 208, FMVSS 214, and FMVSS 301 compliance programs are also used. Additional vehicle measurements are collected in a vehicle parameter database[6] that are not collected in the NHTSA's test programs. This database has a wealth of vehicle measurements, but is limited mostly to passenger cars. The vehicle parameter database can be readily linked to NHTSA test databases.

The correlation between NVPP registration data and the car group codes was last developed for 1996. Car group codes were identified for 93.4% of the 1996 registrations (160,572,388). The 1996 registrations were divided among 402 unique car groups, including both passenger cars and LTV's. The car group coding system developed in Reference 4 also provides the capability to decode vehicle identification numbers (VIN) into the same car group codes. There are 609 frontal NCAP tests with valid 17 digit VIN's that correspond to 275 unique car group codes. Additionally there are 1,236 full scale vehicle crash tests with NHTSA standard pre and post test vehicle measurements available. Correlating the NCAP test data with the 1996 registration data, accounts for approximately 85% of all 1996 registrations. Not all measurements exist for all tests. The fleet coverage for a specific vehicle characteristic will depend upon which tests have valid measurements. Additionally, there are numerous car groups that have multiple corresponding NCAP tests. The results of each NCAP test will be distributed among the registrations for the car group.

EVALUATION OF DESIGN FACTORS

STIFFNESS – One of the principle concerns with the design of an MDB is selecting the appropriate stiffness to use for the honeycomb of the barrier face. The frontal NCAP tests provide force deflection data that can be used to correlate the vehicle's crush with the force measured on the load cell barrier. Reference 3 presents averages, by vehicle category for all of the force-deflection profiles measured in the frontal NCAP program. It is desired to update these averages by weighting the test results according to the corresponding fleet registrations. These measurements are shown in Figure 1. Here the force measured on the load cell barrier is plotted against the displacement measured at the B pillar or the left rear seat, depending upon available sensor data. The data for each of the tests was truncated from time zero until the time of maximum displacement. The force deflection profile was resampled in 1 mm increments, so that multiple tests could be averaged. Since different vehicles have different levels of crush, the right axis is used to plot the percent of registrations, for each vehicle category, that were averaged at each deflection interval. Generally, deflections above 600 mm average the results of a varying number of NCAP tests and their corresponding fleet registrations.

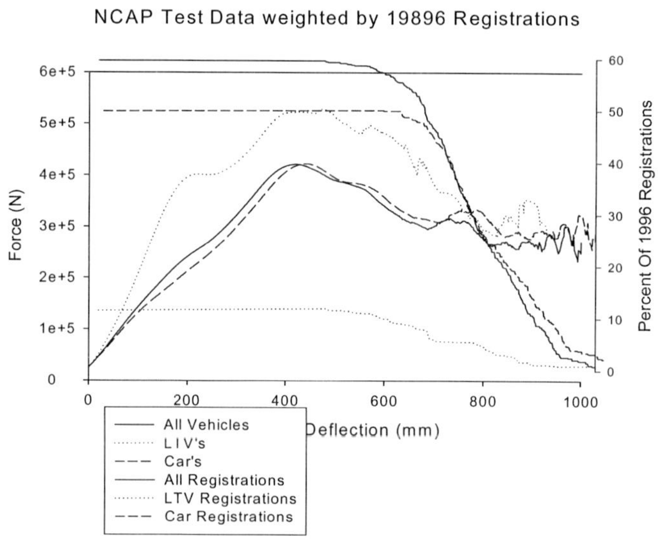

Figure 1: Force Deflection profiles

The NCAP force-deflection data under represents the LTV's, as a percentage of the fleet data. The LTV data represent only 20% of the "All Vehicles" force-deflection data. In 1996, LTV's represented 31% registrations in NVPP. Thus the force-deflection profile for "All Vehicles" significantly under represents the contributions due to LTV's.

A similar analysis was conducted for the average height of force. Here, the height of each load cell is used to compute the average height of force (AHOF) for each time step during the crash for each NCAP test.

$$HOF(t) = \frac{\sum_{1}^{36} Fi \times Hi}{\sum_{1}^{36} Fi}$$

The height of force is then averaged, using the force(t) data as a weighting function. The weighting is intended to insure that the AHOF reflects the height at which the load was transferred to the barrier.

$$AHOF = \frac{\sum_{0}^{t} HOF(t) \times F(t)}{\sum_{0}^{t} F(t)} \qquad t = time\, step$$

The AHOF for each NCAP tested was weighted using the registration data and is shown in figure 2. The median AHOF was 473 mm for cars and 534 mm for trucks. Loadcell barrier testing with the FMVSS 214 barrier produced an AHOF of 506 mm as indicated by the vertical line in figure 2.

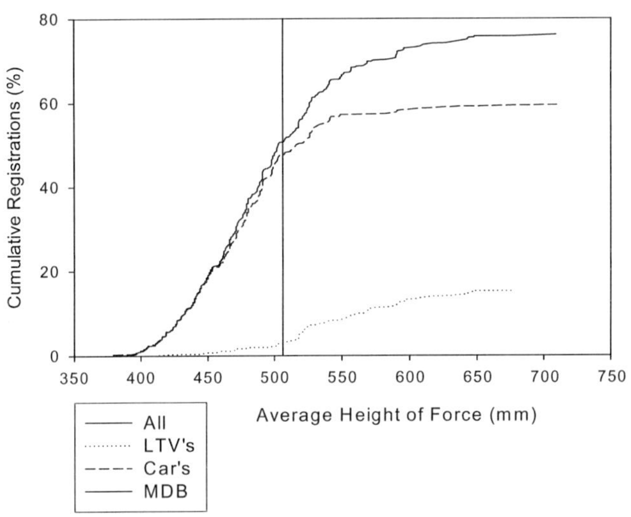

Figure 2: Average Height of Force

One additional measure of interest, particularly in side impact crashes, is the initial stiffness of the force deflection profiles. Visual examination of Figure 1 shows a strong variation in the initial slope of the force deflection profile between cars and LTVs, from 0 to about 400 mm of deflection. To provide a numerical measure of the initial slope, each force deflection profile was linear fit with a straight line that was constrained to start within the first 200 mm and have an R^2 value > 0.95. The slope of the longest straight line, greater than 75 mm in length, that met these criteria was selected as the initial slope for the force deflection curve. The slope measurements, distributed by 1996 registrations, are shown in Figure 3. The vertical line indicates the initial slope of 214 barrier measured from loadcell barrier testing.

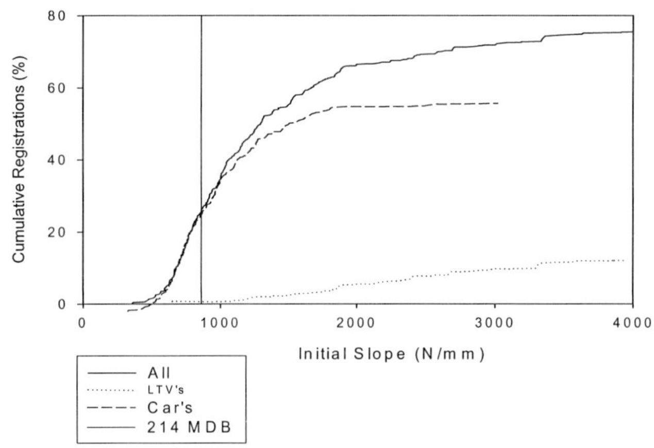

Figure 3: Force Deflection Slope measurements

The median slope for the cars tested is around 1040 N/mm, while for LTV's the median is around 2200 N/mm, or about twice as steep. Several of the older LTV models had extremely steep initial slopes.

WEIGHT – Vehicle weight is one of the most commonly measured vehicle parameters, yet it is generally difficult to find consistent measures for vehicle weight. Polk NVPP reports curb weights for passenger vehicles, but does not report curb weights for LTV's. Additionally, the weights that Polk reports are the base weight for the model and may not reflect the average weight of the vehicles in the fleet. NHTSA test databases record the weight of the vehicle as tested, but this weight includes the weight of dummies, ballast, and stoddard solution. A vehicle's test weight can be 100 to 200 kg heavier than the reported curb weight. This can be clearly seen from Figure 4, which shows curb weight as reported by Polk NVPP and the test weights reported in NHTSA testing. NVPP data has a median car curb weight of 1330 kg while the test data for cars had a median weight of 1494 kg, or an average increase of 164 kg between the NVPP reported curb weights and the NCAP test weight. The vertical line represents the weight of the current FMVSS 214 MDB.

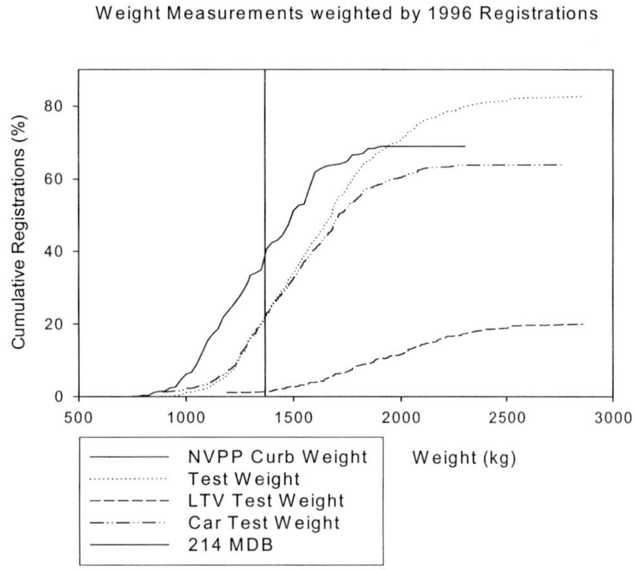

Figure 4: Weight Measurements

While the car test weights are always greater than the curb weight, there is no consistent relationship between them. LTV's represents 23 percent of the test weight distribution, which does not reflect the 31 percent LTV registrations for 1996.

WIDTH – The vehicle width is readily applicable to the design of an MDB. Figure 5 compares the overall vehicle width measured in NCAP testing against the vehicle widths reported in the vehicle parameter database. The vertical line indicates the width of the FMVSS 214 honeycomb face. It is interesting to see that the specifications average about 50 mm wider than was reported in the NHTSA test programs. No explanation was readily apparent for this discrepancy.

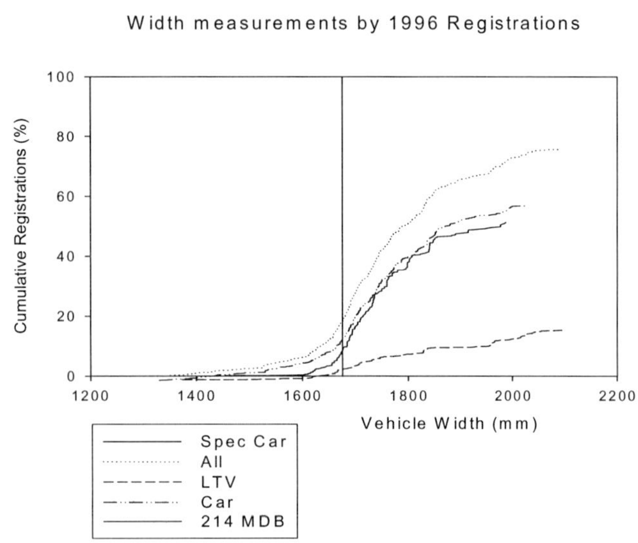

Figure 5: Width Measurements

LENGTH – For a pure frontal crash, the overall vehicle length is not a significant design parameter, however if an oblique or offset test condition is chosen, then the wheelbase can play a significant role. Additionally, the NCAP test procedure includes measurements of vehicle front to the front of the engine and the vehicle front to the firewall. These measures may be useful in determining an appropriate crush depth for the honeycomb face. Figure 6 shows these length measurements for the 1996 fleet as measured in the NCAP test program. The front of engine and firewall measurements are shown for all vehicles only.

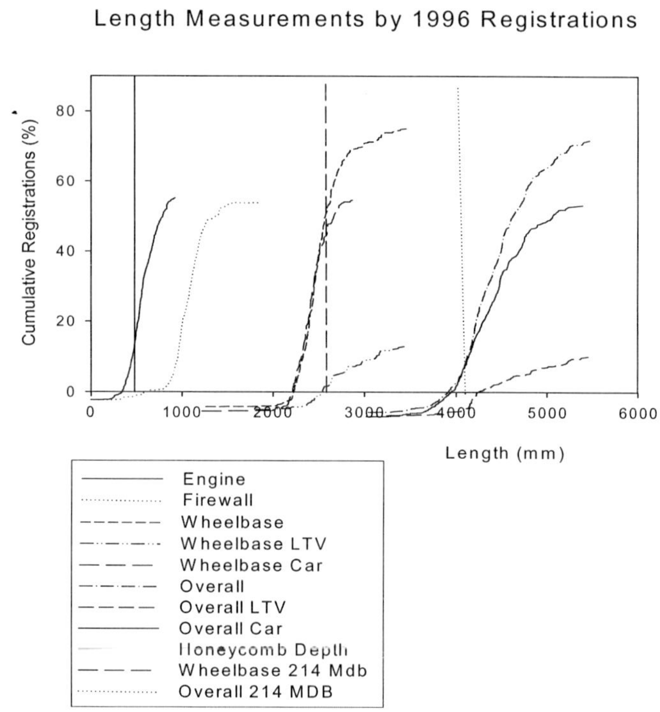

Figure 6: Length Measurements

The three vertical lines in Figure 6 represent measurements from the existing FMVSS 214 MDB. From left to right these are, the length of the honeycomb face, the wheelbase, and the overall length. The wheelbase of the MDB is near the 50%'tile for the car measurements. The honeycomb length is considerably shorter than the typical vehicle front to engine length. This indicates that the MDB may not have sufficient crush depth to represent a typical crash partner.

HEIGHT – Unfortunately, NHTSA's test programs include almost no height measurements. The only data available for height measurements comes from the vehicle parameter database, which is essentially limited to cars. The measurements of interest are the minimum height of the bumper, the height of the cowl or windshield to hood interface, and the minimum rocker panel height. These measurements are plotted in Figure 7. The vertical lines represent the bumper height and the maximum height of the honeycomb face.

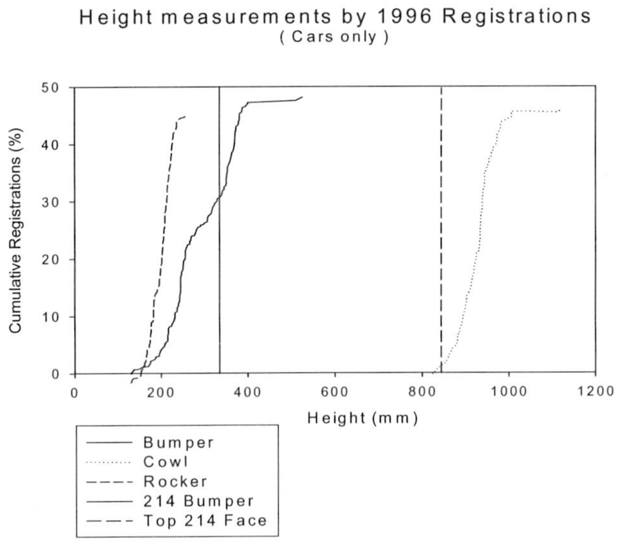

Figure 7: Height Measurements

CONCLUSION

This paper presented a methodology for using vehicle measurements, primarily from the NCAP testing program, to estimate collected fleetwide distributions of vehicle characteristics. These distributions can be used to aid in the development of a MDB for compatibility testing. The optimum vehicle characteristics for an MDB depend upon the measurements, criteria and intent of a compatibility test. Regardless of whether a vulnerability, aggressivity, or combined compatibility test is being developed, these fleet distributions can help in the development of an MDB to represent the existing fleet. It might also be interesting to generate similar distributions based on new vehicle sales rather than registrations. This might provide better predictions of changing fleet characteristics. While there is still considerable work to be done in the development of a meaningful compatibility test, these design charts can provide some insights into the appropriate ranges for many design parameters.

REFERENCES

1. Peter O'Reilly, Status Report of IHRA Vehicle Compatibility Working Group, ESV Paper 337, June 2001
2. Delannoy, P., Diboine, A., "Structural Front Unit Global Approach", 17th International Technical Conference on the Enhanced Safety of Vehicles, Paper number 199, June 2001
3. Summers, Prasad, Hollowell, "NHTSA's Compatibility Research Program Update", SAE 2001-01-1167, March 2001
4. Hans C. Joksch, "Vehicle Design versus Aggressivity", DOT HS 809 194, April 2000
5. Charles Kahane, Relationships between Vehicle Size and Fatality Risk in Model Year 1985-93 Passenger Cars and Light Trucks, DOT HS 808 570, January 1997.
6. McCullough, Catherine A., Hollowell, W.T., Battisti, Peter, "Applications of NHTSA's Vehicle Parameter Database", SAE Paper No. 950360, March 1995

ADDITIONAL SOURCES

See http://www-nrd.nhtsa.dot.gov/departments/nrd-11/aggressivity/ag.html for current information on NHTSA's compatibility research program.

Systems Modeling

2003-06-0153

VEHICLE-TO-VEHICLE FULL FRONTAL CRASH OPTIMIZATION USING A CAE-BASED METHODOLOGY

Saeed Barbat, Xiaowei Li, Phillip Przybylo and Priya Prasad
Safety Research and Development
Ford Motor Company

ABSTRACT

This paper describes a CAE-based methodology used to identify major factors influencing vehicle structural performance and crash energy management in full-frontal vehicle-to-vehicle collisions. Finite element models of an "average" SUV and an "average" full-size passenger vehicle were used in this study. The determining factors of vehicle compatibility in multi-vehicle collisions are relative mass, relative stiffness and relative geometry. Four parameters of the average SUV, mass, fore rail length, fore rail thickness, and fore rail height were selected as design variables. A uniformly spaced Optimal Latin Hypercube sampling technique was employed to probe the design space of these variables using thirteen simulation runs.

Dash intrusions in the passenger vehicle and the absorbed collision energy in both vehicles were selected as response variables. Polynomial response surfaces were constructed, based on the simulation results, and found to fit the results well (R^2= 0.98 for dash intrusion and R^2= 0.85 for absorbed energy). As a result, prediction equations for maximum dash intrusion and absorbed collision energy as a function of the vehicle design variables were obtained. Results indicated that aligning front-end structures (specifically fore rail heights between impacted vehicles) in vehicle-to-vehicle full-frontal collisions has greater effect on reducing dash intrusions and managing crash energy than mass and variables associated with stiffness. An optimal design solution could also be determined with the appropriate introduction of constraint conditions.

INTRODUCTION

Vehicle compatibility has been investigated in many studies using different approaches such as real-world crash statistics, crash testing and computer modeling. Statistics from the General Estimates System (GES) were used to determine the number of vehicle-to-vehicle collisions [1]. One objective of these studies was to identify and demonstrate the extent of the problem of vehicle compatibility. Another objective was to demonstrate, through statistical analysis, the aggressivity metric as a function of vehicle mass, stiffness, and geometry. Other statistical analysis such as that conducted by Evans [2,3] indicated that mass is one of the most significant factors affecting potential risk of occupant injury in vehicle-to-vehicle collisions. His study indicated that the ratio of the injury rate in a lighter vehicle to that in a heavier one can be expressed as the power function of the mass ratio (of the heavier to the lighter vehicle). Accident data in Japan and computer modeling techniques were also used by Mizuno and Kajzer [4] to investigate the compatibility of mini cars in traffic accidents. Barbat, *et al.* [5,6] also investigated factors influencing compatibility in SUV/LTV-to-car crashes. Their study proposed a robust and repeatable vehicle-to-vehicle test procedure to help assess vehicle compatibility, and their preliminary results indicated that geometric compatibility was the dominant factor.

In addition to occupant responses, attempts have been made to characterize frontal crash performance with structural responses (i.e., occupant compartment intrusions and energy distribution) [5-8] in vehicle-to-vehicle collisions. Barbat, *et al.* [5,6] has shown the importance of occupant compartment intrusions in full-frontal SUV/LTV-to-car crashes. Steyer, *et al.* [7] postulated that frontal crash compatibility was dependent upon the distribution of collision energy absorbed in vehicle-to-vehicle crashes.

The consensus is that the significant parameters influencing compatibility in vehicle-to-vehicle crashes are geometric interaction, vehicle stiffness and vehicle mass. The effect of each individual design parameter, however, is not clearly understood. Separating the effects in various crash configurations via experimentation would not only be costly and time consuming, but also would be susceptible to systemic errors due to test-to-test variability. Because of the limitations of the statistical approach and crash testing, math modeling in combination with design of experiments (DOE) methods was deemed necessary to examine the influence of vehicle mass, stiffness, structural interaction and geometry on vehicle compatibility.

Finite element (FE) simulations have been used to study many aspects of vehicle crashworthiness.

Carefully designed experiments can characterize responses over the entire design space using a reduced number of simulations. The development of a FE model-based DOE methodology focused on discerning the effects of a few design variables on structural intrusions in 40% offset, car-to-car collisions has been previously presented by Brown, *et al.* [8] Brown's work revealed some of the economies of combining FE analysis and statistical techniques. The current study, while fundamentally similar, utilizes a deterministic approach that allowed analytical prediction equations for structural responses to be generated. This study combined FE analysis, Latin Hyper Cube Sampling (LHS), and subset selection with sequential replacement to produce a powerful tool that may be used to investigate vehicle compatibility issues. A full-frontal, collinear, SUV-to-passenger vehicle crash configuration (See Figure 1) was used in this study. This technique can also be extended to various crash configurations other than the one used in this study.

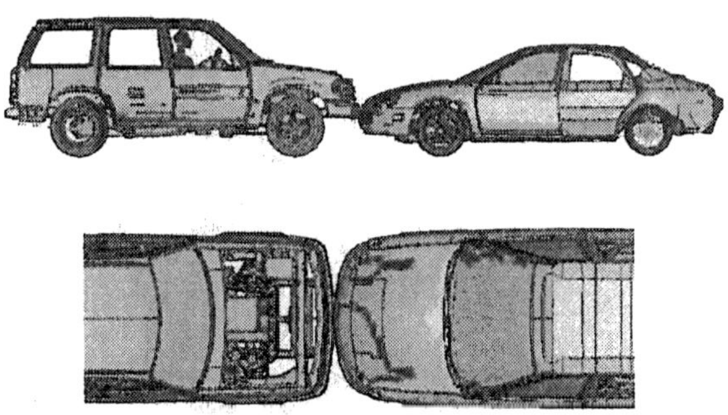

Figure 1. Baseline model of SUV-to-passenger vehicle full-frontal crash simulation.

SYSTEM SETUP

Finite Element Modeling

Reliable finite element models of the vehicles are required to enable reasonable predictions of structural performance. In this study, full-vehicle finite element models validated to rigid barrier tests were used for the "average" SUV and the "average" passenger vehilce. A baseline, full-frontal vehicle-to-vehicle FE model was constructed by combining the two vehicle models and it was correlated to a physical vehicle-to-vehicle full-frontal crash test. As in the physical test, the simulated passenger vehicle was stationary and the simulated SUV was given an initial velocity of 96 km/h.

Full-frontal, collinear SUV-to-passenger vehicle simulations involve many complex and non-linear interactions. The nonlinear, explicit FE crash code, RADIOSS [9], was used for all of the simulations. The appropriate selection of design variables and their levels as well as, the pertinent system responses are basic requirements.

Design Variable Selection

It is generally accepted that the determining factors of vehicle compatibility are relative geometry, relative stiffness and relative mass. All three factors were considered here through four design variables, the SUV fore rail height, fore rail length, fore rail thickness, and mass. Geometric differences between the SUV and the passenger vehicle were modeled as relative vertical alignment between the fore-most structural members in each vehicle. The SUV front-end stiffness was separated into two design variables, fore rail length and fore rail thickness. This separation allows the effects of the design variables to be interpreted independently and to be associated with vehicle characteristics. For each design variable, only the SUV portions of the FE models were changed to reflect the respective ranges. A brief description of how the variables were introduced into the FE models and the levels selected for the design variables follows.

<u>Geometry</u>
The vertical alignment of load-bearing/energy-absorbing structures was varied to allow geometric effects to be observed. The relative vertical offsets were accomplished by changing the ground reference plane of the SUV with respect to that of the passenger vehicle. The four discrete levels selected are depicted in Figure 2.

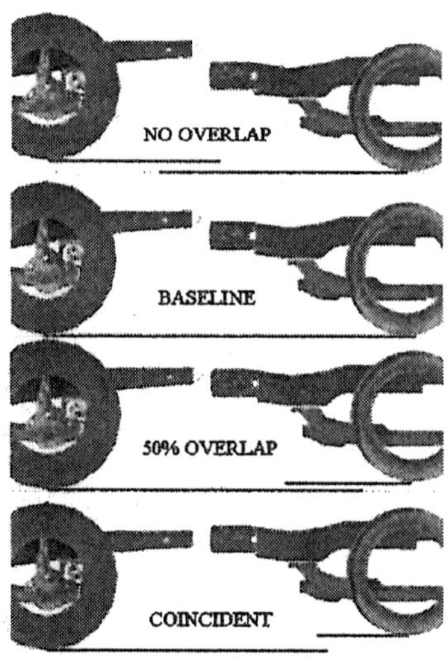

Figure 2. Geometric design variable implementation.

The BASELINE configuration, NO OVERLAP between the fore rails, 50% OVERLAP of the PASSENGER VEHICLE fore rail, and COINCIDENT FORE RAIL centerlines were included in the study. The range of levels encompass from a potentially overriding condition to an aligned configuration.

Modeling the geometric changes in this manner shifts the center of gravity height of the SUV. The effect this amount of the shift would have on the overall responses of the vehicles was considered minimal.

Stiffness

The relative stiffness was modeled with two design variables, SUV fore rail length and SUV fore rail thickness. The length of the fore rail was modeled at four discrete levels, baseline (average SUV), +25 mm, +50 mm and +75 mm. Models associated with each design level are shown in Figure 3. In addition to adjusting the fore rail length, components such as the bumper system, radiator, and light supports were translated in the longitudinal direction by appropriate amounts.

The second stiffness-related deign variable was the SUV fore rail wall thickness. As with the two previous design variables, the thickness was studied at four distinct levels, baseline, -0.5 mm, -1.0 mm, and +0.5 mm.

Mass Ratio

The final design variable studied was the ratio of the SUV mass to the passenger vehicle mass, hereafter referred to as mass ratio. In order to vary the mass ratio, the SUV mass was adjusted by distributing small masses throughout the model such that the center of gravity location remained equivalent to the BASELINE location. The mass of the SUV was varied from BASELINE-20% to BASELINE+20% in 10% increments. The range approximately spans the vehicle segments from small SUV to large SUV. The corresponding mass ratios ranged from 0.9 to 1.38 in five discrete levels (see Table 1). Since the initial velocity of the SUV was the same for all simulations, the velocity change (ΔV) experienced by each vehicle varied with the mass ratio. Table 1 shows the ΔV's experienced by the vehicles at the various mass ratios. The velocity change is an indicator of the crash severity commonly used for vehicle crashes.

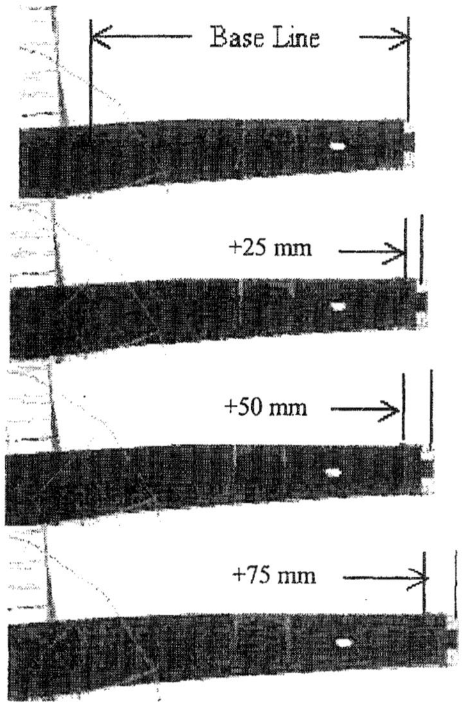

Figure 3. SUV fore rail length.

Table 1.
Passenger Vehicle and SUV Velocity Changes

Mass Ratio	Passenger Vehicle ΔV [km/h]	SUV ΔV [km/h]
0.90	45.7	50.9
1.02	48.8	47.8
1.14	51.5	45.1
1.26	53.9	42.6
1.38	56.0	40.6

Table 2 contains a summary of the design variables and their corresponding levels. The levels have been associated with an integer representation (coded) for simplicity.

Table 2.
Summary of Coded Design Variables and Levels

Fore Rail Height
1 = No overlap
2 = BASELINE
3 = 50% Overlap of PASSENGER VEHICLE fore rail
4 = Coincident
Fore Rail Length
1 = BASELINE
2 = BASELINE + 25 mm
3 = BASELINE + 50 mm
4 = BASELINE + 75 mm
Fore Rail Thickness
1 = BASELINE – 1.0 mm
2 = BASELINE – 0.5 mm
3 = BASELINE
4 = BASELINE + 0.5 mm
Mass
1=BASELINE – 20%
2=BASELINE – 10%
3= BASELINE
4= BASELINE + 10%
5– BASELINE + 20%

System Responses

Occupant compartment intrusion is a metric, which is used in comparative studies of vehicle crash performance (e.g., the Insurance Institute for Highway Safety includes occupant compartment intrusion in their evaluation of vehicles in a 64 km/h 40% offset deformable barrier crash test). In this study, several passenger vehicle intrusions and the vehicles' collision energy management were the responses of interest (see Table 3).

The toe board intrusions were measured along the driver and passenger centerline on the dash, 200 mm up from the floor panel. The windshield intrusions were taken at the intersection between the lower edge of the windshield and the cowl at the windshield centerline. The driver and passenger A-pillar intrusions were monitored near the juncture between the lower windshield, the A-pillar, and the cowl. Bumper to rocker crush was taken as the relative longitudinal displacement between the bumper beam center and rocker at the B-pillar. The maximum dash intrusion was captured from a vertical section along the dash from the cowl to the floor through the driver centerline. The intrusions were calculated as the relative displacement between the point of interest and a reference point located on the rocker at the B-pillar.

Table 3.
Selected Responses

Passenger Vehicle
Driver Toe board Intrusion
Driver A-pillar Intrusion
Lower Windshield Center Intrusion
Passenger A-pillar Intrusion
Passenger Toe board Intrusion
Bumper to Rocker Crush
Driver Centerline Dash Intrusion
Internal Energy Absorbed
SUV
Internal Energy Absorbed

The distribution of collision energy absorbed between the vehicles was also a response monitored in this study. The internal energies absorbed that were calculated from the FE simulations are reported.

EXPERIMENTAL DESIGN

The number of factors and levels included in this study describe a sizable design space. Numerous techniques exist for constructing experimental designs that specify a minimal number of samples throughout the design space to characterize the responses [10,11]. Latin Hypercube Sampling was utilized to select the design levels for the FE models due to its ability to uniformly sample the design space and its flexibility in the number of levels for each design variable. Since no noise factors were introduced into this study, the minimum number of simulations to construct a reasonably accurate response surface was 12 (3N where N is the number of control factors). The outcome of the sampling process is shown in Table 4. One additional run was included in the matrix to represent the baseline simulation that was correlated to a physical test. All 13 simulations were conducted for 150 ms.

Table 4.
Simulation Matrix

	Fore Rail Height	Fore Rail Length	Fore Rail Thickness	Mass
1 (Base line)	*2*	*1*	*3*	*3*
2	4	4	2	2
3	2	4	3	4
4	4	2	3	3
5	1	1	3	1
6	3	1	1	2
7	1	4	1	3
8	4	3	1	5
9	2	2	2	5
10	3	3	4	1
11	2	3	2	1
12	1	2	4	4
13	3	1	4	5

Finite Element Simulation Results

Table 5 contains the responses from the FE simulations. All intrusion responses were normalized with respect to the intrusions associated with the baseline simulation. The energy absorbed in both vehicles was computed and reported as the ratio of the energy absorbed in the passenger vehicle to that absorbed in the SUV. Response surfaces were created to allow interpretation over the design space and efficient (FE model-free) studies to be carried out.

Table 5.
Normalized Simulation Responses

	Driver A-Pillar	Wind-shield	Passenger A-Pillar	Driver Toe board	Passenger Toe board	Bumper To Rocker	Maximum Dash Intrusion	IE Ratio
1	1	1	1	1	1	1	1	2.6
2	0.41	0.31	0.47	0.63	0.78	0.81	0.43	1.5
3	1.17	1.28	1.31	1.17	1.19	0.99	1.16	2.3
4	0.64	0.40	0.68	0.76	0.86	0.83	0.51	1.8
5	0.59	0.71	0.71	1.20	0.86	0.95	0.94	1.9
6	0.57	0.50	0.60	0.78	0.77	0.86	0.50	1.7
7	0.70	1.08	0.97	1.25	0.98	1.07	1.09	1.8
8	1.11	0.68	0.86	0.84	0.78	1.00	0.79	1.5
9	1.07	1.17	1.13	1.23	1.21	1.08	1.10	2.0
10	0.77	0.54	0.71	0.47	0.83	0.80	0.51	2.5
11	0.57	0.62	0.73	0.83	0.81	0.84	0.73	2.0
12	1.26	1.33	1.32	1.39	1.11	1.04	1.28	2.7
13	1.18	1.05	1.22	0.96	1.10	0.96	0.96	2.4

Response Surface Generation

The sample responses obtained from the FE simulations (Table 5) were fit with quadratic polynomials using regression based upon subset selection by sequential replacement. Gu [11] showed that for structural intrusions and internal energy predictions in crash applications, quadratic polynomials are often sufficient. The polynomial basis of the equations allows the response surface dependency on the design variables to be interpreted by observation. Furthermore, the explicit form of the response surfaces lends itself to a variety of design studies (e.g., parameter sensitivity analysis or optimization). The coefficient of determination, referred to symbolically as R^2, is a measure of the model's ability to fit the response data and was used to identify the best-fit equations in the least squares sense.

The response surfaces and R^2 values for the fit polynomials are listed in Table 6. In the response surface expressions, H is the SUV fore rail height, L is the SUV fore rail length, T is the SUV fore rail thickness, M is the mass ratio, and a_1-a_8, b_1-b_8, c_1-c_8, d_1-d_8, and e_1-e_8 are the best fit coefficients. The R^2 values shown in the table indicate that the response surfaces are capable of representing the sampled FE results.

Table 6.
Response Surfaces and R^2

Response Surface	R^2
Driver Toeboard Intrusion $= a_1 - b_1 H M - c_1 T^2 + d_1 T + e_1 M$	0.94
Drive A Pillar Intrusion $= a_2 + b_2 M - c_2 T H - d_2 T M + e_2 T^2$	0.87
Windshield Intrusion $= a_3 - b_3 H^2 + c_3 L T - d_3 M^2 + c_3 M$	0.99
Passenger A Pillar Intrusion $= a_4 + b_4 T M + c_4 L T - d_4 H^2 + e_4 M$	0.97
Passenger Toeboard Intrusion $= a_5 - b_5 H M - c_5 T^2 + d_5 T + e_5 M$	0.9
Bumper to Rocker Crush $= a_6 - b_6 T M + c_6 T - d_6 H + e_6 M$	0.96
Maximum Dash Intrusion $= a_7 - b_7 H^2 + c_7 L M + d_7 T - e_7 L T$	0.98
Energy Ratio $= a_8 + b_8 L M + c_8 T^2 - d_8 H^2 + e_8 H$	0.85

Verification of Equations

Four additional FE simulations were run to confirm the predictive character of the response surfaces. This step is not required in a rigorous mathematical sense, but it was conducted because of the complex nature of the crash mode. The parameters for these additional simulations were selected randomly from within the ranges of each design variable and the levels are shown in Table 7.

Table 7.
Verification Simulation Design Variable Levels

	Fore Rail Height	Fore Rail Length	Fore Rail Thickness	Mass
14	4	1	2	3
15	2	3	1	5
16	1	2	1	1
17	1	1	4	4

Figure 4 shows a bar chart comparing the FE simulation results and the predicted responses for passenger vehicle maximum dash intrusion. The run numbers from one through 13 correspond to the responses that were used to form the response surface. The remaining runs, run numbers 14 through 17, are the verification runs. The figure shows that the passenger vehicle maximum dash intrusion can be described via the prediction equation. The response surface, therefore, can be utilized for additional studies of designs contained within the original design space.

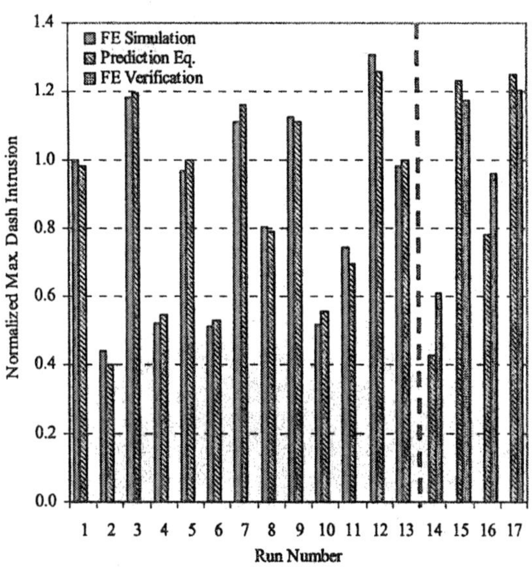

Figure 4. Normalized Passenger Vehicle Maximum Dash Intrusion

Figure 5 compares FE-simulated and predicted internal energy ratio responses for all runs previously described. The internal energy ratio response equation accurately predicted the responses for the additional runs. As with the maximum dash intrusion equation, the internal energy equations can be used to study the design space without conducting additional finite element simulations.

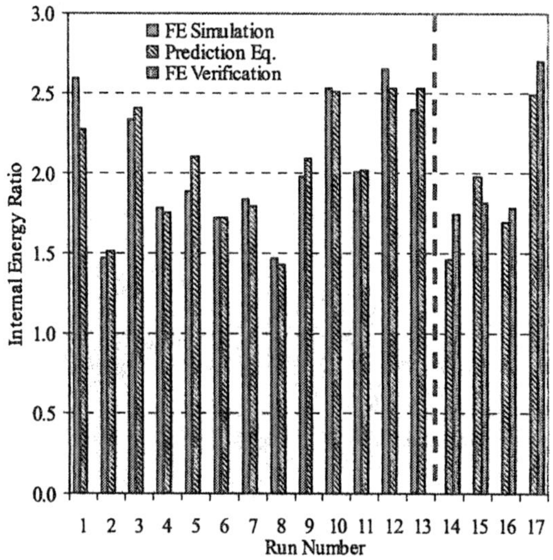

Figure 5. Internal Energy Ratio

RESULTS

Pair-wise Comparison of Design Variable Effects on Passenger Vehicle Maximum Dash Intrusion

Pair-wise comparisons of the predicted effects of the design variables show the relative importance of each factor. For all comparisons described, the variables that are not included were set to their BASELINE levels.

The comparison of the mass effect to the geometric and fore rail thickness effects on the passenger vehicle maximum dash intrusion (MDI) are shown in Figures 6 and 7, respectively. Figure 6 indicates that the MDI increases as the mass ratio increases while it decreases as the overlap percentage of the energy absorbing members of both impacted vehicles increases. Figure 6 also shows that fore rail height influences MDI more than the mass ratio for BASELINE stiffness SUV. That is, for a constant mass, the change in MDI due to changes in SUV fore rail height is greater than the change in MDI due to a change in the mass ratio for a constant fore rail height. This is also apparent from the contours of MDI.

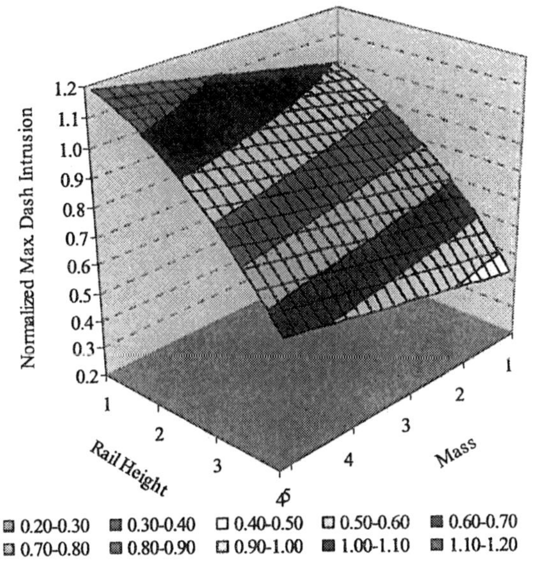

Figure 6. Mass-geometry surface plot of passenger vehicle maximum dash intrusion.

Figure 7 indicates that the MDI increases as the fore rail thickness and mass ratio increases. However, Figure 7 also shows that for BASELINE fore rail height and fore rail length, the fore rail thickness effect on MDI is larger than the mass ratio effect.

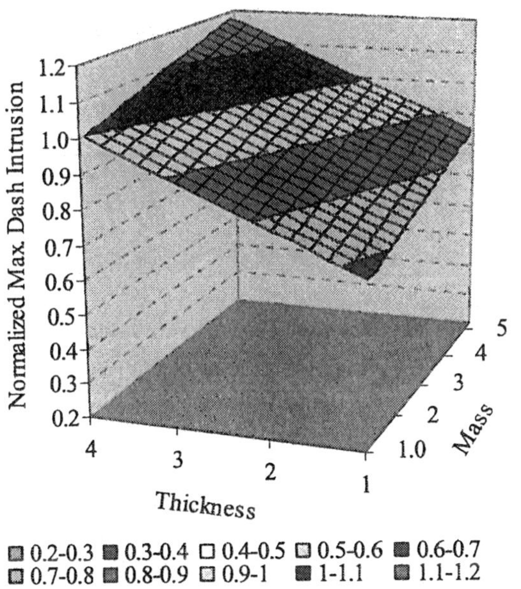

Figure 7. Mass-thickness surface plot of passenger vehicle maximum dash intrusion.

Figure 8 shows the comparison of the geometric effect to the thickness effect for BASELINE mass ratio and SUV fore rail length. The inclination of the surface indicates that the geometry effect is more than the thickness effect for MDI.

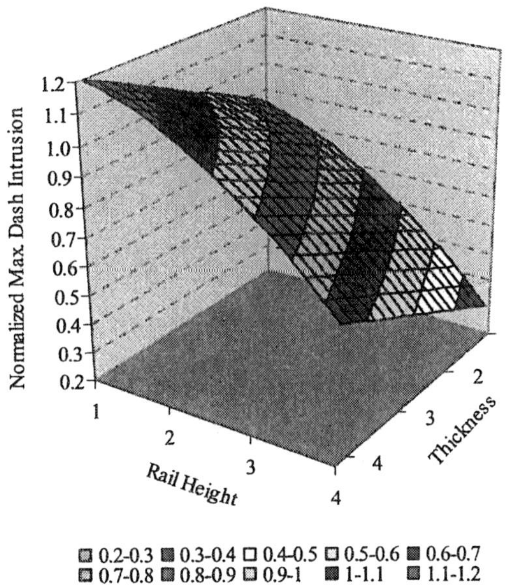

Figure 8. Geometry-thickness surface plot of passenger vehicle maximum dash intrusion.

The response surfaces in Figures 6 – 8 reveal that the geometric factor, relative fore rail height, has a greater effect than the relative mass and stiffness on the passenger vehicle maximum dash intrusion. These figures also showed that the thickness effect is more prominent than the mass effect.

Pair-wise Comparison of Design Variable Effects on Internal Energy Absorbed

Pair-wise comparisons of the effects of the design variables for internal energy absorbed ratio (IEAR) follow in Figures 9 through 11. Figures 9 and 10 contain comparisons of the mass ratio effect to the geometric and thickness effects on the IEAR, respectively. The geometric effect is compared to the thickness effect in Figure 11.

Figure 9 shows that the geometric effect on IEAR is much greater than the mass ratio effect. In fact, for the ranges considered the mass effect on IEAR is small (> 10%). For SUV fore rail height at level 1, the IEAR is lower than at SUV fore rail height level 2. Two possible explanations for this are: 1) structural components other than the fore rails of the vehicles are interacting or 2) the IEAR prediction equation is not accurate for this portion of the design space.

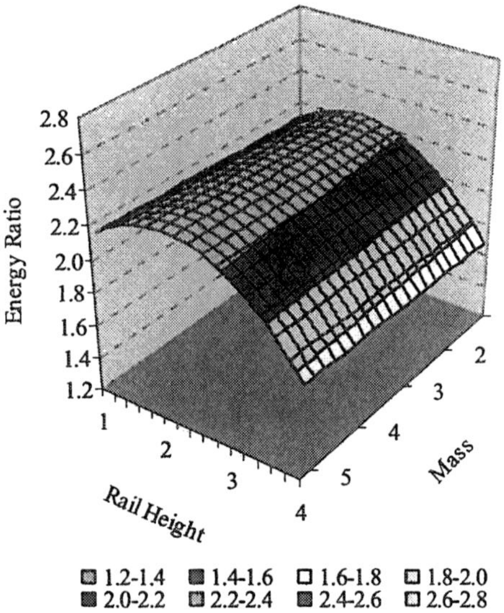

Figure 9. Geometry-mass surface plot of internal energy absorbed ratio.

Figure 10 contains the comparison of the mass ratio effect on IEAR to that of the SUV fore rail thickness. SUV fore rail thickness has a greater effect on IEAR than the mass ratio.

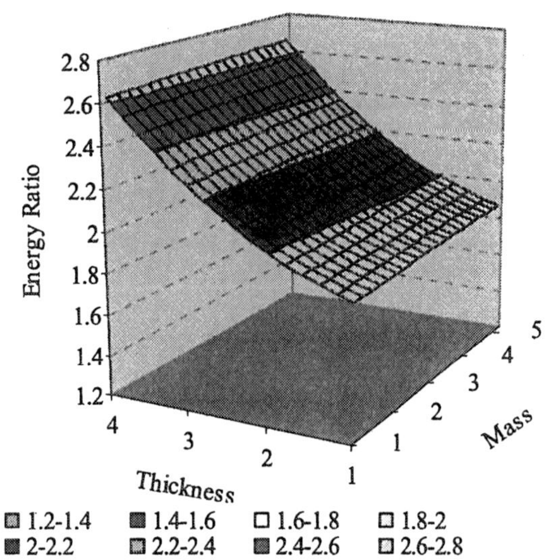

Figure 10. Mass-stiffness surface plot of internal energy absorbed ratio.

From Figure 11, the effects of SUV fore rail height and thickness can be seen. For IEAR the SUV fore rail height has a greater effect than thickness for SUV fore rail height levels 2 and higher. The thickness effect exceeds the height effect for the SUV fore rail height level 1 (i.e., NO OVERLAP/SUV over-ride condition). This switching of effect significance is subject to the explanations issued for Figure 9.

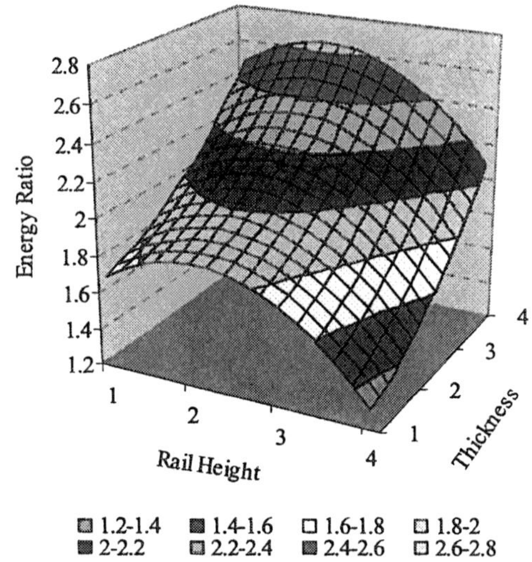

Figure 11. Geometry-stiffness surface plot of internal energy absorbed ratio.

RESPONSE OPTIMIZATION

Due to the predictive capability of the response surfaces, optimization can be performed on this set of surrogate models. The Safety Optimization and Robustness (SOAR) methodology developed by Yang, et al. [10] was the optimization technique applied. The explicit form of the response surfaces simplified the incorporation of the responses as objective or constraint functions. The objective function could be any of the response surfaces, but was chosen to minimize the passenger vehicle maximum dash intrusion subject to the constraints listed in Table 8. Since the optimal configuration depends on the constraint conditions, their magnitudes were based on the FE-simulated passenger vehicle responses taken from a rigid, fixed barrier (RFB) simulation at 56 km/h. That is, the intrusions in the optimal vehicle-to-vehicle full frontal simulation were limited to being no greater than the passenger vehicle striking itself. The internal energy ratio was restricted to being less than or equal to the BASELINE full-frontal vehicle-to-vehicle simulation.

A comparison of the optimized design to the BASELINE design follows. Figure 12 shows the comparison of passenger vehicle intrusions, and Figure 13 shows the internal energy absorption ratios. All intrusions into the passenger vehicle occupant compartment were reduced by at least 35% when compared to the BASELINE simulation. The distribution of absorbed energy between the SUV and passenger vehicle was more balanced in the optimal configuration. These structural metrics indicate that this configuration (fore rail height = 4, fore rail length = 1, fore rail thickness = 1, mass ratio = 1) is more compatible for the full-frontal, collinear crash mode. The SUV described by the optimal configuration was the lightest, least stiff, and most aligned vehicle in the design space. This optimal design is based on only this test mode. However, this is a demonstration of the optimization technique capability and the optimum design needs to be further studied to ensure no degradation in self-protection.

Table 8.
Constraint Conditions

Response	Constraint
Driver A-pillar Intrusion	$\leq RFB$
Windshield Low Intrusion	$\leq RFB$
Passenger A-pillar Intrusion	$\leq RFB$
Driver and Passenger Toe board Intrusion	$\leq RFB$
Bumper to Rocker Deformation	$\leq RFB$
Internal Energy Ratio	$\leq BASELINE$

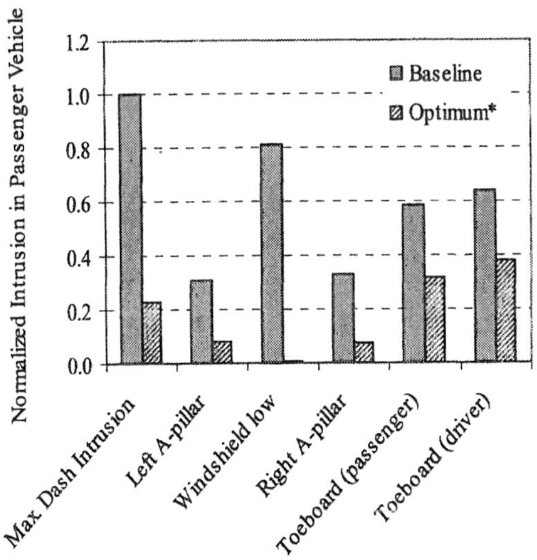

Figure 12. Passenger vehicle maximum dash intrusion – BASELINE vs. optimal.

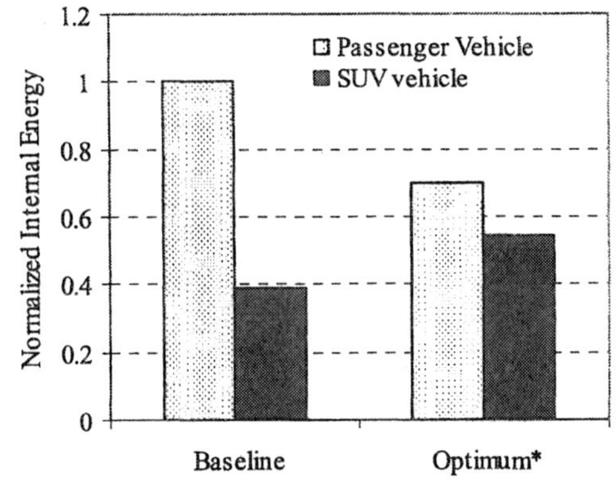

Figure 13. Internal energy absorbed.

SUMMARY

Validated finite element models of an "average" SUV and an "average" passenger vehicle were used to explore the effects of geometry, stiffness and mass in full-frontal collinear vehicle-to-vehicle collisions. Four design variables, the SUV fore rail height, fore rail length, fore rail thickness, and mass were chosen. Since this study was focused on structural performance and did not investigate the relationship between the structural responses and occupant responses, only passenger vehicle intrusions and vehicles' collision energy absorbed were selected as the responses.

A design of experiments methodology involving Latin Hypercube sampling was employed to select the appropriate number of simulations and the design levels of each of the design variables that should be incorporated in each simulation. The responses were characterized by quadratic polynomial surfaces. This characterization of the response surfaces could be done with higher order polynomials or other functions, but was not deemed necessary for this study based on the findings of Gu [11]. Four verification simulations showed that the response surfaces were adequate to capture the responses.

Pair-wise comparisons of the effects of the design variables were used to assess their individual influence on maximum dash intrusion and absorbed internal energy ratio. The pair-wise comparisons were based on the response surfaces generated from the 13 initial FE simulations. When a pair of design variables was compared, the remaining design variables were set to their BASELINE levels. In full-frontal, collinear SUV-to-passenger vehicle collisions, the relative geometric effect on maximum dash intrusion was greater than the mass and thickness effects. In addition, the thickness effect was greater than the mass effect for this response. The test data of Barbat, *et al.* [5] also showed that the geometric effect was the most significant in full-frontal, collinear SUV/LTV-to-car collisions.

For this crash mode and the constraints imposed on the design, the optimal SUV vehicle configuration based upon minimization of passenger vehicle maximum dash intrusion had aligned fore rails, BASELINE mass – 20%, and BASELINE fore rail thickness – 1.0 mm. This configuration represented the best aligned, lightest, and least stiff SUV in the study.

More importantly, the effect of reducing the front-end stiffness of an SUV on self protection is not investigated in this study. This effect may degrade the structural and occupant performances in single vehicle accidents. Therefore, compatibility is a fleet-wide issue and design changes such as stiffness and geometry may involve real-world tradeoffs.

The methodology described can be applied to other crash configurations. In fact, several configurations could be considered simultaneously using this method.

CONCLUSIONS

- In full-frontal, collinear SUV-to-passenger vehicle collisions, the relative geometric effect on maximum dash intrusion was greater than the mass and thickness effects.
- The thickness effect was greater than the mass effect on maximum dash intrusion.
- Quadratic polynomial surfaces can characterize the structural intrusions for full-frontal, collinear SUV-to-passenger vehicle collisions.
- The optimal SUV vehicle configuration based upon minimization of passenger vehicle maximum dash intrusion had aligned fore rails, BASELINE mass – 20%, and BASELINE fore rail thickness – 1.0 mm.

REFERENCES

[1] Gabler, H.C. and W.T. Hollowell, "The Aggressivity of Light Trucks and Vans in Traffic Crashes", SAE Paper No. 980908, Detroit, 1998.

[2] Evans, L. and M.C. Frick, "Mass Ratio and Relative Driver Fatality Risk in Two-Vehicle Crashes", Accident Analysis and Prevention, Volume 25, pp 213-214, 1993.

[3] Evans, L., "Driver Injury and Fatality Risk in Two-Car Crashes versus Mass Ratio Inferred Using Newtonian Mechanics", Accident Analysis and Prevention, Volume 26, No.5, pp 609-616, 1994.

[4] Mizuno, K. and J. Kajzer, "The Compatibility of Mini Cars in Traffic Accidents", 16th ESV Conf., Paper No. 98-S3-O-08, Windsor, Canada, 1998.

[5] Barbat, S., X. Li, and P. Prasad, "A Comparative Analysis of Vehicle-to-Vehicle and Vehicle-to-Rigid Fixed Barrier Frontal Impacts", 17th ESV Conf., Paper No. 01-S7-O-01, Amsterdam, Netherlands, 2001.

[6] Barbat, S., X. Li and P. Prasad, "Evaluation of Vehicle Compatibility in Various Frontal Impact Configurations", 17th ESV Conf., Paper No. 01-S7-O-09, Amsterdam, Netherlands, 2001.

[7] Steyer, C., M. Delhommeau, and P. Delannoy, "Proposal to Improve Compatibility in Head on Collision", 16th ESV Conf., Paper No. 98-S3-O-05, Windsor, Canada, 1998.

[8] Brown, G. , E. Arvelo and A. Strong, "Investigation of the Major Factor Influencing Front Compatibility Design of Vehicles", SAE Paper No. 2001-01-1166, Detroit, 2001.

[9] RADIOSS Crash Guidelines, Mecalog SARL, 1997.

[10] Yang, R.J., A. Akkerman, D.F. Anderson, O.M. Faruque, and L. Gu, "Robustness Optimization for Vehicular Crash Simulations", Computing in Science and Engineering, pp. 8-13, Nov./Dec. 2000.

[11] Gu, L., "A Comparison of Polynomial-based Regression Models in Vehicle Safety Analysis", ASME 2001 DETC, Paper No. DAC-21063, Pittsburgh, 2001.

2001-06-0238

A SYSTEMS MODELING METHODOLOGY FOR ESTIMATION OF HARM IN THE AUTOMOTIVE CRASH ENVIRONMENT

Alexandra C. Kuchar
Volpe National Transportation Systems Center, USDOT
USA
ESV Paper Number 354

ABSTRACT

A systems modeling approach is presented for assessment of harm in the automotive crash environment. The recent surge in light truck sales has highlighted the need to evaluate these vehicles' aggressivity in two-vehicle crashes while also considering potential self protection benefits in single-vehicle crashes. The methodology consists of parametric simulation of several controlled accident variables, with case results weighted by the relative frequency of each specific event. A hierarchy of models is proposed, consisting of a statistical model to define the crash environment and assign weighting factors for each crash situation case, and vehicle models for parametric simulation of crash events. Approximating functions are utilized to estimate occupant harm metrics based on vehicle crash response. Head and chest injury results for each case are converted to harm vectors, in terms of probabilistic Abbreviated Injury Scale (AIS) distributions. These harm vectors are weighted by each case's probability as defined by the statistical model, and summed to obtain a total estimate of harm for the crash environment. The methodology is applied to a subset impact environment consisting of single- and two-vehicle frontal collisions among passenger cars and light trucks. The model is validated against injury field data, and is found to accurately reflect trends in distribution of injury severity. The model is also exercised for variable sensitivity analyses, wherein changes in light truck/car population mix and LTV frontal stiffness are evaluated in terms of their effects on occupant harm within the frontal crash environment.

1.0 Introduction

This paper presents a systems modeling approach for evaluation of overall safety in the automotive fleet. This methodology stands in contrast to typical approaches, where specific safety issues such as air bags are addressed independently. However, the recent surge in light truck sales in the U.S. has led to the advent of a broader problem: how to evaluate the aggressivity of these larger passenger vehicles in two-vehicle accidents while also considering their potential safety benefits in single-vehicle crashes. While light truck vehicles do provide added protection to occupants within the vehicle, one recent statistical study reports that light trucks are so aggressive due to both mass and geometry that in head-on crashes between cars and light trucks, deaths in the cars outnumber those in the light trucks by 70% (Joksch, 1998). The systems model methodology applied here features computational vehicle models to represent cars and light trucks, making it suitable for analysis of aggressivity and compatibility among dissimilar vehicles.

This paper describes a systems modeling methodology for prediction of passenger injuries across the entire accident environment, considering a variety of metrics including vehicle type, impact speed, occupant size, safety belt usage, and other factors which directly affect overall safety. This approach will allow for evaluation of global effects of small changes to the accident environment, so that proposed automotive safety regulations may be evaluated in terms of their total safety benefit. The methodology has been developed as a generalized tool for assessment of a variety of crash-worthiness topics, such as air bags and vehicle design characteristics. The methodology is applied here to study vehicle aggressivity as a function of passenger vehicle fleet mix and light truck frontal stiffness.

History. Several previous studies have considered a systems approach for investigating vehicle safety. During 1975-78, the Ford Motor Company developed the Safety Systems Optimization Model (Ford Motor Co., 1978), featuring a simulation-optimization program for maximizing a single vehicle's safety performance in frontal crashes. The same program was substantially modified by the University of Virginia in the early 1980s (White, et al., 1985) to include updated biomechanical transforms and accident data as well as multivariate analysis capability. This model utilized approximating functions to estimate relationships among crash variables due to limitations in computational power at the time. Other motor vehicle manufacturers, including Fiat and Volkswagen, have also developed programs for optimizing vehicle design for crashworthiness, with emphasis on single-vehicle as opposed to fleet wide performance.

The model presented here differs from these earlier models in several aspects. It predicts total harm over a range of vehicle types rather than a single subject vehicle. While the model estimates injuries over a given set of crashes, it does not include an optimization algorithm for minimization of total harm. The model considers air bags in addition to seat belts, and occupants of varying size. It also incorporates recent accident statistics and more sophisticated biomechanical transforms than earlier approaches. An earlier application of the methodology has been presented to describe the modeling approach and initial validation (Kuchar, 2001). This paper presents an updated application of the model, with more recent crash

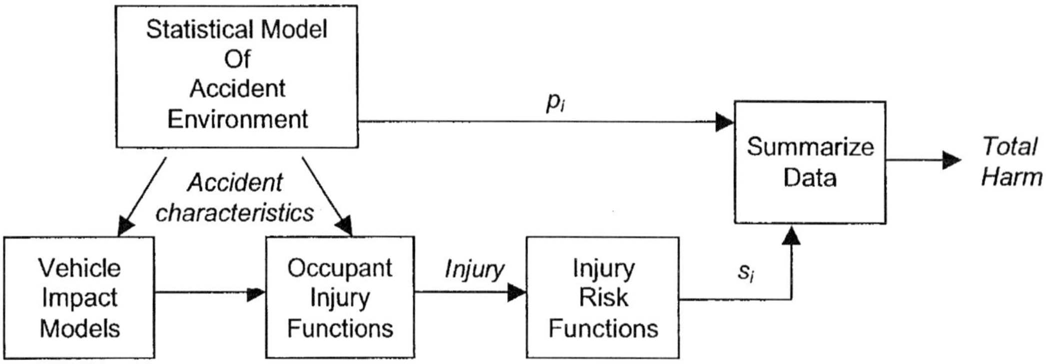

Figure 1. Fleet Systems Model Methodology

statistics, approximating functions to estimate occupant injury as a function of vehicle crash pulse, and consideration of non-air bag equipped vehicles.

Governing Equation and Methodology. The methodology is based upon the following governing equation for estimation of total injuries:

$$Harm = \sum_i p_i s_i \qquad (1)$$

Each i is a specific crash event, defined in terms of assigned values for model variables, such as vehicle type, restraint usage and occupant size. The accident environment is described by the range of i, which may include as few or as many cases as desired. p_i is the probability of each event i, its expected rate of occurrence based on accident statistics. s_i is the expected injury outcome of crash event i, represented in terms of probable levels of harm as measured by the Abbreviated Injury Scale (AIS). Each case's expected harm outcome s_i is determined via computer simulation of vehicle crashes and numerical approximation of occupant kinematics. This formulation allows for consideration of a range of accidents, while assigning a weight to each event based upon field data. Given this methodology, the model's robustness is directly related to three factors: the number and range of accidents considered, the reliability of the accident field data, and the accuracy of the computational models.

Figure 1 graphically depicts the implementation of the methodology. The term "Fleet Systems Model" refers to the whole system, which consists of a family of models. There are statistical models to describe the accident environment, vehicle structural crashworthiness models to predict vehicle behavior, occupant injury functions to estimate dummy motion, and injury risk functions for estimation of harm.

2.0 Statistical Model of Accident Environment

The motive for examination of accident field data for development of a statistical model is threefold: to select and define the boundaries of the model environment, for case weighting (computation of p_i for each case), and to provide a set of validation data against which the model's estimates of total injury are compared.

Subset Environment. A review of past year crash statistics is performed to identify the boundaries of the model environment - the subset of the real crash environment to be represented within the model. By identifying those events that are most frequent and lead to the greatest number of injuries, the model's coverage of the real environment can be maximized for a given number of cases.

The annual distribution of passenger vehicles in single- and two-vehicle towed accidents by impact mode and severity outcome is given in Table 1. Severe crashes are defined as those in which at least one occupant sustains an injury of AIS 3 or higher.

Table 1. Vehicles in Towed Crashes, by Accident Mode, 1994-99[1]

Crash Mode	All Crashes (n=3.45 million vehicles/year)	Severe Crashes (n=93,000 vehicles/year)
Single Vehicle		
Fixed Object, Frontal	6.7%	13.5%
Rollover	6.1%	19.1%
Other	11.3%	12.6%
Two Vehicle		
Head-on	2.7%	11.9%
Side Impact	45.1%	33.6%
Rear	19.8%	4.6%
Sideswipe	5.7%	2.9%
Other	2.5%	1.6%

[1] All accident field data presented are obtained from the NHTSA National Automotive Sampling System (NASS) Crashworthiness Data System (CDS) database, years 1994-99.

Note that vehicles in frontal accidents comprise a large percentage of severe crashes: 13.5% of these vehicles are in frontal single-vehicle impacts, while 11.9% are in two-vehicle head-on accidents for a total of 25.4% of all vehicles in towed accidents. Side impacts are more common, accounting for 45.1% of vehicles in all crashes and 33.6% of vehicles in severe crashes. Single-vehicle rollover accidents also comprise a large percentage of vehicles in severe crashes, at 19.1%. Because current frontal vehicle crashworthiness models are more feasible for parametric simulation than side impact models, the application of the methodology presented here considers only frontal impacts. This includes all single- and two-vehicle frontal impacts among cars and light trucks. Furthermore, while the vehicle crashworthiness models employed here simulate full frontal impacts, they are also assumed to approximate angled and offset crashes.

The same methodology is applied to select other parameters of the subset crash environment, wherein emphasis is placed on the frequency and severity outcome of impact variables, as well as modeling feasibility of impact parameters. For example, vehicle type is limited to passenger vehicles under gross vehicle weight of 4550 kg (10,000 lbs.), while all other vehicle groups such as buses and motorcycles are excluded. The passenger vehicle population is modeled as two separate classes, cars and LTVs (including all light trucks, sport utility vehicles, vans and minivans). Occupant seat position is limited to front seat driver and passenger occupants only, as this group represents over 86% of all occupants in towed crashes. For estimation of harm in each simulated case, only injuries to the head and chest are considered, as these body areas are by far the most common region of serious injury in frontal collisions.

Given this parametric definition of the subset environment, it represents a total of 355,000 vehicles in towed accidents per year, 25,600 of which feature an occupant sustaining a severe injury.

Computation of Case Weights p_i. The p_i term of the governing equation is a function of several accident variables:

$$p_i = fn(mode,\ vehicle,\ speed,\ belt\ usage,\ air\ bag\ deployment,\ seat\ position,\ occupant\ size) \quad (2)$$

Each unique permutation of these variables defines a single case within the methodology, and the sum of all of these cases describes the entire subset environment considered. For each variable, the relative probability of each value is determined from field data. Some interdependencies exist among these six variables, as illustrated in Figure 2. The first and only probabilistically independent simulation parameter is accident mode, which determines the relative probability of each vehicle type. Both impact mode and vehicle type determine weighting of each simulated vehicle impact speed. Vehicle type alone determines occupant seat distribution, which in turn defines the probabilities of the occupant size and seat belt usage variables. For the range of accident years considered (1994-99), air bags were made available in new vehicles in a piecemeal basis: first drivers, then passengers; in cars before LTVs. Therefore, airbag availability is modeled statistically as a function of both vehicle type and seat position. The numbers in parentheses in Figure 2 indicate the number of permutations for each variable. This yields a total of 1,008 cases, obtained via perturbation of 2 vehicle types, 2 impact modes (single- and two-vehicle impacts), 2 partner vehicles (car and LTV) in one of the impact modes (two-vehicle impacts), 7 impact speeds, 3 occupant sizes, 2 occupant locations, 2 seat belt configurations, and 2 air bag availability settings.

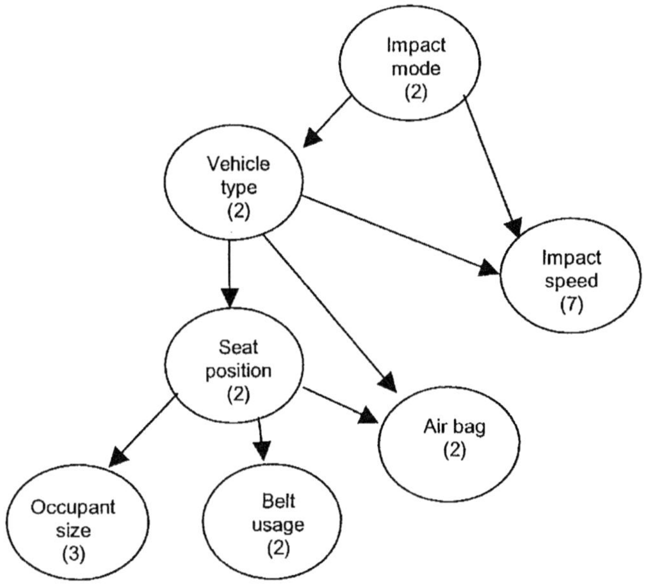

Figure 2. Hierarchy of Dependencies Among Probabilities of Simulation Variables

For variables with discrete values (such as seat belt usage, yes or no), linked probabilities are derived directly from field data. For continuous variables such as vehicle speed and occupant size, probability density functions (PDFs) are defined, and relative probabilities of each value are computed via numerical integration. Impact speeds for case simulation are selected to emphasize severe accidents, but are weighted according to frequency across all accidents.

The statistical model of the subset environment provides the data for computation of p_i for each case in terms of the linked probabilities depicted in Figure 2.

$$p_i = p(mode) \times p(vehicle|mode) \times p(speed|vehicle,\ mode) \times p(seat\ position|vehicle) \times p(occupant\ size|seat\ position) \times p(belt\ usage|seat\ position) \times p(air\ bag|seat\ position,\ vehicle) \quad (3)$$

Note that the computed value of p_i for each case should be very small, given that there are a total of 1,008 cases evaluated for the subset environment, and the sum of all p_i must equal 1.

3.0 Vehicle Crashworthiness Simulation Models

Vehicle response for each case is simulated using computational models of vehicles in frontal impacts. The vehicle models are one dimensional lumped-parameter systems, with three discrete masses representing the Occupant Compartment, Engine, and Wheels. Six non-linear springs represent energy-absorbing load paths in the front end of the vehicle, and are defined to approximate various buckling and crushing modes during frontal impacts. Each spring is described in terms of a segmented force-deflection curve to represent static behavior, plus a dynamic magnifier component, which applies dynamic force as a function of the spring's strain rate and static characteristics.

The models are simulated and developed using the SISAME (Structural Impact Simulation And Model Extraction) program (Mentzer, 1999), which performs forward simulation of one-dimensional lumped-parameter models, and features an optimization tool for the development of these models. This model extraction-optimization approach has been proven for development of highly accurate 1-dimensional models of vehicles in full frontal impacts (Mentzer et al., 1992). More recent developments in the SISAME program enable multiple-event extraction of simulation models, where crash test data from more than one event may be used to identify an optimal simulation model.

The passenger car fleet is represented by a single vehicle model based on a 1995 Chevrolet Lumina and the LTV fleet is represented by a model of a 1995 Ford Explorer. Future development of the methodology will include additional vehicle models to represent multiple weight classes of cars and LTVs, as well as more geometrically detailed vehicle models to capture 3D effects such as bumper mismatch. However, for the current application of this systems approach, the 1-dimensional full frontal crashworthiness models are considered adequate. Each model is extracted from two full frontal crash tests conducted at different speeds (24 and 56 kph for the car, and 48 and 56 kph for the LTV). Both models demonstrate very good correlation with test data at both impact speeds. No test data was available for validation of the models in vehicle-to-vehicle impacts.

The models are simulated in single- and two-vehicle full frontal impacts at 7 different impact speeds to generate occupant compartment response data for input into occupant injury approximating functions.

4.0 Occupant Approximating Functions

A previous application of this methodology consisted of full MADYMO simulation of 504 occupant cases to predict occupant head and chest injury. Because full parametric simulation of such a large number of cases becomes cumbersome for the number of desired occupant cases (1,008), approximating functions were developed to estimate occupant injury from vehicle crash pulse parameters. Linear regression analyses were performed on occupant injury results from NHTSA NCAP and FMVSS208 full vehicle compliance tests as well as occupant simulation results from existing MADYMO models. Input parameters consisted of 5 descriptive crash pulse characteristics, including delta-v, peak acceleration, time of peak acceleration, peak displacement, and time of peak displacement. Three distinct approximating functions for estimating HIC, peak chest acceleration and peak chest deflection were developed for each occupant configuration (characterized by vehicle type, occupant size, occupant position, seat belt usage and air bag availability). The chest injury metrics were treated as purely linear functions across all input observations. Because the known HIC observations clearly reflected an exponential nature with respect to impact speed, ln(HIC) is estimated by the approximating functions.

These functions developed via linear regression on crash pulse parameters demonstrate excellent correlation with known data. It should be noted however that these functions are rather heuristic in nature, generally based on single-point crash tests and computer simulation. Furthermore, in the absence of computer simulation results for non-air bag equipped occupant scenarios, the target results were estimated from scaled injury parameters obtained from air bag simulation cases. Nonetheless, the approximating functions give a good general estimate of head and chest injury for a wide variety of occupant scenarios, and provide accurate trend behavior comparable to that of parametric simulation.

5.0 Biomechanical Models

Because the vast majority of serious injuries in crashes are the result of head and chest trauma, only head and chest injury metrics are used here to measure occupant harm. Further refinement of the methodology may consider other injury mechanisms, such as neck and femur loads.

Head Injury Criterion (HIC) is computed from triaxial head acceleration response from each occupant simulation case. Chest injury for each simulation case is measured in terms of the Combined Thoracic Index (CTI), defined as

$$CTI = \frac{A_{\max}}{A_{\text{int}}} + \frac{D_{\max}}{D_{\text{int}}} \tag{4}$$

where A_{max} and D_{max} are peak values observed during simulation and A_{int} and D_{int} are constants defined for each dummy size. CTI is not currently used as a regulatory criterion, though it is recommended by the National Highway Traffic Safety Administration (NHTSA) for research use (Kleinberger, et al., 1998).

Because both HIC and CTI are computed from a controlled environment - direct measurements of acceleration, deflection, or force during crash tests or computer simulation; they are not obtainable from field data. Injuries in real crashes are recorded in terms of AIS category, which is less precise than the standard injury criteria. Nonetheless, the AIS scale is the only source of injury data available from the field, and provides the validation data for comparison of injury results computed by the Fleet Systems Model.

A series of mathematical models to convert HIC and CTI into the AIS scale are presented in Figures 3 and 4, commonly known as injury risk functions (Eppinger et al., 1999). These functions are proposed by NHTSA based on experimental data and previous research to estimate harm from measured criteria. The HIC injury risk functions are log normal approximations, while the CTI curves are two-parameter Weibull approximations. Both sets of curves are based upon experimental tests within the regulatory range of interest (HIC=1000, A_{max} =60G and D_{max}=76mm), and therefore these approximations are more heuristic for higher injuries. The AIS=6 curve shown below for CTI is not proposed by NHTSA, but has been extrapolated for use within this study.

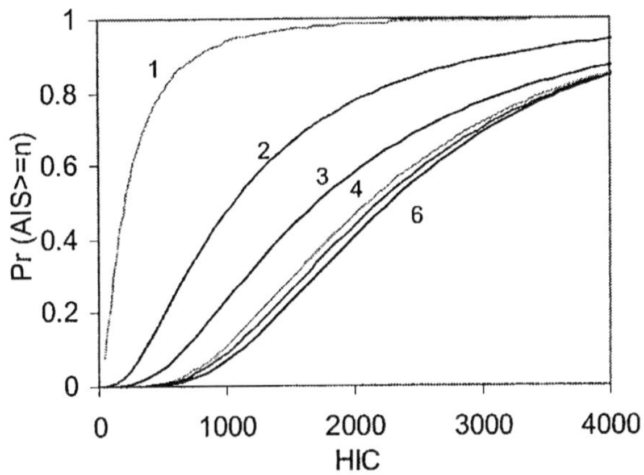

Figure 3. Injury Risk Functions for HIC
(from Eppinger, et al., 1999)

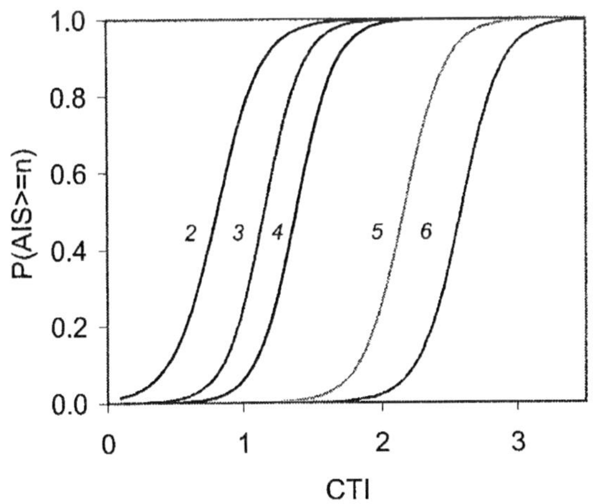

Figure 4. Injury Risk Functions for CTI
(from Eppinger, et al., 1999)

Applying these functions provides a mathematical transform for conversion of HIC or CTI into probabilities of each AIS result. Hence, the injury risk functions can be applied to obtain the outcome probabilities of AIS=0,1,2,3, 4,5, and 6. These values correspond to a vertical "slice" through Figure 3 or 4 at a given injury value. For each HIC and CTI computed from simulation, a vector of AIS probabilities is computed, corresponding to that occupant's probability of sustaining head or chest injuries corresponding to each AIS state. There are therefore two harm vectors obtained from each occupant simulation, one for head injuries (HIC) and the other for chest injuries (CTI). For each occupant simulation, the head and chest harm vectors are each multiplied by a normalized cost function, which quantifies the relative harm of each AIS level (NHTSA, 1999). The vector resulting in greater computed harm is assigned to be the vector s_i, or probabilistic AIS outcome, for that simulation case.

6.0 Fleet Systems Model Results

Parametric simulation of the vehicle models and application of the occupant approximating functions yields a total of 1,008 individual cases. For each of these cases there is a probability value (p_i) obtained from the statistical model and a harm vector (s_i) obtained from vehicle, occupant, and biomechanical models. These quantities are multiplied and summed according to the governing equation (1) to yield an estimated distribution of injuries for the subset environment.

The computed AIS distribution for the entire subset environment is shown in Figure 5, compared against field data. Because non-injuries and minor injuries of AIS 0,1, and 2 comprise the vast majority of the results, and because severe injuries of AIS 3 and higher are of greatest interest, only serious injuries are plotted, with minor injury figures given in text. A total annual number of 427,000 occupants is represented, so 1% of the environment corresponds to roughly 4,270 occupant injuries.

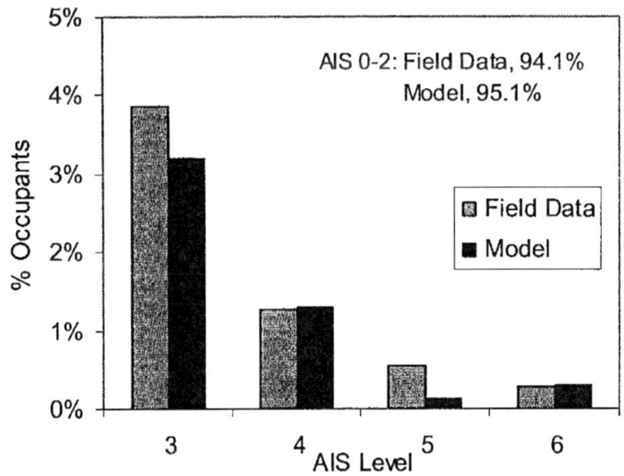

Figure 5. AIS Distribution of Subset Environment Model vs. Field Data Comparison

The model demonstrates very good agreement with field data. The model nearly perfectly predicts the percentage of AIS 4 and 6 injuries within the frontal crash environment. The model underestimates AIS 5 injuries by 0.4% of all occupants, with an underestimation of about 0.7% among

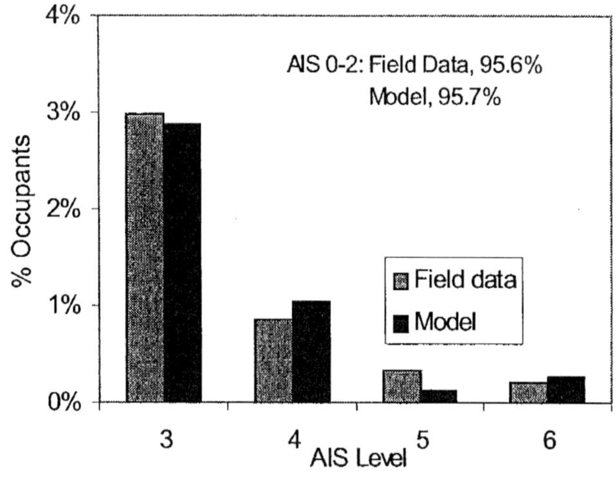

Single-Vehicle Fixed Object Impacts
n=342,000 cases/year

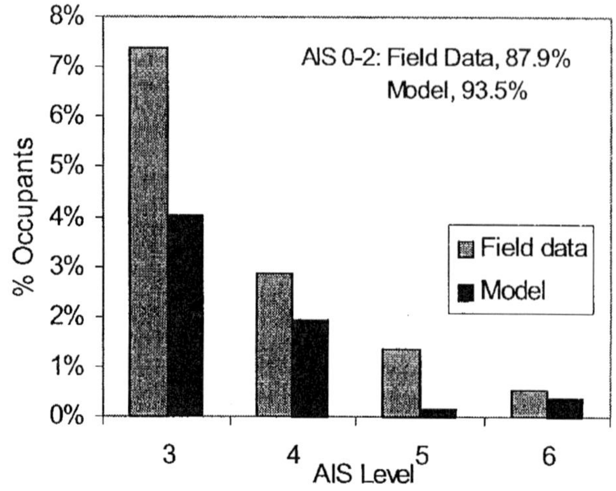

Two-Vehicle Head-on Impacts
n=85,000 cases/year

Figures 6 and 7. AIS Distribution by Impact Mode

AIS 3 injuries. Improvement of the model's accuracy in predicting AIS 5 injuries would likely result in a cumulative shift of injuries across the AIS 4 category. Nonetheless, the data clearly shows very good agreement when compared against field data from crash statistics.

In Figures 6 and 7, the model results and field data are presented by impact mode, so that the model's accuracy within each of these groups may be assessed. Figure 6 shows very good agreement with field data among occupants in single-vehicle fixed object crashes. Figure 7 shows the methodology is fairly accurate for predicting 2-vehicle head-on impacts, though the model's tendency to underrepresent AIS 3 and 5 injuries is evident.

Figures 8 and 9 show the same data sorted by vehicle type, car and LTV. For car impacts (Figure 8), the model appears to closely estimate injuries for AIS 4 and 6, with an underestimation of injuries in the AIS 3 and 5 categories. Among LTV impacts, the model appears to closely predict the severest injuries of AIS 5 and 6, while overestimating AIS 3 and 4 injuries 0.4% and 1.1%, respectively.

Although the model shows less accurate prediction of injury distribution among LTV occupants than car occupants, the number of cases represented in Figure 9 is also much smaller, at n=113,000. Therefore, in terms of the absolute number *n* of injuries, the model results are roughly equally accurate for LTV and car impacts. This suggests that larger percentage errors may be acceptable for smaller subsets of data, when absolute numbers of injuries are considered. For the methodology to accurately reflect

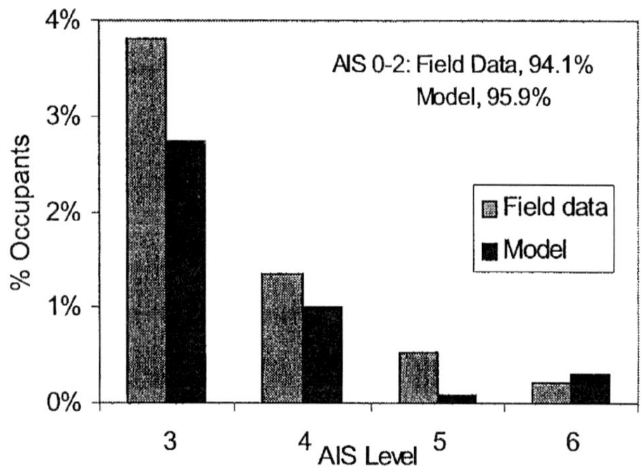

Passenger Cars
n=314,000 cases/year

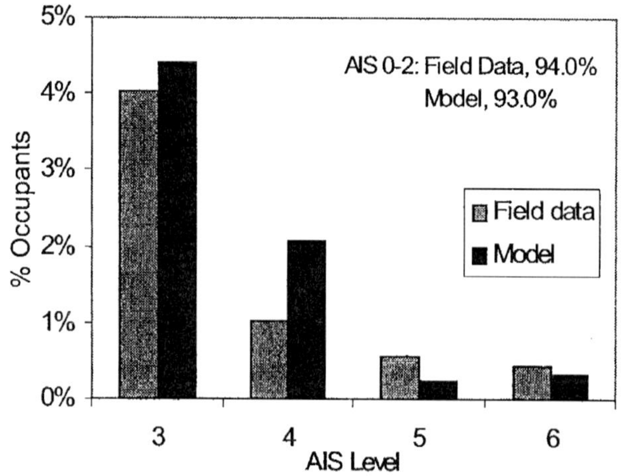

Light Trucks
n=113,000 cases/year

Figures 8 and 9. AIS Distribution by Vehicle Type

the entire fleet, however, it is highly desirable to accurately predict occupant injury in all types of vehicles. The small predictive errors observed in Figures 8 and 9 may therefore highlight some areas for potential improvement of the model results. One source of error may be a systematic tendency within the vehicle models or occupant approximating functions to underestimate severe (AIS 5 and 6) injuries while overestimating moderate (AIS 3 and 4) injuries. Although these vehicle and occupant models have been validated at 48 and 56kph (30 and 35mph), they have been extrapolated to simulate higher initial speeds of up to 80kph (50mph), where the most severe injuries are likely occur. Refinement and validation of the vehicle models and occupant approximating functions at higher speeds may address this issue, although validation data from high speed crash tests are not currently available for cars or LTVs.

Vehicle Aggressivity Analysis. To assess the role of vehicle aggressivity and compatibility within the model environment, serious injury results of only 2-vehicle impacts between LTVs and cars are shown in Figure 10. This subgroup of cases represents roughly 7.8% of the modeled crash environment, or roughly 33,000 occupants annually, 5,000 of which are seriously injured. The y-axis scale in Figure 10 represents only the percentage of occupants involved in LTV-car impacts (where 100% equals 33,000 occupants). The data demonstrates that car occupants undergo significantly more numerous and more severe injuries than their LTV counterparts. AIS 3 and 4 injuries in cars outnumber those in LTVs by roughly 2 to 1 and 3 to 1 in each category, respectively. While small numbers of AIS 5 and 6 injuries are predicted for car occupants, even fewer injuries are predicted for LTV occupants in LTV/car collisions: fatalities in cars during these impacts outnumber those in LTVs by roughly 4 to 1. The data therefore shows a clear disadvantage for car occupants. The mass and stiffness incompatibility between these vehicles is further highlighted by the simulated crashworthiness behavior in the vehicle models. For example, the maximum approach speed simulated for 2-vehicle

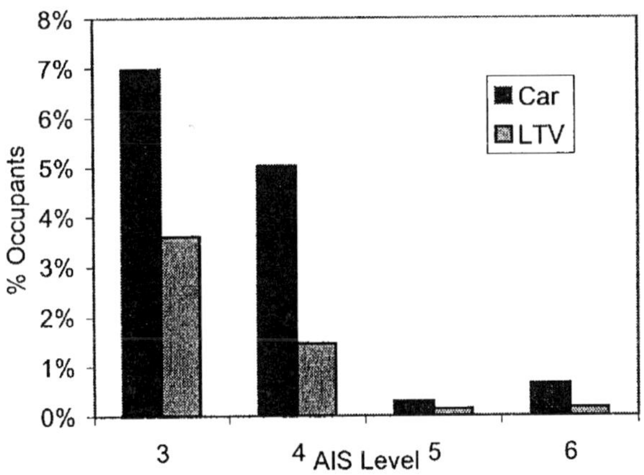

Figure 10. Serious Injuries in LTV/Car Impacts, by Subject Vehicle

collisions is 78 kph per vehicle, or an approach speed of 156 kph. Due to the differences in the two vehicles' mass and stiffness, the LTV undergoes a delta-v of only 90 kph (including rebound effects) at this highest impact speed, whereas the car has a much higher delta-v of 107 kph. Because the vehicle models are 1-dimensional mass/spring representations of vehicles, this simulated behavior reflects vehicle incompatibilities due to vehicle mass and stiffness but not geometry.

As LTVs continue to gain popularity as passenger vehicles, their potential effect on total safety within the accident environment grows in importance. To study the net safety effects of increasing LTV population within the accident environment (or the frontal impact environment, as the methodology covers here), the model was exercised to assess the sensitivity of occupant injuries as a function of LTV/car fleet mix. Occupant injuries were predicted for the hypothetical cases of a 100% car fleet and a 100% LTV fleet, as well as a range of scenarios in between these endpoints, at 10% intervals. The predicted results for serious injuries are shown in Figure 11. This study assumes that the total number of vehicles in the fleet remains fixed. The results reflect all single- and two-vehicle frontal impacts involving cars and LTVs. A reference line indicates the 1994-99 baseline environment, consisting of roughly 27% LTVs, 73% cars.

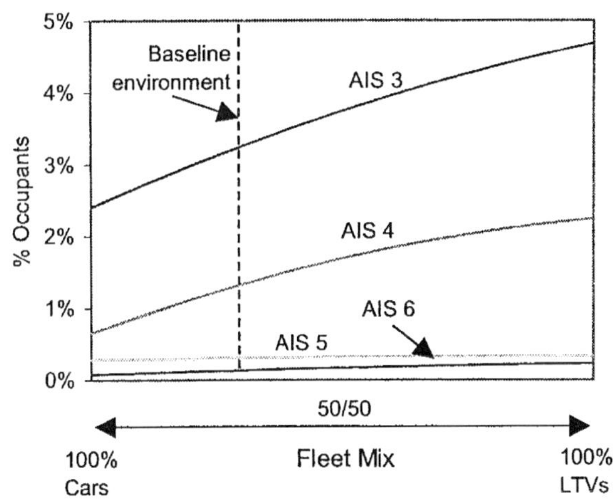

Figure 11. Serious Injuries as a Function of LTV/Car Fleet Mix

The data predict a steady increase in AIS 3 and 4 injuries as LTV share increases. AIS 5 and 6 injuries also demonstrate a smaller but notable climb in injuries as LTV population grows. The general upward trend in all injuries with increasing LTVs in the fleet can be attributed largely to increased injuries in cars during LTV/car impacts, as well as injuries in single-vehicle LTV impacts. Although LTVs are generally assumed to provide greater self-protection than cars, field data shown in Figures 8 and 9 indicate that cars and LTVs demonstrate similar injury distributions in frontal impacts. This may be due to the fact that crash statistics also indicate that LTVs tend to be involved in single-vehicle

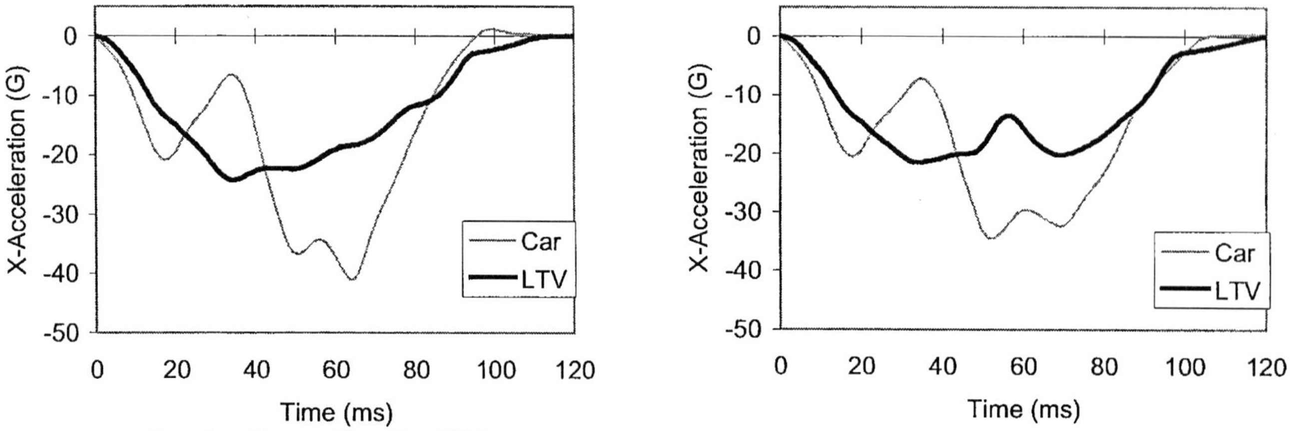

Baseline Car vs. Baseline LTV

Baseline Car vs. Reduced-Stiffness LTV

Figures 12 and 13. Crash Pulse Response in Baseline and Modified Car-LTV Crashes

crashes at slightly higher speeds. Because these results include all single- and two- vehicle collisions, cases resulting from single-vehicle and same-vehicle head-on impacts (car-car and LTV-LTV) simply scale up or down as a function of car or LTV population. Figure 11 demonstrates the net result of fleet mix changes within the accident environment, including aggressivity issues in LTV/car impacts as well as same-vehicle and different-vehicle impacts.

7.0 Future Aggressivity & Compatibility

The methodology can be applied to estimate harm in hypothetical future crash environments, to evaluate potential solutions to the current vehicle aggressivity and compatibility problem. In this approach, the vehicle models are specifically modified to reflect vehicle design changes aimed to reduce aggressivity. Such design changes may be in response to future regulations mandating compatibility among vehicles, or could be the result of independent redesign by automakers to address the LTV aggressivity problem.

In the previous section, the detailed statistical model of the environment was parametrically varied to evaluate injuries as a function of car/LTV fleet mix. Because the methodology also considers physical behavior in the accident environment (via computational modeling of vehicles and occupants), such design changes can be implemented and evaluated in terms of injuries across the entire frontal accident environment. This provides a useful tool for identifying the net effects and tradeoffs incurred when addressing the vehicle aggressivity problem, as opposed to single point simulation, where the broader context of a vehicle design change may be difficult to assess.

For this study, the LTV vehicle model was modified to reduce its overall stiffness by 15%. This was achieved by simply reducing the force-deflection behavior of all of the spring elements in the lumped-parameter LTV vehicle model. Such a reduction in stiffness may be achieved in a real vehicle via a combination of structural modifications, such as material selection and stiffness, as well as part geometry. The LTV vehicle model was only modified to decrease its stiffness – its mass remained the same, and the car model remained unchanged. Furthermore, the baseline comparison scenario, with unchanged vehicle stiffness, consists of 100% dual airbag availability. This is done on the basis that before a more "compatible" fleet (considered here as a less stiff representative LTV model) can be achieved, airbags will be completely phased into the future fleet.

Figures 12 and 13 illustrate the effectiveness of reducing the LTV's stiffness in improving crash pulse behavior in both vehicles during a car/LTV head-on impact. Figure 12 shows the vehicles' baseline acceleration in an impact with a closing speed of 116kph; clearly the car undergoes a far more severe deceleration than the LTV, peaking at 41G. In Figure 13, the car still undergoes a more severe crash pulse than the modified LTV, though not as severe as during the baseline crash case, peaking at less than 35G. The LTV crash response shows similar severity for the

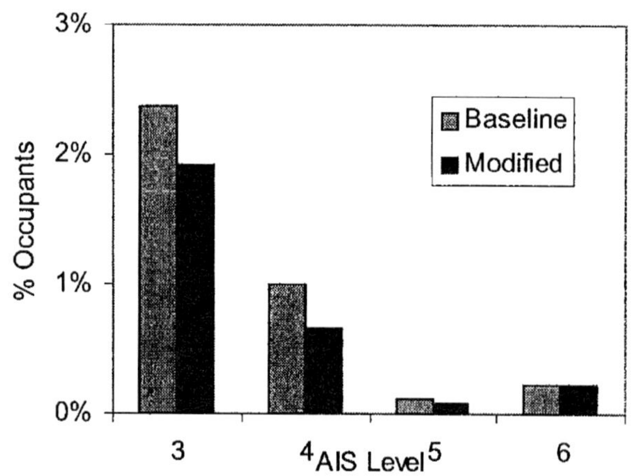

Figure 14. Reduction in Total AIS 3-6 Injuries due to Reduced-Stiffness LTV

Table 2. Annual Change in Injuries due to Reduced-Stiffness LTV Vehicle, Absolute Number of Occupants

AIS	TOTAL Environment	Single-vehicle impacts Cars	Single-vehicle impacts LTVs	Two-vehicle Head-on Impacts Car-Car	LTV-LTV	Car-LTV	LTV-Car
	n=427,260	n=252,091	n=89,742	n=42,922	n=9,318	n=18,940	n=14,248
0, 1, 2	3476	0	2421	0	415	736	-96
3	-1932	0	-1570	0	-265	-96	-1
4	-1446	0	-808	0	-132	-577	71
5	-98	0	-87	0	-4	-22	16
6	0	0	43	0	-13	-41	11

baseline and modified cases. Therefore, it is expected that injuries to car occupants in car-LTV crashes would generally decrease as a result of reducing LTV stiffness. The net effect of this change within the entire frontal crash environment is demonstrated in Figure 14. While total fatalities are left unchanged by the modified LTV, AIS 3, 4, and 5 injuries show a noticeable decrease compared to the baseline scenario. Figure 14 illustrates the net environment effect of decreasing LTV stiffness, and therefore it is difficult to attribute the injury reduction to specific crash modes or vehicles. Table 2 provides a detailed breakdown of this data, in terms of net change in absolute number of injuries as a function of crash mode and vehicle type. The first column describes the entire frontal crash environment, as does Figure 14, but in terms of absolute number of occupants. Note that the reduced-stiffness vehicle leads to the transfer of 3,476 annual AIS 3-5 injuries to the lesser AIS 0-2 categories. The second and fourth columns of Table 2 show a zero gain/loss in all injury categories, as the car model is unchanged for the two environments, leading to identical occupant harm results for single-vehicle car impacts and two-vehicle car-car impacts. The third column shows the results of single-vehicle LTV impacts: note that although the modified LTV model is less stiff than current LTVs, AIS 3-5 injuries are reduced substantially, with a concurrent increase of 43 fatalities. This suggests that reducing frontal stiffness in LTVs increases their self-protection in moderate severity single-vehicle impacts, though it may compromise LTV self-protection in very severe or fatal crashes. In the right half of Table 2, injuries in two-vehicle crashes are described in terms of subject vehicle vs. partner vehicle. In LTV-LTV and Car-LTV collisions, AIS 3-6 injuries are reduced throughout, indicating that reducing the stiffness of LTVs always improves injury results in the partner vehicles, both cars and other LTVs. The rightmost column highlights the negative tradeoff incurred in LTV-car impacts: AIS 4-6 injuries increase within LTVs, including an increase of 11 fatalities. Therefore, while large safety gains can clearly be made when considering the entire frontal crash environment, consequent smaller areas of reduced protection do exist and can be identified. Detailing the data as shown in Table 2 provides a cross-sectional view of the various tradeoffs incurred when considering the aggressivity and compatibility problem.

The topic of vehicle aggressivity and compatibility is typically discussed in terms of three major vehicle parameters: mass, stiffness, and geometry. While the study described here only addresses the stiffness issue and features relatively simple vehicle and occupant models, it serves as a useful tool for estimating injury reductions to be realized from a general reduction in LTV frontal stiffness.

8.0 Conclusions

A systems modeling methodology for estimation of harm has been presented and validated for a subset accident environment consisting of single- and two-vehicle frontal impacts. The model has been applied to study the sensitivity of total harm to fleet mix, wherein an incremental increase in LTV population is linked to a rise in AIS 3-6 injuries.

Components of the methodology include a statistical model providing a probabilistic description of the accident environment, vehicle models and occupant approximation functions for parametric simulation of crashes, and biomechanical transforms for estimation of injury in each case. The application model presented consists of 1,008 occupant cases, representing 427,000 drivers and passengers annually, 25,300 of which sustain a serious injury.

The model demonstrates that overall injury trends are very accurately estimated using the system modeling methodology described. The model accurately predicts distribution of AIS level 3 through 6 injuries in the frontal crash environment. When validated against field data sorted by accident mode and vehicle type, the model demonstrates very close estimation of injuries, with greater percentage accuracy for cars occupants than for LTV occupants.

Observed differences between model results and field data indicate that there exist areas for potential improvement of the application model presented. The occupant approximating functions to estimate occupant injury from vehicle crash pulses may be refined to more accurately predict unrestrained conditions and possibly consider additional crash pulse characteristics as inputs. With regard to the vehicle models, all frontal collisions, including angled and offset frontal collisions, are simulated as full frontal impacts. Hence, some of the more severe injuries resulting from angled and offset impacts that occur in the field may not have been fairly represented by full frontal simulation. The vehicle models could also be improved by representing vehicle geometry, as the 1-dimensional models employed here are adequate for capturing vehicle mass and stiffness behavior, but do not consider geometric effects such as bumper height mismatch. Also, further refinement of the statistical model to include more variable joint dependencies and greater resolution across continuous variables such as impact speed may lead to a more accurate prediction of overall harm. Injury compounding effects of combined injuries are not modeled, and injuries to body regions other than the head and chest are not considered. Finally, the implemented biomechanical models, in terms of injury criteria and risk functions for estimating AIS levels from those criteria, are also subject to known limitations.

To study the problem of vehicle aggressivity and compatibility, LTV/car impacts were separately evaluated to identify serious injury trends within each subject vehicle. Car occupants were found to undergo significantly greater harm than their LTV counterparts, by a factor of 2:1 and 3:1 for AIS 3 and 4 injuries, respectively. Also, the overall sensitivity of total occupant injuries as a function of LTV/car fleet mix was investigated. AIS 3-6 injuries were found to steadily increase with growing share of LTVs in the vehicle fleet. Furthermore, a study was conducted to evaluate changes in injury severity upon reducing LTV stiffness in a hypothetical more compatible future vehicle fleet. Results showed small absolute decreases in LTV self-protection in serious single-vehicle and LTV/car impacts, with significant improvements in occupant protection in LTV partner vehicles in all 2-vehicle crashes.

Future work. This investigation is an ongoing effort to develop methods for evaluation of fleetwide aggressivity and compatibility in support of NHTSA research initiatives. Further studies will include 3-dimensional lumped parameter or hybrid vehicle models to capture occupant compartment response in angled and offset frontal impacts. The scope of the existing model will be expanded to include side impacts in addition to frontal impacts, and include sensitivity analyses to evaluate the relationship between vehicle crashworthiness in frontal collisions and aggressivity in side impacts. Long term developments include addition of optimization capability to the methodology, to identify optimal vehicle features which lead to a minimization of overall harm.

Acknowledgements

The author wishes to thank W. Thomas Hollowell and Stephen Summers of NHTSA for support of systems modeling research at the Volpe Center.

References

Eppinger, R., Sun, E., Bandak, F., Haffner, M., Khaewpong, N., Maltese, M., Kuppa, S. Nguyen, T., Takhounts, E., Tannous, R., Zhang, A., and Saul, R. (1999). "Development of Improved Injury Criteria for the Assessment of Advanced Automotive Restraint Systems – II", Supplement to NHTSA Docket No. 1998-4405-9.

Ford Motor Company. (1978). "Safety Systems Optimization Model, Final Report", DOT Report Number DOT HS 6-01446.

Joksch, H. (1998). "Fatality Risks in Collisions Between Cars and Light Trucks", DOT Report Number DOT HS 808 802.

Kleinberger, M., Sun, E., Eppinger, R., Kuppa, S. and Saul, R. (1998). "Development of Improved Injury Criteria for the Assessment of Advanced Automotive Restraint Systems", NHTSA Docket No. 1998-4405-9.

Kuchar, A.C., Greif, R., and Neat, G.W. (2001). "A Systems Modeling Methodology For Evaluation of Vehicle Aggressivity In The Automotive Accident Environment", SAE Paper 2001-01-1172.

Mentzer, S.G. (1999). "The SISAME Program: Structural Crash Model Extraction and Simulation", U.S. Department of Transportation Report (in preparation).

Mentzer, S.G., Radwan, R.A., and Hollowell, W.T. (1992). "The SISAME Methodology for Extraction of Optimal Structural Crash Models", SAE Paper 920358.

National Highway Traffic Safety Administration. (1999). "National Automotive Sampling System: Crashworthiness Data System, 1994-1996", USDOT Report Number DOT HS 808985.

White Jr., K.P., Gabler III, H.C., Pilkey, W.D., and Hollowell, W.T. (1985). "Simulation Optimization of the Crashworthiness of a Passenger Vehicle in Frontal Collisions using Response Surface Methodology", SAE Paper 850512.

2003-06-0073

PARAMETER OPTIMIZATION FOR VEHICLE TO VEHICLE CRASH COMPATIBILITY USING FINITE ELEMENT METHODS

Intaek Lee
Katsutoshi Ainaka
Junji Saiki
Minoru Yubuchi
Altair Engineering Ltd.
Japan
Paper Number 238

ABSTRACT

Today, many auto makers and national researchers are interested in the compatibility of standard vehicle crash testing to that of real accident conditions. Current standard tests assume like vehicle to vehicle crash events to address the injury or fatality risk to vehicle occupants. Researchers that investigate vehicle to vehicle crash tests struggle to understand the relationship between aggressivity and injury measurements compared to those of accident investigation data like FARS and NASS GES. Such tests, however, can be simulated with finite element methods to re-enact real accident conditions to predict detailed vehicle measurements in order to make study structural improvements. In order to understand the nature of compatibility and to facilitate structural improvements, robust parametric studies and structural optimization methodologies can be employed to manage the complex, coupled design parameters and geometric changes.

In the past, assembling, executing, managing and interpreting the results has prevented this level of parametric study. StudyWizard, developed by Altair Engineering, is a software technology specifically designed to automate and extract meaningful design information from parametric analytical studies. Using StudyWizard, this paper will demonstrate a strong relationship between aggressivity and injury revealed by FARS data set for an oblique offset vehicle to vehicle crash. For this paper, a full-sized sedan compatibility study is performed for impacts with a light truck and van. The aggressivity characteristics; vehicle mass, stiffness and stackup will be discussed. To conclude, compatibility improvements will be examined.

INTRODUCTION

Even though incompatibility has been known since the 1960s, research activities have not been systematic until recently. However, an understanding and improvement of crash performance of vehicle in frontal and side impact was required before addressing compatibility, because they believed this could reduce casualties immediately.

This incompatibility has often been identified in real traffic accidents. For example, the crash between a passenger car and truck may result in unbalanced deformation and one of the vehicles may suffer great injury.

In studying compatibility, there are many aspects, such as crash between cars, car to pedestrian, car to light truck or van. The aim of this compatibility study is to identify how vehicle safety may be improved by structural changes that are designed to interact better during the crash and by restraint systems that are also designed to act properly in the second collision for each. For ideas on pedestrian protection (considering compatibility between car and pedestrian) evaluation methods are undergoing. EEVC and IHRA activities are making progress in this compatibility.

During the past decade, most automakers have done really well for crashworthiness design of the vehicle structure and restraint system to reduce the occupant injury in specific collision situations, in comparison with other cars or with standard car in FMVSS and other safety standards.

Such tests, however, only back up indications that are commonly calculated for broad categories of collisions. By the way, incompatibility in multi-vehicle crashes have been a concern due to promotion activities that have automakers designing cars with better crashworthiness in rigid barrier or stationary deformable barrier crash tests. The reason why better crashworthiness may result in incompatibility is the possibility that the stronger structure makes the vehicle more aggressive in collisions with other vehicles.

At the same time, the mass of vehicle is also observed as an important factor, driven by the laws of conservation of energy and momentum of two objects. The heavier object has small velocity change and the lighter object high with proportional mass ratio.

With respect to the affect on aggressivity, the stiffness of the vehicle is likely to have small influence compared with the mass of the vehicle.

To study and investigate the aggressivity, generic vehicle data were originally developed in 1981 for use in CRASH3 computer program by CALSPAN under contract to NHTSA. This generic data derived from the statistical analysis of a number of vehicles, not specifying any vehicle, could be useful for statistical studies, where vehicle parameter need to be considered as we simulate and analyze traffic accidents and may be a good preview for a comprehensive understanding of crashworthiness and incompatibility in real accidents.

Reference [1] lists the updated vehicle generic data including mass, stiffness and even suspension information. Some of this data is repeated in Table 1 through 3 below for the reader to glance over real incompatibility information.

Table 1. Generic Data for Passenger Cars

			Class1	Class2	Class3	Class4	Class5
Mass	Total Mass		5.398	6.391	7.607	9.528	10.026
	Roll		2200	2300	3085	3822	3834
	Pitch		12383	15845	20001	24960	23985
	Yaw		13489	17286	23989	29294	29279
Stiffness	Front	A	180.25	184.69	206.64	215.40	288.73
		B	72.11	65.38	69.97	66.70	113.45
		K	95.27	84.51	93.40	87.73	147.56
	Rear	A	172.50	162.33	189.62	186.00	292.40
		B	54.40	49.44	51.77	47.00	138.00
		K	80.82	72.57	80.35	70.16	207.47
	Side	A	88.25	100.00	95.75	137.00	137.00
		B	59.75	66.20	77.75	95.00	95.00
		K	88.92	84.06	97.45	119.12	119.12
Susp	Front	R	96.14	102.39	111.15	95.25	96.00
		D	9.90	9.23	8.69	7.02	6.86
	Rear	R	100.59	97.35	114.39	132.50	173.33
		D	7.99	7.01	6.93	6.96	7.64

Table 2. Generic Data for Pickups

			Class1	Class2
Mass	Total Mass		7.324	11.465
	Roll		2427	5430
	Pitch		17835	40008
	Yaw		19979	43081
Stiffness	Front	A	266.08	219.60
		B	108.92	68.40
		K	140.59	89.10
	Rear	A	258.33	290.67
		B	108.83	123.00
		K	151.77	190.30
	Side	A	103.00	78.00
		B	92.00	40.00
		K	110.55	48.65
Susp	Front	R	149.85	184.17
		D	9.6	8.85
	Rear	R	128.00	157.33
		D	7.86	6.84

Table 3. Generic Data for Multi-purpose and Vans

			Multi-Purpose		Van	
			Class1	Class2	Class1	Class2
Mass	Total Mass		9.069	12.536	8.691	13.050
	Roll		3134	6395	5656	9134
	Pitch		19588	38674	24493	47204
	Yaw		21998	40949	26474	51035
Stiffness	Front	A	266.08	219.60	309.00	358.75
		B	108.92	68.40	135.00	154.75
		K	140.59	89.10	170.36	189.74
	Rear	A	258.33	290.67	281.00	312.00
		B	108.83	123.00	118.02	141.73
		K	161.77	190.30	182.02	221.37
	Side	A	103.00	78.00	96.00	137.00
		B	92.00	40.00	78.00	95.00
		K	110.55	48.65	97.00	119.12
Susp	Front	R	156.33	180.08	164.50	180.73
		D	8.03	7.51	9.41	7.36
	Rear	R	134.22	145.48	146.12	158.00
		D	7.33	6.92	7.62	6.57

$\text{mass} = \text{lb}-\text{sec}^2/\text{in}, \quad \text{inertia} = \text{lb}-\text{sec}^2-\text{in}$

$\text{Stiffness } A = \text{lb}/\text{in}, B = \text{lb}/\text{in}^2, K = \text{lb}/\text{in}^2$

$R = \text{ride rate} = \text{ln}/\text{in}, \quad D = \text{damp rate} = \text{lb}-\text{sec}/\text{in}$

From Tables 1 through 3, we can directly imagine incompatibility in mass and stiffness along the vehicle category. Specifically, an incompatibility in crash type can that can be inferred is that there is more fatality risk in side impacted car when the stiffness of front and side are discussed.

Unlike the crashworthiness, the concept of aggressivity is about fatality or injury risk to occupants of other vehicle, applied with measurement as either injuries to occupants of vehicles involved in collisions, of specific vehicle as standards, or the average over all, or representative selection. The terms of compatibility appears in the practical problems, as a vehicle shows a different injury response to occupants of any other vehicles, even that have the same crashworthiness.

The researcher who is interested in accident reconstruction would investigate the reason why injuries are different is due to different stiffness, mass and mismatch of main parts for crash performance. At that time, the generic data would be often used for it. Specifically, this mismatch of main parts require us to review the vehicle structure along the impact types, impact position and angle, resulting in different interactions during the collisions.

Compatibility

Reiterating the concept of compatibility, it is more ideal to use a definition that considers the fatality or injury risk to occupants of the other vehicle with considering the subject vehicle. This is more complicated because both the crashworthiness of the other vehicle, and the aggressivity of the other vehicle influence the injury and fatality risks.

In order to understand the nature of compatibility, many automakers and research organization have conducted vehicle-to-vehicle crash tests and tried to set up the relationship between the injury measures and the aggressivity metric (AM) developed by NHTSA aggressivity research program.

$$\text{Aggressivity} = \frac{\text{Driver Fatalities in collision Partner}}{\text{Number of Crashes of subject Vehicle}}$$

This aggressivity metric is well used to indicate how much any vehicle categories are aggressive in real traffic accident, by their relative aggressivity using FAR and GES, as the overall fleet aggressivity ranking. This is a very typical and easier way to understand vehicle aggressivity by vehicle category from reference [2] repeated in Table 4.

Table 4. Vehicle Aggressivity by vehicle category (FARS/GES 1991-94)

Vehicle Category	Driver fatalities in the struck vehicle per 1000 police-reported crashes
Full size van	2.47
Full size pickup	2.31
Sports-Utility vehicle	1.91
Small Pickups	1.53
Minivans	1.45
Large Cars	1.15
Midsize Cars	0.70
Compact Cars	0.58
Subcompact Cars	0.45

In the same reference, crash compatibility also examined in vehicle-to-vehicle crashes, frontal and side impact types, shows a significant difference in the number of fatalities with respect to impact direction, A side impacted car has more than 3 times the fatalities compared to frontal impacted, and 7 times more at it worst. The results of this reference is also repeated in Table 5.

Table 5. Ratio of Fatally-Injured Drivers in LTV-to-Car Frontal and Side Impacts

Vehicle Category	Frontal impacts	Side impacts
Full size van	1 : 6.0	1 : 23
Full size pickup	1 : 5.3	1 : 17
Utility vehicle	1 : 4.1	1 : 20
Minivan	1 : 3.3	1 : 16
Small pickup	1 : 1.6	1 : 11

These crash statistics also reveal incompatibility between car and LTVs (Light Truck and Van including Sport Utility Vehicles) and impact modes.

A different countermeasure for each crash types is apparent. The engineering is a challenge in that the increase of ride-down efficiency has to be balanced with the restraint systems for frontal impact. The reduction of the effective velocity change is accounted as the exchange of momentum between a car structure accelerated by struck vehicle and occupants from impact to separation for side impact.

For instance of the side impact, the possibility of changing initial impact velocity condition or interactions between the car structure and occupants led by slight change of impact angle or tolerance structural stiffness is so high that different occupant responses could often be seen and complicate the understanding of structural effects, when compared with frontal impacts.

While crashworthiness is focused on structural effects and restraints, the accident reconstructionists are used to saying the basic principals to reconstruct the crash accidents are as follows:

- Impact is to the exchange of momentum.

During the collision a linear impulse is produces between the vehicles is

$$\text{Impulse} = \sqrt{2(E_1 + E_2)(M_1 M_2/(M_1 + M_2)}$$

where E_1, E_2 are the energies absorbed by vehicle 1,2

M_1, M_2 are the mass of vehicle 1,2

- The vehicles are objects strongly influenced by plasticity, but some elasticity works in crash. The restitution is often considered when we calculate the velocity change at relatively low speed impacts showing the small and elastic deformation before the plastic deformation.

- Impact is phenomenon about converting a part of kinetic energy in to plastic deformation energy, and dissipating it. The kinetic energy just before the crash is equal to the plastic deformation energy plus the kinetic energy after crash.
  So, plastic deformation energy is calculated with the following equation involving the restitution coefficient.

$$EA = \frac{M_1 M_2}{2(M_1 + M_2)}(1-\eta^2)(V_{i1} - V_{i2})^2$$

where EA is the energy absorbed by plastic deformation

V_{i1}, V_{i2} are the initial velocity of vehicle 1,2

η is the coefficient of restitution

- The forces engaged in crash are friction force plus impact force. During the crash the total resultant force vector is a function of the reaction force and friction force between vehicles and be calculated with equation as follows:

$$\text{Resultant Force} = Fv\sqrt{1+\mu^2}$$

where Fv = Perpendicular Reaction Force

μ = Friction Coefficient

- Non-central impact is not only exchange of momentum but also converting linear momentum in to angular momentum.
  For instance, the non-central and perpendicular impact case is illustrated graphically in figure 1, and the velocity of each vehicle is calculated as follows:

$$V_{1f,Ydir} = \frac{M_1(L_2^2+\rho_2^2)-\eta I_2}{I_2+M_1(L_2^2+\rho_2^2)} V_{1i}$$

$$V_{2f,Ydir} = \frac{M_1\rho_2^2(1+\eta)}{I_2+M_1(L_2^2+\rho_2^2)} V_{1i}$$

$$\alpha_{2f,center} = \frac{M_1(1+\eta)}{I_2+M_1(L_2^2+\rho_2^2)} L_2 V_{1i}$$

$$\text{Energy Ratio} = \text{Angular Kinetic Energy / Total Kinetic Energy} = 1 \Big/ 1+(\rho_2/L_2)^2$$

where $I_{1,2}$ = Mass of Inertia

$L_{1,2}$ = Distance between CG to impact force position

$\rho_{1,2}$ = radius of gyration

$\alpha_{1,2}$ = Angular Velocity

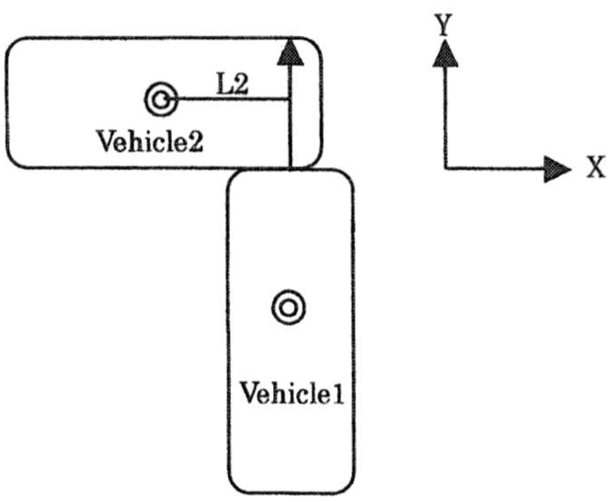

Fig 1. Non-central and perpendicular impact

Consequently, the topic for crash accidents or compatibility study do not involve just one parameter, rather the permutated effects with several parameters. The priority of these parameters may be changed case by case, and have to be studied for the basic incompatibility.

Another consideration regards the injury mechanism to represent the load path to transfer impact forces to the occupant, and what is the major reason to result in fatal injury during the crash.

Comparing side impact and frontal impact, occupants mainly restrained by seat belt and airbag supplemented in frontal impacts, are injured at the second impact time as usual, but in side impacts occupants did not have specific restraint system. Recent side airbag or inflatable tubular system meant to reduce injuries are suffering with the bullet car hitting them directly.

Many aspects of incompatibility are not well understood, but studies focus on engineering spot issues until a comprehensive awareness of compatibility is understood completely, and the methodology for it accepted commonly.

Finite Element vehicle models and optimization

In crash analysis and simulation, the finite element method is regarded as the most powerful tool because of the ability to perform parametric studies requiring iterations in the design of vehicle structure.

For crash simulation, explicit FE method such as LS-DYNA, PAM-CRASH and RADIOSS, ABAQUS-explicit are well used and are common for virtual crash tests.

FHWA/NHTSA National Crash Analysis Center at George Washington have developed various FE vehicle models which are available to download for all researchers of the world who are likely to do crash simulation.

In this study, the 3 FE models extracted from FHWA/NHTSA National Crash Analysis Center-Website are listed on table 6.
(http://www.ncac.gwu.edu/archives/model/index.html)

Table 6. FE models used in LTV-to-Car compatibility study, extracted from FHWA/NHTSA National Crash Analysis Center-Website

Model Name	Model Size
Ford Taurus	Modified Model 28,400 elements
Chevrolet C2500 Pickup	Detailed model 54,800 elements
Dodge Caravan	Detailed model 329,300 elements

Design of experimental and Optimization methods are generally known as the best tools used for design of vehicle structure, because it can drive an optimal conclusion which can be evidenced in a systematic and mathematical sense.

But there are major difficulties as well that have been identified in crash simulation case. The literature mentions that crash simulation models must be accurate to prevent the numerical error affecting the response. Things such as contact routines, rigid linkage, element types/sizes, and sampling frequency may cause degradation of spectral response, more than the effects of parameter changes, and requiring extensive computational resources and time [4].

Another difficult thing that may be said is that "the nature of crash phenomenon" is chaotic as it involves the interaction of hundreds of parts which come into contact causing force oscillations within a very short time. Bifurcation also happens and is known as the qualitative change appearing in the response with respect to change of parameters [5].

It is impossible to assess the crash structure system in view of this bifurcation because bifurcations mean the change of topology in the character of the solution. Contemporary crash simulation techniques are likely to conclude that the optimization of vehicle crashworthiness is possible, because it is always the study about the representative crash case like FMVSS crash tests and simply selecting the best settings of the varied parameters, what designer wants to change. Those who catalogue crash analysis and simulation as chaos, however, may say "must be careful in optimization study". One must take into consideration the sensitivity of initial conditions or uncertainty, such as change of initial velocity, mass, impact angle and interactions between car structure parts, and qualitative leverage from structural improvements or for different crash types.

Ignoring the uncertainty and excluding the possible effects coming from the numerical errors and change of topology, etc, the optimization technique may attempt to solve the crash problems in a relatively easy way. And, it is believed that the optimization theory using sequential approximation method can solve the sensible crash problems if design parameters could be defined and designer wants to get the optimum result within its variations [6][7].

Response Surface optimization for crash problems

The engineering of structural design is a challenge to determine the best one through an iterative process where a series of structural changes regarded as reasonable trials show off the different response. Using optimization and FE simulation tools, engineers might be allowed to change the design parameters and to get the efficient search for right combination of design parameters for a certain design.

The FE codes for analyzing the nonlinear dynamic response of structures have a time discretization which is accomplished through the central difference operator, that expresses velocities at time t+ Δ t/2 and displacements at t+Δt explicitly in terms of present accelerations, velocities, displacements as follows [8][9][10];

$$
\begin{aligned}
{}^{t}\ddot{\mathbf{d}} &= \mathbf{M}^{-1}\left({}^{t}\mathbf{f}\right) \\
{}^{t+\Delta t/2}\dot{\mathbf{d}} &= {}^{t-\Delta t/2}\dot{\mathbf{d}} + {}^{t}\ddot{\mathbf{d}}\Delta t \qquad \ldots\ldots\ldots\ldots\ldots(1) \\
{}^{t+\Delta t}\mathbf{d} &= {}^{t}\mathbf{d} + {}^{t+\Delta t/2}\dot{\mathbf{d}}\Delta t
\end{aligned}
$$

The advantage of this method is only the mass matrix has to be simply inversed.

However this explicit method is conditionally stable with Δ t as the disadvantage to be considered carefully.

From the nonlinear crash experience that looked at plastic folding, thin shell elements are not accommodated by higher order due to plastic hinge line. The fold may be located in the any nodal points of finite elements model where the more dense nodal points are doing well enough to represent the crippling.

There are two distinct optimization schemes. One is the local approximation using most important gradient, design sensitivity analysis. The other is global approximation method which uses higher order analytical expressions to approximate the dependence between structural responses and design parameters with few analyses.

The local approximation scheme is based on the premise that small changes of design parameters occur in each optimization step, and in conjunction with nonlinear dynamic FE code, uses an explicit time integration scheme, where differentiation of the state equations (1) with respect to the design variables p_j yields

$$
\begin{aligned}
\frac{d^{t}\ddot{d}}{dp_j} &= M^{-1}\left(\frac{\partial^{t}f}{\partial p_j} + \frac{\partial^{t}f}{\partial^{t}d}\frac{d^{t}d}{dp_j} + \frac{\partial^{t}f}{\partial^{t}\dot{d}}\frac{d^{t}\dot{d}}{dp_j} - \frac{d^{t}M}{dp_j}\ddot{d}\right) \\
\frac{d^{t+\Delta t/2}\dot{d}}{dp_j} &= \frac{d^{t-\Delta t/2}\dot{d}}{dp_j} + \frac{d^{t}\ddot{d}}{dp_j}\Delta t \qquad \ldots\ldots\ldots\ldots(2) \\
\frac{d^{t+\Delta t}d}{dp_j} &= \frac{d^{t}d}{dp_j} + \frac{d^{t+\Delta t/2}\dot{d}}{dp_j}\Delta t
\end{aligned}
$$

This linear gradient relationship for each optimization step is likely not to be maintained under the significant change of time step during the calculation, even change of material stiffness to keep the constant time step.

In this case, there are notable non-linearities and warring convergence, simply when we think about the significant change of time step and several different element size models, Consequently, the global approximation seems to be a better solution and a more general way of getting the optimized solution with respect to more wide design parameters.

Let $\psi_i(\mathbf{b})$ be the response of interest. Then, a polynomial $\hat{\psi}_i(\mathbf{b})$ of the degree n can be introduced such that

$$\psi_i(b) \approx \hat{\psi}_i(b) = a_{i0} + \sum_{p=1}^{m} a_{ip} b_p + \sum_{p=1}^{m} \sum_{q=1}^{m} a_{ipq} b_p b_q + \sum_{p=1}^{m} \sum_{q=1}^{m} \sum_{r=1}^{m} a_{ipqr} b_p b_q b_r + \dots \qquad \text{.........(3)}$$

where the quantities $a_{i0}, a_{ip}, a_{ipq}, a_{ipqr}$ are constants that are determined from the results of the system analyses by solving a linear system of equations. The number of analyses that are necessary depends on the polynomial degree n of Equation (3). The function of Equation (3) is called the *response surface*.
Using the equation above, an approximation to the classical optimization problem can be established:

$$\hat{\psi}_i(\mathbf{b}) \Rightarrow \min, i = 0$$
$$\hat{\psi}_i(\mathbf{b}) \le 0, i = 1, \dots, n_\psi$$

A modified version of the standard response surface method, as described above, analyzes the designs as the optimization proceeds and will be called the sequential response surface method which is used in *Altair StudyWizard*, (it is to be renamed with *HyperStudy* adding robust and stochastic analysis). This optimization software is interfaced with several FE solvers such as *LS-DYNA* (Hallquist, 1998) or PAM-CRASH. Consider a quadratic response surface as

$$\psi_i(b) \approx \hat{\psi}_i(b) = a_{i0} + \sum_{p=1}^{m} a_{ip} b_p + \sum_{p=1}^{m} \sum_{q=1}^{m} a_{ipq} b_p b_q \qquad \text{............(4)}$$

The coefficients a_{i0}, a_{ip}, a_{ipq} are determined using a least square fit of the function $\hat{\psi}_i$ on the existing designs. In the first step $s=1+m$ designs are analyzed, and the coefficients a_{i0}, a_{ip} are determined. The next m designs are analyzed to determine the coefficients a_{ipp}. Additional designs are used to determine the remaining coefficients as it becomes necessary. After $s=1+m+(1+m)\ m/2$ designs, an algorithm is used to weigh the designs. The optimization procedure is as follows:

1. Analyze the initial design and *m* perturbed designs
2. Use a least square fit to determine the polynomial coefficients for the objective function $\hat{\psi}_o(\mathbf{b})$ and each constraint $\hat{\psi}_i(\mathbf{b})$

   If $s=(1+m)$, then calculate a_{i0}, a_{ip}

   If $1+m < s < 2(1+m)$, then calculate a_{i0}, a_{ip}, a_{ipp}, $i = 1, s-1-m$

   If $1+2m < s < 2+m+(1+m)\ m/2$, then calculate a_{i0}, a_{ip}, a_{ipp} and the successive off diagonal elements a_{ipq}

   If $s > 1+m+(1+m)\ m/2$, then weight the designs and calculated all a_{i0}, a_{ip}, a_{ipq}
3. Solve for the approximate optimum design using mathematical programming
4. Analyze the approximate optimum design
5. If the design has converged, stop
6. Back to 2

The procedure is shown in Figure 2. for a problem with one design variable. The theory of sequential response surface optimization is further explained by Schramm and Thomas in reference [6][11].

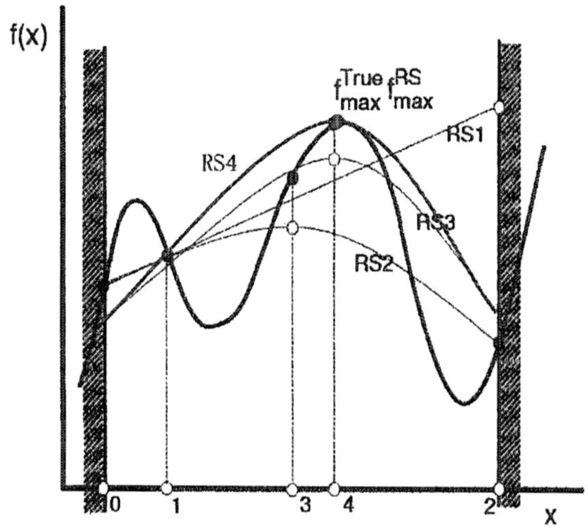

Fig 2. Response Surface Method

In general, crash problems are often defined as follows;

Objective:	$W(\mathbf{p}) \Rightarrow \min$	
Subject to	$\mathbf{M}\,{}^t\ddot{\mathbf{d}} = {}^t\mathbf{f}({}^t\mathbf{d}, {}^t\dot{\mathbf{d}}, \mathbf{p}),\ 0 \le t \le t_e$	
with	${}^0\mathbf{d} = \mathbf{d}_0, {}^0\dot{\mathbf{d}} = \dot{\mathbf{d}}_0$	....(5)
Constraints:	$\mathbf{g}(\mathbf{d}, \dot{\mathbf{d}}, \ddot{\mathbf{d}}, \mathbf{p}) \le 0$	
Side constraints:	$\mathbf{p}^l \le \mathbf{p} \le \mathbf{p}^u$	

The objective function is a structural response like acceleration of vehicle, mass of certain body, deformation of vehicle or whatever engineer could be interested in.
The constraints are also structural responses.
In the structural responses, displacements, velocities, accelerations, stresses are grouped as time dependent

responses, often requiring the noise signal to be filtered, in order to make up a feasible engineering result.
The other responses, like energies, injury criteria, and crash performance taken with average in general may be used directly in definition of response function for objective or constraints because they are integral type.
Using the equations (5), objective and constraints can be calculated and determined by combination of computational analysis programs, ordinary FE and optimization.

Industrial application of structural optimization is divided into the sizing and shape optimization distinguished by the type of design parameters.
Size optimization is about the change of dimensions of structure for instance, thickness of structure.
Shape optimization is about the change of boundary of the structure and its design parameters are to comply with certain design requirements that require moving the nodes of FE mesh, both on the surface and in the inner of mesh, and keeping good shaped elements.
The shape functions that are commonly accepted to prescribe the surface and the inner parts of FE model can be linear or quadratic.
The nodal positions during optimization are computed as follows;

$$x_n = {}^0x_n + \sum_{i=1}^{NDV} \frac{\partial x_n}{\partial b_i} \Delta b_i$$

For this shape optimization, the creation of shape functions is fully supported in HyperMesh, Altair's finite element preprocessor [12].

StudyWizard

Parameter study and optimization can be usefully combined to investigate and optimize the behavior of a structural model. Response surface methods can handle just a few design variables since otherwise the computational effort is too high. Especially in crash analysis, where a single analysis run requires from several hours to days of computer time, any effort to reduce the wait time needs to be made. The number of design variables for an optimization should be limited to about ten.
If the number of design variables is very high, it is advisable to first run a screening Design of Experiment to determine design variables of large influence.
Altair's StudyWizard combines parameter studies using Design of Experiments approaches and sequential response surface optimization [13]. The same parameterized model can be used in both approaches. The StudyWizard is integrated in the post-processing software HyperView [14]. This way all the interfaces to different solvers can be accessed and multi-disciplinary optimization can be performed.

The creation of parameterized models can be automated with custom wizards. For the use of the StudyWizard with LS-DYNA such a parameterization wizard is available. The user can select the type of design variables such as shell thickness, shape, materials, load curves from a wizard window and the parameterized deck is created automatically. Almost no deck editing is involved.
(http://www.altair.com/, http://www.altairjp.co.jp)

Crash Optimization

From the computational crash simulations, a designer could pick up the weak points and make feasible design changes. The procedure of definitive changes depends upon the engineering sense or engineering standard database made by a previous study.

If they have no idea how to measure crash problems and how to attempt crash optimization due to a new study, the process would be as follows;

- Exclude the possible numerical errors with respect to change of design parameters or its instability and extract design parameters considering crash phenomenon that have no change qualitatively.

- Design of Experimental method is used to understand the factorial effects in design parameters and sample the design type. For the sampling design parameters, the Latin square technique is often used.

- Global approximation methods may facilitate optimization for crash problem. A second order Response Surface type model is made up with wide crash design parameters and is implemented to do optimization of the complex structural crash problems.

Frontal Oblique Offset Crash Type

In order to reconstruct the real aggressivity characteristics such as mass, stiffness, geometric profiles, NHTSA has conducted a series of vehicle-to-vehicle crash tests. A typical example of fatality in LTV-to-Car crash types comes from reference [2]. The results are repeated in figure 3 to 4.
According to the research based on the 1988-1996 NASS-CDS files, this crash type is severe in both crash pulse and intrusion. Therefore it is proper to estimate the restraint systems and structural improvements at the same time because restraint systems are generally designed by crash pulse basis and car structure are designed by deformation basis.

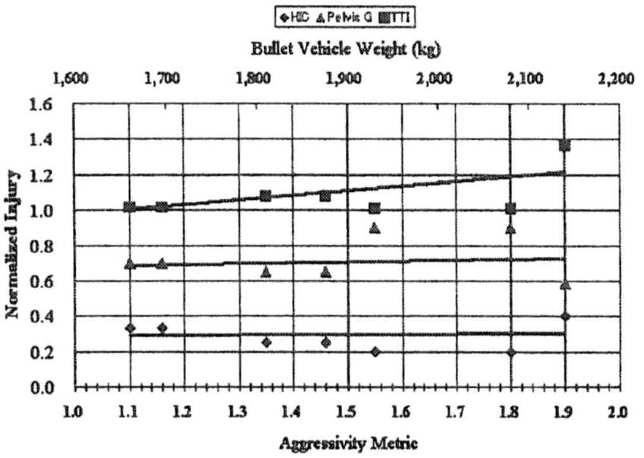

Fig 3. Side Impact Injury Measurements with respect to Aggressivity Metric and Bullet Vehicle Weight.

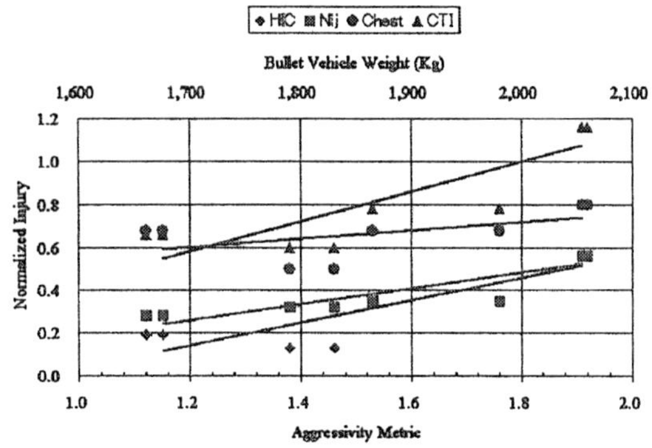

Fig 4. Frontal Impact Injury Measurements with respect to Aggressivity Metric and Bullet Vehicle Weight.

In this study, they concluded the side impact test series did not show a strong relationship between the injury measures of the target vehicle and either the mass of the striking vehicle or the aggessivity metric based on FARS/GES data.

The oblique offset test series, however, shows good correlation of driver injury measure with both the vehicle mass and the aggressivity metric.

And they commented that more research should be done to better understand compatibility. Additional factors may be affecting vehicle aggressivity, even though the frontal offset tests show a good correlation. In the determination of oblique offset test series, analysis of the NASS crash data files of drivers in frontal collision, one representing with the highest frequency and serious-to-fatal injury risk is addressed. For these reasons, the optimization for vehicle to vehicle crash compatibility using FE methods should consider the frontal oblique offset case only, due to the fact that more clarity is seen in the frontal impact than side impact.

As in reference [2], the crash type is shown in figure 5.

To see figure 5, this crash type should be catalogued as the central collinear collision is not so much in rotational effects during the crash and the simulation substantiate this by representing very small amount of rotational acceleration till ending of translational deceleration.

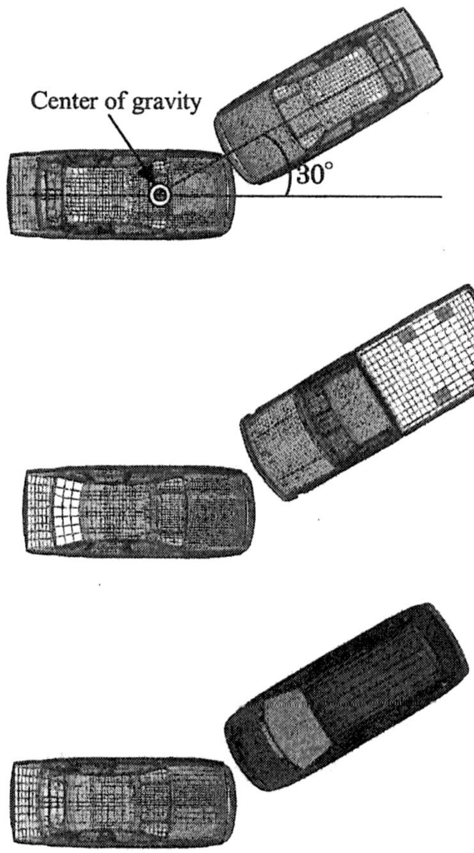

Figure 5. Frontal Oblique Offset Crash configuration

Optimization of car-to-LTV

To represent the LTV-to-Car crash types, the vehicle categories were selected and tuned as the table 7 shows.

Table 7. Frontal Oblique Offset Impact Vehicle Weights

Vehicle Category	Target Vehicle	Striking Vehicle
Passenger Car Class 3	Ford Taurus (W=1,511kg)	Ford Taurus (W=1,511kg)
Pickup Class 2		Chevrolet C2500 (W=2,254Kg)
Van Class 1		Dodge Caravan (W=1,910Kg)

In order to get more accurate body deceleration signal, the original Taurus FE model mesh density (of which the dash and floor was too rough to give us a feasible response) was modified as moderate element size between front and rear mesh density.

The deformable to rigid option of LS-DYNA, allowing deformable parts to be switched to rigid, was used in the Dodge Caravan model that reach 329,300 elements because this is too big to fulfill optimization study.

A parameter study or a DOE (design of experiments) study has been often used for determining which parameters are most influential on thc response.

In this compatibility study, the optimization study follows a DOE study that could give us main parameter effects to be considered at first.

DOE and Optimization study is implemented using StudyWizard, and the crash problem is solved using LS-DYNA. Numerical solutions are performed on DELL Intel Xeon 2.4 GHz (4 CPUs, 4GB Memory100GB free disk Windows 2000). A single analysis of Taurus-to-Taurus impact takes about 7 hours, Taurus-to-C2500 pick up truck about 10 hours and Taurus-to-Caravan about 21 hours.

At the first step, it is necessary to define the parameter that is to be considered or likely to affect the incompatibility problem.

Among the many parameters we could discuss, the following design parameters (as mentioned in table 8) are selected first, because many times crash engineers experience those in crashworthiness.

For a DOE of 4 design parameters for each impact, $L_{16}(2^{15})$ Latin square is used because 4 design parameters of each come to 12 design parameters in 3 impact models.

Table 8. Design Parameters for DOE study

Design Parameters	Description
Engine Position	Movement forward and backward (+/- 30mm)
FRT Member Inner Thickness	Upper Value 3mm Lower Value 1mm
Sub Frame Member Thickness	Upper Value 3.6mm Lower Value 1.5mm
Dash Lower PNL Thickness	Upper Value 3.6mm Lower Value 1.0mm

To extend the commonly accepted idea of head injury criterion (HIC) calculation, the average deceleration means imposing comparative impulse as to different crash conditions. . This average deceleration, however, is not blocked with specific maximum duration time to pick up the burst signal and weighted as like mathematical HIC calculation.

A plot of the relationship between the vehicle weights and average deceleration calculated by integration of deceleration curve with respect to time and then divided by its time duration is shown in fig 6. This substantiates the understanding of aggressivity metric in frontal oblique offset impact case as was done and mentioned in reference 2.

One must carefully consider the fact that structural modifications are included here. The heavier bullet vehicle weight imposes more impulse in general. Thus, the possibility of reducing the impulse with structural modification is found, when we see the bottom case on the heavier bullet vehicle weight. But it should be aggressivity to the other passenger car.

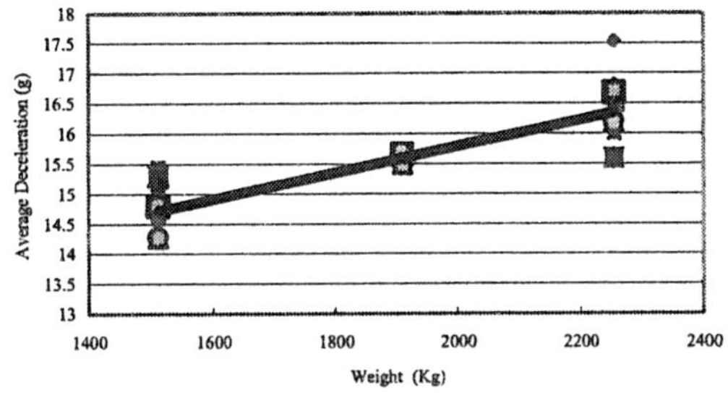

Fig 6. Changing average deceleration with respect to bullet vehicle weight and structural modification.

Finding the main parameter, which most influences the responses of average deceleration and displacement is the goal here. The average deceleration monitored at driver side is influenced more by the sub frame member thickness and Engine position as shown in fig 7.

Meanwhile, the thickness of the front side member and dash does not result in a change in the average deceleration.

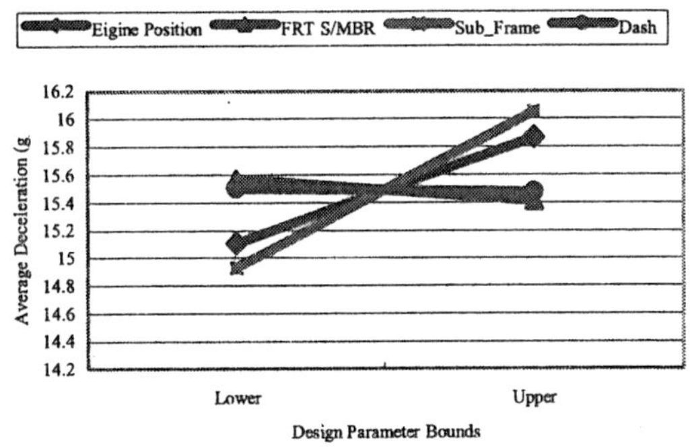

Fig 7. Design Parameters influence in Average Deceleration.

Fig 8 shows that the dash thickness and engine position influenced the other displacement responses monitored at several points of dashboard. Sub frame thickness also shows more influence with an inverse proportional relationship.

A review of fig 9 (which illustrates the deformed shape) shows the reason why this front side member does not affect both deceleration and displacement. The front side member have designed to get axial high deformation energy in the full-lap frontal impact is easy to bend in this frontal oblique offset, because of loading angle changed from perpendicular to 30 degree angled impact condition. But sub frame shows more stable axial deformation due to originally angled layout against oblique impact angle and its support by front crossed member.

Therefore we may say the side member is necessary to change in initial shape instead of its thickness and the shape optimization for this matter of changing initial geometry is efficient to change deformation mode close to axial.

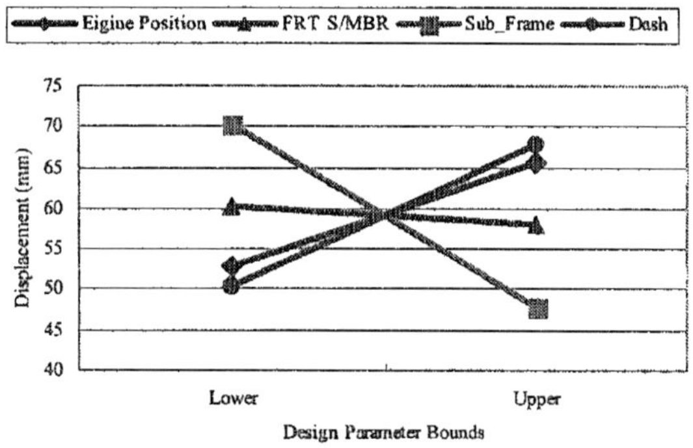

Fig 8. Design Parameters influence in Displacement.

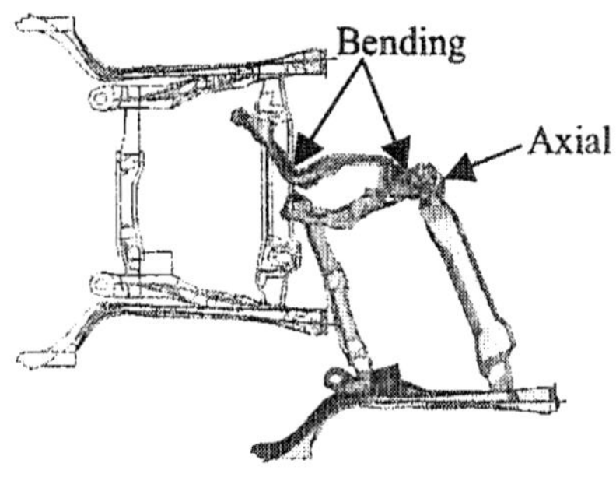

Fig 9. The main members deformed shape during the frontal oblique offset impact

In the deformation of each crash as shown in fig 10, not surprisingly the target Taurus has more deformation than the same bullet Taurus, and that deformation is in the proportion to bullet car weight in general. The deformed shapes of the target car show the more possible injury through the more deformation of compartment as it pinpoint the collapse of center pillar especially in Taurus-to-C2500 due to the high impact position overriding the passenger car.

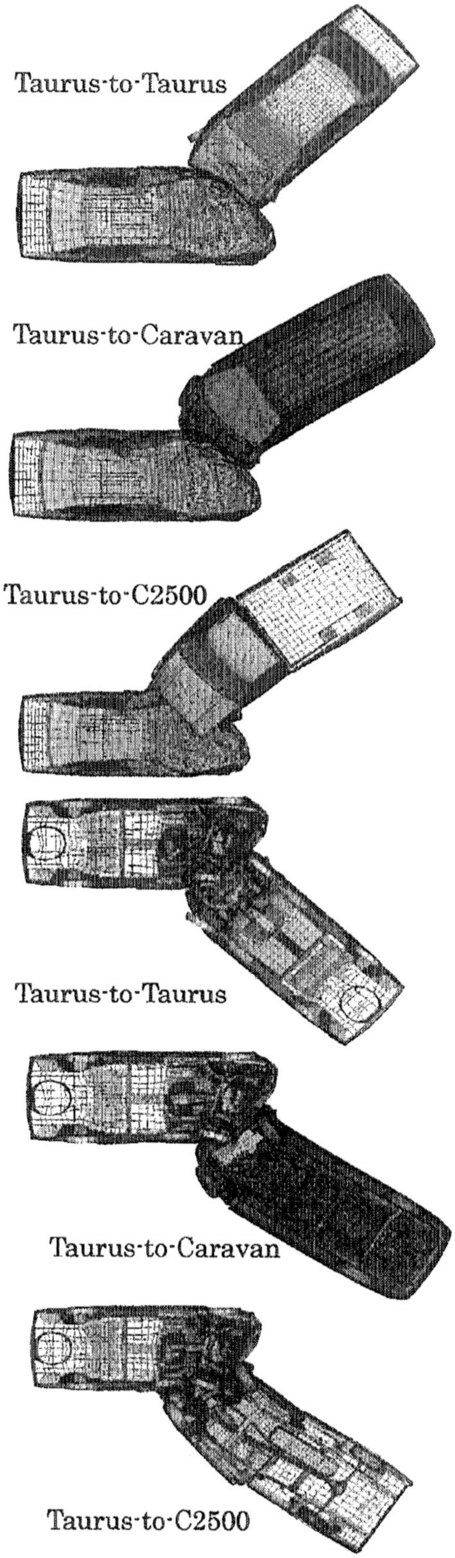

Fig 10. Simulated deformation in the Car-to-LTV impact

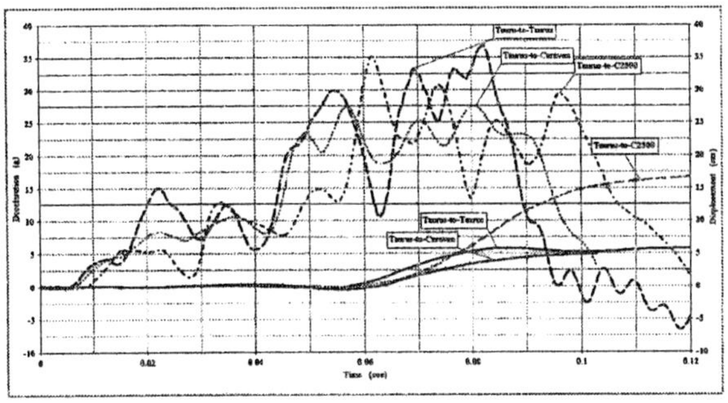

Fig 11. Body Deceleration Curve and Dashboard Intrusion.

Having the DOE results, an optimization study of compatibility follows it, utilizing parameters as listed in table 9 that is distinguished with shape and size variables. Shape optimization refers to the geometric shape changes while size optimization considers the gauge effects of any parts.
In this study, both size and shape optimization have been utilized simultaneously to consider the efficient ways of changing geometry.

Table 9. Design Parameters and its bounds for optimization (Shape and size variables)

Design Parameters	Initial Value	Lower Value	Upper Value
Engine Position	0.0	-30	30
FRT Member Wider Shape	0.0	0	20
FRT Member Inner Thickness	1.5	1	3
FRT Member Reinf Thickness	1.5	1	3
Sub Frame Longitudinal Thickness	1.930	1.5	3.6
Dash LWR Thickness	1.120	1.0	3.6

The shape optimization is used in the engine positioning and shape of front longitudinal side member. To define the shape variable perturbations, nodal domains are defined having vectors specified at the vertices, and internal nodes rounded by that domains are interpolated linearly. This is very similar to finite element shape function algorithm in basic idea. Specifying vectors at the center between vertices, 2nd order shape variables could be represented.
Specifically, the shape optimization of the front side member needs to be considered because the original deformed shape of it was very easy to bend at change of section combined with impact loading direction of frontal oblique offset impact, shown in previous DOE study.

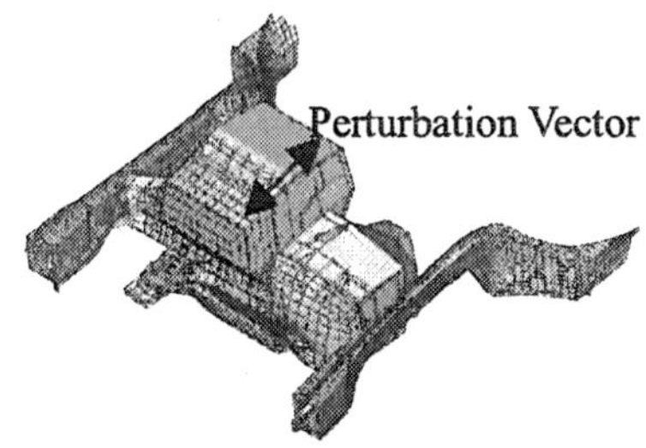

(a) Engine forward and rearward movement

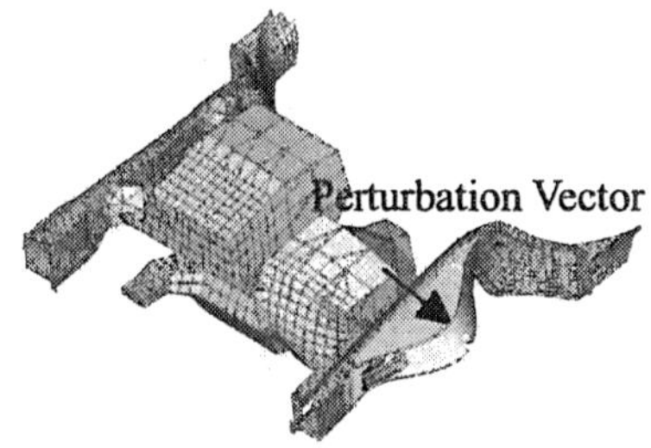

(b) Front Side Member Wider perturbation

Fig 12. perturb FE mesh for shape optimization

The problem is formulated as a typical crashworthiness concept ;
Minimize : intrusion of compartment
Within design range :

$$25g \le \max(\text{Deceleration}) \le 45g$$

$$10g \le \text{mean}(\text{Eng Room Decel}) \le 15g$$

$$13g \le \text{mean}(\text{Compart Decel}) \le 15g$$

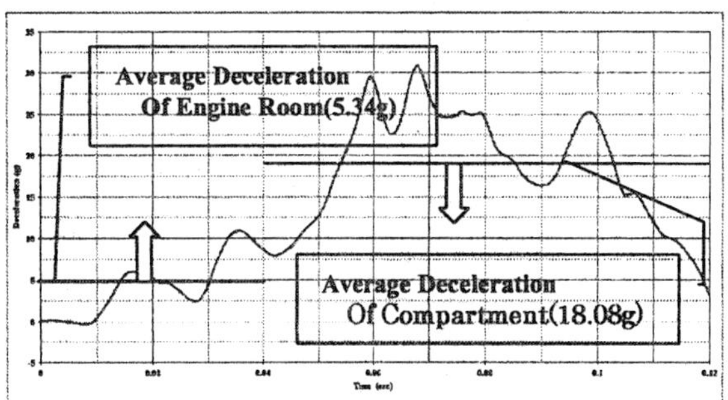

Fig 13. Concept of Design Range

Even though the role of restraint system has not been investigated clearly at this time, the seat belt could be effective to moderate injury level at even high deceleration. The range of deceleration, therefore, may be set to the upper 45g. At that value the seat belt can manage to reduce the injury level without so much venturing. The lower 25g is not thought to be an issue in designing the car structure.

Other design ranges for the engine room average deceleration and the compartment average deceleration, as shown in figure 13, are associated with improving crashworthiness by the increase of front stiffness leading to decrease force level at the second impact stage.

In order to set the threshold, the most moderated impact condition of car-to-car impact was used, and then it were taken out as the increased 10g of lower bound at front stiffness and the moderated 15g of upper bound at second impact stage. In the mean time, the design variables were selected as only rearward of front side member, it may have not related with pedestrian impact.

Minimizing the intrusion into the compartment would make sense if the car structure is going to be strong simultaneously. And it is doable design method to prevent occupant injury suffering from large deformation to the extent that restraint system could not support to moderate injury.

Unfortunately, the number of design parameters are limited because of the computation time, as this full vehicle impact models take about 3 days for one iteration. To think this 6 parameters, sum up to 18 parameters for 3 impact models, it is required to execute 18 runs at least for initialize stage of optimization progress.

The intermediate results are shown in fig 14 and fig 15, the change of objective dashboard intrusion and the change of body deceleration for each.

We hope to show the optimized results at coming 18th ESV conference.

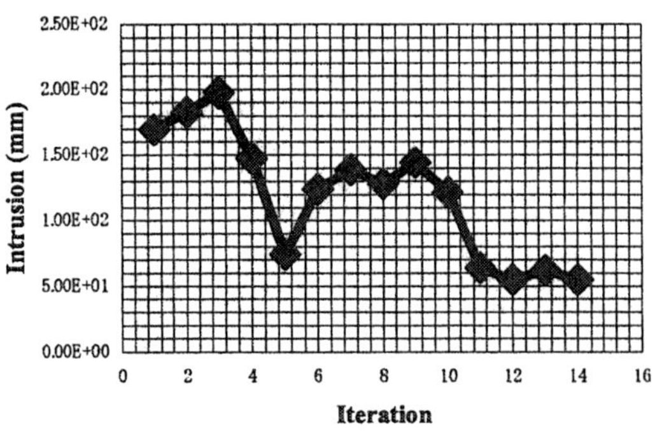

Fig 14. Optimization History for objective function as intrusion minimize

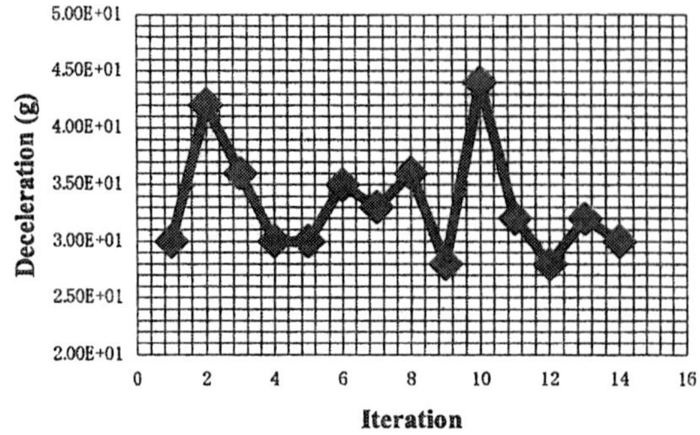

Fig 15. Optimization History for constraint function with body deceleration

Figure 16 compares curves between the initial run and intermediate run at iteration 14, showing some of the changed responses.

In the ideal optimized results to the compatibility problem, the balanced increase of body deceleration and the minimization of intrusion are desired. The intermediate results, therefore, are not sufficient to conclude because deceleration is also decreased along with the decreased intrusion.

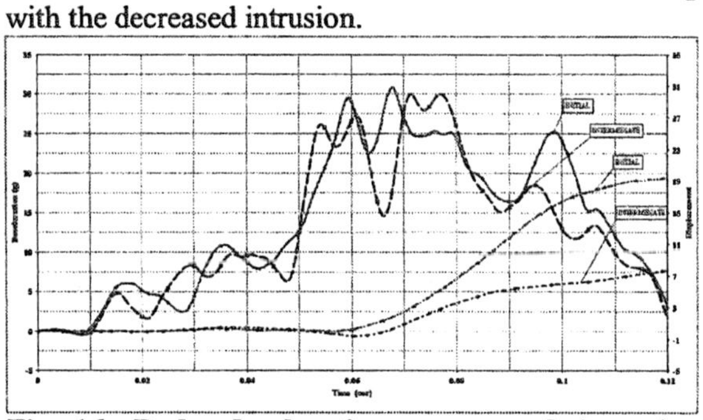

Fig 16. Body deceleration curve and intrusion between initial and intermediate at 14 run.

CONCLUSIONS

The main objective of this study is to show that bullet vehicle weight deepens aggressivity by the DOE study coupled with a optimization method and a nonlinear dynamic FE code. Then, these same methods are demonstrated to propose an optimized passenger car structure against the incompatibility of LTV to car.

As a starting view for compatibility problems, the parameters considered are in the line of crashworthiness, because of the similarity in structural improvement and response analysis.

Combining structural improvement and bullet vehicle weight, the DOE results substantiate that the tendency is close to aggressivity metric, and that structural improvements also can leverage the incompatibility.

The results of the DOE indicate that engine position and sub frame are more influential in both deceleration and displacement responses. In contrast with full-lap or offset case, the front side member is believed to act as the main resistance and crash energy absorbing member having not influenced both responses.

This means front side member need to change its shape to be more effective in resistance force and crash energy absorbing, ideally it should exhibit an axial deformation mode. In this case, shape optimization technique has facilitated the consideration of geometric design space more efficiently.

Even though optimization followed DOE study, the final results are unfortunately unavailable because it is still running and just in initial step.

The optimization for vehicle-to-vehicle crash is a very time consuming job. Of course it depends on the FE model size and the number of design parameters. In this case it took about 2 months just for the initial step with 18 design parameters.

Another problem for FE model size is that enough detail should be included, allowing representation of the large deformation suffered with aggressive bullet vehicle having heavy weight and stiffer structure. The frontal oblique offset crash especially shows more severe deformation of compartment.

The challenge is to reduce the number of elements not affecting the response to be discussed or to seek the way blocked consideration at specific time zone sequentially.

REFERENCES

[1] Donald E. Siddal and Terry D .Day, "Updating the Vehicle Class Categories", Society of Automotive Engineers Paper No. 960897

[2] Summers, S, Prasad, A, .and Hollowel, W.T., "NHTSA`s Vehicle Compatibility Research Program", Society of Automotive Engineers Paper No. 1999-01-0071 March 1999

[3] H. Gabler, W. Hollowell, "NHTSA's vehicle aggressivity and compatibility research program", NHTSA, Paper No. 98-S3-O-01, 16th ESV, 1998

[4] Lars Fredriksson, Christian Alscher, Sensitivity Analysis of an LS-DYNA Crash Model, 19th CAD-FEM Users` Meeting 2001 International Congress on FEM Technology, Berlin, October, 2001.

[5] Marczyk, J., et.al. Uncertainty Management in Automotive Crash: From Analysis to Simulation, ASME 2000 Conference, Baltimore, USA, September 2000.

[6] Schramm, U., Thomas, H., and Schneider, D., Crashworthiness Design using Structural Optimization. Design Optimization, 1 (1999) 374-387.

[7] H. Kurtaran, A. Eskandarian, D. Marzougui, N. E. Bedewi, Crashworthiness Design Optimization using Successive Response Surface Approximations, Computational Mechanics 29 (2002), pp. 409-421.

[8] Hallquist JO (1997) LS-DYNA Theoretical Manual, Livermore Software Technology Corporation, Livermore, CA, USA.

[9] PAM-CRASH(2000) Solver Reference and Note Manual, Pam System International S.A., The software Company of ESI Group.

[10] RADIOSS(1999) Theory Manual 99, MECALOG Les Algorithms, Sophia Antipolis, France

[11] Schramm, U., Thomas, H.(1998), Crashworthiness Design using Structural Optimization. AIAA-paper 98-4729, 7th Symposium on Multidisciplinary Analysis and Optimization, St.Louis, MO.

[12] Altair HyperMesh V5.1, On-line Manual, Altair Engineering Inc., Troy, MI, 2002).

[13] Altair StudyWizard V5.1, On-line Manual, Altair Engineering Inc., Troy, MI, 2002).

[14] Altair HyperView V5.1, On-line Manual, Altair Engineering Inc., Troy, MI, 2002).

[15] Altair OptiStruct V5.1, On-line Manual, Altair Engineering Inc., Troy, MI, 2002).

[16] Altair HyperOpt V5.1, On-line Manual, Altair Engineering Inc., Troy, MI, 2002).

[17] Hans C. Joksch, Vehicle Design versus Aggressivity, DOT HS 809 194, NHTSA, April, 2000

[18] Hans C. Joksch, Dawn Massie, Robert Pichler, Vehicle Aggressivity:Fleet Characterization Using Traffic Collision Data, DOT HS 808 679, DOT-VNTSC-NHTSA-98-1, NHTSA, February, 1998

[19] Sheldon L. Stucki, W. Hollowell, Osvaldo Fessahaie, Determination of frontal offset test conditions based on crash data, Paper Number 98-S1-O-02, The 16th International Technical Conference on the Enhanced Safety of Vehicles (ESV) Proceedings -Windsor, Ontario, Canada, May 31- June 4, 1998

ABOUT THE EDITOR

Stan Backaitis is principal engineer at the National Highway Traffic Safety Administration in Washington DC. His education includes a BSME degree in 1953, MSAE in 1955, and post-graduate studies from 1968-1978, including studies at the Federal Management Institute. He served from 1953 to 1968 in various engineering positions, including program manager, at the Chrysler Corporation's Automotive and Aerospace Divisions. He has worked at NHTSA from 1968 to the present, first as branch chief and then as principal engineer at the Office of Vehicle Crashworthiness Standards. He has authored over 60 technical papers, and edited five books on human biomechanics and crash injury tolerances and two books on accident reconstruction technologies. He organized over 20 technical sessions related to automotive safety at SAE World Congresses.

Mr. Backaitis is a registered professional engineer in Michigan and a certified product safety manager. He was chair of the SAE Truck Ride and Vibrations Committee from 1966-1968, and Chair of the SAE Occupant Protection Committee from 1986-1999, and remains a member of that committee. He is also a member of the Human Use Review Panel at the U.S. Department of Transportation, member of the U.S. delegation to the ISO U.S. Technical Advisory Group TC22/SC 10 and 12, and their technical working groups on biomechanics, crash test methodology, crash test instrumentation, and occupant protection. He is a member of SAE committees and working groups on human biomechanics and simulation, instrumentation, Hybrid III Dummy Family, ATD test procedures, impact, and rollover.

Mr. Backaitis is the recipient of three Secretary of Transportation and eight NHTSA Administration awards for superior achievement, holds an Honorary Doctorate Degree from the Kaunas University of Technology (KTU), and was instrumental in the development of SAE's Lithuania Section.

Besides his daily duties at NHTSA, he is serving as vice president for academic affairs of the American-Lithuanian Community Council, and is a counselor on scientific affairs to a U.S.-Lithuania parliamentary working group.